FEATURES AND BENEFITS

BASIC GEOMETRY

Full coverage of the basic concepts and **methods** provides a secure foundation in the essentials of Euclidean geometry. A new **appendix on** transformational geometry is also included. See pages iii–ix (Contents), 531–546.

An **informal, intuitive approach** includes stimulating *Explorations* that lead students to discover geometric principles for themselves. **Short, readable lessons** make information easy to absorb and give students a genuine sense of progress. Minimal text material and plentiful worked-out examples make concepts easy to understand. See pages 250–251, 293, and 388–389.

There is an **abundance of Classroom Practice and Written Exercises.** *Experiments* often supplement written exercises, reinforcing vital concepts with hands-on activities. *Answers to Selected Exercises* are provided at the back of the textbook. See pages 193, 204–207, 251–254, 333, and 566–583.

Abundant review, including *Skills Reviews, Chapter Reviews, Mixed Reviews,* and *Cumulative Reviews,* strengthens vital skills. **Self-Tests** with vocabulary lists occur after a cluster of related lessons. *Answers to Self-Tests* are supplied at the back of the textbook. A **Chapter Test** is provided at the end of each chapter. See pages 228, 338, 339–340, 341, 342, 384, and 563–565.

Color is used effectively in the text and diagrams to clarify and emphasize important concepts. Postulates, Theorems, and Constructions are displayed in colorful panels. Student notes, providing brief explanations or suggestions, are also highlighted in color. See pages 72, 158, 174, 255, and 289.

Special **Reading Geometry** features help **students effectively** use explanations and diagrams to learn geometric concepts. **Problem Solving Strategies** give students practice in applying general problem solving techniques. **Applications** apply geometric concepts and skills to a variety of areas of interest. See pages xi–xii, 457–459, and 506–507.

Calculator Activities and **Computer Activities** extend and apply geometric concepts by means of calculators and computers. See pages 378–379, 409.

An expanded collection of **supplementary materials** accompanies a **comprehensive Teacher's Edition** that includes lesson-by-lesson commentary, teaching suggestions, chalkboard examples, extensions, quick quizzes, and much, much more. The supplementary materials consist of the following:

Solution Key includes worked-out solutions to all written exercises.

Resource book, on blackline masters, includes algebra and arithmetic reviews, extra testing, practice, mixed reviews, and guided help with proof.

Tests, on duplicating masters, provide cumulative and sectional test coverage different from Resource Book tests).

Practice Masters, on duplicating masters, provide practice on fundamental concepts and skills for slower learners (different from Resource Book practice).

BASIC GEOMETRY

TEACHER'S EDITION

Ray C. Jurgensen
Richard G. Brown

Editorial Adviser: Albert E. Meder, Jr.
Teacher Consultant: Robert J. McMurray

Houghton Mifflin Company • Boston
Atlanta   Dallas   Geneva, Illinois
Palo Alto   Princeton   Toronto

The Authors
Ray C. Jurgensen, formerly Chairman of the Mathematics Department and holder of the Eppley Chair of Mathematics at the Culver Academies, Culver, Indiana. Mr. Jurgensen has been a lecturer at the National Science Foundation institutes for mathematics teachers and meetings of the National Council of Teachers of Mathematics.

Richard G. Brown, Mathematics teacher at the Phillips Exeter Academy, Exeter, New Hampshire. Mr. Brown has taught a wide range of mathematics courses for both students and teachers at several schools and universities, including Newton (Massachusetts) High School, the University of New Hampshire, Arizona State University, and the North Carolina School for Science and Mathematics.

Editorial Adviser
Albert E. Meder, Jr., Professor of Mathematics, Emeritus, Rutgers, The State University of New Jersey.

Teacher Consultant
Robert J. McMurray, Mathematics teacher at the Albuquerque Technical Vocational Institute, Albuquerque, New Mexico. Mr. McMurray has many years of experience teaching geometry in secondary schools.

Copyright © 1990 by Houghton Mifflin Company. All rights reserved.

The pages in this work showing printed notice that they may be reproduced by teachers for use in their classes in which they are using the accompanying Houghton Mifflin material may be reprinted or photocopied in classroom quantities, provided each copy shows the copyright notice. Such copies may not be sold and further distribution is expressly prohibited. No part of this work may be reproduced or transmitted in any form or by any means, electronic or mechanical including photocopying and recording, or by any information storage or retrieval system without prior written permission of Houghton Mifflin Company unless such copying is expressly permitted by federal copyright law. Address inquiries to School Permissions, Houghton Mifflin Company, One Beacon Street, Boston, MA 02108.

Printed in U.S.A.

ISBN: 0-395-50121-0

BCDEFGHIJ–RM–96543

CONTENTS

Teaching the Course — T4
An explanation of the contents of the Teacher's Edition, Resource Book, Solution Key, Tests, and Practice Masters.

Organization of the Textbook — T6
An illustrated introduction to the organization of the text and features.

Chapter Tests — T8
A permission-to-reproduce test for each chapter, followed by complete answers on page T20.

Assignment Guide — T23
A guide that will help you plan the course that best meets the needs of your particular students.

Supplementary Materials Guide — T32
A cross-reference guide to the text and its supplementary Tests, Practice Masters, and Resource Book.

Reading Geometry — T34
Suggestions for helping your students learn how to read a geometry textbook.

Problem Solving Strategies — T38
Some strategies that your students can use to become better problem solvers.

Error Analysis — T41
A discussion of common errors that geometry students make.

Lesson Commentary — T43
A discussion of each lesson, with teaching suggestions, chalkboard examples, extensions, and quick quizzes.

Student Text
A full-sized facsimile of each student page with answers to exercises annotated on the page.

TEACHING THE COURSE

The Teacher's Edition, Resource Book, and Solution Key have been designed to help you teach geometry. For each chapter in the textbook, the Teacher's Edition contains lesson-by-lesson commentary on the student text, with teaching suggestions, chalkboard examples, extensions, and quick quizzes. The Teacher's Edition also includes extra chapter tests, an assignment guide, a guide to supplementary materials, and special Reading Geometry, Problem Solving Strategies, and Error Analysis sections.

The Resource Book offers extra test, practice, and review material—all keyed to the student textbook—in blackline master format. Worksheets designed to help students who are having difficulty with proofs are also included.

The Solution Key provides step-by-step solutions, including diagrams, for all written exercises in the student book.

The supplementary materials on duplicating masters include Tests and Practice Masters. Each set of masters is keyed to the student textbook and has a separate Answer Key with answers annotated on reduced facsimiles of the masters.

The Practice Masters offer concentrated practice on small portions of material in worksheet format. Periodic cumulative reviews are included.

The Tests are a convenient way to keep track of each student's performance. They contain quizzes covering the same groups of lessons as the Self-Tests in the student textbook. There are also four cumulative tests that cover three chapters each.

ORGANIZATION OF THE TEXTBOOK

Lessons

The text is organized into six units, each divided into two chapters. The chapters consist of short lessons with manageable bits of material and a generous number of exercises. Postulates, Theorems, and Constructions are displayed in colorful panels. Illustrative examples are frequently included. Many lessons have **Explorations** that lead students to discover geometric principles. Color is used effectively in the text and the diagrams to clarify and emphasize important concepts.

There is an abundance of classroom and written exercises. Written Exercises are graded by the letters A, B, and C in order of increasing difficulty. Additional review exercises for each chapter begin on page 523.

The Written Exercises are often supplemented by an **Experiments** section. These experiments serve to reinforce concepts already taught and to introduce new ideas for enrichment.

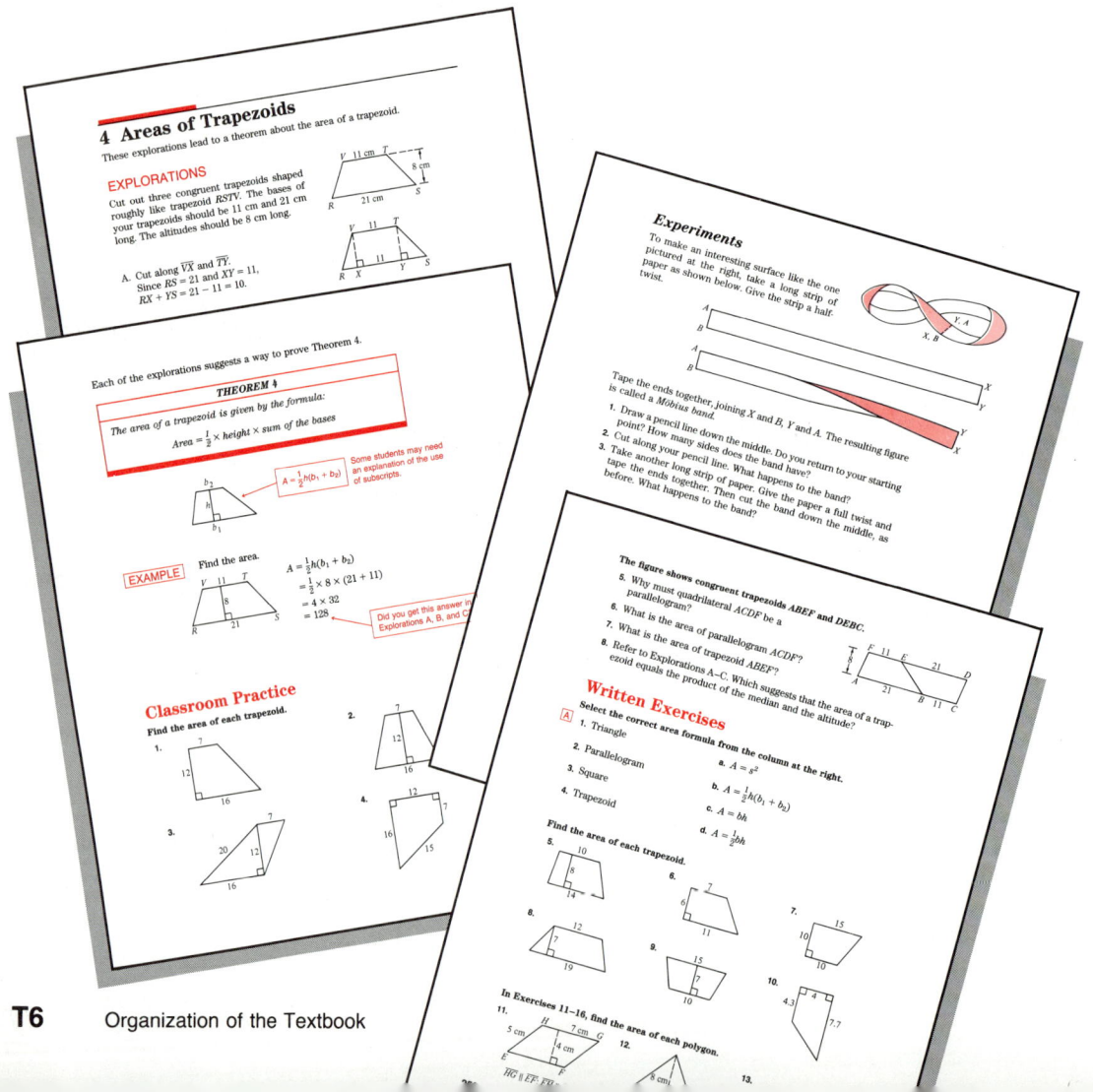

Review/Testing

SELF-TEST

CHAPTER TEST

Self-Tests occur after a cluster of related lessons. Answers to the Self-Tests are supplied at the back of the text. A **Chapter Test** is provided at the end of each chapter.

SKILLS REVIEW

CHAPTER REVIEW

MIXED REVIEW

UNIT E CUMULATIVE REVIEW

Ample review material is also provided. Each chapter has a **Chapter Review** consisting of a summary and review exercises. A **Skills Review,** covering arithmetic and geometry topics precedes each Chapter Review. **Mixed Reviews** at the end of odd-numbered chapters give students practice in solving exercises when they are encountered out of context. **Cumulative Reviews** occur at the end of each unit. They cover the topics in the unit in the order in which they were presented.

Special Features

CALCULATOR ACTIVITIES

COMPUTER ACTIVITIES

PROBLEM SOLVING STRATEGIES

A technology strand, consisting of **Calculator** and **Computer Activities,** indicates ways of using calculators or computers to enhance a geometry course. **Problem Solving Strategies** give students practice in applying general problem solving techniques.

READING GEOMETRY

APPLICATIONS

PUZZLE ♦ PROBLEMS

The **Reading Geometry** features give suggestions for helping students read a geometry textbook. In **Applications** geometric concepts and skills are applied to a variety of areas such as art, nature, sports, architecture and engineering. Many **Puzzle Problems** also occur throughout the text.

CAREER NOTES

CONSUMER APPLICATIONS

Career Notes illustrate many jobs that need a background in mathematics. Practical tips for students are presented in **Consumer Applications.**

CHAPTER TESTS

Chapter 1

Complete each statement.

1. Another name for $\overleftrightarrow{QR}$ is __?__.

2. Another name for $\overrightarrow{SQ}$ is __?__.

3. $PQ = $ __?__ and $QS = $ __?__.

4. If R is the midpoint of $\overline{QS}$, then the coordinate of R is __?__.

5. Another name for $\angle GTA$ is __?__.

6. $\angle 8 + \angle 9 = \angle$ __?__ (use 3 letters)

7. $\angle 4 + \angle 5 = $ __?__ °

8. $\angle 3$ and $\angle$ __?__ are vertical angles.

9. If $\overrightarrow{AN}$ bisects $\angle TAG$ and $\angle 8 = 35°$, then $\angle TAG = $ __?__ °.

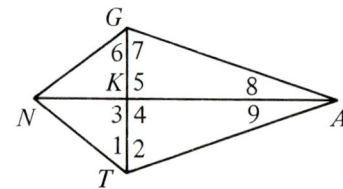

In the diagram above, $\overline{TG} \perp \overline{NA}$. Classify each angle as acute, right, obtuse, or straight.

10. $\angle NTA$ 11. $\angle NKA$ 12. $\angle NKG$ 13. $\angle ANG$

14. Construct a right angle.   15. Construct a 45° angle.
(Use your answer to Exercise 14 if you wish.)

Which postulate can be used to justify each statement?

16. If $3x + 7 = 31$, then $3x = 24$.

17. If $3x = 24$, then $x = 8$.

18. If $\angle 1 = \angle 2$ and $\angle 1 + \angle 5 = 50°$, then $\angle 2 + \angle 5 = 50°$.

19. If $\angle 1 = \angle 2$ and $\angle 3 = \angle 4$, then $\angle ABC = \angle ACB$. (See diagram.)

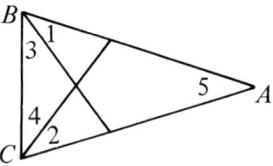

Refer to the rectangular solid shown. Complete.

20. Plane $SBTN$ ∥ plane __?__.

21. J, N, L, and __?__ are coplanar points.

22. $\overleftrightarrow{AJ}$ is formed by the intersection of plane __?__ and plane __?__.

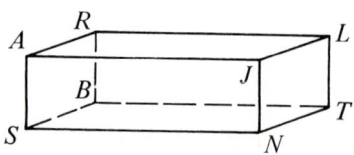

Chapter 2

Given: $\overline{DC} \perp \overline{FC}$

1. Name a complement of ∠ACE.
2. Name *two* pairs of supplementary angles.
3. If $\overline{DE} \parallel \overline{AC}$, then ∠DEC = ∠ ___?___.
4. If $\overline{DE} \parallel \overline{AC}$, then ∠FAE = ∠ ___?___.

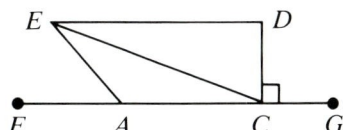

Accept as true: "If $a = 1$, then $a \cdot b = b$."

5. Write the hypothesis and the conclusion.
6. Write the converse.   7. Is the converse true?
8. (Optional) Given that $a \cdot b \neq b$, what, if anything, can you conclude?

Find the values of x and y.

9.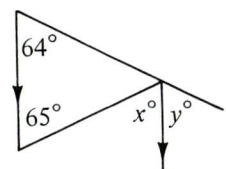
10.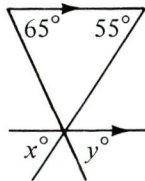
11.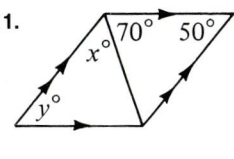

12. Supply the reasons to complete the proof.
    Given: ∠1 = ∠2; ∠3 = ∠4
    Prove: $\overline{AN} \parallel \overline{LB}$

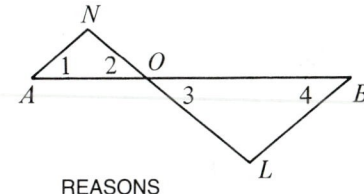

STATEMENTS	REASONS
1. ∠1 = ∠2; ∠3 = ∠4	1. ___?___
2. ∠2 = ∠3	2. ___?___
3. ∠1 = ∠4	3. ___?___
4. $\overline{AN} \parallel \overline{LB}$	4. ___?___

13. Use the figure for Exercise 12.
    Given: ∠1 = ∠2; $\overline{AN} \parallel \overline{LB}$
    Prove: ∠4 = ∠3

14. Draw a triangle ABC. Construct a line through A parallel to $\overline{BC}$.

Chapter 3

Find the value of x.

1.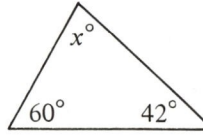
2.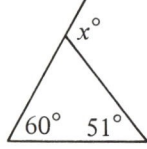
3.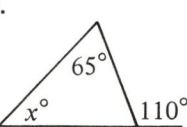

4. Two angles of a triangle have measures 46° and 44°. Is the triangle *acute, right,* or *obtuse?*

5. A triangle has three unequal sides. Is the triangle *isosceles, equilateral,* or *neither* of these?

Complete: If △ TOW ≅ △ RAN, then . . .

6. ∠ W = ∠ __?__   7. TW = __?__

8. Which *two* of the following indicate correct ways of proving triangles congruent?
   SS   HL   SSS   AAA

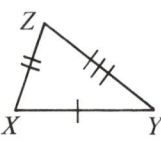

9. Complete the statement correctly.
   △ XYZ ≅ △ __?__ __?__ __?__

10. Supply the reasons to complete the proof.
    Given: $\vec{AB}$ bisects ∠ XAY
            AX = AY
    Prove: △ BAX ≅ △ BAY

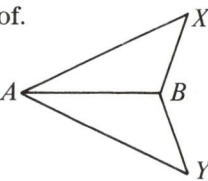

STATEMENTS	REASONS
1. AB = AB	1. __?__
2. ∠ BAX = ∠ BAY	2. __?__
3. AX = AY	3. __?__
4. △ BAX ≅ △ BAY	4. __?__

11. Write a complete proof in two-column form.
    Given: $\overline{VT} \parallel \overline{SR}$
            $\overline{VR} \parallel \overline{TS}$
    Prove: △ RST ≅ △ TVR

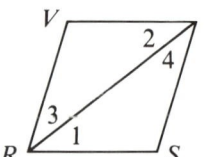

Chapter 4

1. Suppose ∠SAT has been bisected, by construction. Then $\overline{XP}$ and $\overline{YP}$ have been drawn. Supply reasons to complete the proof that ∠SAT really has been bisected.

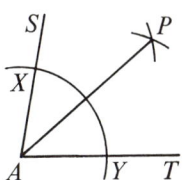

STATEMENTS	REASONS
1. $AX = AY$	1. By construction
2. $XP = YP$	2. ___?___
3. $AP = AP$	3. ___?___
4. $\triangle XAP \cong \triangle YAP$	4. ___?___
5. ∠XAP = ∠YAP	5. ___?___

2. M is the midpoint of $\overline{JK}$.
   a. Name three bisectors of $\overline{JK}$.
   b. $\overleftrightarrow{DE}$ is a perpendicular bisector of $\overline{JK}$ if ∠DMK = ___?___.

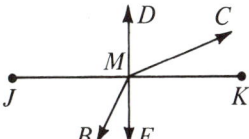

3. Draw a triangle like the one shown, but much larger.
   a. Construct the altitude from T.
   b. Construct the median from T.

4. Draw a △ABC. Circumscribe a circle about △ABC.

5. In △GUM, $GU = MU$ and ∠U = 40°. ∠G = ___?___°

Write a proof in two-column form.

6. **Given:** $\overrightarrow{WX}$ bisects ∠YWZ
   $WY = WZ$
   Prove: $XY = XZ$

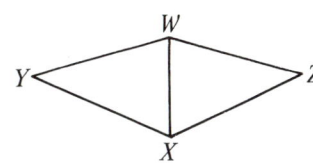

7. **Given:** ∠3 = ∠4
   Prove: $AT = AV$

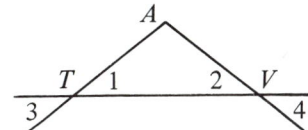

Chapter 5

1. The perimeter of a regular pentagon is 450 cm. How long is a side?
2. Find the interior angle sum of a polygon with 12 sides.
3. Find the exterior angle sum of an octagon.
4. Find the measure of each interior angle of a regular hexagon.

Given: ▱*STAL* and ▱*KBAN*
Find each length and angle measure.

5. *NK*
6. *LS*
7. *LN*
8. ∠*T*
9. ∠*A*
10. ∠*K*

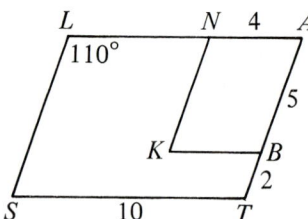

Given: **Rectangle *ABEF* and rhombus *BCDE*.**
Find each length and angle measure.

11. *AE*
12. *BF*
13. *OF*
14. *BC*
15. ∠*BAE*
16. ∠*BCE*

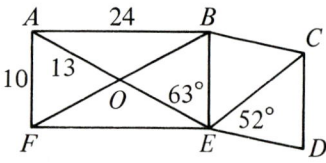

PRIM is an isosceles trapezoid with bases $\overline{PR}$ and $\overline{MI}$.
PR = 30, *MI* = 16, and *RI* = 12.

17. Find the length of the median of the trapezoid.
18. If ∠*P* = 54°, find the measures of ∠*R*, ∠*I*, and ∠*M*.

D, E, and F are midpoints of the sides of △*PQR*.

19. *DE* = ___?___
20. *RQ* = ___?___
21. *PE* = ___?___
22. If ∠*Q* = 90°, what kind of quadrilateral is *FEQD*?

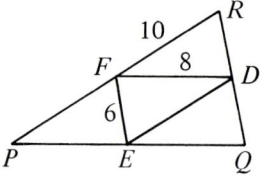

23. State the definition of a parallelogram.
24. State a theorem which can be used to prove that a quadrilateral is a parallelogram.

Chapter 6

Match the expressions or statements in the column at the right with the phrases in the column at the left. *If* the area of a parallelogram were $2\pi r$, for instance, you would write 1j.

1. Area of a parallelogram
2. Area of a square
3. Circumference of a circle
4. Area of a trapezoid
5. Area of a circle
6. Area of a triangle
7. Perimeter of a square
8. Obtuse triangle
9. Acute triangle
10. Perimeter of a rectangle

a. πr^2
b. s^2
c. bh
d. $c^2 > a^2 + b^2$
e. $\frac{1}{2}(bh)$
f. $4s$
g. $\frac{1}{2}h(b_1 + b_2)$
h. $c^2 < a^2 + b^2$
i. $2b + 2h$
j. $2\pi r$

11. The legs of a right triangle have lengths 6 and 8. Find the length of the hypotenuse.

12. The hypotenuse of a right triangle is 13 m long and one leg is 12 m long. Find the length of the other leg.

13. The sides of a rectangle are of length 4 and 13.
    a. perimeter = __?__   b. area = __?__   c. length of a diagonal = __?__

14. The radius of a circle is 6. Find the area (in terms of π).

15. The radius of a circle is 8. Using 3.14 for π, find the area (correct to the nearest integer).

16. The area of a circle is 100π. Find the circumference (in terms of π).

Find the area of each figure.

17.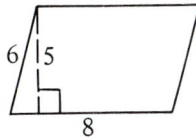
Parallelogram

18.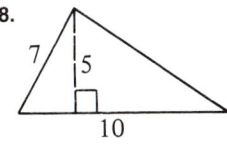
Triangle

19.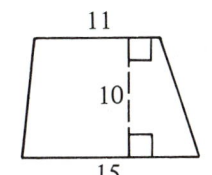
Trapezoid

20.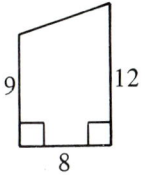
Trapezoid

21. The area of a certain rectangle is 84 cm². Each of the longer sides is 14 cm. How long is each of the shorter sides?

22. The perimeter of a certain square is 28 cm. Find the area.

Chapter 7

Express each ratio in simplest form.

1. $\dfrac{12}{16}$
2. $\dfrac{18}{21}$
3. $\dfrac{15x}{35x}$
4. $30:45$
5. $xy:yz$
6. $48:36:60$
7. $8 \text{ cm}:2 \text{ m}$
8. $400 \text{ m}:4 \text{ km}$
9. $15 \text{ min to } 5 \text{ h}$

Given: $\square ABCD$. **Find each ratio in simplest form.**

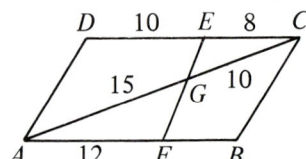

10. $\dfrac{DE}{DC}$
11. $\dfrac{AF}{FB}$
12. $\dfrac{AG}{AC}$
13. $\dfrac{BC}{AD}$

Complete each statement.
All variables represent nonzero numbers.

14. If $\dfrac{a}{b} = \dfrac{5}{7}$, then $\dfrac{a+b}{b} = \underline{}$.

15. If $\dfrac{k}{8} = \dfrac{7}{9}$, then $\dfrac{k}{7} = \underline{}$.

Find the value of x.

16. $\dfrac{x}{6} = \dfrac{3}{2}$
17. $\dfrac{27}{18} = \dfrac{x}{6}$
18. $\dfrac{x-5}{4} = \dfrac{6}{12}$

19. It takes 5 cans of paint to paint a floor 5 m wide and 8 m long. How many cans will it take to paint a floor 4 m wide and 6 m long?

Use the diagram at right.

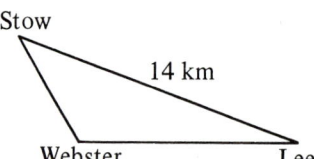

20. Measure the map distance between Stow and Lee to the nearest centimeter. Then give the scale of the map.

21. Find the actual distance between Stow and Webster.

22. Find the actual distance between Lee and Webster.

23. Make an accurate scale drawing of a rectangular plot of land 180 m long and 120 m wide. Give the scale of your drawing.

Chapter 8

The two trapezoids are similar. Complete each statement.

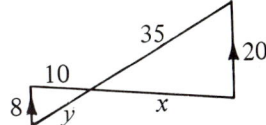

1. $GHIJ \sim$ __?__

2. The scale factor is __?__ : __?__.

3. $\dfrac{12}{8} = \dfrac{x}{?}$

4. $y =$ __?__

5. The ratio of the perimeters of the figures is __?__ : __?__.

6. The ratio of the areas is __?__ : __?__.

Find the values of x and y.

7.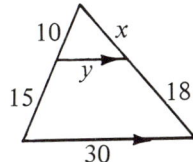

8.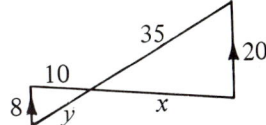

9.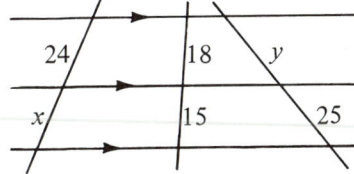

10.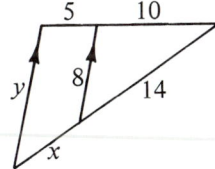

11. Supply the reasons to complete the proof.
    Given: $\square PQRS$
    Prove: $\dfrac{NO}{QO} = \dfrac{RO}{PO}$

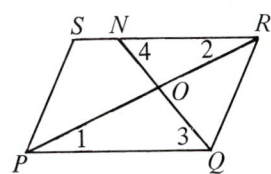

STATEMENTS	REASONS
1. $PQRS$ is a $\square$	1. __?__
2. $\overline{NR} \parallel \overline{PQ}$	2. __?__
3. $\angle 1 = \angle 2$; $\angle 3 = \angle 4$	3. __?__
4. $\triangle NOR \sim \triangle QOP$	4. __?__
5. $\dfrac{NO}{QO} = \dfrac{RO}{PO}$	5. __?__

Chapter 9

Refer to the figure and use letters to name the following:

1. A tangent
2. A radius
3. A chord
4. A secant

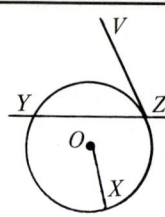

Refer to the diagram and complete the statements.

5. The number of common external tangents that can be drawn to ⊙P and ⊙R is __?__.
6. The number of common internal tangents that can be drawn to ⊙P and ⊙R is __?__.
7. ⊙P and ⊙Q are $\frac{?}{\text{internally/externally}}$ tangent.
8. Suppose a line is drawn perpendicular to $\overline{QX}$ at X. That line will be __?__ to ⊙Q.

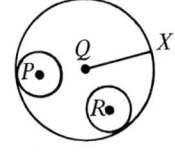

Write numerical values to complete the statements.

9. If $\widehat{BC} = 40°$, $\widehat{AC} = $ __?__ °.
10. If $\widehat{AC} = 150°$, $\widehat{ABC} = $ __?__ °.
11. If $\angle 1 = 36°$, $\widehat{BC} = $ __?__ °.
12. If $\widehat{AC} = 142°$, $\angle 1 = $ __?__ °.

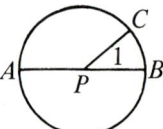

13. If $\widehat{CD} = \widehat{EF}$ and $CD = 13$, then $EF = $ __?__.
14. If $\widehat{CA} = 50°$, then $\widehat{CD} = $ __?__ °.
15. If $OY = OX$ and $EF = 18$, then $CX = $ __?__.
16. If $CD = 16$ and $OX = 6$, then $OC = $ __?__.

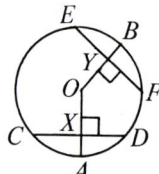

State the numerical value of x.

17.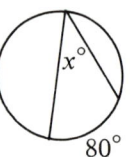
18.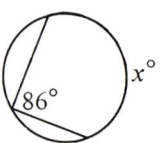
19.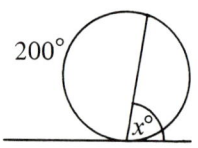
20.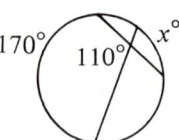
21.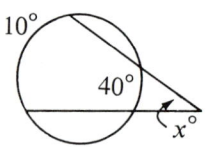
22.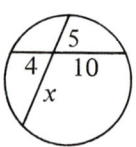

23. Construct a circle and choose a point P on the circle. Construct a line tangent to the circle at point P.

Chapter 10

Tell whether each statement is true or false.

1. Two parallel lines must be coplanar.
2. If two lines are skew, there is one plane that contains both lines.
3. If a line does not lie in a plane, the line must intersect the plane.
4. If $\overleftrightarrow{XY}$ is perpendicular to plane P, then every plane that contains $\overleftrightarrow{XY}$ must be perpendicular to P.

Refer to the right prism shown.

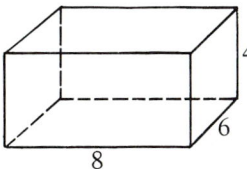

5. $B = $ __?__
6. L.A. = __?__
7. T.A. = __?__
8. $V = $ __?__

Refer to the cylinder shown. Answer in terms of π.

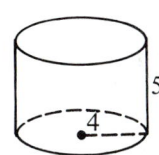

9. $C = $ __?__
10. L.A. = __?__
11. T.A. = __?__
12. $V = $ __?__

Refer to the regular square pyramid shown.

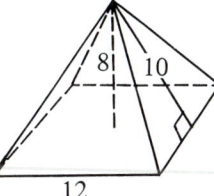

13. $B = $ __?__
14. L.A. = __?__
15. T.A. = __?__
16. $V = $ __?__

Refer to the cone shown. Answer in terms of π.

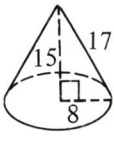

17. $C = $ __?__
18. L.A. = __?__
19. T.A. = __?__
20. $V = $ __?__

21. Find, in terms of π, the area of a sphere in which $r = 8$.
22. The area of a sphere is 36π. Find the volume, to the nearest integer.

(Optional) Two similar prisms have heights of 10 cm and 12 cm, respectively.

23. Find the ratio of their total areas.
24. If the volume of the smaller prism is 750 cm³, find the volume of the larger prism.

Chapter 11

Find the exact values of x and y.

1.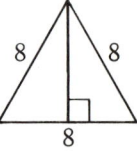

2.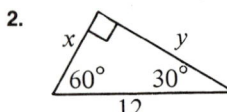

3.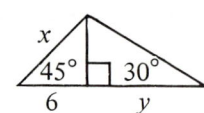

Find the area of each figure.

4.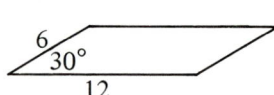

5. (figure with 6, 30°, 12)

6.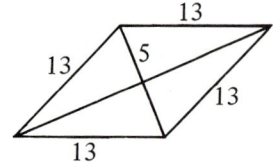

7. The length of a diagonal of a square is $8\sqrt{2}$. What is the area of the square?

8. The dimensions of a rectangular solid are 6 cm, 3 cm, and 2 cm. How long is its diagonal?

Chord $\overline{AB}$ is 16 cm long and is 6 cm from the center of the circle.

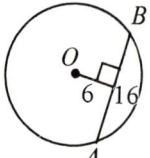

9. Find the radius of the circle.

10. Find the circumference of the circle.

11. Find the lateral area of the regular square pyramid.

12. Find the height.

13. Find the volume.

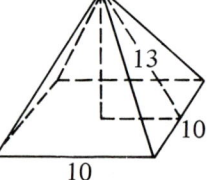

14. Draw a cone with height 9 and slant height 15.

15. Find the radius of the cone.

16. Find the volume of the cone (in terms of π).

Find the value of x to the nearest tenth. Use the table on page 517.

17.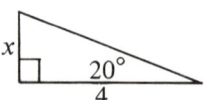

18.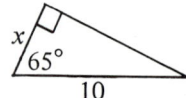

19.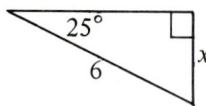

Chapter 12

Complete each statement.

1. Point R lies in Quadrant ___?___.
2. The x-coordinate of point R is ___?___.
3. The y-coordinate of point S is ___?___.
4. If point U is chosen so that quadrilateral $RSTU$ is a rectangle, then U has coordinates (___?___, ___?___).

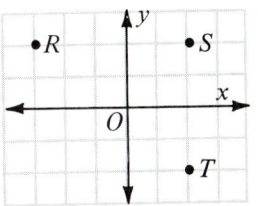

**Find the distance between the points named.
Use graph paper if you wish.**

5. $(4, 1)$ and $(4, 7)$
6. $(0, -3)$ and $(-2, -3)$
7. $(0, 0)$ and $(3, 4)$
8. $(-4, 1)$ and $(2, -7)$

9–12. Find the coordinates of the midpoint of the segment that joins the points named in each of Exercises 5–8.

State the slope of the line. If the slope is not defined for a particular line, write *not defined*.

13. $\overleftrightarrow{OA}$
14. $\overleftrightarrow{AB}$
15. $\overleftrightarrow{BC}$
16. $\overleftrightarrow{DC}$

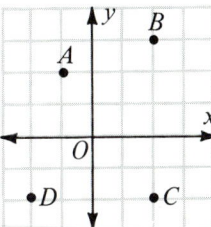

Line t has slope $\frac{5}{2}$.

17. Does line t *rise*, or does it *fall*, to the right?
18. The slope of any line parallel to t equals ___?___.
19. The slope of any line perpendicular to t equals ___?___.

Line k has the equation $3x - 2y = 12$.

20. Does line k pass through point $(5, 1)$?
21. On line k, when $x = 0$, $y = $ ___?___.
22. On line k, when $y = 6$, $x = $ ___?___.

Points E, F, G, and H have the coordinates shown.

23. Prove: $\overline{EF} \perp \overline{GH}$
24. Prove: $EF = GH$

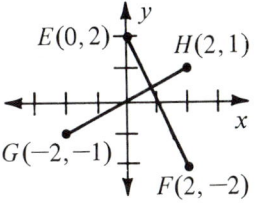

ANSWERS TO CHAPTER TESTS

Chapter 1

1. $\overleftrightarrow{PQ}, \overleftrightarrow{QP}, \overrightarrow{PR}, \overrightarrow{RP}, \overleftrightarrow{QS}, \overleftrightarrow{SQ}, \overleftrightarrow{RS}, \overleftrightarrow{SR}, \overrightarrow{PS}, \overrightarrow{SP},$ or $\overleftrightarrow{RQ}$ 2. $\overrightarrow{SR}$ or $\overrightarrow{SP}$ 3. 5; 10   4. 6
5. $\angle 2, \angle KTA, \angle ATG,$ or $\angle ATK$ 6. GAT or TAG 7. 180   8. 5   9. 70   10. obtuse
11. straight   12. right   13. acute   14. 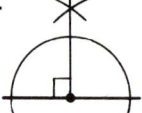   15. 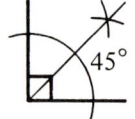   16. subtr.
17. div.   18. subst.   19. add.
20. $ARLJ$ 21. T 22. $ARLJ; AJNS$

Chapter 2

1. $\angle ECD$ 2. $\angle EAF$ and $\angle EAC$; $\angle DCA$ and $\angle DCG$; $\angle ECA$ and $\angle ECG$ 3. ECA
4. AED 5. Hyp.: $a = 1$; Conclu.: $a \cdot b = b$ 6. If $a \cdot b = b$, then $a = 1$.   7. no   8. $a \neq 1$
9. $x = 65; y = 64$ 10. $x = 55; y = 65$ 11. $x = 60; y = 50$
12. 1. Given   2. Vert. $\angle$s are =.   3. Subst. Post. (twice)   4. If 2 lines and a trans. form = alt. int. $\angle$s, then the lines are ∥.
13. 1. $\angle 1 = \angle 2$ (Given)   2. $\angle 2 = \angle 3$ (Vert. $\angle$s are =.)   3. $\angle 1 = \angle 3$ (Subst. Post.)   4. $\overline{AN} \parallel \overline{LB}$ (Given)   5. $\angle 1 = \angle 4$ (If 2 ∥ lines are cut by a trans., then alt. int. $\angle$s are =.)   6. $\angle 4 = \angle 3$ (Subst. Post.)   14.

Chapter 3

1. 78   2. 111   3. 45   4. right   5. neither (scalene)   6. N 7. RN 8. HL; SSS   9. NKP
10. 1. From algebra   2. *Given:* $\overrightarrow{AB}$ bisects $\angle XAY$.   3. Given   4. SAS Post.
11. 1. $\overleftrightarrow{VT} \parallel \overleftrightarrow{SR}; \overleftrightarrow{VR} \parallel \overleftrightarrow{TS}$ (Given)   2. $\angle 1 = \angle 2; \angle 3 = \angle 4$ (If 2 ∥ lines are cut by a trans., then alt. int. $\angle$s are =.)   3. $TR = RT$ (From algebra)   4. $\triangle RST \cong \triangle TVR$ (ASA Post.)

Chapter 4

1. 2. By construction   3. From algebra   4. SSS Post.   5. Corr. parts of $\cong$ $\triangle$s are =.
2. a. $\overrightarrow{MC}; \overrightarrow{MB}; \overleftrightarrow{DE}$ b. 90° (or $\angle DMJ$)
3. 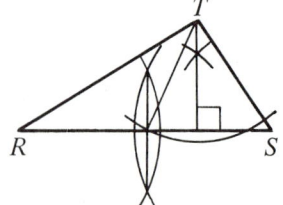   4. 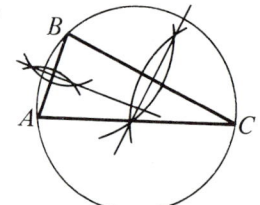   5. 70

6. 1. $WY = WZ$ (Given) 2. $\angle YWX = \angle ZWX$ (*Given:* $\overrightarrow{WX}$ bisects $\angle YWZ$.)
3. $WX = WX$ (From algebra) 4. $\triangle YWX \cong \triangle ZWX$ (SAS Post.)
5. $XY = XZ$ (Corr. parts of $\cong$ ⩟ are =.)

7. 1. $\angle 3 = \angle 1; \angle 4 = \angle 2$ (Vert. ⩟ are =.) 2. $\angle 3 = \angle 4$ (Given)
3. $\angle 1 = \angle 2$ (Subst. Post.—twice) 4. $AT = AV$ (If 2 ⩟ of a △ are =, then the sides opp. those ⩟ are =.)

Chapter 5

1. 90 cm **2.** 1800° **3.** 360° **4.** 120° **5.** 5 **6.** 7 **7.** 6 **8.** 110°
9. 70° **10.** 70° **11.** 26 **12.** 26 **13.** 13 **14.** 10 **15.** 27° **16.** 52°
17. 23 **18.** $R = 54°; \angle I = \angle M = 126°$ **19.** 10 **20.** 12 **21.** 8 **22.** rectangle
23. A ▱ is a quad. with both pairs of opp. sides ∥. **24.** Theorems 10–12, pages 214–215.

Chapter 6

1. c **2.** b **3.** j **4.** g **5.** a **6.** e **7.** f **8.** d **9.** h **10.** i **11.** 10
12. 5 m **13. a.** 34 **b.** 52 **c.** $\sqrt{185} \approx 13.6$ **14.** 36π **15.** 201 **16.** 20π
17. 40 **18.** 25 **19.** 130 **20.** 84 **21.** 6 cm **22.** 49 cm^2

Chapter 7

1. $\frac{3}{4}$ **2.** $\frac{6}{7}$ **3.** $\frac{3}{7}$ **4.** 2:3 **5.** $x:z$ **6.** 4:3:5 **7.** 1:25 **8.** 1:10 **9.** 1:20
10. $\frac{5}{9}$ **11.** $\frac{2}{1}$ **12.** $\frac{3}{5}$ **13.** $\frac{1}{1}$ **14.** $\frac{12}{7}$ **15.** $\frac{8}{9}$ **16.** 9 **17.** 9 **18.** 7 **19.** 3 cans
20. 4 cm; 1 cm:3.5 km **21.** 5.6 km **22.** 10.5 km **23.** (The length:width ratio of the drawing should be 3:2.)

Chapter 8

1. *LMNK* **2.** 3:2 **3.** 14 **4.** 6 **5.** 3:2 **6.** 9:4 **7.** $x = 12; y = 12$
8. $x = 25; y = 14$ **9.** $x = 20; y = 30$ **10.** $x = 7; y = 12$
11. 1. Given 2. Opp. sides of a ▱ are ∥. 3. If 2 ∥ lines are cut by a trans., then alt. int. ⩟ are =. 4. AA Post. 5. Corr. sides of ∼ ⩟ are in proportion.

Chapter 9

1. $\overleftrightarrow{VZ}$ **2.** $\overline{OX}$ **3.** $\overline{YZ}$ **4.** $\overleftrightarrow{YZ}$ **5.** 2 **6.** 2 **7.** internally **8.** tangent
9. 140 **10.** 210 **11.** 36 **12.** 38 **13.** 13 **14.** 100 **15.** 9 **16.** 10
17. 40 **18.** 172 **19.** 80 **20.** 50 **21.** 35 **22.** 8 **23.** See art at right.

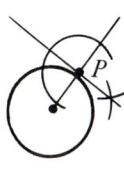

Exercise 23

Chapter 10

1. True **2.** False **3.** False **4.** True **5.** 48 **6.** 112 **7.** 208 **8.** 192
9. 8π **10.** 40π **11.** 72π **12.** 80π **13.** 144 **14.** 240 **15.** 384 **16.** 384
17. 16π **18.** 136π **19.** 200π **20.** 320π **21.** 256π **22.** 113 **23.** $25:36$
24. 1296 cm^3

Chapter 11

1. $x = 6; y = 2\sqrt{41}$ **2.** $x = 6; y = 6\sqrt{3}$ **3.** $x = 6\sqrt{2}; y = 6\sqrt{3}$ **4.** $16\sqrt{3}$
5. 36 **6.** 120 **7.** 64 **8.** 7 cm **9.** 10 cm **10.** 20π cm **11.** 260 **12.** 12
13. 400 **14.** (Drawing omitted) **15.** 12 **16.** 432π **17.** 1.5 **18.** 4.2 **19.** 2.5

Chapter 12

1. II **2.** -3 **3.** 2 **4.** $(-3, -2)$ **5.** 6 **6.** 2 **7.** 5 **8.** 10 **9.** $(4, 4)$
10. $(-1, -3)$ **11.** $(1.5, 2)$ **12.** $(-1, -3)$ **13.** -2 **14.** $\frac{1}{3}$ **15.** not defined
16. 0 **17.** rise **18.** $\frac{5}{2}$ **19.** $-\frac{2}{5}$ **20.** no **21.** -6 **22.** 8
23. 1. Slope of $\overline{EF} = \frac{2-(-2)}{0-2} = -2$; slope of $\overline{GH} = \frac{1-(-1)}{2-(-2)} = \frac{1}{2}$. (Def. of slope)
 2. $(-2)(\frac{1}{2}) = -1$ (From algebra) 3. $\overline{EF} \perp \overline{GH}$ (If the product of the slopes of two lines is -1, then the lines are $\perp$.)
24. 1. $(EF)^2 = 2^2 + 4^2 = 20$; $(GH)^2 = 4^2 + 2^2 = 20$ (Pythagorean Thm.)
 2. $EF = \sqrt{20}; GH = \sqrt{20}$ (From algebra) 3. $EF = GH$ (Subst. Post.)

ASSIGNMENT GUIDE

The following Guide to Individualized Assignments provides suggested assignments for an average course and a comprehensive course. The average course includes all topics that are considered essential and any concepts which are prerequisites for the lessons to follow. The comprehensive course includes all of the topics presented in the text. You will find that the assignments for the comprehensive course provide somewhat less drill than do the assignments for the average course.

Summary Time Schedule for the Assignments

Chapter	1	2	3	4	5	6	7	8	9	10	11	12	Total
Average Course	16	13	15	15	15	17	8	11	15	14	10	11	160
Comprehensive Course	15	15	13	14	14	17	8	10	13	13	14	14	160

AVERAGE COURSE / COMPREHENSIVE COURSE

Day	Pages	Assignment	Pages	Assignment	Day
Chapter 1			**Chapter 1**		
1	5	1–3	5	1–3	1
2	8–9	1–20	8–9	1–20	2
3	9	21–29	9	21–30	3
4	13	1–12	13, 14	1–22; Self-Test	4
5	13, 14	13–22; Self-Test	16–17	1–20	5
6	16–17	1–16	19	1–17	6
7	19	1–16	21	1–13	7
8	21	1–13	22, 24	Experiments; Self-Test Applications	8
9	22, 24	Experiments; Self-Test Applications	27–28	1–15	9
10	27–28	1–15	28	16–24	10
11	31–32	1–14	31–32	1–17	11
12	36	1–17	36–37	1–22	12
13	36–37, 37	18–22; Self-Test	37	23; Self-Test	13

AVERAGE COURSE

Day	Pages	Assignment
14	38 42	Skills Review Mixed Review 1–30
15	39–40 43	Chapter Review Exercises Mixed Review 1–18
16	41 (T8)	Chapter Test (Chapter 1 Test)

Chapter 2

Day	Pages	Assignment
17	48–49	1–18
18	52	1–20
19	53 55	21–24; Self-Test Applications
20	62–64	1–6
21	64–65	7–12
22	69–70	1–25 odd
23	74–75	1–24 odd
24	79	1–10
25	80 81	Self-Test Reading Geometry
26	84	Skills Review
27	85–86	Chapter Review Exercises
28	87 (T9)	Chapter Test (Chapter 2 Test)
29	88	Cumulative Review

Chapter 3

Day	Pages	Assignment
30	94–96	1–27 odd
31	99	1–14
32	100 101	15–19 Self-Test

COMPREHENSIVE COURSE

Pages	Assignment	Day
39–40	Chapter Review Exercises	14
41 (T8)	Chapter Test (Chapter 1 Test)	15

Chapter 2

Pages	Assignment	Day
48–49	1–22	16
52–53	1–24	17
53 55	25–28; Self-Test Applications	18
58	1–11	19
62–65	1–6; 7–11 odd	20
65	13–16	21
69–70	1–25	22
70–71	26–32	23
74–75	1–24	24
75–76	25–30	25
79	1–10	26
79, 80 80 81	11–14 Self-Test Reading Geometry	27
85–86	Chapter Review Exercises	28
87 (T9)	Chapter Test (Chapter 2 Test)	29
88	Cumulative Review	30

Chapter 3

Pages	Assignment	Day
94–96	1–29 odd; 30	31
99	1–14	32

Assignment Guide

	AVERAGE COURSE		COMPREHENSIVE COURSE		
Day	Pages	Assignment	Pages	Assignment	Day
33	104–106	1–12	100 101	15–20 Self-Test	**33**
34	109–110	1–10	104–106	1–21 odd	**34**
35	110–111 112 114	11–15 Self-Test Applications	109–111	1–17	**35**
36	117–118	1–13	111 112 114	18–20 Self-Test Applications	**36**
37	118–119 120	14–17 Experiments	117–118	1–15	**37**
38	122–124	1–9	119 120	16–20 Experiments	**38**
39	124 125	10–11 Self-Test	122–124	1–9	**39**
40	128–130	1–21 odd	124 125	10–15 Self-Test	**40**
41	132 133	Experiments; Self-Test Reading Geometry	128–131	1–30	**41**
42	136 140–141	Skills Review Mixed Review 1–13	132 133 137–138	Experiments; Self-Test Reading Geometry Chapter Review Exercises	**42**
43	137–138 141	Chapter Review Exercises Mixed Review 14–28	139 (T10)	Chapter Test (Chapter 3 Test)	**43**
44	139 (T10)	Chapter Test (Chapter 3 Test)	**Chapter 4**		
			146–148	1–6; 8–22 even	**44**
Chapter 4					
45	146–147	1–14	150–151	1–11	**45**
46	150	1–6	152 153	12–16; Self-Test Reading Geometry	**46**
47	151 152	7–11 Self-Test	156	1–11	**47**
48	156	1–11	156–157	12–17	**48**
49	156–157	12–15	160–161	1–15	**49**

Assignment Guide **T25**

	AVERAGE COURSE		COMPREHENSIVE COURSE		
Day	Pages	Assignment	Pages	Assignment	Day
50	160	1–10	161 162	16–18 Experiments	**50**
51	153 162	Reading Geometry Experiments	165–166 166	1–15 Self-Test	**51**
52	165–166 166	1–10 Self-Test	169–171	1–19 odd; 20–22	**52**
53	169–170	1–20 even	175–177	1–17	**53**
54	175–177	1–14	177 178	18–23 Experiments; Self-Test	**54**
55	178 179	Experiments; Self-Test Applications	179 183–184	Applications Chapter Review Exercises	**55**
56	182	Skills Review	185 (T11)	Chapter Test (Chapter 4 Test)	**56**
57	183–184	Chapter Review Exercises	186	Cumulative Review	**57**

Chapter 5 (Comprehensive)

	AVERAGE COURSE		COMPREHENSIVE COURSE		
58	185 (T11)	Chapter Test (Chapter 4 Test)	192–193 193	1–31 odd Experiments	**58**
59	186	Cumulative Review	196–197 197	1–22 Self-Test	**59**

Chapter 5 (Average)

	AVERAGE COURSE		COMPREHENSIVE COURSE		
60	192	1–19	200–202	1–23 odd	**60**
61	193 195–196	Experiments 1–11	205–206	1–25 odd; 27–32	**61**
62	196–197 197	12–18 Self-Test	206–207 207 208	33–36 Self-Test Applications	**62**
63	200–201	1–15	211–212	1–10; 11–23 odd	**63**
64	205	1–29 odd	213	25–28	**64**
65	206–207 207 208	30–32 Self-Test Applications	216–218	1–12	**65**
66	211–212	1–10; 12–24 even	219	13, 15, 17; Self-Test	**66**
67	216–217	1–6	223–224	1–21 odd; 22–26	**67**
68	218 219	8–12 even Self-Test	226–227	1–21	**68**

AVERAGE COURSE

Day	Pages	Assignment
69	223	1–15
70	226–227	1–21 odd
71	228 229	Self-Test Reading Geometry
72	230 234	Skills Review Mixed Review 1–17
73	231–232 235	Chapter Review Exercises Mixed Review 18–29
74	233 (T12)	Chapter Test (Chapter 5 Test)

Chapter 6

Day	Pages	Assignment
75	241	2–24 even
76	241–242	25–31
77	245	1–9
78	245 246	10–13 Self-Test
79	249	1–14
80	252–253	1–16
81	253 254	17–24 Self-Test
82	257	1–12
83	261 262	1–19 odd Self-Test
84	266	1–23 odd
85	266–267	24–36 even
86	269–270	1–21 odd
87	270 271	22–26 even Self-Test

COMPREHENSIVE COURSE

Pages	Assignment	Day
227–228 228 229	19–25 Self-Test Reading Geometry	**69**
231–232	Chapter Review Exercises	**70**
233 (T12)	Chapter Test (Chapter 5 Test)	**71**

Chapter 6

Pages	Assignment	Day
241–242	1–31 odd	**72**
242	32–35	**73**
245	1–13	**74**
245 246	14–15 Self-Test	**75**
249	1–15	**76**
252–253	1–7; 8–24 even	**77**
253–254 254	25–27 Self-Test	**78**
257–258	2–28 even	**79**
261 262	2–24 even Self-Test	**80**
266	1–35 odd	**81**
267	36–40	**82**
269–270	1–27	**83**
270–271 271	28–33 Self-Test	**84**
273	Applications	**85**
279–280	Chapter Review Exercises	**86**
281 (T13)	Chapter Test (Chapter 6 Test)	**87**

AVERAGE COURSE

Day	Pages	Assignment
88	273 278	Applications Skills Review
89	279–280	Chapter Review Exercises
90	281 (T13)	Chapter Test (Chapter 6 Test)
91	282	Cumulative Review

Chapter 7

Day	Pages	Assignment
92	287–288	1–31 odd
93	291	2–28 even
94	294 295	1–8 Self-Test
95	298	1–12
96	299–300 301 303	1–7 odd Self-Test Applications
97	304 308	Skills Review Mixed Review 1–11
98	305–306 309	Chapter Review Exercises Mixed Review 12–29
99	307 (T14)	Chapter Test (Chapter 7 Test)

Chapter 8

Day	Pages	Assignment
100	314–315	1–16
101	319–320	1–15 odd
102	321 322	16–21 Self-Test
103	323–324	1–9
104	327	1–14
105	331	1–13

COMPREHENSIVE COURSE

Pages	Assignment	Day
282	Cumulative Review	88

Chapter 7

Pages	Assignment	Day
287–288	2–40 even	89
291–292	2–38 even	90
294–295 295	1–11 Self-Test	91
298	1–17	92
299–301	1–9 odd	93
301 303	10–11; Self-Test Applications	94
305–306	Chapter Review Exercises	95
307 (T14)	Chapter Test (Chapter 7 Test)	96

Chapter 8

Pages	Assignment	Day
314–315	1–16	97
315–316	17–26	98
319–321 322	2–24 even Self-Test	99
323–324	1–12	100
327–328	1–23 odd	101
331–332	2–22 even	102
333 335	Experiments; Self-Test Applications	103
339–340	Chapter Review Exercises	104
341 (T15)	Chapter Test (Chapter 8 Test)	105

AVERAGE COURSE

Day	Pages	Assignment
106	331–332 333	14–20 even Experiments; Self-Test
107	335 338	Applications Skills Review
108	339–340	Chapter Review Exercises
109	341 (T15)	Chapter Test (Chapter 8 Test)
110	342	Cumulative Review

Chapter 9

Day	Pages	Assignment
111	347–348	1–18
112	348 349	19–26 Experiments
113	353	1–14
114	357–358 358	2–16 even Self-Test
115	362	1–16
116	363	17–23
117	366	1–14
118	366–367 368	15–23 Self-Test
119	371	1–12
120	372	13–20
121	374–375 376	1–12 Self-Test
122	377 380	Applications Skills Review
123	384–385	Mixed Review

COMPREHENSIVE COURSE

Pages	Assignment	Day
342	Cumulative Review	106

Chapter 9

Pages	Assignment	Day
347–348	1–26	107
348 349	27–32 Experiments	108
353–354	1–25 odd	109
357–358 358	2–26 even Self-Test	110
362–363	2–30 even	111
366–367	1–29 odd	112
367 368	30–33 Self-Test	113
371–372	2–24 even	114
374–375	2–12 even; 13–16	115
375–376 376	17–20 Self-Test	116
377	Applications	117
381–382	Chapter Review Exercises	118
383 (T16)	Chapter Test (Chapter 9 Test)	119

Chapter 10

Pages	Assignment	Day
390–391 392	1–35 odd Experiments	120
396	1–18	121
397	19–27	122
399–400	1–16	123

AVERAGE COURSE

Day	Pages	Assignment
124	381–382	Chapter Review Exercises
125	383 (T16)	Chapter Test (Chapter 9 Test)

Chapter 10

Day	Pages	Assignment
126	390–391	1–24
127	391 392	25–29 Experiments
128	396	1–14
129	396–397	15–23
130	399–400	1–12
131	400 401	17–20 Self-Test
132	404	1–14
133	404 405	15–22 Experiments
134	407–408 409	1–12 Experiments
135	411 412	1–8 Self-Test
136	413 417 420	Applications Reading Geometry Skills Review
137	422	Chapter Review Exercises
138	423 (T17)	Chapter Test (Chapter 10 Test)
139	424	Cumulative Review

Chapter 11

Day	Pages	Assignment
140	429–430 431	2–18 even Applications
141	434–435	1–25 odd

COMPREHENSIVE COURSE

Pages	Assignment	Day
400; 401	17–21; Self-Test	124
404 405	1–20; 21–25 odd Experiments	125
407–408	1–17	126
408 409	18–20 Experiments	127
411–412 412; 413	1–15 odd; 17–18 Self-Test; Applications	128
415–416	1–17 odd	129
417 422	Reading Geometry Chapter Review Exercises	130
423 (T17)	Chapter Test (Chapter 10 Test)	131
424	Cumulative Review	132

Chapter 11

Pages	Assignment	Day
429–430	2–18 even; 19–28	133
430 431	29–36 Applications	134
434–435	1–13 odd; 14–25	135
435–436	26–36	136
437–438	1–9 odd; 10–25	137
438–439 439	26–38 Self-Test	138
442	1–13	139
445	1–15	140
445–446 446	16–23 Self-Test	141

AVERAGE COURSE

Day	Pages	Assignment
142	437–438 439	1–19 odd Self-Test
143	442	1–6
144	445 446	1–12 Self-Test
145	449–450	1–13
146	454–455 456	1–31 odd Self-Test
147	458–459 460	Applications Skills Review
148	461–462 464–465	Chapter Review Exercises Mixed Review
149	463 (T18)	Chapter Test (Chapter 11 Test)

Chapter 12

Day	Pages	Assignment
150	470–471	1–22
151	473	1–12
152	479	1–8; 9–17 odd
153	480 484–485	Self-Test Applications
154	489–490	1–13
155	493 495	1–12 Self-Test
156	499	1–16
157	502–503 504	1–3 Self-Test
158	510 511–512	Skills Review Chapter Review Exercises
159	513 (T19)	Chapter Test (Chapter 12 Test)
160	514	Cumulative Review

COMPREHENSIVE COURSE

Pages	Assignment	Day
449–451	1–21 odd; 22–24	142
454–455	1–39 odd	143
456 458–459	Self-Test Applications	144
461–462	Chapter Review Exercises	145
463 (T18)	Chapter Test (Chapter 11 Test)	146

Chapter 12

Pages	Assignment	Day
470–471	1–22	147
471	23–32	148
473–474	2–24 even	149
479–480 480	2–26 even Self-Test	150
484–485	Applications	151
489–490	1–21 odd; 22–23	152
493–494	1–17 odd	153
494–495 495	19–22 Self-Test	154
499	1–16	155
499–500	17–27	156
502–504 504	1–8 Self-Test	157
505 511–512	Applications Chapter Review Exercises	158
513 (T19)	Chapter Test (Chapter 12 Test)	159
514	Cumulative Review	160

SUPPLEMENTARY MATERIALS GUIDE

The chart below correlates the supplementary materials with the sections and chapters in the student textbook. In addition to tests and practice exercises, the Resource Book contains arithmetic and algebra reviews, mixed reviews, and help with proof worksheets. The front of the Resource Book contains descriptions and correlations of this material.

For use after Section	Tests	Practice Masters	Resource Book		For use after Section	Tests	Practice Masters	Resource Book	
			Tests	Practice Exercises				Tests	Practice Exercises
1–2		Sheet 1			4–1		Sheet 21		
1–3	Test 1				4–2	Test 11			p. 87
1–4		Sheet 2		p. 75	4–3		Sheet 22		
1–6	Test 2	Sheet 3			4–5	Test 12	Sheet 23		p. 88
1–7		Sheet 4		p. 76	4–7	Test 13	Sheet 24		p. 89
1–9	Test 3	Sheet 5		p. 77	Ch. 4		Sheet 25	Tests 8, 9	
Ch. 1		Sheet 6	Tests 1, 2		Cum. Rev. Ch. 3–4		Sheet 26		
2–1		Sheet 7			5–2	Test 14	Sheet 27		p. 90
2–2	Test 4			p. 78	5–4	Test 15	Sheet 28		p. 91
2–3		Sheet 8			5–5		Sheet 29		
2–4		Sheet 9		p. 79	5–6	Test 16	Sheet 30		p. 92
2–5		Sheet 10			5–8	Test 17	Sheet 31		p. 93
2–7	Test 5	Sheet 11		p. 80	Ch. 5		Sheet 32	Tests 10, 11	
Ch. 2		Sheet 12	Tests 3, 4		6–2	Test 18	Sheet 33		p. 94
Cum. Rev. Ch. 1–2		Sheet 13			6–3		Sheet 34		
3–1				p. 81	6–4	Test 19	Sheet 35		p. 95
3–2	Test 6	Sheet 14			6–6	Test 20	Sheet 36		p. 96
3–3				p. 82	6–8	Test 21	Sheet 37		p. 97
3–4	Test 7	Sheet 15			Ch. 6		Sheet 38	Tests 12, 13	
3–5		Sheet 16		p. 83	Cum. Rev. Ch. 5–6				
3–6	Test 8	Sheet 17			Ch. 4–6	Test 22	Sheet 39		
3–7	Test 9	Sheet 18		p. 84					pp. 98–99
Ch. 3		Sheet 19	Tests 5, 6		Ch. 1–6		Sheet 40	Test 14	pp. 100–102
Cum. Rev. Ch. 1–3	Test 10	Sheet 20	Test 7	pp. 85–86					

T32   Supplementary Materials Guide

For use after Section	Tests	Practice Masters	Resource Book		For use after Section	Tests	Practice Masters	Resource Book	
			Tests	Practice Exercises				Tests	Practice Exercises
7–1		Sheet 41			10–5		Sheet 64		
7–2		Sheet 42		p. 103	10–6	Test 32			p. 116
7–3	Test 23	Sheet 43			10–7		Sheet 65		
7–5	Test 24	Sheet 44		p. 104	Ch. 10		Sheet 66	Tests 22, 23	
Ch. 7		Sheet 45	Tests 15, 16						
					Cum. Rev. Ch. 9–10		Sheet 67		
8–1		Sheet 46		p. 105					
8–2	Test 25	Sheet 47		p. 106	11–1		Sheet 68		
8–3		Sheet 48			11–3	Test 33	Sheet 69		p. 117
8–4		Sheet 49		p. 107	11–5	Test 34	Sheet 70		p. 118
8–5	Test 26	Sheet 50		p. 108	11–7	Test 35	Sheet 71		p. 119
Ch. 8		Sheet 51	Tests 17,18		Ch. 11		Sheet 72	Tests 24, 25	
Cum. Rev. Ch. 7–8		Sheet 52			12–1				p. 120
					12–2		Sheet 73		
9–2		Sheet 53		p. 109	12–3	Test 36	Sheet 74		p. 121
9–3	Test 27	Sheet 54			12–5	Test 37	Sheet 75		p. 122
9–4		Sheet 55		p. 110	12–6		Sheet 76		
9–5	Test 28	Sheet 56		p. 111	12–7	Test 38	Sheet 77		p. 123
9–6		Sheet 57			Ch. 12		Sheet 78	Tests 26, 27	
9–7	Test 29	Sheet 58		p. 112					
Ch. 9		Sheet 59	Tests 19, 20		Cum. Rev. Ch. 11–12 Ch. 10–12	Test 39	Sheet 79		pp. 124–125
Cum. Rev. Ch. 7–9	Test 30	Sheet 60	Test 21	pp. 113–114	Ch. 7–12		Sheet 80	Test 28	pp. 126–128
10–2		Sheet 61			Ch. 1–12			Test 29	
10–3	Test 31	Sheet 62		p. 115					
10–4		Sheet 63							

Supplementary Materials Guide **T33**

READING GEOMETRY
Suggestions for Helping Your Students Learn How to Read a Geometry Textbook

The question is sometimes asked, "Why do we need to teach reading in a geometry class?" A study of reading geometry reveals that it is integrally tied to the learning of geometry. Difficulties that students encounter in geometry are often the result of difficulties in reading. To improve reading is to increase the chance of making the student a more independent learner since reading is a learning-to-learn skill. Once a student learns to read proficiently, doors are opened that were previously closed.

When reading geometry, a student has to read symbols, tables, and graphs as well as words, and to understand how these representations of meaning are related. Without adequate preparation, frustration is likely to develop as students try to read quickly such expressions as

$$\text{T.A.} = 2\pi rh + 2\pi r^2$$

or

$$\left(\frac{x_1 + x_2}{2}, \frac{y_1 + y_2}{2}\right) = \left(\frac{1+5}{2}, \frac{2+8}{2}\right) = (3, 5).$$

Many familiar words—for example, *base*, *leg*, and *plane*—take on specialized meanings in geometry. These new meanings need to be pointed out to students, and the importance of learning definitions and using mathematical terms correctly should be emphasized.

Diagrams are of great value in mathematics—particularly in geometry. Students need to realize that diagrams are essential parts of the discussion, the examples, and the proofs and that the diagrams must be read along with the words of the text. Identifying and interpreting the information provided by diagrams are skills that have to be learned.

All the learning activities that have been mentioned are essentially reading activities. Clearly, successful reading calls for practice and patience on the part of both the teacher and students. Consistent work on reading that is integrated into the content of the course seems to be the method most likely to lead to success. This textbook has been written with special concern for students' reading difficulties. It uses language that is as simple as possible without sacrifice of correctness. Long paragraphs of exposition have been avoided; important concepts are highlighted in various ways; and much information is presented in brief but highly visible form. In addition, special devices are used frequently to call students' attention to specific items in diagrams and text.

Six major reading objectives are stressed. These objectives are interrelated; it is hardly possible to teach one without the others. We encourage you to focus on these objectives throughout the year and to emphasize those that best meet the specific needs of your students.

Objective 1: Reading and Communicating Orally

This first objective is to ensure that students have a basic understanding of the relationship between what is spoken and what is written. A mathematics textbook contains many symbols, figures, charts, and graphs that are not commonly used in other books. It is not always possible to express their content in words. Furthermore, the interrelationship between symbols is often complex, and the method of verbalizing may be unclear. It is not uncommon for a set of symbols to be verbalized in several ways within one class lesson. For example, a^2 may be called the square of a, the second power of a, a squared, or a to the second power. Practice is necessary if the relationship between the spoken and the written is to be understood.

Some of your students have probably had very little experience with either oral or written communication of *ideas*. They may have great difficulty in verbalizing even the simpler concepts of geometry. It will be well worth while for students to spend classroom time in reading aloud and discussing the material in the early chapters of the textbook. The Classroom Practice exercises are useful for this purpose. You can help students by asking them to read sentences aloud, repeat, say in other words, summarize, and ask questions of classmates. They should be encouraged to use complete sentences in responding to questions.

Objective 2: Reading Silently

This second objective is to provide students with the necessary skills to be able to read silently, recognizing that one's purpose determines the speed and type of reading. When students read silently it is often beneficial if they do so with a purpose—to find the main idea, to look for a specific detail, to summarize, to answer a question, to study an example.

Different purposes demand different types of reading. When previewing or reviewing a lesson, skimming (rapid reading) is used. By contrast, slow reading is used when the student is trying to learn the main ideas and important details of a lesson.

Your students may need a good deal of help in developing efficient skimming techniques. You can help them by showing how the format of the textbook makes important words and ideas stand out. The learning goals listed at the beginning of each chapter can also be used as aids in skimming. You might want to go through a few chapters with the students, helping them locate the lesson or lessons related to each learning goal.

For slow reading, it is beneficial to students when the teacher provides questions in advance that will be discussed upon completion of the reading. When beginning a section, students may be directed first to skim the page to find new ideas and symbols, then to read carefully, and finally to answer questions or work exercises.

Objective 3: Using Symbols

This third objective is to assist students in recognizing symbols and in associating symbols with words and figures. As was suggested earlier, the verbalization of symbols creates a reading problem for many students. Other difficulties introduced by the use of symbols are their conciseness, the order in which they are read, and their association with concepts not actually stated. This example will demonstrate all three.

To find the length of a, use the Pythagorean Theorem.

$$a = \sqrt{9^2 - 5^2}$$
$$a = \sqrt{81 - 25}$$
$$a = \sqrt{56} = 2\sqrt{14}$$

One way to recognize the conciseness of the symbols above is to write out in words the ideas that appear in symbol form. This becomes a lengthy task. While the compactness demands slower reading and greater concentration, it has the value of showing related structures of ideas. In the example above, the symbols allow us to perceive easily the three sentences together. When we read this example, our eyes do not move from left to right in an orderly fashion. We refer to the figure and move our eyes up and down to relate the three sentences. Words and figures need to be associated with the symbols. In the example above, "length of side a" is associated with a in each sentence and on the figure. Also, "Pythagorean Theorem" is associated with the sentences and the figure.

Objective 4: Using Mathematical Words

This fourth objective is to stress the mathematical meaning of words. A mathematics page contains words of everyday language (*and, when, the*), words that are unique to mathematics (*hypotenuse, rhombus*), and familiar words with specialized meanings in mathematics (*leg, line, table*). The most commonly used words are primarily *structure* words, whereas most of the *content* words are mathematical. This implies that the meaning of the content words needs to be taught in the mathematics classroom.

If your students shy away from geometric terms because they look hard, you might point out that once they have been learned, they are no harder than any other words. Use of correct terminology makes the discussion of geometric concepts easier, not harder. You can help students find clues for words within the words themselves (*coplanar, collinear*), and from content clues on the page, such as charts, figures, and symbols. Understanding definitions of terms is of primary importance. Use of the glossary should be encouraged. In classroom work, it is good practice to emphasize correct spelling and correct pronunciation.

Objective 5: Reading Diagrams, Charts, and Figures

This fifth objective is to make sure that students can relate the reading of diagrams, charts, and figures to the rest of the exposition and exercises. Diagrams, charts, and figures play an integral role in the development of a lesson. Many students fail to use these to their advantage. Through oral reading and questioning you can assist students in relating these visual aids to the words and symbols on the page. The labels, in *red* type within *red* boxes, associated with many of the diagrams have been provided to give students extra help in interpreting the figures.

To help students understand figures for proofs, you may sometimes ask them to draw a diagram from the statement of a theorem or an exercise before looking at the given diagram—or you and your students could discuss the "Given" items and draw a diagram together before beginning the proof.

Objective 6: Understanding the Organization of Proofs

This sixth objective is to assist students to understand the organization of proofs. To read a direct proof of a theorem, students need to see the organization as containing at least these components: a statement of the theorem, a figure, a "given", a "prove", and a proof with statements and reasons listed in logical order. The key to reading the proof is developing an interrelationship among these parts, which involves looking up and down and from side to side many times.

Although many theorems are presented without proof in this textbook, it is important for your students to develop an appreciation of this orderly method of reasoning. The partially proved theorems and exercises will help students gain skill in writing proofs in two-column form. Some of the exercises call for the writing of complete proofs. Indirect proof is presented as an Application. Your better students may enjoy experimenting with this method.

PROBLEM SOLVING STRATEGIES
Some Strategies That Your Students Can Use to Become Better Problem Solvers

A problem solving strategy is simply a plan or technique for solving a problem. There are a number of well-known strategies that relate specifically to geometry. For example, the strategy on page 144 outlines a method for proving that two segments or two angles are equal, and the list on page 216 provides four ways of showing that a quadrilateral is a parallelogram. One of the goals of a geometry course is to familiarize students with such standard techniques and to give students enough practice with these techniques so that they can use them confidently and successfully to solve geometry problems.

These rather specific strategies are not the only ones that students can use in solving geometry problems, however. Other, more general, strategies, such as looking for a pattern or drawing a diagram, can be very effective problem solving tools. These general strategies can help not only with geometry problems, but also with problems in other branches of mathematics and in other subject areas. Since these general strategies provide an approach to solving a problem, rather than a specific method of solution, they are particularly useful for attacking a problem when the method of solution is not obvious.

For example, suppose several geometry students are confronted with a word problem that they do not know how to solve. One student might ask, "Is this problem similar to any other types of problems I have seen before?" Another student might try to organize information in a table or a chart or a diagram. Still another might guess an answer and by checking it with the words of the problem discover a general method of solution. Each of these approaches can be a useful strategy.

The basic five-step guide to problem solving given below can provide students with a general plan of attack that can be used throughout the course:

A GUIDE FOR PROBLEM SOLVING

1. *Read the problem carefully. To do this, it may be necessary to read it more than once.*
2. *Decide what information is known and what information is unknown. Look for the connection between them.*
3. *Devise a plan to solve the problem or carry out the proof.*
4. *Follow the plan step by step.*
5. *Always remember to check your answer.*

For students whose reading skills are limited, reading a problem *with understanding* is in itself a strategy that may need to be specifically taught. Similarly, students may need frequent reminders to check *every* answer for reasonableness in terms of the given conditions.

The Problem Solving Strategies sections in even chapters cover the following topics: *asking questions, diagrams and charts, working backwards, auxiliary lines, counterexamples,* and *patterns*. Opportunities for developing and using these strategies will arise as students deal with various types of word problems and proofs throughout the course. The general problem solving strategies and skills in the following list can also be used to supplement those mentioned above.

- use correct mathematical notation to clarify the situation
- recognize the problem as a standard type
- apply a standard formula to the given information
- write and solve an equation that describes the problem
- make use of known theorems, postulates, corollaries and definitions
- solve a simpler related problem first
- break the problem down into several smaller problems
- use trial and error and the process of elimination
- use geometric models to visualize the problem
- change your perspective by rotating a diagram
- consider alternate ways to do the proof or construction
- recognize the possibility of no solution

You can help your students see cases where one or more of these strategies can be used to good advantage. In this way they can become better problem solvers and may also grow to enjoy problem solving more.

ERROR ANALYSIS
Anticipating Common Errors and Helping Students Avoid Them

There are many different kinds of student errors in geometry, and reasons for making them. Certain types of errors are common enough to be predictable. If you are aware of these, you can help students avoid them. Many of the more common errors fall into one of the following categories.

Formulas

Students may use "formulas" that are simply not correct. Examples include $C = \pi r$ and $A = \pi r$. Another problem occurs when the student chooses the correct formula, but applies it incorrectly. This might involve using the formula $A = \frac{1}{2}bh$ for the area of a triangle (correct), but substituting the lengths of the sides of a triangle *without a right angle* for b and h (incorrect). Likewise, watch out for students who forget that c must be the length of the *hypotenuse* in the formula $c^2 = a^2 + b^2$ for the Pythagorean Theorem.

Definitions

Definitions must be *memorized*, even though students sometimes resist memorizing. The result of such resistance can be the improper use of formulas. For example, some students may apply a formula to the wrong type of figure. They will indiscriminately use the formula $A = bh$ for figures such as trapezoids, failing to realize that trapezoids are not parallelograms.

A more serious difficulty arises when students who have not learned definitions are then not able to apply subsequent theorems. If a student does not know the difference between a rectangle and a rhombus, faulty application of theorems about angles and diagonals may occur. Students who don't know the difference between secants, chords, and tangents will not know whether to add or subtract when calculating the measures of associated angles.

Confusion among alternate interior, corresponding, and same-side interior angles will result in mistaking equal angles for supplementary angles and vice versa. Those who do not learn the correct definitions of complementary and supplementary angles will be unsure whether to use 90° or 180° in solving related problems.

Notation

Geometry has its own special symbols and notation conventions. Students have probably not encountered symbols such as $\perp$, $\parallel$, and $\angle$ before, and need to become comfortable with them. The more subtle differentiation among AB, $\overline{AB}$, $\overleftrightarrow{AB}$, and $\overrightarrow{AB}$ may escape students at first. The proper meanings must be reinforced so that, for example, students don't have lines in mind when segments are intended.

A perennial problem in mathematics, including geometry, is the use of the same symbol, or kind of notation, for several different objects or entities. In algebra we must deal with many uses of the same symbol, for example "−". It can be a minus sign, an opposite sign, or a negative sign. Similarly, in geometry we can find a single capital letter used to refer to a number of different things. We write "A" to refer to a point, a plane, an angle, a variable, and so on. Students must develop the ability to obtain the correct meaning from the *context* of a given situation.

Congruent and Similar Figures

Students may approach problems too casually to notice that the way corresponding parts are paired does make a difference. There are many proportions that can be written for the sides of two similar triangles, but care must be taken to match corresponding parts.

In the diagram, the proportions $\frac{5}{x} = \frac{7}{x}$ and $\frac{5}{y} = \frac{7}{z}$, which may seem as if they should be true, are not. The true proportions are $\frac{5}{x} = \frac{x}{7}$ and $\frac{5}{y} = \frac{x}{z}$. Emphasize the need to identify corresponding sides, which are opposite the equal angles in the triangles.

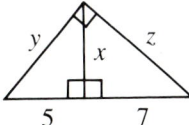

Another very common error is to assume that the areas and volumes of similar figures are in the same ratio as the sides. Students usually recognize the error of this idea readily when given numerical examples.

Coordinate Geometry

Errors in coordinate geometry often involve *inattention to order*. Students may accidentally reverse x- and y-coordinates, either plotting points incorrectly or stating coordinates of given points incorrectly. When computing slopes students sometimes subtract in reverse order, resulting in a slope that is $\frac{y_1 - y_2}{x_2 - x_1}$ or $\frac{y_2 - y_1}{x_1 - x_2}$ rather than $\frac{y_2 - y_1}{x_2 - x_1}$. A more serious error is the reversal of rise and run, such that $\frac{\text{run}}{\text{rise}}$ is given, which is, of course, the reciprocal of the correct slope.

With midpoints there is a tendency to subtract, yielding $\left(\frac{x_2 - x_1}{2}, \frac{y_2 - y_1}{2} \right)$ as the midpoint of a segment with endpoints (x_1, y_1) and (x_2, y_2). This is the result of a thought process in which students take half the difference of two numbers and add this to the smaller number. Applied correctly, this would produce a correct result. Unfortunately the usual effect is simply to replace the "+" in the midpoint formula with a "−" as mentioned above. Numerical examples should make it clear that $\frac{a+b}{2}$ is the number that is halfway between a and b, no matter what the values of a and b.

Error Analysis **T41**

Faulty Logic

A geometry course is often the student's first contact with the notion that logic is a subject with its own formal structure. Piaget's studies suggest that abstract, formal logic becomes comprehensible at an advanced cognitive level, probably not achieved until a relatively advanced age, if at all. Thus we can expect many students will have only the haziest idea of what is "logical" when they enter a geometry course. Two common errors of logic are highlighted below.

In writing a proof, it is not unusual for students, without realizing it, to use the proposition being proved in the proof. If the teacher focuses on the fact that this is what the student has done, the error in logic is easily apparent and readily corrected.

The other logical error that is fairly common and can be clearly categorized is the converse error. Here students reason from "If p, then q" and "q" to conclude "p". For example, they may reason that the theorem stating that the diagonals of a rectangle are equal implies that if $AC = BD$, then $ABCD$ must be a rectangle. This reasoning is "backwards" and not logical. Venn diagrams and specific counterexamples should help convince students that the converse is not equivalent to the given proposition.

LESSON COMMENTARY

Chapter 1. Points, Lines, and Angles

Many students beginning geometry look to the course as a fresh change from algebra. The classroom experiments and constructions in Chapter 1 and throughout the book will help to create a new outlook on mathematics. This does not mean, however, that students may forget algebra during their course in geometry. Algebraic skills are needed to solve many of the exercises in this text. We hope that teachers will frequently assign such exercises.

You will probably want to move students quickly through the first four or five sections of this chapter. The first three sections in particular should not be allowed to drag. The ideas involved are already familiar to many students. Even students who have had no previous experience with the ideas or the notation usually do not find them difficult.

Objectives for Pages 4–13

1. To become acquainted with the types of questions involved in the study of geometry.
2. To name points, lines, rays, and segments.
3. To name angles and use a protractor to measure angles.

1 Beginning Your Study of Geometry *(pp. 4–5)*

Teaching These Pages

We hope the optical illusions that open this text will provide an entertaining "curtain-raiser" for students. They also illustrate the point that one can't always depend on appearances. Sometimes, measurements must be made. Other illusions are presented in Exercise 1 on page 4 and in Chalkboard Examples 1 and 2 on page T44.

In Experiment 1, many students may not know the term *rhombus*. Some may call the figure a "diamond." Try not to let vocabulary problems become a stumbling block. The important thing is for students to see that the sides of the figure inside the rectangle all have the same length. Let students test their conjectures by measuring the sides and angles of the inner figures. Then point out two things: (1) no measurement is completely accurate; (2) it would be impossible to test every rectangle since there are an unlimited number of them. With this in mind, students should appreciate that, later in the course, they will be able to prove without measuring that the figure formed by joining the midpoints of the (consecutive) sides of any rectangle is a rhombus. (See Suggested Extension 2 on page T89.)

Although the results of Experiment 1 may not be surprising, Experiment 2 completely astounds many people. Exercise 3 on page 5 suggests that the space between the hoop and the object will be the same no matter what size object is used. You might do this exercise in class and have students repeat it with different-sized containers as homework. A proof of this surprising result is developed in Written Exercise 40 on page 267. The proof depends on the formula $C = 2\pi r$.

Chalkboard Example 3 below gives students an opportunity to piece together clues to arrive at a "theorem" of a nongeometric nature.

Be sure to remind students that they will need a protractor (for Section 3) and a straightedge and compass (for Section 7).

Chalkboard Examples

1. Are the labeled segments straight? yes

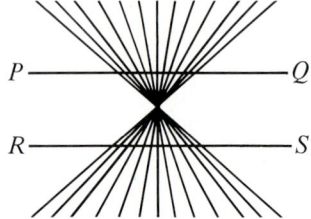

2. Are the labeled segments parts of the same line? yes

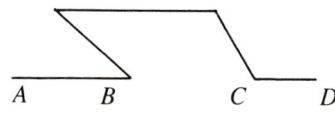

3. In a certain corporation, the positions of sales manager, vice-president, and treasurer are held by Aquino, Chung, and Berman, although not necessarily in that order. The sales manager, who was an only child, earns the least. Chung, who plays chess with Aquino's sister, earns more than the treasurer. What position does each person occupy?
   Chung is vice-president; Aquino is treasurer; Berman is sales manager.

2 Points and Lines (pp. 6–9)

Teaching These Pages

Some students will have been exposed to the basic ideas of this section in previous mathematics courses. You might ask whether anyone in the class can suggest ways of symbolizing line AB, ray AB, segment AB, and distance AB. The Classroom Practice Exercises will help to acquaint students with the notation $\overleftrightarrow{AB}$, $\overrightarrow{AB}$, $\overline{AB}$, and AB.

The introduction of the *midpoint* concept enables students to practice numerical skills while developing geometric intuition. The idea of a number line should be familiar to all students. However, practice with negative coordinates is still advisable. Classroom Practice Exercises 11–15 will help, as will Chalkboard Examples 8–13 on the following page.

T44   Lesson Commentary

Chalkboard Examples

Give the symbol for each:

1. Line PQ $\overleftrightarrow{PQ}$
2. Segment ST $\overline{ST}$
3. Ray ST $\overrightarrow{ST}$
4. Ray TS $\overrightarrow{TS}$
5. Are points P, S, and T collinear? no
6. Name three points which are collinear. *P, S, and Q; R, S, and T*
7. If PS = SQ, then S is called the __?__ of $\overline{PQ}$. midpoint

Use the number line below to find the following:

8. GL 5
9. EJ 5
10. AD 3
11. BL 10
12. The midpoint of $\overline{AK}$ F
13. The midpoint of $\overline{BH}$ E

Suggested Extension

Find the approximate position on the number line of points having these coordinates:

1. $2\frac{1}{2}$
2. $-\frac{3}{4}$
3. 3.9
4. -3.99
5. $\sqrt{2} \approx 1.414$

3 Angles (pp. 10–13)

Teaching These Pages

Familiarize students with the notation used for angles. Begin by asking them to suggest a notation for the first angle shown at the right. Students will probably suggest "∠ ABC" or "∠ B." Some may also suggest "∠ CBA."

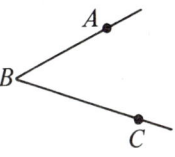

To illustrate that "∠ ABC" is sometimes preferable to "∠ B," draw the second diagram at the right. Here, the symbol "∠ B" is ambiguous. It could refer to ∠ ABC, but also to ∠ ABD or ∠ DBC.

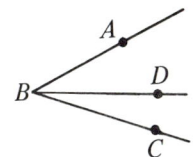

You might illustrate the use of the protractor by drawing angles and having students measure them with a protractor. This can be done on the chalkboard with a large protractor or on an overhead projector with a small transparent protractor. The footnote on page 10 describes our use of the notation "∠ 1 + ∠ 2 = ∠ AOC" instead of "m∠ 1 + m∠ 2 = m∠ AOC." We retain the use of the term "measure" in some situations: for example, "Find the measure of ∠ 3."

Lesson Commentary **T45**

Chalkboard Examples

1. Give other names for each angle. Students' answers may vary.
   - a. ∠RSQ   ∠3
   - b. ∠QTR   ∠8
   - c. ∠SUR   ∠RUS
   - d. ∠7   ∠TQU
   - e. ∠5   ∠QRT
   - f. ∠6   ∠RQS

2. ∠4 + ∠5 = ∠ ?   SRQ

3. ∠6 + ∠7 = ∠ ?   RQT

4. ∠2 + ∠3 = ?°   180

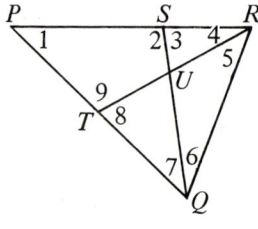

Quick Quiz

Classify as true or false.

1. $\vec{BC}$ is the same as $\vec{BD}$.   true
2. B is the midpoint of $\overline{AC}$.   true
3. $AD = 6$ false
4. There is only one point between B and D.   false

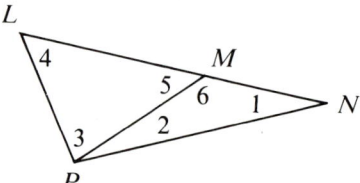

5. Name ∠4 with three letters.
   ∠PLM, ∠MLP, ∠NLP, or ∠PLN

6. Name three angles with vertex P.
   ∠2, ∠3, and ∠LPN

7. ∠ ? + ∠ ? = 180°   5; 6

8. If ∠3 = 80° and ∠LPN = 100°, then ∠2 = ?°.   20

Objectives for Pages 15–22

1. To classify angles as *acute*, *right*, *obtuse*, or *straight*.
2. To name and calculate the measures of vertical angles.
3. To use angle bisectors in calculating the measures of angles.

4 Classifying Angles (pp. 15–17)

Teaching These Pages

You may want to introduce the terms *right angle* and *perpendicular* first, since they are more familiar than *acute* and *obtuse*. One method is to ask students to estimate the measures of the angles formed by intersecting horizontal and vertical lines. In addition, be sure to give examples of right angles whose sides are not horizontal and vertical. Then present the special names given for angles smaller than right angles and angles larger than right angles.

T46   Lesson Commentary

To describe 180° angles, we have decided to use the term *straight angle*. Note that we use the notation ∠ B = ∠ C ("angle B equals angle C") to state that angles have equal measures. "Equal angles have equal measures" is an idea students find easy, so you need not belabor it.

Chalkboard Examples

Draw several angles on the board and have students classify them as *acute*, *right*, *obtuse*, or *straight* angles. You might also have them estimate the measure of each angle.

5 Vertical Angles *(pp. 18–19)*

Teaching These Pages

The figure at the top of page 18 illustrates the idea of vertical angles. To develop the relationship between vertical angles, let ∠ 1 = 30° and have students find the measures of ∠ 2, ∠ 3, and ∠ 4. You may want to repeat this exercise, letting ∠ 1 = 20°, ∠ 2 = 40°, and so on, recording your answers in a table like that shown on page 18. Have students make a conjecture about vertical angles (vertical angles are equal). You might challenge students to prove their conjecture, letting the class decide which arguments should be accepted as "proofs." Students may enjoy naming the result after the discoverer and prover. It appears as Theorem 3 in Chapter 2 (page 47).

Chalkboard Examples

Suppose $\overline{QT} \perp \overline{UR}$ and ∠ POQ = 35°.
Find the measure of each angle.

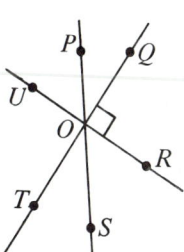

1. ∠ SOT   35°
2. ∠ POU   55°
3. ∠ ROS   55°
4. ∠ POR   125°
5. ∠ SOU   125°
6. ∠ TOU   90°

7–12. Repeat Examples 1–6 using ∠ POQ = 27°.

7. 27°   8. 63°   9. 63°   10. 117°   11. 117°   12. 90°

6 Angle Bisectors *(pp. 20–22)*

Teaching These Pages

You might begin this lesson by holding up a piece of string and having a student "bisect" it with a pair of scissors to demonstrate the "cutting in half" concept. Then ask a student to draw a ray which will bisect an angle you've drawn on the board. When students have grasped the basic concept, move on to the numerical exercises.

Lesson Commentary   **T47**

In Classroom Practice Exercise 5, students show that the line bisecting an angle is unique. Since many students are confused by detailed considerations of uniqueness and existence issues, the text omits such a discussion. Some of your more advanced students might be interested in using Postulate 6 on page 35 to justify the uniqueness of the angle bisector.

Experiment 2 offers students a chance to propose a theorem concerning paper folding, and then to argue the correctness of the theorem. The Suggested Extension for this lesson uses paper folding to introduce the concept of a circle inscribed in a triangle.

Chalkboard Examples

Draw $\angle ABC$ with the given measure. Using a protractor, draw $\overrightarrow{BD}$ which bisects the angle. Name the measures of $\angle ABD$ and $\angle DBC$.

1. $\angle ABC = 70°$ 35°, 35°
2. $\angle ABC = 90°$ 45°, 45°
3. $\angle ABC = 150°$ 75°, 75°
4. $\angle ABC = 180°$ 90°, 90°

Suggested Extension

1. Draw a large triangle on a sheet of paper and cut the triangle out.
2. Can you discover a way of folding the paper so that the fold will bisect $\angle A$?
3. Now unfold the paper and refold it to bisect $\angle B$.
4. Finally, bisect $\angle C$ with a fold.
5. If you have worked accurately, your three folds meet in a point. Do they?
6. Put the point of a compass where the three folds meet. Can you use the compass to draw a circle that will just touch the three sides of the triangle?

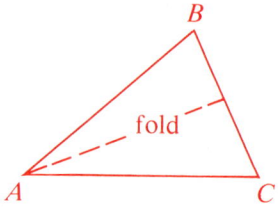

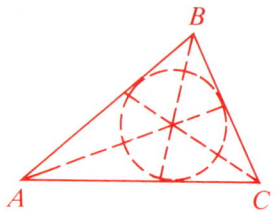

Quick Quiz

In the diagram, $\overline{BF} \perp \overline{AC}$.
Classify each angle as acute, right, obtuse, or straight.

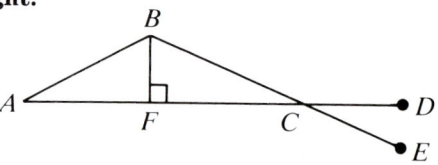

1. $\angle BAF$ acute
2. $\angle FCE$ obtuse
3. $\angle BFC$ right
4. $\angle DCF$ straight
5. Name two pairs of vertical angles.
   $\angle BCD$ and $\angle ECF$, $\angle DCE$ and $\angle BCF$
6. If $\overrightarrow{BF}$ bisects $\angle ABC$, what two angles must be equal?
   $\angle ABF$ and $\angle CBF$

Objectives for Pages 25-37

1. To construct bisectors of angles and perpendiculars to segments.
2. To use postulates of equality to draw conclusions about geometric figures.
3. To use postulates of geometry.

7 Three Constructions (pp. 25-28)

Teaching These Pages

Demonstrate these constructions one at a time, allowing students to practice each before you show the next one. After students have learned all three constructions, you might try the Classroom Practice Exercises and then the Chalkboard Examples below, which use the three constructions in a different order. Proofs that these constructions are valid are developed in the text and exercises of Section 2 of Chapter 4 (pages 149–152).

Chalkboard Examples

Draw a figure similar to, but larger than, the one shown.

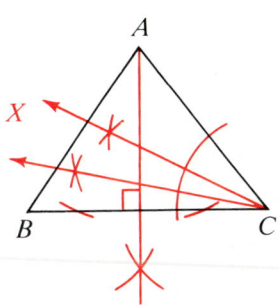

1. Construct a line through A perpendicular to $\overline{BC}$.
2. Bisect $\angle C$. Call the bisector $\overrightarrow{CX}$.
3. Bisect $\angle BCX$.
4. Construct a line perpendicular to $\overleftrightarrow{PQ}$ at P. (*Note:* $\overline{PQ}$ must be extended to the left.)

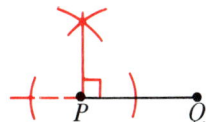

Suggested Extension

The designs below were made using only a compass. Have students experiment to make designs of their own.

8 Postulates of Equality *(pp. 29–32)*

Teaching These Pages

Your students may not be able to quote the postulates of equality, but they have used them in their study of algebra. Before taking up the geometric examples in the text, you may wish to illustrate the postulates with the algebraic examples below.

Addition Postulate
If $x - 5 = 7$
and $\underline{5 = 5}$
then $(x - 5) + 5 = 7 + 5$.
Simplify: $x = 12$.

Subtraction Postulate
If $x + 9 = 25$
and $\underline{9 = 9}$
then $(x + 9) - 9 = 25 - 9$.
Simplify: $x = 16$.

Multiplication Postulate
If $\frac{1}{3}x = 21$,
then $3\left(\frac{1}{3}x\right) = 3(21)$.
Simplify: $x = 63$.

Division Postulate
If $13x = 52$,
then $\frac{13x}{13} = \frac{52}{13}$.
Simplify: $x = 4$.

Substitution Postulate
If $y = 2x + 7$ and $x + y = 10$,
then $x + (2x + 7) = 10$.

Written Exercises 13–17 offer students a chance to draw their own conclusions. You probably won't want to insist on a written formal proof to support these conclusions: a student's oral argument to the class is sufficient. You may wish to partially formalize the argument by writing the steps on the chalkboard. Having students make a list of steps may help to prepare them for the transition to writing formal proofs in Chapter 2.

Chalkboard Examples

Name the postulate of equality used in each step.

1. $3x - 7 = 8$ Given
   $3x = 15$ _?_ Addition
   $x = 5$ _?_ Division

2. $2x + 18 = 5x$ Given
   $18 = 3x$ _?_ Subtraction
   $6 = x$ _?_ Division

3. $\frac{x + 5}{4} = 9$ Given
   $x + 5 = 36$ _?_ Multiplication
   $x = 31$ _?_ Subtraction

4. $y = 3x - 5$
   and $x = 10$ Given
   $y = 3(10) - 5$ _?_ Substitution
   $y = 25$ _?_ Substitution

5. Suppose $AB = CD$. Which postulate allows you to conclude that
$\frac{1}{2}AB = \frac{1}{2}CD$?   Multiplication

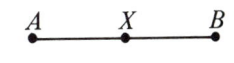

Suppose you also know that $AX = \frac{1}{2}AB$ and that $CY = \frac{1}{2}CD$.
Which postulate allows you to conclude that $AX = CY$?   Substitution

9 Postulates of Geometry *(pp. 33–37)*

Teaching These Pages

Use your classroom to illustrate some of the terminology and postulates of this section:

> *plane*—desk top; floor; walls
> *collinear points*—points along one edge of a desk
> *coplanar points*—points on one wall
> *non–coplanar points*—two points in the middle of the floor, one in the middle of the ceiling, and one in a side wall
> *parallel planes*—floor and ceiling; opposite walls

It is probably unwise to prolong the presentation of the postulates in this section. Deal with them matter-of-factly, and move on to the exercises, which reinforce the learning of the postulates.

Some students may note that the Ruler Postulate is true only after a coordinate system has been selected for the line. Most students, however, feel satisfied with the postulate as it is stated in the text. Similarly, some students may note that there are two numbering schemes for rays, corresponding to the two scales on a standard protractor.

Chalkboard Exercises

1. Mark two points on the chalkboard and have a student indicate a third point "in space." Have another student indicate the plane determined by these points (Postulate 2).

2. Mark three collinear points on the chalkboard. Have a student indicate several planes which contain the points.

3. Describe the intersection of: a. a side wall of the classroom and the floor   line
   b. a side wall and the front wall   line   c. the plane containing a desk top and the floor
   no intersection

Suggested Extension

A carpenter sometimes checks the flatness of a board by laying a straightedge across the board. Suppose there is a part of the board where the endpoints of the straightedge lie on the board but the middle of the straightedge does not. The carpenter then concludes that the board is not flat. State the postulate to which this conclusion is related.   Postulate 3:
If two points lie in a plane, then the line joining them lies in that plane.

Quick Quiz

1. Draw a figure similar to the one shown. Then construct a perpendicular to $\overleftrightarrow{AB}$ through C.
   Use Construction 3 (shown in red on the diagram).

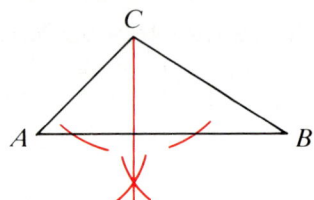

Which postulate can be used to justify each statement?

2. If $\angle 1 + \angle 2 = 90°$ and $\angle 2 = 60°$, then $\angle 1 + 60° = 90°$.   Substitution
3. If $\angle 1 + 60° = 90°$, then $\angle 1 = 30°$.   Subtraction

Questions 4 and 5 refer to the figure.

4. $\angle TOR = \underline{}°$ 140
5. Name the bisector of $\angle QOP$. $\overrightarrow{OT}$

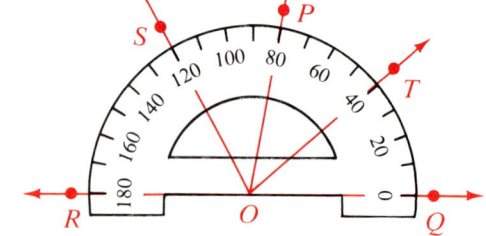

Classify as true or false.

6. The intersection of two planes can be a single point.   false
7. Three points must lie in exactly one plane.   false
8. The two rays which are the sides of an angle are coplanar.   true

Chapter 2. Introduction to Proof

This chapter begins with proofs of three simple theorems. The informal, step-by-step presentation of these proofs paves the way to the usual two-column format. In their first exposure to two-column proof, students supply "Reasons" to complete proofs. Then they construct their own proofs concerning parallel lines. The chapter closes with a construction of parallel lines using straightedge and compass.

Objectives for Pages 46–53

1. To use theorems about complementary, supplementary, and vertical angles.
2. To write "if . . . then" statements and their converses.

1 Three Theorems (pp. 46–49)

Teaching These Pages

You might begin the introduction of proof by supplying students with the *Given* information from the Example on page 46. Ask them to draw their own conclusion. Then have a student present an oral argument to justify the conclusion. Summarize the argument in writing, then compare this with the five steps given on page 46.

The same strategy could be used for Theorems 2 and 3. For Theorem 3, a proof could be based on Theorem 2 (as on page 47) or could follow the steps suggested in Chapter 1 on page 18.

The proofs of the three theorems in this section are written in a style which is intended to provide a transition between oral proof and the more formal two-column proof to be introduced in Section 4.

Chalkboard Examples

Find the complement and supplement of ∠A if:

1. ∠A = 10° 80°, 170°
2. ∠A = 80° 10°, 100°
3. ∠A = 45° 45°, 135°
4. ∠A = 16° 74°, 164°

5. Given: $\overline{AB} \perp \overline{CD}$
   a. Name the complement of ∠DBE. ∠ABE
   b. Name the supplement of ∠DBE. ∠CBE

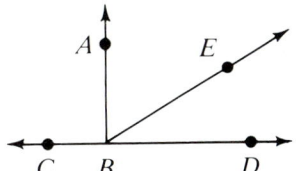

6. a. Name two supplements of ∠2. ∠1, ∠7
   b. Name two supplements of ∠4. ∠3, ∠5
   c. If ∠2 = ∠4, tell why ∠1 = ∠3.
      If two angles are supplements of equal angles, then the two angles are equal. (Theorem 2)
   d. Why does ∠3 = ∠5? Vertical angles are equal.

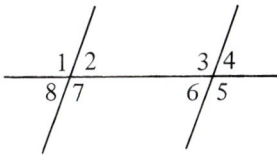

Suggested Extension

1. a. If ∠A = x°, what is the measure of a complement of ∠A?
      $(90 - x)$°
   b. What is the measure of a supplement of ∠A? $(180 - x)$°
   c. Write an equation, in terms of x, which says that a supplement of ∠A is four times as large as a complement of ∠A. $180 - x = 4(90 - x)$
   d. Solve your equation for x. $x = 60$

2. A supplement of ∠B is six times as large as a complement of ∠B. Find the measure of ∠B. 72°

2 "If ... Then" Statements *(pp. 50–53)*

Teaching These Pages

You might like to supply examples of "if ... then" statements from a current newspaper or magazine. Your students can supply other examples from their daily lives. Examples 1 and 2 on page 50 illustrate the key ideas of the section: (1) An "if ... then" idea is often expressed without the words "if" and "then"; (2) The converse of a true statement may be false.

Chalkboard Examples

State the hypothesis and the conclusion of each statement.

1. If he remembers, then we're sunk. Hyp: he remembers   Conc: we're sunk
2. She should stay home if she is sick. Hyp: she is sick   Conc: she should stay home

State the converse of each statement.

3. If n is an integer, then $n + 1$ is an integer. If $n + 1$ is an integer, then n is an integer.
4. All cats are animals. All animals are cats.
5. Decide if the statements in 3 and 4 above are true. **3.** true   **4.** true
   Then decide if their converses are true. **3.** true   **4.** false

3 Using "If . . . Then" Statements (Optional) *(pp. 56–59)*

Teaching These Pages

Type 1 reasoning is the kind most commonly used in geometry. Most students view this kind of reasoning as common sense. Unfortunately, students too often employ Type 3 and Type 4 reasoning in their geometry proofs. The fallacy committed in Type 3 reasoning is called "arguing from the converse." The fallacy committed in Type 4 reasoning is called "arguing from the inverse." Type 2 reasoning ("arguing from the contrapositive") is valid. It is the logical basis of indirect proof. (See Applications, pages 334–335.)

If time is a problem, consider omitting all of the text, but including the Classroom Practice Exercises. As the class discusses the exercises, you can illustrate the correct and incorrect types of reasoning.

Chalkboard Examples

Given: "**If Nan is in Nebraska, then Ryan is in Alaska.**" **From each statement below, draw a conclusion or state "no conclusion possible."**

1. Ryan is in Alaska. no conclusion
2. Nan is in Nebraska. Ryan is in Alaska.
3. Ryan is not in Alaska. Nan is not in Nebraska.
4. Nan is not in Nebraska. no conclusion

Decide whether the reasoning is correct.

5. (1) All persons 18 and over are qualified to vote.
   (2) Cecily is 18.
   (3) *Conclusion:* Cecily is qualified to vote. correct

6. (1) Those who brush their teeth with Dentashine have no cavities.
   (2) Jared has no cavities.
   (3) *Conclusion:* Jared brushes his teeth with Dentashine. not correct

Quick Quiz

In the diagram, $\overrightarrow{AD} \perp \overleftrightarrow{BE}$.

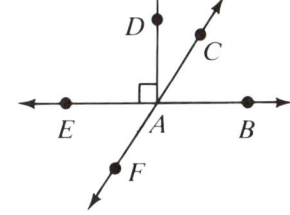

1. Name a complement of ∠ CAB. ∠ CAD
2. Name two supplements of ∠ EAC. ∠ CAB, ∠ EAF
3. Name an angle equal to ∠ FAB. ∠ EAC
4. If ∠ Q = 45°, what is the measure of a supplement of ∠ Q? 135°

Consider the statement "If $4x = 28$, then $x = 7$."

5. Write the conclusion of the statement. $x = 7$
6. Write the converse of the statement. If $x = 7$, then $4x = 28$
7. Is the converse true? yes

Objectives for Pages 60–80

1. To understand and write basic geometry proofs.
2. To use the properties of parallel lines cut by a transversal.
3. To use facts about transversals to prove lines parallel.
4. To construct an angle equal to a given angle.
5. To construct parallel lines.

4 Writing Proofs *(pp. 60–65)*

Teaching These Pages

Teaching students to write proofs is one of the toughest jobs of a geometry teacher. Most students find it easier to write a proof after understanding the broad outline of the reasoning. One way to help students to develop this kind of understanding is by having them present oral arguments, and allowing them to make heavy use of a figure.

Once the outline of a proof is given orally, help students refine it into a written proof. You may wish to use this procedure with the Classroom Practice Exercises and the Chalkboard Examples on page T56.

The order of the steps, even the basic strategy, may vary from one student's proof to another's. This is as it should be. One of the beauties of geometry is that there is so much room for individuality in constructing proofs.

We suggest that, at this early stage, you not risk destroying students' interest in proof by being overly fussy about details. With more experience, many students will come to appreciate the finer points of logical argument.

Chalkboard Examples

Write a proof in two-column form.

1. Given: $\angle 1 = \angle 4$
   Prove: $\angle 2 = \angle 3$

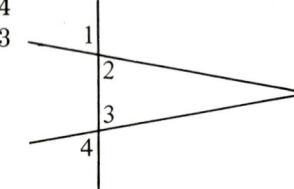

   1. $\angle 1 = \angle 2$ (Vertical angles are equal.)
   2. $\angle 1 = \angle 4$ (Given)
   3. $\angle 2 = \angle 4$ (Substitution)
   4. $\angle 3 = \angle 4$ (Vertical angles are equal.)
   5. $\angle 2 = \angle 3$ (Substitution)

2. Given: $\angle 1 = \angle 2$
   $\overrightarrow{AX}$ bisects $\angle CAB$
   $\overrightarrow{BX}$ bisects $\angle ABC$
   Prove: $\angle 5 = \angle 6$

3. Given: $\angle 3 = \angle 4$
   $\overrightarrow{AX}$ bisects $\angle CAB$
   $\overrightarrow{BX}$ bisects $\angle ABC$
   Prove: $\angle 5 = \angle 6$

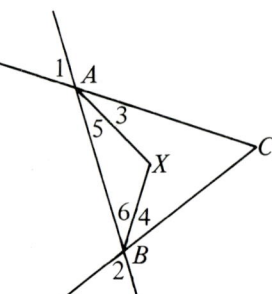

Ex. 2
1. $\angle 1 = \angle CAB$; $\angle 2 = \angle CBA$ (Vertical angles are equal.)
2. $\angle 1 = \angle 2$ (Given)
3. $\frac{1}{2}\angle 1 = \frac{1}{2}\angle 2$ (Mult. Post.)
4. $\frac{1}{2}\angle CAB = \frac{1}{2}\angle CBA$ (Subst.)
5. $\angle 5 = \frac{1}{2}\angle CAB$; $\angle 6 = \frac{1}{2}\angle CBA$ (*Given:* angle bisectors)
6. $\angle 5 = \angle 6$ (Subst.)

Ex. 3
1. $\angle 3 = \frac{1}{2}\angle CAB$; $\angle 4 = \frac{1}{2}\angle CBA$;
   $\angle 5 = \frac{1}{2}\angle CAB$; $\angle 6 = \frac{1}{2}\angle CBA$
   (*Given:* angle bisectors)
2. $\angle 3 = \angle 5$; $\angle 4 = \angle 6$ (Subst.)
3. $\angle 3 = \angle 4$ (Given)
4. $\angle 5 = \angle 4$ (Subst.)
5. $\angle 5 = \angle 6$ (Subst.)

5 Parallel Lines *(pp. 66–71)*

Teaching These Pages

You can illustrate skew and parallel lines using your classroom. The intersection lines of ceiling, walls, and floor provide good examples of both of these concepts.

The word *transversal* may have more meaning for students if compared with *reversal:*

     reversal—a turning back   transversal—a turning across

Point out that the prefix *trans* means "across," as in *transfer, transport,* and *transmit.*

You may wish to teach students exterior-angle terminology.
A suggested sequence follows.

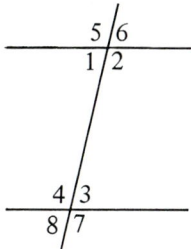

  interior angles: ∠ 1, ∠ 2, ∠ 3, and ∠ 4
  exterior angles: ∠ 5, ∠ 6, ∠ 7, and ∠ 8

  alternate interior angles: ∠ 1 and ∠ 3; ∠ 2 and ∠ 4
  same-side interior angles: ∠ 2 and ∠ 3; ∠ 1 and ∠ 4

  alternate exterior angles: ∠ 5 and ∠ 7; ∠ 6 and ∠ 8
  same-side exterior angles: ∠ 6 and ∠ 7; ∠ 5 and ∠ 8

Note that Postulate 7 and Theorem 4 are sometimes worded this way:

  Postulate 7: Corresponding angles of parallel lines are equal.
  Theorem 4: Alternate interior angles of parallel lines are equal.

In this text, we have often chosen an "if . . . then" phrasing of a theorem or postulate instead of another, shorter phrasing. We have found that learning "if . . . then" forms enables students to understand better the relationships among statements and their converses.

Postulate 7 is roughly equivalent to Euclid's "Parallel Postulate." Some texts use the following instead of Postulate 7: "Through any point outside a line, there is exactly one line parallel to the given line."

Chalkboard Examples

Classify each pair of angles as (1) *alternate interior*, (2) *same-side interior*, (3) *corresponding*, or (4) *none of these*.

(You may wish to add: (5) *alternate exterior angles* and (6) *same-side exterior angles*.)

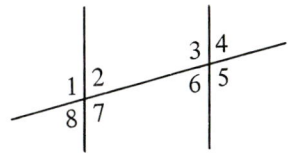

1. ∠ 2 and ∠ 6 (1)
2. ∠ 3 and ∠ 1 (3)
3. ∠ 1 and ∠ 5 (4) or (5)
4. ∠ 2 and ∠ 4 (3)
5. ∠ 7 and ∠ 6 (2)
6. ∠ 5 and ∠ 8 (4) or (6)
7. ∠ 8 and ∠ 6 (3)
8. ∠ 2 and ∠ 5 (4)

Find the values of *x* and *y* in each diagram.

9.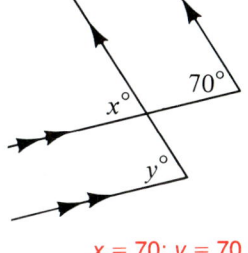

   x = 70; *y* = 70

10.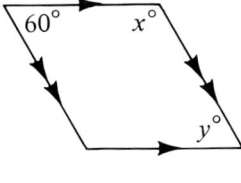

    x = 120; *y* = 60

11.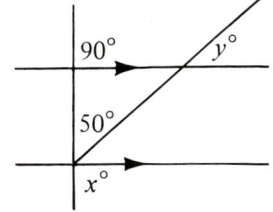

    x = 90; *y* = 40

Lesson Commentary **T57**

Suggested Extension

If you decide to teach exterior-angle terminology, have students discover two additional theorems:

> *Theorem:* If two parallel lines are cut by a transversal, then alternate exterior angles are equal.
>
> *Theorem:* If two parallel lines are cut by a transversal, then same-side exterior angles are supplementary.

You may wish to have some students confirm the converses of these theorems as an extension of their work in Section 6.

6 Proving Lines Parallel *(pp. 72–76)*

Teaching These Pages

Postulates 7 and 8 are converse statements, as are Theorems 4 and 6 and Theorems 5 and 7. It is important that students realize that Theorems 6 and 7 are not automatically true just because their converses have been proved. Theorems 6 and 7 do not follow from Theorems 4 and 5, but rather from Postulate 8.

You may wish to use the following example to remind students that the converse of a true statement is not necessarily true.

> TRUE: If two lines are parallel, then they do not intersect.
>
> FALSE: If two lines do not intersect, then they are parallel.

The converse is true in a plane, but in three dimensions, lines may be skew.

Chalkboard Examples

In each figure, tell which lines are parallel. Give reasons for your answers.

1.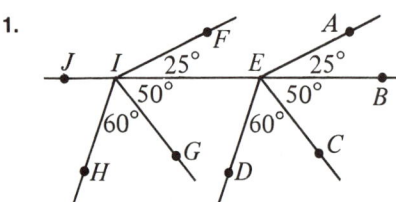

2.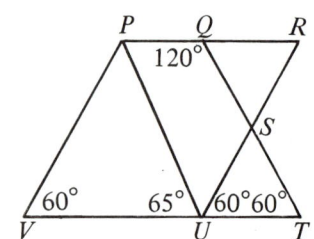

$\overleftrightarrow{IF}$ and $\overleftrightarrow{EA}$ (corr. angles)
$\overleftrightarrow{IG}$ and $\overleftrightarrow{EC}$ (corr. angles)
$\overleftrightarrow{HI}$ and $\overleftrightarrow{DE}$ (corr. angles)

$\overleftrightarrow{PV}$ and $\overleftrightarrow{RU}$ (corr. angles)
$\overleftrightarrow{PQ}$ and $\overleftrightarrow{TU}$ (same-side int. angles)

Find the value of x which makes lines s and t parallel.

3.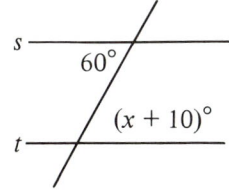

    $x = 50$

4.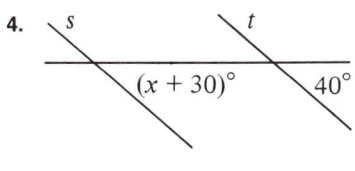

    $x = 10$

5.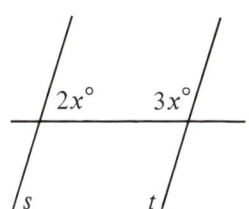

    $x = 36$

Suggested Extension

If $l \parallel m$ and $l \perp n$, what are the possible relationships between m and n? Draw a diagram to illustrate each possibility. (Diagrams omitted)

If l, m, and n are coplanar, then $m \perp n$. Otherwise, m and n are skew.

7 Constructing Parallel Lines *(pp. 77–80)*

Teaching These Pages

You may wish to challenge students to copy an angle using straightedge and compass only (no protractor). After explaining Construction 4, let students practice copying angles of various sizes. Then see if any student can describe a strategy for Construction 5. The method given in the text is based on Postulate 8. Methods based on Theorems 6, 7, or 8, or on exterior angles are equally possible. (See Written Exercises 7 and 10.) Students will use the SSS Postulate to prove the validity of Construction 4 in Written Exercise 9 on page 151.

The proof that the lines constructed in Written Exercise 9 on page 79 are perpendicular depends on a fact that some students may not yet know: The angle sum of a triangle is 180°.

The Chalkboard Examples below preview the Midpoints Theorem. (See Chapter 5, Section 8.)

Chalkboard Examples

Draw a triangle similar to, but larger than, the one shown.

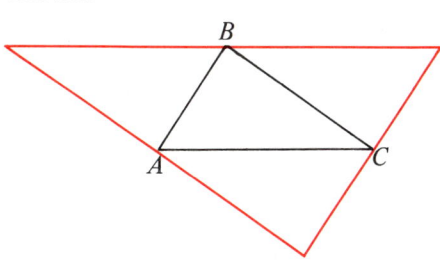

1. Draw a line through A parallel to $\overleftrightarrow{BC}$.
2. Draw a line through B parallel to $\overleftrightarrow{AC}$.
3. Draw a line through C parallel to $\overleftrightarrow{AB}$.
4. Measure $\overline{AB}$, $\overline{BC}$, $\overline{AC}$, and the sides of the new triangle just constructed. What relationships do you find? The sides of the new triangle are twice as long as the sides of the original triangle.

Suggested Extension

Written Exercises 7 and 10 offer other methods of constructing a line parallel to a given line. Those methods are based on Theorems 6 and 8. Challenge your students to find a method based on Theorem 7.

Let *l* be the given line. Draw a point *P* not on *l*. Draw a line through *P* that intersects *l*; label the line *m*. Both ∠ 2 and ∠ 3 are supplementary to ∠ 1. At *P*, construct ∠ 4 equal to ∠ 2 (or ∠ 3). By substitution ∠ 1 is supplementary to ∠ 4. Then *n* ∥ *l* by Theorem 7.

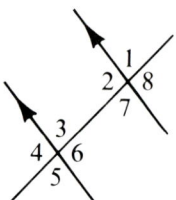

Some of your students may be interested in finding out the methods that commercial artists use to draw parallel lines (drafting board with T-squares and triangles, folding parallel rules, rolling rules).

Quick Quiz

1. A statement which is accepted without proof is called a __?__. (definition/postulate/theorem) postulate

2. ∠ 2 and ∠ __?__ are corresponding angles. 4

3. ∠ 2 and ∠ __?__ are alternate interior angles. 6

4. If ∠ 7 = 80°, then ∠ 5 = __?__ and ∠ 6 = __?__.
   80°, 100°

Use the given information to tell which lines, if any, are parallel.

5. ∠ 1 = ∠ 2   c ∥ d
6. ∠ 3 + ∠ 4 = 180°   a ∥ b
7. ∠ 4 = ∠ 5   none

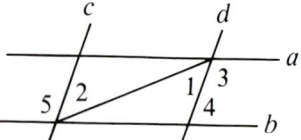

8. Draw an acute angle *A*. Then construct an angle whose measure is twice that of ∠ *A*.

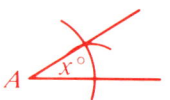
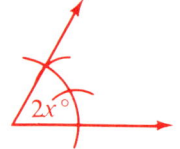

Chapter 3. Triangles

This chapter opens with a proof of a property that many students already know: The angle sum of a triangle is 180°. Terminology is presented which enables students to classify triangles according to their sides and their angles. After the meaning of congruent triangles is developed, the SSS, SAS, and ASA Postulates are considered in separate sections. The final section is devoted to the AAS and HL Theorems.

Objectives for Pages 92–100

1. To use the angle sum theorem to find angle measures.
2. To use the exterior angle corollary to find angle measures.
3. To classify triangles by sides.
4. To classify triangles by angles.

1 The Angle Sum of a Triangle *(pp. 92–96)*

Teaching These Pages

Encourage students to draw diagrams for the investigation of the angle sum of a triangle. Using large diagrams will heighten the impact of the lesson and help students recall it better. Some students may suggest measuring angles with a protractor. They shouldn't expect to get an angle sum of exactly 180° on each trial, but they shouldn't miss by much if they work carefully and use large triangles. A proof of the corollary to Theorem 1 is outlined in Written Exercise 25 on page 96. Exercises 26 and 27 anticipate the theorem concerning the angle sum of a convex polygon (Theorem 1, page 194).

Chalkboard Examples

Given: $l \parallel \overline{AC}$

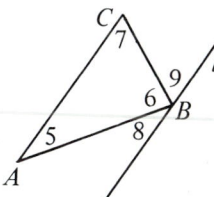

1. What angle of $\triangle ABC$ equals $\angle 8$? $\angle 5$
2. What angle of $\triangle ABC$ equals $\angle 9$? $\angle 7$
3. $\angle 8 + \angle 6 + \angle 9 = \underline{}°$ 180
4. $\angle 5 + \angle 6 + \angle 7 = \underline{}°$ 180

Complete the table, referring to the diagram below.

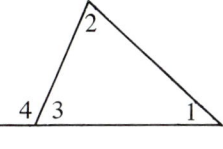

	$\angle 1$	$\angle 2$	$\angle 3$	$\angle 4$	
5.	40°	80°	?	?	60°, 120°
6.	40°	?	65°	?	75°, 115°
7.	?	70°	60°	?	50°, 120°
8.	40°	?	?	100°	60°, 80°

Suggested Extension

Can two angles of a triangle have the measures listed?

1. 90° and 80°   yes
2. 90° and 90°   no
3. 90° and 100° no
4. 81° and 93°   yes

Is it possible for a triangle to have the types of angles listed?

5. two obtuse angles   no
6. two right angles   no
7. one right angle and one obtuse angle   no
8. two acute angles and one non-acute angle   yes

2 Classifying Triangles *(pp. 97–100)*

Teaching These Pages

Group activity as well as individual study may make mastery of new vocabulary more pleasant. You might like to conduct an oral drill in which students classify different triangles as you draw them on the chalkboard.

Students should realize that more than one term may apply to a figure. If, in $\triangle ABC$, $AB = 10$ and $AC = 10$, then one can conclude that $\triangle ABC$ is isosceles. It is possible that additional information, acquired later, will permit a further conclusion: for instance, it may develop that $\triangle ABC$ is acute, obtuse, or even equilateral. Be sure students understand that it remains true that $\triangle ABC$ is isosceles, because $AB = AC$.

Chalkboard Examples

Draw a triangle that fits the description. Drawings are omitted.
If the condition cannot be met, write *not possible*.

1. An acute triangle that is isosceles
   Any $\triangle$ whose $\angle$s are all less than 90° with 2 sides equal.
2. A right triangle that is isosceles   Any right $\triangle$ with equal legs.
3. An obtuse triangle that is isosceles
   Any $\triangle$ with one angle greater than 90° with 2 sides equal.
4. A scalene triangle that is isosceles   not possible
5. A right triangle that is equilateral   not possible
6. An isosceles triangle that is equilateral   Any equilateral $\triangle$.
7. A right triangle in which the hypotenuse equals a leg   not possible

Quick Quiz

Find the value of *x*.

1. 52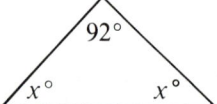
2. 44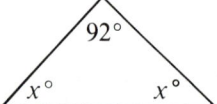
3. 50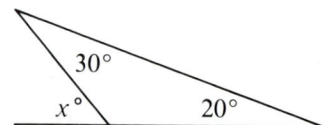

4.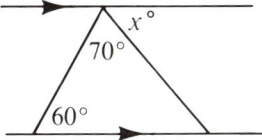
50

5.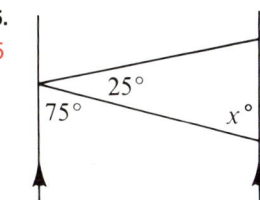
75

In Questions 6–8, classify each triangle as *acute*, *right*, or *obtuse*. Then classify it as *scalene*, *isosceles*, or *equilateral*.

6.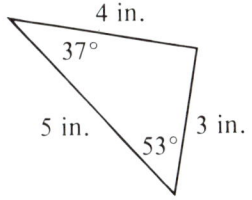

right; scalene

7.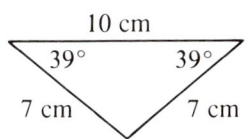

obtuse; isosceles

8.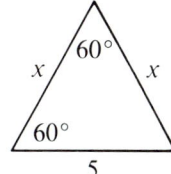

acute; equilateral

Objectives for Pages 102–111

1. To name congruent triangles and their corresponding parts.
2. To use the SSS Postulate to prove that two triangles are congruent.

3 Defining Congruent Triangles *(pp. 102–106)*

Teaching These Pages

In some situations, it is convenient to make informal statements such as "The triangles are congruent" or "Triangle I is congruent to Triangle II." But students must realize that in a statement of the form △ ___ ≅ △ ___, the order in which the letters appear is crucial. An advantage of the more formal statement is that we can identify corresponding parts of congruent triangles without needing to consult a diagram.

Written Exercises 5–8 involve graphing skills. You may wish to assign Skills Review, page 84, to students who need extra practice using the coordinate plane.

Chalkboard Examples

1. Name the six parts of △ *RAT*.
   $\overline{RA}, \overline{AT}, \overline{TR}, \angle 2, \angle A, \angle 3$
2. Name the six parts of △ *ERT*.
   $\overline{ER}, \overline{RT}, \overline{TE}, \angle E, \angle 1, \angle 4$

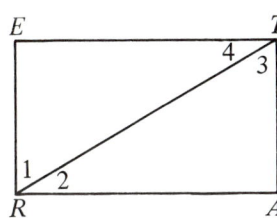

Lesson Commentary **T63**

In the diagram at the bottom of page T63, ∠ 1 = 60°, ∠ 2 = 30°, ∠ 3 = 60°, and ∠ 4 = 30°. Tell whether the following statements are correct or incorrect.

3. △RAT ≅ △RTE incorrect
4. △RAT ≅ △TER correct
5. △TAR ≅ △RET correct
6. △ERT ≅ △ART incorrect

Suggested Extension

Have students check to see that these two statements are both correct: △ABC ≅ △RST   △ABC ≅ △XYZ

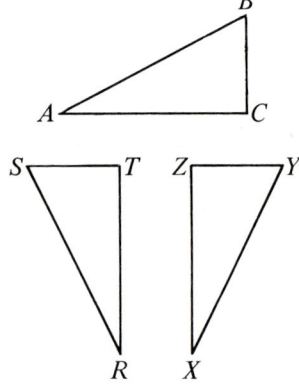

The first statement seems easier, however, because we can slide △ABC—mentally—and get a fit with △RST. To fit △ABC over △XYZ we must both *flip* and slide.

Point out that congruence in solid geometry is more complicated. The two solids shown below are *oppositely congruent*. One cannot be made to fit over the other, no matter how much flipping and sliding is done.

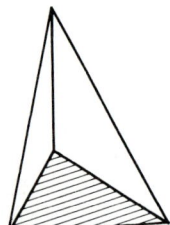 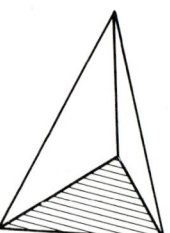

You may ask students to name pairs of oppositely congruent objects that can be found in everyday life. gloves, shoes

4 The SSS Postulate *(pp. 107–111)*

Teaching These Pages

Some students will be convinced of the validity of the SSS Postulate by the diagram on page 107. Other students may need to handle some sticks like those in the diagram. Students who like woodworking may enjoy preparing several sets of three sticks 20 cm, 30 cm, and 40 cm long.

Chalkboard Examples

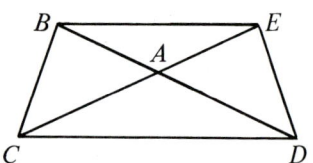

1. Suppose you want to prove △BAC ≅ △EAD by using the SSS Postulate. List the three pairs of sides of these triangles which you must prove equal.
   $\overline{BA}$ and $\overline{EA}$; $\overline{AC}$ and $\overline{AD}$; $\overline{CB}$ and $\overline{DE}$

2. Repeat Exercise 1, but work with △BCD and △EDC. $\overline{BC}$ and $\overline{ED}$; $\overline{CD}$ and $\overline{DC}$; $\overline{DB}$ and $\overline{CE}$

3. Given: $GX = JX$
   E is the midpoint of $\overline{GJ}$
   Prove: $\triangle XEG \cong \triangle XEJ$

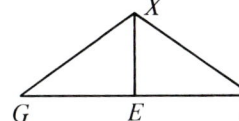

1. $GX = JX$ (Given)
2. $GE = JE$ (Given: E is midpt. of $\overline{GJ}$.)
3. $XE = XE$ (From algebra)
4. $\triangle XEG \cong \triangle XEJ$ (SSS Post.)

4. Suppose you want to prove that $\triangle ABC \cong \triangle DEF$. You find that $BC = 17$ and $EF = 18$. Should you continue your proof? no

5. Suppose you want to prove that $\triangle RST \cong \triangle XYZ$. You find that $RS = 102$ and $YZ = 103$. Should you continue your proof? yes

Suggested Extension

The Application feature, Rigidity of Triangles, on pages 113–114 discusses a very important practical use of the SSS Postulate. The exercises in this feature encourage students to experiment with one of the fundamental concepts of architecture.

Quick Quiz

The two triangles shown are congruent. Complete each statement based on the appearance of the figure.

1. $\triangle BCA \cong \triangle \underline{} DAC$
2. $DC = \underline{} BA$
3. $\angle DCA = \angle \underline{} BAC$
4. $AC = \underline{} CA$

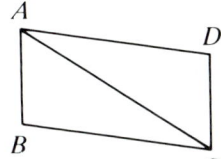

5. Supply the reasons to complete the proof.
   Given: $QP = PS$; $QR = SR$
   Prove: $\triangle RPQ \cong \triangle RPS$

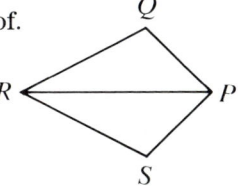

STATEMENTS	REASONS
1. $QP = PS$	1. __?__ Given
2. $QR = SR$	2. __?__ Given
3. $RP = RP$	3. __?__ From algebra
4. $\triangle RPQ \cong \triangle RPS$	4. __?__ SSS Postulate

Objectives for Pages 115–124

1. To use the SAS Postulate to prove that two triangles are congruent.
2. To use the ASA Postulate to prove that two triangles are congruent.

Lesson Commentary **T65**

5 The SAS Postulate *(pp. 115–120)*

Teaching These Pages

If some students prepared sticks for use with Section 4, the sticks can be adapted for use here. An angle indicator like the one pictured at right should be made for each set of sticks.

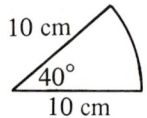

Chalkboard Examples

State whether the SSS Postulate or the SAS Postulate could be used to prove △ GRU ≅ △ GTU.

1. Given: ∠ RGU = ∠ TGU; RG = TG   **SAS**
2. Given: ∠ GUR and ∠ GUT are right angles; RU = TU   **SAS**
3. Given: RG = TG; RU = TU   **SSS**
4. Given: ∠ UGR = 25°; ∠ UGT = 25°; RG = TG   **SAS**
5. Given: U is the midpoint of $\overline{RT}$; RG = TG   **SSS**
6. Given: RU = TU; $\overline{GU} \perp \overline{RT}$ **SAS**

Suggested Extension

Supply reasons to complete the proof.

Given: XY = XZ

Prove: △ XYZ ≅ △ XZY

STATEMENTS	REASONS	
1. XY = XZ	1. _?_	Given
2. ∠ X = ∠ X	2. _?_	From algebra
3. XZ = XY	3. _?_	Given
4. △ XYZ ≅ △ XZY	4. _?_	SAS Postulate

This strategy can be used in an alternate proof of Theorem 3 on page 126.

6 The ASA Postulate *(pp. 121–124)*

Teaching These Pages

You may wish to have students make stick models similar to the one pictured on page 121 to demonstrate the ASA Postulate. Two angle indicators should be made for each set of sticks. Such models help students to understand the activity described on page 121.

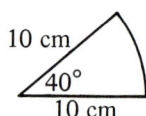

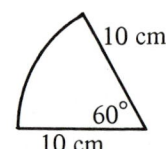

The constructions in Written Exercises 7–9 on page 124 also serve to convince students of the validity of the ASA Postulate.

Chalkboard Examples

State whether the SAS Postulate or the ASA Postulate could be used to prove $\triangle RED \cong \triangle REN$.

1. *Given:* $RD = RN$
   $\overrightarrow{RE}$ bisects $\angle DRN$.
   SAS

2. *Given:* $\angle DRE = \angle NRE$
   $RD = RN$
   SAS

3. *Given:* $\overline{RE} \perp \overline{DN}$
   $\angle D = \angle N$
   $DE = NE$
   ASA

4. *Given:* $\overrightarrow{RE}$ bisects $\angle DRN$.
   $\overline{RE} \perp \overline{DN}$
   ASA

5. *Given:* $DE = NE$
   $\angle RED = \angle REN$
   SAS

6. *Given:* $\angle RDE = 65°$; $\angle RNE = 65°$
   $DR = NR$; $\overrightarrow{RE}$ bisects $\angle DRN$.
   ASA

Exercises 1–6

7. *Given:* $\angle F = \angle G$; $\angle 1 = \angle 2$; $\overline{DE} \perp \overline{FG}$; E is the midpoint of $\overline{FG}$.
   Prove: $\triangle FEH \cong \triangle GEK$

   1. $\angle DEF = \angle DEG$ (*Given:* $\overline{DE} \perp \overline{FG}$)
   2. $\angle 1 = \angle 2$ (*Given*)
   3. $\angle HEF = \angle KEG$ (Subtr. Post.)
   4. $FE = GE$ (*Given:* E is the midpt. of $\overline{FG}$.)
   5. $\angle F = \angle G$ (*Given*)
   6. $\triangle FEH \cong \triangle GEK$ (ASA Post.)

Suggested Extension

Supply reasons to complete the proof.

Given: $\angle A = \angle B$
Prove: $\triangle ACB \cong \triangle BCA$

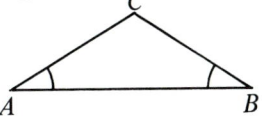

STATEMENTS		REASONS
1. $\angle A = \angle B$	1. __?__	Given
2. $AB = BA$	2. __?__	From algebra
3. $\angle B = \angle A$	3. __?__	Given
4. $\triangle ACB \cong \triangle BCA$	4. __?__	ASA Postulate

This strategy can be used in an alternate proof of Theorem 4 on page 173.

Lesson Commentary **T67**

Quick Quiz

State which postulate, if any, could be used to show that the triangles are congruent. If none applies, write *none*.

1.
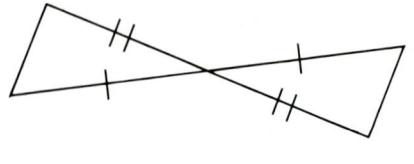
SAS

2.
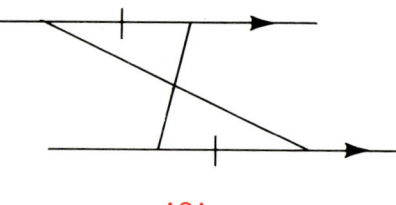
ASA

3.
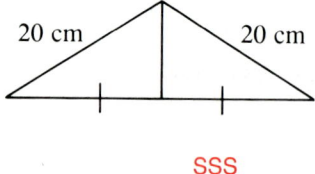
SSS

4.
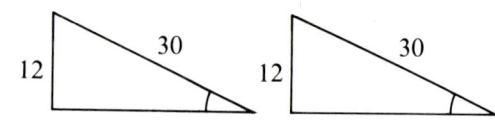
none

5. Write a proof in two-column form.

   Given: $\overline{AD} \perp \overline{CB}$
           $\overline{AD}$ bisects $\angle CAB$.

   Prove: $\triangle ACD \cong \triangle ABD$

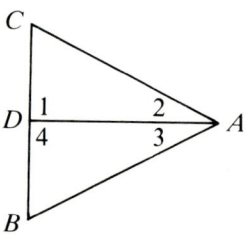

1. $\angle 1$ and $\angle 4$ are right angles. (*Given:* $\overline{AD} \perp \overline{CB}$)
2. $\angle 2 = \angle 3$ (*Given:* $\overline{AD}$ bisects $\angle CAB$.)
3. $AD = AD$ (From algebra)
4. $\triangle ACD \cong \triangle ABD$ (ASA Post.)

Objectives for Pages 126–132

1. To use the AAS Theorem to prove that two triangles are congruent.
2. To use the HL Theorem to prove that two triangles are congruent.

7 The AAS and HL Theorems (pp. 126–132)

Teaching These Pages

Point out that these two congruence statements are theorems, not postulates. In introducing Theorem 2, the text uses one numerical case to suggest a strategy for a general proof.

The proof of Theorem 3 is long and hard, and has been omitted from the text. The Class Experiment on page 132 is convincing at an informal level.

Chalkboard Examples

Given: ∠ T and ∠ X are right angles; AT = AX; ∠ 1 = ∠ 2

1. Use the AAS theorem to prove △ ATE ≅ △ AXE.
   1. ∠ T = ∠ X (Given: ∠ T and ∠ X are rt. ⦟.)
   2. ∠ 1 = ∠ 2; AT = AX (Given)
   3. △ ATE ≅ △ AXE (AAS Thm.)

2. Use the HL theorem to prove △ ATE ≅ △ AXE.
   1. △ ATE and △ AXE are rt. ⦟.
      (Given: ∠ T and ∠ X are rt. ⦟.)
   2. AT = AX (Given)
   3. AE = AE (From algebra)
   4. △ ATE ≅ △ AXE (HL Thm.)

3. Use the HL Theorem to prove that a diagonal of a rectangle divides it into two congruent triangles.

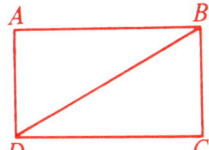

   1. ∠ A, ∠ C are rt. ⦟ (Def. of rectangle)
   2. △ ABD, △ CDB are rt. ⦟ (Def. of rt. △)
   3. AD = CB (Opp. sides of a rect. are =.)
   4. DB = DB (From algebra)
   5. △ ABD ≅ △ CDB (HL Thm.)

Suggested Extension

1. Some books prove an HA theorem (hypotenuse and acute angle theorem) that could be used for the triangles shown at right. What method have you studied that you could use to prove the triangles at the right congruent? **AAS Theorem**

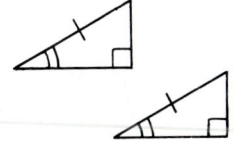

2. Some books prove an LA theorem (leg–acute angle theorem) that could be used for the triangles shown at right. What method have you studied that you could use to prove the triangles at the right congruent?
AAS Theorem

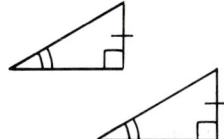

Quick Quiz

Indicate, by abbreviation, the postulate or theorem which could be used to prove that the triangles are congruent. If none applies, write *none*.

1.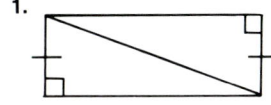

   HL Theorem

2.

   none

3.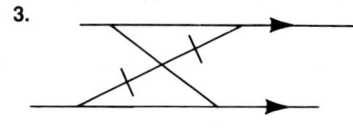

   AAS Theorem or ASA Postulate

Indicate, by abbreviation, the postulate or theorem which could be used to prove that the triangles are congruent. If none applies, write *none*.

4. SSS Postulate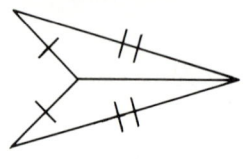

5. AAS Theorem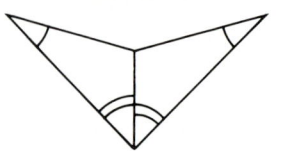

6. none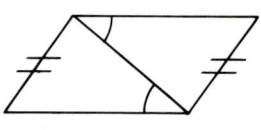

7. Write a proof in two-column form.
Given: $TS = TQ$; $\overline{TR} \perp \overline{QS}$
Prove: $\triangle TRS \cong \triangle TRQ$
1. $TS = TQ$ (Given)
2. $TR = TR$ (From algebra)
3. $\angle 1$ and $\angle 2$ are rt. $\angle$s. (Given: $\overline{TR} \perp \overline{QS}$)
4. $\triangle TRS \cong \triangle TRQ$ (HL Theorem)

Chapter 4. Using Congruent Triangles

This chapter opens by calling attention to six things that must be true when two triangles are congruent. After a section which introduces the use of the reason *Corresponding parts of congruent triangles are equal*, there is a section dealing with proofs that Constructions 1–4 are logically sound.

Three new constructions are presented and justified. Altitudes and medians of triangles are defined and constructed. Important theorems concerning isosceles and equilateral triangles are used in proofs, in computational work, and in justifying the construction of a 60° angle.

Objectives for Pages 144–152

1. To prove segments and angles equal by using the reason: Corresponding parts of congruent triangles are equal.

2. To show that the following constructions are logically sound:
   a. Bisecting an angle
   b. Constructing a perpendicular at a point on a line
   c. Constructing a perpendicular to a line from an outside point (Exercise 7, page 151)
   d. Constructing an angle equal to a given angle (Exercise 9, page 151)

1 Proving Corresponding Parts Equal *(pp. 144–148)*

Teaching These Pages

Encourage students to use the time-honored wording *Corresponding parts of congruent triangles are equal*. Quoting the definition of congruent triangles doesn't appeal to most students, and many find writing "Definition of congruent triangles" unconvincing. The strategy which appears in the panel on page 144 is a very useful tool for the geometry student. Urge students to learn these steps.

Chalkboard Examples

Supply reasons to complete the following proofs:

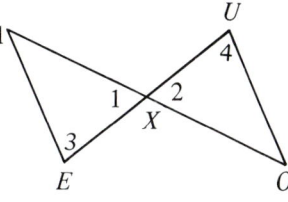

1. Given: $\overline{AO}$ bisects $\overline{EU}$; $\overline{EU}$ bisects $\overline{AO}$.
   Prove: $\angle 3 = \angle 4$

STATEMENTS	REASONS
1. $XE = XU$	1. __?__ Given: $\overline{AO}$ bisects $\overline{EU}$.
2. $XA = XO$	2. __?__ Given: $\overline{EU}$ bisects $\overline{AO}$.
3. $\angle 1 = \angle 2$	3. __?__ Vertical angles are equal.
4. $\triangle AXE \cong \triangle OXU$	4. __?__ SAS Postulate.
5. $\angle 3 = \angle 4$	5. __?__ Corr. parts of $\cong \triangle$ are $=$.

2. Given: $AB = AC$; $BD = CE$
   Prove: $BE = CD$

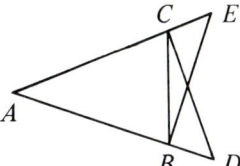

STATEMENTS	REASONS
1. $BD = CE$	1. __?__ Given
2. $AB = AC$	2. __?__ Given
3. $AD = AE$	3. __?__ Addition Postulate
4. $\angle A = \angle A$	4. __?__ From algebra
5. $\triangle ABE \cong \triangle ACD$	5. __?__ SAS Postulate
6. $BE = CD$	6. __?__ Corr. parts of $\cong \triangle$ are $=$.

(These steps correspond to the first steps of Euclid's proof of Theorem 3, page 167.)

Suggested Extension

Have students prove the *transitive* property of congruence: If two triangles are congruent to the same triangle, they are congruent to each other.

Given: $\triangle ABC \cong \triangle XYZ$
$\triangle DEF \cong \triangle XYZ$
Prove: $\triangle ABC \cong \triangle DEF$

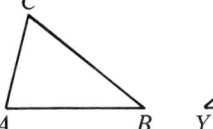 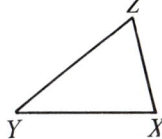 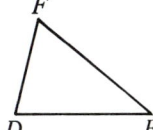

Proof on page T72

Lesson Commentary **T71**

1. $\triangle ABC \cong \triangle XYZ$; $\triangle DEF \cong \triangle XYZ$ (Given)
2. $AB = XY$; $DE = XY$
   $BC = YZ$; $EF = YZ$ (Corr. parts of $\cong \triangle$ are =.)
   $AC = XZ$; $DF = XZ$
3. $AB = DE$; $BC = EF$; $AC = DF$ (Subst. Post.)
4. $\triangle ABC \cong \triangle DEF$ (SSS Post.)

2 Congruent Triangles and Constructions (pp. 149–152)

Teaching These Pages

Point out to students that in showing Constructions 1 and 2 to be correct, we use the fact that corresponding parts of congruent triangles are equal. In each case, the construction involves locating the vertices of congruent triangles, even if the procedure does not, itself, require drawing complete triangles. In Construction 1, for instance, we must complete $\triangle OXP$ and $\triangle OYP$ in order to prove that the construction is correct. The justifications of Constructions 3 and 4 are left as Written Exercises 7 and 9, respectively, on page 151.

Chalkboard Examples

1. Draw an $\angle AOB$ and bisect the angle in this particular way: After you draw an arc with point O as center, use the *same* compass setting to draw the arcs that intersect at P.

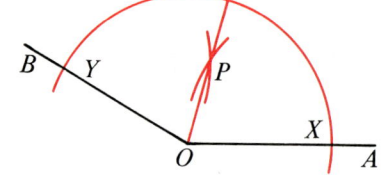

2. Are the five steps of the proof of Construction 1 on page 149 true for your figure? yes

3. What special kind of triangle is $\triangle OXP$? isosceles

Suggested Extension

Prove that:
   (1) Any point on the bisector of an angle is equidistant from the sides of the angle.
   (2) Any point that is equidistant from the sides of an angle is on the bisector of the angle.

Note the similarity between these two facts and Theorems 1 and 2 on pages 154–155.

Quick Quiz

1. Supply reasons to complete the proof.

   Given: $\overline{OM}$ bisects $\angle NOL$.
   $ON = OL$

   Prove: $\angle 3 = \angle 4$

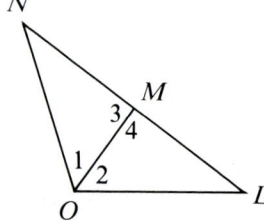

STATEMENTS	REASONS
1. $ON = OL$	1. __?__ Given
2. $\angle 1 = \angle 2$	2. __?__ Given: $\overline{OM}$ bisects $\angle NOL$
3. $OM = OM$	3. __?__ From algebra
4. $\triangle MON \cong \triangle MOL$	4. __?__ SAS Postulate
5. $\angle 3 = \angle 4$	5. __?__ Corr. parts of $\cong \triangle$ are $=$.

2. Write a proof in two-column form.
   Given: M is the midpoint of $\overline{AB}$.
   $\angle C$ and $\angle D$ are right angles.
   $AD = BC$
   Prove: $\angle A = \angle B$

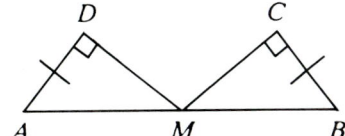

   1. $AM = MB$ (Given: M is the midpt. of $\overline{AB}$.)
   2. $AD = BC$ (Given)
   3. $\angle C$ and $\angle D$ are rt. $\triangle$. (Given)
   4. $\triangle AMD \cong \triangle BMC$ (HL Theorem)
   5. $\angle A = \angle B$ (Corr. parts of $\cong \triangle$ are $=$.)

3. In the diagram, $\angle P$ was given.
   $\angle Q$ was constructed using Construction 4.
   Give the reasons for each statement below.
   a. $\triangle PLM \cong \triangle QRS$ SSS Postulate
   b. $\angle P = \angle Q$ Corr. parts of $\cong \triangle$ are $=$.

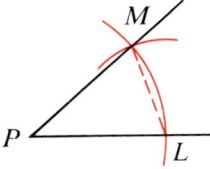 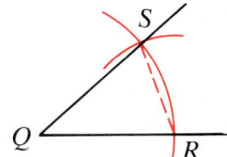

Objectives for Pages 154–166

1. To construct the **perpendicular bisector** of a segment.
2. To construct the **altitudes** of a triangle.
3. To construct the **medians** of a triangle.
4. To **inscribe** a circle in a triangle.
5. To **circumscribe** a circle about a triangle.

3 Segment Bisectors (pp. 154–157)

Teaching These Pages

To most students, it is quite clear that we can always construct a line which is a perpendicular bisector of a segment, and that there is only one such line. As in the case of angle bisectors (Chapter 1, Section 6), the text omits a discussion of existence and uniqueness issues. Students may refer either to *the* perpendicular bisector or to *a* perpendicular bisector of a segment.

Lesson Commentary **T73**

You may want to help the students who appreciate these issues to use Theorems 1 and 2 to show that, in a plane, every segment has one, and only one, perpendicular bisector. Some students may be interested in showing that, in space, the perpendicular bisector of a segment is the *plane* of points equidistant from the endpoints of the segment.

Written Exercise 17 outlines a justification for Construction 6, page 154.

Chalkboard Examples

Begin each example by drawing a segment AB.

1. Draw a line that bisects $\overline{AB}$ but is not perpendicular to $\overline{AB}$.
2. Draw a line that is perpendicular to $\overline{AB}$ but does not bisect $\overline{AB}$.
3. Draw a line that is the perpendicular bisector of $\overline{AB}$.
4. Draw a line that is perpendicular to $\overleftrightarrow{AB}$ but does not intersect $\overline{AB}$.

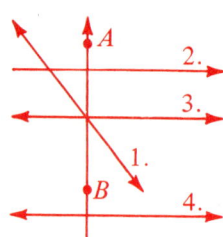

Suggested Extension

1. Draw a segment $\overline{AB}$. Construct the perpendicular bisector of $\overline{AB}$ using the procedure suggested by the diagram at right.
   See black portion of art.

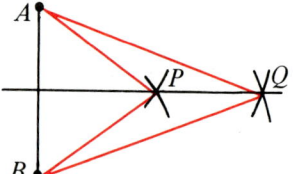

2. Draw the appropriate segments in your construction.
   See red portion of art.
   Prove that $\angle PAQ = \angle PBQ$.
   1. $PA = PB$; $QA = QB$ (Construction)
   2. $PQ = PQ$ (From algebra)
   3. $\triangle PAQ \cong \triangle PBQ$ (SSS Postulate)
   4. $\angle PAQ = \angle PBQ$ (Corr. parts of $\cong \triangle$ are =.)

4 Altitudes and Medians of a Triangle
(pp. 158–162)

Teaching These Pages

Some students find it a formidable challenge to draw the altitude from X in $\triangle XYZ$ pictured at right. You might suggest that they draw line YZ and point X very prominently. In this way, they can focus their attention, properly, on a line and a point. Some students may need to physically rotate their papers so that $\overleftrightarrow{YZ}$ is horizontal with X above it.

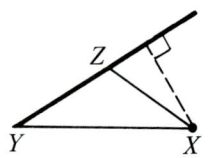

Experiment 1 on page 162 and the Suggested Extension on page T75 provide two demonstrations that the altitude and median to the base of an isosceles triangle are the same segment.

Chalkboard Examples

1. In each triangle below, decide which altitude is the shortest, the one from X, Y, or Z.

   a.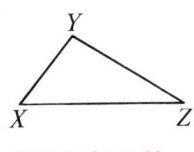
   altitude from Y

   b.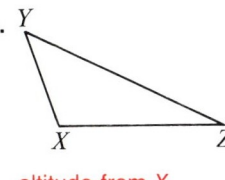
   altitude from X

   c.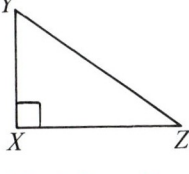
   altitude from X

2. In each triangle above, decide which median is the shortest, the one from X, Y, or Z. **a.** median from Y **b.** median from X **c.** median from X

3. $\triangle RST$ is an isosceles triangle, with $RS = RT$. Which two altitudes of $\triangle RST$ are equal? altitudes from S and T

4. Why must the medians of a triangle meet *inside* the triangle?
   The medians themselves are entirely inside the triangle.

Suggested Extension

1. Draw a $\triangle ABC$ in which the median from A is longer than the altitude from A. Can $\triangle ABC$ be an isosceles triangle, with $AB = AC$? no

2. Draw a $\triangle ABC$ in which the median from A is the same segment as the altitude from A.
   (isosceles, with vertex A)

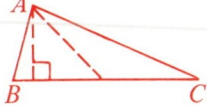
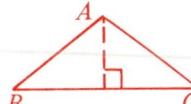

3. Can you draw a $\triangle ABC$ in which the median from A is shorter than the altitude from A? no

4. In Exercise 2 above, note that the altitude from A is the perpendicular bisector of $\overline{BC}$. From Theorem 1 on page 154, you can conclude that ___?___ = ___?___. AB; AC

5 Inscribed and Circumscribed Circles *(pp. 163–166)*

Teaching These Pages

Students should realize that all three perpendicular bisectors of the sides of a triangle intersect at one point. (See Classroom Practice Exercise 5.) Students may select any pair of sides to bisect in order to locate that point, the *circumcenter*. You may want students to construct all the perpendicular bisectors as a check on their mechanical accuracy.

Chalkboard Examples

1. Draw an acute $\triangle ABC$ with the shortest side at least 10 cm long. Inscribe a circle, M, in $\triangle ABC$.

2. Use your work from Example 1. Label as R, S, and T the points where the circle touches $\overline{AB}$, $\overline{BC}$, and $\overline{AC}$. Draw $\triangle RST$ and inscribe a circle, N, in $\triangle RST$.

3. Do the circles in Examples 1 and 2 have the same center? no

4. Which of the circles in Examples 1 and 2 can be called a circumscribed circle? the circle through points R, S, and T

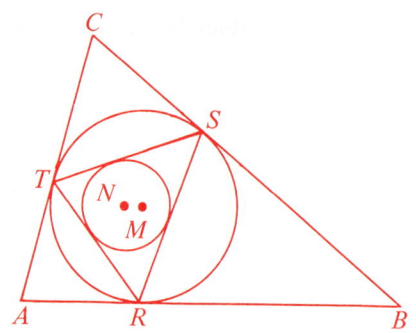

Suggested Extension

The steps below are an attempt to adapt Construction 7 on page 163 to construct a circle that passes through three collinear points.

1. Draw three collinear points A, B, and C so that B is between A and C.

2. Construct the perpendicular bisectors of $\overline{AB}$ and $\overline{BC}$.

3. What can you say about the intersection of the perpendicular bisectors? They do not intersect (Theorem 8, page 72).

4. Is there any circle that contains points A, B, and C? no

Quick Quiz

Constructions are shown in red in the diagrams.

Draw four triangles roughly like, but much larger than, the ones shown below.

1. In $\triangle ABC$, construct the altitude from A. Use Construction 3.

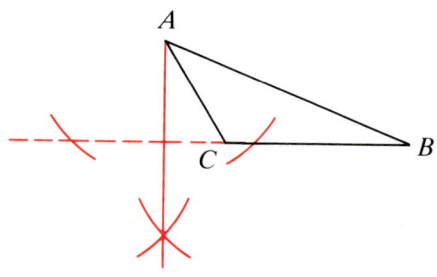 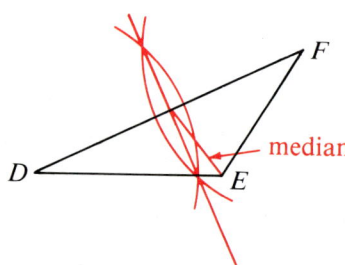

2. In $\triangle DEF$, construct the median from E. Use Construction 6.

Lesson Commentary

3. Inscribe, by construction, a circle in △ GHI. Construction 8

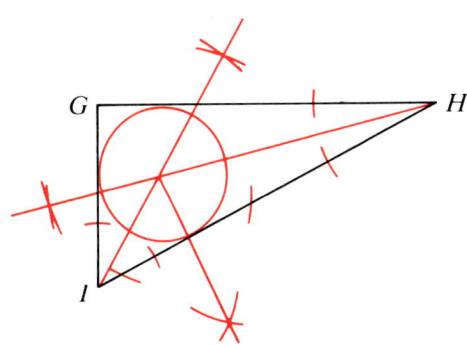
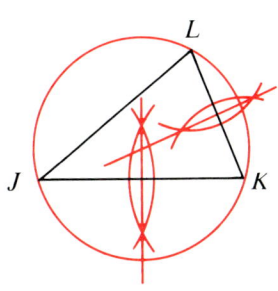

4. Circumscribe, by construction, a circle about △ JKL. Construction 7

Objectives for Pages 167–178

1. To use the theorem: If two sides of a triangle are equal, then the angles opposite those sides are equal.
2. To use the theorem: If two angles of a triangle are equal, then the sides opposite those angles are equal.
3. To construct a 60° angle.

6 Triangles with Two Equal Sides
   (pp. 167–171)

Teaching These Pages

We have chosen a somewhat lengthy phrasing for Theorem 3 on page 167. Some teachers prefer to have students use a simpler statement of this theorem: Base angles of an isosceles triangle are equal. A disadvantage of the latter phrasing, however, is that many students find it difficult to formulate the converse (Theorem 4, page 173).

The corollary on page 167, along with Theorem 1 on page 92, provides the logical basis for the construction of a 60° angle (Construction 9, page 174).

Chalkboard Examples

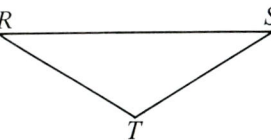

Given: △ RST with RT = ST. Classify each statement as *true*, *false*, or *cannot be determined*.

1. △ RST is isosceles.   true
2. △ RST is equilateral.   can't be det.
3. △ RST is equiangular.   can't be det.
4. RT > RS   can't be det.
5. ∠ R = ∠ S   true
6. ∠ S = ∠ T   can't be det.

Find the value of x.

7.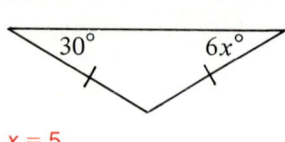
   $x = 5$

8.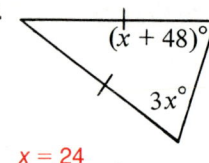
   $x = 24$

9.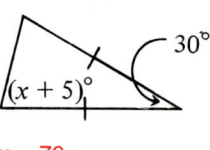
   $x = 70$

Suggested Extension

In the diagram $OA = OB$, and $\angle O = 20°$. Triangles ABC, BCD, and CDE are isosceles. What is the measure of $\angle ODE$? 20°

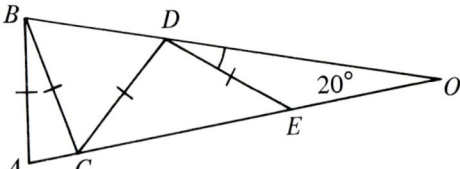

7 Triangles with Two Equal Sides
(pp. 173–178)

Teaching These Pages

Point out that we construct a 60° angle as part of an equilateral (and therefore equiangular) triangle. The outline of a proof justifying Construction 9 appears on page 174 following the construction.

Chalkboard Examples

Equal angles are marked. Name the sides that must be equal. Then find the value of x.

1.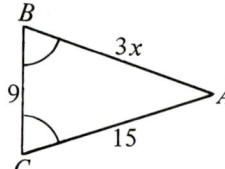
   $AB = AC$; $x = 5$

2.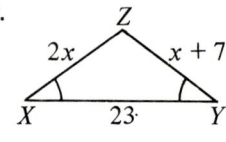
   $XZ = YZ$; $x = 7$

3.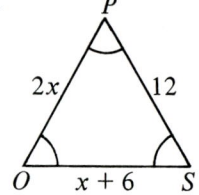
   $PO = OS = PS$; $x = 6$

4.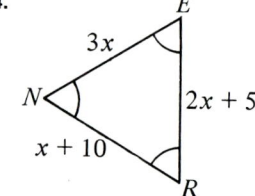
   $NE = ER = NR$; $x = 5$

5.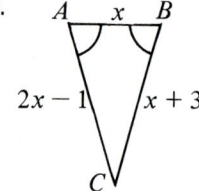
   $AC = BC$; $x = 4$

6.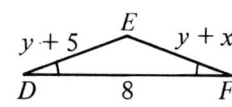
   $DE = FE$; $x = 5$

Suggested Extension

Steps 1–3 describe the construction of a 75° angle.

1. Construct an equilateral $\triangle ABC$.
2. Bisect $\angle ABC$, letting the bisector intersect $\overline{AC}$ in X.
3. Bisect $\angle ABX$, letting the bisector intersect $\overline{AX}$ in Y.
4. State the measures of the following angles:
   a. $\angle ABC$ 60° **b.** $\angle ABY$ 15° **c.** $\angle YBC$ 45° **d.** $\angle BYC$ 75°

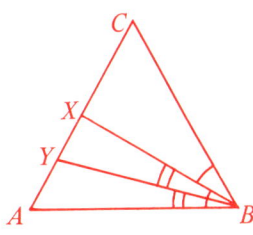

Quick Quiz

In each diagram, there are two equal angles. Find the value of x.

1. 11
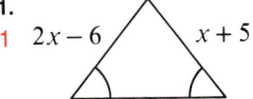

2. 25
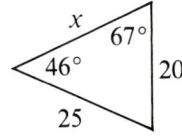

3. 6
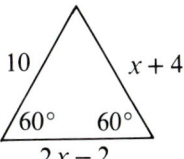

In each diagram, equal sides are marked. Find the value of y.

4. 70
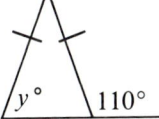

5. 81
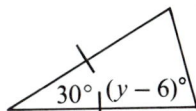

6. 12
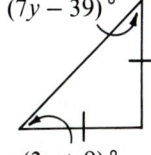

7. Construct an angle measuring 120°.

Construct two 60° $\angle$s as shown, or construct a 60° $\angle$ and extend one side through the vertex, thus creating a supplementary $\angle$ of 120°.

Chapter 5. Polygons

In the first two sections of this chapter, polygons and their angle sums are discussed. In the remaining sections, the properties of special quadrilaterals are investigated. Included in the latter sections of the chapter are several Explorations. You might have students try these Explorations in class or as homework, encouraging them to hypothesize the properties of special quadrilaterals.

Allowing students to be a part of the discovery process is especially important in the study of geometry. Proving one's own conjectures is much more rewarding and motivating for a student than proving a theorem listed in a book or stated by a teacher.

Lesson Commentary

Objectives for Pages 190–197

1. To classify polygons by the number of sides, and by the terms *equilateral*, *equiangular*, *regular*, and *convex*.

2. To find the interior and exterior angle sums of a convex polygon.

3. To find the measures of the interior and exterior angles of a regular polygon.

1 Introducing Polygons *(pp. 190–193)*

Teaching These Pages

The easiest way to teach the polygon concept is to give examples. A formal definition of *polygon* can be complicated and not very helpful to beginning geometry students.

> A polygon is a figure formed by three or more coplanar segments such that:
> 1. No two segments with a common endpoint are collinear.
> 2. Each segment intersects exactly two other segments, but only in endpoints.

You may wish to present this only as an illustration of how difficult a formal definition can be. Some students might have fun trying to make their own precise definitions of "polygon" or "convex polygon."

> A convex polygon is a polygon with the property that every segment which joins two points of the polygon's interior lies wholly in the interior. *Example:* The figure at right is not a convex polygon because segment *PQ* does not lie wholly in the interior of the figure.

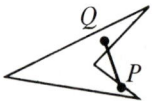

In the study of triangles, the terms *equilateral*, *equiangular*, and *regular* all denote the same figure. You might use a nonsquare rhombus, a rectangle, and a square to demonstrate that the terms are not equivalent in general. You can provide further reinforcement with the three hexagons pictured below.

 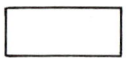 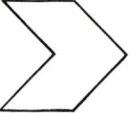  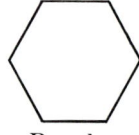

Equilateral   Equiangular   Regular   Equilateral   Equiangular   Regular

Chalkboard Examples

Draw each of the following figures. (Drawings are omitted.)

1. A pentagon
2. An octagon
3. A nonconvex quadrilateral
4. An equilateral triangle
5. An equiangular triangle
6. A regular quadrilateral

7. Is it possible to draw an equiangular triangle that is not equilateral? no

8. What is a regular quadrilateral usually called? square

Suggested Extension

Challenge students to find a pattern in the table given for Written Exercises 28–32 on page 193. They can use this pattern to predict the number of diagonals if the number of sides is 9, 10, 11, and so on. Some students may like to find a formula that gives the number of diagonals, d, in terms of the number of sides, s.

$$d = \frac{s(s-3)}{2}$$

From each of the s vertices, $s - 3$ diagonals can be drawn, for a total of $s(s - 3)$. But each diagonal has been counted twice in this scheme. Divide by 2 to get the correct count.

2 Angle Sums of Polygons (pp. 194–197)

Teaching These Pages

Remind students that one case has already been settled: all triangles have an angle sum of 180°. You may wish to begin further investigation with the "simplest" quadrilaterals: squares and rectangles. Students will quickly see that the interior angle sum is always $4 \times 90° = 360°$. Then try another quadrilateral (perhaps one with just 2 right angles): will its angle sum also be 360°? Suggest to students that they "break" the quadrilateral into triangles and use what they already know about a triangle's angle sum.

You might get more than one proof of the 360° angle sum for the quadrilateral. Here are two possibilities, both of which suggest the general formula, $S = (n - 2) \times 180°$.

Sum = $2 \times 180°$

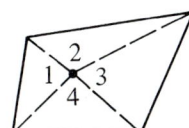

Sum = $4 \times 180° - (\angle 1 + \angle 2 + \angle 3 + \angle 4)$
   = $720° - 360° = 360°$

Generalize:
  Sum = $(n - 2) \times 180°$

Sum = $n \times 180° - 360°$
   = $n \times 180° - 2 \times 180°$
   = $(n - 2) \times 180°$

The measure of each angle of a regular polygon is given by the formula $A = \frac{(n-2) \times 180°}{n}$, as illustrated in the Example on page 195. Students, however, usually prefer the formula $A = 180° - \frac{360°}{n}$, which follows from the fact that each interior angle is the supplement of an exterior angle whose measure is $\frac{360°}{n}$.

Chalkboard Examples

Complete the table for convex polygons.

                                 1.   2.   3.

Number of sides	5	7	102
Interior angle sum	?	?	?
Exterior angle sum	?	?	?

                   540° 900° 18,000°
                   360° 360°   360°

Complete the table for regular polygons.

                           4.   5.   6.

Number of sides	6	10	20
Each exterior angle	?	?	?
Each interior angle	?	?	?

             60°   36°   18°
           120° 144° 162°

Suggested Extension

Some students may be interested in the art of M. C. Escher. Many of his works involve filling the plane with interesting repeated figures. An excellent reference is: Sheila Haak, "Transformation Geometry and the Artwork of M. C. Escher," *The Mathematics Teacher*, 69:647–652 (December, 1976).

Quick Quiz   1. g   2. d   3. b   4. e   5. a   6. h

Match each figure with the name that best describes it.

    a. equiangular quadrilateral   **e.** convex hexagon
    b. regular pentagon   **f.** regular hexagon
    c. regular quadrilateral   **g.** non-convex hexagon
    d. equilateral octagon   **h.** equilateral quadrilateral

1. 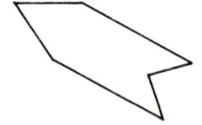   **2.** 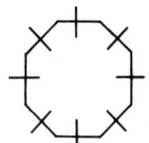   **3.**

4. 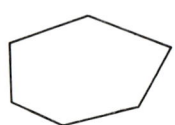   **5.**    **6.**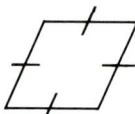

7. Find the measure of each interior angle of a regular 12-sided polygon. **150°**

8. Find the sum of the exterior angles, one at each vertex, of a convex octagon. **360°**

Objectives for Pages 198–207

1. To classify quadrilaterals as parallelograms, rectangles, rhombuses, squares, and trapezoids.

2. To use the properties of parallelograms.

3 Special Quadrilaterals (pp. 198–202)

Teaching These Pages

We suggest that you present examples of the five special quadrilaterals one at a time, in the order implied by the chart on page 198; trapezoid, parallelogram, rectangle, rhombus, and square. You may wish to have students suggest definitions for the figures. Listed below are some possible definitions.

> Rectangle: A parallelogram with 4 right angles
> A parallelogram with 1 right angle

> Rhombus: An equilateral parallelogram
> A parallelogram with two adjacent sides equal

> Square: A rhombus which is a rectangle
> A rectangle which is a rhombus
> A regular quadrilateral

Note that the text defines a trapezoid as having *exactly* one pair of parallel sides. Hence, a parallelogram is not a special case of a trapezoid. More advanced students might like to contrast this situation with that for isosceles and equilateral triangles.

Chalkboard Examples

1. Draw a parallelogram and one of its diagonals. What appears to be true about the two triangles formed? congruent

2. $PQRS$ is a rhombus.

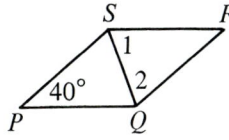

$\angle 1 = $ __?__ 70°
$\angle 2 = $ __?__ 70°

3. $ABCD$ is a rectangle.

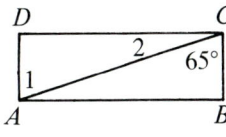

$\angle 1 = $ __?__ 65°
$\angle 2 = $ __?__ 25°

4. $PQRS$ is a square.

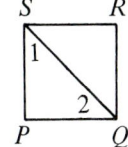

$\angle 1 = $ __?__ 45°
$\angle 2 = $ __?__ 45°

4 Properties of Parallelograms (pp. 203–207)

Teaching These Pages

We suggest that you teach this section by following the steps of the Explorations in the text. Allow plenty of time for students to form and test their conclusions. Students should be able to express the theorems on page 204 in "if . . . then" form. For example: "Theorem 3: If a polygon is a parallelogram, then opposite sides of the polygon are equal." The converses of the theorems will be considered in Section 6.

Lesson Commentary **T83**

Chalkboard Examples

ABCD is a parallelogram. Find:

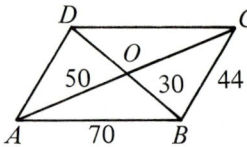

1. *CO* 50
2. *BD* 60
3. *CD* 70
4. *AD* 44

LOAB is a parallelogram. *SAOT* is a rhombus. Find each length and angle measure.

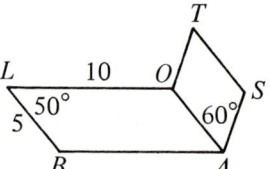

5. *AO* 5
6. *OT* 5
7. *AB* 10
8. ∠*LBA* 130°
9. ∠*BAS* 110°
10. ∠*TSA* 120°

Quick Quiz

Classify as true or false.

1. Every rhombus is a parallelogram. true
2. A trapezoid is a parallelogram. false
3. A square is both a rectangle and a rhombus. true
4. A quadrilateral with four right angles must be a square. false

PQRS is a parallelogram.

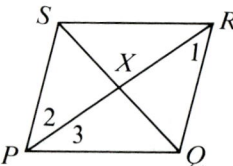

5. *SR* = __?__ PQ
6. *SX* = __?__ XQ
7. *SP* = __?__ RQ
8. ∠1 = ∠ __?__ ∠2
9. ∠*RQP* = ∠ __?__ ∠*PSR*

Objectives for Pages 209–219

1. To use the properties of rectangles, rhombuses, and squares.
2. To prove that quadrilaterals with particular properties are parallelograms.

5 Properties of Special Parallelograms (pp. 209–213)

Teaching These Pages

We suggest that you teach this section by following the steps of the Explorations in the text. Theorem 9, a corollary of Theorems 6 and 7, appears unexpectedly, providing information about right triangles. The Suggested Extension on page T85 shows how this result sheds light on the construction of circumscribed circles.

Chalkboard Examples

**TRAK is a rectangle. BARS is a square.
Find the measure of each angle.**

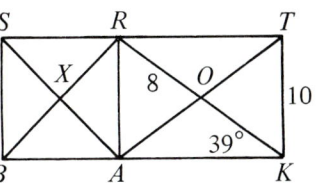

1. ∠KAT 39°
2. ∠TRO 39°
3. ∠RXS 90°
4. ∠XRS 45°

Find each length in the figure above.

5. KO 8
6. OT 8
7. AT 16
8. AB 10

BLUR is a rhombus. Find the measure of each angle.

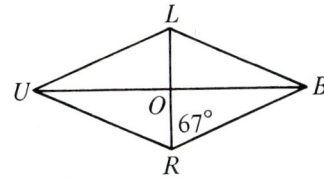

9. ∠LRU 67°
10. ∠BOR 90°
11. ∠RBO 23°
12. ∠LBO 23°

Suggested Extension

Theorem 9 implies that the midpoint of the hypotenuse of a right triangle is also the center of the circumscribed circle. Some students might like to verify these related facts:

> If a triangle is acute, the center of the circumscribed circle is inside the triangle.

> If a triangle is obtuse, the circumcenter is outside the triangle.

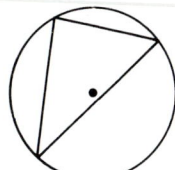
Acute triangle

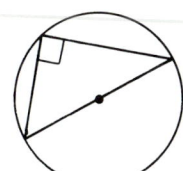
Right triangle

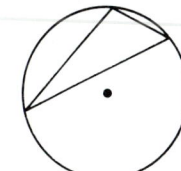
Obtuse triangle

6 Proving Figures Are Parallelograms (pp. 214–219)

Teaching These Pages

As a sequel (or an alternative) to the Explorations in the text, you might have students review the properties of a parallelogram (Theorems 3–6, page 204). Then ask whether or not they think the converse of each theorem is true. (They all are!)

The converse of Theorem 4 is not listed on page 216 as a method for proving that a figure is a parallelogram, because this method is not used very often (but see Suggested Extension 4 for this lesson). Students can prove the validity of this method in Written Exercise 15 on page 219.

Lesson Commentary   **T85**

In Written Exercises 16 and 17 on page 219, students use conditions on diagonals to prove that figures are special parallelograms. Point out to students that if *both* conditions are met, then the figure must be a square.

We have not included in this text the theorem that *if two lines are parallel to a third line, then they are parallel to each other.* This useful principle is mentioned and used in Written Exercise 8 on page 218 and in Chalkboard Example 3 below.

Chalkboard Examples

1. *Given:* PQRS is a parallelogram
TR = UQ

Prove: (a) RTUQ is a parallelogram.
(b) STUP is a parallelogram.

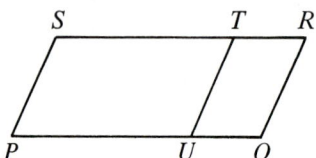

1. $\overline{TR} = \overline{UQ}$ (Given)
2. $\overline{TR} \parallel \overline{UQ}$ (Opp. sides of a ▱ are ∥.)
3. RTUQ is a ▱ (Thm. 10)
4. SR = PQ (Opp. sides of a ▱ are =.)
5. ST = PU (Subtraction Post.)
6. $\overline{ST} \parallel \overline{PU}$ (Opp. sides of a ▱ are ∥.)
7. STUP is a ▱ (Thm. 10)

2. *Given:* LANK is a parallelogram.
$\overrightarrow{KX}$ bisects ∠ LKN.
$\overrightarrow{AY}$ bisects ∠ NAL.

Prove: △ XLK ≅ △ YNA

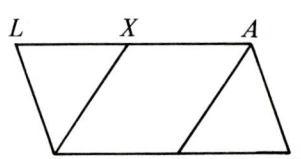

1. ∠ LKN = ∠ NAL; ∠ L = ∠ N
(Opp. ∠s of a ▱ are =.)
2. ∠ LKX = $\frac{1}{2}$ ∠ LKN; ∠ NAY = $\frac{1}{2}$ ∠ NAL
(*Given:* angle bisectors)
3. ∠ LKX = ∠ NAY (Mult. and Subst. Post.)
4. LK = NA (Opp. sides of a ▱ are =.)
5. △ XLK ≅ △ YNA (ASA Post.)

3. Supply the reasons to complete the proof. Use the fact that *if two lines are parallel to the same line, then they are parallel to each other.*

Given: TONG is a parallelogram.
KANG is a parallelogram.
Prove: TOAK is a parallelogram.

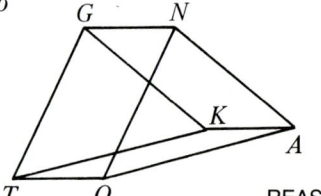

STATEMENTS		REASONS
1. TO = GN; GN = KA	1.	__?__ Opp. sides of a ▱ are equal
2. TO = KA	2.	__?__ Substitution Postulate
3. $\overline{TO} \parallel \overline{GN}$; $\overline{GN} \parallel \overline{KA}$	3.	__?__ Definition of parallelogram
4. $\overline{TO} \parallel \overline{KA}$	4.	__?__ If two lines are ∥ to the same line, then they are ∥ to each other.
5. TOAK is a parallelogram.	5.	__?__ If a quad. has one pair of opp. sides both ∥ and =, then it is a ▱.

Lesson Commentary

Suggested Extension   1–3. Methods of construction may vary.

1. Construct a rhombus whose diagonals have lengths 4 cm and 7 cm.
2. Construct a rhombus which has a 60° angle and sides with length 5 cm.
3. Construct a rectangle having one side with length 6 cm and a diagonal with length 8 cm.

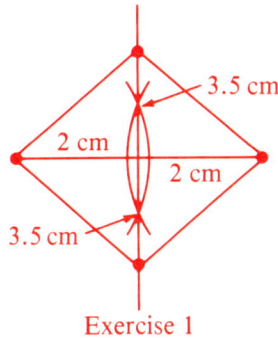

Exercise 1 / Exercise 2 / Exercise 3

4. Use the converse of Theorem 4 in a proof of the following theorem.
   Given: PQRST is a regular pentagon.
   Prove: STOR is a rhombus.

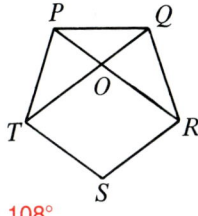

Strategy:
(1) Regular pentagon: ∠ TPQ = ∠ PQR = ∠ S = 108°.
(2) In isos. △ TPQ and QPR, show ∠ PTQ = ∠ QRP = 36°.
(3) Show ∠ OTS = ∠ ORS = 72°.
(4) Use angle sum of quad. to show ∠ TOR = 108°.
(5) Opp. angles in STOR are =. Hence it is a ▱.
(6) Since TS = SR, ▱STOR is a rhombus.

Quick Quiz

QRST is a rhombus, RS = 13, QS = 10, and ∠ RSQ = 67°.

1. RQ = __?__ 13
2. ∠ TSQ = __?__ ° 67
3. ∠ RXQ = __?__ ° 90
4. SX = __?__ 5
5. ∠ SRX = __?__ ° 23
6. ∠ SRQ = __?__ ° 46

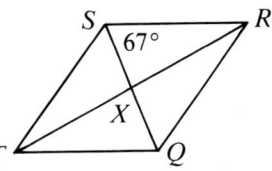

Given LMNO is a rectangle, OM = 20, and ∠ 1 = 22°.

7. LN = __?__ 20
8. XM = __?__ 10
9. ∠ 2 = __?__ ° 22
10. ∠ 3 = __?__ ° 68

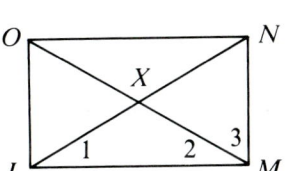

Lesson Commentary   **T87**

Is the information enough to prove that *QUAD* is a parallelogram?

11. X is the midpoint of $\overline{AQ}$ and of $\overline{DU}$. yes
12. $\overline{AD} \parallel \overline{QU}$ and $QD = AU$. no
13. $\triangle AUQ \cong \triangle QDA$ yes

Objectives for Pages 220–228

1. To use the properties of trapezoids.   **2.** To apply the Midpoints Theorem.

7 Properties of Trapezoids *(pp. 220–224)*

Teaching These Pages

You may wish to introduce the isosceles trapezoid by slicing off the vertex of an isosceles triangle with a cut parallel to the base. If you adopt this approach, students should be able to reason that the base angles of an isosceles trapezoid are equal (Theorem 13, page 221). An outline of a formal proof appears on page 221 after the statement of the theorem.

Students should note that, unlike a median of a triangle, the median of a trapezoid does not contain a vertex of the polygon. The properties of the median developed in Exploration B are listed in Theorem 14. A formal proof of this theorem is postponed until Chapter 12 (see page 501), but Classroom Practice Exercises 8 and 9 make the result seem very reasonable. (Better students may note that Theorem 5 on page 68 is needed to prove that the two trapezoids "fit together" to form a parallelogram.)

Chalkboard Examples

For each trapezoid, find the length of its median $\overline{XY}$ and the measures of its angles.

1.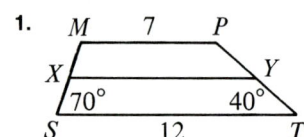

   $XY = 9.5$
   $\angle M = 110°$
   $\angle P = 140°$

2.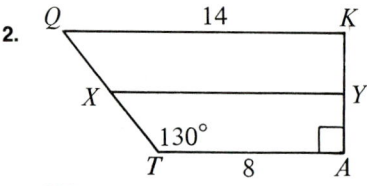

   $XY = 11$
   $\angle Q = 50°$
   $\angle K = 90°$

3.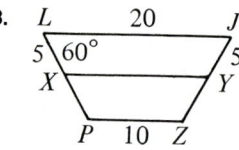

   $XY = 15$
   $\angle P = \angle Z = 120°$
   $\angle J = 60°$

Suggested Extension

The Puzzle Problems feature on page 162 reviews several skills which students have learned in this course: the angle sum of a triangle (Chapter 3), the measure of a straight angle (Chapter 1), vertical angles (Chapter 1), and inscribed circles (Chapter 4).

8 The Midpoints Theorem *(pp. 225–228)*

Teaching These Pages

You might begin this lesson by drawing figures (a), (b), and (c) of page 225 on the chalkboard. Ask students what they know about the median $\overline{MN}$. $\overline{MN} \parallel \overline{AB}$ and $MN = \frac{1}{2}(AB + DC)$ Then draw figure (d) and ask for their conjectures about $\overline{MN}$. (Remind students that in a triangle, $\overline{MN}$ is *not* called a median.)

The strategy of considering a triangle to be a quadrilateral with one side of zero length often yields theorems about triangles from theorems about quadrilaterals. (See the Suggested Extension for this lesson.) Here, Theorem 15 becomes a special case of Theorem 14. The formal proof of Theorem 15 has been postponed until Chapter 12. (See Classroom Practice, page 502.)

Chalkboard Examples

1. $PQ = \underline{\ ?\ }$ 2
2. $XY = \underline{\ ?\ }$ 8
3. Why is $\overline{MN} \parallel \overline{XY}$? Midpoints Thm.
4. Why is $\overline{PQ} \parallel \overline{MN}$? Midpoints Thm.

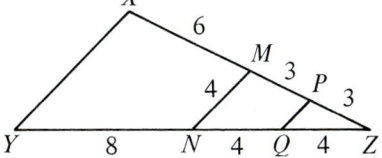

5. Draw equilateral $\triangle ABC$ and $\overline{DE}$, the segment joining the midpoints of sides $\overline{AB}$ and $\overline{AC}$. What kind of triangle is $\triangle ADE$? Why? equilateral; $AD = \frac{1}{2}AB$ and $AE = \frac{1}{2}AC$ by construction; $DE = \frac{1}{2}BC$ by the Midpoints Thm.; $AB = AC = BC$ by hypothesis, so $\frac{1}{2}AB = \frac{1}{2}AC = \frac{1}{2}BC$ by the Mult. Post. of Equality; $AD = AE = DE$ by substitution.

Suggested Extension

1. Derive a theorem about triangles from the following theorem about trapezoids. To do this, let one base of the trapezoid shrink to zero length, turning the figure into a triangle.

   Theorem: The area of a trapezoid with base lengths a and b and height h is $A = \frac{1}{2}h(a + b)$.

   Let $a = 0$ in the formula $A = \frac{1}{2}h(a + b)$: $A = \frac{1}{2}h(0 + b) = \frac{1}{2}h(b) = \frac{1}{2}bh$.
   Hence, the area of a triangle is given by the formula $A = \frac{1}{2}bh$.

2. Prove that the polygon formed by joining the midpoints of the sides of a rectangle is a rhombus. (At long last, we can prove the result of the activity on page 4!) See page T90 for a strategy for proof.

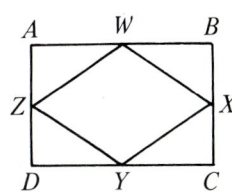

Lesson Commentary **T89**

2. *Strategy:*
   (1) Show AW = BW = CY = DY.
   (2) Show AZ = DZ = BX = CX.
   (3) In rectangle ABCD, ∠A = ∠B = ∠C = ∠D.
   (4) SAS Postulate: △AZW ≅ △BXW ≅ △CXY ≅ △DZY.
   (5) Corr. parts: ZW = XW = XY = ZY.
   (6) By definition, WXYZ is a rhombus.

Quick Quiz

$\overline{MD}$ is the median of trapezoid *TRAP*.

1. MD = __?__ 17
2. MT = __?__ 4
3. ∠T = __?__° 116
4. ∠1 = __?__° 64
5. If *TRAP* is an isosceles trapezoid, then ∠A = __?__. 64°

M, *I*, and *D* are midpoints, *MI* = 4, *AC* = 9, and *CB* = 12.

6. AB = __?__ 8
7. ID = __?__ 4.5
8. MD = __?__ 6
9. Name an angle equal to ∠IDM.
   ∠AMD, ∠MCI, or ∠DIB

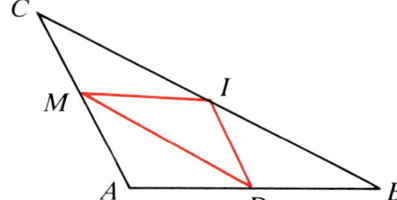

Chapter 6. Areas

The formula for the area of a rectangle is postulated and used. In succeeding sections, area formulas for parallelograms, triangles, and trapezoids are derived and used. The Pythagorean Theorem is developed by reasoning with areas. The last two sections of the chapter deal with circumferences and areas of circles. Every section provides opportunities for students to sharpen their computational skills, a change of pace from the deductive work of the three preceding chapters.

Objectives for Pages 238–245

1. To find the perimeter and area of a rectangle and a square.
2. To find the perimeter and area of a parallelogram.

1 Areas of Rectangles (pp. 238–242)

Teaching These Pages

Make sure students see that whenever a rectangle has sides with integral lengths, segments can be drawn to divide the rectangle into unit squares. Students will soon learn that after they count the squares in the first row, they automatically know the number of squares in each of the other rows. Hence, the total number of unit squares can be found by multiplying. For rectangles whose sides have lengths involving irrational numbers, the counting method breaks down. For this reason, we postulate the formula for the area of a rectangle: $A = bh$.

Note: In Postulate 12 on page 238, the term *base* is used to mean *base length*. This abbreviation is convenient and widely accepted. We occasionally shall use *base* to denote *base length* and *side* to denote *side length* when no confusion is likely.

Chalkboard Examples

1. Find the area of each of the five non-overlapping rectangles. (See the figure.)

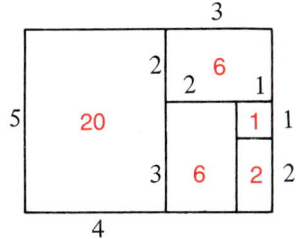

2. The sum of the areas of the five rectangles is __?__. 35

3. In the complete rectangle, $b = $ __?__ 7 and $h = $ __?__. 5

4. Find the area of the large rectangle by using the values of b and h found in Example 3. 35 Compare this with your answer to Example 2. They are equal.

5. Find the area of a square with sides 3 cm long. 9 cm²

Suggested Extension

Each side of a square is 4 m long.

1. The number of meters in the perimeter is __?__. 16

2. The number of square meters in the area is __?__. 16

3. The number of centimeters in the perimeter is __?__. 1600

4. The number of square centimeters in the area is __?__. 160,000

5. Is it proper to say: "The perimeter of this square is equal to its area"? No; the numbers depend on the unit of measure used.

Lesson Commentary   **T91**

2 Areas of Parallelograms *(pp. 243–245)*

Teaching These Pages

Plan to reinforce repeatedly the statement on page 243: The area of a parallelogram is not equal to the product of two consecutive sides (unless the parallelogram is a rectangle). Sometimes you will merely want to give a verbal reminder. As a stronger reminder, you might suggest that students study the three parallelograms above Classroom Practice Exercise 5 on page 244. They should see that these parallelograms cannot all have areas of (15 × 8) square units.

Some students may benefit from handling an actual stick model like that described on page 243.

Chalkboard Examples

For each parallelogram find (a) the area and (b) the perimeter. When the area or the perimeter cannot be computed, write *cannot be found*.

1.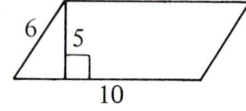
   a. 50 **b.** 32

2.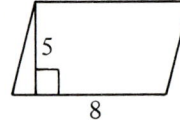
   a. 40 **b.** can't be found

3.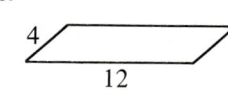
   a. can't be found **b.** 32

Suggested Extension

1. Find the areas of the three parallelograms.
   I. 200   II. 200   III. 200

   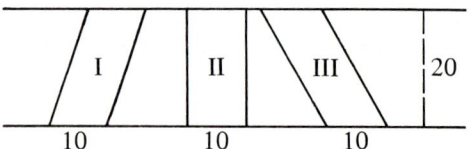

2. Can a parallelogram be drawn with
   $b = 12$ cm, $A = 24$ cm^2, and $p = 50$ cm? yes

3. Can a parallelogram be drawn with
   $b = 12$ cm, $A = 24$ cm^2, and $p = 25$ cm? no

Quick Quiz

1. Find the perimeter and area of the figure.
   (All angles are right angles.) $p = 22$ m; $A = 18$ m^2

   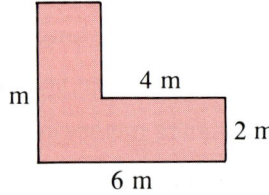

2. The perimeter of a square is 36 cm.
   The length of a side is __?__, 9 cm
   and the area is __?__. 81 cm^2

3. Find the perimeter and area of the
   parallelogram shown at the right.
   $p = 36$ cm, $A = 50$ cm^2

   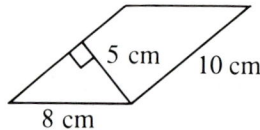

Objectives for Pages 247–254

1. To find the area of a triangle.
2. To find the area of a trapezoid.

3 Areas of Triangles *(pp. 247–249)*

Teaching These Pages

To develop the idea that the area of a triangle is half the area of a parallelogram, you might begin by having a student draw any triangle on the chalkboard. Complete a parallelogram and point out that the triangles are congruent, so that the original triangle has half the area of the parallelogram. Draw an altitude and emphasize that the triangle and the parallelogram have the same base length and height. The area formula follows easily.

Chalkboard Examples

1. Area of $\triangle ATS =$ __?__ 96
2. Area of $\triangle BTS =$ __?__ 54
3. Add the areas found in Examples 1 and 2.
   Area of $\triangle ABS =$ __?__ 150
4. Use AB as b and ST as h. Area of $\triangle ABS =$ __?__. $\frac{1}{2}(25)(12) = 150$
5. $\angle ASB$ is a right angle. Use SB as b and SA as h.
   Area of $\triangle ABS =$ __?__. $\frac{1}{2}(15)(20) = 150$

Find the area of the shaded triangle.

6.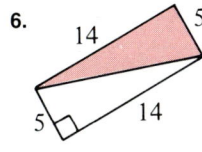

   35

7.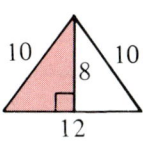

   24

8.

   41.5

Suggested Extension

The exercises below illustrate the idea of indirect proof. See Applications, pages 334–335.

1. Using DE as b, find the area of $\triangle DUE$. 504
2. Assume that $\angle DUE$ is a right angle. Using UE as b and DU as h, find the area of $\triangle DUE$. 520
3. Compare the results of the first two exercises. Can $\angle DUE$ be a right angle? no

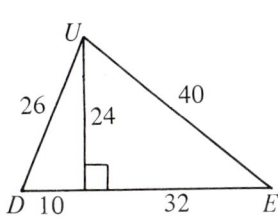

Lesson Commentary   **T93**

4 Areas of Trapezoids (pp. 250–254)

Teaching These Pages

You may wish to split the class into groups, each doing one of the Explorations on page 250. Then the class as a whole can focus on just one method, explained at the chalkboard.

Chalkboard Examples

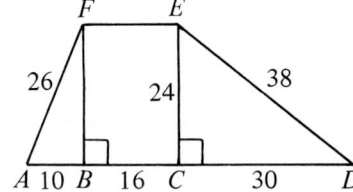

1. Area of $\triangle ABF$ = __?__ 120
2. Area of rectangle $BCEF$ = __?__ 384
3. Area of $\triangle DCE$ = __?__ 360
4. Add the areas found in Examples 1–3.
   Area of trapezoid $ADEF$ = __?__ 864
5. In trapezoid $ADEF$: b_1 = __?__; b_2 = __?__; h = __?__ 56; 16; 24
6. Using the formula $A = \frac{1}{2}h(b_1 + b_2)$, find the area of trapezoid $ADEF$. 864

Suggested Extension

In trapezoid $QRST$, $\overline{QX}$ has been drawn parallel to $\overline{TS}$.

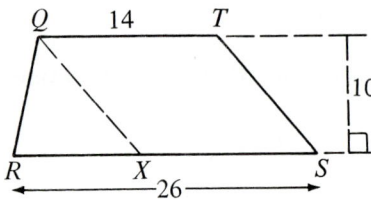

1. XS = __?__ 14
2. RX = __?__ 12
3. Area of $\triangle RXQ$ = __?__ 60
4. Area of $\square XSTQ$ = __?__ 140
5. Add the areas found in Exercises 3 and 4.
   Area of trapezoid $QRST$ = __?__ 200
6. Using the formula $\frac{1}{2}h(b_1 + b_2)$, find the area of trapezoid $QRST$. 200

Quick Quiz

Find the area of each polygon.

1.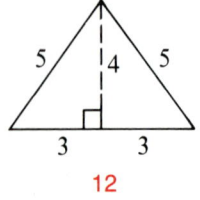
   12

2.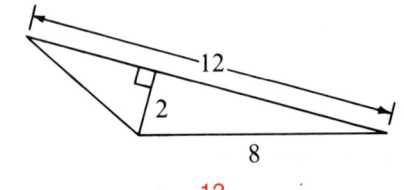
   12

3.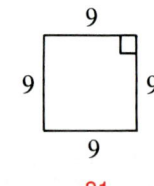
   81

T94   Lesson Commentary

4. 160

5. 150

6. 144

Objectives for Pages 255–261

1. To find the hypotenuse of a right triangle, given the legs.
2. To find a leg of a right triangle, given the hypotenuse and the other leg.
3. To determine, given the three sides of a triangle, whether the triangle is acute, right, or obtuse.

5 The Pythagorean Theorem *(pp. 255–258)*

Teaching These Pages

You might begin by having a student draw any right triangle on the chalkboard. Then you can build the rest of the figure shown at the bottom of page 255 step-by-step. Use ruler and protractor to draw the diagram so that students aren't distracted by straightedge and compass constructions.

The text uses the symbol "≈" to mean "is approximately equal to" in situations involving irrational square roots. Another symbol with the same meaning is "≐".

The Puzzle Problems feature on page 459 provides another area-based demonstration of the Pythagorean Theorem.

Chalkboard Examples

Find the indicated values. When necessary, use the square root table on page 516 and round to the nearest tenth.

1. *OB*   5
2. *OC*   13
3. *OD*   10.2
4. *PR*   1.4
5. *PS*   1.7
6. *PT*   2

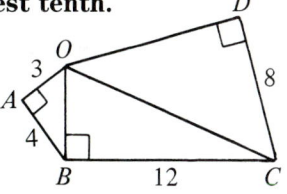
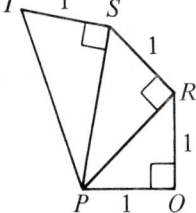

Suggested Extension

Have students use the Pythagorean Theorem to prove the validity of the HL Theorem (page 126). *Strategy:*
(1) *Given: AB = DE* and *AC = DF*
(2) $(BC)^2 = (AB)^2 - (AC)^2$; $(EF)^2 = (DE)^2 - (DF)^2$
(3) Subst. Postulate: $(BC)^2 = (EF)^2$, so *BC = EF*
(4) SSS Postulate: △*ABC* ≅ △*DEF*

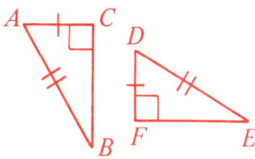

6 Converse of the Pythagorean Theorem (pp. 259–261)

Teaching These Pages

The text does not prove the theorem which underlies the strategy on page 260. You might use a 4-unit stick and a 3-unit stick fastened together to offer the class some visual evidence in support of the theorem. By holding one stick stationary and rotating the other, you can show how the distance between the free ends determines the angle which the sticks make. Only when the free ends are 5 units apart do we have a right angle.

Many students will see that the angle between the sticks determines the distance between the free ends (SAS Postulate). Hence, the converse strategy can be used to determine whether $c^2 < a^2 + b^2$, $c^2 = a^2 + b^2$, or $c^2 > a^2 + b^2$.

Chalkboard Examples

The lengths of the three sides of a triangle are given. Is the triangle a *right* triangle, an *acute* triangle, or an *obtuse* triangle?

1. 9, 12, 16   obtuse
2. 5, 12, 13   right
3. 4, 4, 6   obtuse
4. 11, 60, 61   right
5. 10, 10, 14   acute
6. 15, 112, 113   right

Quick Quiz

Find the value of x.

1.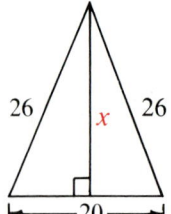

2.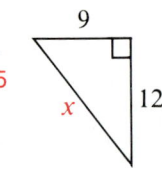

3.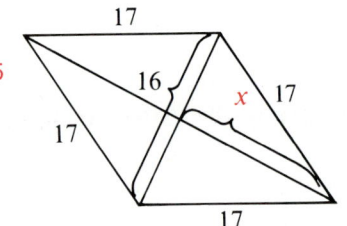

4.

The lengths of three sides of a triangle are given. Is the triangle right, acute, or obtuse?

5. $a = 1.5, b = 2, c = 2.5$ right
6. $a = 2, b = 4, c = 5$ obtuse
7. $a = 10, b = 15, c = 17$ acute
8. $a = 10, b = 24, c = 26$ right

Objectives for Pages 263–271

1. To find the circumference, radius, or diameter of a circle.
2. To find the area, radius, or diameter of a circle.
3. To state an approximation for π, correct to the nearest hundredth.

T96   Lesson Commentary

7 Circumferences of Circles (pp. 263–267)

Teaching These Pages

You might like to have some student volunteers measure the diameter and circumference of circular objects. When different students report that their measurements of a basketball hoop, a tractor tire, a pie tin, a coffee can, and so on, all support the statement $\frac{C}{d} \approx 3.1$, all students should be convinced that $\frac{C}{d}$ really does have just one value—whatever circles are used.

Chalkboard Examples

Complete the table about circles. When necessary, express answers in terms of π.

	1.	2.	3.	4.	5.	6.	
r	5	?	?	12	?	?	6, 9, 4.5, 5.5
d	?	12	?	?	9	?	10, 18, 24, 11
C	?	?	18π	?	?	11π	$10\pi, 12\pi, 24\pi, 9\pi$

Complete. Use 3.14 for π. Round answers to the nearest tenth.

7. When $r = 7.21$, $C =$ __?__ 45.3   8. When $C = 10$, $r =$ __?__ 1.6

8 Areas of Circles (pp. 268–271)

Teaching These Pages

Some students enjoy subjecting the formula $A = \pi r^2$ to a test. Have them draw circles of various radii on graph paper. They can estimate the areas of the circles by counting the unit squares that lie entirely inside the circle or appear to have at least half their area inside the circle. Then they can compare the counts to the values obtained for A using the formula $A = \pi r^2$.

Chalkboard Examples

Copy and complete the table about circles. When necessary, express answers in terms of π.

	1.	2.	3.	4.	5.	6.	
r	3	?	?	$\frac{3}{4}$	?	?	$7, 4, 2.5, \frac{4}{7}$
d	?	14	?	?	5	?	$6, 8, 1\frac{1}{2}, 1\frac{1}{7}$
A	?	?	16π	?	?	$\frac{16\pi}{49}$	$9\pi, 49\pi, \frac{9\pi}{16}, 6.25\pi$

7. Find the area of a circle whose circumference is 10π. 25π

8. Using 3.14 for π, find the area—correct to the nearest integer—of a circle whose radius is 7. 154

Quick Quiz

Complete these formulas about circles:

1. $A = \underline{\quad?\quad}$ πr^2 2. $C = \underline{\quad?\quad}$ or $\underline{\quad?\quad}$ $2\pi r;\ \pi d$

Complete the table about circles. Answers may be given in terms of π.

	3.	4.	5.	6.	
r	3 cm	?	?	?	10 in., 4 mi, 9 km
d	?	20 in.	?	?	6 cm, 8 mi, 18 km
C	?	?	8π mi	?	6π cm, 20π in., 18π km
A	?	?	?	81π km²	9π cm², 100π in.², 16π mi²

Chapter 7. Ratios and Proportions

In this chapter, we have two broad goals. First, we seek to teach soundly the fundamental ideas of ratio and proportion which are crucial to the study of similarity in Chapter 8. Second, we seek to show applications of ratio and proportion which are not part of the traditional study of similarity. These applications include the reading and making of maps, scale drawings, and mechanical drawings. Students should enjoy these ideas and find them helpful in their lives.

Objectives for Pages 286–295

1. To find the ratios of numbers and the ratios of lengths of segments.
2. To write and solve proportions.
3. To solve problems using proportions.

1 Ratios *(pp. 286–288)*

Teaching These Pages

Even if your students are good at algebra, it would be wise to review the concept of ratio and to run through both methods of Example 2 on page 286. Classroom Practice Exercises 13–15 and Written Exercises 17–28 are particularly important preparation for work with similar figures. The Consumer Applications on page 148 discusses an important ratio in daily life:

Consumer Price Index = (current total price of selection of typical consumer purchases) / (total price during "base year")

Chalkboard Examples

Express each ratio in simplest form.

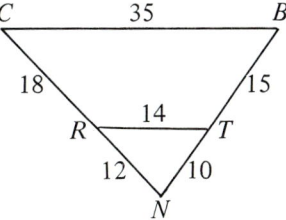

1. $\dfrac{RT}{CB}$ $\dfrac{2}{5}$
2. $\dfrac{BT}{TN}$ $\dfrac{3}{2}$
3. $\dfrac{BT}{BN}$ $\dfrac{3}{5}$
4. $\dfrac{RN}{CR}$ $\dfrac{2}{3}$
5. $\dfrac{RN}{CN}$ $\dfrac{2}{5}$
6. $\dfrac{TN}{RN}$ $\dfrac{5}{6}$
7. $\dfrac{\text{perimeter of } \triangle RTN}{\text{perimeter of } \triangle CBN}$ $\dfrac{2}{5}$

Express in simplest form:
a. the ratio of the area of the shaded part to the area of the unshaded part;
b. the ratio of the area of the shaded part to the total area.

8. 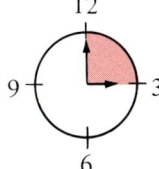 a. $\dfrac{1}{3}$ b. $\dfrac{1}{4}$
9. 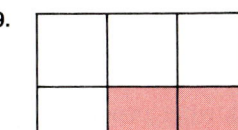 a. $\dfrac{1}{2}$ b. $\dfrac{1}{3}$

2 Proportions (pp. 289–292)

Teaching These Pages

After explaining that a proportion is an equation that states two ratios are equal, try Classroom Practice Exercises 1–4. Then discuss the four properties of proportions. (These properties are given names for reference purposes, but it is not necessary that students learn these names.) Classroom Practice Exercises 5–18 and the Chalkboard Examples below drill these four properties.

Chalkboard Examples

Complete each statement.

1. If $\dfrac{4}{12} = \dfrac{1}{3}$, then $\dfrac{4}{1} = \dfrac{?}{?}$ and $\dfrac{3}{12} = \dfrac{?}{?}$. $\dfrac{12}{3}; \dfrac{1}{4}$

2. If $\dfrac{x}{y} = \dfrac{3}{5}$, then $\dfrac{x+y}{y} = \dfrac{\frac{8}{5}?}{?}$.

3. If $\dfrac{2}{12} = \dfrac{3}{18}$, then $2 \times 18 = \underline{\ ?\ } \times \underline{\ ?\ }$. 12; 3

4. If $\dfrac{a}{b} = \dfrac{4}{5}$, then $\dfrac{b}{a} = \dfrac{?}{?}$ and $5a = \dfrac{\frac{5}{4}; 4b}{?}$.

5. If $\dfrac{13}{2} = \dfrac{x}{y}$, then $\dfrac{x+y}{y} = \dfrac{?}{?}$ and $\dfrac{x-y}{y} = \dfrac{?}{?}$. $\dfrac{15}{2}; \dfrac{11}{2}$

6. If $\dfrac{3}{a} = \dfrac{4}{b}$, then $\dfrac{3}{4} = \dfrac{?}{?}$ and $\dfrac{b}{a} = \dfrac{?}{?}$. $\dfrac{a}{b}; \dfrac{4}{3}$

7. If $\dfrac{2}{x} = \dfrac{5}{15}$, then $5x = \underline{\ ?\ }$ and $x = \underline{\ ?\ }$. 30; 6

Lesson Commentary T99

3 Using Proportions *(pp. 293–295)*

Teaching These Pages

This section applies proportions to the algebraic solution of word problems. Several exercises reinforce students' work with area in Chapter 6. The Suggested Extension for this lesson presents two very *non*standard proportion problems. Students should find them both entertaining and valuable in showing the use of proportions in estimation.

Chalkboard Examples

1. Amy Goldberg used 30 liters of gasoline for driving to and from work last month. Estimate how many liters she uses in a year of driving to and from work.   **360 liters**

2. In a certain school with 1000 students, 10 students chosen at random are asked for whom they are going to vote for student council president. Seven of these students say "Vasquez." Estimate the number of votes that Vasquez will receive in the election.   **700 votes**

Suggested Extension

1. Guide students through the following procedure, which estimates the number of ping-pong balls needed to fill your classroom.
   a. Fill a small box with ping-pong balls. Count the balls.
   b. Find the volume of the box and of your classroom by determining the length, width, and height and using the formula $V = lwh$.
   c. Solve the proportion $\dfrac{\text{number of balls in box}}{x} = \dfrac{\text{volume of box}}{\text{volume of classroom}}$.
2. Have students discuss a procedure for estimating the number of hot dogs eaten in the entire nation in one day.

Quick Quiz

Express each ratio in simplest form.

1. $\dfrac{6}{81}$ $\dfrac{2}{27}$
2. $24:32$ $3:4$

Complete each statement.

3. If $\dfrac{t}{m} = \dfrac{9}{5}$, then $5t = \underline{\ ?\ }$. $9m$
4. If $\dfrac{a}{7} = \dfrac{b}{12}$, then $\dfrac{a+7}{7} = \underline{\ ?\ }$. $\dfrac{b+12}{12}$
5. If $\dfrac{2}{3} = \dfrac{x}{y}$, then $\dfrac{y}{x} = \underline{\ ?\ }$. $\dfrac{3}{2}$
6. If $\dfrac{4}{r} = \dfrac{s}{9}$, then $\dfrac{4}{s} = \underline{\ ?\ }$. $\dfrac{r}{9}$

Find the value of x.

7. $2:3 = x:15$ 10
8. $\dfrac{7}{2x} = \dfrac{5}{20}$ 14

9. If 3 pounds of rice cost $2.91, how much do 5 pounds cost? $4.85

10. Express in simplest form the ratio of the area of the shaded part to the area of the unshaded part. $\frac{3}{5}$

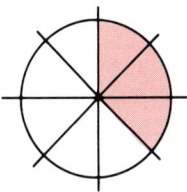

Objectives for Pages 296–301

1. To use maps and scale drawings.   2. To make maps and scale drawings.

4 Maps and Scale Drawings *(pp. 296–298)*

Teaching These Pages

The map of Lakeville and the floor plan of the Briggs home (pages 296–297) are appropriate classroom examples for this topic. Note that both the Classroom Practice Exercises and the Written Exercises refer to the map and floor plan on pages 296 and 297. Emphasize to students that different scales are used to interpret the same diagrams. Hence, for instance, the kitchen has different "real" dimensions in the Written Exercises than it has in the Classroom Practice Exercises.

Chalkboard Example

Have students name 3 or 4 fictitious cities and place them as dots at random on the chalkboard. Then arbitrarily set the distance in kilometers (or miles) between any two of the cities. The diagram at right shows one possibility.

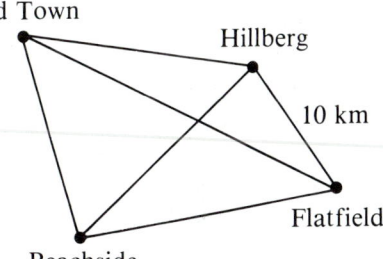

Then have students:
1. Measure the distances on the chalkboard.
2. Calculate the scale factor.
3. Determine the "real" distances between all pairs of cities.

5 Making Maps and Scale Drawings *(pp. 299–301)*

Teaching These Pages

An excellent project for teaching the ideas of scale drawings is to have small groups of students work together to make scale drawings of your classroom. If there are not enough tape measures or measuring sticks available, students can measure dimensions in paces or shoe lengths. For example, your classroom might be "30 shoe lengths by 20 shoe lengths." You may wish to point out that no matter what unit of measurement is used, the dimensions of the classroom will be in the same ratio.

The Applications feature on page 431 illustrates how the concepts presented in this section can be applied to mechanical drawing.

Lesson Commentary   **T101**

Chalkboard Example

Make a map using the following information.
  a. Johnstown is 40 km north of Sayville.
  b. El Placido is 20 km east of Sayville.
  c. Everton is 32 km northeast of Johnstown.
  d. Sherborn is 32 km west of Everton.

(At the chalkboard, use the scale 2 cm : 1 km. Students may use the scale 0.25 cm : 1 km to produce maps which fit on standard size notebook paper.)

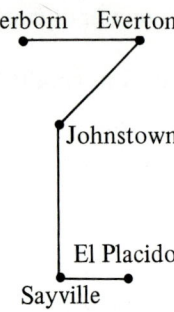

Quick Quiz

1. Measure the map distance between Harrisburg and Lancaster to the nearest half centimeter. Then give the scale of the map.
   2.5 cm; 1 : 2,000,000

2. Find the actual distance from Lancaster to Allentown.   100 km

3. Find the actual distance from Allentown to Harrisburg.   120 km

4. Make a scale drawing of a rectangular lot which is 125 m long and 60 m wide.

   (Suggested scale is 1 cm : 10 m or $\frac{1}{1000}$.)

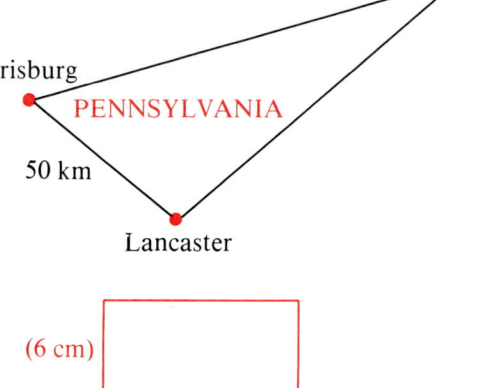

Chapter 8. Similar Polygons

Students have already been introduced to similar figures in their study of scale drawings and maps (Chapter 7, Sections 4 and 5). In this chapter, students first learn how to decide whether two figures are similar. Then they use proportions to calculate lengths in similar figures. Students learn the important case of a line parallel to one side of a triangle cutting off a similar triangle. Finally, they use the relationships involving the perimeters and areas of similar figures.

Objectives for Pages 312–321

1. To recognize similar polygons and use their properties.
2. To use the AA Postulate.
3. To find missing parts of similar triangles.

1 Defining Similar Polygons (pp. 312–316)

Teaching These Pages

You might begin this lesson by drawing the pairs of figures shown in the Chalkboard Examples below, and asking which pairs are similar. Then ask the same question for the pairs of figures in Classroom Practice Exercises 1–3. Students usually can recognize similar figures even if they can't formulate a definition.

Such a lesson might lead to students making up a definition of similar figures. The examples they have discussed should make clear to them why *both* of the following conditions are necessary:

(1) Corresponding angles are equal.
(2) Corresponding sides are in proportion.

This definition applies to polygons, but not to figures like circles. You might ask students which of several circles are similar. If you remind them that "similar figures have the same shape," they should conclude successfully that *all* circles are similar.

Chalkboard Examples

State whether the polygons are, or are not, similar.

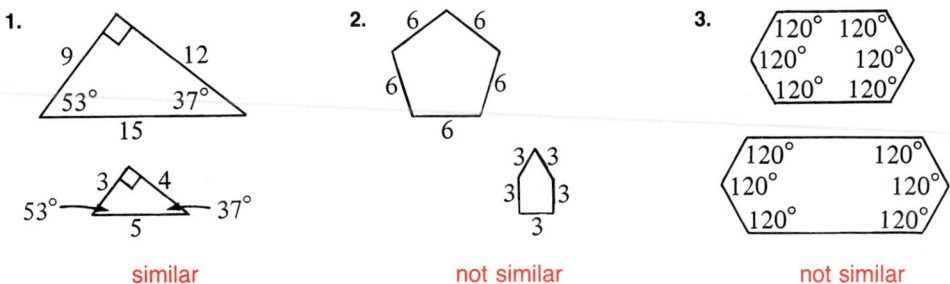

1. similar
2. not similar
3. not similar

In each example, the two polygons are similar. Find the values of x and y.

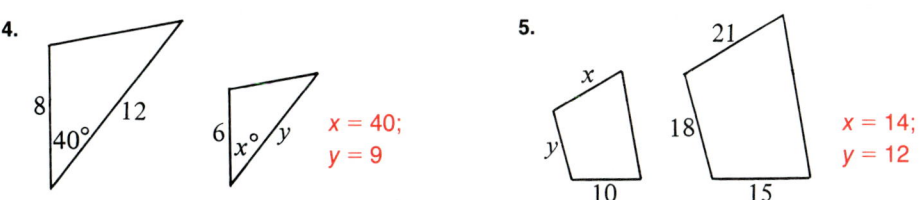

4. $x = 40$; $y = 9$
5. $x = 14$; $y = 12$

6. What is the scale factor of two congruent polygons?   1

7. If the symbol ~ is used for similar polygons, why is the symbol ≅ used for congruent polygons?   Congruent figures are both similar and of equal size.

Lesson Commentary   **T103**

8. The sides of a quadrilateral are 6 cm, 9 cm, 12 cm, and 15 cm long. The longest side of a similar quadrilateral is 20 cm long. How long are its other sides? 8 cm, 12 cm, 16 cm

Suggested Extension

Students may find the game Encounter (page 456) challenging and entertaining. To "solve" the game, a student must realize that player B traces a figure which is similar (with scale factor 2:1) to that of player A. You might vary the rules of the game as follows:

1. Player B draws a segment three times as long as player A's segment.
2. Player B draws a segment in the *opposite* direction and twice as long as player A's segment.

(*Warning:* In the second variation, the "meeting point" is the point X on $\overline{AB}$ such that $BX = 2AX$. If this point X does not have integral coordinates, it will be impossible to force player B to move to the same point as player A.)

2 The AA Postulate (pp. 317–321)

Teaching These Pages

We suggest that you begin this lesson with the remarks and the Explorations on page 317. These Explorations should make the AA Postulate seem reasonable to students. The AA method of showing triangles to be similar can be proved as a theorem instead of being postulated. However, the proof is so involved that it seems wise to postulate the result and have students learn it by *using* it.

Note that Written Exercise 16 proves the special case of similar triangles treated in the next section. Written Exercise 24 proves a result which will help students understand the area ratio of similar triangles. (See Section 5.)

Chalkboard Examples

1. Find the measures of all angles in the three triangles. (See the figure.)

2. Which of the triangles are similar?
   △ ABE ~ △ GEF

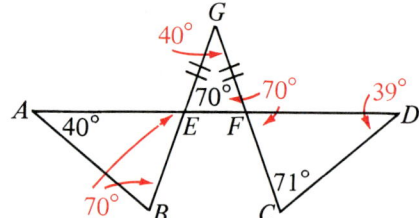

Supply the reasons to complete the proof.

3. Given: $\overline{PQ} \parallel \overline{ST}$

   Prove: $\dfrac{PR}{TR} = \dfrac{QR}{SR}$

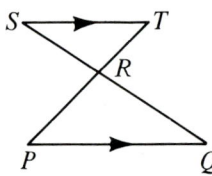

STATEMENTS	REASONS
1. $\overline{PQ} \parallel \overline{ST}$	1. __?__ Given
2. $\angle P = \angle T;\ \angle Q = \angle S$	2. __?__ Alt. int. angles
3. $\triangle PQR \sim \triangle TSR$	3. __?__ AA Postulate
4. $\dfrac{PR}{TR} = \dfrac{QR}{SR}$	4. __?__ Def. of similar triangles

Suggested Extension

To prove that two polygons are similar, one must show two things:

       (1) Corresponding angles are equal.
       (2) Corresponding sides are in proportion.

The AA postulate asserts that (1) alone suffices to prove that two *triangles* are similar. The activity outlined below should convince students that (2) alone also suffices to prove similarity in the case of triangles.

1. Draw any triangle *ABC*.   Drawings may vary.
2. Draw or construct $\triangle DEF$ with sides twice as long.
3. Draw or construct $\triangle GHI$ with sides three times as long.
4. Draw or construct $\triangle JKL$ with sides 3.5 times as long.
5. Use a protractor to measure the angles of all the triangles.
   Are corresponding angles equal? yes
   Are the triangles similar? yes

Quick Quiz

The two polygons are similar. Complete each statement.

1. $SIML \sim$ __?__   PLAR

2. $\dfrac{SI}{?} = \dfrac{IM}{LA} = \dfrac{?}{AR} = \dfrac{?}{?}$
   $\dfrac{SI}{PL} = \dfrac{IM}{LA} = \dfrac{ML}{AR} = \dfrac{LS}{RP}$

3. $\angle L = \angle$ __?__ and $\angle I = \angle$ __?__. R; L

4. The scale factor is __?__ : __?__. 6 : 1   5. $x=$ __?__, $y=$ __?__, and $z=$ __?__. 5; 18; 24

State whether or not the triangles can be proved similar.

6. yes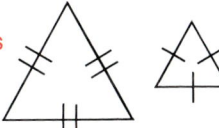

7. no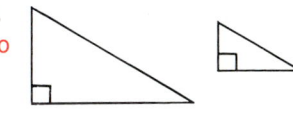

8. yes

9. yes

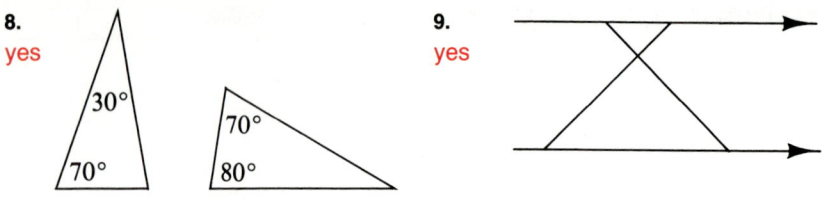

Objectives for Pages 323–332

1. To use similar triangles involving parallel segments.
2. To use the Triangle Proportionality Theorem.
3. To find the perimeters and areas of similar polygons.

3 A Special Case of Similar Triangles (pp. 323–324)

Teaching These Pages

Of all the circumstances in which similar triangles appear, the most common is the one discussed in this section. Make a point to assign or discuss Written Exercise 16 of the previous section, since this exercise proves the key result needed in this section:

> If a line through two sides of a triangle is parallel to the third side, then it cuts off a triangle similar to the original triangle.

When using this result, students commonly make the error of using segments which are not sides of triangles, but only parts of sides.

Common Student Error: $\dfrac{10}{5} = \dfrac{8}{y}$, so $y = 4$

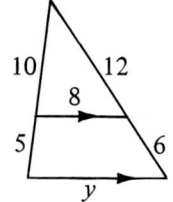

The Triangle Proportionality Theorem (Section 4) will enable us to draw conclusions about parts of sides of triangles. For now, it is best to emphasize to students that we can make proportions only with the lengths of *complete* sides of similar triangles. Some students may need to make additional diagrams in order to identify correctly the corresponding sides.

Chalkboard Examples

Complete each statement. Then find the values of x and y.

1. $x = 7.2$; $y = 12$

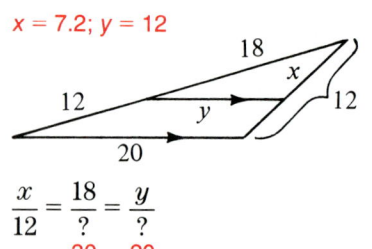

$\dfrac{x}{12} = \dfrac{18}{?\ 30} = \dfrac{y}{?\ 20}$

2. $x = 8$; $y = 15$

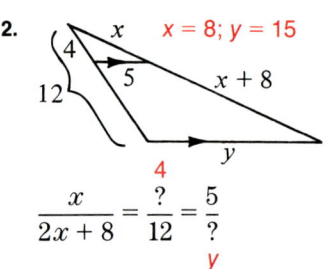

$\dfrac{x}{2x+8} = \dfrac{?\ 4}{12} = \dfrac{5}{?\ y}$

3. $x = 15$; $y = 30$

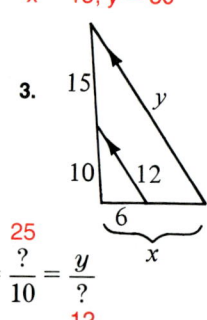

$\dfrac{x}{6} = \dfrac{?\ 25}{10} = \dfrac{y}{?\ 12}$

T106   Lesson Commentary

4 The Triangle Proportionality Theorem
(pp. 325–328)

Teaching These Pages

In the previous sections, students have worked only with complete sides of similar polygons. Theorem 1 in this section deals with *parts* of two different sides of a triangle. You might summarize the situation with the diagram and discussion below.

A. When dealing with $\overline{XY}$ and $\overline{XZ}$, the two sides intersected by the parallel $\overline{PQ}$, we may form proportions using whole sides or parts of sides.

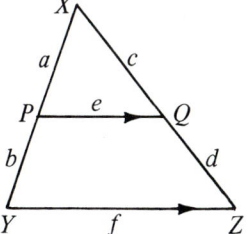

1. The top-to-whole ratios are equal:
$$\frac{a}{a+b} = \frac{c}{c+d}$$

2. The top-to-bottom ratios are equal:
$$\frac{a}{b} = \frac{c}{d}$$

B. When dealing with the parallel sides, we must not form proportions with parts of sides. We must form ratios of complete sides of similar triangles.
$$\frac{e}{f} = \frac{a}{a+b} = \frac{c}{c+d}$$

Chalkboard Examples

Find the values of x and y.

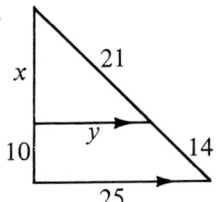

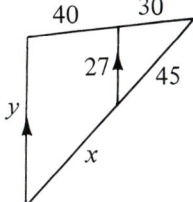

1. $x = 15;\ y = 15$

2. $x = 60;\ y = 63$

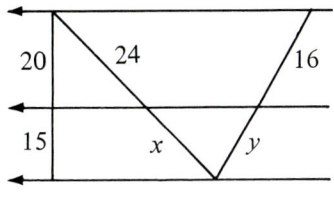

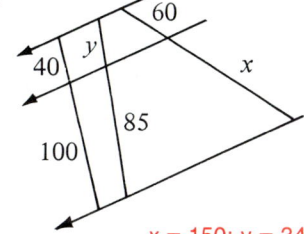

3. $x = 18;\ y = 12$

4. $x = 150;\ y = 34$

Lesson Commentary **T107**

Suggested Extension

1. An extension of the corollary on page 326 asserts:

    If three or more parallel lines cut off equal segments on one transversal, then they cut off equal segments on any transversal.

    A page of lined notebook paper has a set of lines meeting this description. Challenge students to use notebook paper to divide a segment 10 cm long into 3, 4, or 5 equal segments.

2. Have students supply reasons to complete the following proof of the converse of the Triangle Proportionality Theorem.

    Given: $\dfrac{b}{a} = \dfrac{d}{c}$

    Prove: $\overline{PQ} \parallel \overline{YZ}$

    1. Draw $\overline{YN} \parallel \overline{PQ}$, as shown. (We wish to show that point N is the same as point Z so that "$\overline{YN} \parallel \overline{PQ}$" becomes "$\overline{YZ} \parallel \overline{PQ}$.") Construction 5, page 78

    2. $\dfrac{b}{a} = \dfrac{d}{c+e}$ Triangle Proportionality Thm.

    3. $\dfrac{b}{a} = \dfrac{d}{c}$ Given

    4. $e = 0$ Subst., Subtr., Mult. Post.

    5. N coincides with Z and $\overline{YZ} \parallel \overline{PQ}$ Ruler Postulate

Point out that the Midpoints Theorem (page 225) is a special case of this theorem. In the Midpoints Theorem, $a = b$ and $c = d$.

5 Perimeters and Areas (pp. 329–332)

Teaching These Pages

You may wish to let students discover for themselves the relationships described in this section. To start, simply draw the following pairs of rectangles and have students calculate **(a)** the ratio of the perimeters and **(b)** the ratio of the areas.

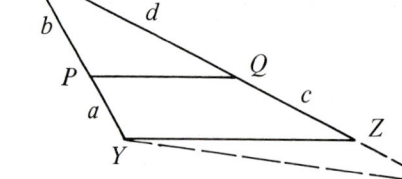

$\dfrac{\text{perimeter of I}}{\text{perimeter of II}} = \dfrac{8}{16} = \dfrac{1}{2}$

$\dfrac{\text{area of I}}{\text{area of II}} = \dfrac{3}{12} = \dfrac{1}{4}$

$\dfrac{\text{perimeter of III}}{\text{perimeter of IV}} = \dfrac{2(a + ka)}{2(b + kb)} = \dfrac{(2k + 2)a}{(2k + 2)b} = \dfrac{a}{b}$

$\dfrac{\text{area of III}}{\text{area of IV}} = \dfrac{a(ka)}{b(kb)} = \dfrac{ka^2}{kb^2} = \dfrac{a^2}{b^2}$

T108 Lesson Commentary

The work with the second pair of rectangles proves Theorem 2 for the special case of rectangles.

Students may continue their investigation by guessing (or calculating) the perimeter and area ratios of two similar rectangles with scale factor 1:3, or 1:4, or 2:3. Try different scale factors until students get the idea. The billboard illustrated on page 329 gives another example of similar rectangles.

You might want to have students investigate other polygons, or simply tell them that the principles they've discovered are true for all pairs of similar polygons. The Calculator Activities on page 322 provide practice with perimeters and areas of equilateral triangles, any two of which are similar. Optional Section 7 of Chapter 10 extends the discussion of similar figures to solids.

Chalkboard Examples

**In each example, there is a pair of similar figures.
Find (a) the ratio of the perimeters and (b) the ratio of the areas.**

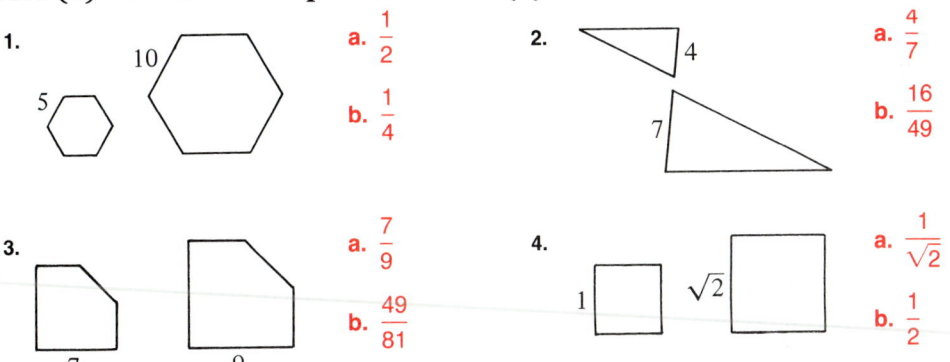

1. a. $\dfrac{1}{2}$ b. $\dfrac{1}{4}$

2. a. $\dfrac{4}{7}$ b. $\dfrac{16}{49}$

3. a. $\dfrac{7}{9}$ b. $\dfrac{49}{81}$

4. a. $\dfrac{1}{\sqrt{2}}$ b. $\dfrac{1}{2}$

5. Given: $\triangle PST \sim \triangle PQR$.
   Find (a) the ratio of the perimeters of $\triangle PST$ and $\triangle PQR$ and
   (b) the ratio of the areas.

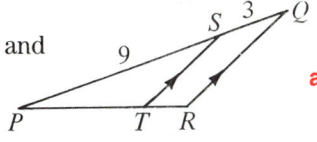

a. $\dfrac{3}{4}$

b. $\dfrac{9}{16}$

Suggested Extension

Have students sketch a chest of drawers, closet, bookcase, suitcase, or any other familiar item whose shape is that of a rectangular solid. Then have them make an enlarged sketch with dimensions twice that of the first sketch. Point out that if we calculated the volumes of the two objects, the ratio would be 1:8. In general, we can add a clause to Theorem 1:

> If two similar solids have a scale factor $a:b$, then the ratio of their volumes is $a^3:b^3$.

Some students might like to verify this fact using pairs of similar rectangular solids and the formula $V = lwh$.

Lesson Commentary   **T109**

Quick Quiz

Find the values of x and y.

1. [triangle diagram with sides 2, 3, 7.5, 6 and segments x, y]

   $x = 3,\ y = 4$

2. [triangle diagram with sides 8, 15, 6, 5 and segments x, y]

   $x = 12,\ y = 3\frac{1}{3}$

3. [diagram with parallel lines cut by transversals, segments $x+3$, 10, x, 8]

   $x = 12$

4. What is the ratio of the perimeters of the two similar triangles shown?   **5:3**

5. The larger triangle has perimeter 30 m. What is the perimeter of the smaller triangle?   **18 m**

6. What is the ratio of the areas of the triangles?   **25:9**

7. The smaller triangle has area 10.8 m². What is the area of the larger triangle?   **30 m²**

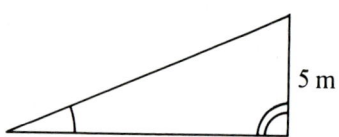

Chapter 9. Circles

Circumferences and areas of circles were treated in Chapter 6. In this chapter, other aspects of circles are studied. The measures of angles are calculated using the measures of intercepted arcs. Relationships among radii, chords, and tangents are developed. Because the theorem about the products of the lengths of the segments of two intersecting chords seems contrary to the intuition of most people, it is treated in detail.

Objectives for Pages 346–358

1. To use terms related to circles.
2. To construct a tangent to a circle at a given point on the circle.
3. To apply theorems relating tangents and radii.
4. To find measures of minor arcs, major arcs, and central angles.

1 Basic Terms (pp. 346–349)

Teaching These Pages

Brief, fast-moving drills on vocabulary will ease the learning process for students and promote retention. You might draw a circle and then add a radius, a chord, and so on, one at at time. Remind students that we use the term radius in two correct senses, to denote a certain segment and to denote a length.

Chalkboard Examples

State whether the figure named (though not necessarily drawn in the diagram) is a chord of circle O. If the figure has another name, give that name.

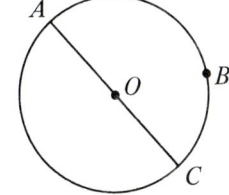

1. $\overline{OA}$ no; radius
2. $\overleftrightarrow{AB}$ no; secant
3. $\overline{AC}$ yes; diameter
4. $\overleftrightarrow{OA}$ no; secant
5. $\overrightarrow{AB}$ no
6. $\overline{BC}$ yes

Suggested Extension

Point X lies on a sphere.

1. How many chords of the sphere contain point X?   infinitely many
2. How many diameters of the sphere contain point X?   one

2 Tangents *(pp. 350–354)*

Teaching These Pages

The displayed statements on page 350 are referred to as theorems, since they can be proved. We recommend that you not use class time for proofs. Most students accept these statements as being quite reasonable without needing a formal argument. Caution students to distinguish carefully between a common internal tangent and the common tangent of internally tangent circles. As the figures on page 351 illustrate, these phrases apply to quite different situations.

Chalkboard Examples

$\overleftrightarrow{AB}$ **is externally tangent to $\odot X$ and $\odot Y$.**

1. Copy the diagram. Draw $\overline{XA}$ and $\overline{YB}$.
2. $\overline{XA}$ __?__ $\overline{AB}$; $\overline{YB}$ __?__ $\overline{AB}$ ⊥ ; ⊥
3. $\overline{XA}$ __?__ $\overline{YB}$ ∥

4. Draw $\overline{XY}$. What kind of quadrilateral is $AXYB$?   trapezoid
5. Suppose $\odot X$ and $\odot Y$ were intersecting circles. Would your answers to Examples 2–4 be different?   no

Suggested Extension

Construct a circle O and draw a diameter CD. Construct a diameter UV that is perpendicular to $\overline{CD}$. Construct lines tangent to the circle at points C, U, D, and V. What kind of quadrilateral is formed?   square

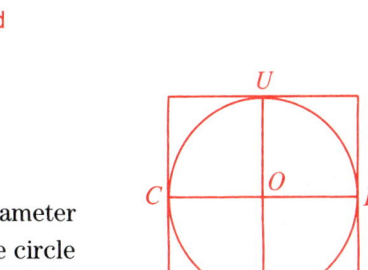

Lesson Commentary   **T111**

3 Arcs and Central Angles (pp. 355–358)

Teaching These Pages

You might want to show students that the use of a protractor is based on the idea that a minor arc and its central angle have the same degree measure. Explain that when we use a protractor to draw a 40° angle, we are really drawing a central angle which cuts off an arc with measure 40° in a certain circle.

Chalkboard Examples

$\overline{AC}$ is a diameter of $\odot O$. Find the measure of:

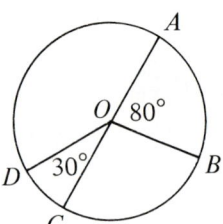

1. $\overset{\frown}{BC}$ 100°
2. $\overset{\frown}{DA}$ 150°
3. $\overset{\frown}{BD}$ 130°
4. $\overset{\frown}{BAD}$ 230°
5. $\angle AOD$ 150°
6. $\angle BOC$ 100°

Quick Quiz

Match each item with an example from the diagram.

1. a diameter g
2. a secant f
3. a point of tangency b
4. a minor arc a
5. a major arc h
6. a radius e
7. a central angle j
8. a semicircle d

a. $\overset{\frown}{CR}$
b. R
c. C
d. $\overset{\frown}{CEL}$
e. $\overline{OL}$
f. $\overleftrightarrow{CE}$
g. $\overline{CL}$
h. $\overset{\frown}{CLR}$
i. $\angle LCE$
j. $\angle ROC$
k. $\overline{CE}$

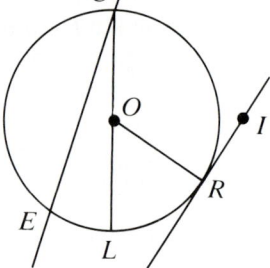

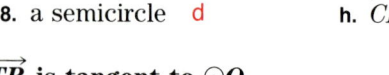

$\overleftrightarrow{TR}$ is tangent to $\odot O$.

9. $\overset{\frown}{AC} = \underline{\ ?\ }°$ 40
10. $\angle BOA = \underline{\ ?\ }°$ 60
11. $\overset{\frown}{BT} = \underline{\ ?\ }°$ 120
12. $\overset{\frown}{BAT} = \underline{\ ?\ }°$ 240
13. $\angle RTO = \underline{\ ?\ }°$ 90

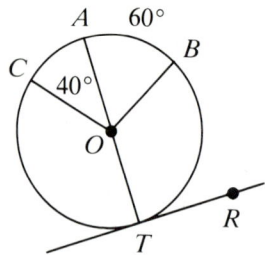

14. Construct a circle with diameter $\overline{AB}$. Construct tangents to the circle at A and B. Do these tangents intersect?   no

Use Construction 6.
Then use Construction 2 twice.

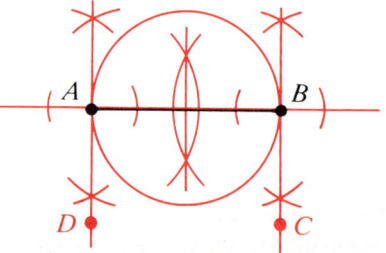

Objectives for Pages 359–367

1. To apply theorems about chords and arcs.

2. To find the measure of an inscribed angle, given the measure of its intercepted arc.

3. To find the measure of an angle formed by a chord and a tangent, given the measure of its intercepted arc.

4 Chords (pp. 359–363)

Teaching These Pages

We suggest that you have students discover the properties described in Theorems 5, 6, and 7 by having them do the Explorations on page 359. You might challenge more advanced students to devise proofs of these theorems. Written Exercise 30 outlines a proof of Theorem 7.

The construction in the Puzzle Problems feature on page 376 can be justified using the converse of Theorem 7:

> The perpendicular bisector of a chord passes through the center of the circle.

Chalkboard Examples

In $\odot O$, $JY = KY$. Copy the figure, then draw $\overline{OJ}$, $\overline{OY}$, and $\overline{OK}$.

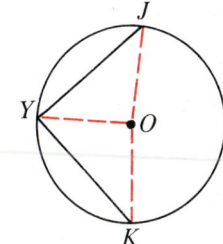

1. Why is $\triangle JOY \cong \triangle KOY$?   SSS postulate

2. Why is $\angle JOY = \angle KOY$?   Corr. parts of $\cong \triangle$ are =.

3. Why is $\widehat{JY} = \widehat{KY}$?   Thm. 3 (In a $\odot$, equal central $\angle$ have equal minor arcs.) or Thm. 5 (In a $\odot$, equal chords have equal arcs.)

Suggested Extension

Suppose you wish to circumscribe a circle about a square. How would you find the center of the circumscribed circle?
Draw the diagonals of the square as shown at the right.
How would you find the center of the inscribed circle?
Use the same method described above.
What can you say about these two circles?
The circles have the same center. The larger circle has twice the area of the smaller circle.

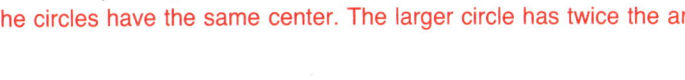

5 Inscribed Angles (pp. 364–367)

Teaching These Pages

You may want to review what it means for a triangle to be inscribed in a circle. Students should realize that when $\triangle ABC$ is inscribed in a circle, each of the

Lesson Commentary   T113

angles A, B, and C is inscribed in the circle. For example, if side $\overline{BC}$ is erased—except for endpoints B and C—then inscribed ∠A remains.

Note that the omission of the term "measure" simplifies the statements of Theorems 8 and 9 in this section and Theorems 10 and 11 in Section 6.

Chalkboard Examples

Given: $\overleftrightarrow{YZ}$ is tangent to $\odot O$; $\overset{\frown}{XB} = 80°$; ∠AXZ = 70°

1. ∠YXB = __?__° 40
2. $\overset{\frown}{XA}$ = __?__° 140
3. $\overset{\frown}{AB}$ = __?__° 140
4. ∠AXB = __?__° 70
5. $\overset{\frown}{XA} + \overset{\frown}{AB} + \overset{\frown}{BX}$ = __?__° 360
6. ∠AXZ + ∠AXB + ∠YXB = __?__° 180

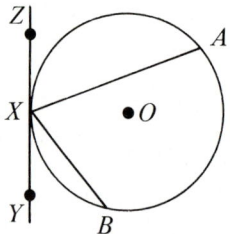

Suggested Extension

In the diagram, ∠TEY intercepts $\overset{\frown}{TY}$. We say that ∠TEY is inscribed in $\overset{\frown}{TEY}$. What can you say about the measure of:

1. An angle inscribed in a semicircle? right angle
2. An angle inscribed in a major arc? obtuse angle
3. An angle inscribed in a minor arc? acute angle

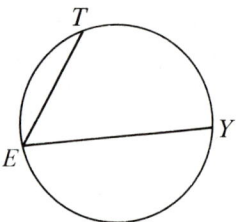

Quick Quiz

In $\odot O$, $LN = RP$.

1. $\overset{\frown}{LN}$ = __?__   RP
2. OS = __?__   OT
3. $\overset{\frown}{LM}$ = __?__   MN
4. SP = __?__   SR, LT, or TN

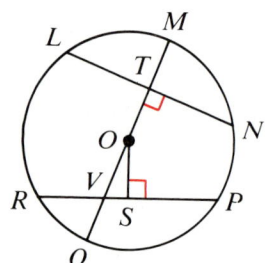

Given: $\overleftrightarrow{TD}$ is tangent to $\odot O$.

5. ∠1 = __?__° 45
6. ∠2 = __?__° 70
7. $\overset{\frown}{TC}$ = __?__° 130
8. ∠3 = __?__° 65
9. ∠4 = __?__° 45

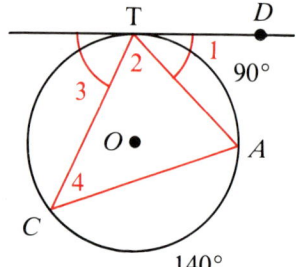

10. Draw a trapezoid inscribed in a circle.

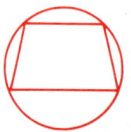

Objectives for Pages 369–376

1. To find the measure of an angle formed by two chords.
2. To find the measure of an angle formed by two secants.
3. To apply the theorem about the lengths of the segments of two intersecting chords.

6 Other Angles (pp. 369–372)

Teaching These Pages

We suggest that you have students discover the properties described in Theorems 10 and 11 by having them do the Explorations on page 369. You might like to lead students through the following verification of a special case of Theorem 10: the two chords are diameters.

Suppose $\angle 1 = 30°$.
Since $\angle 1$ is a central angle, $\widehat{AC} = 30°$.
Since $\angle 1$ and $\angle 2$ are vertical angles, $\angle 2 = 30°$.
Since $\angle 2$ is a central angle, $\widehat{BD} = 30°$.
Now test: $\angle 1 \stackrel{?}{=} \frac{1}{2}(\widehat{AC} + \widehat{BD})$

$$30° \stackrel{?}{=} \frac{1}{2}(30° + 30°)$$

$$30° \stackrel{?}{=} \frac{1}{2}(60°)$$

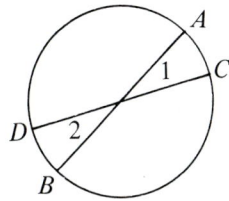

Thus the definition of central angle and Theorem 10 both lead to the same result.

Chalkboard Examples

Draw a circle. Place points R, S, T, and U, in that order, on the circle so that $\widehat{RS} = 120°$, $\widehat{ST} = 30°$, and $\widehat{TU} = 70°$.

1. $\widehat{UR} = $ __?__ °   140
2. Let the intersection of $\overleftrightarrow{RT}$ and $\overleftrightarrow{SU}$ be A. $\angle SAT = $ __?__ °   85
3. Let the intersection of $\overleftrightarrow{RS}$ and $\overleftrightarrow{UT}$ be B. $\angle RBU = $ __?__ °   55
4. Let the intersection of $\overleftrightarrow{RU}$ and $\overleftrightarrow{ST}$ be C. $\angle RCS = $ __?__ °   25

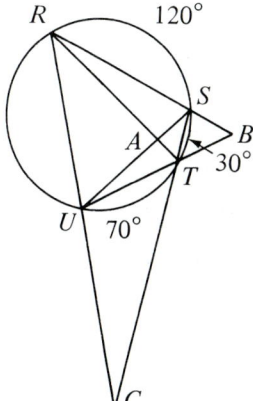

Suggested Extension

In $\odot O$, $\widehat{AB} = 100°$. Point X lies on the same side of $\overleftrightarrow{AB}$ as O does. Does X lie inside, outside, or on the circle if:

1. $\angle AXB = 50°$? on the circle   2. $\angle AXB = 40°$? outside   3. $\angle AXB = 52°$? inside

7 Segments of Chords (pp. 373–376)

Teaching These Pages

Point out to students that almost all the theorems they have encountered in geometry "look true." But in Theorem 12, we have a theorem which "looks false" to many people.

Chalkboard Examples

Draw a diagram for each exercise. Chords $\overline{RS}$ and $\overline{TV}$ intersect at P, inside a circle. Find PV.

1. $RP = 3$, $PS = 4$, and $TP = 6$
   $PV = 2$

2. $RP = 6$, $PS = 10$, and $TP = 5$
   $PV = 12$

3. $RP = 12$, $PS = 12$, and $TP = 16$
   $PV = 9$

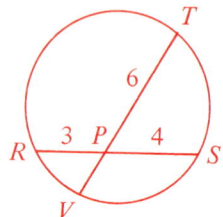
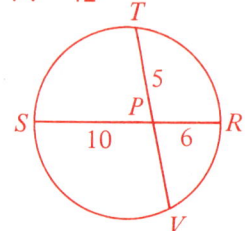
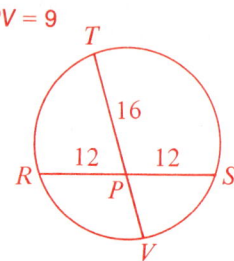

Suggested Extension

In the Chalkboard Examples above, it is difficult to draw accurate diagrams. A method for drawing an accurate figure for Chalkboard Example 1 is given below.

1. Draw a segment RS that is 7 cm long.
2. Take P on $\overline{RS}$ so that $RP = 3$ cm and $PS = 4$ cm.
3. Take a point T, not on $\overline{RS}$, so that $TP = 6$ cm.
4. Draw $\triangle RTS$ and circumscribe a circle about $\triangle RTS$.
5. Extend $\overline{TP}$ to the point where it intersects the circle and call that point V.

The resulting figure should be quite accurate. If another point T had been chosen 6 cm away from P, would the circumscribed circle be the same? **no**

Quick Quiz

Find the measures.

1. $\angle 1 = $ __?__° 35
2. $\angle 2 = $ __?__° 60
3. $\angle 3 = $ __?__° 95
4. $x° = $ __?__° 70
5. $\angle 4 = $ __?__° 85

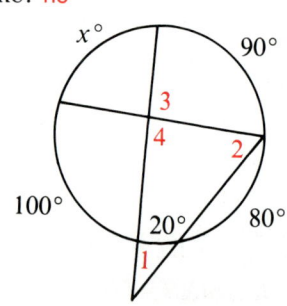

Find the value of y.

6.
8
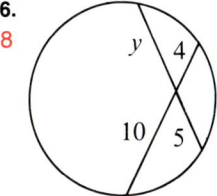

7.
8
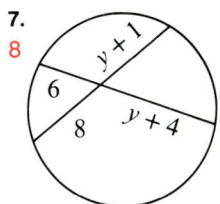

8.
9
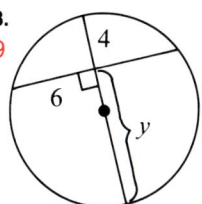

Chapter 10. Areas and Volumes of Solids

This chapter opens with a section that develops students' awareness of geometric relationships in three dimensions. The possible relative positions of two lines, two planes, and a line and a plane are explored. The rest of the chapter deals with the most common solids. The material on prisms emphasizes rectangular solids. The cylinders and cones studied are the types that students see most often, right circular cylinders and cones. The pyramids studied are regular pyramids. After a section on spheres, there is an optional section on similar solids. The chapter as a whole helps students to develop the ability to picture three-dimensional figures while reinforcing their arithmetic and algebraic skills.

Objectives for Pages 388–400

1. To draw and to interpret drawings of solid figures.
2. To find the lateral area, total area, and volume of a right prism.
3. To find the lateral area, total area, and volume of a right circular cylinder.

1 Lines and Planes in Space *(pp. 388–391)*

Teaching These Pages

The classroom itself provides examples of lines and planes in space: for instance, the floor, ceiling, and walls, and the lines in which they intersect. You might use a chalkboard pointer and a desk top to illustrate possible relative positions of a line and plane.

Encourage students to make drawings and to use handy objects (desk top, pencils, notebooks, and so on) in making judgments about three-dimensional relationships. Remind students that lines and planes extend indefinitely: The plane suggested by the ceiling, for instance, doesn't end at the walls of the classroom.

Chalkboard Examples

Classify as true or false.

1. If two lines are parallel, there must be a plane that contains both lines. true
2. If two lines are skew, there must be a plane that contains both lines. false
3. Any line in the ceiling must be parallel to the floor. true
4. There are many lines in the floor that intersect the front wall. true
5. There are many lines in the floor that are parallel to the front wall. true
6. If a pencil is held parallel to a desk top, then any plane that contains the pencil must be parallel to the desk top. false
7. A line and a plane may be skew. false
8. A volleyball net and the gym floor form perpendicular planes. true
9. A wall and an open door along that wall are skew. false

Suggested Extension

The exercises below make use of a cylinder and a cone. Some students may not be familiar with these figures.

1. Draw a plane. Show two points, A and B, in the plane. Does $\overleftrightarrow{AB}$ have to lie in the plane? yes (Postulate 3)
2. Draw a cylinder. Show two points, A and B, such that $\overline{AB}$ lies in the cylindrical (the curved) surface. Also show two points, X and Y, both in the cylindrical surface but such that $\overline{XY}$ does not lie in the cylindrical surface.

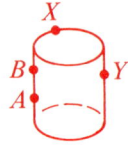

3. Repeat Exercise 2, using a cone and its conical surface.

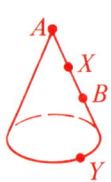

2 Right Prisms (pp. 393–397)

Teaching These Pages

Use the classroom or a cardboard packing box to illustrate a rectangular solid. If you don't have a set of models, you can find objects around the classroom to illustrate other prisms. A pencil can be cut twice to form a right prism with a hexagonal base. It could be cut at a slant to form a prism that is not a right prism. (Round pencils can be used to illustrate cylinders, which the class will be studying next.) An open textbook, standing upright, suggests a right prism with a triangular base.

Chalkboard Examples

A rectangular solid has edges that are 10, 8, and 3 units long.

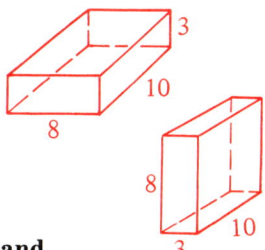

1. Draw the solid, showing the 10 × 8 face as the base. Find:
   a. the lateral area 108 **b.** the total area 268 **c.** the volume 240

2. Repeat Exercise 1, but show the 10 × 3 face as the base.
   a. 208 **b.** 268 **c.** 240

For each right prism, find (a) the lateral area, (b) the total area, and (c) the volume.

3.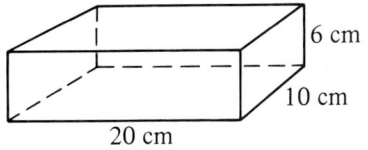
   a. 360 cm² **b.** 760 cm²
   c. 1200 cm³

4.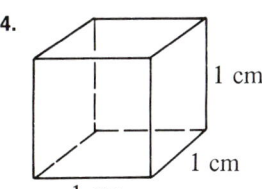
   a. 4 cm² **b.** 6 cm²
   c. 1 cm³

5.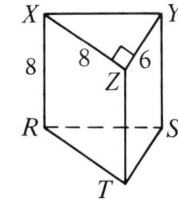
   a. 192 **b.** 240
   c. 192

Suggested Extension

Think of a cube whose edges are 6 cm long.

1. The volume of the cube is __?__ cm³. 216
2. The total area of the cube is __?__ cm². 216

Now use the fact that the edges of the cube are 60 mm long.

3. The volume of the cube is __?__ mm³. 216,000
4. The total area of the cube is __?__ mm². 21,600
5. Look at your results in Exercises 1 and 2; then at your results in Exercises 3 and 4. Is it proper to say that the volume of the cube is numerically equal to the total area? No, since the numbers depend on the unit of measure used.

3 Right Circular Cylinders *(pp. 398–400)*

Teaching These Pages

Use a few minutes to show the class that not all cylinders are circular. You might place a large oval paper cutout on the classroom floor, and then trace its boundary with a chalkboard pointer held upright. A round pencil with both ends cut off at an angle provides another example. You could demonstrate that a circular cylinder need not be a right cylinder by tracing a circular cutout with a chalkboard pointer kept at an angle to the floor. After such demonstrations, point out that right circular cylinders are the most common ones; for simplicity, the class may use the term *cylinder* for *right circular cylinder*.

Lesson Commentary   **T119**

Chalkboard Examples

Express answers in terms of π.

1. The radius of a cylinder is 10; the height is 2. Find:

   a. circumference of the base 20π **b.** area of the base 100π

   c. L.A. 40π **d.** T.A. 240π **e.** V 200π

2. Repeat Exercise 1, using a cylinder in which $r = 2$ and $h = 10$.
   a. 4π **b.** 4π **c.** 40π **d.** 48π **e.** 40π

Use 3.14 for π.

3. A 20 cm by 30 cm sheet of paper is rolled, without overlap, to form the lateral surface of a cylinder with height 30 cm. Find the volume to the nearest tenth of a cubic centimeter.   955.4 cm³

4. Repeat Example 3, but roll the paper so that the height of the cylinder is 20 cm.   1433.1 cm³

Suggested Extension

1. Describe a cylinder with height 10 cm whose lateral area (in cm²) is many times as great as its volume (in cm³).
   It has a radius much smaller than 2 cm.

2. Describe a cylinder with height 10 cm whose volume (in cm³) is many times as great as its lateral area (in cm²).
   It has a radius much greater than 2 cm.

Quick Quiz

The base of the right prism shown is a right triangle.

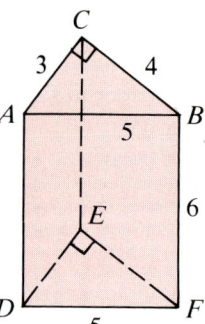

1. Name two lines parallel to $\overleftrightarrow{AD}$. $\overleftrightarrow{BF}$ and $\overleftrightarrow{CE}$
2. Name a line skew to $\overleftrightarrow{DE}$. $\overleftrightarrow{BF}$, $\overleftrightarrow{AB}$, or $\overleftrightarrow{CB}$
3. State the number of base edges.   6
4. State the number of lateral faces.   3
5. Find the base area.   6
6. Find the lateral area.   72
7. Find the volume.   36

In a cylinder, $r = 4$ and $h = 7$.
Find the indicated values in terms of π.

8. L.A. = __?__ 56π **9.** T.A. = __?__ 88π **10.** V = __?__ 112π

Objectives for Pages 402–412

1. To find the lateral area, the total area, and the volume of a pyramid.
2. To find the lateral area, the total area, and the volume of a cone.
3. To find the area and the volume of a sphere.

4 Regular Pyramids (pp. 402–404)

Teaching These Pages

It is helpful to have a model pyramid on which you can show students the parts of a pyramid. Some students might enjoy making cardboard models based on the pattern at right.

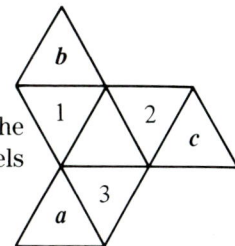

Chalkboard Examples

1. Draw a pyramid so that the base is a 10 × 10 square and the height is 12. The slant height of this pyramid is 13.

2. Find: **a.** the area of the base   100
   b. the area of one lateral face   65
   c. the total area   360
   d. the volume   400

3–4. Repeat Examples 1 and 2 using a regular pyramid with base edges of length 3, a height of 2, and a slant height of 2.5.
   4. a. 9   **b.** 3.75   **c.** 24   **d.** 6

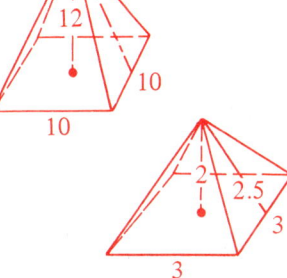

Suggested Extension

The four diagonals of a cube meet at a point V. Consider the six congruent pyramids $VABCD$, $VBCTS$, and so on, that have vertex V.

1. Use the formula $V = \frac{1}{3}Bh$ to find the volume of one of these pyramids.   288

2. Does your answer in Exercise 1 indicate that each of the congruent pyramids "fills" one-sixth of the cube?
   Yes, $288 = \frac{1}{6}(12 \times 12 \times 12)$.

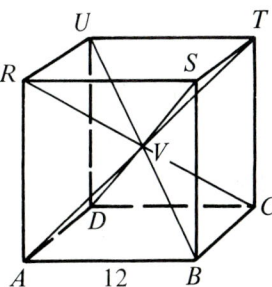

5 Right Circular Cones (pp. 406–408)

Teaching These Pages

Explain that just as not all cylinders are right circular cylinders, not all cones are right circular cones. You might draw a figure like the one shown at right as an example of a cone in the broad sense of the word. Recalling the agreement concerning cylinder terminology, tell students that *cone* will mean *right circular cone* in this course.

Lesson Commentary

The activity described in Classroom Practice Exercises 2–10 justifies the formula $A = \pi r l$ for the lateral area of a cone by showing that this area is two-thirds the area of a certain circle.

Chalkboard Examples

Express answers in terms of π.

1. In a certain cone: $r = 9$; $h = 12$; $l = 15$. Find:
   a. L.A. 135π b. T.A. 216π c. V 324π

2. Repeat Exercise 1, using: $r = 12$; $h = 9$; $l = 15$.
   a. 180π b. 324π c. 432π

3. A cone and a cylinder both have a radius of 5 and a height of 2. Find the ratio of their volumes. $1:3$

6 Spheres *(pp. 410–412)*

Teaching These Pages

The formulas for the area and volume of a sphere are very difficult to prove. Most students are quite willing to accept these formulas without proof. You may want to point out that the area formula involves "r^2" and the volume formula involves "r^3," corresponding to the fact that area is measured in square units and volume is measured in cubic units.

Chalkboard Examples

Find, in terms of π, the area of a sphere whose radius is:

1. 3 cm 36π cm²   2. 4 cm 64π cm²   3. 5 cm 100π cm²

4. Does the area of the largest of the three spheres equal the sum of the areas of the other two? yes

5–8. Repeat Examples 1–4, calculating volumes instead of areas.

   5. 36π cm³   6. $85\tfrac{1}{3}\pi$ cm³   7. $166\tfrac{2}{3}\pi$ cm³   8. no

Suggested Extension

Show that the volume of the cylinder is three times that of the cone, and that the volume of the sphere is twice that of the cone.

$V_{CY} = \pi r^2(2r) = 2\pi r^3$; $V_{SP} = \dfrac{4}{3}\pi r^3$;

$V_{CO} = \dfrac{1}{3}\pi r^2(2r) = \dfrac{2}{3}\pi r^3$

$V_{CY} = 2\pi r^3 = 3\left(\dfrac{2}{3}\pi r^3\right) = 3(V_{CO})$

$V_{SP} = \dfrac{4}{3}\pi r^3 = 2\left(\dfrac{2}{3}\pi r^3\right) = 2(V_{CO})$

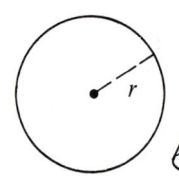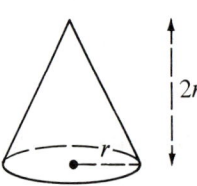

Quick Quiz

The pyramid shown has height 12 and slant height 13.

1. L.A. = __?__ 260
2. B = __?__ 100
3. V = __?__ 400

A cone has radius 15 cm, height 8 cm, and slant height 17 cm.

4. C = __?__ 30π cm
5. B = __?__ 225π cm^2
6. L.A. = __?__ 255π cm^2
7. T.A. = __?__ 480π cm^2
8. V = __?__ 600π cm^3

The radius of a sphere is 3 cm.

9. A = __?__ 36π cm^2
10. V = __?__ 36π cm^3

7 Similar Solids (Optional) *(pp. 414–416)*

Teaching These Pages

This section extends the treatment of measurements of similar figures begun in Section 5 of Chapter 8. We suggest that you have students do the Explorations on page 414. Explain that Theorem 1 on page 414 is an extension of Theorem 2 on page 330. Students may find Theorem 1 easier to remember if they note that the exponent (1, 2, or 3) in the ratio of measurements corresponds to the unit of measure involved (for example: m, m^2, or m^3).

Chalkboard Examples

1. A photograph is enlarged to form a new one that is twice as long and twice as wide as the original.
   a. Are angle measures changed in the enlarging process? no
   b. Suppose the picture shows a bridge. What is the ratio of the length of the bridge in the enlargement to the length in the original photograph? 2:1
   c. What is the ratio of the area of the enlarged photograph to the area of the original? 4:1

2. One ball bearing has one-third the diameter of another. There are 10 g of steel in the smaller one. How many grams of steel are there in the larger one? 270 g

3. Two swimming pools have the same shape, but one is twice as wide.
   a. How many times larger must a cover for the larger pool be? 4 times
   b. How many times as much water will the larger pool hold? 8 times

4. Two soup cans are similar solids. The ratio of their diameters is 3:5.
   a. The smaller can has a label with area 450 cm^2. What is the area of the larger can's label? 1250 cm^2
   b. The larger can has a capacity of 2500 cm^3. What is the capacity of the smaller can? 540 cm^3

Lesson Commentary

Chapter 11. Right Triangles

This chapter offers both a review and an extension of important ideas presented in the course. The Pythagorean Theorem is applied to derive special relationships in 45°-45°-90° and 30°-60°-90° triangles. These relationships are used in exercises which also review a wide variety of topics: properties of circles, triangles, and special quadrilaterals; area and volume formulas; properties of similar figures; inscribed figures. The Pythagorean Theorem is further applied to find the diagonals of rectangular solids and to extend the study of pyramids and cones.

The chapter ends with an introduction to trigonometry. A complete table of trigonometric ratios is given on page 517.

Objectives for Pages 428–439

1. To solve problems involving some common right triangle lengths.
2. To use the relationships among the sides of 45°-45°-90° triangles and 30°-60°-90° triangles.
3. To apply theorems about 45°-45°-90° triangles and 30°-60°-90° triangles to other figures.

1 Reviewing Right Triangles *(pp. 428–430)*

Teaching These Pages

You might begin by reviewing the Pythagorean Theorem with students, having them recall the common right triangles with integral side lengths. The exercises in the text review many of the situations in geometry in which right triangles occur.

You may wish to move right to Section 2, assigning problems from both of the first two sections for two or three days. You might try the same tactic with Sections 3 and 4, discussing and assigning problems from both sections concurrently.

Chalkboard Examples

Find the value of each variable.

1.

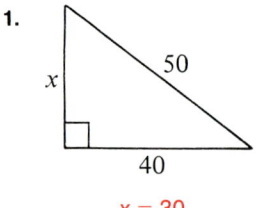

$x = 30$

2.

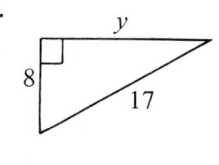

$y = 15$

3.

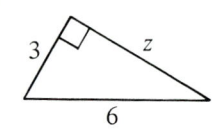

$z = 3\sqrt{3}$ or $\sqrt{27}$

4.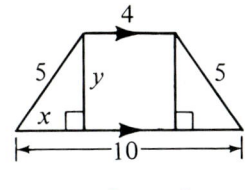

$x = 3; y = 4$

5.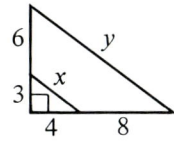

$x = 5; y = 15$

6.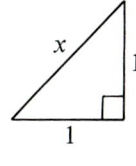

$x = \sqrt{2}$

Suggested Extension

Some students may be interested in using whole-number values of r and s in the formulas below to generate "Pythagorean triples," sets of whole numbers which are the lengths of sides of right triangles.

For any values of r and s:

If $a = r^2 - s^2$, $b = 2rs$, and $c = r^2 + s^2$, then $a^2 + b^2 = c^2$.

You might challenge more advanced students to prove the validity of this method.
$$\begin{aligned} a^2 + b^2 &= (r^2 - s^2)^2 + (2rs)^2 \\ &= r^4 - 2r^2s^2 + s^4 + 4r^2s^2 \\ &= r^4 + 2r^2s^2 + s^4 \\ &= (r^2 + s^2)^2 = c^2 \end{aligned}$$

2 Special Right Triangles (pp. 432–436)

Teaching These Pages

To motivate students to use Theorem 1 on page 432, you might use the Pythagorean Theorem to find the hypotenuse of each of the following isosceles right triangles.

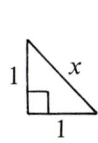
$x^2 = 1^2 + 1^2$
$x^2 = 2$
$x = \sqrt{2}$

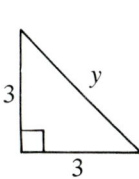
$y^2 = 3^2 + 3^2$
$y^2 = 3^2 \cdot 2$
$y = 3\sqrt{2}$

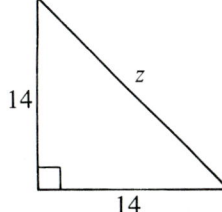
$z^2 = 14^2 + 14^2$
$z^2 = 14^2 \cdot 2$
$z = 14\sqrt{2}$

Note that the method above eliminates the need to reduce the radicals $\sqrt{18}$ and $\sqrt{392}$.

By this time, several students should be ready to state a conjecture about 45°-45°-90° triangles. (Theorem 1) The proof of this theorem in the text uses the algebraic equivalent of the method shown above.

To motivate Theorem 2 on page 432, have students find the altitude of an equilateral triangle with side length 2. Point out that the altitude forms two 30°-60°-90° triangles.

$x = 1$
$y^2 = 2^2 - 1^2 = 3$
$y = \sqrt{3}$

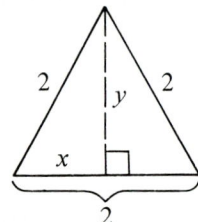
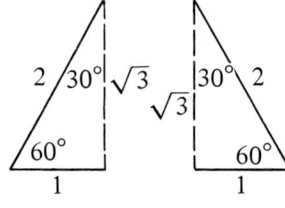

Have students form other 30°-60°-90° triangles in the same way. After working several such examples, students should be ready to suggest Theorem 2 on their own.

Chalkboard Examples

In $\triangle ABC$, $\angle A = 90°$, $\angle B = 60°$, and $\angle C = 30°$.

1. If $AB = 6$, $BC =$ __?__.   12
2. If $AB = 10$, $BC =$ __?__.   20
3. If $AC = 4\sqrt{3}$, $BC =$ __?__.   8
4. If $BC = 2$, $AC =$ __?__. $\sqrt{3}$

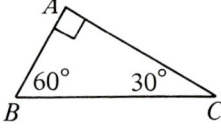

In $\triangle PQR$, $\angle P = 30°$, $\angle Q = 90°$, and $\angle R = 60°$.

5. If $PR = 8$, $QR =$ __?__ and $PQ =$ __?__.   4; $4\sqrt{3}$
6. If $PR = 12$, $QR =$ __?__ and $PQ =$ __?__.   6; $6\sqrt{3}$
7. If $QR = 5$, $PR =$ __?__ and $PQ =$ __?__.   10; $5\sqrt{3}$
8. If $PQ = 7\sqrt{3}$, $QR =$ __?__ and $PR =$ __?__.   7; 14

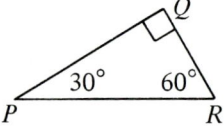

3 Using Special Right Triangles (pp. 437–439)

Teaching These Pages

Explain to students that the special right triangles they have just studied occur often in geometry problems. The theorems concerning these triangles frequently provide the key to solving such problems.

Work the Example on page 437 with students, pointing out how a knowledge of 30°-60°-90° triangles enables us to find the altitude of the parallelogram and, hence, the area. You may want to work the problem a second time, letting $\angle A = 45°$, and a third time, letting $\angle A = 30°$.

Chalkboard Examples

1. Figure $ABCD$ is a square with $BC = 7$.
   a. Find BD.
      $7\sqrt{2}$
   b. Find CO.
      $\dfrac{7\sqrt{2}}{2}$

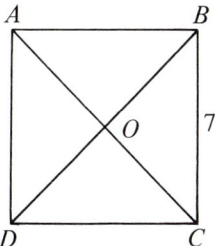

2. Figure $ABCD$ is a rhombus with $\angle DAC = 30°$.
   a. Find the measures of $\angle 1$, $\angle 2$, and $\angle 3$.
      $\angle 1 = 60°$; $\angle 2 = 90°$; $\angle 3 = 30°$
   b. Find BO and AO.
      $BO = 5$; $AO = 5\sqrt{3}$

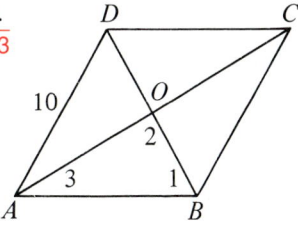

3. a. Find PR. 3
   b. Find QR. 3
   c. Find PQ. $3\sqrt{2}$

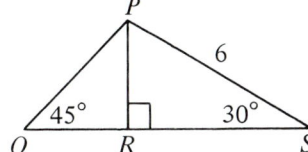

4. The altitude of an equilateral triangle is $6\sqrt{3}$.
   a. What is the perimeter of the triangle? 36
   b. What is the area of the triangle? $36\sqrt{3}$

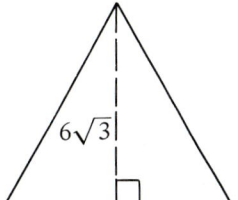

Quick Quiz

Find the exact values of x and y.

1.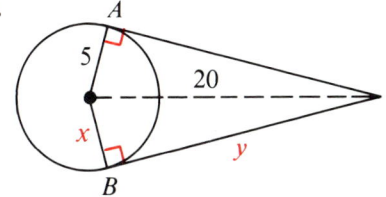
   $x = 5$, $y = 5\sqrt{15}$

2.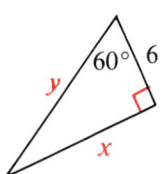
   $x = 6\sqrt{3}$, $y = 12$

3.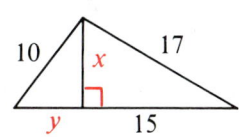
   $x = 8$, $y = 6$

Find the area of each polygon.

4.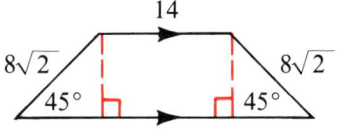
   176

5.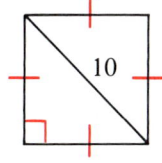
   50

6.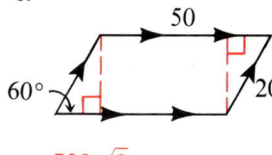
   $500\sqrt{3}$

Objectives for Pages 440–446

1. To find the length of the diagonal of a rectangular solid.
2. To use right triangles to solve problems involving pyramids and cones.

4 Diagonals of Rectangular Solids (pp. 440–442)

Teaching These Pages

Try *leading* your students to the formula $d = \sqrt{a^2 + b^2 + c^2}$ rather than presenting it without justification. This can be done by analyzing several rectangular solids using the three steps illustrated on page 440. These same three steps are used in Classroom Practice Exercises 1–3 and 5–7 and in the Chalkboard Examples below. A few such examples should prepare students for the derivation of the formula.

Chalkboard Examples

1. $x^2 = (\underline{})^2 + (\underline{})^2$ 3; 4
2. $d^2 = x^2 + (\underline{})^2$ 2

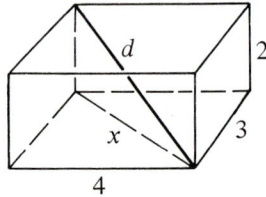

3. Combining the results of Examples 1 and 2:

   $d^2 = (\underline{})^2 + (\underline{})^2 + (\underline{})^2 = \underline{}$ 3; 4; 2; 29

   $d = \sqrt{\underline{}}$ 29

4. $x^2 = (\underline{})^2 + (\underline{})^2$ 2; 5
5. $d^2 = x^2 + (\underline{})^2$ 3

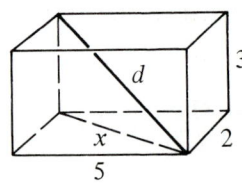

6. Combining the results of Examples 4 and 5:

   $d^2 = (\underline{})^2 + (\underline{})^2 + (\underline{})^2 = \underline{}$ 2; 5; 3; 38

   $d = \sqrt{\underline{}}$ 38

Suggested Extension

Many of your students will enjoy the problem about the ant and the sugar in the Puzzle Problems on page 446. The hint will help students see that the ant crawls a distance of $\sqrt{8^2 + 6^2} = 10$. But what if the box is unfolded differently, suggesting another path for the ant? The work below shows that another path is of length $\sqrt{130}$, which is larger than 10.

You can also vary the problem by placing the ant and the sugar at different places.

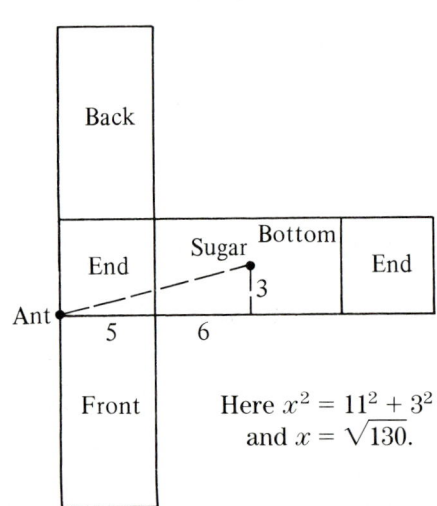

Here $x^2 = 11^2 + 3^2$ and $x = \sqrt{130}$.

5 Right Triangles in Pyramids and Cones
(pp. 443–446)

Teaching These Pages

The work in this section is an extension of the area and volume work in Chapter 10. In that chapter, students were *given* the slant height or the height of a pyramid in order to calculate its area or volume. Now students must use right triangles to *calculate* the slant height or height.

Perhaps the best way to teach these pages is to pose to the class the problems given in Examples 1 and 2 on pages 443–444. After their work with rectangular solids, most students should have little difficulty visualizing right triangles in space. Some students, however, may need to work with models of pyramids and cones.

Chalkboard Examples

A cone has radius 6 and slant height 10.

1. What is its height?   8
2. What is its volume? 96π
3. What is its lateral area? 60π

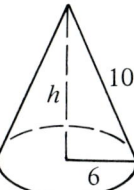

A square pyramid has base edges of length 6 and a height of 4.

4. $MB =$ __?__   3
5. $OM =$ __?__   3
6. $VM =$ __?__   5
7. Area of $\triangle VBC =$ __?__   15
8. Lateral area of pyramid = __?__   60
9. Total area of pyramid = __?__   96

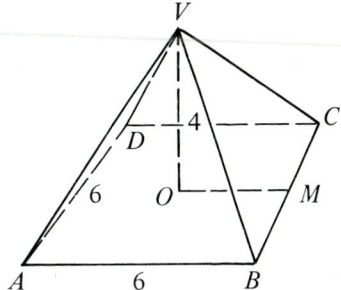

Suggested Extension

A plane slices a sphere, passing 4 cm from the center O. It forms a circle with radius 3.

1. What is the radius of the sphere?   5
2. What is the area of the sphere? 100π
3. What is the volume of the sphere? $166\frac{2}{3}\pi$

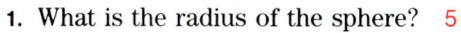

Lesson Commentary   **T129**

4. a. Carefully draw a larger copy of the diagram.
   b. Make folds along each side of the square.
   c. Tape together the edges with the same lengths in order to form a pyramid.

5. The pyramid in Exercise 4 is not regular. Why?
   The top does not lie directly over the base (the square).

6. Make two more pyramids like the one in Exercise 4. Combine all three pyramids to form a cube.

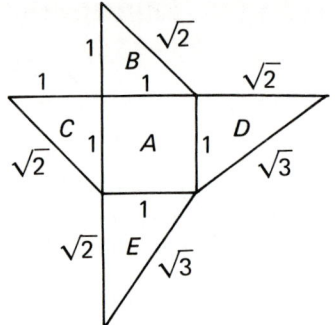

Quick Quiz

1. A rectangular solid is 3 m wide, 8 m long, and 6 m high. Find the length of a diagonal. $\sqrt{109} \approx 10.4$ m

Complete the table about right circular cones.

	r	h	l	L.A.	V
2.	15 cm	?	25 cm	?	?
3.	5 m	8 m	?	?	?

2. 20 cm, 375π cm², 1500π cm³
3. $\sqrt{89}$ m, $5\sqrt{89}\pi$ m², $66\frac{2}{3}\pi$ m³

A regular square pyramid has base edges 42 and slant height 35.

4. Find the lateral area. 2940   5. Find the volume. 16,464

Objective for Pages 447–455

To use the tangent, sine, and cosine ratios to find the missing parts of right triangles.

6 The Tangent Ratio *(pp. 447–451)*

Teaching These Pages

If you have a limited time in which to teach trigonometry, you may wish to combine Sections 6 and 7. Explain the terminology "opposite leg" and "adjacent leg," and then define the sine, cosine, and tangent ratios. Then move immediately to the Classroom Practice Exercises for Section 7.

Of course, it is more desirable to do Section 6 by itself before proceeding to Section 7. After you define the tangent ratio, you might work through Example 1 in the text and Chalkboard Examples 1–3 below. To do Example 2 in the text and Chalkboard Examples 4–6 below, students must be introduced to the Table of Trigonometric Ratios on page 517.

Point out to students that in Chalkboard Example 5, we may write either

$$\tan 35° = \frac{x}{2} \quad \text{so that} \quad x = 2(\tan 35°)$$
$$\text{or} \quad \tan 55° = \frac{2}{x} \quad \text{so that} \quad x = \frac{2}{\tan 55°}.$$

For those having a calculator, both expressions for x are equally easy to evaluate. For those without a calculator, $2(\tan 35°)$ is easier to evaluate than $\frac{2}{\tan 55°}$.

Chalkboard Examples

Find tan A and tan B for each right triangle.

1.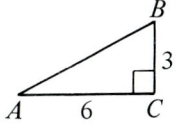

   $\tan A = \frac{1}{2}$; $\tan B = 2$

2.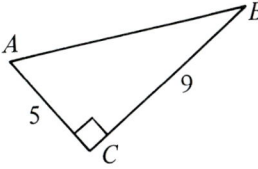

   $\tan A = \frac{9}{5}$; $\tan B = \frac{5}{9}$

3.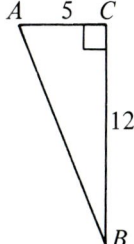

   $\tan A = \frac{12}{5}$; $\tan B = \frac{5}{12}$

Find the value of x to the nearest tenth. Use the table on page 517.

4.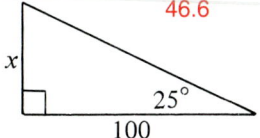

5.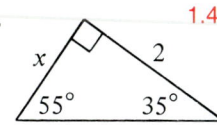

6. **a.** Express tan A as a fraction. $\frac{1}{2}$

   b. Express tan A as a decimal. 0.5

   c. Use the table on page 517 to find the measure of $\angle A$ to the nearest 5°. 25°

   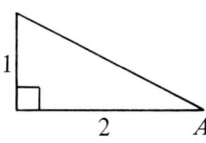

7 The Sine and Cosine Ratios (pp. 452–455)

Teaching These Pages

After you define the sine and cosine ratios, work through Examples 1, 2, and 3 on pages 452–453. Be sure that students understand these ratios thoroughly before you assign the Written Exercises.

Chalkboard Examples

Express the sine, cosine, and tangent of ∠A and ∠B as decimals.

1.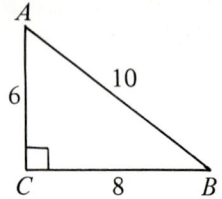

   sin A = 0.8000; sin B = 0.6000
   cos A = 0.6000; cos B = 0.8000
   tan A = 1.3333; tan B = 0.7500

2.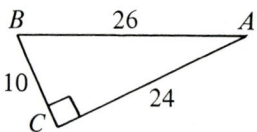

   sin A = 0.3846; sin B = 0.9231
   cos A = 0.9231; cos B = 0.3846
   tan A = 0.4167; tan B = 2.4000

3.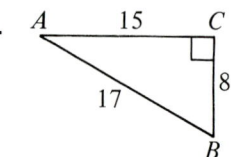

   sin A = 0.4706; sin B = 0.8824
   cos A = 0.8824; cos B = 0.4706
   tan A = 0.5333; tan B = 1.8750

Suggested Extension

The Calculator Activities feature on page 481 provides a nice introduction to the use of trigonometric ratios in oblique triangles. Advanced students might like to tackle the problems below.

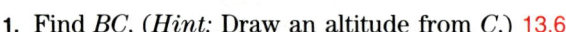

1. Find *BC*. (*Hint:* Draw an altitude from *C*.) 13.6
2. Find *AB*. 8.19 + 12.32 ≈ 20.5   3. Find the area. 58.8
4. Find the length of the altitude from *B*. 11.8

5. Show that $\dfrac{\sin A}{\cos A} = \tan A$.

   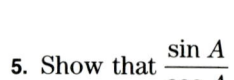

   $\dfrac{\sin A}{\cos A} = \dfrac{\frac{a}{c}}{\frac{b}{c}} = \dfrac{a}{b} = \tan A$

6. Show that $a \sin B = b \sin A$.
   (*Hint:* Calculate the length of altitude *CD* in two ways.)

   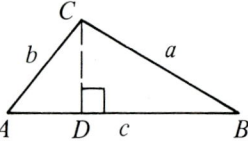

   $\sin A = \dfrac{CD}{b}$; $\sin B = \dfrac{CD}{a}$; $a \sin B = CD = b \sin A$

7. Show that $c = b \cos A + a \cos B$.
   (*Hint:* Find *AD* and *BD*.)

   $\cos A = \dfrac{AD}{b}$, so $AD = b \cos A$

   $\cos B = \dfrac{BD}{a}$, so $BD = a \cos B$

   $c = AD + BD = b \cos A + a \cos B$

T132   Lesson Commentary

Quick Quiz

Complete.

1. $\sin M = \underline{\quad?\quad} \; \frac{21}{29}$
2. $\cos M = \underline{\quad?\quad} \; \frac{20}{29}$
3. $\tan F = \underline{\quad?\quad} \; \frac{20}{21}$
4. $\cos F = \underline{\quad?\quad} \; \frac{21}{29}$

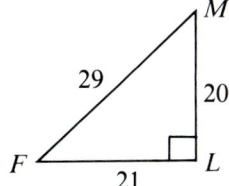

Solve for x, y, and z to the nearest tenth.

5.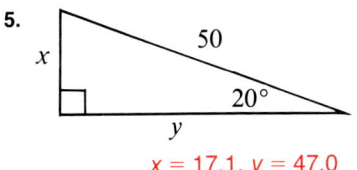

$x = 17.1, y = 47.0$

6.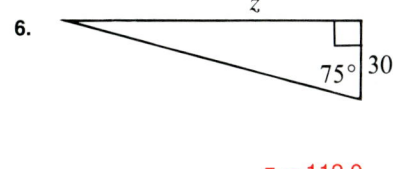

$z = 112.0$

Chapter 12. Coordinate Geometry

This final chapter begins with drill on basic vocabulary and plotting points. After students have had experience finding the distance between two points on horizontal or vertical lines, they move on to finding distances by applying the Pythagorean Theorem. The midpoint formula grows out of an averaging approach. The idea of slope is developed and applied to parallel and perpendicular lines. Linear equations are graphed. In the final section of the book two theorems that lend themselves to coordinate treatment are proved. Students first saw these theorems in Chapter 5, but the proofs were postponed until the introduction of coordinates.

Objectives for Pages 468–480

1. To specify points by their coordinates in the coordinate plane.
2. To work with simple figures in the coordinate plane.
3. To find the distance between two points in the coordinate plane.
4. To apply the midpoint formula.

1 Points and Coordinates (pp. 468–471)

Teaching These Pages

Drill vocabulary according to needs of the class; what is review to some students may be new material to others. Even experienced students should do most of the exercises, for the ability to make judgments about simple figures is developed along with proficiency in plotting points.

Lesson Commentary   **T133**

Chalkboard Examples

1. State the coordinates of each of the labeled points.
   A(3, 2); B(1, 2); C(−3, 2); D(−2, 0); E(−2, −3); F(0, −3);
   G(2, −1); O(0, 0)
2. Find the distance between points:
   a. A and B 2
   b. B and C 4
   c. D and E 3
   d. E and F 2

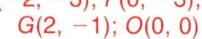

Suggested Extension

1. Points (−1, −1) and (3, 3) are two of the vertices of a square.
   Find three pairs of points that could be the other two vertices.
   (7, −1) and (3, −5); (−1, 7) and (−5, 3); (3, −1) and (−1, 3)

2. Repeat Exercise 1, using points (0, 2) and (4, 8).
   (6, −2) and (10, 4); (−6, 6) and (−2, 12); (5, 3) and (−1, 7)

2 Distance Between Two Points *(pp. 472–474)*

Teaching These Pages

Encourage students to recognize the fact that they don't need a special formula to find the distance between two given points. When a segment is horizontal or vertical its length can be determined by subtraction or by counting squares. For any other segment, students may draw horizontal and vertical segments to form a right triangle, then apply the Pythagorean Theorem.

Chalkboard Examples

Plot points and draw segments as necessary. Then find the distance between the two points. (Drawings are omitted.)

1. (−3, 0) and (−5, 0) 2
2. (2, −4) and (2, 1) 5
3. (1, 2) and (5, 5) 5
4. (−3, 0) and (5, 6) 10
5. (0, 2) and (5, 14) 13
6. (1, −1) and (−11, −10) 15

Suggested Extension

Plot point $A(2, 1)$. What points lie above the x-axis exactly 5 units from A?
One coordinate of several such points is given. Find the other coordinate.

1. (7, __?__) 1
2. (6, __?__) 4
3. (5, __?__) 5
4. (2, __?__) 6
5. (−1, __?__) 5
6. (−2, __?__) 4

7. There is one other point that lies above the x-axis, has integer coordinates, and is 5 units from A. Find the coordinates of that point. (−3, 1)

8. There are five points that lie below the x-axis, have integer coordinates, and are 5 units from A. Find the coordinates of the five points.
   (−2, −2); (−1, −3); (2, −4); (5, −3); (6, −2)

3 Midpoint of a Segment *(pp. 476–480)*

Teaching These Pages

We suggest that you have students carry out Part A of the Explorations individually. Then do Part B—step by step—as a group, using a large figure like that on page 477. In Steps 4 and 5, be sure students see that we find a midpoint coordinate by averaging the endpoint coordinates.

Some students may need a careful explanation of the use of subscripts in the midpoint formula.

Chalkboard Examples

M is the midpoint of $\overline{RS}$. State the coordinates of M, using:

1. $R(0, 0)$ and $S(8, -6)$ $(4, -3)$
2. $R(5, 0)$ and $S(11, 2)$ $(8, 1)$
3. $R(-3, 2)$ and $S(5, -10)$ $(1, -4)$
4. $R(0, 7)$ and $S(8, 0)$ $(4, 3.5)$

Suggested Extension

Have students plot points $A(0, 0)$, $B(8, 0)$, $C(10, 6)$, and $D(2, 6)$. Then have them complete the parallelogram and find the midpoint of each diagonal. $(5, 3)$

The fact that the same point is the midpoint of both diagonals confirms Theorem 6 of Chapter 5 (page 204):
Diagonals of a parallelogram bisect each other.

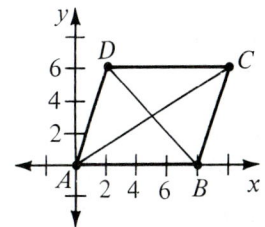

Quick Quiz

The sides of rectangle $QRST$ are parallel to the coordinate axes.

Give the coordinates of:

1. R $(1, -3)$
2. T $(-4, 2)$
3. S $(-4, -3)$
4. The midpoint of $\overline{TR}$ has coordinates $(\underline{\ ?\ }, \underline{\ ?\ })$. $-\frac{3}{2}$ $-\frac{1}{2}$
5. $TR = \underline{\ ?\ }$ $5\sqrt{2}$

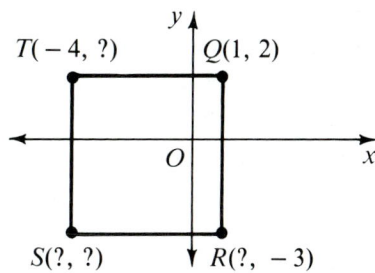

In Questions 6–8, use points $L(-35, 6)$ and $R(25, -5)$.

6. $LR = \underline{\ ?\ }$ 61
7. The midpoint of $\overline{LR}$ has coordinates $(\underline{\ ?\ }, \underline{\ ?\ })$. -5 $\frac{1}{2}$
8. L is in Quadrant $\underline{\ ?\ }$ and R is in Quadrant $\underline{\ ?\ }$.   II; IV

Objectives for Pages 486–495

1. To find the slope of a line.
2. To apply the fact that parallel lines have equal slopes.
3. To apply the fact that the product of the slopes of two (nonvertical) perpendicular lines equals -1.

4 Slope of a Line *(pp. 486–490)*

Teaching These Pages

Encourage students to go from left to right when they compute slopes. Then all lines that rise have positive slopes and all that fall have negative slopes. Use language carefully when you talk about vertical lines, but don't be too critical of students who think correctly but express themselves ambiguously. Note how easy it is to go from "That part of the road doesn't slope" to "That part of the road has no slope" to "A horizontal line doesn't have a slope." However, students must learn that a horizontal line does have a slope; the slope is the number zero. Use the phrasing, "If a line is vertical, slope is not defined for the line." Some students may form confusing mental images if you say "A vertical line has no slope."

Chalkboard Examples

Use points $A(1, 2)$, $B(5, 4)$, $C(4, 2)$, and $D(4, 0)$. State the slope of the line named. Write *not defined* if the line named is vertical.

1. $\overleftrightarrow{AB}$ $\frac{1}{2}$
2. $\overleftrightarrow{AC}$ 0
3. $\overleftrightarrow{AD}$ $-\frac{2}{3}$
4. $\overleftrightarrow{BC}$ 2
5. $\overleftrightarrow{BD}$ 4
6. $\overleftrightarrow{CD}$ not defined

Suggested Extension

In the diagram, $\overline{AR}$ and $\overline{CS}$ are horizontal segments; $\overline{RB}$ and $\overline{SD}$ are vertical segments. Note that $\overleftrightarrow{AR}$ and $\overleftrightarrow{CS}$ are parallel lines cut by transversal k. Tell why:

1. $\angle BAR = \angle DCS$ Postulate 7 (p. 67)
2. $\angle R = \angle S$ Hor. and vert. lines make rt. $\triangle$.
3. $\triangle BAR \sim \triangle DCS$ AA Postulate
4. $\dfrac{BR}{DS} = \dfrac{AR}{CS}$ Corr. parts of $\sim \triangle$ are in proportion
5. $\dfrac{BR}{AR} = \dfrac{DS}{CS}$ Switching Property

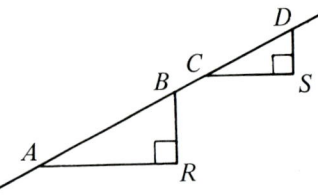

Note that $\dfrac{BR}{AR}$ and $\dfrac{DS}{CS}$ both represent the slope of line k.

5 Parallel and Perpendicular Lines
(pp. 491–495)

Teaching These Pages

You should use, without fanfare, the idea that vertical lines are parallel and horizontal lines are parallel. The coordinate grid itself implies this. It is better to say to beginning students, "If two lines are parallel, they have equal slopes" than "If two nonvertical lines are parallel, they have equal slopes." The precision you would achieve by using the second form would come at the expense of simplicity and clarity.

Chalkboard Examples

Use points $R(2, 2)$, $S(5, 4)$, and $T(5, 6)$. Find:

1. The slope of $\overleftrightarrow{RS}$; of $\overleftrightarrow{RT}$ $\quad \frac{2}{3}; \frac{4}{3}$

2. The slope of any line parallel to $\overleftrightarrow{RS}$; of any line parallel to $\overleftrightarrow{RT}$ $\quad \frac{2}{3}; \frac{4}{3}$

3. The slope of any line perpendicular to $\overleftrightarrow{RS}$; of any line perpendicular to $\overleftrightarrow{RT}$ $\quad -\frac{3}{2}; -\frac{3}{4}$

4. The slope of any line parallel to $\overleftrightarrow{ST}$ not defined

5. The slope of any line perpendicular to $\overleftrightarrow{ST}$ 0

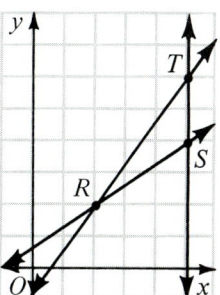

Suggested Extension

Plot points $A(-2, 0)$, $B(2, 3)$, $C(2, -2)$, and $D(-2, -5)$. Complete.

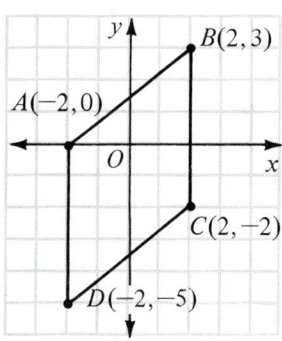

1. Because $\overline{AD}$ and $\overline{BC}$ are vertical segments, they are __?__ to each other.   parallel

2. $\overline{AB}$ and $\overline{CD}$ are parallel segments because the slope of each equals __?__. $\frac{3}{4}$

3. Because opposite sides are parallel, quadrilateral $ABCD$ is a __?__.   parallelogram

4. The slope of $\overline{AC}$ equals __?__, and the slope of $\overline{BD}$ equals __?__. $-\frac{1}{2}; 2$

5. $\overline{AC}$ and $\overline{BD}$ are __?__ to each other.   perpendicular

6. From Exercises 3 and 4 you see that quadrilateral $ABCD$ must be a __?__.   rhombus

7. Confirm your conclusion in Exercise 6 by finding the length of each side of quadrilateral $ABCD$.   All sides are of length 5.

Quick Quiz

Find the slope of:

1. $\overline{CD}$ $\frac{3}{2}$
2. $\overline{DA}$ undefined
3. $\overline{CB}$ $-\frac{1}{2}$
4. $\overline{AB}$ $\frac{3}{2}$
5. any line perpendicular to $\overline{AD}$ 0
6. Is $\overline{CB} \perp \overline{CD}$?   no
7. Is $\overline{CD} \parallel \overline{AB}$?   yes

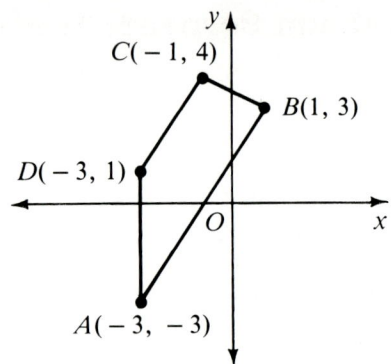

Objectives for Pages 496–504

1. To make a table of values for a linear equation.
2. To graph the line specified by a given equation.
3. To prove theorems using coordinate geometry.

6 Equations and Lines *(pp. 496–500)*

Teaching These Pages

Remember that some students have difficulty graphing equations like $x = 3$. When you drill equations of the type $x = a$, all goes speedily. But later on, when students encounter $x = 3$ in the midst of equations like $x + y = 5$, confusion arises. There is no magic remedy. Patience on your part, and willingness to spread the graphing of lines over a period of several days will help. Encourage students to use simple numbers, especially 0, when they make tables of values, and to show fortitude when they need to plot such points as $(0, 2\frac{1}{3})$.

Chalkboard Examples

Prepare tables of values for the equations.

1. $x + y = 6$

x	y
0	6
2	4
4	2
6	0

2. $2x - y = 6$

x	y
0	−6
1	−4
2	−2
3	0

3. $2x + 3y = 12$

x	y
0	4
3	2
6	0

Lesson Commentary

4–6. Draw graphs of the equations in Examples 1–3.

4.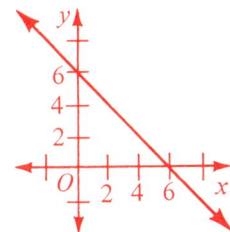
5.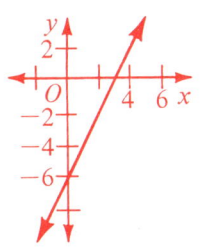
6.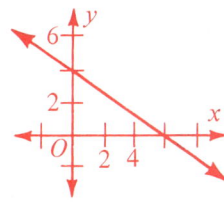

Suggested Extension

Graph the equations: $2x - 3y = -4$
$7x - 4y = 12$
$3x + 2y = -6$

What kind of triangle is formed by the lines? right triangle
Support your answer using two different arguments.
Prove this by (1) using the converse to the Pythagorean Theorem, or
(2) showing that the line with equation $2x - 3y = -4$ is perpendicular to the line with equation $3x + 2y = -6$.

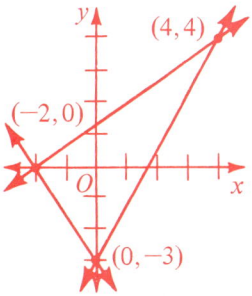

7 Coordinate Geometry Proofs (pp. 501–504)

Teaching These Pages

Allow students plenty of time to study page 501. You may want to develop the theorem as a class project. Begin with a large figure drawn on the chalkboard. Assign coordinates to the vertices, explaining that we use the coefficient "2" in the coordinates to simplify the calculations. Then find the coordinates of M and N. Have students state the slope of $\overline{MN}$, then of $\overline{AB}$ and $\overline{DC}$, leading to the deduction that $\overline{MN} \parallel \overline{AB} \parallel \overline{DC}$. Finally, have students find the lengths of $\overline{MN}, \overline{AB}$, and $\overline{DC}$, showing that $MN = \frac{1}{2}(AB + DC)$.

Chalkboard Examples

Coordinates have been assigned to three vertices of an isosceles trapezoid.

1. Find the coordinates of vertex C. (5, 5)

2. Check your work by showing that $AD = BC$.
   $AD = \sqrt{2^2 + 5^2} = \sqrt{29} = BC$

3. Prove that the diagonals of this isosceles trapezoid are equal.
   $BD = \sqrt{12^2 + 5^2} = 13$
   $AC = \sqrt{12^2 + 5^2} = 13$

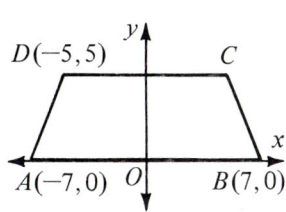

Suggested Extension

Given points $P(a, b)$ and $Q(c, d)$:

1. Prove that point $J\left(\dfrac{2a + c}{3}, \dfrac{2b + d}{3}\right)$ is one of the trisection points of $\overline{PQ}$ by proving that $PJ = \dfrac{1}{3}PQ$ and $JQ = \dfrac{2}{3}PQ$.

$(PJ)^2 = \left(\dfrac{2a + c}{3} - a\right)^2 + \left(\dfrac{2b + d}{3} - b\right)^2$

$= \left(\dfrac{c - a}{3}\right)^2 + \left(\dfrac{d - b}{3}\right)^2$

$= \dfrac{1}{9}[(c - a)^2 + (d - b)^2] = \dfrac{1}{9}(PQ)^2$

Hence $PJ = \dfrac{1}{3}PQ$.

$(JQ)^2 = \left(\dfrac{2a + c}{3} - c\right)^2 + \left(\dfrac{2b + d}{3} - d\right)^2$

$= \left[\dfrac{2}{3}(a - c)\right]^2 + \left[\dfrac{2}{3}(b - d)\right]^2$

$= \dfrac{4}{9}[(a - c)^2 + (b - d)^2] = \dfrac{4}{9}(PQ)^2$

Hence $JQ = \dfrac{2}{3}PQ$.

2. Use the triangle at right to prove that the medians of a triangle meet at a point $\dfrac{2}{3}$ of the way along each median. *Strategy:*
   1. Find the coordinates of the midpoints of the sides.
   2. Use the formula from the exercise above to show that $(2a + 2b, 2c)$ is a trisection point of all three medians.

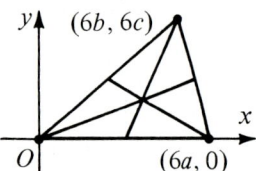

Quick Quiz

1. Is the point with coordinates $(3, -8)$ on the graph of the line $5x - 2y = 6$? **no**

2. Given the equation $9x - 6y = 15$, **(a)** make a table of values and **(b)** plot points and draw the graph of the equation.

   a.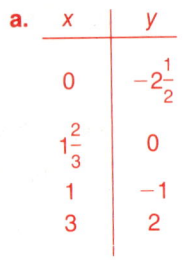

   b.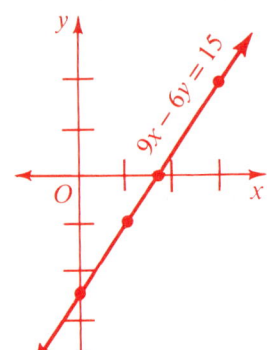

3. You are given the points $A(-1, -1)$, $B(6, 0)$, and $C(2, 3)$. Prove that $\triangle ABC$ is a right triangle.

   The slope of $\overline{AC}$ is $\dfrac{4}{3}$; the slope of $\overline{BC}$ is $-\dfrac{3}{4}$. Since $\dfrac{4}{3} \cdot -\dfrac{3}{4} = -1$, $\overline{AC} \perp \overline{BC}$.

   Therefore, $\angle C$ is a right angle and $\triangle ABC$ is a right triangle.

BASIC
GEOMETRY

Ray C. Jurgensen
Richard G. Brown

Editorial Adviser: Albert E. Meder, Jr.
Teacher Consultant: Robert J. McMurray

Houghton Mifflin Company • Boston
Atlanta   Dallas   Geneva, Illinois
Palo Alto   Princeton   Toronto

The Authors
Ray C. Jurgensen, formerly Chairman of the Mathematics Department and holder of the Eppley Chair of Mathematics at the Culver Academies, Culver, Indiana. Mr. Jurgensen has been a lecturer at the National Science Foundation institutes for mathematics teachers and meetings of the National Council of Teachers of Mathematics.

Richard G. Brown, Mathematics teacher at the Phillips Exeter Academy, Exeter, New Hampshire. Mr. Brown has taught a wide range of mathematics courses for both students and teachers at several schools and universities, including Newton (Massachusetts) High School, the University of New Hampshire, Arizona State University, and the North Carolina School for Science and Mathematics.

Editorial Adviser
Albert E. Meder, Jr., Professor of Mathematics, Emeritus, Rutgers, The State University of New Jersey.

Teacher Consultant
Robert J. McMurray, Mathematics teacher at the Albuquerque Technical Vocational Institute, Albuquerque, New Mexico. Mr. McMurray has many years of experience teaching geometry in secondary schools.

Copyright © 1990 by Houghton Mifflin Company. All rights reserved.

No part of this work may be reproduced or transmitted in any form or by any means, electronic or mechanical, including photocopying and recording, or by any information storage or retrieval system without the prior written permission of Houghton Mifflin Company unless such copying is expressly permitted by federal copyright law. Address inquiries to School Permissions, Houghton Mifflin Company, 222 Berkeley Street, Boston, MA 02116.

Printed in the U.S.A.

ISBN: 0-395-50120-2

CONTENTS

UNIT A

CHAPTER 1 Points, Lines, and Angles
1 Beginning Your Study of Geometry 4
2 Points and Lines 6
3 Angles 10
4 Classifying Angles 15
5 Vertical Angles 18
6 Angle Bisectors 20
7 Three Constructions 25
8 Postulates of Equality 29
9 Postulates of Geometry 33

FEATURES
Consumer Applications 14   Puzzle Problems 17   Experiments 22
Applications 23   Career Notes 32

REVIEWS AND TESTS
Self-Tests 14, 22, 37   Skills Review 38   Chapter Review 39
Chapter Test 41   Mixed Review 42

CHAPTER 2 Introducing Proof
1 Three Theorems 46
2 "If . . . then" Statements 50
3 Reaching Conclusions (Optional) 56
4 Writing Proofs 60
5 Parallel Lines 66
6 Proving Lines Parallel 72
7 Constructing Parallel Lines 77

FEATURES
Applications 54   Career Notes 59   Calculator Activities 71
Consumer Applications 76   Reading Geometry 81   Problem Solving Strategies 82

REVIEWS AND TESTS
Self-Tests 53, 80   Skills Review 84   Chapter Review 85   Chapter Test 87
Cumulative Review 88

UNIT B

CHAPTER 3 Triangles
1 The Angle Sum of a Triangle 92
2 Classifying Triangles 97
3 Defining Congruent Triangles 102
4 The SSS Postulate 107
5 The SAS Postulate 115
6 The ASA Postulate 121
7 The AAS and HL Theorems 126

FEATURES
Puzzle Problems 101, 112, 132   Applications 113   Experiments 120, 132
Consumer Applications 120   Career Notes 125   Reading Geometry 133
Computer Activities 135

REVIEWS AND TESTS
Self-Tests 101, 112, 125, 132   Skills Review 136   Chapter Review 137
Chapter Test 139   Mixed Review 140

CHAPTER 4 Using Congruent Triangles
1 Proving Corresponding Parts Equal 144
2 Congruent Triangles and Constructions 149
3 Segment Bisectors 154
4 Altitudes and Medians of Triangles 158
5 Inscribed and Circumscribed Circles 163
6 Triangles with Two Equal Sides 167
7 Triangles with Two Equal Angles 173

FEATURES
Consumer Applications 148   Reading Geometry 153   Puzzle Problems 157, 162
Experiments 162, 178   Career Notes 172   Applications 179
Problem Solving Strategies 180

REVIEWS AND TESTS
Self-Tests 152, 166, 178   Skills Review 182   Chapter Review 183
Chapter Test 185   Cumulative Review 186

UNIT C

CHAPTER 5 Polygons
1 Introducing Polygons 190
2 Angle Sums of Polygons 194
3 Special Quadrilaterals 198
4 Properties of Parallelograms 203
5 Properties of Special Parallelograms 209
6 Proving Figures Are Parallelograms 214
7 Properties of Trapezoids 220
8 The Midpoints Theorem 225

FEATURES
Experiments 193   Career Notes 202   Applications 208   Puzzle Problems 224
Reading Geometry 229

REVIEWS AND TESTS
Self-Tests 197, 207, 219, 228   Skills Review 230   Chapter Review 231
Chapter Test 233   Mixed Review 234

CHAPTER 6 Areas
1 Areas of Rectangles 238
2 Areas of Parallelograms 243
3 Areas of Triangles 247
4 Areas of Trapezoids 250
5 The Pythagorean Theorem 255
6 Converse of Pythagorean Theorem 259
7 Circumferences of Circles 263
8 Areas of Circles 268

FEATURES
Consumer Applications 246   Puzzle Problems 254, 261   Career Notes 262
Applications 272   Problem Solving Strategies 274   Computer Activities 276

REVIEWS AND TESTS
Self-Tests 246, 254, 262, 271   Skills Review 278   Chapter Review 279
Chapter Test 281   Cumulative Review 282

UNIT D

CHAPTER 7 Ratios and Proportions
1 Ratios 286
2 Proportions 289
3 Using Proportions 293
4 Maps and Scale Drawings 296
5 Making Maps and Scale Drawings 299

FEATURES
Calculator Activities 292   Puzzle Problems 295   Applications 302

REVIEWS AND TESTS
Self-Tests 295, 301   Skills Review 304   Chapter Review 305
Chapter Test 307   Mixed Review 308

CHAPTER 8 Similar Polygons
1 Defining Similar Polygons 312
2 The AA Postulate 317
3 A Special Case of Similar Triangles 323
4 The Triangle Proportionality Theorem 325
5 Perimeters and Areas 329

FEATURES
Career Notes 316   Calculator Activities 322, 332   Puzzle Problems 324
Experiments 333   Applications 334   Problem Solving Strategies 336

REVIEWS AND TESTS
Self-Tests 322, 333   Skills Review 338   Chapter Review 339
Chapter Test 341   Cumulative Review 342

UNIT E

CHAPTER 9 Circles
1 Basic Terms 346
2 Tangents 350
3 Arcs and Central Angles 355
4 Chords 359
5 Inscribed Angles 364
6 Other Angles 369
7 Segments of Chords 373

FEATURES
Experiments 349   Consumer Applications 349   Puzzle Problems 354, 368, 376
Applications 377   Computer Activities 378

REVIEWS AND TESTS
Self-Tests 358, 368, 376   Skills Review 380   Chapter Review 381
Chapter Test 383   Mixed Review 384

CHAPTER 10 Areas and Volumes of Solids
1 Lines and Planes in Space 388
2 Right Prisms 393
3 Right Circular Cylinders 398
4 Regular Pyramids 402
5 Right Circular Cones 406
6 Spheres 410
7 Similar Solids (Optional) 414

FEATURES
Experiments 392, 405, 409   Consumer Applications 392   Puzzle Problems 401, 405
Calculator Activities 409   Applications 413   Reading Geometry 417
Problem Solving Strategies 418

REVIEWS AND TESTS
Self-Tests 401, 412   Skills Review 420   Chapter Review 421   Chapter Test 423
Cumulative Review 424

UNIT F

CHAPTER 11 Right Triangles
1 Reviewing Right Triangles 428
2 Special Right Triangles 432
3 Using Special Right Triangles 437
4 Diagonals of Rectangular Solids 440
5 Right Triangles in Pyramids and Cones 443
6 The Tangent Ratio 447
7 The Sine and Cosine Ratios 452

FEATURES
Applications 431, 457   Calculator Activities 436   Puzzle Problems 442, 446, 456, 459
Career Notes 451

REVIEWS AND TESTS
Self-Tests 439, 446, 456   Skills Review 460   Chapter Review 461
Chapter Test 463   Mixed Review 464

CHAPTER 12 Coordinate Geometry
1 Points and Coordinates 468
2 Distance Between Two Points 472
3 Midpoint of a Segment 476
4 Slope of a Line 486
5 Parallel and Perpendicular Lines 491
6 Equations and Lines 496
7 Coordinate Geometry Proofs 501

FEATURES
Calculator Activities 475, 481   Applications 482, 505   Career Notes 500
Problem Solving Strategies 506   Computer Activities 508

REVIEWS AND TESTS
Self-Tests 480, 495, 504   Skills Review 510   Chapter Review 511
Chapter Test 513   Cumulative Review 514

TABLE OF SQUARES 515
TABLE OF SQUARE ROOTS 516
TABLE OF TRIGONOMETRIC RATIOS 517

POSTULATES 518
THEOREMS 519
CONSTRUCTIONS 522

REVIEW EXERCISES 523
APPENDIX: TRANSFORMATIONAL GEOMETRY 531

GLOSSARY 547
INDEX 555

ANSWERS TO SELF-TESTS 563
ANSWERS TO SELECTED EXERCISES 566

ACKNOWLEDGMENTS 584

SYMBOLS

Symbol	Meaning	Page
∠, ∠s	angle, angles	10
∠ABC	angle ABC, measure of ∠ABC	10
$\widehat{CD}$	arc with endpoints C and D, measure of $\widehat{CD}$	355
A	area	238
B	area of base	394
⊙P	circle with center P	263
C	circumference	263
≅	congruent, is congruent to	102
cos	cosine	452
°	degrees	10
d	diameter,	263
	length of diagonal,	440
	distance	472
=	equals, is equal to	7
≈	is approximately equal to	264
≠	is not equal to	59
>	is greater than	34
<	is less than	53
L.A.	lateral area	394
h	length of altitude, height	238
b	length of base	238
s	length of a side of a square	239
l	length of slant height	402
XY	length of $\overline{XY}$	7
$\overleftrightarrow{CD}$	line through points C and D	6
(x, y)	point with coordinates x and y	84
∥	parallel, is parallel to	34
▱	parallelogram	198
p	perimeter	240
⊥	perpendicular, is perpendicular to	15
π	pi	264
r	radius	264
$\frac{a}{b}$, $a:b$	ratio of a to b	286
$\overrightarrow{AB}$	ray with endpoint A	6
$\overline{AB}$	segment with endpoints A and B	6
~	similar, is similar to	312
sin	sine	452
$\sqrt{x}$	positive square root of x	256
tan	tangent	447
T.A.	total area	394
△, △s	triangle, triangles	15
V	volume	394

METRIC SYMBOLS

mm	millimeter
cm	centimeter
m	meter
km	kilometer
cm²	square centimeter
m²	square meter
km²	square kilometer
cm³	cubic centimeter
m³	cubic meter
L	liter
g	gram
g/cm³	grams per cubic centimeter
min	minute
h	hour

READING GEOMETRY

HOW TO READ YOUR GEOMETRY TEXTBOOK

When you read a story or a newspaper article, you probably read rather quickly, without giving full attention to every sentence. To get the most out of a textbook, you need to read differently. A geometry textbook calls for slow, thoughtful reading. Here are some suggestions to help you comprehend more when you read about geometry.

VOCABULARY

You will find many new words in this book. Some of them are not much used except in geometry—for example, *collinear, vertices,* and *transversal.* Others are everyday words that have special meanings in geometry—for example, *ray, acute,* and *supplement.* As you turn the pages, you will see words printed in **heavy type.** These are important words and phrases, whose meanings you will learn as you study the lessons. They are defined or explained when they first appear, and are often illustrated by diagrams. For example, see the definition and illustration of *vertical angles* on page 18. Later, if you need to refresh your memory, you can look up "vertical angles" in the Glossary at the back of the book, or in the Index. The Glossary repeats the definition and tells you the page where it is first stated in the text. The Index may give other page references if you are looking for additional information.

SYMBOLS

Almost as soon as you began to study arithmetic, you learned to use symbols for the operations. Later you became acquainted with other mathematical symbols. Symbols are especially useful in geometry because they provide a lot of information in a small amount of space. For example, look at Statement 1 in the proof of Theorem 1 on page 154, "$\overleftrightarrow{RM} \perp \overline{CD}$." In words this says, "The line through points R and M is perpendicular to the segment with endpoints C and D." Be sure to look for *all* the information given by statements in condensed form. If you have trouble reading a statement expressed by symbols, refer to the list of symbols on the facing page. Then reread the statement slowly. Saying it aloud or writing it out in words may help you. Exercises such as Classroom Practice Exercises 6–10 on page 8 will give you practice in interpreting symbols.

DIAGRAMS

Open this book almost anywhere and you will see diagrams. They are there to help you understand the explanations and the exercises, and to present additional information. You need to read the diagrams along with the words of the text. Be careful not to miss any of the information that is given. Sometimes, as on page 6, you will find red boxes and arrows below or beside the diagrams. These often provide a word statement of information given in a diagram. They may also call your attention to a particular part of the figure.

In reading a diagram, be careful not to take anything for granted that isn't actually shown. Look again at the second diagram on page 7. It wouldn't be safe to say that *AM* and *MB* are equal if the little red marks weren't there to tell you that this is true.

Sometimes you may find it helpful to make a quick sketch of your own to illustrate an explanation that you are studying. Be sure to label your drawing to show the relationships correctly.

Exercises such as Written Exercises 3–20 on page 9 and Written Exercises 8–14 on page 49 will help you gain skill in reading diagrams.

ORGANIZATION OF THE TEXTBOOK

Your textbook has been planned to make reading and learning geometry as easy as possible. New words and phrases are printed in **heavy type** and carefully explained when they first occur. Brief explanations in red boxes often appear along with diagrams and other important items. Definitions are repeated in the Glossary, with page references, and the Index lists all pages where you can find information about a particular item. Large red boxes are used to indicate theorems, postulates, and constructions. Other important information is also shown in red. All the postulates, theorems, and constructions are listed at the back of the book, with page references to where they first occur.

Each chapter contains one or more Self-Tests that you can use to check your progress. Answers are given at the back of the book. At the end of every chapter you will find a summary, review exercises, and a review test for the chapter. There is either a cumulative test or a mixed review at the end of every chapter. Reviews of arithmetic and algebraic skills are also included. You will have many opportunities to check your progress and strengthen your skills.

UNIT A

Here's what you'll learn in this chapter:

To name points, lines, rays, segments, angles, and planes.

To measure and classify angles.

To do basic constructions with a compass and a straightedge.

To use postulates of equality and postulates of geometry to justify statements.

Many examples of the geometric figures discussed in this chapter and later in the book can be found in the photograph above. Look for parts of the building that suggest points, lines and line segments, planes, and angles.

Chapter 1

Points, Lines, and Angles

1 Beginning Your Study of Geometry

For each diagram below, tell which line looks longer—the solid red one or the solid black one.

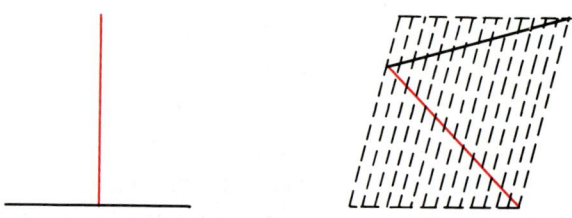

Check your answers by measuring.

Were you fooled by these optical illusions? They show that you cannot always depend on appearances. Sometimes you need to measure. In this course you will need three tools for measuring: a ruler, a protractor, and a compass.

Not all geometry problems can be solved by measuring. Many are solved by reasoning logically, starting with known geometric facts. This is the kind of reasoning you will study in this course. The following activities preview some things that you will study. Try them.

1. Draw a rectangle.
   Join the midpoints of the sides.
   What kind of figure is formed?

   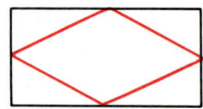

   Now join the midpoints of the "inside" figure to form a third four-sided figure. What kind of figure is formed? How do its dimensions compare with the dimensions of the original rectangle?

   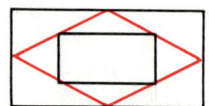

   Try this experiment again, beginning with a different-looking rectangle.

2. Imagine that the earth is a perfectly round ball.
   A giant hoop fits exactly around the earth's equator. The hoop is then stretched so that it is 2 m longer. (Two meters is about the height of a tall person.)

   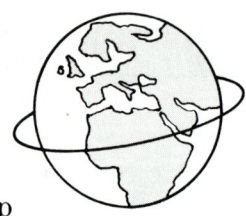

   The hoop now floats evenly around the earth.
   Is there enough room between the earth and the hoop for a flea to crawl under? a fly? a mouse? a cat?

The answers are at the top of the next page.

Chapter 1

Activity 1: When you join the midpoints of a rectangle, you form a parallelogram with all sides equal. This figure is called a *rhombus*.

When you join the midpoints of this rhombus, you form a rectangle. The length and width of this rectangle are half those of the original rectangle.

Activity 2: All of them could crawl under the hoop. Later in the course you will be able to prove that there are about 32 cm between the earth and the hoop. See Exercise 40 on page 267.

The activities should suggest that some geometric problems have surprising results. To get these results, we begin our study with some very *un*surprising ones. We will piece these together to prove more substantial results, called *theorems*. In a way, this process is like detective work. We will fit together geometric clues to arrive at conclusions. The exercises will give you a chance to do some detective work.

Exercises

1. Which line at the right of the rectangle is the continuation of the line at the left? the bottom line

2. Draw several four-sided figures. For each, join the midpoints of the sides. In each case, what kind of figure is formed?
   parallelogram

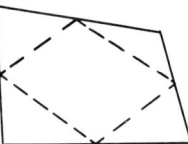 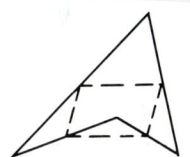 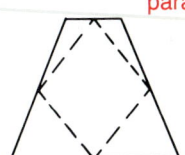

3. Wrap a piece of string tightly around a large wastebasket. Wrap another piece around a tin can. Then add two meters of string to each piece. Finally, spread the lengthened strings evenly around the wastebasket and the can as shown.

Compare the space between the string and the wastebasket with the space between the string and the can. What do you notice? The spaces are the same; for any diameter, the distance will be $\frac{1}{\pi}$ meters.

Points, Lines, and Angles 5

2 Points and Lines

Do you know that a color television picture is composed of thousands of dots? You don't notice the spaces between the dots because the dots are so close to each other.

The idea of tightly-packed dots is sometimes used in mathematics. For example, the picture below suggests that a **line** contains very many **points.**

○○

Each dot represents a point of the line.

Actually, a line contains an unlimited number of points. We usually picture a line by drawing along the edge of a ruler, like this.

←——————————————→

The arrowheads in the drawing suggest that a line extends indefinitely far in both directions.

The drawings below explain how we name points and lines.

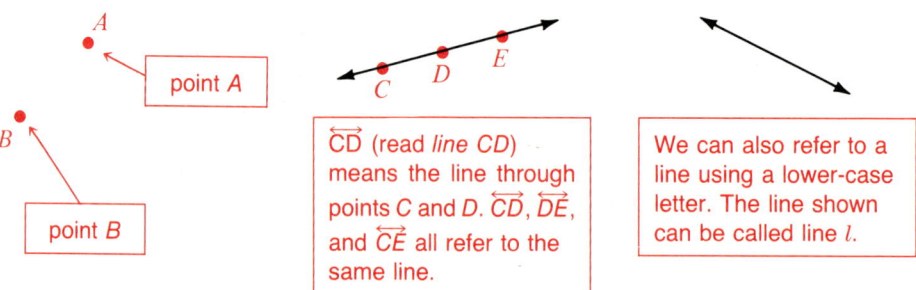

point A

point B

$\overleftrightarrow{CD}$ (read *line CD*) means the line through points C and D. $\overleftrightarrow{CD}$, $\overleftrightarrow{DE}$, and $\overleftrightarrow{CE}$ all refer to the same line.

We can also refer to a line using a lower-case letter. The line shown can be called line *l*.

The points *C*, *D*, and *E* shown above are called **collinear points.** This means that they are all on one line. On the other hand, the points *X*, *Y*, and *Z* pictured at the right are not collinear points. They do not lie on one line.

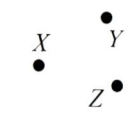

A *segment* and a *ray* are parts of a line as shown below.

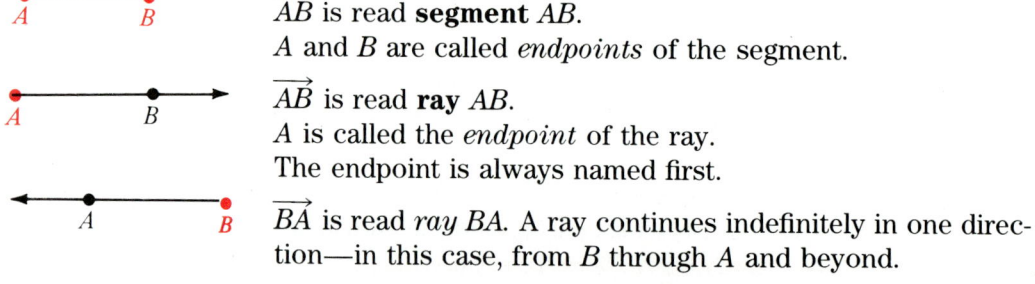

$\overline{AB}$ is read **segment** *AB*.
A and *B* are called *endpoints* of the segment.

$\overrightarrow{AB}$ is read **ray** *AB*.
A is called the *endpoint* of the ray.
The endpoint is always named first.

$\overrightarrow{BA}$ is read *ray BA*. A ray continues indefinitely in one direction—in this case, from *B* through *A* and beyond.

Notice that $\vec{BA}$ and $\vec{AB}$ refer to different rays. On the other hand, $\overline{BA}$ and $\overline{AB}$ refer to the same segment.

Each point of a line can be paired with a real number. See the Ruler Postulate Similarly, each real number can be paired with a point. on page 34.

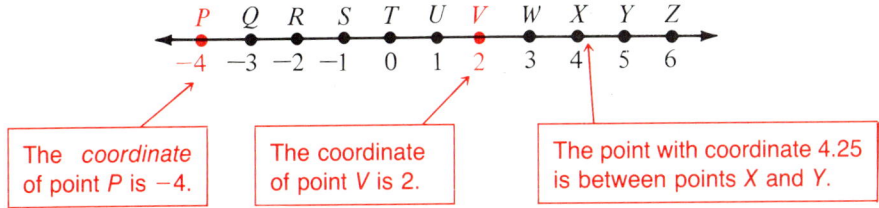

The *coordinate* of point P is -4.

The coordinate of point V is 2.

The point with coordinate 4.25 is between points X and Y.

The symbol XY stands for the length of $\overline{XY}$. This length is just the distance between X and Y. It can be found by counting the number of units from X to Y. It can also be found by subtracting the smaller coordinate (the one on the left) from the larger coordinate (the one on the right). Using the figure above, we find:

$$XY = 5 - 4 = 1$$
$$SU = 1 - (-1) = 1 + 1 = 2$$
$$PS = -1 - (-4) = -1 + 4 = 3$$

In the diagram above, the *midpoint* of $\overline{SU}$ is the point T. It is exactly halfway between S and U. In general, the **midpoint** of a segment is defined in this way:

M is the midpoint of $\overline{AB}$ means that

(1) M is on $\overline{AB}$ and

(2) $AM = MB$.

The little marks show that $AM = MB$.

From the definition of midpoint it is clear that

$$AM = \tfrac{1}{2}AB \text{ and } MB = \tfrac{1}{2}AB.$$

EXAMPLE   In the diagram, Q is the midpoint of $\overline{PS}$, $RS = 2$, and $QR = 3$. Find PS.

$QS = QR + RS = 3 + 2 = 5$
Since $QS = 5$, PQ also equals 5.
Then $PS = PQ + QS = 10$.

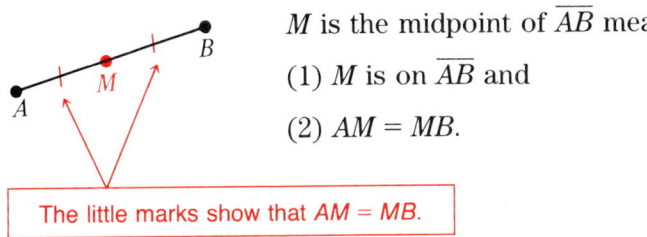

Points, Lines, and Angles 7

Classroom Practice

Two lines meet at O as shown. Tell whether each statement is true or false.

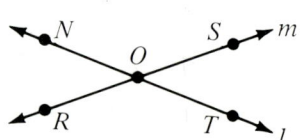

1. $\overleftrightarrow{OS}$ is the same as $\overleftrightarrow{OR}$. **true**
2. $\overline{OS}$ is the same as $\overline{OR}$. **false**
3. Line m contains just three points. **false**
4. Points R, O, and S are collinear. **true**
5. Points N, O, and S are collinear. **false**

Exercises 6–10 refer to the diagram at right.

6. Read the symbol $\overline{QR}$ and explain what it means. **(See below.)**
7. Name the endpoints of: **a.** $\overline{PQ}$; **b.** $\overline{QP}$.   **a.** P and Q **b.** P and Q
8. Read the symbol $\overrightarrow{QP}$ and explain what it means. **(See below.)**
9. Name the endpoint of: **a.** $\overrightarrow{PQ}$; **b.** $\overrightarrow{QP}$.   **a.** P **b.** Q
10. Which points are in both $\overrightarrow{QR}$ and $\overrightarrow{PQ}$? $\overrightarrow{QR}$

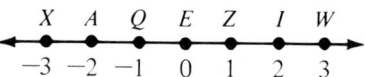

Exercises 11–15 refer to the diagram.

11. Name the coordinate of X. -3
12. Name the point with coordinate 3. W
13. Name the midpoint of $\overline{ZW}$. I

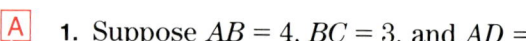

14. Find each distance: **a.** ZW; **b.** EI; **c.** XZ.   **a.** 2   **b.** 2   **c.** 4
15. How many real numbers are there between 1 and 2—a limited number or an unlimited number? How many points are there between Z and I?   **unlimited number; unlimited number**

6. Segment QR; it is points Q and R and all the points on $\overleftrightarrow{QR}$ between Q and R.
8. Ray QP; it is point Q and all points on $\overleftrightarrow{QR}$ on the same side of Q as P.

Written Exercises

A   1. Suppose $AB = 4$, $BC = 3$, and $AD = 14$.
     a. Then $CD = \underline{\ ?\ }$.   7
     b. Is C the midpoint of $\overline{AD}$?   yes

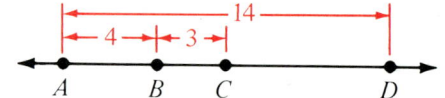

2. Suppose O is the midpoint of $\overline{HL}$.
     a. If $HL = 16$, then $OL = \underline{\ ?\ }$.   8
     b. Suppose you also know that W is the midpoint of $\overline{OL}$. Then $OW = \underline{\ ?\ }$.   4

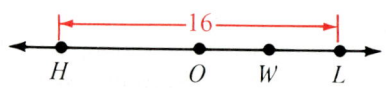

Exercises 3–10 refer to the figure.

3. The coordinate of point U is __?__. −3
4. If I is the midpoint of $\overline{TO}$, then the coordinate of I is __?__. 5
5. If U is the midpoint of $\overline{QE}$, then the coordinate of E is __?__. −1
6. If $ON = 2$, then the coordinate of N is __?__. 9

Exercises 3–10

Find each distance.

7. TO 4
8. TU 6
9. QU 2
10. OQ 12

The coordinates of I and A are −10 and 20, as shown. O is the midpoint of $\overline{IA}$, $IE = 6$, and $IU = 23$. Find each of the following.

11. The coordinate of O 5
12. The coordinate of E −4
13. The coordinate of U 13
14. IO 15
15. EU 17
16. EA 24

Exercises 11–20

Using the diagram above, tell whether each statement is true or false.

17. There are an unlimited number of points between E and O. true
18. $\overline{EU}$ and $\overline{UE}$ refer to the same segment. true
19. $\overrightarrow{EU}$ and $\overrightarrow{UE}$ refer to the same ray. false
20. $\overleftrightarrow{EU}$ and $\overleftrightarrow{AO}$ refer to the same line. true

Suppose S is the midpoint of $\overline{RT}$. Complete the table.

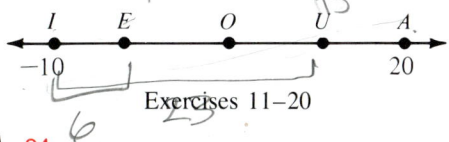

	21.	22.	23.	24.	25.	26.	27.	28.
coordinate of R	8	−4	−18	5	17	−18	a	$b - 10$
coordinate of T	12	12	−2	? 13	? 41	? −2	$a + 6$	6
length of $\overline{RT}$	? 4	16	? 16	? 8	? 24	? 16	? 6	16 ? −b
coordinate of S	10	? 4	? −10	9	29	−10	a ? +3	$\frac{b}{2}$? −2

29. Look up the meaning of the prefix "geo-." Relate this meaning to the words "geometry," "geography," and "geology."

30. Points X, Y, and Z have coordinates 2.001, $\frac{21}{8}$, and $\sqrt{5}$. Which of these points are between points C and D? All three

3 Angles

As you may know, an **angle** is a figure formed by two rays or segments with a common endpoint. The rays or segments are called the **sides** of the angle and the endpoint is called the **vertex** of the angle. The diagrams below show how angles are named.

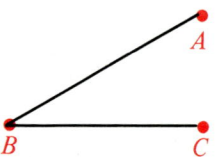

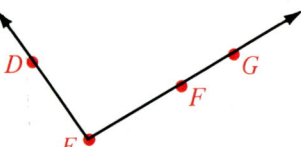

Notice that the vertex is always named in the middle.

∠ABC means angle ABC.
∠CBA is another name for ∠ABC.

∠DEF means angle DEF.
∠DEF is the same as ∠DEG.

$\overrightarrow{BA}$ and $\overrightarrow{BC}$ are the sides.
B is the vertex.

$\overrightarrow{ED}$ and $\overrightarrow{EF}$ are the sides.
E is the vertex.

Sometimes we use just one letter or one number to name an angle.

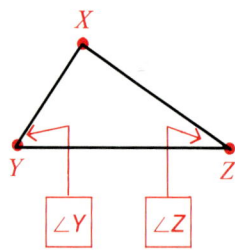

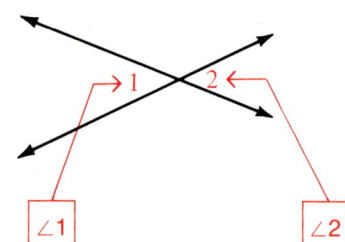

∠Y   ∠Z

∠1   ∠2

To find the *measure in degrees* of an angle, we use an instrument called a *protractor*. The diagram below shows how to do this. Can you see that the measure of ∠AOB is 40°? This fact is written: ∠AOB = 40°.*

See the Protractor Postulate on page 35.

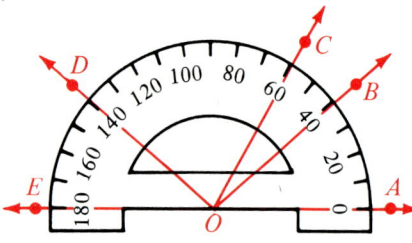

Notice that:
∠AOB = 40°
∠AOE = 180°
∠BOC = 60° − 40° = 20°
∠DOB = 140° − 40° = 100°
∠EOC = 180° − 60° = 120°

*Note: Some books use the notation "m∠AOB = 40°" and say, "the measure of ∠AOB is 40°." In this book, we shall use the simpler notation "∠AOB = 40°" and say, "∠AOB equals 40°."

Many protractors have two scales, one reading from left to right, and the other reading from right to left. To use this kind of protractor, follow these steps:

1. Estimate: Is the angle measure less than or greater than 90°?

2. Carefully line up your protractor as shown below.
   If the sides of the angle do not reach the scale, extend them.

 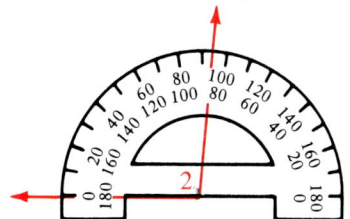

3. Choose the scale that has 0° at one side of the angle.
   Read the measure of the angle.
   Does the measure agree with your estimate?

EXAMPLE   What is the measure of ∠3?

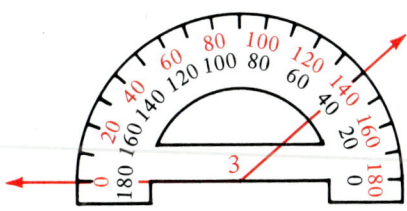

∠3 is greater than 90°.

Choose the scale on top.

∠3 = 140°

The diagram at the bottom of page 10 suggests that we can add and subtract angle measures. Here are some other examples.

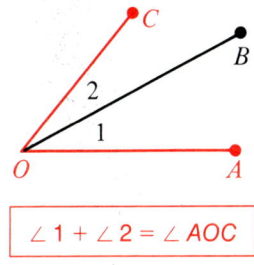 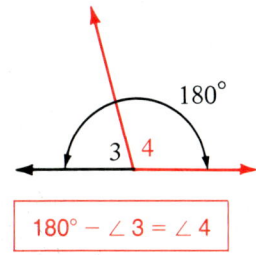

∠1 + ∠2 = ∠AOC   180° − ∠3 = ∠4

Points, Lines, and Angles   **11**

Classroom Practice

Exercises 1–7 refer to the diagram.

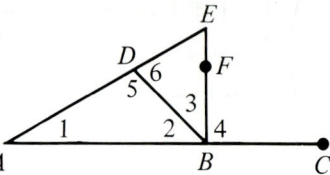

1. Name the vertex of ∠ 4. *B*

2. Name the sides of ∠ 2.
   $\overline{BD}$ and $\overline{BA}$

3. ∠ BAE and ∠ 1 are names for the same angle. Give another name for each of the following angles:
   ∠ ABD, ∠ AEB, ∠ BDE, ∠ 3, ∠ 5, ∠ 4. ∠ 2; ∠ E, ∠ DEB, ∠ AEF, or ∠ DEF;
   ∠ 6; ∠ DBE or ∠ DBF; ∠ ADB;
4. ∠ 5 + ∠ 6 = __?__ ° 180   ∠ EBC or ∠ FBC

5. ∠ ABE − ∠ 3 = ∠ __?__   2 or ABD

6. Explain why you should not refer to ∠ B. Because it would not be clear whether you meant ∠ 2, ∠ 3, or ∠ 4.

7. **a.** Is it correct to refer to ∠ A? yes
   b. Is it correct to refer to ∠ D? no

In Exercises 8–10, use the diagram shown.

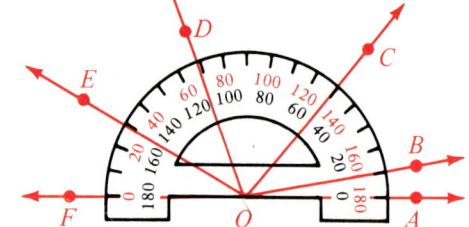

8. Using three letters for each angle, name five angles that have $\overrightarrow{OA}$ as one side.
   ∠ AOB, ∠ AOC, ∠ AOD, ∠ AOE, ∠ AOF

9. Using three letters for each angle, name five angles that have $\overrightarrow{OC}$ as a side.
   ∠ AOC, ∠ BOC, ∠ DOC, ∠ EOC, ∠ FOC

10. State the measure of each of the following:
    ∠ AOB, ∠ AOC, ∠ AOD, ∠ AOE, ∠ AOF.
      10°   50°   110°   150°   180°

11. Estimate the measure of each angle shown.

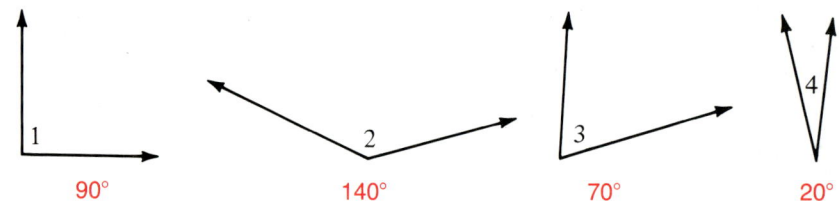

    90°   140°   70°   20°

12. Without measuring, sketch each angle. Make your sketches large enough so that you can use a protractor to check the accuracy of your work.

    a. 90° angle   **b.** 60° angle   **c.** 10° angle   **d.** 170° angle

Written Exercises

Complete each statement.

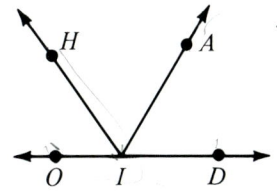

A 1. The sides of ∠AIH are $\vec{IA}$? and $\vec{IH}$?.
2. The vertex of ∠AIH is ? . I
3. Another name for ∠OIH is ∠ ? . HIO
4. ∠DIA + ∠AIH = ∠ ? DIH
5. ∠DIH + ∠HIO = ∠ ? = ? ° DIO; 180
6. 180° − ∠DIA = ∠ ? OIA

Find the measure of each angle named.

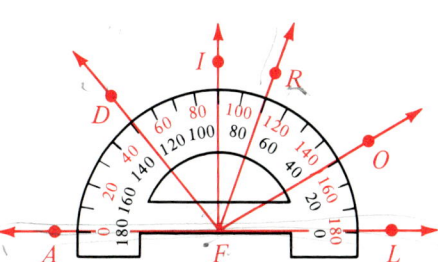

7. ∠LFO 30°
8. ∠RFA 110°
9. ∠DFO 100°
10. ∠AFL 180°
11. ∠DFR 60°
12. ∠IFL 90°

Use your protractor to draw an angle with the given measure.

13. 20°   14. 65°   15. 90°   16. 140°

17. Estimate the measure of each angle shown.

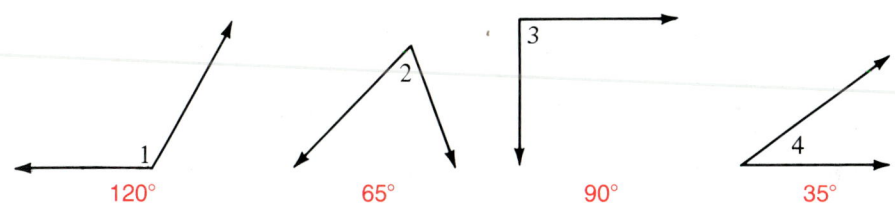

120°   65°   90°   35°

18. Without measuring, sketch each angle. Then use a protractor to check the accuracy of your work.
   a. 45° angle   b. 15° angle   c. 85° angle   d. 160° angle

Complete each statement.

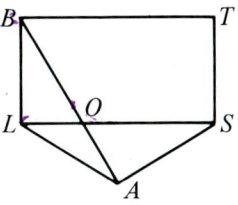

SAMPLE   ∠BLO + ∠OLA = ?
         ∠BLO + ∠OLA = ∠BLA

 B 19. ∠BAS + ∠BAL = ∠ ? LAS
20. ∠TSO + ∠OSA = ∠ ? TSA
21. ∠LAS − ∠LAB = ∠ ? BAS
22. 180° − ∠BOS = ∠ ? = ∠ ? SOA; BOL

Points, Lines, and Angles

SELF-TEST

Vocabulary

line (p. 6)
point (p. 6)
collinear points (p. 6)
segment (p. 6)
ray (p. 6)
midpoint (p. 7)
angle (p. 10)
sides of an angle (p. 10)
vertex of an angle (p. 10)

Tell whether each statement is true or false.

(1-2)

1. $\overleftrightarrow{AB}$ is the same as $\overleftrightarrow{BC}$.   true

2. $\overrightarrow{AB}$ is the same as $\overrightarrow{BA}$.   false

3. C is the midpoint of $\overline{BD}$.   true

4. $AB = 6$ true

(1-3)

5. What is the vertex of $\angle PRQ$?   R

6. Name $\angle 2$ using three letters. $\angle PQS$

7. What angle has a measure of about 90°? Use three letters to name the angle. $\angle PQR$

8. $\angle 5 + \angle 6 = \underline{\ ?\ }°$ 180

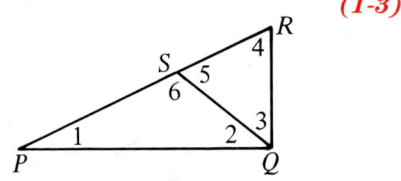

CONSUMER APPLICATIONS

TIME IS MONEY

People use the money they earn in their jobs to buy things they want and need: food, clothing, housing, books, and recreation. How much time must a person work to make these common purchases?

Suppose a worker takes home $5.40 per hour. That is equivalent to 9¢ per minute. (5.40 ÷ 60 = 0.09) How much time will it take the worker to earn the money to pay for the following items?

3 h 7 min	a $16.83 toaster	a monthly rent of $360	66 h 40 min
40 h	a $216 television set	a $48.60 grocery bill	9 h
42 min	a $3.78 paperback book	a 5¢ piece of gum	$\frac{5}{9}$ min

14   Chapter 1

4 Classifying Angles

Angles are often classified by their measures.

An **acute angle** has a measure between 0° and 90°.

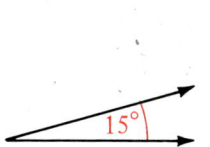 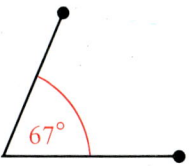 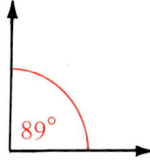

A **right angle** has measure 90°. We indicate a right angle by drawing a small square inside the angle as shown at the right.

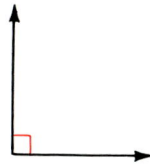

If an angle formed by two lines is a right angle, then the lines are **perpendicular.** Also, if two lines are perpendicular, they form right angles. In the diagram, line AB is perpendicular to line CD. This is written:

$$\overleftrightarrow{AB} \perp \overleftrightarrow{CD}.$$

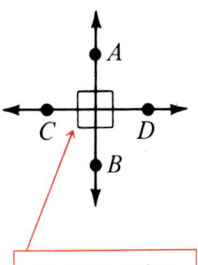

4 right angles

An **obtuse angle** has a measure between 90° and 180°.

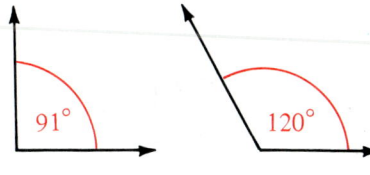

A **straight angle** has measure 180°.

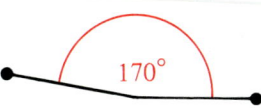

∠ AOB is a straight angle.

The diagram at the right shows a **triangle** with **vertices** A, B, and C. Vertices is the plural of *vertex*. The triangle pictured is called *triangle ABC* and is represented by the symbol $\triangle ABC$. Since $\angle B$ and $\angle C$ have equal measures, we say that the angles are equal and write

$$\angle B = \angle C.$$

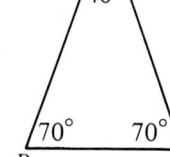

Also, since the measures of $\angle A$, $\angle B$, and $\angle C$ total 180°, we write

$$\angle A + \angle B + \angle C = 180°.$$

Points, Lines, and Angles **15**

EXAMPLE In the diagram, $\overline{AC} \perp \overline{EB}$ and $\angle EAD = 40°$.
Find the measure of **a.** $\angle DAC$ and **b.** $\angle DAB$.

a. $\angle DAC = \angle EAC - \angle EAD$
$= 90° - 40°$
$= 50°$

b. $\angle DAB = \angle EAB - \angle EAD$
$= 180° - 40°$
$= 140°$

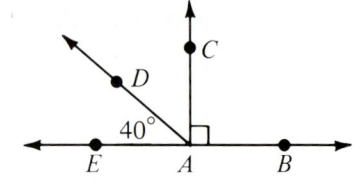

Classroom Practice

Classify each angle as acute, right, obtuse, or straight.

1. $\angle 1$ acute
2. $\angle 2$ right
3. $\angle DAB$ obtuse
4. $\angle ACB$ right
5. $\angle CAB$ acute
6. $\angle DCB$ straight
7. An angle that is equal to $\angle 2$ is $\angle$ __?__. ACB
8. You know that $\overline{AC} \perp \overline{BD}$ because $\angle 2$ is __?__. a right angle
9. The three triangles pictured are $\triangle$ __?__, $\triangle$ __?__, and $\triangle$ __?__.
   ACB   ACD   ABD
10. Name the three vertices of $\triangle ZOS$. Z, O, S
11. Name the three angles of $\triangle XOZ$. $\angle XOZ$, $\angle OZX$, $\angle ZXO$
12. How many angles are shown in the diagram? (Do not count straight angles.) 15
13. How many triangles are shown in the diagram? 8

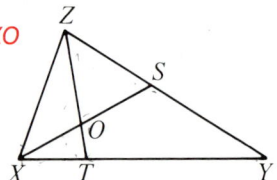

Written Exercises

**The measures of several angles are given below.
Classify each angle as acute, right, obtuse, or straight.**

A 1. 18° acute 2. 157° obtuse 3. 128° obtuse 4. 90° right 5. 180° straight 6. 89° acute

7. Name four obtuse angles. $\angle 3, \angle 5, \angle 7, \angle 10$
8. Name six acute angles. $\angle 1, \angle 2, \angle 4, \angle 6, \angle 8, \angle 9$
   ($\angle JON$ is also acute)
9. Name four pairs of angles whose sum is a straight angle. $\angle 2, \angle 3; \angle 5, \angle 6; \angle 6, \angle 7; \angle 8, \angle 10$
10. Name four triangles.
    $\triangle JOK, \triangle MON, \triangle JML, \triangle NKL$

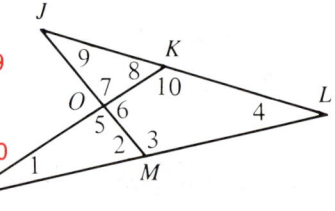

Exercises 7–10

In the figure, $\overline{AB} \perp \overline{CD}$ and $\angle SRD = 30°$.

11. Name four right angles. ∠CRA, ∠DRA, ∠DRB, ∠CRB
12. Find the measure of each angle named.
    a. ∠ARS 60°   b. ∠CRS 150°   c. ∠BRS 120°

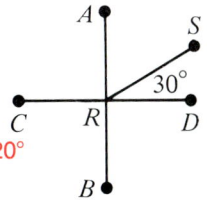

In the figure, $\overline{TQ} \perp \overline{VR}$ and $\overline{UQ} \perp \overline{QS}$.

13. Name three right angles. ∠VQT, ∠TQR, ∠UQS
14. Name two obtuse angles. ∠UQR, ∠VQS
15. Find the measure of each angle named.
    a. ∠VQU 65°   b. ∠TQS 65°   c. ∠SQR 25°

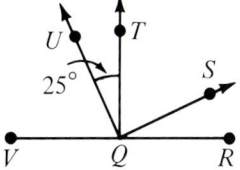

B 16. a. Draw a triangle with three acute angles. Art may vary.
       b. Draw a triangle with just two acute angles. Art may vary.
       c. Can you draw a triangle with just one acute angle? no

17. If the angles between the panes of the stained glass window shown are all equal, find the measure of each angle. 60°

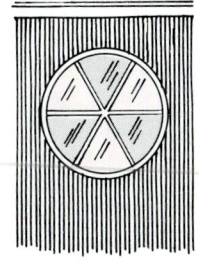
Exercise 17

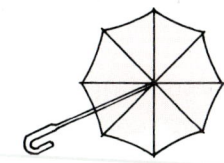
Exercise 18

18. If the angles between the spokes of an umbrella are all equal, find the measure of each angle. 45°

19. Use a dictionary to find the meanings of an "acute" person and an "obtuse" person. An acute person is one who is keenly perceptive or discerning; an obtuse person is one who is slow to perceive.

20. Study a corner of a room where two walls and the floor meet. How many right angles are formed at the corner? 3

PUZZLE ♦ PROBLEMS

Here are six lines.
Add five more lines to make NINE!

Points, Lines, and Angles **17**

5 Vertical Angles

When two lines meet, they form four angles. Two angles that are opposite each other, such as $\angle 1$ and $\angle 3$ in the diagram, are called **vertical angles.** $\angle 2$ and $\angle 4$ are also vertical angles.

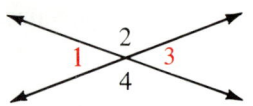

Suppose $\angle 1 = 30°$. Do you see that $\angle 2 = 150°$? (Remember that $\angle 1 + \angle 2 = 180°$.) Now do you see that $\angle 3 = 30°$? How large is $\angle 4$? This information is given in the first row of the table at the right. Copy and complete rows 2 and 3.

$\angle 1$	$\angle 2$	$\angle 3$	$\angle 4$
30°	150°	30°	150°
20°	?	?	?
25°	?	?	?

You can probably guess that vertical angles are always equal. Before reading further, see if you can explain why this is so.

160°   20°   160°
155°   25°   155°

The following steps show that *vertical angles are always equal.* Refer to the diagram above.

Step 1 $\angle 1 = 180° - \angle 2$ because $\angle 1 + \angle 2 = 180°$.
Step 2 $\angle 3 = 180° - \angle 2$ because $\angle 2 + \angle 3 = 180°$.
Step 3 $\angle 1 = \angle 3$ because both angles equal $180° - \angle 2$.

See Theorem 3 on page 47.

In the same way, you can show that $\angle 2 = \angle 4$.

| EXAMPLE | Study the diagram and find the measure of
a. $\angle DOE$ **b.** $\angle BOC$ **c.** $\angle COD$.

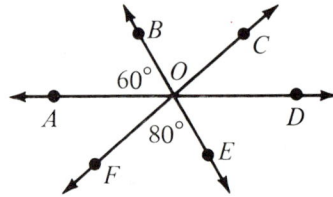

a. $\angle DOE$ and $\angle AOB$ are vertical angles. Therefore, both equal 60°.

b. $\angle BOC$ and $\angle FOE$ are vertical angles. Therefore, both equal 80°.

c. $\angle AOB + \angle BOC + \angle COD = 180°$
   $60° + 80° + \angle COD = 180°$
   $140° + \angle COD = 180°$
   $\angle COD = 40°$

Classroom Practice

Find the measure of each angle.

1. ∠ POU = __?__°   20
2. ∠ QOR = __?__°   85
3. ∠ POQ = __?__°   75
4. ∠ SOT = __?__°   75

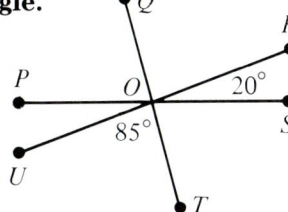

5. There are six pairs of vertical angles in the diagram. Can you find all six pairs? ∠ RAP, ∠ TAS; ∠ NAT, ∠ PAG; ∠ RAN, ∠ GAS; ∠ RAT, ∠ SAP; ∠ RAG, ∠ NAS; ∠ NAP, ∠ TAG

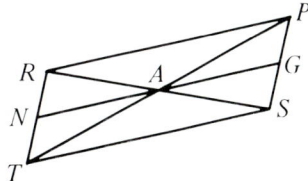

Written Exercises

Name an angle equal to the given angle.

A
1. ∠ BOA   2. ∠ BOC   3. ∠ AOC   4. ∠ DOC
    ∠ DOE   ∠ FOE   ∠ FOD   ∠ AOF

Suppose ∠ BOA = 30° and ∠ DOF = 140°. Find the measure of each angle.

5. ∠ COD   6. ∠ BOC   7. ∠ COE   8. ∠ BOF
    40°   110°   70°   70°

Exercises 1–8

In the diagram, $\overline{AE} \perp \overline{GC}$, ∠ GOH = 25°, and ∠ AOB = 35°. Find the measure of each angle.

9. ∠ AOH   65°   10. ∠ BOC   55°   11. ∠ COD   25°
12. ∠ DOE   65°   13. ∠ EOF   35°   14. ∠ FOG   55°

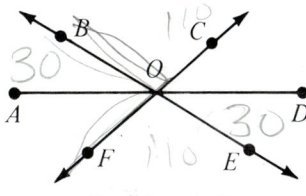

Exercises 9–14

Find the values of x and y.

B
15. $x = 20$
    $y = 9$

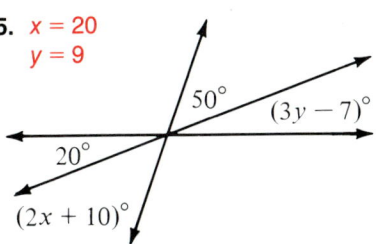

16. $x = 14$
    $y = 8$

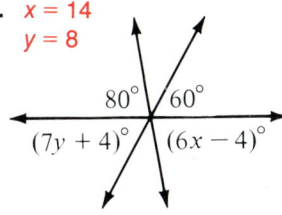

C
17. Draw a diagram showing $\overleftrightarrow{AB}$ and $\overleftrightarrow{CD}$ intersecting at point O so that ∠ AOD = ∠ BOD. Why must $\overleftrightarrow{AB}$ be perpendicular to $\overleftrightarrow{CD}$?
Since ∠ AOD + ∠ BOD = 180° and ∠ AOD = ∠ BOD, each is 90°. Thus $\overleftrightarrow{AB}$ is perpendicular to $\overleftrightarrow{CD}$.

6 Angle Bisectors

A **bisector of an angle** is a ray or line which divides the angle into two equal angles. In the diagram below, $\overrightarrow{OX}$ bisects $\angle AOB$.

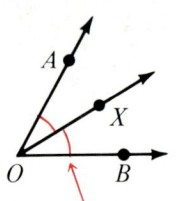

When $\overrightarrow{OX}$ bisects $\angle AOB$, $\angle AOX = \angle BOX$.

Thus $\angle AOX = \frac{1}{2} \angle AOB$

and $\angle BOX = \frac{1}{2} \angle AOB$.

The red marks show that $\angle AOX = \angle BOX$.

EXAMPLE $\overrightarrow{OB}$ bisects $\angle AOC$.

Find the measure of $\angle AOB$.

Then find the number on the protractor that corresponds to $\overrightarrow{OB}$.

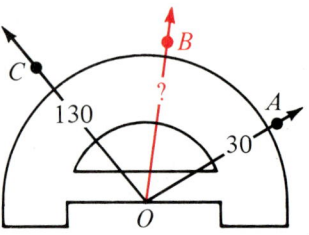

$\angle AOB = \frac{1}{2} \angle AOC$

$= \frac{1}{2}(130° - 30°)$

$= \frac{1}{2}(100°) = 50°$

$\overrightarrow{OB}$ corresponds to $30° + 50°$, or $80°$.

Classroom Practice

In each diagram, $\overrightarrow{SZ}$ bisects $\angle RST$.

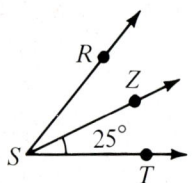   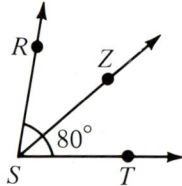

1. $\angle RSZ = \underline{\quad?\quad}°$ 25
2. $\angle RST = \underline{\quad?\quad}°$ 50

3. $\angle RSZ = \underline{\quad?\quad}°$ 40
4. $\angle ZST = \underline{\quad?\quad}°$ 40

5. How many lines can bisect a given angle?   one

Written Exercises

A 1. Draw an acute angle. Label it ∠ JOG.
Draw a bisector of ∠ JOG and call it $\overrightarrow{OP}$.
Now draw $\overrightarrow{ON}$ bisecting ∠ JOP.
Suppose ∠ JOG = 40°.
Then ∠ JOP = __?__° and ∠ JON = __?__°. 20; 10

In each exercise, $\overrightarrow{OB}$ bisects ∠ AOC. Find
a. the measure of ∠ AOB and
b. the number on the protractor that corresponds to $\overrightarrow{OB}$.

2. a. 25° b. 55°
3. a. 45° b. 65°
4. a. 70° b. 82°

2.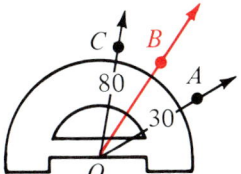

3.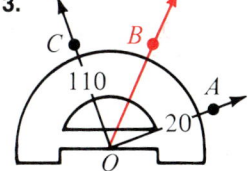

4.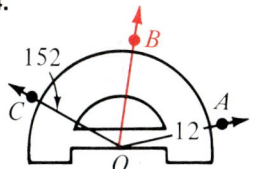

LFAE is a four-sided figure and $\overline{EF} \perp \overline{AL}$.

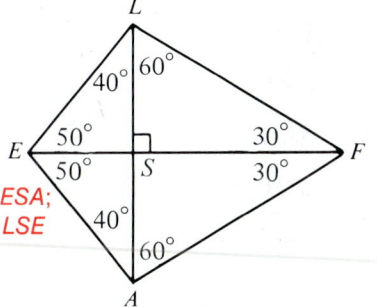

5. How many right angles are shown? 4
6. How many obtuse angles are shown? 3
7. How many acute angles are shown? 9
8. Name two pairs of vertical angles. ∠ FSL, ∠ ESA; ∠ FSA, ∠ LSE
9. Name the bisector of ∠ LFA. $\overline{FE}$
10. Name the bisector of ∠ LEA. $\overline{FE}$

B 11. Make a paper airplane. Then unfold it and mark on the paper the measures of the angles between the folds.   Answers will vary; the △ will be bisected.

12. Fold a rectangular sheet of paper to form a square.

13. a. Draw any angle and its bisector like that at the left below.

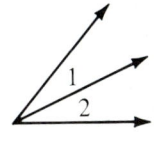   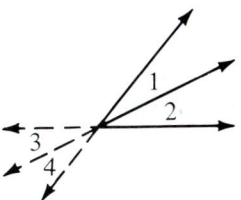

b. Extend each ray in the opposite direction as shown above.
c. What can you conclude about ∠ 3 and ∠ 4? ∠ 3 = ∠ 4
d. Explain why you think your conclusion is correct.
∠ 2 = ∠ 3, and ∠ 1 = ∠ 4; since ∠ 1 = ∠ 2, ∠ 3 = ∠ 4

Points, Lines, and Angles   21

Experiments

Experiment 1

1. Fold a rectangular sheet of paper to form a 45° angle.
   Fold any side up to an adjacent side.
2. Can you discover how to form a $22\frac{1}{2}°$ angle?
   Bring the two sides of the 45° angle together.
3. Could an $11\frac{1}{4}°$ angle be formed by the folding process?
   Yes; bring the two sides of the $22\frac{1}{2}°$ angle together.

Experiment 2

1. Fold over a corner of a rectangular sheet of paper as in Figure 1 below.
2. Now fold the next corner so the edges touch as in Figure 2.

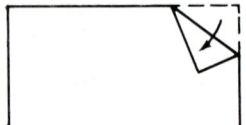

Figure 1   Figure 2

3. Open the paper and measure the angle between the creases. 90°
4. Can you explain why your classmates' angles have the same measure as your angle? By folding, we bisect two angles whose sum is 180°. Hence the angle between the creases will be $\frac{1}{2} \cdot 180°$ or 90°.

SELF-TEST

Vocabulary

acute angle (p. 15)
right angle (p. 15)
perpendicular lines (p. 15)
obtuse angle (p. 15)
straight angle (p. 15)

triangle (p. 15)
vertices of a triangle (p. 15)
vertical angles (p. 18)
bisector of an angle (p. 20)

In the diagram, $\overline{AD} \perp \overline{BC}$. Classify each angle as acute, right, obtuse, or straight.

1. $\angle ADC$ right
2. $\angle ABC$ acute
3. $\angle BEA$ obtuse
4. $\angle AFC$ straight

(1-4)

5. Name two pairs of vertical angles.
   $\angle BED, \angle AEF; \angle BEA, \angle DEF$

(1-5)

6. If $\overrightarrow{BF}$ bisects $\angle ABC$, what two angles must be equal?
   $\angle ABF, \angle FBC$

(1-6)

Applications

LATITUDE AND LONGITUDE

When traveling over land, you can sometimes tell your location by noticing landmarks that are near you. In the middle of the ocean, this would be very difficult to do. One of the ways that air and sea navigators describe their exact location is by using the system of *latitude* and *longitude*. This system consists of a grid of imaginary lines, east-west lines of latitude and north-south lines of longitude. Each point on the earth is at the crossing of a particular line of latitude and a particular line of longitude. Most globes and large maps show some of these lines.

The *equator*, a line that divides the earth in half, is used as the line of zero latitude. Other lines of latitude, also called parallels of latitude, are marked to indicate degrees north and south of the equator. We measure 90° of latitude north of the equator and 90° of latitude south.

Lines of longitude, also called meridians, run from the North Pole to the South Pole. These lines indicate degrees east or west of a zero line, called the prime meridian. We measure 180° of longitude west of the prime meridian and 180° of longitude east.

On navigational charts, points are drawn showing a ship's position at various times. Lines are drawn to connect these points. The woman shown is using parallel rulers to transfer such lines to another part of a chart. Navigators also need to know how to use a compass and dividers.

Points, Lines, and Angles

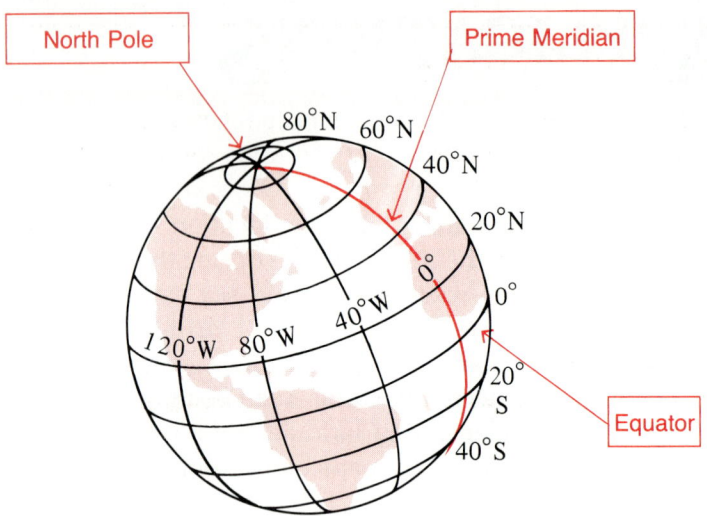

| EXAMPLE | Look at the illustration of a globe shown above. Do you see that 40°N, 40°W is near the middle of the North Atlantic Ocean?

Find the location of 20°S, 80°W. It is near the west coast of South America.

There is no natural location for the prime meridian as there is for the equator. Originally each nation used the line of longitude passing through its own national observatory. This led to some confusion, and in 1884, most nations agreed to use the meridian passing through Greenwich, England, as the prime meridian.

EXERCISES

Use a map or globe showing latitude and longitude.

1. Name the ocean in which the following locations are found.
   a. 80°N, 160°W Arctic   **b.** 5°S, 60°E Indian   **c.** 30°S, 10°W So. Atlantic
   d. 40°N, 150°W No. Pacific   **e.** 50°N, 20°W No. Atlantic   **f.** 50°S, 160°W So. Pacific

2. Find the approximate latitude and longitude for the place where you live. Answers will vary.

3. The location 0° latitude, 0° longitude is near what continent? Africa

4. Name the countries in Africa through which the equator passes. (See below.)

5. Name the countries in Europe through which the prime meridian passes. England, France, Spain

4. Gabon, Congo, Uganda, Kenya, Somali Republic, Zaire

7 Three Constructions

Making a construction is a mathematical game in which figures are drawn using only a *straightedge* and *compass*. The use of a ruler or protractor for measuring purposes is not allowed.

CONSTRUCTION 1

Given: *An angle*
Construct: *A bisector of the angle*

1.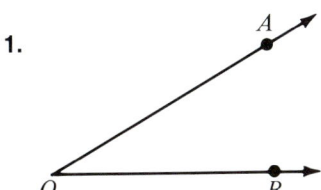

You are given ∠AOB.

2.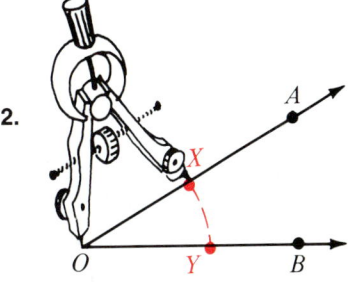

Put the point of your compass at O and swing an arc which cuts the sides of ∠AOB at X and Y.

3.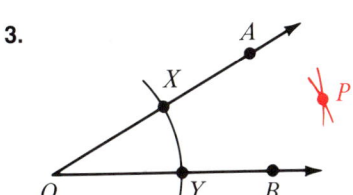

Now draw two arcs, one with the compass point at X and the other with the point at Y. Use the same compass setting for both arcs. The arcs meet at P.

4.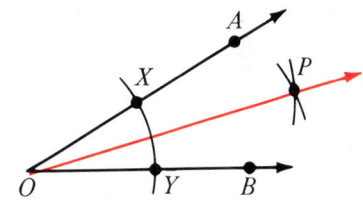

Join points P and O.
$\overrightarrow{OP}$ is a bisector of ∠AOB.

Suppose that ∠AOB given in Construction 1 is a straight angle. When you bisect this straight angle with measure 180°, you form two angles with measure 90°. In other words, bisecting a straight angle produces perpendicular lines. This idea is used in the next construction.

Points, Lines, and Angles

CONSTRUCTION 2

Given: *A line and a point on the line*
Construct: *A perpendicular to the line through the point*

Turn back to Construction 1. You will notice that the steps in Construction 2 are almost the same as those of Construction 1.

1, 2.

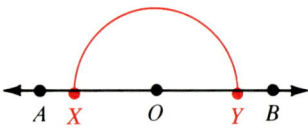

3. 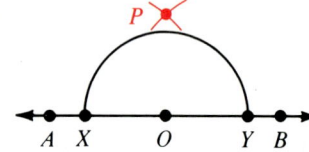   **4.**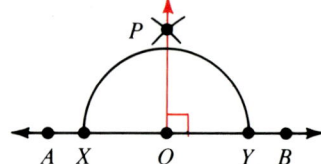

In Construction 2, the perpendicular passes through a point on the given line. In the next construction, the perpendicular passes through a point outside the given line.

CONSTRUCTION 3

Given: *A line and a point not on the line*
Construct: *A perpendicular to the line through the point*

Compare the steps in this construction with those of Constructions 1 and 2.

1. 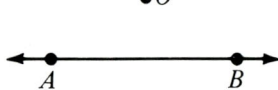   **2.**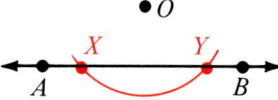

3. 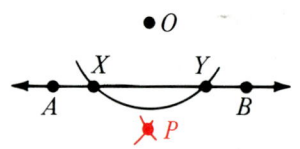   **4.**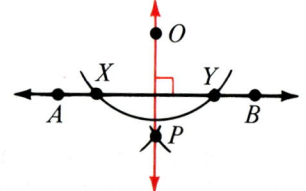

26   Chapter 1

Classroom Practice

1. Draw an obtuse angle. Const. 1
   Then bisect it, using a straightedge and compass.

2. Use Construction 2 to construct a 90° angle.
   Then construct a 45° angle by bisecting the 90° angle.

3. Draw a line l and choose a point O not on l.
   Use Construction 3 to construct a line, through O,
   which is perpendicular to line l.

Written Exercises

**Draw an angle similar to, but larger than, the one shown.
Then bisect it.**

A 1. Const. 1   2. Const. 1   3. Const. 1

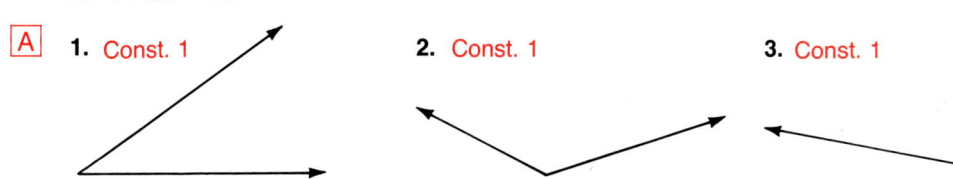

**Draw a figure similar to, but larger than, the one shown.
Then construct a perpendicular to $\overleftrightarrow{AB}$ through point O.**

4. Const. 2   5. Const. 2   6. Const. 2

7. Const. 3   8. Const. 3   9. Const. 3

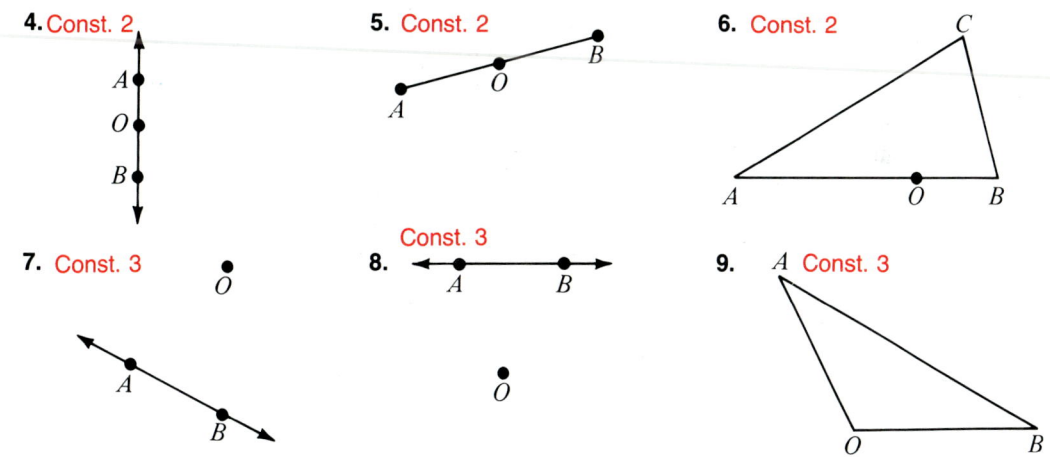

10. Draw an acute angle. Construct the bisector of the angle. Const. 1

11. Draw an obtuse angle. Construct the bisector of the angle. Const. 1

12. **a.** Draw a large triangle.
    b. Construct the bisectors of the three angles. Const. 1
       The bisectors should meet in a point. Do they? yes

13. **a.** Draw a line m and choose a point K on the line.
    b. Construct four right angles, each of which has K as vertex.

14. Use your construction from Exercise 13. Construct a 45° angle.

15. Use your construction from Exercises 13 and 14. Construct a $22\frac{1}{2}°$ angle.

**Draw a figure similar to, but larger than, the one shown.
Then construct perpendiculars to $\overleftrightarrow{AB}$ from C and D.**

B 16. 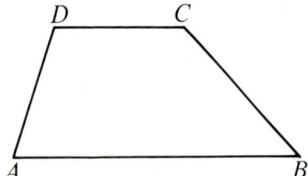   17.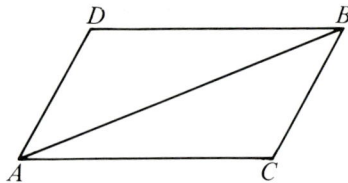

**Draw a figure similar to, but larger than, the one shown.
Construct a perpendicular to $\overleftrightarrow{RS}$ at S and a perpendicular to $\overleftrightarrow{RT}$ at T.**

18. 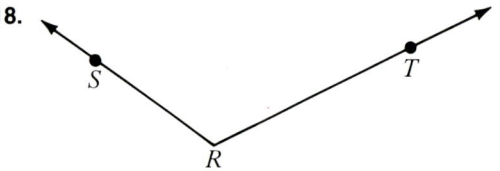   19.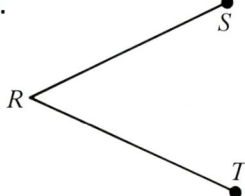

**Draw a figure similar to, but larger than, the one shown.
Then construct a bisector of $\angle UVW$ and a bisector of $\angle WVZ$.**

20. 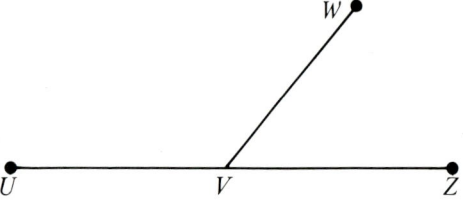   21.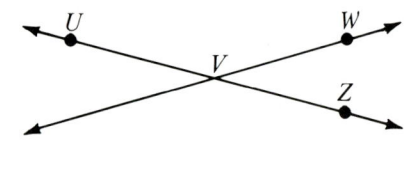

22. In Exercises 20 and 21, how do the two angle bisectors you constructed seem to be related? They are perpendicular.

C 23. Construct a 135° angle. (*Hint:* 135° = 180° − 45°.)

24. Construct a square. Remember: You may not use your ruler to measure distances; you may use it only as a straightedge.

8 Postulates of Equality

The following facts about real numbers are called **postulates.** These are statements which are accepted without proof. Postulates are used to prove other statements called *theorems*.

THE ADDITION POSTULATE

If $a = b$ and $c = d$, then $a + c = b + d$.

EXAMPLE 1

In the diagram, $AB = CD$.

But we know that $BC = BC$.

Conclusion: $AB + BC = CD + BC$.

In other words, $AC = BD$.

THE SUBTRACTION POSTULATE

If $a = b$ and $c = d$, then $a - c = b - d$.

EXAMPLE 2

In the diagram, $\angle ABC = \angle ACB$

and also $\angle 1 = \angle 2$.

Conclusion: $\angle ABC - \angle 1 = \angle ACB - \angle 2$.

In other words, $\angle 3 = \angle 4$.

THE MULTIPLICATION POSTULATE

If $a = b$, then $ac = bc$.

EXAMPLE 3

In the diagram, M and N are midpoints of $\overline{AB}$ and $\overline{CD}$, and also $AM = CN$.

Conclusion: $2 \times AM = 2 \times CN$.

In other words, $AB = CD$.

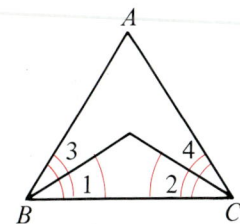

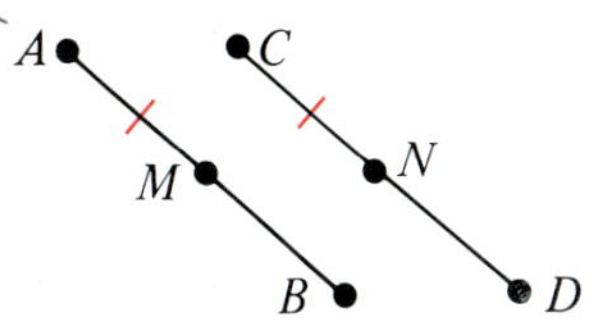

Points, Lines, and Angles

> **THE DIVISION POSTULATE**
>
> If $a = b$ and $c \neq 0$, then $\dfrac{a}{c} = \dfrac{b}{c}$.

> **THE SUBSTITUTION POSTULATE**
>
> If $a = b$, then a can be substituted for b in any equation or inequality.

EXAMPLE 4 In the diagram, $BC = BD$.
Since $AB + BC = AC$,
we can substitute
and get $AB + BD = AC$.

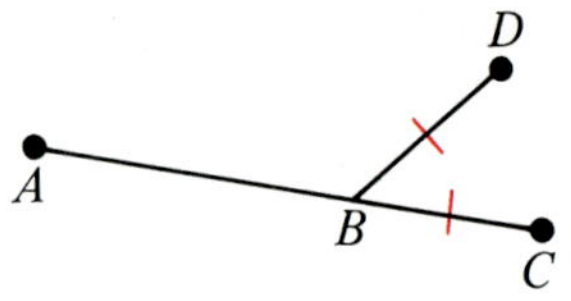

EXAMPLE 5 In the diagram, $\angle 1 = \angle 2$
and $\angle 3 = \angle 4$.
Since $\angle 2 = \angle 3$, (Why?) Vertical $\angle$s are $=$.
we can substitute
and get $\angle 1 = \angle 4$.

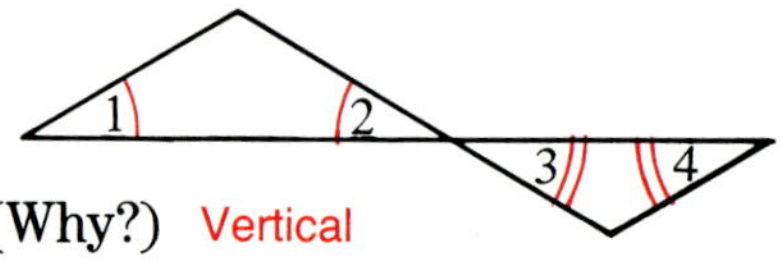

Classroom Practice

Name the postulate that can be used to justify each statement.

1. If $x - 5 = 21$, then $x = 26$. addition
2. If $\tfrac{1}{2}x = 5$, then $x = 10$. multiplication
3. If $3x + 4 = 19$, then $3x = 15$. subtraction
4. If $3x = 15$, then $x = 5$. division
5. If $x + y = z$ and $z = 24$, then $x + y = 24$. substitution

6. If $CL = AM$, then $CA = LM$. addition
7. If $CA = LM$, then $CL = AM$. subtraction

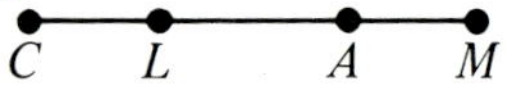

8. If $\angle 1 = \angle 3$ and $\angle 2 = \angle 4$,
   then $\angle ABC = \angle XYZ$. addition
9. If $\angle XYZ = \angle ABC$ and $\angle 1 = \angle 3$,
   then $\angle 2 = \angle 4$. subtraction

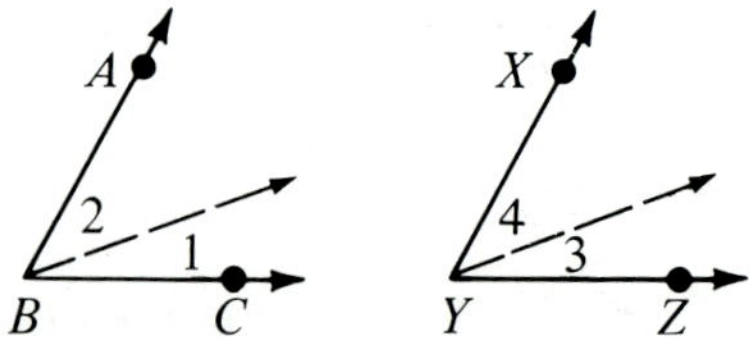

Written Exercises

Which postulate can you use to justify each statement below?

A 1. If $3x = 18$, then $x = 6$. division

2. If $x + 9 = 2$, then $x = -7$. subtraction

3. If $\frac{x}{3} = 9$, then $x = 27$. multiplication

4. **a.** If $2x - 20 = 22$, then $2x = 42$. addition
   b. If $2x = 42$, then $x = 21$. division

5. If $x + y = 8$ and $x - y = 4$, then $2x = 12$. addition

6. **a.** If $x + y = 180$ and $y = 2x$, then $3x = 180$. substitution
   b. If $3x = 180$, then $x = 60$. division

7. If $\angle 2 = \angle 4$ and $\angle 1 = \angle 3$, then $\angle ABC = \angle ACB$. addition

8. If $\angle ABC = \angle ACB$ and $\angle 1 = \angle 3$, then $\angle 2 = \angle 4$. subtraction

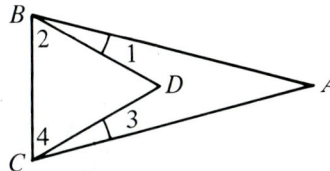

9. If $\angle 1 = \angle 3$, then $\angle POR = \angle QOS$. addition

10. If $\angle 1 = \angle 2$ and $\angle 2 = \angle 3$, then $\angle 1 = \angle 3$. substitution

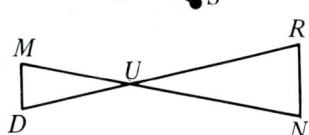

addition
11. If $MU = DU$ and $UN = UR$, then $MN = DR$.

12. If $MN = DR$ and $MU = DU$, then $UN = UR$. subtraction

Study the given information. Then complete the conclusion. Be sure that you understand how the postulates justify your conclusion.

13. Given information: $\angle 2 = \angle 3$

    Conclusion: $\angle 1 = \angle \underline{}_2 = \angle \underline{}_3 = \angle \underline{}_5$

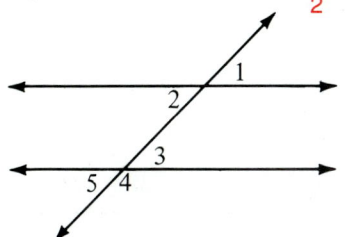

Points, Lines, and Angles

Study the given information. Then complete each conclusion.

14. Given information: $\overline{AB} \perp \overline{BC}$
 $\overline{PQ} \perp \overline{QR}$
 $\angle 1 = \angle 3$

 Conclusion: $\dfrac{?}{\angle 2} = \dfrac{?}{\angle 4}$

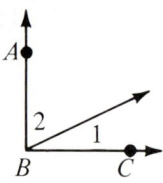

B 15. Given information: $\overrightarrow{AK}$ bisects $\angle RAC$.
 $\overline{AK} \perp \overline{TN}$

 Conclusions: $\dfrac{\angle 1\ ?}{?} = \dfrac{?\ \angle 2}{?}$
 $\dfrac{?}{\angle 3} = \dfrac{?}{\angle 4}$

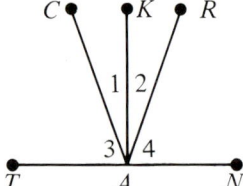

16. Given information: $\angle 2 = \angle 3$
 $\angle 5 = \angle 6$

 Conclusions: $\angle 2 = \angle 3 = \dfrac{?}{\angle 5} = \dfrac{?}{\angle 6}$
 $\angle 1 = \dfrac{?}{\angle 4} = \dfrac{?}{\angle 8} = \dfrac{?}{\angle 7}$

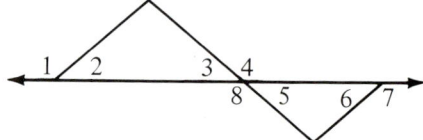

17. Given information: M is the midpoint of $\overline{AB}$.
 N is the midpoint of $\overline{PQ}$.

 What additional information would permit you to conclude that $AM = PN$? $AB = PQ$ or $BM = QN$

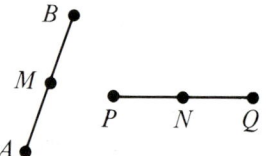

CAREER NOTES

ASTRONOMER

Did you ever watch the stars and planets on a clear night? Did you wonder how and where the planets move?

Astronomers who specialize in celestial mechanics deal with the locations and motions of objects in the solar system. These astronomers compute the positions of planets and chart the orbits of comets, meteors, asteroids, and artificial satellites. They use optical devices on telescopes to make measurements. To process the data, they use computers and other electronic equipment.

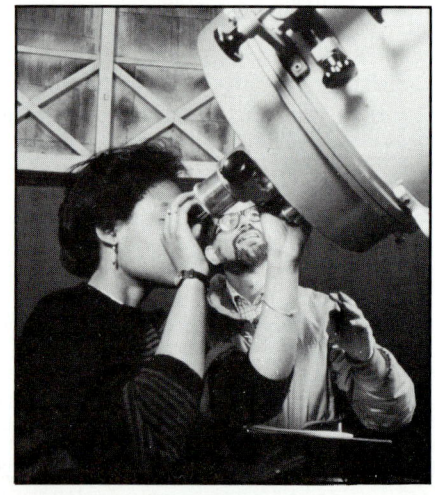

9 Postulates of Geometry

A **plane** is a flat surface that extends indefinitely in all directions. A floor suggests part of a plane. So does a wall.

Usually a plane is represented by a four-sided figure. The plane can be named by the four vertices of the figure or by a single capital letter.

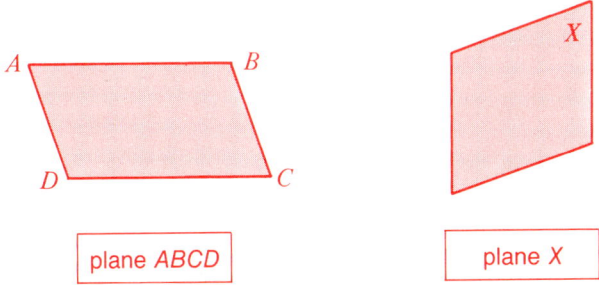

plane ABCD

plane X

You should realize that drawings like those above only *suggest* planes. There is no good way of showing in a drawing that a plane extends indefinitely. An actual plane does *not* have sides or vertices.

Remember that points all on one line are called *collinear* points. Similarly, points all in one plane are called **coplanar** points.

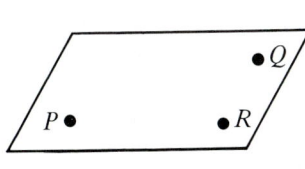
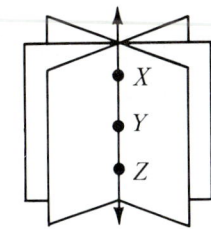

P, Q, and R are *not* collinear. But they are coplanar.

X, Y, and Z are collinear. They are also coplanar.

This drawing suggests that there is exactly one plane containing three noncollinear points.

This drawing suggests that there are an unlimited number of planes containing three collinear points.

A summary of the relationships between points, lines, and planes is given by the postulates on the next page. Remember that postulates are statements which we assume to be true without proof.

Points, Lines, and Angles

POSTULATE 1

Through any two points there is exactly one line.

POSTULATE 2

Through any three noncollinear points there is exactly one plane.

POSTULATE 3

If two points lie in a plane, then the line joining them lies in that plane.

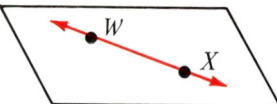

POSTULATE 4

If two planes intersect, then their intersection is a line.

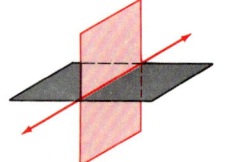

If two planes do not intersect, we say that they are parallel. Planes *M* and *N*, shown at the right, are parallel planes. We indicate this by writing "plane *M* ∥ plane *N*."

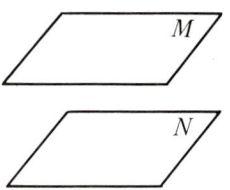

Postulates 1–4 deal with the geometric ideas of *points*, *lines*, and *planes*. Postulates 5 and 6 connect geometric and algebraic ideas. You do not need to memorize the statements of these two postulates. Just be sure that you understand them.

POSTULATE 5 (The Ruler Postulate)

Each point on a line can be paired with exactly one real number called its coordinate. The distance between two points is the positive difference of their coordinates.

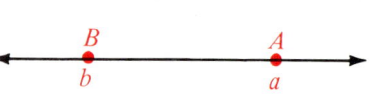

If $a > b$, then $AB = a - b$.

34 Chapter 1

POSTULATE 6 (The Protractor Postulate)

Suppose O is a point of $\overleftrightarrow{XY}$. Consider all rays with endpoint O which lie on one side of $\overleftrightarrow{XY}$. Each ray can be paired with exactly one real number between 0 and 180, as shown.

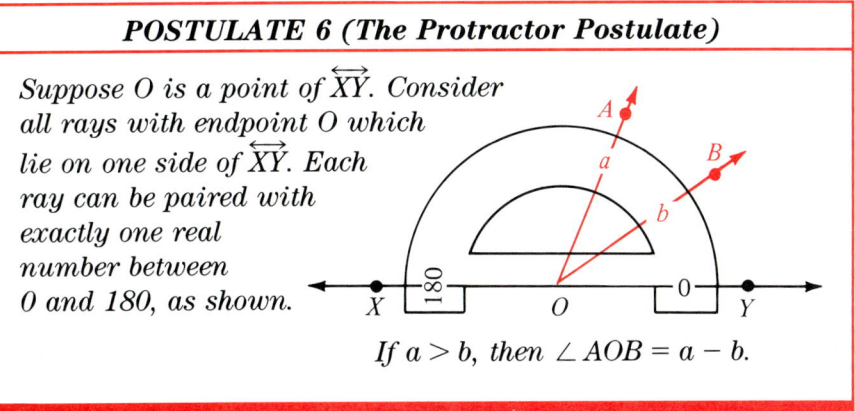

If $a > b$, then $\angle AOB = a - b$.

Do the ideas of the Ruler and Protractor Postulates seem familiar? When we found lengths of segments and measures of angles earlier, we were informally using the ideas expressed in these postulates.

Classroom Practice

Find each length.

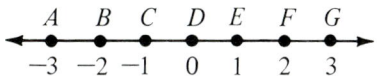

1. CF 3
2. EG 2
3. EA 4
4. BG 5

Find the measure of each angle.

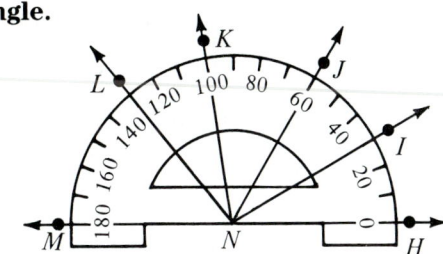

5. $\angle INL$ 100°
6. $\angle KNJ$ 40°
7. $\angle HNM$ 180°
8. $\angle JNM$ 120°

The diagram shows a figure called a *rectangular solid*. Decide if the given points are coplanar.

9. $P, Q, R,$ and S coplanar
10. $P, Q, B,$ and A coplanar
11. $B, C, R,$ and S not coplanar
12. $B, Q, S,$ and D coplanar

Name another point coplanar with the given points.

13. $A, B,$ and D C
14. $A, D,$ and S P
15. $A, B,$ and R S
16. $A, P,$ and R C

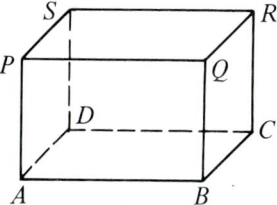

Exercises 9–18

Complete each statement.

17. Plane $ABCD$ ∥ plane _?_. PQRS
18. Plane $ABQP$ ∥ plane _?_. DCRS

Points, Lines, and Angles

Written Exercises

The diagram shows a rectangular solid. Complete each statement.

[A] 1. Plane ABCD ∥ plane __?__. EFGH

2. Plane ADHE ∥ plane __?__. BCGF

3. Plane ABFE ∥ plane __?__. DCGH

4. Plane EFGH intersects plane BCGF in the line __?__. FG

5. Plane ABFE intersects plane BCGF in the line __?__. BF

6. Points E, A, D, and __?__ are coplanar. H

7. Points B, F, H, and __?__ are coplanar. D

8. Points C, G, H, and __?__ are coplanar. D

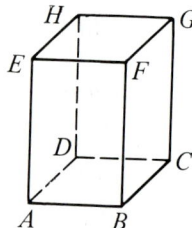

Draw each of the following.

9. Two parallel planes   10. Two intersecting planes

11. A corner of a room where two walls and the floor meet

12. Plane M and plane N both contain point P.
    a. Do the planes have any other points in common? yes
    b. State the postulate that answers this question. Post. 4

13. The diagram suggests what would happen if two different "lines" both contained points A and B. State the postulate that says this cannot happen. Postulate 1

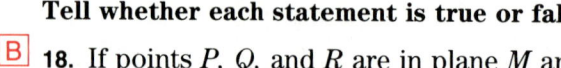

14. A point is given. Is the point contained in just one line or in more than one line? more than one line

15. A line is given. Is the line contained in just one plane or in more than one plane? more than one plane

16. A line and a point not on the line are given. Are the line and the point both contained in one plane or in more than one plane? one plane (Postulate 2)

17. A rectangle is given. Is the rectangle contained in just one plane or in more than one plane? one plane (Postulate 2)

Tell whether each statement is true or false.

[B] 18. If points P, Q, and R are in plane M and also in plane N, then the planes must be the same. false (true only if the points are not collinear)

19. A triangle is contained in exactly one plane. true (Postulate 2)

20. Any four points are contained in exactly one plane. false

21. If points A and B are in plane X, then every point of $\overline{AB}$ is in X. **true (Post. 4)**

22. If points A and B are on a soup can, then every point of $\overline{AB}$ must be on the can. **false**

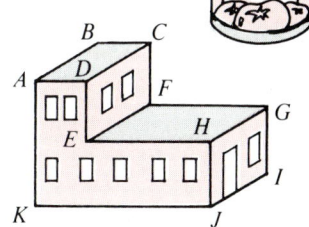

C 23. When a three-dimensional figure is represented on a flat surface, it is usually drawn in perspective. For example, angles are drawn larger or smaller than they actually are. In the diagram, ∠HJI is an acute angle but it represents a right angle in the actual building. Complete the table.

	∠HJI	∠EDC	∠BAD	∠AKJ	∠ABC
In the diagram	acute	?	?	?	?
In the actual building	right	?	?	?	?
		obtuse	acute	right	obtuse
		right	right	right	right

SELF-TEST

Vocabulary

straightedge (p. 25)   plane (p. 33)
compass (p. 25)   coplanar points (p. 33)
postulate (p. 29)   rectangular solid (p. 35)

1. Draw an acute angle. Bisect it using straightedge and compass.   *(1-7)*

Which postulate can be used to justify each statement?

2. If ∠1 + ∠2 = 180° and ∠2 = ∠3, then ∠1 + ∠3 = 180°. **substitution**   *(1-8)*

3. If $x - 7 = 3$, then $x = 10$. **addition**

4. Use the diagram at the right.
   Given information: $\overrightarrow{BD}$ bisects ∠CBE.
   Conclusion: ∠1 = ∠ ?/2 = ∠ ?/5

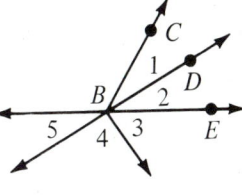

The diagram shows a rectangular solid. Complete each statement.

5. S, V, Z, and ? are coplanar points. **W**   *(1-9)*

6. Plane STUV ∥ plane ? . **WXYZ**

7. Plane STXW intersects plane WXYZ in the line ? . **WX**

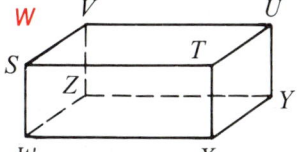

Points, Lines, and Angles   37

SKILLS REVIEW

ARITHMETIC OPERATIONS

Add. The samples will help you remember.

SAMPLES
$7 + 4 = 11 \qquad -7 + 4 = -3$
$7 + (-4) = 3 \qquad -7 + (-4) = -11$

1. $-3 + 12$ 9
2. $0 + (-5)$ -5
3. $-4 + (-8)$ -12
4. $9 + (-2)$ 7
5. $-15 + (-6)$ -21
6. $-11 + 19$ 8
7. $3 + (-14)$ -11
8. $-23 + 23$ 0
9. $-5.7 + 13.8$ 8.1
10. $6.2 + (-10.4)$ -4.2
11. $-7.6 + (-8.5)$ -16.1
12. $0 + (-9.5)$ -9.5

Subtract.

SAMPLES
$4 - 7 = 4 + (-7) = -3 \qquad -4 - 7 = -4 + (-7) = -11$
$4 - (-7) = 4 + 7 = 11 \qquad -4 - (-7) = -4 + 7 = 3$

13. $13 - 17$ -4
14. $-2 - 8$ -10
15. $9 - (-2)$ 11
16. $10 - (-10)$ 20
17. $-7 - 7$ -14
18. $20 - 34$ -14
19. $-16 - (-9)$ -7
20. $-14 - 30$ -44
21. $6.4 - 9.8$ -3.4
22. $-5.7 - 4.9$ -10.6
23. $7.2 - (-7.2)$ 14.4
24. $-5.3 - (-8.6)$ 3.3

Multiply.

SAMPLES
$3(8) = 24 \qquad 3(-8) = -24$
$-3(-8) = 24 \qquad -3(8) = -24$

25. $5(12)$ 60
26. $-4(20)$ -80
27. $13(-7)$ -91
28. $-8(-9)$ 72
29. $-29(0)$ 0
30. $-3(-50)$ 150
31. $8(-11)$ -88
32. $-6(25)$ -150
33. $8(11.2)$ 89.6
34. $-5.1(9)$ -45.9
35. $7(-10.5)$ -73.5
36. $-0.2(-0.6)$ 0.12

Divide.

SAMPLES
$40 \div 8 = 5 \qquad -40 \div 8 = -5$
$-40 \div (-8) = 5 \qquad 40 \div (-8) = -5$

37. $-63 \div 9$ -7
38. $42 \div (-6)$ -7
39. $-52 \div (-4)$ 13
40. $-17 \div 17$ -1
41. $-44 \div (-11)$ 4
42. $-72 \div 8$ -9
43. $39 \div (-13)$ -3
44. $-125 \div 25$ -5
45. $69.3 \div (-3)$ -23.1
46. $-4.8 \div 12$ -0.4
47. $-63 \div (-0.7)$ 90
48. $8.6 \div -8.6$ -1

Compare the numbers. Write <, =, or >.

49. $17 + (-10)$ __?__ 5 $>$
50. $(-5)^2$ __?__ 10 $>$
51. $-2(-9)$ __?__ $0 + 18$ $=$
52. $-(-3)^2$ __?__ 9 $<$
53. $-13 - (-4)$ __?__ -7 $<$
54. $0(-10)$ __?__ $5 - 8$ $>$

Chapter 1

CHAPTER REVIEW

CHAPTER SUMMARY

1. The basic figures studied in this chapter are lines, rays, segments, and angles. Each of these is a set of points.
   Line AB is written $\overleftrightarrow{AB}$.
   Ray AB is written $\overrightarrow{AB}$. A is the endpoint.
   Segment AB is written $\overline{AB}$. A and B are endpoints.
   Angle AOB is written $\angle AOB$. O is the vertex.

2. Angles are classified as acute, right, obtuse, and straight according to their measures. Perpendicular lines form right angles.

3. Vertical angles are equal. $\angle 1 = \angle 2$ $\angle 3 = \angle 4$

4. If M is the midpoint of $\overline{AB}$, then $AM = MB = \frac{1}{2}AB$.
   If $\overrightarrow{OM}$ bisects $\angle AOB$, then $\angle AOM = \angle MOB = \frac{1}{2}\angle AOB$.

5. In constructing a geometric figure, the only instruments which may be used are a compass and a straightedge. Three basic constructions are shown on pages 25–26:
   (1) a bisector of an angle;
   (2) a line $\perp$ to a given line at a point on the line;
   (3) a line $\perp$ to a given line through a point not on the line.

6. Postulates are statements which are accepted without proof. They are used to prove other statements and theorems. You should review the postulates of equality on pages 29–30 and the geometric postulates on pages 34–35.

REVIEW EXERCISES

Complete. *(See pp. 6–9.)*

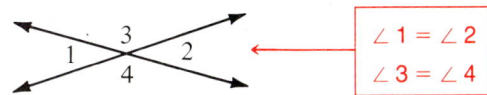

1. Another name for $\overleftrightarrow{AB}$ is __?__. $\overleftrightarrow{AC}$, for example
2. Another name for $\overrightarrow{BD}$ is __?__. $\overrightarrow{BC}$ or $\overrightarrow{BE}$
3. $BC =$ __?__   4
4. If $AB = 5$, then the coordinate of A is __?__. -6
5. If D is the midpoint of $\overline{CE}$, then the coordinate of E is __?__.   11

Points, Lines, and Angles

Complete. *(See pp. 10–13.)*

6. The vertex of ∠1 is __?__. Q
7. ∠7 + ∠8 = __?__°. 180
8. The sides of ∠7 are __?__ and __?__. $\overline{BN}$ and $\overline{BQ}$ (or $\overline{BK}$)
9. Another name for ∠5 is __?__. (Use three letters.) ∠KQR
10. ∠SQB − ∠NQB = ∠__?__ 3 or SQN

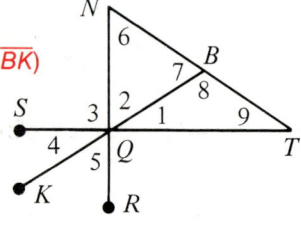

Exercises 6–16

In the diagram above, $\overline{NQ} \perp \overline{ST}$. Classify each angle as acute, right, obtuse, or straight. *(See pp. 15–17.)*

11. ∠RQT right
12. ∠QBN acute
13. ∠QBT obtuse
14. ∠KQB straight

Complete. Use the diagram above. *(See pp. 18–19.)*

15. ∠2 and ∠__?__ are vertical. 5
16. If ∠KQN = 122°, ∠__?__ = 122°. RQB

Solve. *(See pp. 20–22.)*

17. When you bisect a right angle, you divide the angle into two smaller angles. What is the measure of each? 45°

For each exercise, draw a large triangle like △RST. Then construct the required figure. *(See pp. 25–28.)*

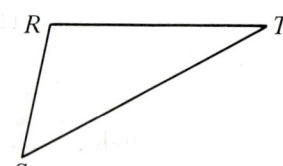

18. Construct a bisector of ∠T.
19. Construct a line perpendicular to $\overline{ST}$ through R.

Which postulate can be used to justify each statement? *(See pp. 29–32.)*

20. If $5x = 35$, then $x = 7$. division
21. If ∠1 = ∠2 and ∠4 = ∠5, then ∠1 + ∠4 = ∠2 + ∠5. addition
22. If ∠1 + ∠2 = 180° and ∠2 = ∠4, then ∠1 + ∠4 = 180°. substitution

Refer to the rectangular solid shown. Complete. *(See pp. 33–37.)*

23. T, P, Q, and __?__ are coplanar points. A
24. A, T, R, and __?__ are coplanar points. S
25. Plane RKPT ∥ plane __?__. SLQA
26. Plane PKLQ intersects plane RKLS in the line __?__. KL
27. How many planes contain three noncollinear points? exactly one

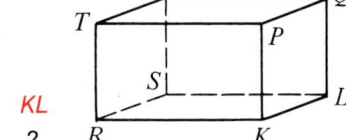

CHAPTER TEST

On a number line, point *R* has coordinate −7 and point *S* has coordinate −1.

1. $RS = $ __?__ 6
2. Find the coordinate of the midpoint of $\overline{RS}$. −4 (1-2)
3. If *X* lies on $\overrightarrow{RS}$, must *X* have a negative coordinate? no

In Questions 4–10, refer to the diagram below.

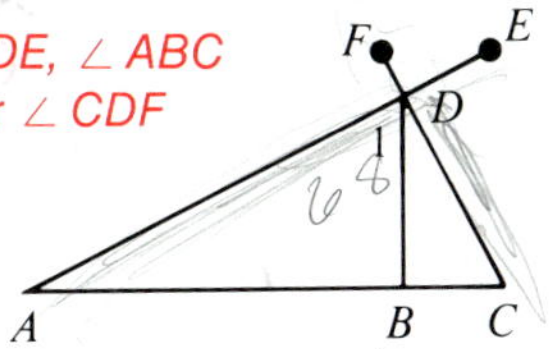

4. Name an angle with measure 180°. ∠ADE, ∠ABC or ∠CDF (1-3)
5. Use your protractor to find the approximate measure of ∠A. 25°
6. ∠ADC − ∠1 = ∠ __?__   CDB
7. If $\overline{CD} \perp \overline{AE}$, name **(a)** a right angle with vertex *D* and **(b)** an obtuse angle with vertex *D*. **a.** ∠ADC, ∠EDC, ∠EDF, or ∠FDA **b.** ∠FDB or ∠BDE (1-4)
8. If $\overline{DA} \perp \overline{DC}$ and ∠1 = 68°, find the measure of ∠BDC. 22°
9. Can a triangle contain two obtuse angles? no
10. Explain why ∠ADF and ∠CDE must be equal. Vertical ∠s are =. (1-5)
11. Find the value of *x*. 40
12. Find the value of *y*. 60

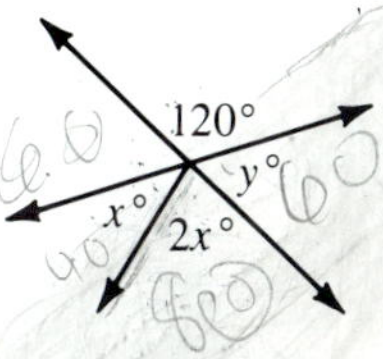

Exercises 11–12

13. If you bisect a right angle forming ∠1 and ∠2, then ∠1 = __?__°. 45 (1-6)
14. If $\overrightarrow{OE}$ bisects ∠DOF, ∠DOE = (y + 6)°, and ∠DOF = 40°, then y = __?__. 14
15. Draw a large obtuse angle, ∠XYZ. Construct $\overrightarrow{YW}$, a bisector of ∠XYZ. (1-7)
16. Draw $\overleftrightarrow{QR}$. Construct a right angle, ∠PQR.

Which postulate can you use to justify each statement?

17. If AB + BC = BC + CD, then AB = CD. subtraction (1-8)
18. If ∠1 + ∠2 = 90° and ∠3 = 90°, then ∠1 + ∠2 = ∠3. substitution
19. If x = y and x = z, then 2x = y + z. addition
20. Two perpendicular lines are given. How many planes contain the lines? one
21. Can three planes intersect in a line? in a point? yes; yes (1-9)
22. If *A* and *B* lie in plane *X*, what can you conclude about $\overleftrightarrow{AB}$? $\overleftrightarrow{AB}$ lies in plane X.

MIXED REVIEW

ALGEBRA PROBLEMS

Solve each equation. Show your work.

SAMPLES

$$3x - 1 = 17$$
$$3x - 1 + 1 = 17 + 1$$
$$3x = 18$$
$$\frac{3x}{3} = \frac{18}{3}$$
$$x = 6$$

$$m + 4m = 90$$
$$5m = 90$$
$$\frac{5m}{5} = \frac{90}{5}$$
$$m = 18$$

1. $x + 120 = 180$ $x = 60$
2. $z - 10 = 25$ $z = 35$
3. $7 - k = 1$ $k = 6$
4. $5 - c = 8$ $c = -3$
5. $3x = 60$ $x = 20$
6. $7n = -14$ $n = -2$
7. $2a - 1 = 7$ $a = 4$
8. $5b + 2 = 2$ $b = 0$
9. $8q + 12 = 4$ $q = -1$
10. $4d - 15 = -3$ $d = 3$
11. $19 - 10j = -11$ $j = 3$
12. $-6s - 70 = 20$ $s = -15$
13. $2x - 3x = -4$ $x = 4$
14. $7y + 2y = 3$ $y = \frac{1}{3}$
15. $t - 3t = 8$ $t = -4$
16. $6(z - 7) = 3$ $z = 7\frac{1}{2}$
17. $-(2 + w) = 2$ $w = -4$
18. $4(9p + 2) = 0$ $p = -\frac{2}{9}$

SAMPLES

$$2(r - 1) = 1 + 2r$$
$$2r - 2 = 1 + 2r$$
$$2r = 3 + 2r$$
$$0 = 3$$
No solution

$$7r - 1 = 5 - (6 - 7r)$$
$$7r - 1 = 5 - 6 + 7r$$
$$7r - 1 = 7r - 1$$
$$-1 = -1$$
All real numbers are solutions.

19. $11g + 7 = 5(3 - g)$ $g = \frac{1}{2}$
20. $2(6c - 1) = 3(4c + 7)$ No solution
21. $-(3 - 2v) = v + v$ No solution
22. $5(4 + x) = -2(2x - 1)$ $x = -2$
23. $1 - 2(7 - h) = -4$ $h = 4\frac{1}{2}$
24. $7(y + 5) - 2y = 5(y + 7)$
25. $-8(2k - 1) = -3k - 31$ $k = 3$
26. $2l + 2(4 - l) = 12$ No solution
27. $h + 2h + 4h = -7h$ $h = 0$
28. $(9 - k) - (2 - k) = 7$
29. $5n - (2 + n) = 2(2n - 1)$
30. $\frac{1}{2}(4x - 8) = -3x + 11$ $x = 3$

24., 28., 29. All real numbers are solutions.

GEOMETRY PROBLEMS

Draw a diagram similar to, but larger than, the one shown.

1. Construct a line perpendicular to l through P.

2. Construct a ray perpendicular to l through Q.

3. How many lines contain both P and Q? 1
   State the postulate that answers this question.
   Through any 2 pts. there is exactly 1 line.

4. Name an obtuse angle shown. ∠ACB or ∠DCE

5. ∠ACB + ∠BCE = ∠ _?_ ACE = _?_° 180

6. Given that ∠ACD = 85°, find the measures of ∠ECB and ∠DCE.
   85°   95°

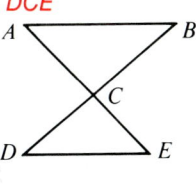

Tell whether each statement is true or false.

7. R lies on $\vec{ST}$. true

8. If A is the midpoint of $\overline{SR}$ and T is the midpoint of $\overline{SA}$, then $ST = \frac{1}{3} SR$. false

9. The plane of the paper is the only plane containing $\overleftrightarrow{ST}$. false

10. If $ST = TA$ and $SA = AR$, then $TA = AR - ST$. true

11. Which segment is longer, $\overline{RS}$ or $\overline{XY}$?
    Neither is longer; $RS = XY$.

12. Use your protractor to draw an angle with measure 130°. Construct the bisector of the angle.

13. Find the values of x and y. $x = 6$
    $y = 40$

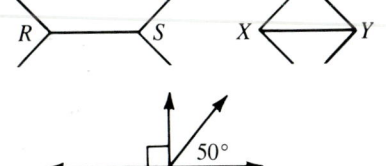

Exercise 13

Write the name of the postulate used.

14. If $5x - 1 = y$ and $y = 9$, then $5x - 1 = 9$. substitution

15. If $5x - 1 = 9$, then $5x = 10$. add.

16. If $5x = 10$, then $x = 2$. division

Complete each sentence with *always*, *sometimes*, or *never*.

17. Three different points are _?_ collinear. sometimes

18. Three different points are _?_ coplanar. always

Here's what you'll learn in this chapter:

To use theorems about complementary, supplementary, and vertical angles.

To write "If . . . then" statements and their converses.

To understand and write basic geometric proofs.

To use the properties of parallel lines cut by a transversal.

To prove lines parallel.

To construct parallel lines.

This curving fence is used to hold sand dunes in place and prevent them from drifting. The upright stakes that make up the fence suggest parallel lines, discussed in this chapter.

Chapter 2

Introducing Proof

1 Three Theorems

Logical reasoning is an important part of geometry. You are given certain facts about a geometric figure and you use these facts to reach conclusions. So far, your conclusions have been based on definitions of figures and on the postulates of equality. Here is an example of logical reasoning.

EXAMPLE *Given information:* $\overline{AB} \perp \overline{BC}$
$\overline{PQ} \perp \overline{QR}$
$\angle 1 = \angle 2$

Conclusion: $\angle 3 = \angle 4$

These are the steps used in reaching the conclusion:

Step 1 $\angle ABC = 90°$ because $\overline{AB} \perp \overline{BC}$, and perpendicular lines (by definition) form 90° angles.

Step 2 $\angle PQR = 90°$ because $\overline{PQ} \perp \overline{QR}$.

Step 3 $\angle 1 + \angle 3 = \angle 2 + \angle 4$ because both sums equal 90°.

Step 4 $\angle 1 = \angle 2$ because this is given information.

Step 5 $\angle 3 = \angle 4$ by using the Subtraction Postulate.

In the example above, $\angle 1$ and $\angle 3$ are called *complementary angles* or just *complements*. **Complementary angles** are two angles whose measures total 90°. In the diagram above, $\angle 2$ and $\angle 4$ are also complementary angles. So are $\angle A$ and $\angle B$ in the diagram at the right.

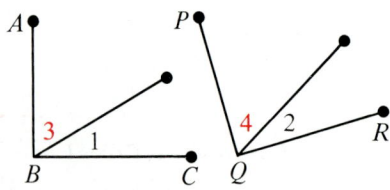

The conclusion reached in the example is really a statement about complementary angles:

> If two angles are equal (such as $\angle 1 = \angle 2$), then their complements are equal ($\angle 3 = \angle 4$).

Let us list this statement as Theorem 1. Remember that a **theorem** is nothing more than a statement which has been proved. Of course, we won't list every proved statement as a theorem. We'll list only the ones which will be most helpful to us in our later work.

THEOREM 1

If two angles are complements of equal angles (or of the same angle), then the two angles are equal.
---

46   Chapter 2

The next theorem refers to **supplementary angles** (sometimes called **supplements**). These are two angles whose measures total 180°.

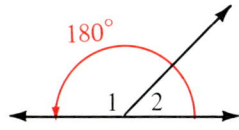

∠ 1 and ∠ 2 are supplementary angles.   ∠ A and ∠ B are supplements.

THEOREM 2

If two angles are supplements of equal angles (or of the same angle), then the two angles are equal.

Here is a diagram and proof of Theorem 2.

Given information: ∠ 1 and ∠ 2 are supplements.
∠ 3 and ∠ 4 are supplements.
∠ 1 = ∠ 3

Conclusion: ∠ 2 = ∠ 4

Step 1 ∠ 1 + ∠ 2 = 180°   because ∠ 1 and ∠ 2 are supplements, and supplements (by definition) total 180°.

Step 2 ∠ 3 + ∠ 4 = 180° because ∠ 3 and ∠ 4 are supplements.

Step 3 ∠ 1 + ∠ 2 = ∠ 3 + ∠ 4 because both sums are 180°.

Step 4 ∠ 1 = ∠ 3 because this is given information.

Step 5 ∠ 2 = ∠ 4 by using the Subtraction Postulate.

Our next theorem is the familiar statement that vertical angles are equal. Of course, we introduced this result on page 18.

THEOREM 3

Vertical angles are equal.

Step 1 ∠ 1 and ∠ 2 are supplements.

Step 2 ∠ 2 and ∠ 3 are supplements.

Step 3 ∠ 1 = ∠ 3 because supplements of the same angle are equal.

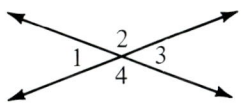

Classroom Practice

1. If ∠A = 70°, what is the measure of a complement of ∠A? 20°
   a supplement of ∠A? 110°

2. If ∠B = 89°, what is the measure of a complement of ∠B? 1°
   a supplement of ∠B? 91°

3. If ∠C = x°, what is the measure of a complement of ∠C? 90° − x°
   a supplement of ∠C? 180° − x°

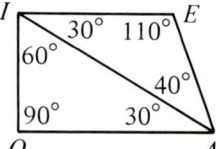

4. Name two pairs of complementary angles. ∠OIA, ∠EIA; ∠OIA, ∠OAI

5. Name a supplement of ∠OAE. ∠IEA

6. Name another pair of supplementary angles. ∠EIO, ∠IOA

7. Name three pairs of supplementary angles.
   ∠DHE, ∠EHG; ∠DHF, ∠FHG; ∠EFH, ∠HFG

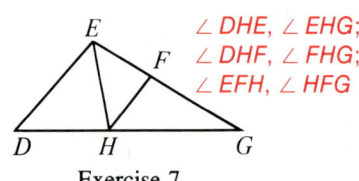

Exercise 7

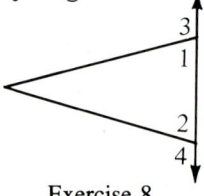

Exercise 8

8. Suppose that ∠1 = ∠2. State the theorem that allows you to conclude that ∠3 = ∠4. If two angles are supplements of equal angles, then the two angles are equal.

Suppose $\overline{RY} \perp \overline{SY}$ and $\overline{UY} \perp \overline{TY}$.

9. Name a complement of ∠1. ∠2

10. Name a complement of ∠3. ∠2

11. State the theorem that allows you to conclude that ∠1 = ∠3. If two angles are complements of the same angle, then the two angles are equal.

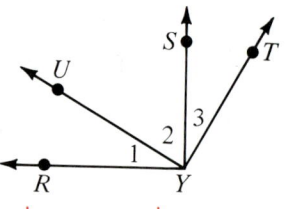

Written Exercises

Copy and complete the table.

	1.	2.	3.	4.	5.	6.	7.
∠A	50°	62°	18°	?	?	?	?
Complement of ∠A	?	?	?	70°	48°	?	?
Supplement of ∠A	?	?	?	?	?	100°	135°

Answers:
- 1: 40°, 130°
- 2: 28°, 118°
- 3: 72°, 162°
- 4: 20°, 160°
- 5: 42°, 138°
- 6: 80°, 10°
- 7: 45°, 45°

In the diagram, $\overline{AB} \perp \overline{BC}$ and $\overline{DC} \perp \overline{BC}$.

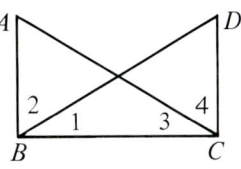

8. Name a complement of ∠1. ∠2
9. Name a complement of ∠3. ∠4
10. If ∠1 = ∠3, state the theorem that allows you to conclude that ∠2 = ∠4. If two angles are complements of equal angles, then the two angles are equal.

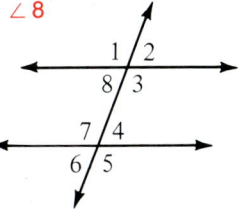

11. In the diagram, name two different supplements of ∠3. ∠2, ∠8
12. If ∠3 and ∠4 are supplements, state the theorem that allows you to conclude that ∠2 = ∠4. (See below.)
13. From Exercise 12, you know that ∠2 = ∠4. State the theorem that allows you to conclude that ∠1 = ∠7.
    If two angles are supplements of equal angles, then the two angles are equal.

14. *Given information:*   ∠ABC = 90°
                                     ∠ADC = 90°
                                     ∠1 = ∠2

    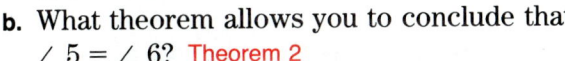

    a. What theorem allows you to conclude that ∠3 = ∠4? Theorem 1
    b. What theorem allows you to conclude that ∠5 = ∠6? Theorem 2

15. Draw an acute angle, ∠ABC. Then use Construction 2, page 26, to construct the complement of ∠ABC.

12. If two angles are supplements of the same angle, then the two angles are equal.

Name the postulate of equality that justifies each statement.

16. If $\frac{5x+3}{4} = 7$, then $5x + 3 = 28$. multiplication
17. If $5x + 3 = 28$, then $5x = 25$. subtraction
18. If $5x = 25$, then $x = 5$. division

B 19. If $3x - 4y = 5$ and $2x + 4y = 10$, then $5x = 15$. addition

20. If $\frac{2x}{3} = \frac{7}{5}$, then $10x = 21$. multiplication

21. Two angles are complementary and equal. What is the measure of each? 45°

22. Two angles are supplementary and equal. What is the measure of each? 90°

Introducing Proof   **49**

2 "If...then" Statements

"If...then" statements are common in everyday speech and in mathematics. Here are some examples.

1. **If** Jolynda is 16, **then** she can apply for a driver's license.

2. **If** we win, **then** we'll be in first place.

3. **If** two angles are equal, **then** their supplements are equal.

These "if...then" statements all have this form:

If A, **then** B.

A is called the **hypothesis;** B is called the **conclusion.**

Sometimes you can express an "if...then" idea without even using the words *if* and *then*. Note that the form of a statement does not affect the truth or falsity of a statement.

EXAMPLE 1   All owls are birds.
   Restatement: **If** X is an owl,
   then X is a bird.

   This is a true statement.

A **true statement** is *always* true.

EXAMPLE 2   All birds are owls.
   Restatement: **If** X is a bird,
   then X is an owl.

   This is a false statement.

A **false statement** is *not always* true.

The statements in Examples 1 and 2 are *converses*, and they say very different things. The **converse** of a statement is formed by exchanging the hypothesis and the conclusion of the statement.

Statement: If A, then B.   Switch
Converse: If B, then A.

Some true statements have converses that are true. Other true statements have converses that are false.

1. True statement: If two lines are ⊥, then they form 90° angles.
   True converse: If two lines meet at a 90° angle, then they are ⊥.

2. True statement: False converse:
   If Kai lives in Texas,   If Kai lives in North America,
   then he lives in North America.   then he lives in Texas.

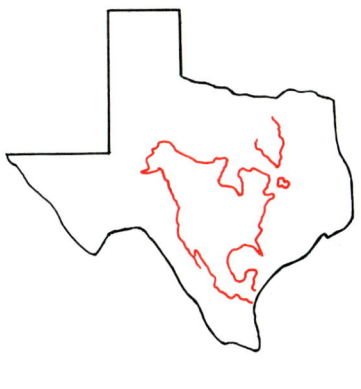

True   False!

Classroom Practice

State the hypothesis and the conclusion of each statement.

————— hypothesis
- - - - - conclusion

1. If <u>the anchor gets loose</u>, then <u>the boat drifts</u>.

2. If <u>I am smart</u>, then <u>you are a genius</u>.

3. If <u>$3x - 7 = 23$</u>, then <u>$x = 10$</u>.

4. I will go if you will. (*Warning:* This exercise shows that the hypothesis is not always mentioned first.) Hyp.: You go   Concl.: I will go

5. <u>We shall succeed</u> if <u>we try</u>.

6. If <u>you want to play lead guitar</u>, then <u>you must practice</u>.

7–12. State the converse of each statement in Exercises 1–6.

Express each statement in "If . . . then" form.

13. The integer n is even when $n + 1$ is odd. If $n + 1$ is odd, then n is even.

14. All Olympic athletes like competition. If a person is an Olympic athlete,
    (*Hint:* Begin with "If a person is . . .") then he or she likes competition.

15. All record collectors are interested in music. If a person is a record collector,
    then he or she is interested in music.

16. Too many sweets spoil the appetite.
    If there are too many sweets,
    then the appetite is spoiled.

Introducing Proof

Express each statement in "If . . . then" form.

17. $y = 14$ when $y - 8 = 6$. If $y - 8 = 6$, then $y = 14$.

18. Every positive number less than 100 has a square root less than 10.
    If a number is positive and less than 100, then it has a square root which is less than 10.

In Exercises 19–22: a. decide if the statement is true;
  b. state the converse;
  c. decide if the converse is true.

19. If $t + 4 = 8$, then $8 = t + 4$. a. true b. If $8 = t + 4$, then $t + 4 = 8$. c. true

20. If $a = 2$ and $b = 3$, then $ab = 6$. a. true b. If $ab = 6$, then $a = 2$ and $b = 3$. c. false

21. If x is an even integer, then $2x$ is an even integer. a. true b. If $2x$ is an even integer, then x is an even integer. c. false

22. All right angles have the same measure.
    a. true b. If angles have the same measure, then they are right angles.
    c. false

Written Exercises

Write the hypothesis and the conclusion of each statement.

A 1. If the sum of two angles is 90°, then the angles are complements.

2. If it rains, it pours.

3. If $5x + 7 = 27$, then $x = 4$.

―――――― hypothesis
- - - - - conclusion

4. They'll be here at noon if they take the train.

5. We can go to the beach Saturday, provided it's a nice day.

6. If a number is greater than 4, it is greater than 3.

7–12. Write the converse of each statement in Exercises 1–6.

Write each statement in "If . . . then" form. Note that true statements below remain true when stated in "if . . . then" form; false statements remain false.

13. All spiders are insects. (*Hint:* Begin with "If X is . . .") If X is a spider, then X is an insect.

14. All snakes are poisonous. If X is a snake, then X is poisonous.

15. Vertical angles are equal. If two angles are vertical angles, then they are equal.
    (*Hint:* Begin with "If two angles . . .")

16. $x^2 = 4$ when $x = 2$ or $x = -2$.
    If $x = 2$ or $x = -2$, then $x^2 = 4$.
17. Every turtle is a reptile.
    If X is a turtle, then X is a reptile.
18. Every history student knows about Paul Revere.
    If a person is a history student, then he or she knows about Paul Revere.
19. People who live in glass houses should not throw stones.
    If a person lives in a glass house, then he or she should not throw stones.
20. An obtuse angle has a measure that is greater than 90°.
    If an angle is obtuse, then it has a measure greater than 90°.

In Exercises 21–28: a. decide if the statement is true;
b. write the converse;
c. decide if the converse is true.

B 21. If two angles are equal, then they are vertical angles. **a.** false **c.** true

22. If $x = -3$, then $x^2 = 9$. **a.** true **c.** false

23. If a figure is a square, then all its sides have the same length. **a.** true **c.** false

24. If $AM = MB$, then M is the midpoint of $\overline{AB}$. **a.** false **c.** true

C 25. Every rectangle is a square. **a.** false **c.** true

26. If $a < 0$, then $a^2 > 0$. **a.** true **c.** false

27. Every number divisible by five is also divisible by ten. **a.** false **c.** true

28. Whenever each of two numbers is negative, the sum of the numbers is negative. **a.** true **c.** false

SELF-TEST

Vocabulary

complementary angles (p. 46)
theorem (p. 46)
supplementary angles (p. 47)
"If . . . then" statement (p. 50)
hypothesis (p. 50)

conclusion (p. 50)
true statement (p. 50)
false statement (p. 50)
converse (p. 50)

1. If $\angle A = 20°$, what is the measure of a complement of $\angle A$? 70° (2-1)

2. If $\angle B = 38°$, what is the measure of a supplement of $\angle B$? 142°

In the diagram, $\overline{AL} \perp \overline{RK}$.   ∠ SAL

3. Name a complement of ∠ RAS.

4. Name a supplement of ∠ RAS.
∠ SAK

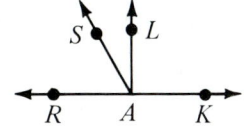

Consider the statement "If you read the newspaper every day, then you are well informed."

5. Write the hypothesis and the conclusion of the statement. (2-2)

6. Write the converse of the statement.

APPLICATIONS

ART AND GEOMETRY

In the period between 1400 and 1700, artists began to use geometric ideas to make their paintings more realistic. The sketch below illustrates some of their techniques.

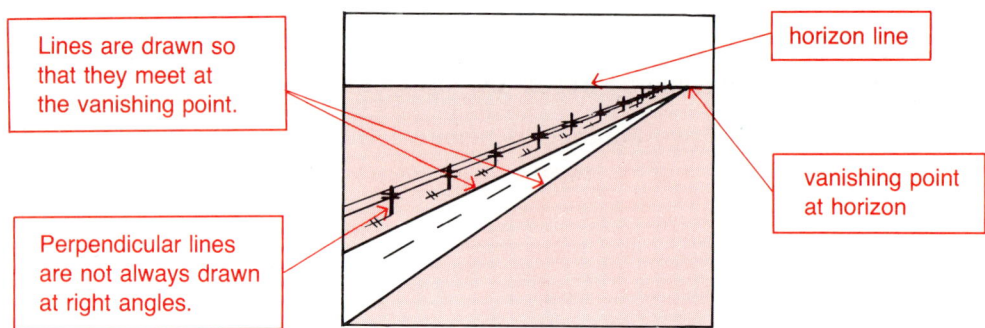

Lines are drawn so that they meet at the vanishing point.

Perpendicular lines are not always drawn at right angles.

horizon line

vanishing point at horizon

Baldassare Peruzzi, a sixteenth century Italian architect and painter, drew this picture of ancient Roman monuments. Notice his use of the vanishing point.

EXERCISES

1. Draw the tiled hallway shown by following these steps:

   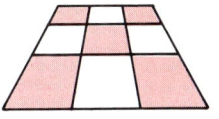

   (1) Draw the horizon and three vanishing points A, B, and C with $AB = BC$.

   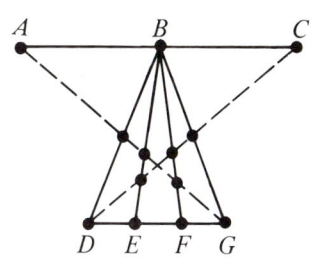

   (2) Draw $DEFG$ with $DE = EF = FG$. Draw $\overline{BD}$, $\overline{BE}$, $\overline{BF}$, and $\overline{BG}$.

   (3) Draw $\overline{DC}$ and $\overline{GA}$, marking the points in which they intersect $\overline{BD}$, $\overline{BE}$, $\overline{BF}$, and $\overline{BG}$.

   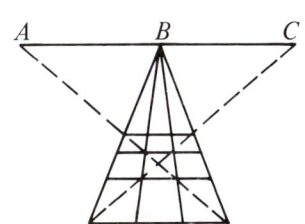

   (4) Draw horizontal lines through the intersection points.

   (5) Erase the construction lines and shade every other square.

2. The humorous picture at the right is called "False Perspective." It was created in 1753 by the famous British artist William Hogarth. Describe some of the false perspectives in this work.

3. Cut a picture from a magazine or newspaper and enlarge it by the method described below.
   (1) Draw a rectangular grid on the picture.
   (2) Draw a larger rectangular grid.
   (3) Copy each square of the picture in the corresponding square of the larger grid.

ORIGINAL PICTURE

ENLARGEMENT

Introducing Proof

3 Reaching Conclusions (Optional)

When we try to prove something to others, we often use "If ... then" statements. In this section, we shall use these statements to reach conclusions. Four kinds of simple reasoning are shown below. The first two are correct, and the last two are incorrect.

CORRECT REASONING

Type 1 *Given information:*
(1) If a figure is a square, then it has four right angles.
(2) *ABCD* is a square.

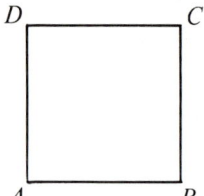

CORRECT CONCLUSION: *ABCD* has four right angles.

Type 2 *Given information:*
(1) If a figure is a square, then it has four right angles.
(2) *EFGH* does not have four right angles.

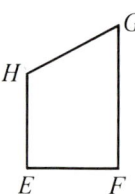

CORRECT CONCLUSION: *EFGH* is not a square.

INCORRECT REASONING

Type 3 *Given information:*
(1) If a figure is a square, then it has four right angles.
(2) *PQRS* has four right angles.

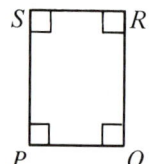

INCORRECT CONCLUSION: *PQRS* is a square.

Type 4 *Given information:*
(1) If a figure is a square, then it has four right angles.
(2) *WXYZ* is not a square.

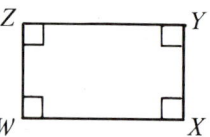

INCORRECT CONCLUSION: *WXYZ* does not have four right angles.

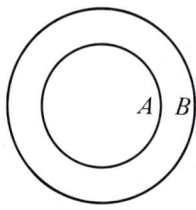

The basic form of each type of reasoning follows, along with a diagram. The diagram at the left illustrates the statement "If *A*, then *B*." When *A* is true, we place a point inside circle *A*. When *A* is not true, we place a point outside circle *A*.

Type 1
(1) If A, then B.
(2) A is true.
CORRECT CONCLUSION: B is true.

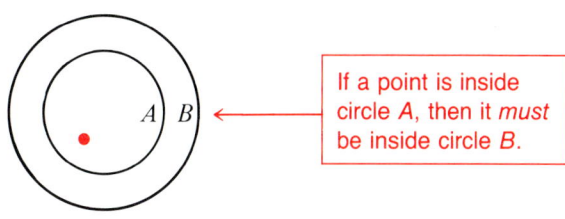

If a point is inside circle A, then it *must* be inside circle B.

Type 2
(1) If A, then B.
(2) B is not true.
CORRECT CONCLUSION: A is not true.

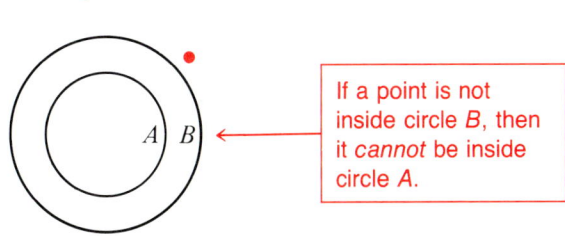

If a point is not inside circle B, then it *cannot* be inside circle A.

Type 3
(1) If A, then B.
(2) B is true.
NO CONCLUSION POSSIBLE.

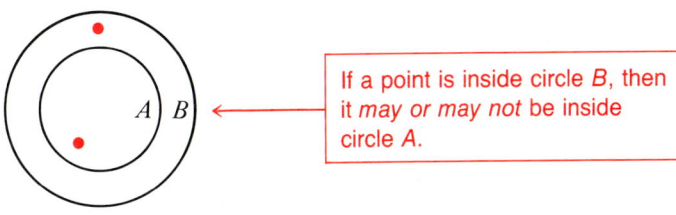

If a point is inside circle B, then it *may or may not* be inside circle A.

Type 4
(1) If A, then B.
(2) A is not true.
NO CONCLUSION POSSIBLE.

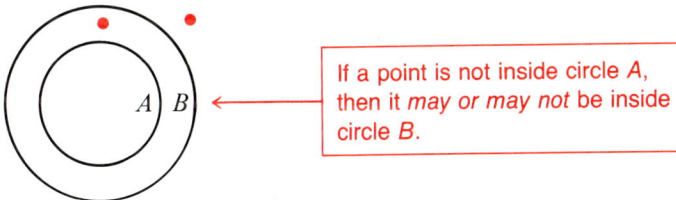

If a point is not inside circle A, then it *may or may not* be inside circle B.

Classroom Practice

Accept this statement as true:

"All basketball players are tall."

1. Reword this statement in "If . . . then" form.

2. Make a circle diagram of this statement.

3. For each statement below, tell what you can conclude. If no conclusion is possible, say so.
   a. Julia is tall. no conclusion
   b. A giraffe is not a basketball player. no conclu.
   c. My cousins play basketball. My cousins are tall.
   d. Irving is short. Irving is not a basketball player.

Introducing Proof   **57**

In Exercises 4–5: a. decide whether the reasoning is Type 1, 2, 3, or 4;
b. decide whether the reasoning is correct.
(*Hint:* Rewrite (1) in "If . . . then" form.)

4. (1) All comedians love laughter.
   (2) Larry Larynx loves laughter.
   Conclusion: Larry Larynx is a comedian. **a.** Type 3 **b.** not correct

5. (1) When the product of two numbers is zero, one of the numbers must be zero.
   (2) $(x - 2) \cdot (x - 3) = 0$
   Conclusion: $x - 2 = 0$ or $x - 3 = 0$. **a.** Type 1 **b.** correct

6. *Given information:* (1) $\angle 1 = \angle 2$
   (2) $\angle 3 = \angle 4$
   What, if anything, can you conclude about $\angle 1$ and $\angle 3$?
   Nothing can be concluded.

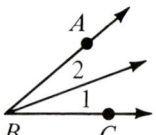 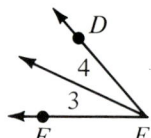

Written Exercises

Accept this statement as true: "All spiders have hairy legs."

A 1. Express the statement in "If . . . then" form.
   If *X* is a spider, then *X* has hairy legs.
2. Make a circle diagram of this statement.
3. For each statement, tell what you can conclude.
   a. Daddy Long-Legs is a spider. Daddy Long-Legs has hairy legs.
   b. A tarantula has hairy legs. no conclusion
   c. Charlotte is not a spider. no conclusion
   d. An octopus does not have hairy legs.
      An octopus is not a spider.
4. Accept this statement as true: "If a figure is a rectangle, then the opposite sides have the same length." For each statement, tell what, if anything, you can conclude.
   a. Opposite sides of figure *ABCD* have the same length.
   b. Opposite sides of figure *PQRS* do not have the same length.
   c. Figure *WXYZ* is a rectangle.

rectangle

a. no conclusion **b.** *PQRS* is not a rectangle.
c. The opposite sides in figure *WXYZ* have the same length.

Accept the numbered steps as true. For each exercise:
a. decide whether the reasoning is Type 1, 2, 3, or 4;
b. decide whether the reasoning is correct.

5. (1) If $x = 5$, then $y = 0$.
   (2) $x = 5$
   Conclusion: $y = 0$
   a. Type 1 **b.** correct

6. (1) If $x = 5$, then $y = 0$.
   (2) $y = 0$
   Conclusion: $x = 5$
   a. Type 3 **b.** not correct

7. (1) Vertical angles are equal.
   (2) ∠1 and ∠2 are not vertical angles.
   Conclusion: ∠1 ≠ ∠2
   a. Type 4 **b.** not correct

8. (1) Vertical angles are equal.
   (2) ∠3 ≠ ∠4
   Conclusion: ∠3 and ∠4 are not vertical angles.
   a. Type 2 **b.** correct

9. (1) If ∠D is larger than ∠E, then $\overline{EF}$ is longer than $\overline{DF}$.
   (2) $\overline{EF}$ is shorter than $\overline{DF}$.
   Conclusion: ∠D is not larger than ∠E.
   a. Type 2 **b.** correct

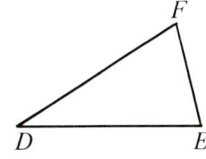

In Exercises 10–11, try to draw a conclusion from the given information. If none is possible, say so.

 10. *Given information:* (1) $\overrightarrow{EN}$ bisects ∠AEG.
   (2) $\overrightarrow{EG}$ bisects ∠NEL.

   What, if anything, can you conclude about ∠1 and ∠3?
   ∠1 = ∠3 since ∠1 = ∠2 and ∠2 = ∠3.

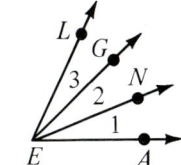

11. *Given information:* (1) AE = EG = GH
   (2) NO = OS = ST

   What, if anything, can you conclude about $\overline{AE}$ and $\overline{NO}$?
   Nothing can be concluded.

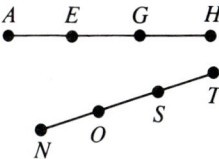

CAREER NOTES

MARINE DIETICIAN

Have you ever tried tuna hot dogs or fish luncheon rolls? These are two examples of food products that marine dieticians have developed. Marine dieticians are both nutrition experts and oceanographers. They analyze possible food sources in the oceans.

Most of the marine life that can be harvested is not considered desirable as human food. One of the most challenging problems facing marine dieticians is the conversion of such marine life into acceptable food products.

Science and math courses will help prepare you for this field. Marine dieticians often begin their careers as lab assistants.

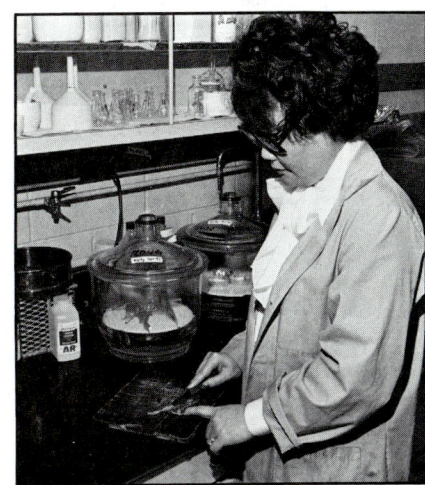

Introducing Proof

4 Writing Proofs

A geometric proof consists of steps that show how a conclusion follows logically from other statements. As your proofs become longer than just a few steps, it helps to organize them in two columns as shown below.

Given: ∠1 and ∠4 are supplements.

Prove: ∠2 = ∠3

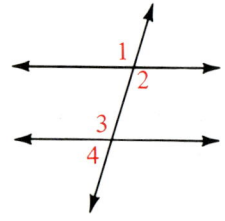

Here is the proof:

STATEMENTS	REASONS
1. ∠1 and ∠4 are supplements.	1. Given
2. ∠3 and ∠4 are supplements.	2. If the measures of two angles total 180°, the angles are supplements. (A definition)
3. ∠1 = ∠3	3. Supplements of the same angle are equal. (A theorem)
4. ∠1 = ∠2	4. Vertical angles are equal. (A theorem)
5. ∠2 = ∠3	5. Substitution Postulate (Using Steps 3 and 4)

The comments in parentheses are not really necessary. They are included to show you that there are four kinds of reasons which can be used to justify a step in a proof.

REASONS USED IN A PROOF

1. *Given information*
2. *Definitions*
3. *Postulates (These are statements accepted without proof.)*
4. *Theorems (These are statements which have been proved.)*

These four kinds of reasons are used in the following proof. Notice that Steps 5 and 6 use both the given information and the definition of angle bisector.

Given: ∠1 = ∠2
BX bisects ∠ABC.
CX bisects ∠ACB.

Prove: ∠3 = ∠4

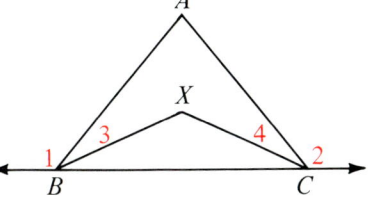

STATEMENTS	REASONS
1. ∠1 and ∠ABC are supplements. ∠2 and ∠ACB are supplements.	1. Definition of supplementary angles
2. ∠1 = ∠2	2. Given
3. ∠ABC = ∠ACB	3. Supplements of equal angles are equal.
4. ½∠ABC = ½∠ACB	4. Multiplication Postulate
5. ½∠ABC = ∠3	5. Given: BX bisects ∠ABC.
6. ½∠ACB = ∠4	6. Given: CX bisects ∠ACB.
7. ∠3 = ∠4	7. Substitution Postulate

Classroom Practice

Supply the missing reasons in the proofs.

1. Given: ∠1 = ∠2
   Prove: $\overleftrightarrow{LI} \perp \overleftrightarrow{NE}$

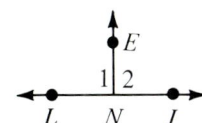

STATEMENTS	REASONS
1. ∠1 + ∠2 = 180°	1. Their sum is a __straight__ angle.
2. ∠1 = ∠2	2. __?__ Given
3. ∠1 + ∠1 = 180°, or 2 · ∠1 = 180°	3. __?__ Substitution Postulate
4. ∠1 = 90°	4. __?__ Division Postulate
5. $\overleftrightarrow{LI} \perp \overleftrightarrow{NE}$	5. __?__ ⊥ lines form right ∠.

Introducing Proof 61

2. *Given:* ∠1 and ∠3 are supplementary.

Prove: ∠2 = ∠4

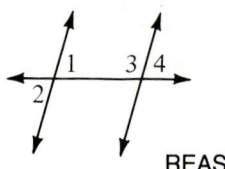

STATEMENTS	REASONS
1. ∠1 and ∠3 are supplementary.	1. __?__ Given
2. ∠4 and ∠3 are supplementary.	2. __?__ Their sum is a straight ∠.
3. ∠1 = ∠4	3. __?__ Supplements of the same ∠ are =.
4. ∠1 = ∠2	4. __?__ Vertical ∠s are =.
5. ∠2 = ∠4	5. __?__ Substitution Postulate

Written Exercises

Supply the missing statements and reasons in the proofs.

 1. *Given:* ∠1 = ∠2
∠3 = ∠4

Prove: ∠1 = ∠4

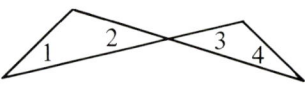

STATEMENTS	REASONS
1. ∠1 = ∠2 ∠3 = ∠4	1. __?__ Given
2. ∠2 = ∠3	2. __?__ Vertical ∠s are =.
3. ∠1 = ∠4	3. __?__ Substitution Postulate

2. *Given:* ∠1 and ∠2 are complements.
∠3 and ∠4 are complements.

Prove: ∠1 = ∠4

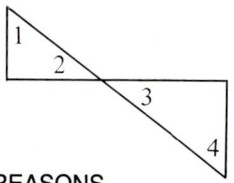

STATEMENTS	REASONS
1. ∠1 and ∠2 are complements. ∠3 and ∠4 are complements.	1. __?__ Given
2. ∠2 = ∠3	2. __?__ Vertical ∠s are =.
3. ∠1 = ∠4	3. __?__ Complements of = ∠s are =.

3. *Given:* ∠1 = ∠4
Prove: ∠3 and ∠2 are supplements.

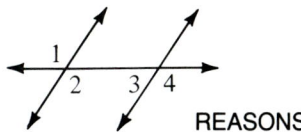

STATEMENTS	REASONS
1. ∠1 = ∠4	1. __?__ Given
2. ∠2 = ∠1	2. __?__ Vertical ∠s are =.
3. ∠2 = ∠4	3. __?__ Subst. Post.
4. ∠3 + ∠4 = 180°	4. Their sum is a __?__ straight angle.
5. ∠3 + ∠2 = 180°	5. __?__ Subst. Post.
6. __?__ ∠3 and ∠2 are supplementary.	6. Def. of supplementary angles

4. *Given:* $\overline{AB} \perp \overline{BD}$
$\overrightarrow{BD}$ bisects ∠EBC.
Prove: ∠1 and ∠3 are complements.

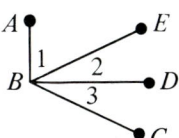

STATEMENTS	REASONS
1. $\overline{AB} \perp \overline{BD}$	1. __?__ Given
2. ∠ABD = 90°	2. __?__ ⊥ lines form right ∠s.
3. ∠1 + ∠__?__ 2 = 90°	3. Subst. Post.: ∠ABD = ∠1 + ∠2
4. ∠2 = ∠__?__ 3	4. *Given:* $\overrightarrow{BD}$ bisects ∠EBC.
5. ∠1 + ∠3 = 90°	5. __?__ Substitution Postulate
6. ∠1 and ∠3 are complements.	6. __?__ Def. of complementary ∠s

5. *Given:* $\overleftrightarrow{XY}$ bisects ∠AOC.
Prove: $\overleftrightarrow{XY}$ bisects ∠DOB.

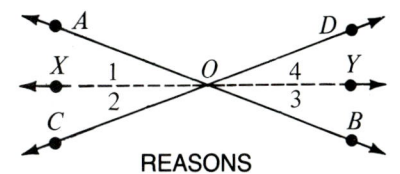

STATEMENTS	REASONS
1. ∠1 = ∠2	1. *Given:* __?__ Def. of ∠ bisector
2. ∠1 = ∠3	2. __?__ Vertical ∠s are =.
3. ∠2 = ∠__?__ 4	3. __?__ Vertical ∠s are =.
4. ∠3 = ∠4	4. __?__ Substitution Postulate
5. __?__ $\overleftrightarrow{XY}$ bisects ∠DOB.	5. Definition of angle bisector

Introducing Proof

6. Given: $AC = BD$
O is the midpoint of $\overline{AC}$.
O is the midpoint of $\overline{BD}$.

Prove: $AO = DO$

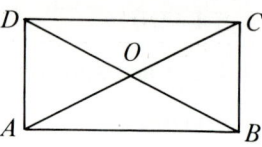

STATEMENTS	REASONS
1. $AC = BD$	1. <u>?</u> Given
2. $\frac{1}{2}AC = \frac{1}{2}BD$	2. <u>?</u> Multiplication Postulate
3. $AO = \frac{1}{2}AC$	3. Given: <u>?</u> O is the midpt. of $\overline{AC}$.
4. $DO = \frac{1}{2}BD$	4. <u>?</u> Given: O is the midpt. of $\overline{BD}$.
5. $\underset{AO}{\underline{?}} = \underset{DO}{\underline{?}}$	5. Substitution Postulate (Steps 2, 3, and 4)

B **7.** Given: $\overrightarrow{RO}$ bisects $\angle YRS$.
$\overrightarrow{SO}$ bisects $\angle YSR$.
$\angle 1 = \angle 2$

Prove: $\angle 5 = \angle 6$

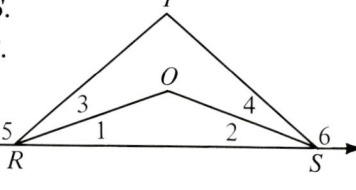

STATEMENTS	REASONS
1. $\angle 1 = \angle 3$	1. <u>?</u> Given: $\overrightarrow{RO}$ bisects $\angle YRS$.
2. $\angle 2 = \angle 4$	2. <u>?</u> Given: $\overrightarrow{SO}$ bisects $\angle YSR$.
3. $\angle 1 = \angle 2$	3. <u>?</u> Given
4. $\angle 3 = \angle 4$	4. <u>?</u> (Steps 1, 2, and 3) Subst. Post.
5. $\angle 1 + \angle 3 = \angle 2 + \angle 4$, or $\angle YRS = \angle YSR$	5. <u>?</u> Addition Postulate
6. $\angle YRS$ and $\angle 5$ are supplements. $\angle YSR$ and $\angle 6$ are supplements.	6. <u>?</u> Def. of supp. $\angle$s
7. $\angle 5 = \angle 6$	7. <u>?</u> Supplements of = $\angle$s are =.

8. Given: $\angle ABC = 90°$
$\angle ADC = 90°$
$\angle 1 = \angle 4$

Prove: $\angle 2 = \angle 5$

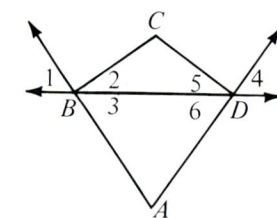

STATEMENTS	REASONS
1. ∠1 = ∠3	1. __?__ Vertical ∠s are =.
2. ∠4 = ∠6	2. __?__ Vertical ∠s are =.
3. ∠1 = ∠4	3. __?__ Given
4. ∠3 = ∠6	4. __?__ Substitution Postulate
5. ∠2 and ∠ __?__ are complements. (3)	5. Given: ∠ABC = 90°
6. ∠5 and ∠ __?__ are complements. (6)	6. __?__ Given: ∠ADC = 90°
7. ∠2 = ∠5	7. __?__ Complements of = ∠s are =.

For each exercise, copy what is given and what is to be proved. Write a proof in two-column form. Refer to the diagrams below.

9. Given: ∠3 = ∠5
   Prove: ∠4 = ∠6

10. Given: ∠2 = ∠8
    Prove: ∠4 = ∠6

11. Given: ∠3 and ∠6 are supplements.
    Prove: ∠2 = ∠6

12. Given: ∠1 = ∠5
    Prove: ∠4 and ∠5 are supplements.

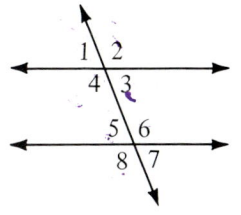

Exercises 9–12

C 13. Given: ∠2 = ∠3
       ∠3 = ∠5
    Prove: ∠1 = ∠6

14. Given: ∠4 and ∠5 are supplements.
    ∠2 = ∠3
    Prove: ∠2 = ∠5

15. Given: ∠1 = ∠2
    ∠3 and ∠5 are complements.
    ∠4 and ∠6 are complements.
    Prove: ∠5 = ∠6

16. Given: $\overline{AB} \perp \overline{BD}$
    $\overline{CD} \perp \overline{BD}$
    ∠5 = ∠6
    Prove: ∠1 = ∠2

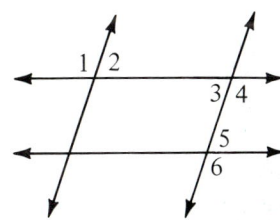

Exercises 13, 14

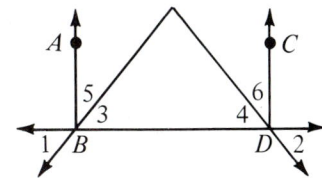

Exercises 15, 16

Introducing Proof **65**

5 Parallel Lines

If two lines do not intersect, they are either *parallel* or *skew*.

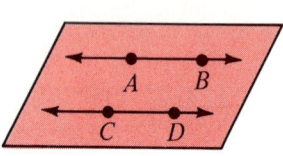 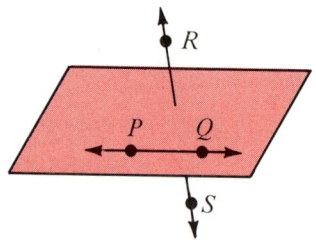

$\overleftrightarrow{AB}$ and $\overleftrightarrow{CD}$ are **parallel lines.**
(1) They are coplanar lines.
(2) They do not intersect.

$\overleftrightarrow{PQ}$ and $\overleftrightarrow{RS}$ are **skew lines.**
(1) They are *not* coplanar lines.
(2) They do not intersect.

We abbreviate the statement "$\overleftrightarrow{AB}$ is parallel to $\overleftrightarrow{CD}$" by writing $\overleftrightarrow{AB} \parallel \overleftrightarrow{CD}$. We shall also say that $\overline{AB} \parallel \overline{CD}$ because the segments are contained in parallel lines. In a drawing, we indicate parallel lines and segments by arrowheads and, if necessary, double arrowheads.

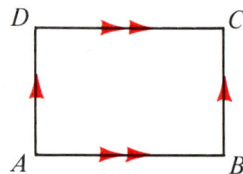

The single arrowheads show that $\overline{AD} \parallel \overline{BC}$.

The double arrowheads show that $\overline{AB} \parallel \overline{CD}$.

The diagram at the right shows two parallel lines cut by a third line. This cutting line is called a **transversal.** Angles 1, 2, 3, and 4 formed on the inside of the parallel lines are called **interior angles.** Angles formed by parallel lines are given special names as follows:

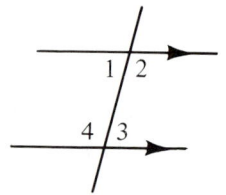

Alternate interior angles are interior angles on alternate sides of the transversal. (Examples above: ∠ 1 and ∠ 3; also ∠ 2 and ∠ 4) To spot alternate interior angles, look for a "Z-shaped" figure in various positions.

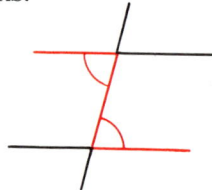 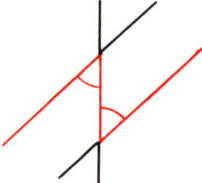 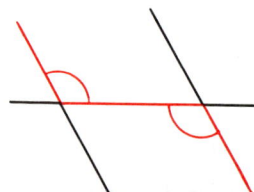

Same-side interior angles are interior angles on the same side of the transversal. (Examples on page 66: ∠ 1 and ∠ 4; also ∠ 2 and ∠ 3) To spot same-side interior angles, look for a "U-shaped" figure in various positions.

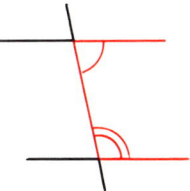 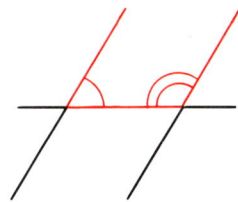 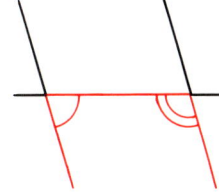

Corresponding angles are so named because they appear to be in corresponding positions in relation to the two lines. To spot corresponding angles, look for an "F-shaped" figure in various positions.

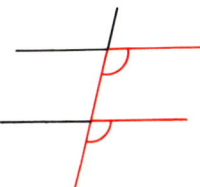 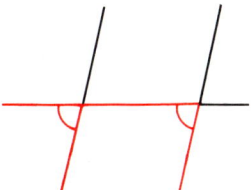 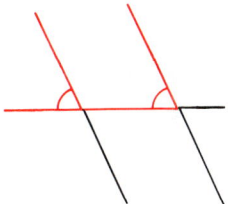

Even when lines cut by a transversal are not parallel, we still use the same vocabulary. For example:

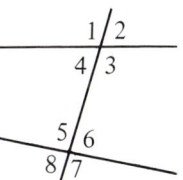

∠ 4 and ∠ 8 are corresponding angles;

∠ 4 and ∠ 6 are alternate interior angles;

∠ 4 and ∠ 5 are same-side interior angles.

The diagram at the right suggests that when two parallel lines are cut by a transversal, corresponding angles are equal. You can convince yourself of this by drawing and measuring many pairs of corresponding angles formed by different transversals. You may suspect that we can prove this property. However, there is just not enough information in our previous postulates and theorems to deduce this result as a theorem. Let us therefore accept it without proof as a postulate. This is a version of the "Parallel Postulate."

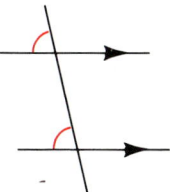

POSTULATE 7

If two parallel lines are cut by a transversal, then corresponding angles are equal.

Introducing Proof

Once we have this postulate, we can immediately deduce the following theorems. The proofs are left as Exercises 31 and 32 on page 71.

THEOREM 4

If two parallel lines are cut by a transversal, then alternate interior angles are equal.

THEOREM 5

If two parallel lines are cut by a transversal, then same-side interior angles are supplementary.

Classroom Practice

1. ∠4, ∠6; ∠3, ∠5
2. ∠3, ∠6; ∠4, ∠5
3. ∠1, ∠5; ∠4, ∠8; ∠2, ∠6; ∠3, ∠7

In the diagram, $l \parallel m$.

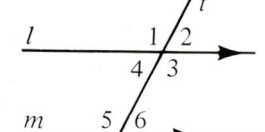

1. Name two pairs of alternate interior angles.
2. Name two pairs of same-side interior angles.
3. Name four pairs of corresponding angles.
4. What is the special name given to line t? transversal
5. Suppose ∠4 = 60°. Find the measures of the other seven angles. (See below.)
6. Suppose ∠4 = x°. Find the measures of ∠5 and ∠6. ∠5 = 180° − x°, ∠6 = x°
7. Although we have not discussed *alternate exterior angles*, you may be able to guess what they are. Name two pairs of them. ∠1, ∠7; ∠2, ∠8

5. ∠1 = ∠3 = ∠5 = ∠7 = 120°, ∠2 = ∠6 = ∠8 = 60°

State the special name for each pair of angles.

8. ∠3 and ∠5   alt. int. angles
9. ∠2 and ∠6   corr. ∠s
10. ∠3 and ∠6   same-side interior angles
11. ∠1 and ∠5   corresponding angles

Find the values of x and y in each diagram.

12.
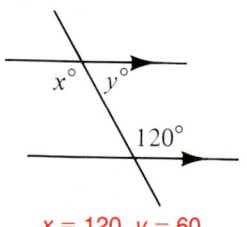
$x = 120, y = 60$

13.
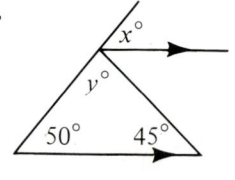
$x = 50, y = 85$

14.
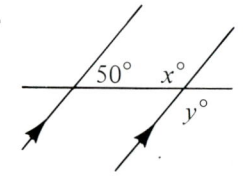
$x = 130, y = 130$

Written Exercises

Classify each pair of angles as (1) alternate interior angles, (2) same-side interior angles, (3) corresponding angles, or (4) none of these.

A
1. ∠1 and ∠5   (3)
2. ∠3 and ∠8   (2)
3. ∠2 and ∠8   (1)
4. ∠4 and ∠8   (3)
5. ∠2 and ∠7   (4)
6. ∠3 and ∠5   (1)
7. ∠2 and ∠5   (2)
8. ∠3 and ∠7   (3)

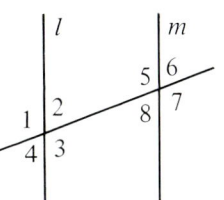

In the diagram above, suppose $l \parallel m$. ∠2 = ∠4 = ∠6 = ∠8 = 70°,

9. If ∠1 = 110°, find the measures of the other seven angles.   ∠3 = ∠5 = ∠7 = 110°
10. If ∠2 = x°, find the measures of ∠5 and ∠8.   ∠5 = 180° − x°, ∠8 = x°

Complete each statement.

11. ∠1 and ∠2 are formed by lines __?__ AD and __?__ BC and the transversal __?__. AC

    ∠3 and ∠4 are formed by lines __?__ AB and __?__ CD and the transversal __?__. AC

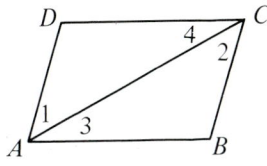

12. ∠5 and ∠6 are formed by lines __?__ PQ and __?__ RS and the transversal __?__. QS

    ∠7 and ∠8 are formed by lines __?__ QR and __?__ PT and the transversal __?__. RS

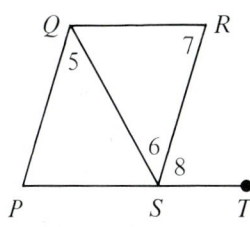

In the diagram, $a \parallel b$ and $c \parallel d$. State the special name for each pair of angles. Then tell if the angles are equal or supplementary.

13. ∠6 and ∠10   corr.; =
14. ∠7 and ∠9   s. − s. int., supp.
15. ∠3 and ∠14   alt. int.; =
16. ∠2 and ∠14   corr.; =
17. ∠9 and ∠16   alt. int.; =
18. ∠8 and ∠10   s. − s. int.; supp.
19. ∠9 and ∠13   corr.; =
20. ∠8 and ∠12   corr.; =

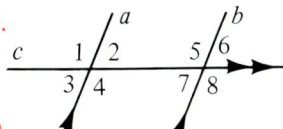

21. Refer to the diagram above. Name seven angles equal to ∠4.   ∠1, ∠5, ∠8, ∠9, ∠12, ∠13, ∠16
22. Name eight angles supplementary to ∠4.
    ∠2, ∠3, ∠6, ∠7, ∠10, ∠11, ∠14, ∠15

Introducing Proof

Find the values of x and y in each diagram.

23.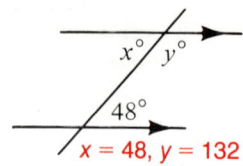
$x = 48, y = 132$

24.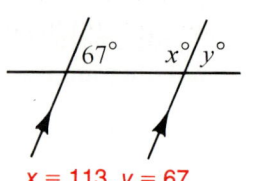
$x = 113, y = 67$

25.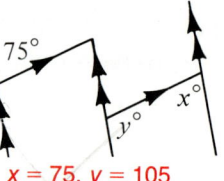
$x = 75, y = 105$

B 26.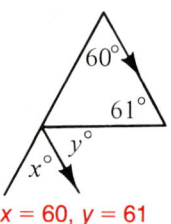
$x = 60, y = 61$

27.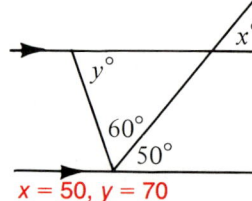
$x = 50, y = 70$

28.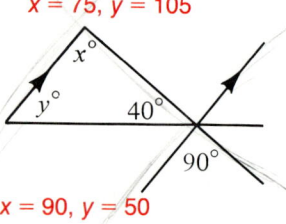
$x = 90, y = 50$

Supply the missing statements and reasons in each proof.

29. Given: $\overline{RU} \parallel \overline{ST}$
    $\overline{RS} \parallel \overline{UT}$

    Prove: $\angle R = \angle T$

STATEMENTS	REASONS
1. $\overline{RU} \parallel \overline{ST}$	1. __?__ Given
2. $\angle R$ and $\angle S$ are supplements.	2. If 2 $\parallel$ lines are cut by a trans., then __?__. s. – s. int. $\angle$s are supp.
3. $\overline{RS} \parallel \overline{UT}$	3. __?__ Given
4. $\angle S$ and $\angle T$ are supplements.	4. __?__ If 2 $\parallel$ lines are cut by a trans., then s. – s. int. $\angle$s are supp.
5. $\angle R = \angle T$	5. __?__ Supplements of the same $\angle$ are =.

30. Given: $\angle B = \angle C$
    $\overline{AD} \parallel \overline{BC}$

    Prove: $\angle 1 = \angle 2$

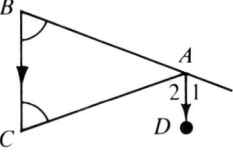

STATEMENTS	REASONS
1. $\overline{AD} \parallel \overline{BC}$	1. __?__ Given
2. $\angle 1 = \angle B$	2. If 2 $\parallel$ lines are cut by a trans., then __?__. corr. $\angle$s are =
3. $\angle 2 = \angle C$	3. __?__ If 2 $\parallel$ lines are cut by a trans., then alt. int. $\angle$s are =.
4. $\angle B = \angle C$	4. __?__ Given
5. $\angle 1 = \angle$ __?__ 2	5. Subst. Post. (Steps 2, 3, 4)

C 31. Prove Theorem 4: If two parallel lines are cut by a transversal, then alternate interior angles are equal.

Given: $r \parallel s$
Prove: $\angle 1 = \angle 2$ (*Hint:* Use $\angle 3$.)

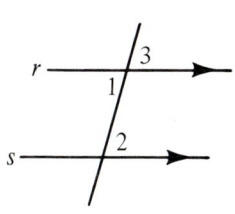

32. Prove Theorem 5: If two parallel lines are cut by a transversal, then same-side interior angles are supplementary.

Given: $t \parallel u$
Prove: $\angle 4$ and $\angle 5$ are supplementary. (*Hint:* Use $\angle 6$.)

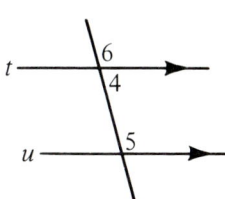

CALCULATOR ACTIVITIES

5. 109.201
6. 81.588
11. 2750.019

Perform the indicated operations using a calculator. Round your answers to three decimal places.

1. 42.31 + 13.26 + 8.79   64.360
2. 2.3 + 9.41 + 0.00048   11.710
3. 107.236 + 2.0007 + 44   153.237
4. 69.1 + 332 + 0.4445   401.545
5. 99 + 0.11 + 10.0909
6. 3.3333 + 77.7 + 0.555
7. 153.2 − 89.378   63.822
8. 1436 − 725.23   710.770
9. 5.698 − 0.7531   4.945
10. 0.74963 − 0.54296   0.207
11. 3647.02 − 897.001
12. 24.27 − 19.0005   5.270
13. 6.7 × 418   2800.600
14. 0.38 × 0.57   0.217
15. 46.12 × 0.9   41.508
16. 0.763 × 97.2   74.164
17. 1.1111 × 26   28.889
18. 15.6 × 89.21   1391.676
19. 437 ÷ 33   13.242
20. 56 ÷ 8.2   6.829
21. 115.3 ÷ 0.3   384.333
22. 8.426 ÷ 0.002   4213.000
23. 0.567 ÷ 40   0.014
24. 215 ÷ 5.375   40.000

Use the division key to find the *terminating* decimal *equal* to each fraction. Write your answer exactly as it appears on the calculator.

25. $\frac{15}{75}$ 0.2
26. $\frac{17}{16}$ 1.0625
27. $\frac{31}{32}$ 0.96875
28. $\frac{12}{25}$ 0.48
29. $\frac{43}{64}$ 0.671875
30. $\frac{23}{92}$ 0.25

The fractions below can be expressed as *repeating* decimals. Use the division key to find a decimal *approximation* for each fraction by rounding to four decimal places.

31. $\frac{5}{9}$ 0.5556
32. $\frac{2}{3}$ 0.6667
33. $\frac{4}{9}$ 0.4444
34. $\frac{6}{11}$ 0.5455
35. $\frac{5}{6}$ 0.8333
36. $\frac{2}{11}$ 0.1818

6 Proving Lines Parallel

In this section, we shall study ways of proving lines parallel. The first way is given in Postulate 8. Postulate 7 is listed with Postulate 8 so that you can see that these two postulates are converses.

POSTULATE 7

If two parallel lines are cut by a transversal, then corresponding angles are equal.

POSTULATE 8

If two lines and a transversal form equal corresponding angles, then the lines are parallel.

Postulates 7 and 8 can be written in convenient picture statements as follows:

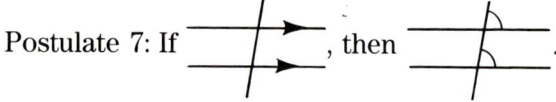

From Postulate 8, we can easily determine other ways to prove lines parallel. These are given in Theorems 6, 7, and 8 below.

THEOREM 6

If two lines and a transversal form equal alternate interior angles, then the lines are parallel.

THEOREM 7

If two lines and a transversal form supplementary same-side interior angles, then the lines are parallel.

THEOREM 8

In a plane, if two lines are each perpendicular to a third line, then the two lines are parallel.

WAYS TO PROVE TWO LINES PARALLEL

1. *Show that corresponding angles are equal.*
2. *Show that alternate interior angles are equal.*
3. *Show that same-side interior angles are supplementary.*
4. *Show that the lines are both perpendicular to a third line.*

EXAMPLE Which lines are parallel?

$\overleftrightarrow{BP}$ and $\overleftrightarrow{CR}$ are not parallel because two corresponding angles are not equal (50 ≠ 51).

$\overleftrightarrow{BQ}$ and $\overleftrightarrow{CS}$ are parallel because two corresponding angles are equal (50 + 30 = 80 and 51 + 29 = 80).

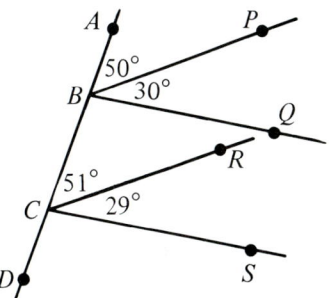

Classroom Practice

For each figure, tell whether you can correctly conclude that $l \parallel m$.

1. yes
2. no
3. yes

In each figure, tell which lines are parallel.

4.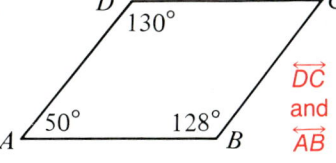
   $\overleftrightarrow{DC}$ and $\overleftrightarrow{AB}$

5.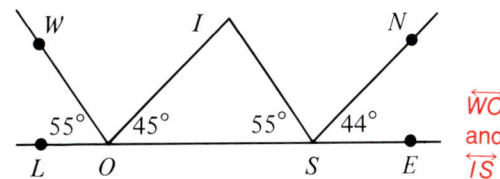
   $\overleftrightarrow{WO}$ and $\overleftrightarrow{IS}$

6. Suppose ∠1 = ∠2. Tell which lines are parallel in each figure.

   a.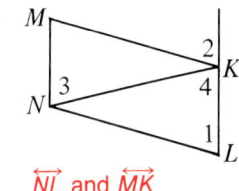
      $\overleftrightarrow{NL}$ and $\overleftrightarrow{MK}$

   b.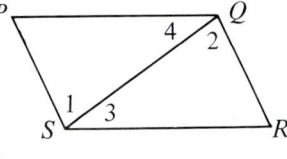
      $\overleftrightarrow{PS}$ and $\overleftrightarrow{QR}$

Introducing Proof

7. Draw a picture statement for Theorem 7 on page 72.

8. Suppose ∠1 = ∠2.
   a. Tell why $\overline{AB} \parallel \overline{CD}$.
   b. Tell why ∠3 = ∠4.

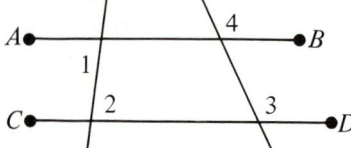

   a. If 2 lines and a trans. form = alt. int. ∠s, then the lines are ∥.
   b. If 2 ∥ lines are cut by a trans., then corr. ∠s are =.

Written Exercises

For each figure, tell if you can correctly conclude that $u \parallel v$.

A 1. yes 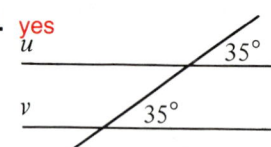 2. no 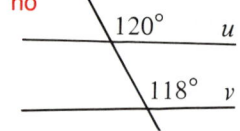 3. yes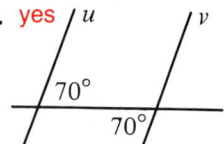

4. yes 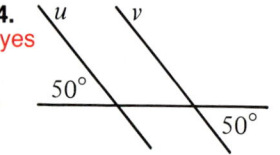 5. no 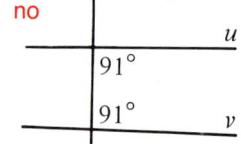 6. yes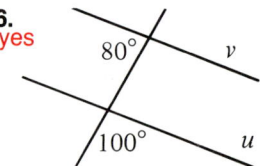

Draw a picture statement for each theorem.

7. Theorem 4 (page 68)
8. Theorem 6 (page 72)
9. Theorem 8 (page 72)

For each figure, name two pairs of parallel lines.

10.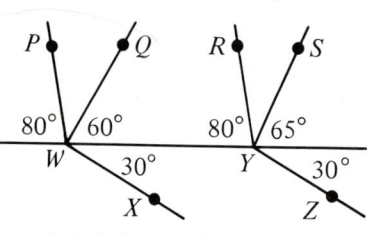

11.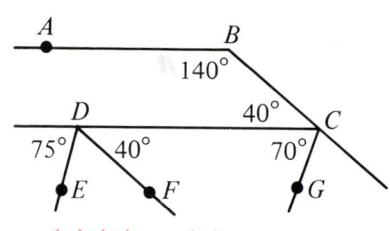

$\overleftrightarrow{PW}$ and $\overleftrightarrow{RY}$; $\overleftrightarrow{WX}$ and $\overleftrightarrow{YZ}$ $\overleftrightarrow{BC}$ and $\overleftrightarrow{DF}$; $\overleftrightarrow{AB}$ and $\overleftrightarrow{CD}$

Find the value of x which will make lines a and b parallel.

12. 35 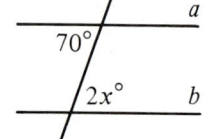 13. 120 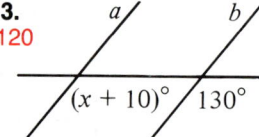 14. 40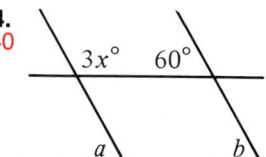

Use the given information to tell which lines, if any, must be parallel.

SAMPLE   ∠1 = ∠3

Since ∠1 and ∠3 are equal corresponding angles, e ∥ f.

15. ∠2 = ∠4   e, f
16. ∠4 = ∠14   c, d
17. ∠8 = ∠11   e, f
18. ∠7 = ∠13   none
19. ∠5 + ∠10 = 180°   c, d
20. ∠2 + ∠3 = 180°   e, f
21. ∠1 = ∠10   c, d
22. ∠1 = ∠9   e, f
23. ∠9 + ∠14 = 180°   c, d

24. Complete the following proof of Theorem 6.

Given: ∠1 = ∠2
Prove: r ∥ s

STATEMENTS	REASONS
1. ∠1 = ∠2	1. __?__ Given
2. ∠1 = ∠3	2. __?__ Vertical ∠s are =.
3. ∠2 = ∠3	3. __?__ Substitution Postulate
4. __?__ r ∥ s	4. If 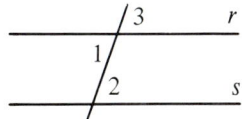, then __?__. the lines are ∥

25. Complete the following proof of Theorem 7.

Given: ∠1 and ∠2 are supplements.
Prove: l ∥ m

STATEMENTS	REASONS
1. ∠1 and ∠2 are supplements.	1. __?__ Given
2. ∠3 and ∠2 are supplements.	2. Definition of __?__ supp. ∠s
3. ∠1 = ∠3	3. Supplements of __?__. the same ∠ are =
4. l ∥ m	4. __?__ (See below.)

4. If 2 lines and a trans. form = corr. ∠s, then the lines are ∥.

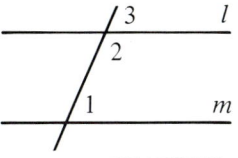

Introducing Proof

B 26. A technical artist sets a plastic triangle in position *ABC* and draws $\overline{AB}$. Then the artist slides the triangle along the T-square to position *XYZ* and draws $\overline{XY}$.

Why is $\overline{AB} \parallel \overline{XY}$? ∠ ABC = ∠ XYZ and if 2 lines and a trans. form = corr. ∠s, then the lines are ∥.

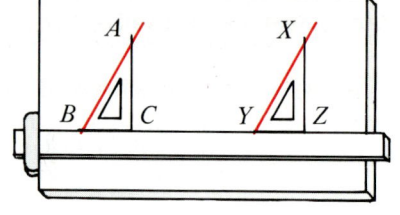

27. Explain why the phrase *In a plane* is needed in Theorem 8.

    Draw a diagram to illustrate your answer.

28. Given: ∠ 1 = ∠ 2

    Prove: ∠ 3 = ∠ 4

    (*Hint:* This proof requires just three steps.)

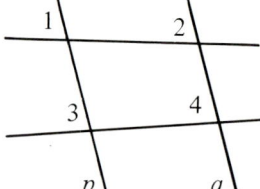

C 29. Given: ∠ 5 = ∠ 6
    ∠ 7 = ∠ 8

    Prove: $\overline{AB} \parallel \overline{CD}$

30. Prove Theorem 8.

    Given: $a \perp b$; $a \perp c$

    Prove: $b \parallel c$ (*Hint:* Use Postulate 8.)

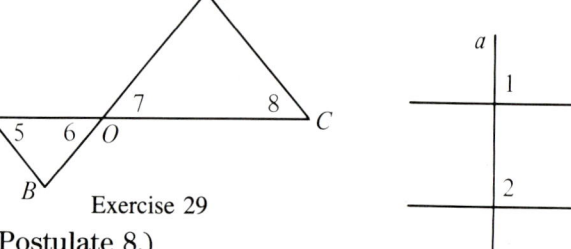

Exercise 29

Exercise 30

CONSUMER APPLICATIONS

ENERGY USAGE

Every time you toast a piece of bread, use a hair dryer, or play the radio, you use energy. Some machines use more energy than others. The chart at the right lists this usage in terms of the length of time the machine would run if supplied with the amount of energy in a liter of gasoline.

electric clock	31 weeks
black & white television	$2\frac{3}{4}$ days
color television	$1\frac{1}{2}$ days
toaster	$8\frac{1}{2}$ hours
oven	3 hours

1. Which of the machines in the chart uses the most energy?   oven

2. Estimate the cost of running an electric clock for one year. (Assume that gasoline costs 32¢ per liter.)   about 54¢

76 Chapter 2

7 Constructing Parallel Lines

Our goal in this section is to construct a line that passes through a given point P and is parallel to a given line l. To do this, we must first be able to copy an angle with a compass and straightedge.

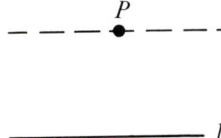

CONSTRUCTION 4

Given: *An angle*

Construct: *An angle equal to the given angle*

Students are asked to justify this construction in Wr. Ex. 9, on page 151.

1. Draw a ray with endpoint O.

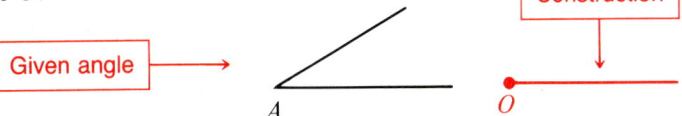

2. Construct an arc with center at A.
3. Without changing your compass, construct an arc with center at O.

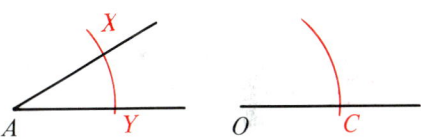

4. Put the compass point at Y. Then adjust your compass so that you can draw an arc passing through X.
5. Without changing your compass, construct an arc with center at C.

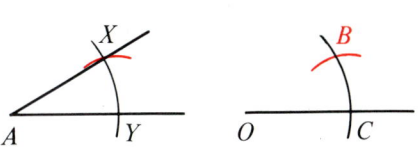

6. Draw $\overrightarrow{OB}$. $\angle BOC = \angle A$.

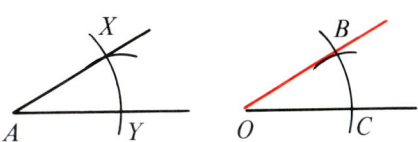

Introducing Proof

CONSTRUCTION 5

Given: *A line and a point not on the line*

Construct: *A line that passes through the point and is parallel to the given line*

Follow these steps:

1.

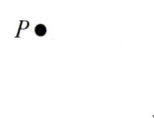

You are given *l* and *P*.

2.

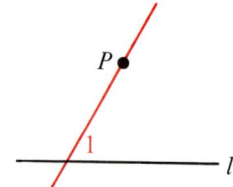

Draw any transversal through *P*. Label ∠ 1.

3.

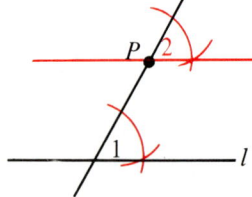

At *P*, construct ∠ 2, a corresponding angle equal to ∠ 1.

4.

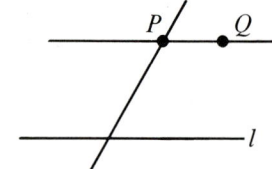

$\overleftrightarrow{PQ} \parallel l$. Do you know why?

Classroom Practice

1. Draw an acute angle.
   Then copy the angle using compass and straightedge.

2. Draw an obtuse angle.
   Then copy the angle using compass and straightedge.

3. Draw a line *l* and choose a point *P* not on *l*.
   Construct a line through *P* and parallel to *l*.

4. Draw a triangle similar to, but larger than, the one shown.
   Then construct a line through *O* parallel to $\overleftrightarrow{JY}$.

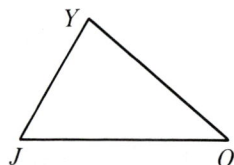

78   Chapter 2

Written Exercises

**Draw an angle similar to, but larger than, the one shown.
Then copy the angle using compass and straightedge.**

A 1.    2.    3.

**Draw a figure similar to, but larger than, the one shown.
Then construct a parallel to $\overleftrightarrow{RS}$ through point P.**

4. 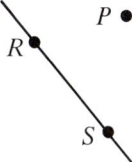   5. 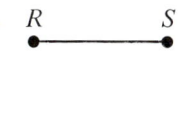   6.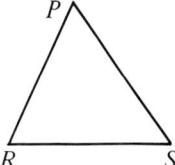

7. Draw a line l and a point P not on l.
   Construct a line which passes through P and is parallel to l by constructing equal alternate interior angles.

8. a. Using lined paper, draw two parallel lines and a transversal as shown. Then construct the bisectors of two alternate interior angles. **b. They are parallel.**
   b. What appears to be true about the bisectors?
   c. Explain why your answer in part (b) is true. (See below.)

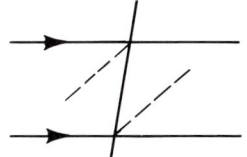

9. a. Using lined paper, draw two parallel lines and a transversal. Then construct the bisectors of two same-side interior angles.
   b. Measure an angle formed by the bisectors. **90°**
   c. Repeat the experiment with a different transversal. **90°**
   d. What appears to be true about the angles formed by the bisectors? **They are right angles.**

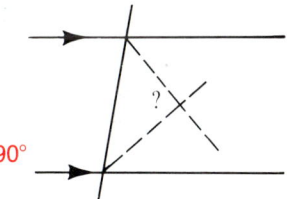

10. a. Construct a line through P and parallel to l by these steps:
    (1) Use Construction 3 on page 26 to construct m, a line perpendicular to l through P.
    (2) Use Construction 2 on page 26 to construct n, a line perpendicular to m through P.
    b. Tell why $l \parallel n$. **In a plane, if 2 lines are each ⊥ to a third line, then the 2 lines are ∥.**

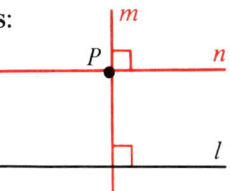

B 11. Draw an acute angle. Call it $\angle B$.
    Draw a line t and choose a point Q on t.
    At Q, construct an angle that is supplementary to $\angle B$.

12. Repeat Exercise 11, using an obtuse $\angle B$.

  8. c. Since ∠1 and ∠2 are each half the measure of = ∠s, they are =.
     Thus the bisectors are ∥ by Thm. 6.

Introducing Proof

| SAMPLE | Given ∠A and ∠B, construct ∠A + ∠B.

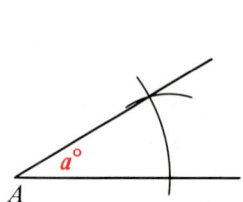 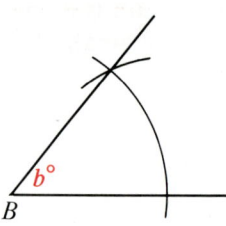 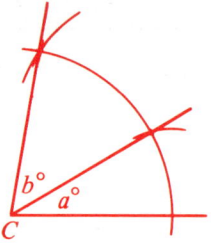

13. Draw two acute angles, ∠X and ∠Y. Construct ∠X + ∠Y.

14. Draw an acute angle, ∠Z. Construct an angle with measure 2 · ∠Z.

SELF-TEST

Vocabulary

parallel lines (p. 66) alternate interior angles (p. 66)
skew lines (p. 66) same-side interior angles (p. 67)
transversal (p. 66) corresponding angles (p. 67)
interior angles (p. 66)

1. What is the difference between a postulate and a theorem? *A postulate is accepted without proof; a theorem is proved.* (2-4)

2. Name two pairs of alternate interior angles. (2-5)
   ∠2, ∠7; ∠3, ∠6

3. Name two pairs of same-side interior angles.
   ∠2, ∠3; ∠6, ∠7

4. Suppose ∠6 = 130°. Find the measures of the other numbered angles. *∠1 = 130°, ∠2 = 50°, ∠3 = 130°, ∠4 = 50°, ∠5 = 50°, ∠7 = 50°, ∠8 = 130°*

5. Find the values of x and y in the diagram. *x = 70*
   y = 50

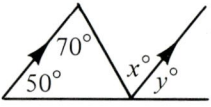

Exercise 5

6. List four ways to prove that two lines are parallel. (2-6)
   See page 73.

7. Given: $l \parallel m$
             ∠1 = ∠2
   Prove: ∠1 = ∠3

8. Draw a line m. Choose a point A that is several centimeters from m. Construct a line through A parallel to m. (2-7)

READING GEOMETRY

DIAGRAMS

Have you ever thought what this book would be like if everything had to be described in words, without the help of diagrams? Diagrams are very useful in geometry because they provide a great deal of information in a form that can be easily understood.

Most of the explanations in this book are illustrated by diagrams. You need to read the text and the diagrams *together* in order to get the most out of your reading. Next to a diagram you may sometimes see a label in a red box. Be sure to look at all the labels when you read the diagrams. Some have arrows that point to important parts of the figure, and some state facts that you need to know about the diagram.

Many of the exercises include diagrams. When you work an exercise, you are allowed to use *all* the information given in the diagram, but you are not allowed to take anything for granted that isn't actually given. For example, you may not say that two sides of a figure are parallel just because they *look* parallel to you.

Sometimes you may want to draw a diagram of your own to help you understand an explanation or a problem. In such a case, you don't need to make a careful construction. A reasonably accurate sketch is usually all you need. Be sure to letter the points correctly and to show only what is given in the description of the figure.

EXERCISES

Use the information given in the diagrams. Tell whether each statement is true or false.

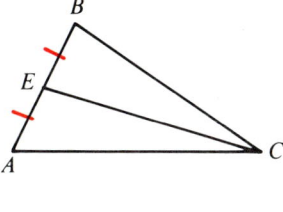

1. E is the midpoint of $\overline{AB}$. true

2. $\angle BCE = \angle ACE$ false

3. $\angle BEC = \angle AEC$ false

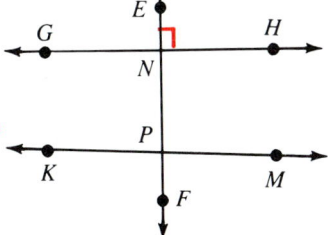

4. $\overleftrightarrow{EF} \perp \overleftrightarrow{GH}$ true

5. $\overrightarrow{KM} \parallel \overleftrightarrow{GH}$ false

6. $\angle GNE$ is a right angle. true

7. $\angle MPE$ is a right angle. false

8. $\angle KPF = \angle MPE$ true

Introducing Proof 81

PROBLEM SOLVING STRATEGIES

ASKING QUESTIONS

Here are some questions to ask when solving geometry problems.

1. DO I UNDERSTAND THE PROBLEM?
   Did I read the problem carefully?
   Do I know what all the words mean?

2. WHAT DO I KNOW?
   What is the given information?
   What conditions are part of the problem?
   Can I draw a picture of the information?
   Can I make a chart?

3. WHAT DON'T I KNOW?
   What information is unknown?
   What is the problem asking for?
   Do I have to find something else first?
   Do I have to prove something?

4. LOOK AT THE WHOLE PROBLEM.
   How can I use the information I have?
   How can I tie together the known and the unknown?

EXERCISES

Answer the questions that follow each problem.

1. Gail brings some herbs to sell at the Farmers' Market.
   In the morning, she sells them for $1.50 each and receives $16.50.
   In the afternoon, Gail lowers the price to $1.00 each.
   She sells three times as many as in the morning.
   What is her gross income for the entire day?

a. You know Gail's gross income from morning sales is $16.50.
b. What is the problem asking for? Gail's gross income for the entire day
c. Find Gail's gross income from afternoon sales:

You need to know how many herbs Gail sold in the afternoon. To do this you need to find out how many herbs she sold in the morning. Use the information about her morning sales to find the number of herbs she sold then.

How many herbs did Gail sell in the afternoon? 33 herbs

What was her gross income for the afternoon? $33

d. What was Gail's gross income for the entire day? $49.50

2. b. It must be divisible by 6, 20, and 33; it must be as small as possible.
2. Find the smallest integer that will be divisible by each of the following: 6, 20, 33
   a. What do you have to find in this problem? an integer
   b. List two conditions this integer must satisfy. (See above.)
   c. Write out the prime factors of 6, 20, and 33. 6 = 2·3; 20 = 2·2·5; 33 = 3·11
   d. How can you find an integer that is divisible by all of these factors? Multiply these factors together: (2·3)(2·2·5)(3·11).
   e. What is the smallest such integer? The least common multiple is formed by eliminating factors repeated in one of the other numbers. Thus we get 2·2·3·5·11 = 660.

3. Prove that if the same-side interior angles of two parallel lines and a transversal are bisected, then the two angle bisectors will form a 90° angle.
   a. Here is a diagram of the situation.
      Copy the diagram and mark in the information that is given.

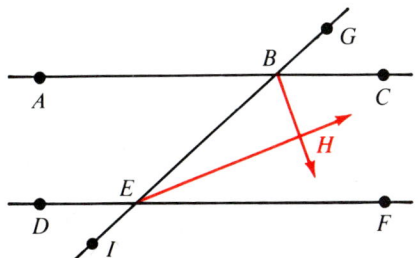

   b. Using the letters in the diagram, state what you are being asked to prove. ∠EHB = 90°

Introducing Proof   83

SKILLS REVIEW

COORDINATES IN THE PLANE

Give the coordinates of each point.

SAMPLE   Point A, $(-3, 3)$

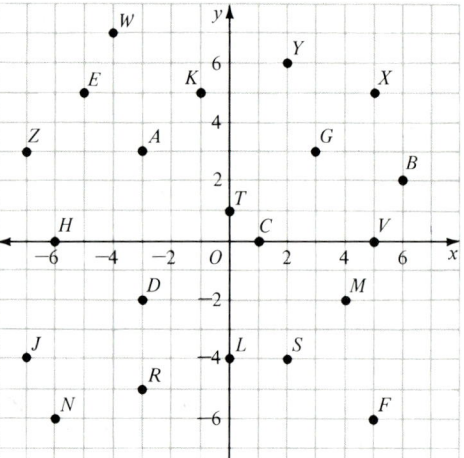

1. B (6, 2)   2. C (1, 0)   3. D $(-3, -2)$
4. E $(-5, 5)$ 5. F $(5, -6)$ 6. G (3, 3)
7. H $(-6, 0)$ 8. J $(-7, -4)$ 9. K $(-1, 5)$

Name the point by its letter.

10. $(-6, -6)$ N   11. $(0, -4)$ L
12. $(-4, 7)$ W   13. $(5, 0)$ V
14. $(-3, -5)$ R   15. $(4, -2)$ M
16. $(2, 6)$ Y   17. $(-7, 3)$ Z
18. $(0, 1)$ T   19. $(2, -4)$ S

Plot the points in Exercises 20–31. Then draw line segments to connect the points in order. You should now have the outline of half a fir tree. (See the Solution Key.)

20. $(0, 6)$ 21. $(-2, 4)$ 22. $(-1, 4)$ 23. $(-3, 2)$
24. $(-2, 2)$ 25. $(-5, -1)$ 26. $(-3, -1)$ 27. $(-6, -4)$
28. $(-1, -4)$ 29. $(-1, -6)$ 30. $(-2, -7)$ 31. $(0, -7)$

Complete the ordered pairs in Exercises 32–43 to tell how to draw the other half of the fir tree. Start at the top again.

32. $(0, \underline{6})$ 33. $(2, \underline{4})$ 34. $(\underline{1}, 4)$ 35. $(3, \underline{2})$
36. $(\underline{2}, 2)$ 37. $(\underline{5}, -1)$ 38. $(\underline{3}, -1)$ 39. $(6, \underline{-4})$
40. $(\underline{1}, -4)$ 41. $(\underline{1}, \underline{-6})$ 42. $(\underline{2}, -7)$ 43. $(\underline{0}, \underline{-7})$

44. Plot and connect the points $(2, -1)$ and $(2, -5)$. Name the vertices of the two squares that can be drawn with this segment as a side.
    $(2, -1), (2, -5), (-2, -1), (-2, -5); (2, -1), (2, -5), (6, -1), (6, -5)$

CHAPTER REVIEW

CHAPTER SUMMARY

1. Complementary angles are two angles whose measures total 90°.
   Supplementary angles are two angles whose measures total 180°.

2. "If . . . then" statements are important in logical reasoning.
       Statement: If A, then B.   Converse statement: If B, then A.

3. There are four kinds of reasons used in a geometric proof.
       Given information   Definition   Postulate   Theorem

4. Parallel lines are coplanar lines which do not intersect.
   Skew lines are not coplanar lines, and they do not intersect.

5. When two parallel lines are cut by a transversal:
   a. corresponding angles are equal;
   b. alternate interior angles are equal;
   c. same-side interior angles are supplementary.

6. Ways to prove that two lines cut by a transversal are parallel:
   a. show that corresponding angles are equal;
   b. show that alternate interior angles are equal;
   c. show that same-side interior angles are supplementary;
   d. show that both lines are perpendicular to the transversal.

7. It is possible to construct an angle equal to a given angle. (Construction 4, page 77)
   Given a line and a point not on the line, you can construct a line that passes through the point and is parallel to the given line. (Construction 5, page 78)

REVIEW EXERCISES

In the diagram shown, $\overline{TR} \perp \overline{RO}$ and $\overline{SO} \perp \overline{RO}$. (See pp. 46–49.)

1. If $\angle 2 = 30°$, then a complement of $\angle 2$ has measure ___?___°. 60

2. If $\angle 2 = 30°$, then a supplement of $\angle 2$ has measure ___?___°. 150

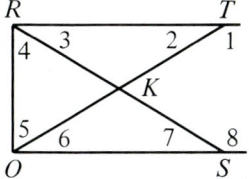

3. Name a complement of $\angle 3$. $\angle 4$

4. If $\angle 3 = \angle 6$, which theorem lets you conclude $\angle 4 = \angle 5$?
       Complements of $=$ ⚋ are $=$.

Introducing Proof   **85**

Use the statement, "Every acute angle has a measure less than 90°."
(See pp. 50–53.)

5. Write the statement in "if . . . then" form. If an ∠ is acute, then it has a measure less than 90°.

6. State the hypothesis and the conclusion. Hyp.: An ∠ is acute.; Concl.: It has a measure less than 90°.

7. State the converse of the given statement. If an ∠ has a measure less than 90°, then it is acute.

(Optional) Accept as true: "All cats have sharp claws." *(See pp. 56–59.)*

8. For each statement, tell what you can conclude. **a., b.** No conclusion possible.
   a. Cyrus is not a cat.   b. Catarina has sharp claws.
   c. Cinnamon is a cat.   d. Cleo does not have sharp claws.
   Cinnamon has sharp claws.   Cleo is not a cat.

Write a proof in two-column form. *(See pp. 60–65.)*

9. Given: ∠1 = ∠4
   Prove: ∠2 = ∠5

Classify each pair of angles as alternate interior angles, same-side interior angles, or corresponding angles. *(See pp. 66–71.)*

10. ∠1 and ∠3 corr. ∠s
11. ∠2 and ∠6 alt. int. ∠s
12. ∠7 and ∠6 s. – s. int. ∠s
13. ∠8 and ∠6 corr. ∠s

Find the values of *x* and *y* in each diagram. *(See pp. 66–71.)*

14.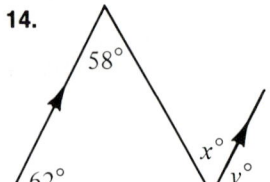
    $x = 58, y = 62$

15.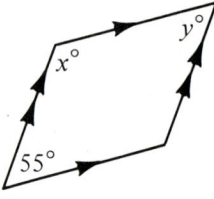
    $x = 125, y = 55$

16.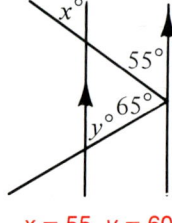
    $x = 55, y = 60$

Use the given information to tell which lines must be parallel.
(See pp. 72–76.)

17. ∠1 = ∠10 a, b
18. ∠7 = ∠12 a, b
19. ∠8 = ∠11 c, d
20. ∠2 + ∠3 = 180° c, d

Construct the required figure. *(See pp. 77–80.)*

21. Draw a line *l* and choose a point *P* not on *l*. Construct a line through *P* and parallel to *l*.

CHAPTER TEST

1. If ∠A = 57°, find the measures of (**a**) a supplement of ∠A and (**b**) a complement of ∠A. **a.** 123°; **b.** 33°   *(2-1)*

2. If ∠1 = ∠2, what do you know about ∠3 and ∠4?
   ∠3 = ∠4

3. Write the theorem that justifies your conclusion in Exercise 2.
   Supplements of = ≜ are =.

In Exercises 4–9 accept as true: "If $x = 0$, then $x + x = 0$."

4. Write the hypothesis and the conclusion. Hyp.: $x = 0$; Concl.: $x + x = 0$ *(2-2)*

5. Write the converse.   6. Is the converse true? yes
   If $x + x = 0$, then $x = 0$.

(Optional) Tell what, if anything, you can conclude.

7. $x + x = 0$ 8. $x + x \neq 0$ 9. $x \neq 0$ no conclusion   *(2-3)*
   no conclusion $x \neq 0$

10. Given: ∠1 and ∠2 are complements.   *(2-4)*
    ∠3 and ∠4 are complements.
    $\overrightarrow{BD}$ bisects ∠EBF.

    Prove: ∠1 = ∠4

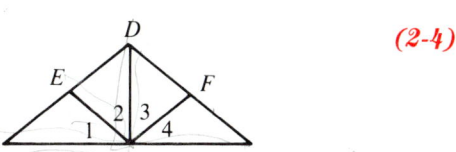

11. ∠3 and ∠ __?__ are corresponding angles.   7   *(2-5)*

12. ∠1 is supplementary to which four angles? ∠2, ∠3, ∠6, ∠7

13. ∠6 and ∠ __?__ are same-side interior angles.   4
    Is it possible for these angles to be equal angles?
    to be complementary? Yes, if they both = 90°; no.

14. Name two pairs of alternate interior angles. ∠3 and ∠6,
    ∠4 and ∠5

Use the given information to tell which segments must be parallel.

15. ∠1 = ∠2 $\overline{RZ} \parallel \overline{XT}$ *(2-6)*

16. $\overline{RZ} \perp \overline{RS}$; $\overline{RS} \perp \overline{SY}$ $\overline{RZ} \parallel \overline{SY}$

17. ∠TRZ + ∠XZR = 180°
    $\overline{RT} \parallel \overline{ZX}$

18. ∠RSY = ∠STX
    $\overline{SY} \parallel \overline{TX}$

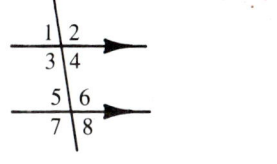

Draw a large acute angle, ∠1.

19. Construct an angle ∠RST equal to ∠1.   *(2-7)*

20. Construct a line through T and parallel to $\overleftrightarrow{RS}$.

UNIT A CUMULATIVE REVIEW

Find the following:

L I Z A R D
−8 −6 −4 −2 0 2

1. AD 4
2. LR 8
3. ZR 4
4. The coordinate of Z   −4
5. The midpoint of $\overline{IA}$ Z

Tell whether each statement is true or false.

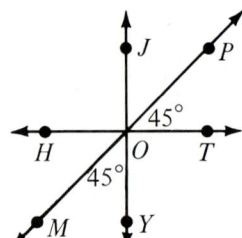

6. ∠HOJ + ∠JOP = ∠HOP true
7. ∠MOY is an obtuse angle. false
8. ∠HOM = 45° true
9. $\overrightarrow{OM}$ bisects ∠HOY. true
10. ∠HOY is an acute angle. false

11. It is possible to draw two different lines containing both points P and T. false

12. Exactly one plane contains points P, O, and Y. true

13. ∠TOP and ∠HOM are complementary angles. true

14. Complete the following proof.

   Given: ∠1 and ∠3 are supplements.
   s ∥ t

   Prove: l ∥ m

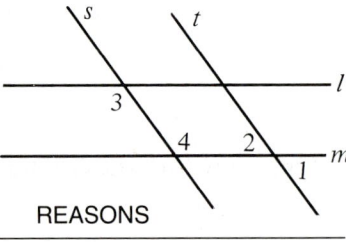

STATEMENTS	REASONS
1. ∠1 and ∠3 are supplements.	1. __?__ Given
2. ∠2 and ∠4 are supplements.	2. __?__ (See below.)
3. ∠1 = ∠2	3. __?__ Vertical ⦞ are =.
4. ∠3 = ∠4	4. __?__ Supplements of = ⦞ are =.
5. __?__ l ∥ m	5. __?__ If 2 lines and a trans. form = alt. int. ⦞, then the lines are ∥.

15. Write the converse of the statement: "If two angles are vertical angles, then they are equal." Is the converse true? If 2 ⦞ are =, then they are vertical ⦞; not true.

16. Draw a line l and points P and Q not on l.
    a. Construct a perpendicular to l through P.
    b. Construct a line through Q parallel to l.

14. 2. If 2 ∥ lines are cut by a trans., then s. − s. int. ⦞ are supp.

UNIT B

Here's what you'll learn in this chapter:

To use the properties of the angles of a triangle.

To classify triangles according to their sides or angles.

To name congruent triangles and their corresponding parts.

To use postulates and theorems to prove that triangles are congruent.

Because triangles are rigid figures, they are often used in the construction of bridges and other buildings. The frames of the bridges shown above contain many congruent triangles. (See the Application on pages 113–114.)

Chapter 3

Triangles

1 The Angle Sum of a Triangle

Draw a large triangle on a sheet of paper and cut the triangle out. Then tear off the three corners of the triangle and arrange them as shown. What appears to be true about the angles of the triangle? Try this experiment with a few other triangles and see if your results are the same.

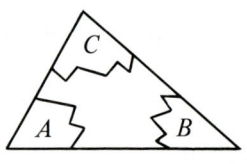

The property illustrated by this experiment is one of the best known and most important theorems in geometry.

THEOREM 1

The angle sum of a triangle is 180°.

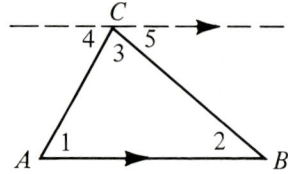

Given: △ABC

Prove: ∠1 + ∠2 + ∠3 = 180°

STATEMENTS	REASONS
1. Through C, draw a line parallel to $\overline{AB}$.	1. Through any point not on a line, a parallel can be drawn to the line. (Recall Construction 5, page 78.)
2. ∠4 = ∠1 and ∠5 = ∠2	2. If 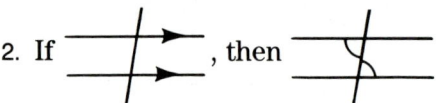, then
3. ∠4 + ∠5 + ∠3 = 180°	3. Their sum is a straight angle.
4. ∠1 + ∠2 + ∠3 = 180°	4. Substitution Postulate

EXAMPLE Find the measures of ∠1 and ∠2.

∠1 + 60° + 70° = 180°

∠1 +   130°   = 180°

∠1 = 50°

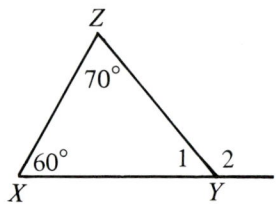

Since ∠1 and ∠2 are supplements,
∠2 = 180° − 50° = 130°.

In the preceding example, ∠ 2 is called an **exterior angle** of △ XYZ. Notice that ∠ 2 = 130° = ∠ X + ∠ Z.

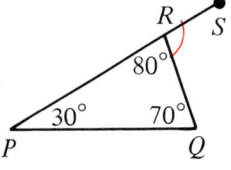

Likewise, in the diagram at the right, ∠ QRS is an exterior angle of △ PQR. Do you see that ∠ QRS = 100° = ∠ P + ∠ Q?

These observations lead to an important *corollary* of Theorem 1. (A **corollary** of a theorem is a result which follows from the theorem with very little extra work.) The proof of the corollary is left as an exercise.

COROLLARY

An exterior angle of a triangle is equal to the sum of the two opposite angles of the triangle.

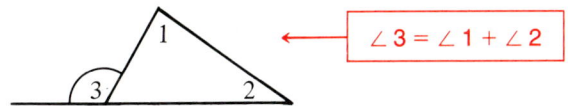

Classroom Practice

Find the measure of each numbered angle.

1.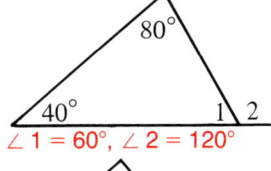
∠ 1 = 60°, ∠ 2 = 120°

2.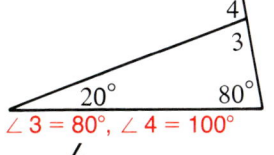
∠ 3 = 80°, ∠ 4 = 100°

3.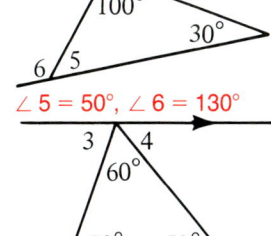
∠ 5 = 50°, ∠ 6 = 130°

4.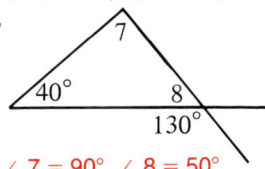
∠ 7 = 90°, ∠ 8 = 50°

5.
∠ 1 = 90°, ∠ 2 = 90°

6.
∠ 3 = 70°, ∠ 4 = 50°

In △ ABC, ∠ C = 90°.

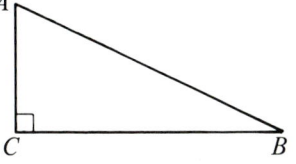

7. Find ∠ B if ∠ A = 60°. 30°

8. Find ∠ B if ∠ A = 70°. 20°

9. Find ∠ B if ∠ A = 65°. 25°

10. From Exercises 7–9, we see that if a triangle has a right angle, then the acute angles are __?__ complementary

Triangles

11. In the diagram, ∠1 + ∠2 = __?__°; 90
∠1 + ∠3 = __?__°. 90

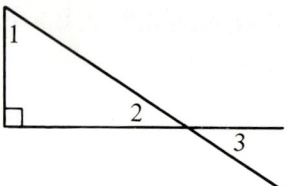

Find the value of x in each diagram.

12. 45 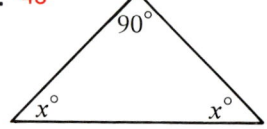   **13.** 60 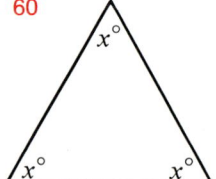   **14.** 90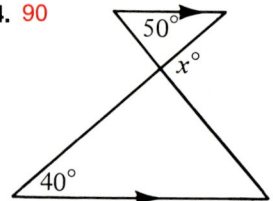

15. How many right angles can a triangle have? Explain your answer. One; if it had two, the sum would be 180° and the third angle would have to be 0°.

16. How many obtuse angles can a triangle have? Explain your answer. One; when the measure of one angle is greater than 90°, the sum of the two remaining angles is less than 90°, so neither is obtuse.

Written Exercises

Find the measure of each numbered angle.

[A] **1.**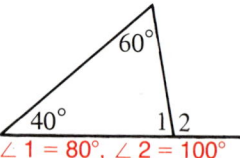
∠1 = 80°, ∠2 = 100°

2.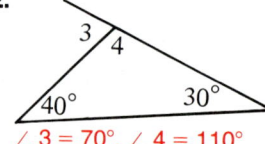
∠3 = 70°, ∠4 = 110°

3.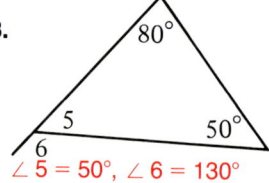
∠5 = 50°, ∠6 = 130°

4.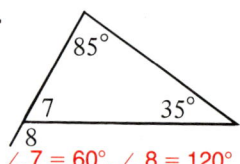
∠7 = 60°, ∠8 = 120°

5.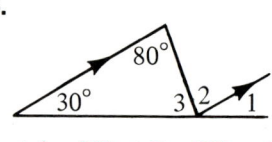
∠1 = 30°, ∠2 = 80°, ∠3 = 70°

6.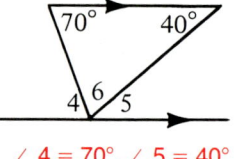
∠4 = 70°, ∠5 = 40°, ∠6 = 70°

In Exercises 7–10, the measures of two angles are given. Find the measure of the third angle using the Corollary of Theorem 1.

	7.	8.	9.	10.
∠1	25°	40°	30°	?
∠2	65°	?	70°	65°
∠3	?	110°	?	115°
	90°	70°	100°	50°

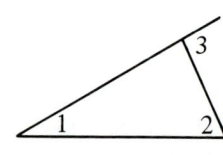

Find the measures of ∠1 and ∠2.

11. ∠1 = 30°, ∠2 = 50°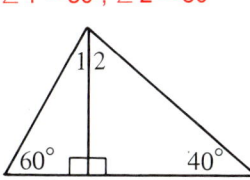

12. ∠1 = 60°
    ∠2 = 50°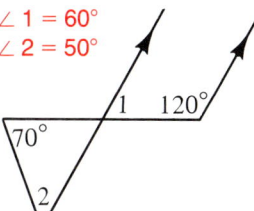

13. ∠1 = 45°
    ∠2 = 85°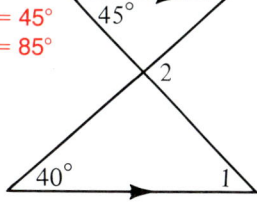

14. The three angles of a certain triangle are all equal. Find the measure of each angle. 60°

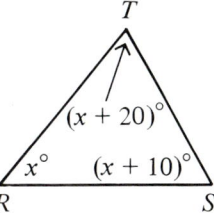

15. Find the numerical measure of each angle of △RST.
    ∠R = 50°, ∠S = 60°, ∠T = 70°

Find the value of x in each diagram.

16. 20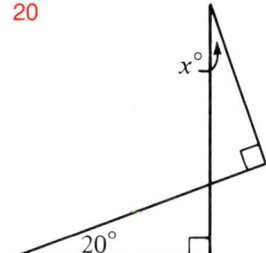

17. 30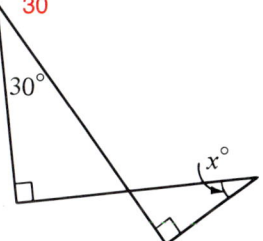

18. 41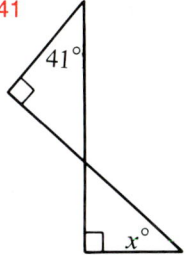

B 19. If two angles of one triangle are equal to two angles of another triangle, what can you say about the third angles of the triangles? Explain your answer. They must also be equal, since each is 180° − (the equal sum of 2 ∠s).

In each diagram, $\overrightarrow{RT}$ and $\overrightarrow{ST}$ are angle bisectors. Find the value of x.

20. 130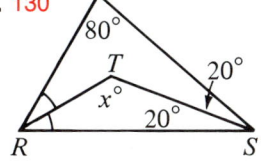

21. 130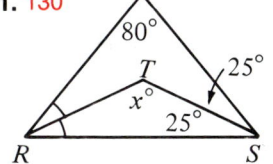

22. 130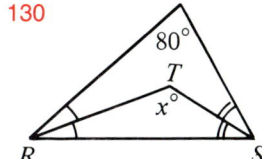

23. In △DEF, ∠E is 15° larger than ∠D. Also, ∠F is 45° larger than ∠D. Find the measure of each angle. 55   85
    (*Hint:* Let ∠D = x°. Then ∠E = __?__° and ∠F = __?__°.)
    40

24. In △ABC, ∠B is twice as large as ∠A. Also, ∠B = ∠C. Find the measure of each angle. ∠A = 36°, ∠B = 72°, ∠C = 72°

Triangles **95**

25. Complete this proof of the Corollary of Theorem 1.

Given: △ABC
Prove: ∠3 = ∠1 + ∠2

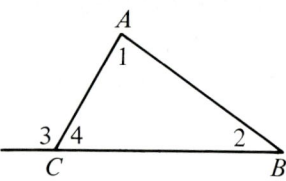

STATEMENTS	REASONS
1. ∠1 + ∠2 + ∠4 = 180°	1. __?__ The ∠ sum of a △ is 180°.
2. ∠1 + ∠2 = 180° − ∠4	2. __?__ Subtraction Postulate
3. ∠4 + ∠3 = 180°	3. Their sum is a __?__ angle. straight
4. ∠3 = 180° − ∠4	4. __?__ Subtraction Postulate
5. ∠3 = ∠1 + ∠2	5. __?__ Substitution Postulate

In each exercise, find the indicated sum.

26.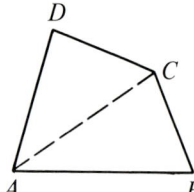

∠A + ∠B + ∠C + ∠D = __?__ °   360

27.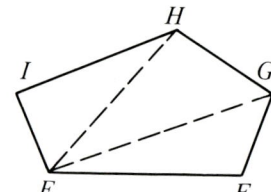

∠E + ∠F + ∠G + ∠H + ∠I = __?__ °   540

In each exercise, find the value of x.

[C] **28.** 120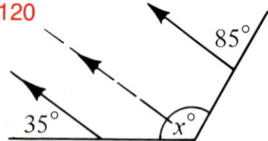

29. 90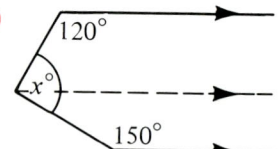

30. Given: ∠B = ∠C
　　　　　$\overrightarrow{AE}$ bisects ∠DAC.
　　Prove: $\overline{AE} \parallel \overline{BC}$

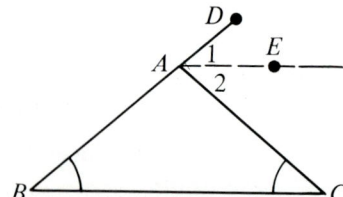

2 Classifying Triangles

△ABC is shown at the right. The segments, $\overline{AB}$, $\overline{BC}$, and $\overline{AC}$ are called the **sides** of the triangle. Notice that two of these sides have equal lengths. We say that the two sides are equal. (This is like saying that two angles are equal if they have equal measures.)

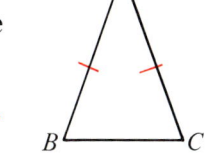

AB = AC

Classification of Triangles by Sides:

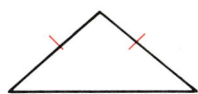
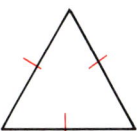

Scalene triangle   Isosceles triangle   Equilateral triangle

No sides are equal.   At least *two* sides are equal.   All *three* sides are equal.

Notice that an equilateral triangle is also isosceles.

Classification of Triangles by Angles:

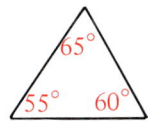

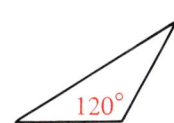
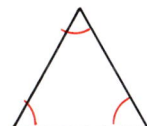

Acute triangle   Right triangle   Obtuse triangle   Equiangular triangle

All angles are acute.   One angle is right.   One angle is obtuse.   All angles are equal.

Special Names for Sides:

The two equal sides of an isosceles triangle are called **legs**. The third side is the **base**.

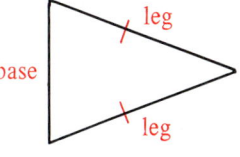

The side opposite the right angle of a right triangle is called the **hypotenuse**. The other sides are called **legs**.

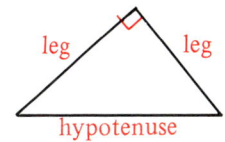

Triangles

Classroom Practice

Classify each triangle shown as scalene, isosceles, or equilateral.

1.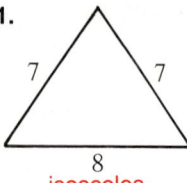
isosceles

2.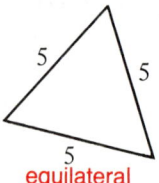
equilateral

3.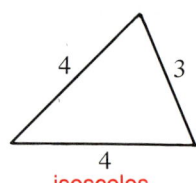
isosceles

4.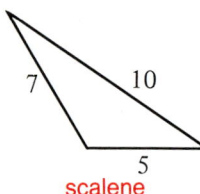
scalene

Classify each triangle shown as acute, right, obtuse, or equiangular.

5.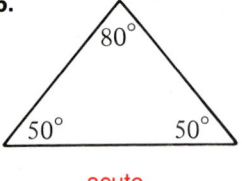
acute

6.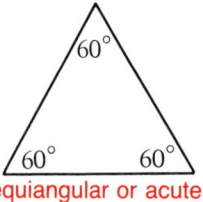
equiangular or acute

7.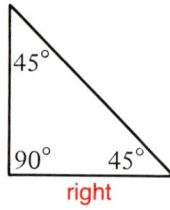
right

8.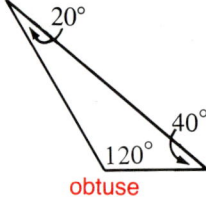
obtuse

9. △ABC is a right triangle.
    Sides $\overline{AC}$ and $\overline{BC}$ are called __?__. legs
    Side $\overline{AB}$ is called the __?__. hypotenuse

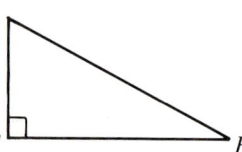

10. △PQR is an isosceles triangle.
    Sides $\overline{PQ}$ and $\overline{PR}$ are called __?__. legs
    Side $\overline{QR}$ is called the __?__. base

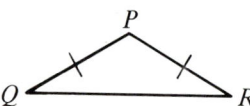

Given: $\overline{BE} \perp \overline{AC}$.

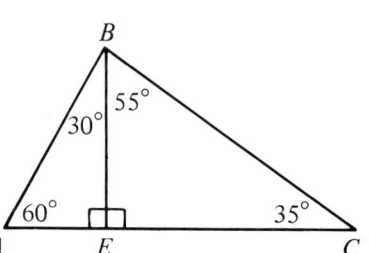

11. Name two right triangles. △ABE, △BEC
12. Name an acute triangle. △ABC
13. Name an exterior angle of △BEC. ∠BEA
14. Name a side common to △ABE and △BEC. $\overline{BE}$
15. Name a vertex common to △ABE and △BEC. E
16. Name an angle common to △ABE and △ABC. ∠A

17. If a triangle is equilateral, is it also isosceles? yes
18. If a triangle is isosceles, is it also equilateral? no
19. The *perimeter* of a triangle is the sum of the lengths of its sides. If the perimeter of an equilateral triangle is 30 cm, find the length of each side. 10 cm

Chapter 3

Written Exercises

Classify each triangle shown as scalene, isosceles, or equilateral.

A 1. 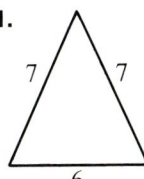   2. 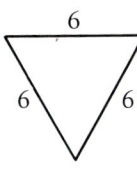   3. 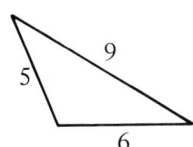   4.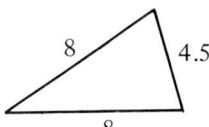

isosceles   equilateral   scalene   isosceles

Classify each triangle shown as acute, right, obtuse, or equiangular.

5. 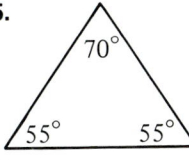   6. 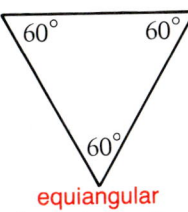   7. 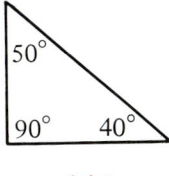   8.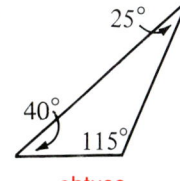

acute   equiangular   right   obtuse

9. **a.** Draw an acute scalene triangle.   **b.** Draw an acute isosceles triangle.
10. **a.** Draw an obtuse scalene triangle.   **b.** Draw an obtuse isosceles triangle.
11. **a.** Draw a right scalene triangle.   **b.** Draw a right isosceles triangle.

12. **a.** Fold a sheet of paper in half.

    b. Use scissors to cut from any point on the fold to a corner point.

    c. What kind of triangle is formed when you unfold the paper? isosceles

   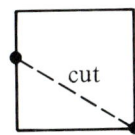   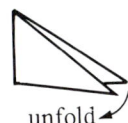

13. *Given:* △ ABC is equiangular.
    $\overrightarrow{AX}$ bisects ∠ BAC.

    Then: ∠ C = ___?___° 60
    ∠ CAX = ___?___° 30
    ∠ AXC = ___?___° 90

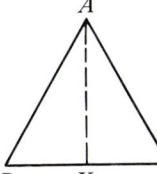

14. Figure *ABCD* has four right angles.
    (In other words, it is a rectangle.)

    △ ABC, △ ACD

    a. Name two right triangles with hypotenuse $\overline{AC}$.

    b. Name two right triangles with hypotenuse $\overline{BD}$.

    △ ABD, △ BDC

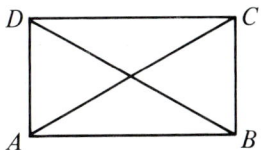

Triangles   **99**

In Exercises 15–17, find the lengths of the sides of the triangle.

B 15.

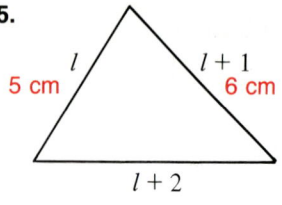

5 cm, 6 cm, 7 cm

The perimeter is 18 cm.

16.

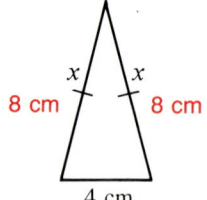

8 cm, 8 cm, 4 cm

The perimeter is 20 cm.

17.

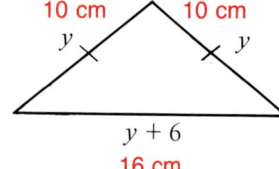

10 cm, 10 cm, 16 cm

The perimeter is 36 cm.

Exercises 18–20 suggest some important properties of triangles. Proofs of these properties will be postponed until Chapter 4.

18. **a.** Construct a triangle with two equal sides. Using a protractor, carefully measure its three angles.

   b. Repeat this process for other triangles with two equal sides.

   c. What do you discover about the angles of such triangles?
   The angles opposite the equal sides are equal.

19. **a.** Construct a triangle with two equal angles. Using a ruler, carefully measure its three sides.

   b. Repeat this process for other triangles with two equal angles.

   c. What do you discover about the sides of such triangles?
   The sides opposite the equal angles are equal.

C 20. **a.** Express your discoveries of Exercises 18 and 19 in "If . . . then" form. If 2 sides of a △ are =, then the ⦞ opp. these sides are =.
   If 2 ⦞ in a △ are =, then the sides opp. these ⦞ are =.

   b. As a pair, the statements can be called __?__. converses

100   Chapter 3

SELF-TEST

Vocabulary

exterior angle (p. 93)
corollary (p. 93)
scalene triangle (p. 97)
isosceles triangle (p. 97)
equilateral triangle (p. 97)
acute triangle (p. 97)

right triangle (p. 97)
obtuse triangle (p. 97)
equiangular triangle (p. 97)
base (p. 97)
legs (p. 97)
hypotenuse (p. 97)

Find the value of x.

1. 40   2. 120   (3-1)

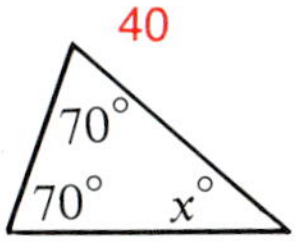

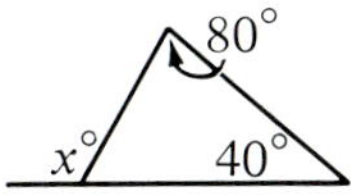

3. 70 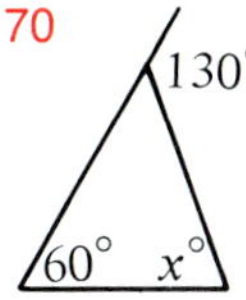   4. 74

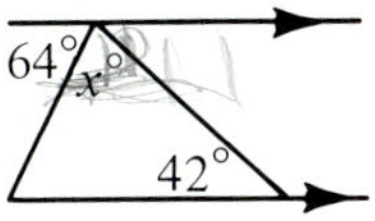

Write a, b, c, or d to identify a triangle of the type named.

5. Scalene triangle c or d
6. Obtuse triangle a
7. Right triangle c
8. Equilateral triangle b

(3-2)

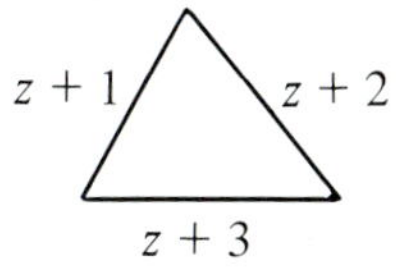

PUZZLE ◆ PROBLEMS

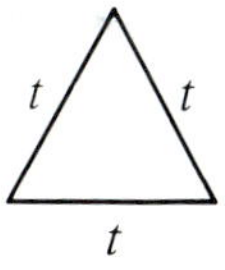

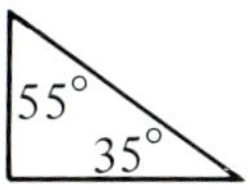

Suppose you don't like obtuse triangles. How can you divide the triangle shown into acute triangles, by using only straight line cuts?

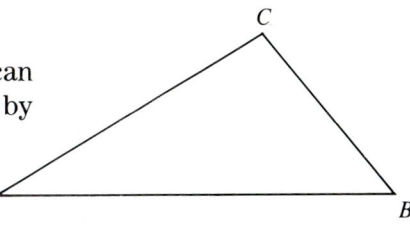

Triangles   **101**

3 Defining Congruent Triangles

When you buy a package of notebook paper, you get sheets that are all alike. Any two sheets have the same size and the same shape. One sheet will fit directly over another sheet. In the language of geometry, one sheet is *congruent* to another sheet.

Likewise △ EFG and △ PQR are congruent. They have the same size and shape.

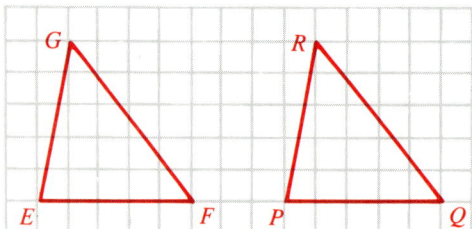

Imagine sliding one triangle over to fit on the other triangle. You put
   point E on point P;   point F on point Q;   point G on point R.

When the vertices are matched in this way:

      ∠ E and ∠ P are called **corresponding angles**;
      $\overline{EG}$ and $\overline{PR}$ are called **corresponding sides.**

We often refer to corresponding angles and corresponding sides as corresponding **parts** of the triangles. In congruent triangles, *corresponding parts are always equal.*

When △ EFG **is congruent to** △ PQR we write

$$\triangle EFG \cong \triangle PQR$$

△ EFG ≅ △ PQR means that all of the following are true:

    ∠ E = ∠ P   ∠ F = ∠ Q   ∠ G = ∠ R
    EF = PQ   FG = QR   EG = PR

Suppose that instead of writing △ EFG ≅ △ PQR you started to write:

$$\triangle GEF \cong$$

Having started with G to name one triangle, you must start with the corresponding letter, R, to name the other triangle. Corresponding parts are named in the same order. When you begin with △ GEF, the complete correct statement is:

$$\triangle GEF \cong \triangle RPQ$$

EXAMPLE The two triangles shown are congruent. Name the
(a) corresponding angles and
(b) corresponding sides.

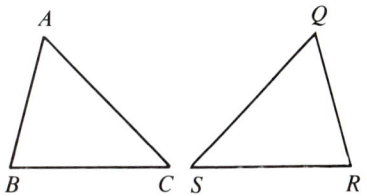

a. *Corresponding angles:*
∠ A and ∠ Q, ∠ B and ∠ R, ∠ C and ∠ S

b. *Corresponding sides:*
$\overline{AB}$ and $\overline{QR}$, $\overline{BC}$ and $\overline{RS}$, $\overline{CA}$ and $\overline{SQ}$

Classroom Practice

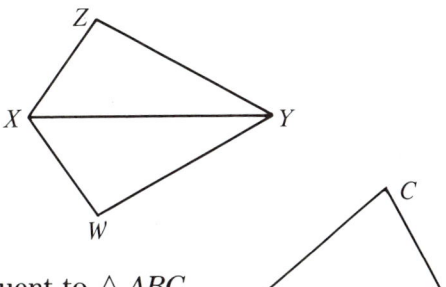

1. Name the six parts of △ XYZ.
   ∠ ZXY, ∠ ZYX, ∠ Z, $\overline{XZ}$, $\overline{ZY}$, $\overline{XY}$

2. Name the six parts of △ XYW.
   ∠ WXY, ∠ WYX, ∠ W, $\overline{WX}$, $\overline{WY}$, $\overline{XY}$

3. Kim drew a triangle that is congruent to △ ABC.
   Is Kim's triangle acute, right, or obtuse? acute

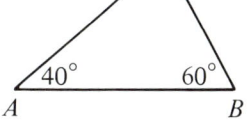

4. Erin drew a triangle that is congruent to △ DEF.
   Is Erin's triangle acute, right, or obtuse? right

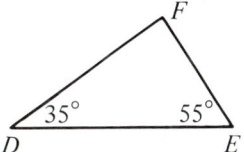

5. Do the triangles appear to be congruent?
   If you can't decide, trace △ ABC and slide
   the tracing over △ XYZ. Does one triangle
   fit exactly over the other? yes; yes

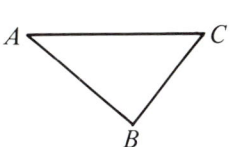 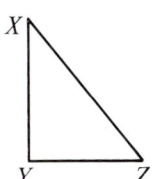

6. Do the triangles appear to be congruent? If you can't decide, trace
   △ DEF and see if it will fit exactly over △ RST. yes

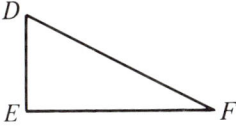 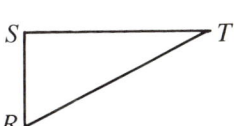

Triangles   **103**

7. Given: △ABC ≅ △RST.

∠R corresponds to __?__. ∠A
∠T corresponds to __?__. ∠C
∠S corresponds to __?__. ∠B

$\overline{RS}$ corresponds to __?__. $\overline{AB}$
$\overline{RT}$ corresponds to __?__. $\overline{AC}$
$\overline{TS}$ corresponds to __?__. $\overline{CB}$

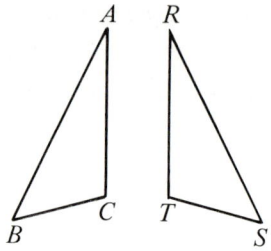

8. Given: △DEF ≅ △VWZ.
State the angle measures and the segment lengths.
∠W = __?__ 90°   ∠V = __?__ 30°   ∠Z = __?__ 60°
DF = __?__ 10   ZW = __?__ 5   WV = __?__ 5√3

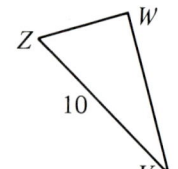

9. Given: △ABC ≅ △RST.
Complete each statement.

a. △ACB ≅ △ __?__ RTS
b. △RST ≅ △ __?__ ABC
c. △BAC ≅ △ __?__ SRT
d. △RTS ≅ △ __?__ ACB
e. △CAB ≅ △ __?__ TRS
f. △TSR ≅ △ __?__ CBA
g. △CBA ≅ △ __?__ TSR
h. △STR ≅ △ __?__ BCA

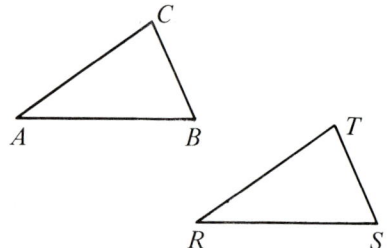

10. Sandy said: "Take any three points X, Y, and Z. Draw $\overline{XY}$, $\overline{YZ}$, and $\overline{XZ}$ and you form a triangle." Was Sandy correct? no

11. Carlos said: "I know a way to pick out three points so you don't get a triangle. Take one point in the middle of the front wall, another point in the ceiling, and a third point on a side wall." Was Carlos correct? no

Written Exercises

In Exercises 1 and 2, copy and complete each statement.

A 1. Suppose you know that △EJM ≅ △IPT.

a. ∠J = __?__ ∠P
b. ∠M = __?__ ∠T
c. ∠E = __?__ ∠I
d. JE = __?__ PI
e. __?__ EJ = IP
f. __?__ JM = PT

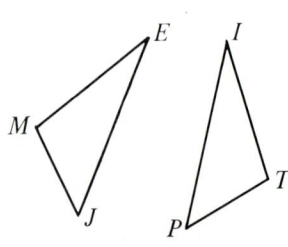

2. The statement △ABC ≅ △DEF can be a true statement only if:

   a. ∠A = __?__ ∠D
   b. __?__ ∠B ≅ ∠E
   c. __?__ ∠C ≅ ∠F
   d. __?__ AB ≅ DE
   e. __?__ AC ≅ DF
   f. BC = __?__ EF

3. Given: △RST ≅ △XYZ. Write six statements, about equal parts, that follow from the definition of congruent triangles. ∠R = ∠X, ∠S = ∠Y, ∠T = ∠Z, RS = XY, RT = XZ, ST = YZ

4. List the six requirements that must be met for the statement △JKM ≅ △PUG to be true. ∠J = ∠P, ∠K = ∠U, ∠M = ∠G, JK = PU, JM = PG, KM = UG

In Exercises 5–8, do the following:

a. Plot points A, B, and C on graph paper. Draw △ABC.

b. Plot points R, E, and W. Draw △REW. (See the Solution Key for diagrams.)

c. If possible, complete the statement △ABC ≅ △__?__.
   Otherwise, write *not congruent*.

SAMPLE

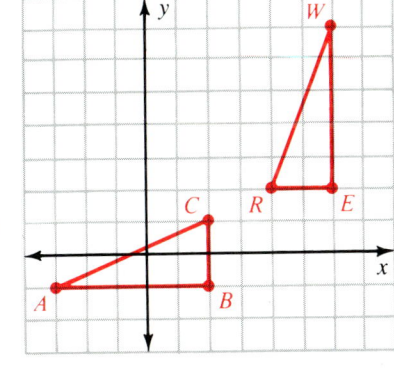

A(−3, −1)
B(2, −1)
C(2, 1)

R(4, 2)
E(6, 2)
W(6, 7)

Answer: △ABC ≅ △WER

	A	B	C	R	E	W
5.	(−3, 1)	(2, 1)	(0, 3)	(4, 2)	(9, 2)	(7, 4)
6.	(−1, 2)	(4, 2)	(2, 4)	(5, −1)	(7, 1)	(10, −1)
7.	(−6, −3)	(−2, −3)	(−4, 0)	(0, 1)	(4, 1)	(2, 2)
8.	(1, 1)	(8, 1)	(4, 3)	(3, −7)	(5, −3)	(3, 0)

5. △ABC ≅ △REW   6. △ABC ≅ △WRE   7. not congruent   8. △ABC ≅ △WRE

Triangles

Basing your decision on the appearance of the figure, judge whether the statement is correctly written.

9. △ ABC ≅ △ OND no
10. △ ABC ≅ △ DON yes
11. △ BAC ≅ △ NOD no
12. △ CAB ≅ △ NDO yes

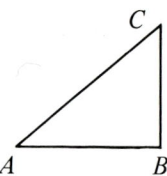 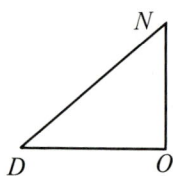

B 13. △ RSJ ≅ △ CNE yes
14. △ JRS ≅ △ ECN yes
15. △ RJS ≅ △ NEC no
16. △ SRJ ≅ △ ENC no

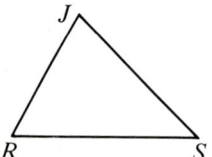 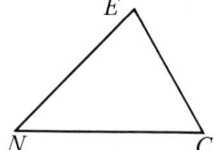

In each exercise it is given that two triangles are congruent. Judge, from the appearance of the figure, which triangles are congruent. List, in pairs, the corresponding parts.

17.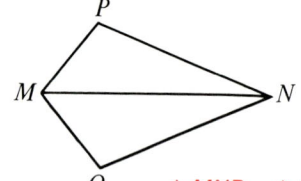

△ MNP ≅ △ MNQ;
∠ PMN = ∠ QMN, ∠ P = ∠ Q, ∠ PNM = ∠ QNM,
MP = MQ, PN = QN, MN = MN

18.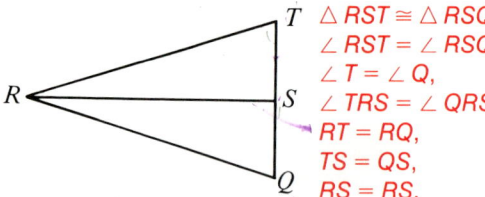

△ RST ≅ △ RSQ;
∠ RST = ∠ RSQ,
∠ T = ∠ Q,
∠ TRS = ∠ QRS,
RT = RQ,
TS = QS,
RS = RS,

19.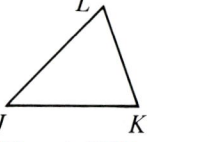

△ JKL ≅ △ RZW,
∠ L = ∠ W, ∠ K = ∠ Z, ∠ J = ∠ R,
LK = WZ, LJ = WR, JK = RZ

20.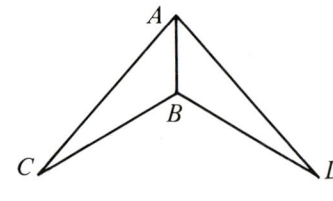

△ ABC ≅ △ ABD;
∠ C = ∠ D,
∠ CAB = ∠ DAB,
∠ CBA = ∠ DBA,
CB = DB,
CA = DA,
AB = AB

21.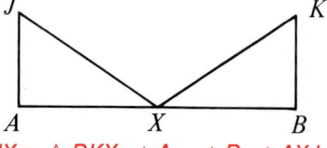

△ AJX ≅ △ BKX; ∠ A = ∠ B, ∠ AXJ = ∠ BXK, ∠ J = ∠ K,
AJ = BK, AX = BX, JX = KX

22.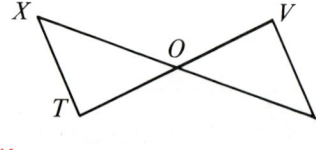

△ XOT ≅ △ YOV;
∠ X = ∠ Y,
∠ T = ∠ V,
∠ XOT = ∠ YOV,
XT = YV
TO = VO
YO = XO

4 The SSS Postulate

Suppose that you are handed the three sticks shown. You fasten the sticks together at the ends to form a triangle.

   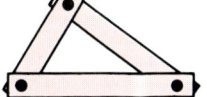

Now suppose someone has a duplicate set of sticks and tries to form a different-looking triangle by joining the sticks at their ends. It is impossible to do this. Only one kind of triangle can be formed from three particular sticks.

The idea suggested by the sticks is a postulate of geometry.

> **POSTULATE 9 (SSS Postulate)**
>
> *If three sides of one triangle are equal to the corresponding parts of another triangle, the triangles are congruent.*

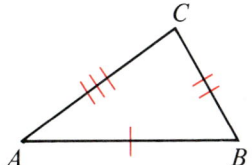 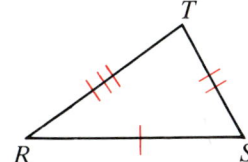

Postulate 9 tells us that $\triangle ABC \cong \triangle RST$.

It follows from the definition of congruent triangles that all six parts of $\triangle ABC$ must be equal to the corresponding parts of $\triangle RST$. Besides the three given pairs of equal sides, we have three pairs of equal angles:

$$\angle A = \angle R \qquad \angle B = \angle S \qquad \angle C = \angle T.$$

Notice in the diagram above that equal sides of the triangles are indicated with marks. $\overline{AB}$ and $\overline{RS}$ each have one mark, $\overline{BC}$ and $\overline{ST}$ each have two, and so on. It is often convenient when writing a proof to mark corresponding sides and angles of congruent triangles.

Suppose someone cannot see that the triangles below are congruent. You can point out:

$AC = RT$ (each length is 10);
$AB = RS$ (each length is 12);
$BC = ST$ (each length is 15).

Then $\triangle ABC \cong \triangle RST$ by the SSS Postulate.

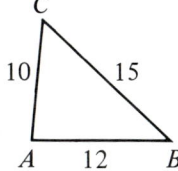 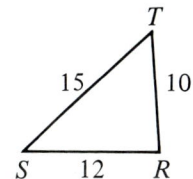

Triangles

Can you prove that $\triangle XAZ \cong \triangle XBZ$? You can do so if you notice that $\overline{XZ}$ is a side of each of the triangles. Recall, from algebra, that $XZ = XZ$. The length of any segment is equal to itself. We use this idea in the following proof.

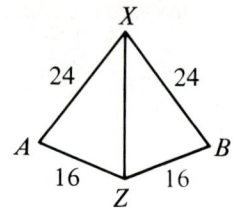

Given: $AX = 24$; $BX = 24$; $AZ = 16$; $BZ = 16$

Prove: $\triangle XAZ \cong \triangle XBZ$

STATEMENTS	REASONS
1. $AX = BX$	1. *Given:* each length is 24.
2. $AZ = BZ$	2. *Given:* each length is 16.
3. $XZ = XZ$	3. From algebra
4. $\triangle XAZ \cong \triangle XBZ$	4. SSS Postulate

Classroom Practice

1. To use the SSS Postulate to prove $\triangle JKV \cong \triangle RNC$, you must first show that:

   $JK = \underset{RN}{\underline{\ ?\ }}$; $KV = \underset{NC}{\underline{\ ?\ }}$; $JV = \underset{RC}{\underline{\ ?\ }}$.

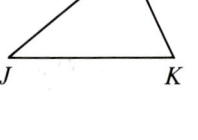
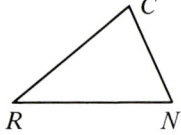

2. To use the SSS Postulate to prove $\triangle ABC \cong \triangle VTS$, you must first show that:

   $AC = \underset{VS}{\underline{\ ?\ }}$; $AB = \underset{VT}{\underline{\ ?\ }}$; $BC = \underset{TS}{\underline{\ ?\ }}$.

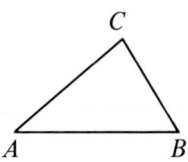
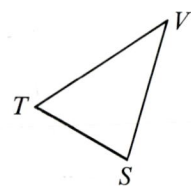

In Exercises 3–8, the goal is to prove that the two triangles are congruent.

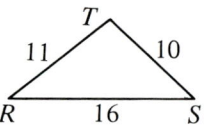

3. Vertex R should be matched with vertex $\underline{\ ?\ }$. H

4. Vertex T should be matched with vertex $\underline{\ ?\ }$. X

5. $\overline{TS}$ and $\underset{\overline{XK}}{\underline{\ ?\ }}$ are corresponding sides.

6. $\underset{\overline{TR}}{\underline{\ ?\ }}$ and $\overline{XH}$ are corresponding sides.

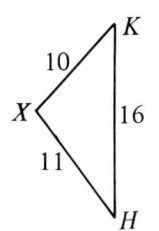

7. Is $\triangle RST \cong \triangle KHX$? no

8. Is $\triangle RTS \cong \triangle HXK$? yes

Written Exercises

Copy what is shown. Complete the proofs by supplying the reasons.

A **1.** **Given:** X is the midpoint of $\overline{AB}$ and $\overline{CD}$.
AC = BD

Prove: △AXC ≅ △BXD

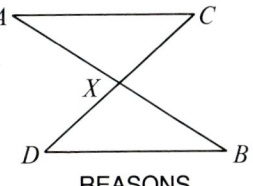

STATEMENTS	REASONS
1. CX = DX	1. *Given:* X is the midpoint of __?__ $\overline{CD}$.
2. AX = BX	2. __?__ *Given:* X is the midpoint of $\overline{AB}$.
3. AC = BD	3. __?__ Given
4. △AXC ≅ △BXD	4. __?__ SSS Postulate

2. **Given:** S is the midpoint of $\overline{TV}$.
TR = VR

Prove: △TSR ≅ △VSR

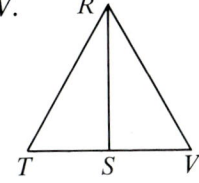

STATEMENTS	REASONS
1. TS = VS	1. __?__ *Given:* S is the midpoint of $\overline{TV}$.
2. RS = RS	2. __?__ From algebra
3. TR = VR	3. __?__ Given
4. △TSR ≅ △VSR	4. __?__ SSS Postulate

3. **Given:** ZY = WX
ZW = YX

Prove: △WZY ≅ △YXW

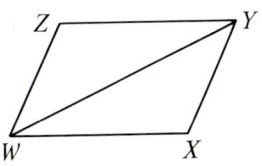

STATEMENTS	REASONS
1. ZY = WX; ZW = YX	1. __?__ Given
2. WY = WY	2. __?__ From algebra
3. △WZY ≅ △YXW	3. __?__ SSS Postulate

Triangles   **109**

Suppose you wish to use the SSS Postulate to prove that $\triangle DEF \cong \triangle JKM$. What value must x have?

4. 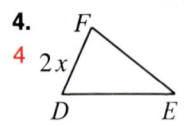 5.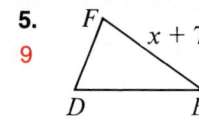
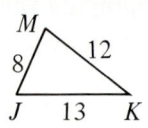 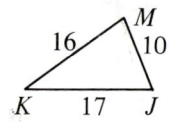

6. 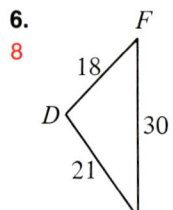 7.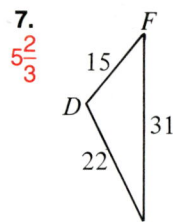
8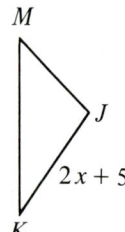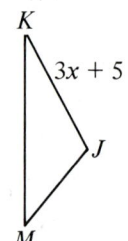

Given: $\triangle ABC$ and $\triangle RST$, with $AB = RS$, $BC = ST$, and $AC = RT$.
Copy and complete each row of angle measures.

	∠A	∠B	∠C	∠R	∠S	∠T
8.	70°	80°	? 30°	? 70°	? 80°	? 30°
9.	100°	? 60°	20°	? 100°	? 60°	? 20°
10.	? 65°	52°	? 63°	? 65°	? 52°	63°
B 11.	? y°	?	x°	y°	?	? x°

11. $\angle B = 180° - (x + y)°$, $\angle S = 180° - (x + y)°$

Draw on your paper three segments roughly like those shown.

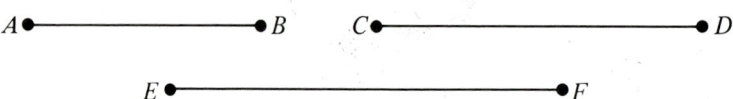

12. Construct a segment $\overline{XY}$ so that $XY = AB$.
    With X as center, draw an arc with radius $= CD$.
    With Y as center, draw an arc with radius $= EF$.
    Use Z to label the point where the arcs intersect.
    Draw $\overline{XZ}$ and $\overline{YZ}$.

13. Construct a segment $\overline{RU}$ so that $RU = AB$.
    With R as center, draw an arc with radius $= EF$.
    With U as center, draw an arc with radius $= CD$.
    Use N to label the point where the arcs intersect.
    Draw $\overline{RN}$ and $\overline{UN}$.

14. a. Refer to Exercises 12 and 13 to complete the statement:

$\triangle XYZ \cong \triangle \underline{\quad ? \quad}$.   URN

  b. What postulate supports your answer?   SSS Postulate

15. a. Plot points $A(-7, 1)$, $B(-3, 1)$, and $C(-6, 3)$ on a graph. Draw $\triangle ABC$.

  b. Plot points $R(2, 0)$ and $S(6, 0)$ on the same graph. Draw $\overline{RS}$.

  c. On your graph, locate a point T so that the statement $\triangle ABC \cong \triangle RST$ will be true. (*Note:* There is more than one correct point. How many can you find?)   There are 2 possible points: $(3, 2)$, $(3, -2)$.

16. a. Plot points $D(2, 0)$, $E(2, 7)$, and $F(5, 7)$ on a graph. Draw $\triangle DEF$.

  b. Plot points $V(-6, 0)$ and $W(-6, 7)$ on the same graph. Draw $\overline{VW}$.

  c. Locate a point K so that the statement $\triangle DEF \cong \triangle VWK$ will be true. (*Note:* There is more than one correct point.)   2 points: $(-3, 7)$, $(-9, 7)$

  d. Locate a point L so that the statement $\triangle DEF \cong \triangle WVL$ will be true. (*Note:* There is more than one correct point.)   2 points: $(-3, 0)$, $(-9, 0)$

17. To strengthen a triangular frame, Myra fastened three additional sticks as shown. How many triangles were formed?   16 triangles

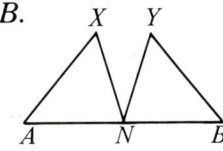

For each exercise, copy what is shown. Then write a complete proof in two-column form.

C **18.** *Given:* N is the midpoint of $\overline{AB}$.
$AX = BY$
$NX = NY$

  Prove: $\triangle AXN \cong \triangle BYN$

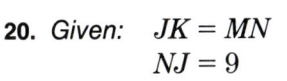

19. *Given:* $RT = RV$
$TS = VS$

  Prove: $\triangle RST \cong \triangle RSV$

20. *Given:* $JK = MN$
$NJ = 9$
$KM = 9$

  Prove: $\triangle NJK \cong \triangle KMN$

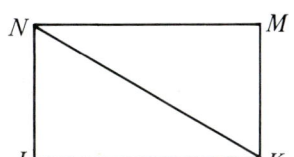

SELF-TEST

Vocabulary

corresponding angles (p. 102)
corresponding sides (p. 102)
congruent (p. 102)

You are given: △ENP ≅ △OBJ. Complete the statements.

1. △NEP ≅ △ _?_ BOJ
2. ∠E = _?_ ∠O
3. OJ = _?_ EP   (3-3)

4. Supply reasons to complete the proof.   (3-4)

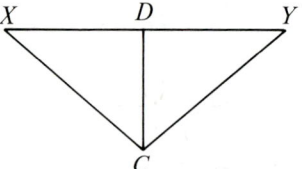

Given: CX = CY
D is the midpoint of $\overline{XY}$.

Prove: △CDX ≅ △CDY

STATEMENTS	REASONS
1. CX = CY	1. _?_ Given
2. CD = CD	2. _?_ From algebra
3. DX = DY	3. _?_ Given: D is the midpoint of $\overline{XY}$.
4. △CDX ≅ △CDY	4. _?_ SSS Postulate

PUZZLE ♦ PROBLEMS

1. Arrange twelve toothpicks as in Figure 1. Now can you rearrange three toothpicks to form three squares of the same size?

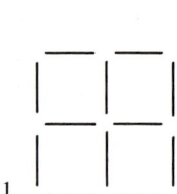

Figure 1

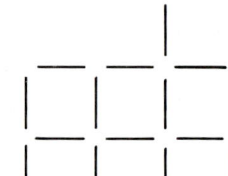

Figure 2

2. In Figure 2, rearrange three toothpicks to form five squares of the same size.

3. Form four equilateral triangles with six toothpicks. (This puzzle is solved in a way completely different from the last two puzzles. Can you see how?) This puzzle is solved in 3 dimensions rather than on a plane.

Applications

RIGIDITY OF TRIANGLES

The SSS Postulate states that if three sides of one triangle are equal to the corresponding sides of another triangle, the triangles are congruent. We have seen that to change the shape of a triangle we must change the length of at least one of its sides. Thus, we say that a triangle is a *rigid* figure.

Architects and builders use the fact that triangles are rigid when they want to build a structure that is strong enough to hold its shape when acted upon by external and internal forces.

Many roofs have a triangular cross-section. In a properly designed triangular roof, all the force is exerted down. None of the force pushes out. As a result, the walls do not need to be reinforced to hold their shape.

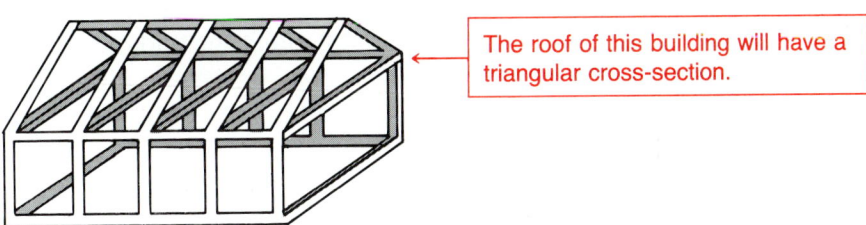

The roof of this building will have a triangular cross-section.

Another example of the use of triangles in construction is the truss bridge. Builders often combine several triangles to make a strong bridge with a wide span. Many covered bridges built in the late 18th century and early 19th century are of truss-type construction.

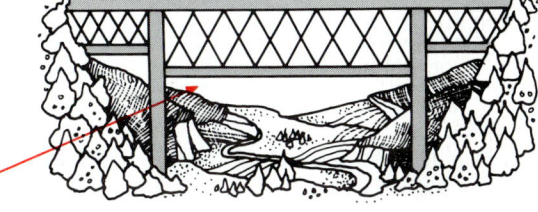

Notice the overlapping triangles in this truss bridge.

Triangles   113

R. Buckminster Fuller used the rigidity of triangles in designing his famous geodesic dome. His design combines many triangular surfaces to form a dome that is strong, but lightweight. Large geodesic domes form pavilions for fairs, sporting events, and horticultural exhibits. Smaller domes make interesting houses.

EXERCISES

Cut some cardboard into 1 cm strips of various lengths. Punch holes in the strips 1 cm from each end. Use metal fasteners to attach the ends and form models like those pictured here.

Which of the models are rigid figures? Models 3 and 6 are rigid figures.

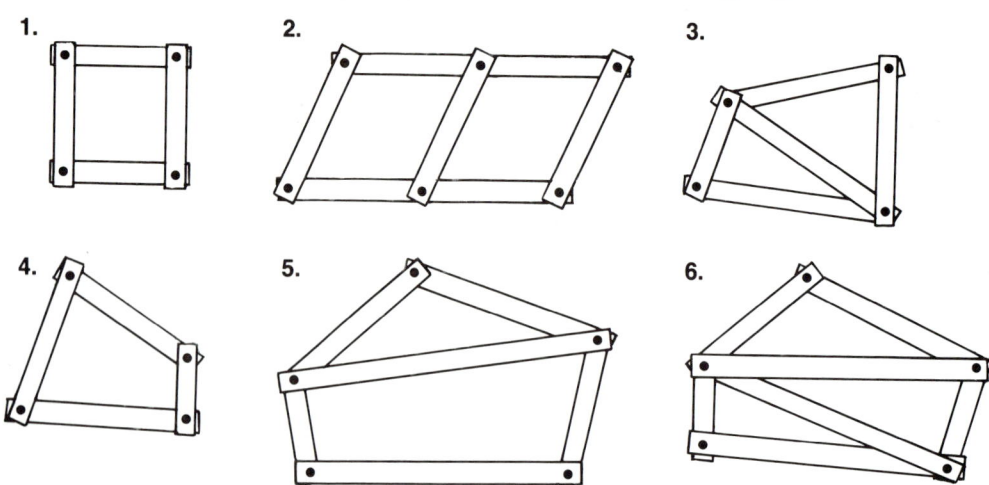

See if you can combine your strips into other models that form rigid figures.

5 The SAS Postulate

Suppose you have two sticks. You fasten them together at a 40° angle.

There is only one way to finish forming a triangle. A third stick must be of just the right length.

Think again of the two sticks. What would happen if the sticks were fastened like this?

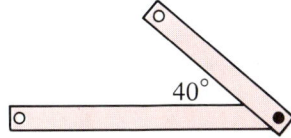

Would the triangle you could form be a different-looking triangle? To decide, you can perform an experiment in your mind.

Flip 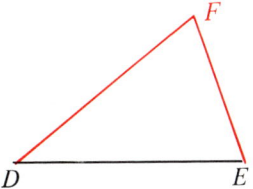 to get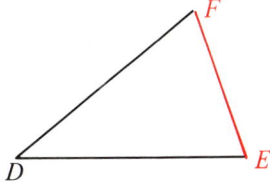

The models are identical.

When two sticks are joined at a certain angle, a third stick must be of just the right length to form a triangle. The triangle can have only one size and shape.

The 40° angle is included between the sides formed by the two sticks. Look at △ DEF.

| ∠ F is included between sides $\overline{FD}$ and $\overline{FE}$. ∠ F is opposite side $\overline{DE}$. | Side $\overline{EF}$ is included between ∠ E and ∠ F. $\overline{EF}$ is opposite ∠ D. |

Triangles 115

> **POSTULATE 10 (SAS Postulate)**
>
> *If two sides and the included angle of one triangle are equal to the corresponding parts of another triangle, the triangles are congruent.*

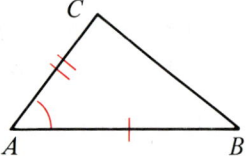 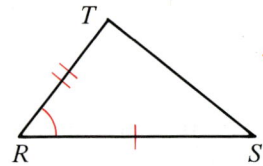

Postulate 10 tells us that △ ABC ≅ △ RST.

Classroom Practice

Complete each statement.

1. ∠ E lies opposite side ___?___ . **$\overline{PN}$**
2. ∠ E is included between sides ___?___ and ___?___ . **$\overline{PE}$**, **$\overline{NE}$**
3. ∠ ___?___ lies opposite side $\overline{PE}$. **N**
4. ∠ ___?___ is included between sides $\overline{PN}$ and $\overline{PE}$. **P**
5. Side $\overline{NE}$ is included between ∠ ___?___ and ∠ ___?___ . **N**, **E**
6. Side ___?___ is included between ∠ P and ∠ N. **$\overline{PN}$**

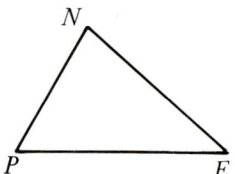

In Exercises 7–12, two triangles are to be proved congruent.

7. Vertex A should be paired with vertex ___?___ . **V**
8. Vertex ___?___ should be paired with vertex J. **C**
9. $\overline{AB}$ and ___?___ are corresponding sides. **$\overline{VP}$**
10. ___?___ and $\overline{VJ}$ are corresponding sides. **$\overline{AC}$**
11. Is △ ACB ≅ △ VJP? **yes**   12. Is △ CAB ≅ △ JVP? **yes**

**In each exercise, two triangles are congruent.
In order to fit one triangle over the other could you:
a. slide the first triangle over the second?
b. flip the first triangle and then slide it over the second?**

13. 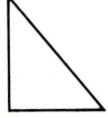 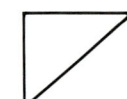 **b**

14. 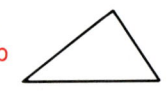 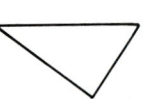 **b**

15. b 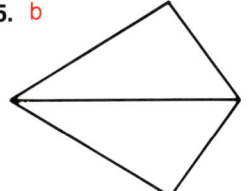 16. b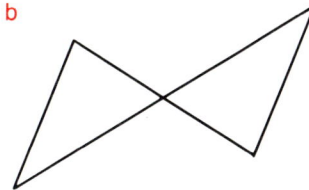

17. b 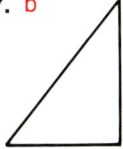 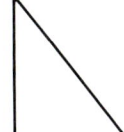 18. a

Written Exercises

In Exercises 1–4, we will prove that $\triangle ABC \cong \triangle JKN$.

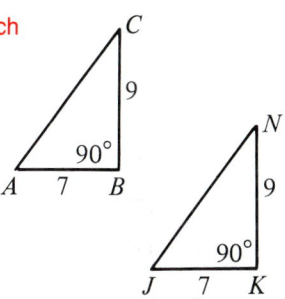

A 1. Is $AB = JK$? Is $BC = KN$? Why? Yes; yes; the lengths of each pair of sides are equal.

2. Is $\angle B = \angle K$? Why? Yes; each is 90°.

3. Do you know that two sides and the included angle of $\triangle ABC$ are equal to the corresponding parts of $\triangle JKN$? yes

4. State the postulate that supports the statement $\triangle ABC \cong \triangle JKN$. SAS Postulate

In Exercises 5–8, we will prove that $\triangle PXQ \cong \triangle RXS$.

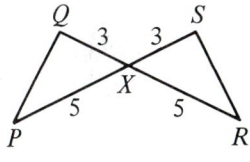

5. In $\triangle PXQ$, what angle is included between sides $\overline{XQ}$ and $\overline{XP}$? In $\triangle RXS$, what angle is included between sides $\overline{XS}$ and $\overline{XR}$? $\angle QXP$, $\angle SXR$ 6. Yes; vert. $\angle$s are =.

6. Is $\angle PXQ = \angle RXS$? Why?   7. Is $QX = SX$? Is $PX = RX$? yes; yes

8. To support the statement $\triangle PXQ \cong \triangle RXS$, would you use the SSS Postulate or the SAS Postulate? SAS Postulate

State whether the SSS Postulate or the SAS Postulate could be used to prove the triangles congruent.

9.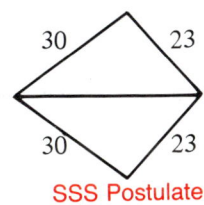
SSS Postulate

10.
SAS Postulate

11.
SAS Postulate

Triangles   **117**

Complete the proofs by supplying the reasons.

12. Given: $\angle A = 40°$
    $\angle R = 40°$
    $AB = RS$
    $AC = RT$
    Prove: $\triangle ABC \cong \triangle RST$

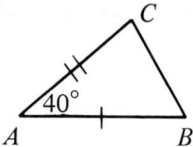

STATEMENTS	REASONS
1. $\angle A = \angle R$	1. *Given:* Each angle has measure __?__. 40°
2. $AB = RS$; $AC = RT$	2. __?__ Given
3. $\triangle ABC \cong \triangle RST$	3. __?__ SAS Postulate

13. Given: $\overrightarrow{ZW}$ bisects $\angle XZY$.
    $XZ = YZ$
    Prove: $\triangle XWZ \cong \triangle YWZ$

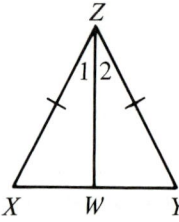

STATEMENTS	REASONS
1. $\angle 1 = \angle 2$	1. __?__ Given: $\overrightarrow{ZW}$ bisects $\angle XZY$.
2. $ZW = ZW$	2. __?__ From algebra
3. $XZ = YZ$	3. __?__ Given
4. $\triangle XWZ \cong \triangle YWZ$	4. __?__ SAS Postulate

Draw, on your paper, an angle and two segments like those shown.

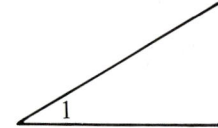
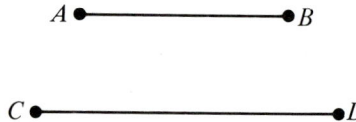

B 14. Construct an angle equal to $\angle 1$. Call it $\angle R$.
    On one side of $\angle R$, construct $\overline{RS}$ so that $RS = AB$.
    On the other side of $\angle R$, construct $\overline{RT}$ so that $RT = CD$.
    Draw $\overline{ST}$.

15. Repeat Exercise 14, using X, Y, and Z instead of R, S, and T.
    What postulate supports the statement: $\triangle RST \cong \triangle XYZ$? SAS Postulate

16. Complete the proof by supplying the reasons.

    Given: X is the midpoint of $\overline{AB}$ and $\overline{CD}$.
    Prove: $\triangle AXC \cong \triangle BXD$

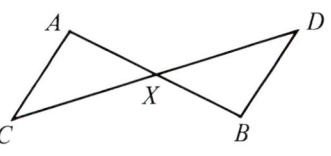

STATEMENTS	REASONS
1. $CX = DX$	1. Given: X is __?__ the midpoint of $\overline{CD}$
2. $AX = BX$	2. __?__ Given: X is the midpoint of $\overline{AB}$.
3. $\angle AXC = \angle BXD$	3. __?__ Vertical angles are equal.
4. $\triangle AXC \cong \triangle BXD$	4. __?__ SAS Postulate

17. Given: $RS = VT$
    $\overline{RS} \parallel \overline{VT}$

    Prove: $\triangle VRS \cong \triangle STV$

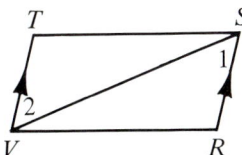

    Copy what is shown.
    Then write a proof in two-column form.

    Suggested strategy:
    A. Use the parallel lines to prove two angles equal.
    B. Use the fact that $\overline{SV}$ is a side of each triangle.
    C. List enough equal parts so that you can use the SSS Postulate or the SAS Postulate.

Copy what is shown. Then write a proof in two-column form.

C 18. Given: $AC = AD$
    $\overrightarrow{AB}$ bisects $\angle CAD$.

    Prove: $\triangle ABC \cong \triangle ABD$

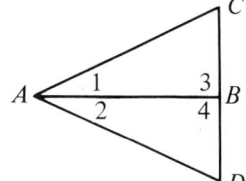

19. Given: $\overline{AB} \perp \overline{CD}$
    B is the midpoint of $\overline{CD}$.

    Prove: $\triangle ABC \cong \triangle ABD$

20. Refer to the figure shown. We wish to prove the following: If the two legs of one right triangle are equal to the two legs of another right triangle, then the triangles are congruent.

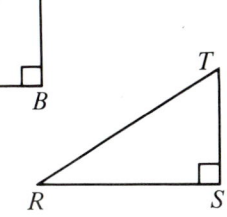

    a. List what is given and what is to be proved, in terms of the diagram. Given: $\triangle ABC$ and $\triangle RST$ are right $\triangle$; $AB = RS$; $BC = ST$.
    b. Write a proof in two-column form. Prove: $\triangle ABC \cong \triangle RST$

Triangles 119

Experiments

Each student needs paper, scissors, a pencil, a ruler, and a protractor.

1. Draw a segment 20 cm long.
2. Choosing either endpoint of the segment, draw a ray so that a 35° angle is formed.
3. From the endpoint of that ray, mark off a segment 15 cm long.
4. Connect the two free endpoints to form a triangle.
5. Shade the triangular region and cut out the triangle.

Now students should compare their models until they agree on answers to the following questions.

a. Are there some pairs of triangles that fit exactly, one over the other, each with the shaded side up?

b. Are there some pairs that fit exactly when the shaded side of one triangle faces up and that of the other triangle faces down?

c. Are there any two models that don't fit at all? Should there be?
All the triangles should fit exactly with the shaded side up.
(SAS Postulate)

CONSUMER APPLICATIONS

DO YOU GET WHAT YOU PAY FOR?

Consumers often have to decide, "Which one should I buy?" For example, you may have to decide among two, three, or a dozen different radios. Several different winter coats may look attractive. You may find a bargain and be able to buy the best product at the lowest price. Often, however, you have to pay more for a better quality item.

Is the extra expense worth it? The answer depends on your own needs, tastes, and financial resources. The radio which costs an extra $15 may produce a much better sound. The camera which costs twice as much may last three times longer. Sometimes, however, extra dollars only buy extra features or frills which are of no use.

Here are some common-sense suggestions for wise purchasing:
(1) Decide exactly what you want and how long it must last.
(2) Talk with people who have bought a similar item.
(3) Consult a trade magazine or a consumer magazine.
(4) Compare prices in several different stores before you buy.

6 The ASA Postulate

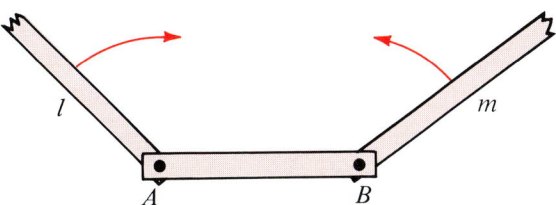

The diagram shows three sticks. The fasteners at *A* and *B* allow sticks *l* and *m* to turn.

1. Stick *l* is rotated until ∠ *A* = 40°.

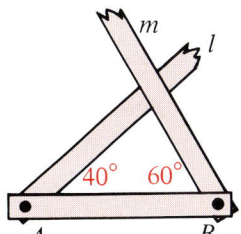

2. Stick *m* is rotated until ∠ *B* = 60°.

3. The extra wood is sawed off and a fastener is attached at *C*.

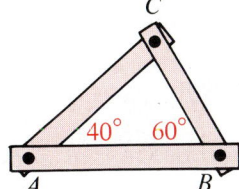

Do you see that for a particular length *AB*, a 40° angle at *A*, and a 60° angle at *B*, only one kind of triangle is possible?

POSTULATE 11 (ASA Postulate)

If two angles and the included side of one triangle are equal to the corresponding parts of another triangle, the triangles are congruent.

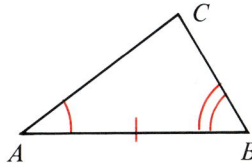
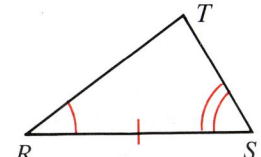

Postulate 11 tells us that △ *ABC* ≅ △ *RST*.

Triangles   **121**

Classroom Practice

Name the side included between the two angles.

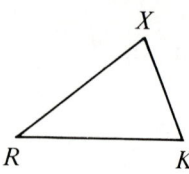 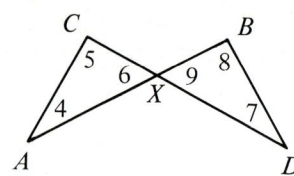 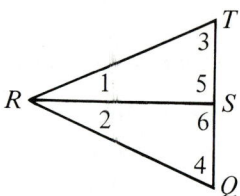

1. ∠R and ∠K $\overline{RK}$
2. ∠X and ∠R $\overline{XR}$
3. ∠5 and ∠6 $\overline{CX}$
4. ∠7 and ∠8 $\overline{BD}$
5. ∠1 and ∠5 $\overline{RS}$
6. ∠6 and ∠4 $\overline{SQ}$

Suppose you know that ∠K = ∠X and KL = XO.

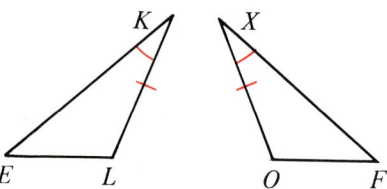

7. To use the ASA Postulate to prove that △KEL ≅ △XFO, you must also show that ___?___. ∠L = ∠O

8. To use the SAS Postulate to prove that △KEL ≅ △XFO, you must also show that ___?___. KE = XF

9. Given: $\overrightarrow{RS}$ bisects ∠XRY.
   S is the midpoint of $\overline{XY}$.
   $\overline{RS} \perp \overline{XY}$

   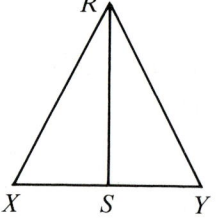

   Could you prove that △RSX ≅ △RSY
   a. by using the SSS Postulate? no
   b. by using the SAS Postulate? yes
   c. by using the ASA Postulate? yes

Written Exercises

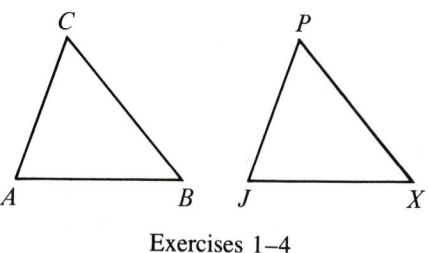

Exercises 1–4

A
1. If ∠A = ∠J, ∠B = ∠X, and AB = JX, then you can use the ASA Postulate to conclude △ABC ≅ △___?___. JXP

2. You intend to use the ASA Postulate to prove △ABC ≅ △JXP. You have stated that ∠C = ∠P and ∠B = ∠X. You also need the statement $\dfrac{?}{CB} = \dfrac{?}{PX}$.

3. You intend to use the SAS Postulate to prove △ABC ≅ △JXP. You have stated that ∠A = ∠J and AB = JX. You also need the statement $\dfrac{?}{CA} = \dfrac{?}{PJ}$.

4. You intend to use the ASA Postulate to prove △ABC ≅ △JXP. You have stated that AC = JP and ∠C = ∠P. You also need the statement __?__ = __?__.
   ∠A   ∠J

In Exercises 5–6, state the reasons needed to complete the proofs.

5. Given: ∠D = ∠R; ∠F = ∠T
         DF = 8; RT = 8

   Prove: △DEF ≅ △RST

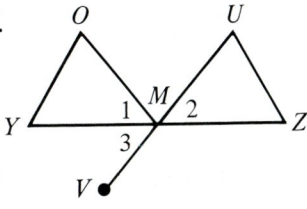

STATEMENTS	REASONS
1. ∠D = ∠R; ∠F = ∠T	1. __?__ Given
2. DF = RT	2. *Given:* each length __?__ equals 8
3. △DEF ≅ △RST	3. __?__ ASA Postulate

6. Given: M is the midpoint of $\overline{YZ}$.
         $\overrightarrow{MY}$ bisects ∠OMV.
         ∠Y = ∠Z

   Prove: △YOM ≅ △ZUM

STATEMENTS	REASONS
1. ∠1 = ∠3	1. __?__ *Given:* $\overrightarrow{MY}$ bisects ∠OMV.
2. ∠2 = ∠3	2. __?__ Vertical angles are equal.
3. ∠1 = ∠2	3. __?__ Substitution Postulate
4. YM = ZM	4. __?__ *Given:* M is the midpoint of $\overline{YZ}$.
5. ∠Y = ∠Z	5. __?__ Given
6. △YOM ≅ △ZUM	6. __?__ ASA Postulate

Draw, on your paper, two angles and a segment roughly like those shown. Use your drawings for Exercises 7–9 on page 124.

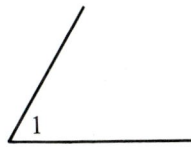
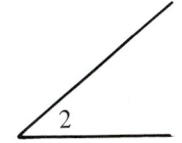
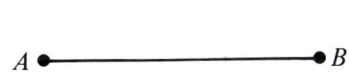

Triangles

For Exercises 7–9, use your drawings from page 123.

B 7. Construct a segment $\overline{RS}$ so that $RS = AB$.
Using $\overrightarrow{RS}$ as one side, construct $\angle R = \angle 1$.
Using $\overrightarrow{SR}$ as one side, construct $\angle S = \angle 2$.
Use T to label the point where two rays intersect.

8. Construct a segment $\overline{EZ}$ so that $EZ = AB$.
Using $\overrightarrow{EZ}$ as one side, construct $\angle E = \angle 2$.
Using $\overrightarrow{ZE}$ as one side, construct $\angle Z = \angle 1$.
Use N to label the point where two rays intersect.

9. **a.** Exercises 7 and 8 show that $\triangle RTS \cong \triangle \underline{\ ?\ }$. ZNE
   b. What postulate supports your answer? ASA Postulate

Copy what is shown. Write complete proofs in two-column form.

10. Given: $AC = AD$
    $BC = BD$

    Prove: $\triangle ACB \cong \triangle ADB$

11. Given: $CB = DB$
    $\angle ABC = \angle ABD$

    Prove: $\triangle ACB \cong \triangle ADB$

C 12. Given: $\overleftrightarrow{AB}$ bisects $\angle CAD$.
    $\overleftrightarrow{AB}$ bisects $\angle CBD$.

    Prove: $\triangle ACB \cong \triangle ADB$

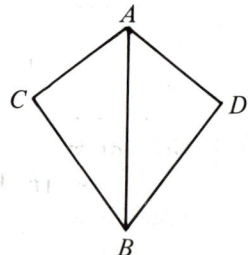

Exercises 10–13

13. Given: $AC = AD$
    $\overrightarrow{AB}$ bisects $\angle CAD$.

    Prove: $\triangle ACB \cong \triangle ADB$

14. Given: $\overline{RT} \parallel \overline{VS}$
    $RT = VS$

    Prove: $\triangle RMT \cong \triangle SMV$

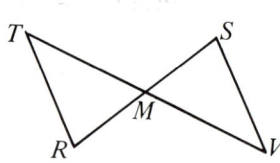

Exercise 14

15. **a.** Draw a large right triangle, $\triangle ABC$, with $\angle B = 90°$.
    b. Use Construction 2, page 26, to construct a 90° angle. Label it $\angle K$.
    c. Construct $KL = BC$.
    d. Construct $\angle L = \angle C$. Label the third vertex as J.
    e. Explain why $\triangle ABC \cong \triangle JKL$. ASA Postulate

SELF-TEST

Write proofs in two-column form.

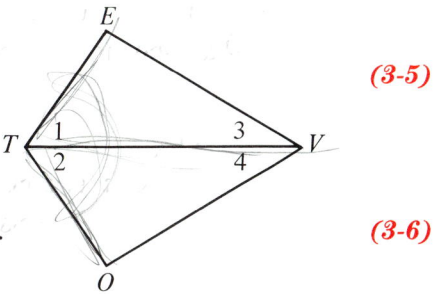

1. Given: $\overrightarrow{TV}$ bisects $\angle ETO$.
   $TE = TO$

   Prove: $\triangle TEV \cong \triangle TOV$ *(3-5)*

2. Given: $\overleftrightarrow{TV}$ bisects both $\angle ETO$ and $\angle EVO$.

   Prove: $\triangle TEV \cong \triangle TOV$ *(3-6)*

CAREER NOTES

REALTOR

MARSTON, 8 rm. older home in excel. cond., 2 full baths, porch, walking distance to shopping and bus, $108,900. Call 555-6789.

Do you ever read the newspaper advertisements for houses? Chances are that this ad was placed by a realtor.

Realtors examine the property to help determine the value of the real estate. They collect information about tax rates, insurance, and financing. A realtor often has to calculate percentages when working with any of these figures.

If the property is residential, realtors find the locations of schools, public transportation, and nearby shopping areas. For commercial property, the realtor may have to determine the price of office rental space. This is often based upon the number of square feet of floor space available. When his or her research is complete, the realtor uses advertisements to locate buyers or renters.

All realtors must obtain licenses by passing written tests. Then they may represent people who wish to sell or rent their property. Realtors may obtain part or all of their income from a commission based upon a percentage of their dollar sales.

Triangles

7 The AAS and HL Theorems

In the diagram below, two angles and a *non*-included side of △ RST are equal to the corresponding parts of △ XYZ. Can we prove that the triangles are congruent?

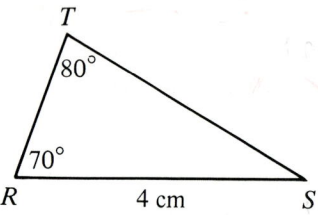 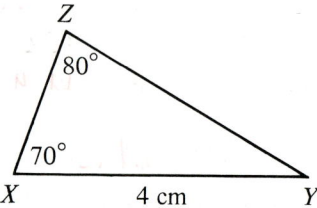

If we use the fact that ∠ R + ∠ S + ∠ T = 180°, we can show that ∠ S = 30°. We can likewise show that ∠ Y = 30°. Then ∠ S = ∠ Y, and △ RST ≅ △ XYZ by the ASA Postulate.

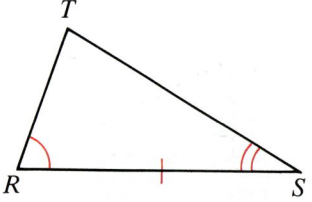 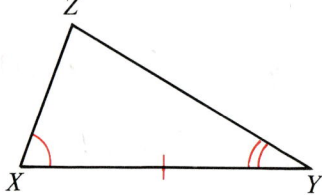

The discussion above suggests our next theorem. You will be asked to prove this theorem in Exercise 30, page 131.

THEOREM 2 (AAS Theorem)

If two angles and a non-included side of one triangle are equal to the corresponding parts of another triangle, the triangles are congruent.

A theorem about right triangles is stated here without proof.

THEOREM 3 (HL Theorem)

If the hypotenuse and a leg of one right triangle are equal to the corresponding parts of another right triangle, the triangles are congruent.

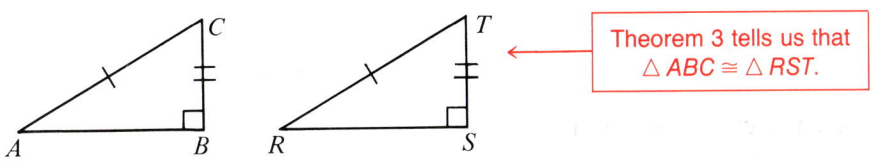

WAYS TO PROVE TRIANGLES CONGRUENT				
SSS	SAS	ASA	AAS	HL

Classroom Practice

What is the measure of the angle marked in color?

SAMPLE  Answer: $180° - c° - d°$
Also correct: $180° - (c° + d°)$

1. 58° 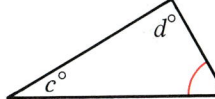   **2.** 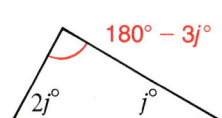   **3.**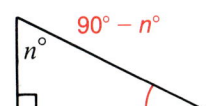

In Exercises 4–7, identify all right triangles shown in each figure. Name the hypotenuse of each right triangle.

4. 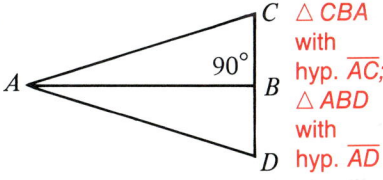 △CBA with hyp. $\overline{AC}$; △ABD with hyp. $\overline{AD}$

5. 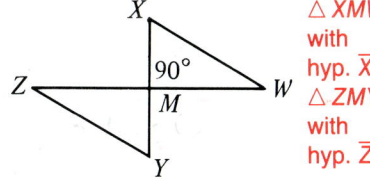 △XMW with hyp. $\overline{XW}$; △ZMY with hyp. $\overline{ZY}$

6. 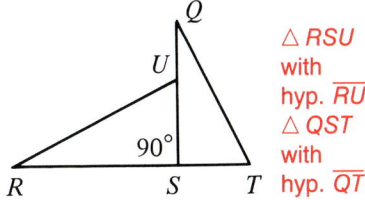 △RSU with hyp. $\overline{RU}$; △QST with hyp. $\overline{QT}$

7. 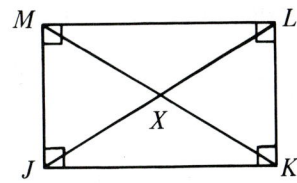 △MJK with hyp. $\overline{MK}$; △JKL with hyp. $\overline{JL}$; △KLM with hyp. $\overline{MK}$; △LMJ with hyp. $\overline{JL}$

In Exercises 8–17, state which postulate or theorem you would use to prove the two triangles congruent. If no method applies, say so.

8.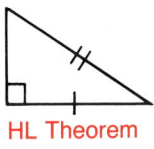
HL Theorem

9.
ASA Postulate

Triangles   **127**

State the postulate or theorem you would use to prove the two triangles congruent. If no method applies, say so.

10. 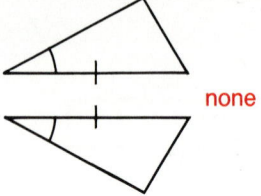 none

11. 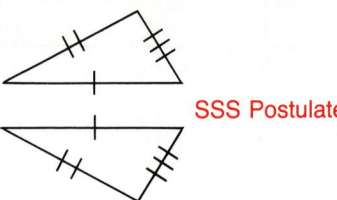 SSS Postulate

12.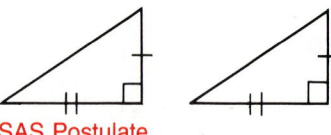
SAS Postulate

13.

AAS Theorem

14.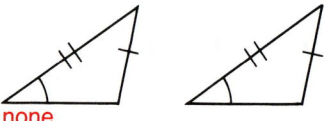
none

15.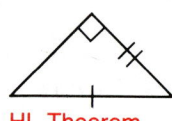
HL Theorem

16.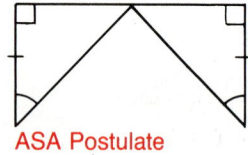
ASA Postulate

17.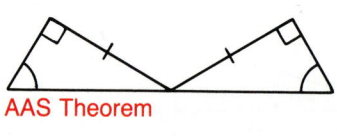
AAS Theorem

Written Exercises

Write SSS, SAS, ASA, AAS, or HL to state the method needed to prove the two triangles congruent. If no method applies, write *none*.

A **1.** HL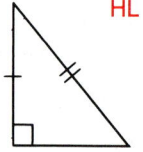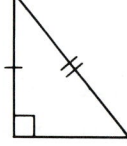

2. none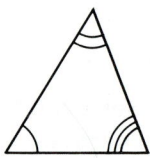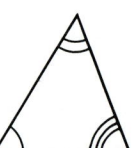

3. ASA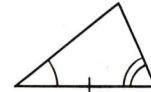

4. AAS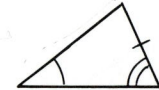

5. HL (or AAS)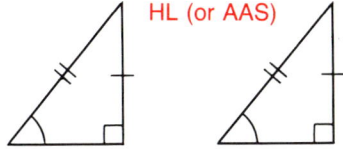
6. SAS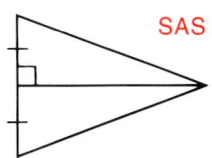
7. AAS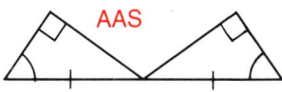
8. HL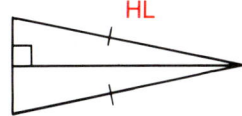
9. SSS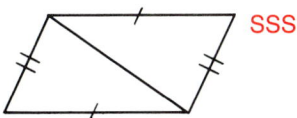
10. ASA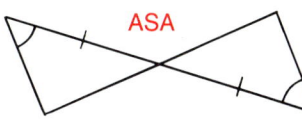
11. none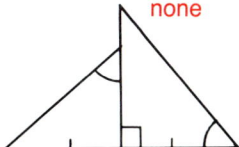
12. AAS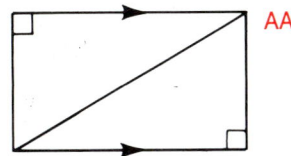

In Exercises 13 and 14, list *all* possible methods that could be used to prove the two triangles congruent.

13.
SAS, ASA, AAS

14.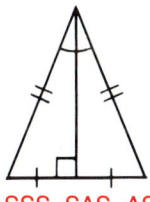
SSS, SAS, ASA, AAS, HL

In Exercises 15–17, supply the reasons to complete the proofs.

15. *Given:* $\angle B$ and $\angle X$ are right angles.
$BY = AX$

Prove: $\triangle ABY \cong \triangle YXA$

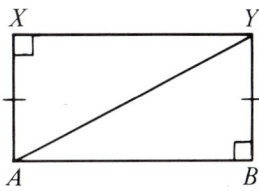

STATEMENTS	REASONS
1. $\angle B$ and $\angle X$ are right angles.	1. Given
2. $AY = AY$	2. From algebra
3. $BY = AX$	3. Given
4. $\triangle ABY \cong \triangle YXA$	4. HL Theorem

Triangles 129

16. Given: $\overrightarrow{BD}$ bisects $\angle ABC$.
$\angle A = \angle C$

Prove: $\triangle ABD \cong \triangle CBD$

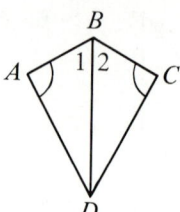

STATEMENTS	REASONS
1. $\angle 1 = \angle 2$	1. __?__ Given: $\overrightarrow{BD}$ bisects $\angle ABC$.
2. $\angle A = \angle C$	2. __?__ Given
3. $BD = BD$	3. __?__ From algebra
4. $\triangle ABD \cong \triangle CBD$	4. __?__ AAS Theorem

17. Given: $\angle 1 = \angle 2$
$\angle 5 = \angle 6$

Prove: $\triangle WXY \cong \triangle WXZ$

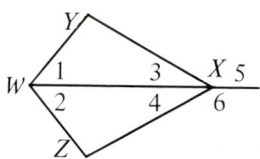

STATEMENTS	REASONS
1. $\angle 3$ and $\angle 5$ are supplements. $\angle 4$ and $\angle 6$ are supplements.	1. __?__ Def. of supplementary angles
2. $\angle 5 = \angle 6$	2. __?__ Given
3. $\angle 3 = \angle 4$	3. __?__ Supplements of $=\angle$s are $=$.
4. $\angle 1 = \angle 2$	4. __?__ Given
5. $WX = WX$	5. __?__ From algebra
6. $\triangle WXY \cong \triangle WXZ$	6. __?__ ASA Postulate

**Each diagram shows two congruent triangles.
Find the values of x, y, and z.**

18.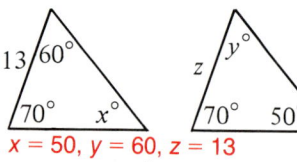
$x = 50, y = 60, z = 13$

19.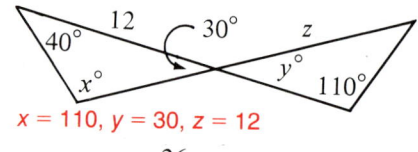
$x = 110, y = 30, z = 12$

B 20.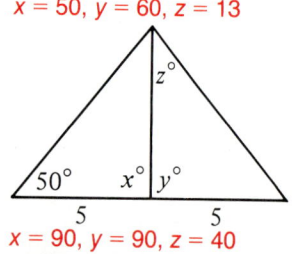
$x = 90, y = 90, z = 40$

21.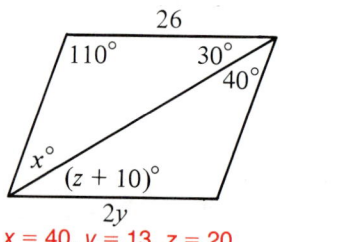
$x = 40, y = 13, z = 20$

A geometry class is asked to write this proof:

Given: $\overline{AB} \parallel \overline{XY}$
M is the midpoint of $\overline{AY}$.

Prove: $\triangle AMB \cong \triangle YMX$

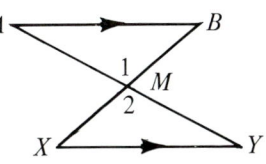

22. Jason uses the ASA Postulate. Name the three parts of $\triangle AMB$ that he uses in his proof. ∠A, ∠1, $\overline{MA}$

23. Jasmine writes ∠1 = ∠2, and then decides to use the AAS Theorem. List the two parts of $\triangle YMX$, in addition to ∠2, that she uses in her proof. ∠X, $\overline{YM}$

24. Miguel uses the AAS Theorem, but doesn't use ∠1. List the three parts of $\triangle AMB$ that he uses. ∠A, ∠B, $\overline{MA}$

25. Leora remarks: "If I knew that M was the midpoint of $\overline{BX}$, then I could use something other than the ASA Postulate or the AAS Theorem." What theorem or postulate could she use? SAS Postulate

26. Given: M is the midpoint of $\overline{YZ}$.
M is the midpoint of $\overline{AB}$.
∠Y and ∠Z are right angles.

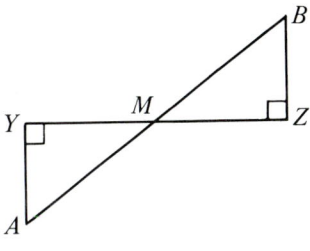

a. List four methods you could use to prove
$\triangle AYM \cong \triangle BZM$. HL, SAS, ASA, AAS

b. Choose one method and write a complete proof.
Answers may vary.

In each exercise, equal parts are indicated. Can the two triangles be proved congruent? If so, what postulate or theorem would be simplest to use?

27.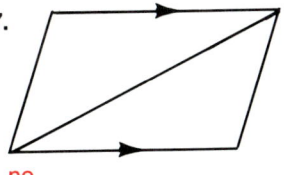
no

28.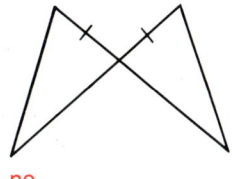
no

29.
yes; SAS Postulate

 30. Prove the AAS Theorem.

Given: ∠R = ∠X; ∠T = ∠Z
RS = XY

Prove: $\triangle RST \cong \triangle XYZ$
(*Hint:* Let ∠R = j° and ∠T = k°.
Show that ∠S = ∠Y.)

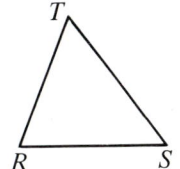 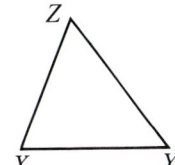

Triangles **131**

Experiments

Each student needs paper, scissors, a pencil, a ruler, and a compass.

1. Draw a segment 8 cm long.
2. Choosing either endpoint of the segment, construct a ray so that a 90° angle is formed.
3. Set your compass for a radius of 12 cm. Using the free endpoint of the original segment as center, draw an arc intersecting the ray.
4. Connect the two points to form a triangle.
5. Cut out your triangle and compare it with triangles that other students have made. If each student works carefully, all the triangles should be congruent.

SELF-TEST

1. Given: $\overline{AC} \parallel \overline{DB}$; $AX = BX$; $CX = DX$. (3-7)

   Indicate, by abbreviation, three different postulates or theorems you could use to prove $\triangle AXC \cong \triangle BXD$. **ASA, SAS, AAS**

   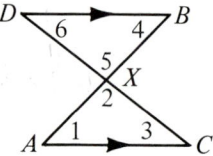

2. Write a proof in two-column form.

   Given: $\overline{RT} \perp \overline{SV}$
   $RS = TV$
   $RM = TM$

   Prove: $\triangle RMS \cong \triangle TMV$

   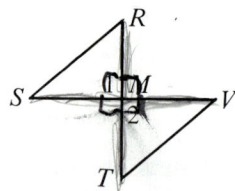

PUZZLE ◆ PROBLEMS

Sid and Gregory were hiking in the woods when they came to the Red Hawk River. Sid challenged Gregory to guess the width of the river. This is what Gregory did:

He stood at B and looked out across the river. He then adjusted his cap until the tip of his visor was in line with his eye and point X. Keeping his neck stiff, he turned and noted the point Z, on the ground, that was in line with his eye and the tip of the visor. By pacing, he found that the distance BZ was about 12 m. He claimed, "The river is about 12 meters wide." Explain why Gregory was right.

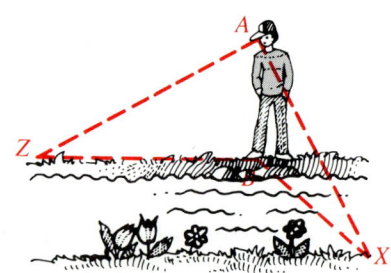

$\triangle AZB \cong \triangle AXB$ since $AB = AB$, $\angle ZAB = \angle XAB$, and $\angle ABZ = \angle ABX$ (ASA Post.).
Thus $BX = ZX = 12$ m.

READING GEOMETRY

SYMBOLS

The symbols used in this book tell you a lot about geometry without using many words. You find symbols in the explanations, in theorems and proofs, in exercises, and even in diagrams. In order to get the message, you must pay just as much attention to the symbols as you do to the words of the text. See the list following the Table of Contents if you have forgotten the meaning of any symbol.

Look at the diagram at the right. The symbols $\overline{AB}$, $\overleftrightarrow{AB}$, $\overrightarrow{AB}$, $\overrightarrow{BA}$, and AB can all be used in referring to this simple diagram. Each symbol has a different meaning. Can you tell what each one means? (The answers are printed upside down at the bottom of the page.)

Symbols can be parts of diagrams. Examples are the right-angle mark ⌐, the little marks, often in color, that tell you lines or angles are equal, and the heavy arrowheads used to show that lines are parallel. Look for such symbols as you study the lessons.

1. No; ∠ABC has vertex B and ∠BAC has vertex A.
2. No; the vertices must be listed in corresponding order: △MNO ≅ △PQR.

EXERCISES

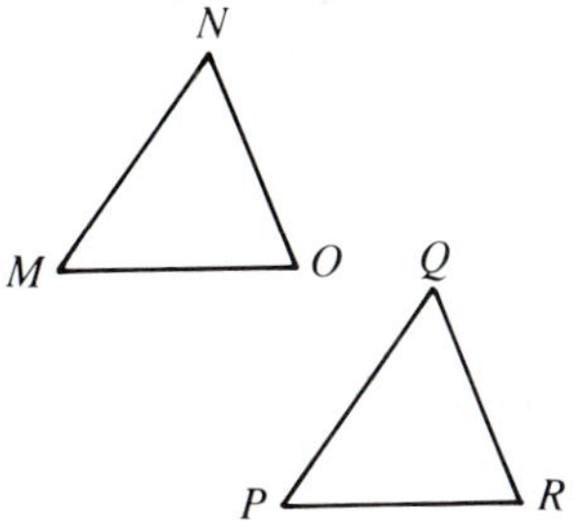

1. Do the symbols ∠ABC and ∠BAC stand for the same angle? Why or why not?

2. The triangles shown at the right are congruent. Is "△MNO ≅ △QRP" a correct way of saying this in symbols? Why or why not?

3. How would you mark the diagram at the right to show that the triangles are congruent by the SAS Postulate?

4. Using a different inequality symbol, write an expression with the same meaning as "∠FGH > ∠JKL".   ∠JKL < ∠FGH

5. On page 72, Postulates 7 and 8 are stated in words and then written in a short form that uses diagrams and symbols. Do Exercises 7–9 on page 74 if you have not already done so.

In order as above: segment with endpoints A and B; line through points A and B; ray with endpoint A through point B; ray with endpoint B through point A; length of segment AB

Triangles   133

COMPUTER ACTIVITIES

COMPLEMENTARY AND SUPPLEMENTARY ANGLES

In the exercises that follow, you must find the complement or supplement of many angles. A computer is an excellent tool for doing repeated or tiresome calculations. If you know what kinds of calculations are needed to solve a problem, you can write a computer program for it. The program is a plan of the procedure to be followed in solving the problem. It functions as a set of instructions for the computer. The programs in this book are written in BASIC, one of many computer programming languages.

You may already know some BASIC, or you may learn some in these computer activities. But you do not need to master BASIC or programming to be able to use a computer successfully. Every day millions of people who know little or nothing about computer languages or programming use computers at school, at home, or on the job.

The program below is a very simple one. You are asked to complete it in the Exercises. The purpose is to compute the measure of the complement of a given angle. Note that each line has a line number. This tells the computer the order in which its instructions are to be followed. The INPUT statement in line 40 allows the person using the computer to supply information. Statements beginning with REM are ignored by the computer. Their purpose is to explain the program itself.

```
10   REM ∗∗∗ ANGLE PROBLEMS
20   PRINT "WHAT IS THE ANGLE"
30   PRINT "(TYPE 0 WHEN DONE)";
40   INPUT A
50   IF A = 0 THEN 200
60   REM ∗∗∗ FIGURING COMPLEMENT
70   C = __?__     90 − A
80   REM ∗∗∗ CHECK FOR IMPOSSIBLE ANSWERS
90   REM ∗∗∗ AND PRINT RESULTS
100  PRINT "ANGLE:   "; A
110  IF A > = 90 THEN 140
120  PRINT "COMPLEMENT:   "; C
130  GOTO 10
140  PRINT "NO COMPLEMENT EXISTS"
150  GOTO 10
200  END
```

EXERCISES

1. Line 70 is not complete. After the equals sign, what expression would you use to represent the complement of angle A? Use this expression to complete line 70 when you type it into the computer. 90 − A

2. RUN the program to find the complement of each angle. All numbers must be entered into the computer as decimals. The computer will continue until you signal you've finished by giving 0, an impossible angle measure.
   - a. 22°   b. 46°   c. 90°   d. 78.9°   e. $12\frac{1}{2}°$ f. 150°
     68°   44°   none   11.1°   77.5°   none

3. Using trial and error (trying many values until you find the correct one), RUN the program to find an angle satisfying the given condition.
   - a. The angle measures 20° more than its complement.   55°
   - b. The angle measures 15° less than its complement.   37.5°
   - c. The measure of the angle is twice the measure of its complement.   60°
   - d. The measure of the angle is the same as that of its complement.   45°

If you want the program to figure the supplement also, you need to add to the program. Type in the following lines to expand the program. The computer will put these lines in order by line number, no matter what order you have typed them in.

```
74   REM ***FIGURING SUPPLEMENT
78   S = __?__   180 − A
105  IF A > = 180 THEN 135
107  PRINT "SUPPLEMENT:   ";S
135  PRINT "NO SUPPLEMENT EXISTS"
```

4. Complete line 78 with an expression for the supplement of an angle whose measure is A.   180 − A

5. RUN the expanded program. Find an angle satisfying the condition.
   - a. The angle is equal to its supplement.   90°
   - b. The angle is triple its supplement.   135°
   - c. The supplement of the angle is triple the complement of the angle.   45°
   - d. The supplement of the angle is 20° more than twice the complement of the angle.   20°

Triangles   **135**

SKILLS REVIEW

WHOLE NUMBERS AND FRACTIONS

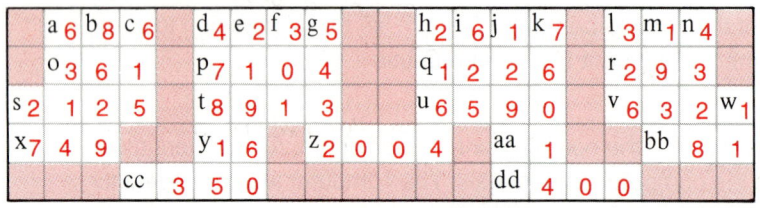

box aa is 3
box bb is 6

Copy the pattern above on squared paper. Then work the exercises and fill in the blanks just as you would in a crossword puzzle.

Across

- **a.** $905 - 219$
- **o.** $(19)^2$
- **s.** 25×85
- **x.** $5992 \div 8$
- **bb.** 3×227
- **d.** 5×847
- **p.** 96×74
- **t.** $995 + 7918$
- **y.** 2^4
- **cc.** $1833 + 2517$
- **h.** $6421 - 3804$
- **q.** $709 + 517$
- **u.** $7309 - 719$
- **z.** $2823 - 819$
- **dd.** 200×47
- **l.** $2826 \div 9$
- **r.** $1172 \div 4$
- **v.** 21×301
- **aa.** $2139 \div 69$

Down

- **a.** $5406 + 908$
- **e.** 72×305
- **i.** $2500 \div 4$
- **m.** $(44)^2$
- **b.** $9303 - 674$
- **f.** $3913 \div 13$
- **j.** $5347 + 7592$
- **n.** $8001 - 3673$
- **c.** 15×41
- **g.** 8×679
- **k.** 618×123
- **s.** 3^3
- **d.** $53{,}913 - 6098$
- **h.** $968 + 1196$
- **l.** $49{,}552 \div 152$
- **w.** $528 \div 48$

Simplify.

1. $\frac{7}{8} + \frac{5}{8}$ $1\frac{1}{2}$
2. $\frac{9}{16} - \frac{5}{16}$ $\frac{1}{4}$
3. $12 - \frac{5}{6}$ $11\frac{1}{6}$
4. $\frac{7}{16} + \frac{1}{2}$ $\frac{15}{16}$
5. $\frac{11}{12} - \frac{3}{4}$ $\frac{1}{6}$
6. $8\frac{5}{9} - 2\frac{1}{3}$ $6\frac{2}{9}$
7. $3\frac{2}{5} + 4\frac{3}{10}$ $7\frac{7}{10}$
8. $4\frac{3}{8} - 2\frac{1}{2}$ $1\frac{7}{8}$
9. $\frac{5}{6} \times \frac{2}{3}$ $\frac{5}{9}$
10. $8 \div \frac{2}{7}$ 28
11. $\frac{3}{4} \div \frac{9}{10}$ $\frac{5}{6}$
12. $\frac{1}{3} \times \frac{4}{5} \times \frac{3}{8}$ $\frac{1}{10}$
13. $10\frac{2}{3} \div \frac{1}{3}$ 32
14. $4 \times 2\frac{5}{16}$ $9\frac{1}{4}$
15. $24 \div 1\frac{1}{5}$ 20
16. $3\frac{3}{4} \times \frac{1}{10} \times \frac{2}{9}$ $\frac{1}{12}$
17. $6\frac{1}{4} - \frac{7}{8}$ $5\frac{3}{8}$
18. $\frac{9}{16} \div 1\frac{1}{2}$ $\frac{3}{8}$
19. $3\frac{1}{7} \times 56$ 176
20. $8\frac{1}{2} + \frac{9}{10}$ $9\frac{2}{5}$
21. $\frac{5}{12} \div \frac{5}{8}$ $\frac{2}{3}$
22. $\frac{19}{20} + \frac{4}{5}$ $1\frac{3}{4}$
23. $9\frac{5}{6} - 1\frac{2}{3}$ $8\frac{1}{6}$
24. $\frac{1}{6} \times 1\frac{3}{5} \times \frac{3}{16}$ $\frac{1}{20}$

CHAPTER REVIEW

CHAPTER SUMMARY

1. **a.** In any triangle, the sum of the angles is 180°.
   b. In any triangle, an exterior angle is equal to the sum of the two opposite angles.

2. Corollaries, like theorems, are proved. A corollary follows directly from a theorem.

3. Triangles can be classified as scalene, isosceles, or equilateral according to their sides.

4. Triangles can be classified as acute, right, obtuse, or equiangular according to their angles.

5. $\triangle ABC \cong \triangle DEF$ means that each part of $\triangle ABC$ is equal to the corresponding part of $\triangle DEF$. Corresponding parts are named in the same order. That is, $\angle A = \angle D$, $\angle B = \angle E$, $\angle C = \angle F$, $AB = DE$, $BC = EF$, and $AC = DF$.

6. Postulates and theorems used to prove triangles congruent are abbreviated by: SSS SAS ASA AAS HL

REVIEW EXERCISES

Find the value of x in each diagram. *(See pp. 92–96.)*

1. 80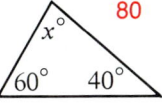

2. 70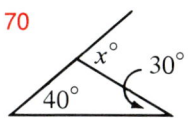

3. 35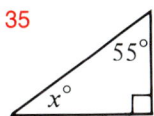

4. 132

5. 15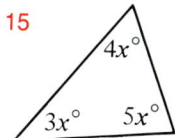

6. 40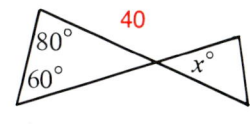

Complete. *(See pp. 97–100.)*

7. If the perimeter of an equilateral triangle is 42 cm, the length of each side is __?__ cm. 14

8. The base of an isosceles triangle is 8 cm long, and a leg is 7 cm long. The perimeter of the triangle is __?__ cm. 22

9. An isosceles triangle cannot be ___?___ scalene.
   scalene/equilateral

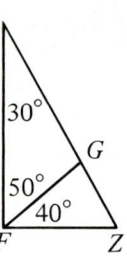

10. A right triangle found in the diagram is ___?___. △ SFZ
11. An acute triangle found in the diagram is ___?___. △ GFZ
12. An obtuse triangle found in the diagram is ___?___. △ SFG

Complete. *(See pp. 102–106.)*

If △ CAP ≅ △ DIR, then:

13. ∠ P = ∠ ___?___ R   14. PC = ___?___ RD   15. △ PCA ≅ △ ___?___ RDI

Write SSS, SAS, ASA, AAS, or HL to indicate the method you would use to prove the two triangles congruent. Do not write the proof.
(See pp. 107–132.)

16. Given: AY = BY and AX = BX
    Prove: △ AXY ≅ △ BXY   SSS

17. Given: ∠ C = ∠ X and CO = XO
    Prove: △ BOC ≅ △ YOX   ASA

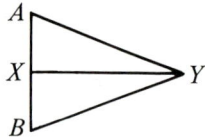
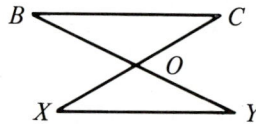

18. Given: $\overline{AB} \perp \overline{CD}$ and AC = AD
    Prove: △ ABC ≅ △ ABD   HL

19. Given: ∠ S and ∠ V are rt. ∠s.
    M is the midpoint of $\overline{SV}$.
    RS = TV
    Prove: △ RSM ≅ △ TVM   SAS

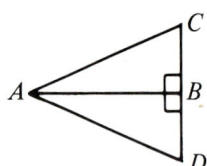
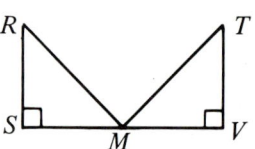

20. Write a complete proof in two-column form.
    Given: ∠ R = ∠ S and $\overrightarrow{GH}$ bisects ∠ RGS.
    Prove: △ RHG ≅ △ SHG

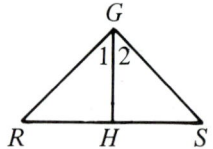

CHAPTER TEST

9. If 3 sides of one △ are = to the corr. parts of another △, the ▲ are ≅.

1. If $x + 3$, $2x - 6$, and $4x + 1$ are the measures of the angles of a triangle, find the numerical measure of the largest angle. 105° *(3-1)*

2. Find the measures of ∠1 and ∠2. ∠1 = 140°, ∠2 = 70°

3. Write the property that you used in Exercise 2 to find ∠1. An ext. ∠ of a △ is = to the sum of the 2 opp. ▲ of the △.

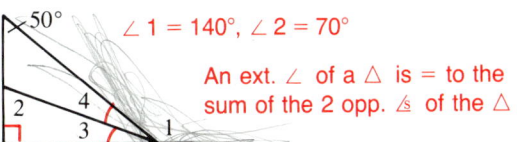

4. Name a scalene right triangle. △ABD *(3-2)*

5. Name an obtuse triangle. △ACD

6. Name the base of an isosceles triangle. $\overline{AC}$

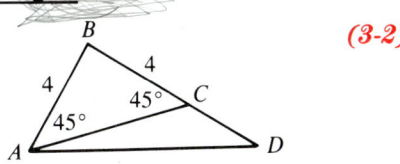

Given: △FLO ≅ △REW **7.** ∠F = ∠R, ∠L = ∠E, ∠O = ∠W, FL = RE, LO = EW, OF = WR

7. Write six statements about equal parts that must be true. *(3-3)*

8. Complete: △EWR ≅ △ __?__ . △LOF

9. Write the SSS Postulate. (See above.) *(3-4)*

10. To prove that △WXY ≅ △WZY by the SSS Postulate, what given information is needed? WX = WZ and XY = ZY

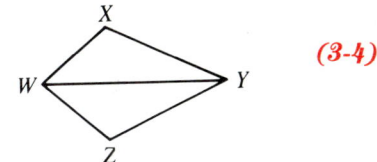

11. Name two pairs of equal sides. $\overline{LQ}$ and $\overline{MQ}$, $\overline{TQ}$ and $\overline{RQ}$ *(3-5)*

12. Name a pair of equal angles and explain why they must be equal. ∠LQT and ∠MQR Vert. ▲ are =.

13. Name two congruent triangles and tell why they must be congruent. ∠LQT and ∠MQR; SAS Post.

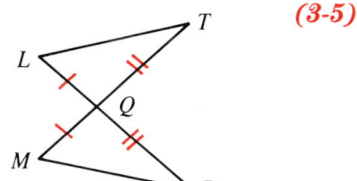

14. Since $\overline{EF} \parallel \overline{GH}$, ∠ __?__ = ∠ __?__   1, 2 *(3-6)*

15. Since $\overline{FG} \parallel \overline{HE}$, ∠ __?__ = ∠ __?__   3, 4

16. Write a reason to support your answers to Exercises 14 and 15. If 2 ∥ lines are cut by a trans., then alt. int. ▲ are =.

17. By the __?__ Postulate, △EFG ≅ △ __?__ . ASA, GHE

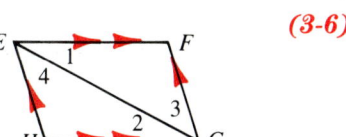

In Exercises 18–20, list *all* methods that could be used to prove the two triangles congruent. If no method exists, write "none."

18.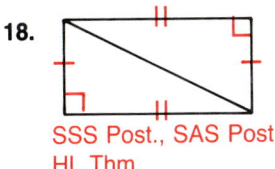
SSS Post., SAS Post., HL Thm.

19.
ASA Post., AAS Thm.

20.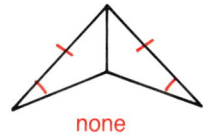
none

(3-7)

Triangles   **139**

MIXED REVIEW

Consider the statement "Perpendicular lines form right angles."

1. Write the statement in "If ... then" form. If 2 lines are ⊥, then they form right ⚞.
2. Write the converse of the statement.
   If 2 lines form right ⚞, then they are ⊥.
3. Supply the reasons to complete the proof.

   Given: $AR = BS$; $\angle 1 = \angle 2$; $\angle A = \angle B$
   Prove: $\triangle ACR \cong \triangle BCS$

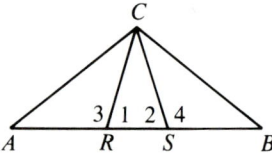

STATEMENTS	REASONS
1. $\angle 1 = \angle 2$	1. __?__ Given
2. $\angle 3 = \angle 4$	2. __?__ Supplements of = ⚞ are =.
3. $AR = BS$; $\angle A = \angle B$	3. __?__ Given
4. $\triangle ACR \cong \triangle BCS$	4. __?__ ASA Postulate

4. If $\angle 1 = x°$, give the measure of a complement of $\angle 1$ in terms of x. $(90 - x)°$

Suppose B is the midpoint of $\overline{AC}$ and C is the midpoint of $\overline{BD}$.

5. Explain why $AB = BC$ and $BC = CD$. Def. of midpoint
6. Which postulate justifies the statement $AB = CD$? substitution
7. Which postulate justifies the statement $AC = BD$? addition
8. If $\overrightarrow{OJ}$ bisects $\angle AOB$, $\angle AOJ = (2x - 3)°$, and $\angle AOB = (3x + 10)°$, then $x =$ __?__ and the numerical measure of $\angle BOJ =$ __?__ °.
   16   29
9. If A has coordinate -5 and C has coordinate 1, find the coordinate of D. 4
10. Does the given information tell you that $r \parallel s$?
    yes **a.** $\angle 1 + \angle 2 = 180°$ **b.** $\angle 3 = \angle 4$ yes
    c. $\angle 1 = \angle 2$ no   **d.** $r \perp t$ and $s \perp t$ yes
11. Is it possible for four points to have the stated relationships? Draw a diagram to support each answer.
    yes **a.** The points are collinear.   **b.** The points are not coplanar. yes
    c. The points are coplanar but not collinear. yes

12. Draw a line l and a point A not on l. Construct a line m that is parallel to l and contains point A. Construction 5

13. Refer to your construction in Exercise 12. Write the statement that supports the claim that $l \parallel m$. If 2 lines and a trans. form = corr. ∠s, then the lines are ∥.

Find the values of x and y.

14. 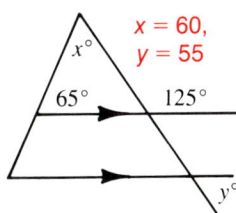 $x = 60$, $y = 55$

15. 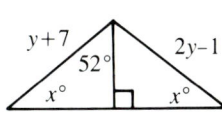 $x = 38$, $y = 8$

16. 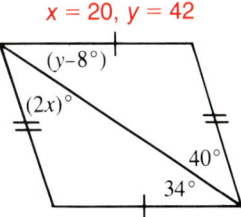 $x = 20$, $y = 42$

Suppose that $\triangle IJK \cong \triangle RST$. Which statements *must* be true?

17. $IK = TR$ true

18. $\triangle KJI \cong \triangle TSR$ true

19. $\angle K = \angle S$ false

Write SSS, SAS, ASA, AAS, or HL to indicate the method you would use to prove two triangles congruent.

20. 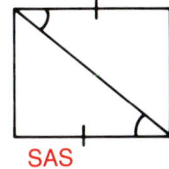 SAS

21. 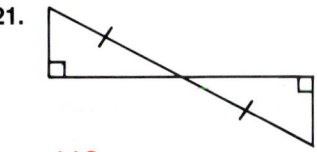 AAS

22. 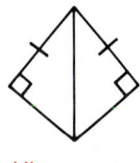 HL

23. Draw a large $\triangle PQR$. Then use the SSS Postulate to construct a triangle congruent to $\triangle PQR$.

24. Construct a right angle and its bisector.

25. If two parallel lines are cut by a transversal and two same-side interior angles are equal, what is the measure of each? 90°

26. If $\triangle ABC$ is an acute triangle, then each exterior angle of $\triangle ABC$ is a(n) __?__ angle. obtuse

27. True or false? If line j lies in plane P, line k lies in plane Q, and $j \parallel k$, then plane $P \parallel$ plane Q. false

28. Write a two-column proof.
    Given: X is the midpoint of $\overline{AC}$ and of $\overline{BD}$.
    Prove: $\triangle AXB \cong \triangle CXD$

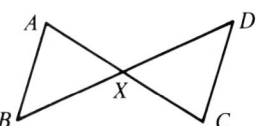

Triangles 141

Here's what you'll learn in this chapter:

To use congruent triangles to prove that two segments or two angles are equal.

To use congruent triangles to justify constructions.

To name the altitudes and the medians of a triangle.

To inscribe a circle in a triangle and circumscribe a circle about a triangle.

To use the properties of isosceles triangles.

Patchwork is made from many small pieces of material sewed together to form patterns. Notice the congruent triangles in the patchwork quilt shown above.

Chapter 4

Using Congruent Triangles

1 Proving Corresponding Parts Equal

Suppose you are told that $\triangle ABC \cong \triangle RST$. Then you know that six things must be true. You should be able to complete the statements below:

$\angle A = \angle R \quad\quad AB = RS$
$\angle B = \underline{\ ?\ } \quad\quad \angle S\,BC = \underline{\ ?\ }\ ST$
$\angle C = \underline{\ ?\ } \quad\quad \angle T\,AC = \underline{\ ?\ }\ RT$

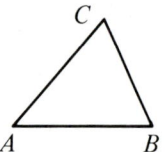
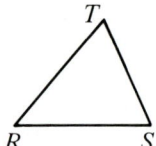

The six statements are true because of the definition of congruent triangles. *Corresponding parts of congruent triangles are equal.*

A STRATEGY FOR PROVING THAT TWO SEGMENTS OR TWO ANGLES ARE EQUAL

1. *Find two triangles in which the two sides or the two angles are corresponding parts.*
2. *Prove that the two triangles are congruent.*
3. *State that the two parts are equal, using as the reason, "Corr. parts of $\cong \triangle$ are $=$."*

This strategy will be used in the examples that follow.

EXAMPLE 1

Given: $JP = JQ$
$\quad\quad\quad PK = QK$

Prove: $\angle P = \angle Q$

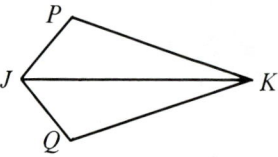

STATEMENTS	REASONS
1. $JP = JQ$	1. Given
2. $PK = QK$	2. Given
3. $JK = JK$	3. From algebra
4. $\triangle PJK \cong \triangle QJK$	4. SSS Postulate
5. $\angle P = \angle Q$	5. Corr. parts of $\cong \triangle$ are $=$.

EXAMPLE 2

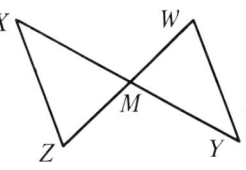

Given: M is the midpoint of $\overline{XY}$.
M is the midpoint of $\overline{ZW}$.

Prove: $XZ = YW$

STATEMENTS	REASONS
1. $XM = YM$	1. *Given:* M is the midpoint of $\overline{XY}$.
2. $ZM = WM$	2. *Given:* M is the midpoint of $\overline{ZW}$.
3. $\angle XMZ = \angle YMW$	3. __?__ Vertical angles are =.
4. $\triangle XMZ \cong \triangle YMW$	4. SAS Postulate
5. $XZ = YW$	5. Corr. parts of $\cong$ △ are =.

Classroom Practice

In Exercises 1–8, it is known that $\triangle ABC \cong \triangle JTN$.

1. $\angle A =$ __?__ $\angle J$ 　　2. $\angle B =$ __?__ $\angle T$ 　　3. $\angle C =$ __?__ $\angle N$

4. What reason supports each of statements 1, 2, and 3 above? Corr. parts of $\cong$ △ are =.

5. $AB =$ __?__ JT 　　6. $BC =$ __?__ TN 　　7. $AC =$ __?__ JN

8. What reason supports each of statements 5, 6, and 7 above? Corr. parts of $\cong$ △ are =.

Suppose, in Exercises 9–14, that you want to prove that $\angle 1 = \angle 2$. Does the figure suggest that congruent triangles might be used?

9.
yes
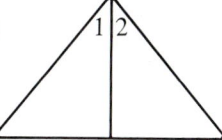

10.
yes
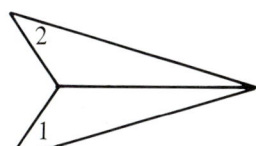

11.
yes
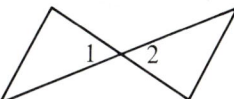

12.
yes
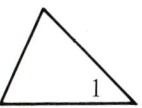 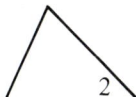

13.
yes

14.
no
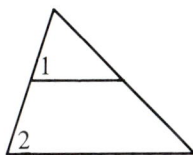

Using Congruent Triangles   **145**

Written Exercises

1–6. *Given:* △ HIJ ≅ △ UVW

Write the six statements that can be supported by the reason:
Corr. parts of ≅ △ are =.

∠ H = ∠ U, ∠ I = ∠ V, ∠ J = ∠ W, HI = UV, IJ = VW, HJ = UW

Name the triangles you might try to prove congruent in order to prove each of the following.

7. NM = NO   △ MEN ≅ △ OLN
8. EO = LM   △ OEL ≅ △ MLE
9. ∠ 1 = ∠ 2   △ OEL ≅ △ MLE
10. EN = LN   △ ENM ≅ △ LNO

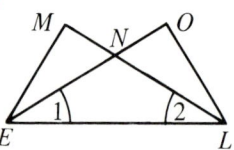

In Exercises 11–14, supply the reasons to complete the proofs.

11. *Given:* AC = BC
    $\overrightarrow{CD}$ bisects ∠ ACB.

    Prove: AD = BD

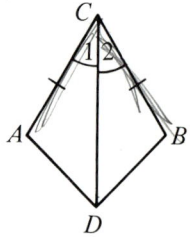

STATEMENTS	REASONS
1. AC = BC	1. __?__ Given
2. ∠ 1 = ∠ 2	2. __?__ Given: $\overrightarrow{CD}$ bisects ∠ ACB
3. CD = CD	3. __?__ From algebra
4. △ ACD ≅ △ BCD	4. __?__ SAS Postulate
5. AD = BD	5. __?__ Corr. parts of ≅ △ are =.

12. *Given:* ∠ O and ∠ E are right angles.
    RO = ES

    Prove: $\overline{RO} \parallel \overline{ES}$

    Strategy for proof:

    A. Prove that two triangles are congruent.
    B. Think of a transversal. Use corresponding parts to show that angles are equal.
    C. Use the equal angles to prove that two lines are parallel.

STATEMENTS	REASONS
1. $\angle O$ and $\angle E$ are right angles.	1. __?__ Given
2. $RO = ES$	2. __?__ Given
3. $RS = RS$	3. __?__ From algebra
4. $\triangle ROS \cong \triangle SER$	4. __?__ HL Theorem
5. $\angle 1 = \angle 2$	5. __?__ Corr. parts of $\cong$ △ are =.
6. $\overline{RO} \parallel \overline{ES}$	6. __?__ If 2 lines and a trans. form = alt. int. ∡, then the lines are $\parallel$.

 13. Given: $\angle A = \angle B$
$\angle 1 = \angle 2$

Prove: $AX = BY$

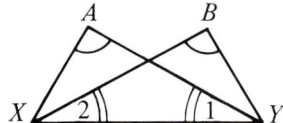

STATEMENTS	REASONS
1. $\angle A = \angle B$; $\angle 1 = \angle 2$	1. __?__ Given
2. $XY = XY$	2. __?__ From algebra
3. $\triangle AXY \cong \triangle BYX$	3. __?__ AAS Theorem
4. $AX = BY$	4. __?__ Corr. parts of $\cong$ △ are =.

14. Given: $\overline{EB} \perp \overline{AC}$
$\angle E = \angle C$
$EB = BC$

Prove: $AB = BD$

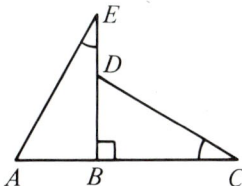

STATEMENTS	REASONS
1. $\angle EBA$ and $\angle DBC$ are right angles.	1. __?__ Given: $\overline{EB} \perp \overline{AC}$
2. $\angle EBA = \angle DBC$	2. __?__ Substitution Postulate
3. $\angle E = \angle C$; $EB = BC$	3. __?__ Given
4. $\triangle EBA \cong \triangle CBD$	4. __?__ ASA Postulate
5. $AB = BD$	5. __?__ Corr. parts of $\cong$ △ are =.

Using Congruent Triangles

Copy what is shown. Then write a proof in two-column form.

15. Given: $AX = BX$; $\angle A = \angle B$
    Prove: $CX = DX$

16. Given: $AX = BX$; $\angle C = \angle D$
    Prove: $AC = BD$

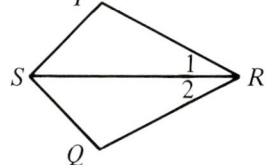

17. Given: $RT = RQ$; $ST = SQ$
    Prove: $\angle T = \angle Q$

18. Given: $\angle 1 = \angle 2$; $RT = RQ$
    Prove: $ST = SQ$

C 19. Given: $\overline{JK} \perp \overline{MN}$; $MJ = NJ$
    Prove: $MK = NK$

20. Given: $\overline{JK} \perp \overline{MN}$; $\angle 3 = \angle 4$
    Prove: $MJ = NJ$

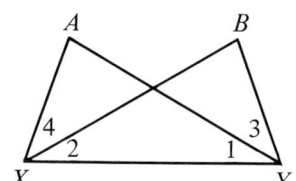

21. Given: $\angle 1 = \angle 2$; $AY = BX$
    Prove: $\angle A = \angle B$

22. Given: $\angle 1 = \angle 2$; $\angle 3 = \angle 4$
    Prove: $AX = BY$

CONSUMER APPLICATIONS

THE CONSUMER PRICE INDEX

The Consumer Price Index (CPI) is a number which keeps track of inflation. It takes into account the cost of food, clothing, housing, transportation, medical care, and other goods and services used by consumers. The CPI compares the *total price now* for a selection of typical consumer purchases to the *former price* during a base year.

If the CPI is 175, then the current total price of a selection of typical consumer purchases is 175% of the total price during the base year.

Consult a newspaper or your library to find out the current CPI. What year is used as the base year? Get a copy of your family's last grocery bill. How much would the bill have been during the base year?

2 Congruent Triangles and Constructions

In this section, we shall use congruent triangles to show that our construction methods are correct.

Bisecting an Angle (Construction 1, page 25)

To show that $\angle AOB$ really is bisected, draw $\overline{XP}$ and $\overline{YP}$. Then:

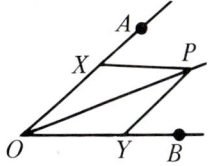

1. $OX = OY$ By construction
2. $XP = YP$ By construction
3. $OP = OP$ From algebra
4. $\triangle XOP \cong \triangle YOP$ SSS Postulate
5. $\angle XOP = \angle YOP$ Corr. parts of $\cong \triangle$ are =.

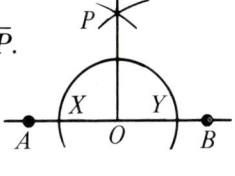

Therefore, $\angle AOB$ is bisected.

Constructing a Perpendicular at a Point on a Line
(Construction 2, page 26)

To show that $\overrightarrow{OP}$ really is perpendicular to $\overleftrightarrow{AB}$, draw $\overline{XP}$ and $\overline{YP}$. Use the SSS Postulate to show that $\triangle XOP \cong \triangle YOP$. Then you know that $\angle XOP = \angle YOP$. You can show that $\angle XOP$ and $\angle YOP$ are right angles (see Written Exercise 8). Therefore, $\overrightarrow{OP} \perp \overleftrightarrow{AB}$.

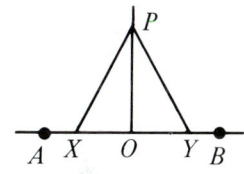

After a construction has been proved to be correct, the construction may be used in proofs. See, for example, Step 1 of Theorem 3 on page 167.

Classroom Practice

Elka started out, in the usual way, to bisect $\angle AOB$. But Elka likes to experiment. So she set her compass for a good-sized radius, used X and Y as centers, and drew arcs to get point P as shown. Finally she drew $\overline{OP}$ and claimed: $\overline{OP}$ bisects $\angle AOB$.

1. Was Elka's statement correct? yes
   Explain. (See the Solution Key.)

2. How could Elka add to her diagram, without using a compass, so that she would clearly show a bisector of $\angle AOB$? Draw $\overrightarrow{PO}$

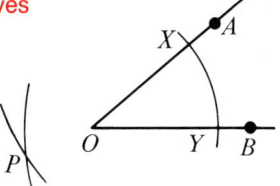
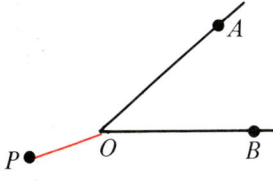

Using Congruent Triangles

Written Exercises

On your paper, draw an acute $\triangle XYZ$ roughly like, but much larger than, the one shown. Use your triangle in Exercises 1–3.

A 1. a. Draw a line. On it choose a point D.
   b. Set your compass for a radius equal to XY. Then with D as center, draw an arc that intersects the line at a point E.
   c. At D, copy $\angle X$.
   d. At E, copy $\angle Y$.
   e. Use F to label the point at which the new sides of $\angle D$ and $\angle E$ intersect.
   f. Explain why $\triangle DEF \cong \triangle XYZ$. **ASA Postulate**

2. a. Draw a line. On it choose a point G.
   b. Set your compass for a radius equal to XY. Then with G as center, draw an arc that intersects the line at a point H.
   c. At G, copy $\angle X$.
   d. Set your compass for a radius equal to XZ. On the new side of $\angle G$, draw an arc.
   e. Use J to label the point at which the arc intersects the side of $\angle G$. Draw $\overline{JH}$.
   f. Explain why $\triangle GHJ \cong \triangle XYZ$. **SAS Postulate**

3. a. Draw a line. On it choose a point K.
   b. Set your compass for a radius equal to XY. Then with K as center, draw an arc that intersects the line at a point M.
   c. Set your compass for a radius equal to XZ. Use K as center and draw an arc.
   d. Set your compass for a radius equal to YZ. Use M as center and draw an arc.
   e. Use N to label the point at which the arcs intersect.
   f. Draw $\overline{KN}$ and $\overline{MN}$.
   g. Explain why $\triangle KMN \cong \triangle XYZ$. **SSS Postulate**

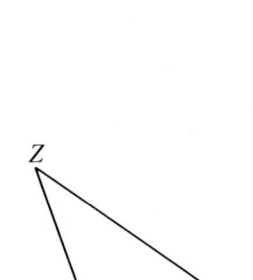

4–6. On your paper, draw an obtuse $\triangle XYZ$ roughly like, but much larger than, the one shown. Using $\triangle XYZ$, repeat Exercises 1–3. **4f. ASA Postulate**
   5f. SAS Postulate
   6g. SSS Postulate

B 7. The diagram at the left, below, illustrates Construction 3 on page 26. When $\overline{RS}, \overline{RT}, \overline{US},$ and $\overline{UT}$ are drawn, we have the diagram at the right, below.

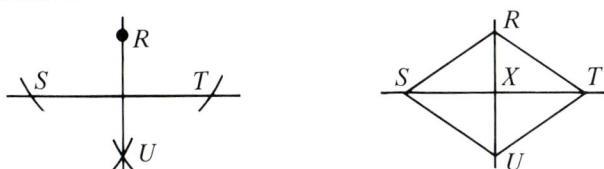

The steps below are the main ones in the proof that $\overleftrightarrow{RU} \perp \overleftrightarrow{ST}$. Complete each statement.

  a. $\triangle RSU \cong \triangle RTU$ by ___?___ (SSS/SAS/ASA/AAS). SSS
  b. Then $\angle SRU = \angle TRU$ because ___?___. Corr. parts of $\cong \triangle$ are =.
  c. Using $\triangle RXS$ and $\triangle RXT$:
    $RS = RT$ by ___?___ (algebra/construction); construction
    $\angle SRX = \angle TRX$ because $\angle SRU = \angle TRU$ (see step **b**);
    $RX = RX$ by ___?___; $\triangle RXS \cong \triangle RXT$ by ___?___. algebra; SAS
  d. $\angle RXS = \angle RXT$ because ___?___. Corr. parts of $\cong \triangle$ are =.
  e. Since $\angle RXS = \angle RXT$ and $\angle RXS + \angle RXT = 180°$, both $\angle RXS$ and $\angle RXT$ have measure ___?___°. 90
    Then $\overleftrightarrow{RU} \perp \overleftrightarrow{ST}$.

8. Refer to the construction at the right, which illustrates Construction 2 on page 26. Explain why $\overrightarrow{OP} \perp \overleftrightarrow{XY}$. (*Hint:* See step **e**, above.) $\triangle XOP \cong \triangle YOP$ by SSS Post.; hence $\angle XOP = \angle YOP$; from step e above, $\overrightarrow{OP} \perp \overleftrightarrow{XY}$.

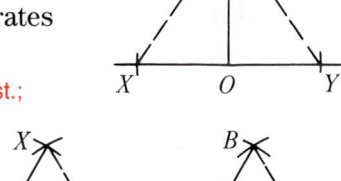

9. In the diagram at the right, $\angle A$ was given. $\angle O$ was constructed using Construction 4 on page 77. Explain why $\angle O = \angle A$. $\triangle AXY \cong \triangle OBC$ by SSS Post.; $\angle O = \angle A$ since corr. parts of $\cong \triangle$ are =.

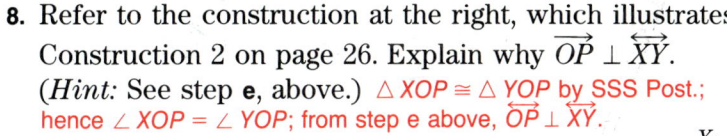

10. On your paper, draw a $\triangle DEZ$ with DZ clearly shorter than EZ. Construct the bisector of $\angle Z$. Use V to label the point where the bisector meets $\overline{DE}$. Is $\triangle DVZ \cong \triangle EVZ$? no Explain. We have only two pairs of corr. parts.

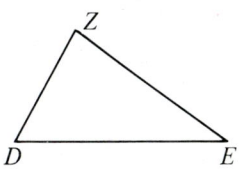

11. Draw a segment $\overline{DE}$. Using one setting of your compass, construct two arcs, one with center at D and the other with center at E. Z is the point where the arcs meet. isosceles;
  a. What kind of triangle is $\triangle DEZ$? Explain.   by construction
  b. Construct the bisector of $\angle Z$. Use V to label the point where the bisector meets $\overline{DE}$.
  c. Is $\triangle DVZ \cong \triangle EVZ$? Explain. yes; SAS Postulate

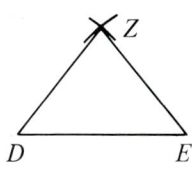

Using Congruent Triangles   **151**

C 12. Using a straightedge and lined paper, draw (you need not construct) a rectangle *ABCD* that is clearly longer than it is wide. Construct the bisector of ∠*A* and the bisector of ∠*B*. Use *M* to label the point where the bisectors intersect. Explain why △*AMB* is a right triangle. (*Hint:* Each angle of a rectangle is a right angle.) ∠*BMA* = 180° − $\frac{1}{2}$∠*A* − $\frac{1}{2}$∠*B* = 180° − 45° − 45° = 90°

13. As in Exercise 12, draw a rectangle *ABCD* that is clearly longer than it is wide. Construct the bisector of ∠*A* and the bisector of ∠*C*. Explain why the bisectors are parallel. (*Hint:* Label the angle measures in your diagram. If necessary, glance back at Section 6 of Chapter 2.) Using $\overline{DC}$ as a trans., we have corr. ≜ =, so the bisectors are ∥.

14. Draw a quadrilateral *ABCD* roughly like the one shown, but much larger. Draw $\overline{AC}$. Construct a △*A'B'C'* that is congruent to △*ABC*. Building on $\overline{A'C'}$, construct a △*A'C'D'* that is congruent to △*ACD*. Does each side and each angle of quadrilateral *A'B'C'D'* equal the corresponding part of quadrilateral *ABCD*? yes

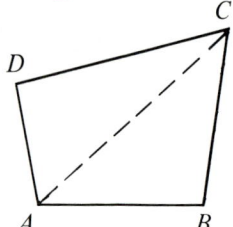

15. Draw a pentagon *ABCDE*. Draw diagonals $\overline{AC}$ and $\overline{AD}$. Construct a pentagon *A'B'C'D'E'* that is congruent to pentagon *ABCDE*. yes

16. Draw a pentagon *RSTVW*. Choose any point *P* inside the pentagon. Draw $\overline{PR}, \overline{PS}, \overline{PT}, \overline{PV},$ and $\overline{PW}$. Construct a pentagon *R'S'T'V'W'* that is congruent to pentagon *RSTVW*.

SELF-TEST

Write proofs in two-column form.

1. Given: *RM* = *TM*
   UM = *SM*

   Prove: *RU* = *TS*   (4-1)

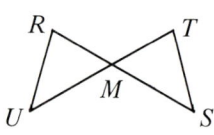

2. Given: *AX* = *AY*
   ∠*X* = ∠*Y* = 90°

   Prove: ∠1 = ∠2

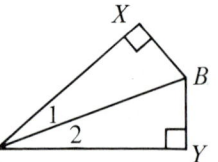

3. Suppose you want to prove that your construction of a perpendicular from *P* to *m* is correct. What postulate supports the statement: △*TEP* ≅ △*TIP*?   (4-2)
   SSS Postulate

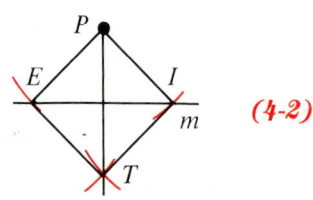

READING GEOMETRY

INDEPENDENT STUDY

When traveling, you usually have a destination in mind. Similarly, when studying geometry, you want to know where you are headed. At the beginning of every chapter in this book, you will find a list of goals to guide you through the chapter. Keeping these goals in mind, you can take a quick look through the chapter to get a preview of it before you begin to study the first lesson. Pay attention to the lesson titles and to statements that are highlighted by red frames or pointers.

As you study each lesson, pay special attention to new words and phrases, in heavy type, and to any constructions, postulates, or theorems that may be included. Read the explanations slowly and think as you read. If the lesson describes a construction, follow the directions and do the construction yourself. If the lesson includes a proof, follow it step by step. Study any worked-out examples in the same way. Try solving some examples before you look at the answers in the book.

After carefully reading a lesson, check your understanding by trying a few of the exercises. You may find that you need to reread some parts of the lesson. If there is anything that you still don't understand, make sure to discuss it with your teacher.

The Self-Tests generally cover several lessons. Before taking one of them, look back through the lessons that it covers and review important words and ideas. Answers to the Self-Tests are printed at the back of the book so that you can use them to check your progress. The Chapter Reviews and Cumulative Reviews will also help you.

EXERCISES

2. bisects, bisector, ⊥ bisector, equidistant; any segment, ray, or line that passes through the midpoint of the segment.

Look at pages 144–164 to find answers to the following questions.

1. Which lesson shows how to use congruent triangles to justify constructions? Lesson 4 – 2

2. What new words appear on page 154? What is a segment bisector?

3. Any point on the perpendicular bisector of a segment is __?__ from the endpoints of the segment. equidistant

4. Is a median of a triangle perpendicular to one of the sides? not necessarily

5. A circle that fits exactly inside a triangle is a(n) __?__ circle. inscribed

3 Segment Bisectors

Suppose that M is the midpoint of $\overline{YZ}$. Any segment, ray, or line that passes through M bisects $\overline{YZ}$ and is a **bisector** of $\overline{YZ}$. The diagram suggests that a segment has an unlimited number of bisectors.

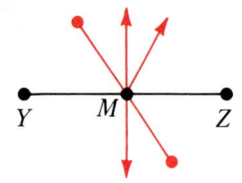

A bisector of a segment that is perpendicular to the segment is called a **perpendicular bisector** of the segment. As shown at the right, $YM = MZ$ and $\overline{LM} \perp \overline{YZ}$, so $\overline{LM}$ is a perpendicular bisector of $\overline{YZ}$. You may wish to refer to $\overline{LM}$ as *the* perpendicular bisector of $\overline{YZ}$.

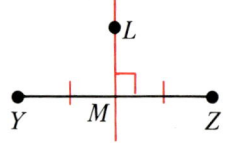

CONSTRUCTION 6

Given: *A segment*
Construct: *A perpendicular bisector of the segment*

Follow these steps:

1.

![segment AB]

2.

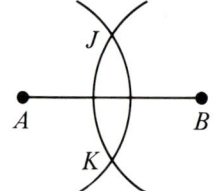

3.

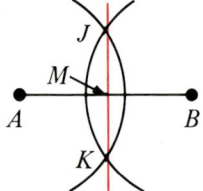

You are given $\overline{AB}$.

Set your compass for a convenient radius. Using A and B as centers, draw arcs that meet at J and K.

Draw $\overleftrightarrow{JK}$. $\overleftrightarrow{JK}$ is a perpendicular bisector of $\overline{AB}$. M is the midpoint of $\overline{AB}$.

A segment, $\overline{CD}$, and its perpendicular bisector are shown at the right. Is point R closer to point C or to point D? Neither one; $RC = RD$, and we say that R is *equidistant* from C and D. Point S is also equidistant from C and D because $SC = SD$.

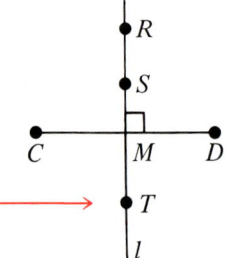

> Any point on ℓ is equidistant from C and D.

THEOREM 1

Any point on the perpendicular bisector of a segment is equidistant from the endpoints of the segment.

Given: $\overleftrightarrow{RM}$ is the perpendicular bisector of $\overline{CD}$.

Prove: $RC = RD$

See if you can supply the reasons for the proof.

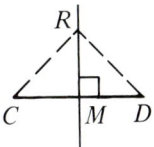

STATEMENTS	REASONS
1. $\overleftrightarrow{RM} \perp \overline{CD}$	1. __?__ Given
2. $\angle RMC$ and $\angle RMD$ are right angles.	2. __?__ Def. of perpendicular
3. $\angle RMC = \angle RMD$	3. __?__ All right angles are =.
4. $CM = DM$	4. __?__ Given: $\overleftrightarrow{RM}$ bisects $\overline{CD}$
5. $RM = RM$	5. __?__ From algebra
6. $\triangle RMC \cong \triangle RMD$	6. __?__ SAS Postulate
7. $RC = RD$	7. __?__ Corr. parts of $\cong$ $\triangle$ are =.

Theorem 2 is the converse of Theorem 1.
The proof is left as Exercise 16 on page 157.

THEOREM 2

Any point that is equidistant from the endpoints of a segment is on the perpendicular bisector of the segment.

Classroom Practice

In Exercises 1–7, line l is a perpendicular bisector of $\overline{AB}$. Classify each statement as true or false.

1. M is the midpoint of $\overline{AB}$.   true
2. $\angle 1$ is an acute angle.   false
3. $l \perp \overline{AB}$ true
4. $AD = BD$ true
5. $AB = CD$ false
6. $\angle 1 = \angle 2$ true
7. Any point on $\overline{CD}$ is equidistant from A and B.   true

8. Think of the perpendicular bisector of $\overline{XY}$ and the perpendicular bisector of $\overline{YZ}$. Do the two perpendicular bisectors intersect?   no
Explain.   The two $\perp$ bisectors are $\perp$ to $\overline{XZ}$, hence the two $\perp$ bisectors are $\parallel$.

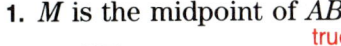

9. A student tried to construct a perpendicular bisector as shown. What went wrong?   The arcs do not intersect. The student must use a larger radius.

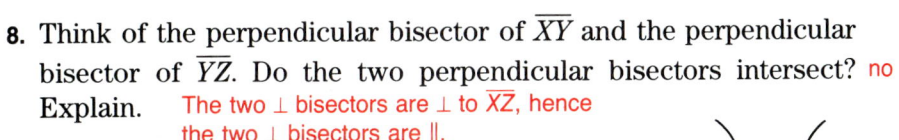
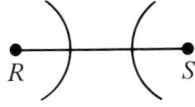

Written Exercises

In Exercises 1–5, *JK = KL*. Complete each statement.

A 1. *K* is the __?__ of *JL*. midpoint

2. Since $\overleftrightarrow{RS}$ passes through *K*, $\overleftrightarrow{RS}$ is a __?__ of $\overline{JL}$. bisector

3. If $\overline{RK} \perp \overline{JL}$, then $\overline{RK}$ is a __?__ of $\overline{JL}$. ⊥ bisector

4. If *SJ = SL*, then *S* is __?__ from *J* and *L*. equidistant

5. Suppose $\overleftrightarrow{RS}$ is the perpendicular bisector of $\overline{JL}$.
State the theorem that allows you to conclude that *RJ = RL*.
Any pt. on the ⊥ bisector of a segment is equidistant from the endpts. of the segment.

In Exercises 6–11, draw a figure roughly like the one shown.
Then construct the indicated perpendicular.

6. The perpendicular bisector of $\overline{AB}$

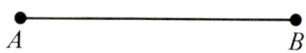

7. The perpendicular bisector of $\overline{XY}$

8. The perpendicular to $\overline{CD}$ at *P*

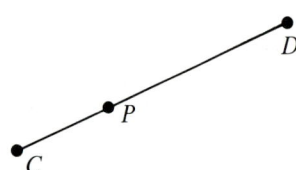

9. The perpendicular to $\overline{JK}$ through *R*

10. The perpendicular bisector of $\overline{PQ}$

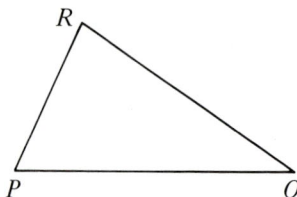

11. The perpendicular bisector of $\overline{BD}$

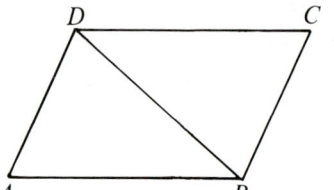

B 12. Construct a 90° angle. Then draw a segment to form a right triangle. Construct the perpendicular bisector of each leg of the right triangle. Where do the perpendicular bisectors appear to meet? the midpoint of the hypotenuse

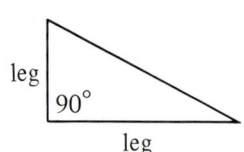

13. Repeat Exercise 12, using a different-looking right triangle. The perpendicular bisectors of the legs of a right triangle appear to meet at __?__. the midpoint of the hypotenuse

14. Construct a large isosceles triangle, using the method outlined in Exercise 11 on page 151. Then construct the perpendicular bisector of the base. Through what point does the perpendicular bisector seem to pass? *the vertex of the ∠ opposite the base*

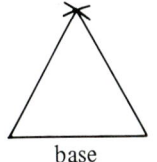
base

15. Repeat Exercise 14, using a different-looking isosceles triangle. The perpendicular bisector of the base of an isosceles triangle appears to pass through ___?___. *the vertex of the ∠ opposite the base*

C 16. One way to prove Theorem 2 is suggested below. Write the proof.

Given: $RC = RD$
M is the midpoint of $\overline{CD}$.

Prove: $\overleftrightarrow{RM}$ is the perpendicular bisector of $\overline{CD}$.

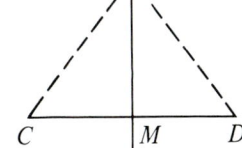

Strategy for proof:
A. Prove that $\triangle CRM \cong \triangle DRM$.
B. Prove that $\angle RMC$ and $\angle RMD$ are equal and are right angles.
C. Use the right angles to prove that $\overline{RM} \perp \overline{CD}$.

17. Use congruent triangles to show that Construction 6 does produce the perpendicular bisector of a segment.

Given: $JA = JB$; $KA = KB$

Prove: $AM = MB$; $\overleftrightarrow{JK} \perp \overline{AB}$

Strategy for proof:
A. Prove that $\triangle AJK \cong \triangle BJK$.
B. Prove that $\angle AJM = \angle BJM$.
C. Prove that $\triangle AJM \cong \triangle BJM$.
D. Prove that $AM = MB$ and $\angle AMJ = \angle BMJ$.
E. Use right angles to prove that $\overleftrightarrow{JK} \perp \overline{AB}$.

PUZZLE ◆ PROBLEMS

Draw a large triangle on cardboard. Construct the midpoints of the three sides. Draw the medians (see page 158) and let P be the intersection point. Cut the triangle out.

Hold the model horizontally and place your finger tip under it at P. The model should balance. Point P is called the *center of gravity* of the triangle.

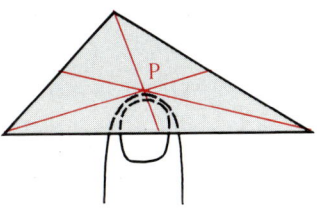

Try this experiment with other triangles.

4 Altitudes and Medians of a Triangle

The diagram at the right shows how Construction 3 can be used to construct a perpendicular to side $\overline{AB}$ through vertex C. We call $\overline{CR}$ the *altitude* of $\triangle ABC$ drawn to side $\overline{AB}$. In general, an **altitude** of a triangle is a segment, drawn from any vertex, perpendicular to the line that contains the opposite side.

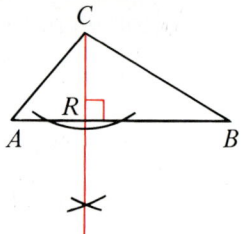

Every triangle has three altitudes, one from each vertex.

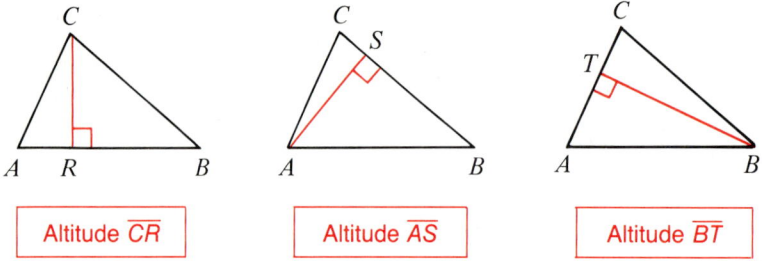

Sometimes it isn't so easy to draw an altitude. Here the altitude from F doesn't intersect *segment DE*. But it does intersect *line DE*.

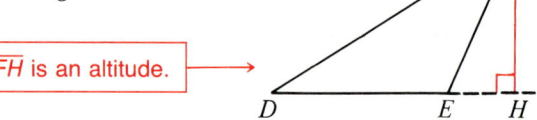

$\overline{FH}$ is an altitude.

In any triangle, the three lines containing the altitudes meet in one point.

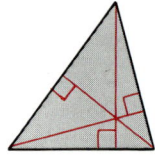
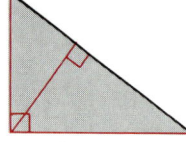
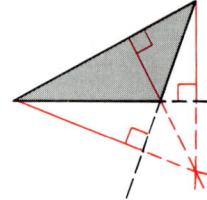

Acute triangle — The lines meet inside the triangle.

Right triangle — The lines meet at the vertex of the right angle.

Obtuse triangle — The lines meet outside the triangle.

Let's look at $\triangle ABC$ once again. Since M is the midpoint of side $\overline{AB}$, $\overline{CM}$ is a *median* of the triangle. A **median** is a segment that joins a vertex of the triangle to the midpoint of the opposite side.

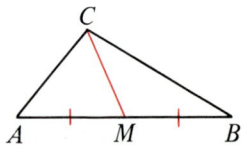

158   Chapter 4

Every triangle has three medians, and they always meet inside the triangle.

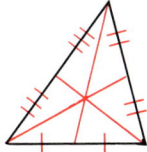 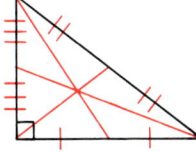 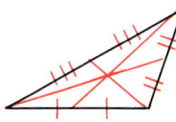

Acute triangle   **Right triangle**   **Obtuse triangle**

Compare the altitudes and the medians of the triangles shown. Notice that, in general, the altitude and the median to a side of a triangle are *different* segments.

Classroom Practice

1. Point K is the midpoint of $\overline{XY}$, and $\overline{ZJ} \perp \overline{XY}$.
   a. The altitude shown is segment __?__. ZJ
   b. The median shown is segment __?__. ZK

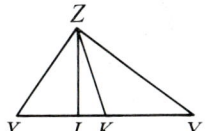

2. Point T is the midpoint of leg $\overline{SQ}$ in right $\triangle QRS$.
   a. The altitude to side $\overline{SQ}$ is segment __?__. RS
   b. The median to side $\overline{SQ}$ is segment __?__. RT

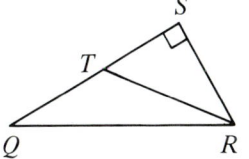

For each exercise, think of the three altitudes of the triangle shown. (Do not draw in your book.) Tell whether each altitude lies inside, on, or outside the triangle.

3. all inside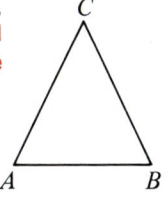
4. only alt. to $\overline{EF}$ inside; others outside
5. all inside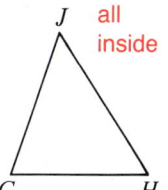
6. only alt. to $\overline{KM}$ inside; others lie on triangle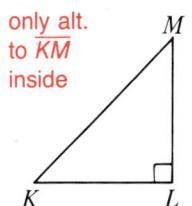

7–10. Use the figures for Exercises 3–6. Think of the three medians of the triangle shown. Tell whether each median lies inside, on, or outside the triangle. All medians lie inside.

11. You know that, in general, the altitude and the median to a side of a triangle are different segments. Describe a triangle in which an altitude and a median are the very same segment. The altitude and median from the vertex of an isosceles triangle are the same segment.

Using Congruent Triangles   **159**

Written Exercises

Draw a triangle roughly like the one shown, but larger. Draw and label the altitude to side $\overline{AB}$.

A 1. 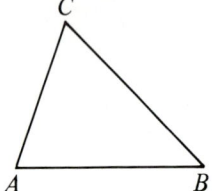 2. 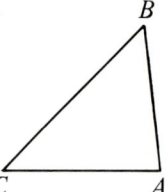 3.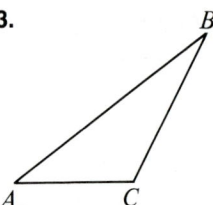

Draw a triangle roughly like the one shown, but larger. Construct and label the midpoint of side $\overline{YZ}$. Then draw the median to $\overline{YZ}$.

4. 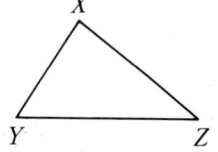 5. 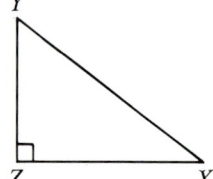 6.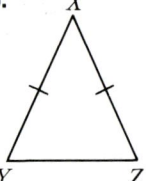

7. Draw a triangle roughly like △ RST, but larger. Then draw each of the following (you need not construct):
   a. The altitude from R
   b. The median from R
   c. The perpendicular bisector of $\overline{ST}$

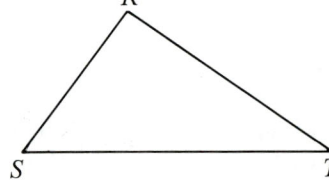

8. Repeat Exercise 7, using an isosceles △ RST with RS = RT. What do you notice? *The altitude, median, and ⊥ bisector of the base of an isosceles △ lie on the same line.*

Draw a triangle roughly like the one shown, but larger. Construct the three altitudes of the triangle.

B 9. 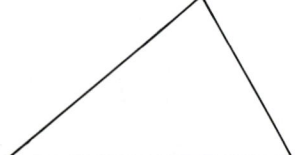 10.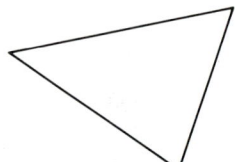

11. a. Draw an acute triangle.
    b. Use Construction 6 to locate the midpoint of each side.
    c. Draw the three medians of the triangle.

160 Chapter 4

12. **a.** Construct a right angle.
    b. Draw a segment to form a right triangle.
    c. Use Construction 6 to locate the midpoint of each side.
    d. Draw the three medians of the triangle.

13. **a.** Draw a large acute $\triangle ABC$.
    b. Construct: the midpoint M of $\overline{BC}$;
       the midpoint N of $\overline{AC}$;
       the midpoint K of $\overline{AB}$.
    c. Draw the three medians.
       Use P to label the point of intersection.
    d.

By measurement		By arithmetic	
$AP =$ __?__	$AM =$ __?__	$\frac{2}{3}(AM) =$ __?__	Measurements will vary; first and third columns should be the same.
$BP =$ __?__	$BN =$ __?__	$\frac{2}{3}(BN) =$ __?__	
$CP =$ __?__	$CK =$ __?__	$\frac{2}{3}(CK) =$ __?__	

14. Repeat Exercise 13 using an obtuse triangle. same relationships

15. Examine part **d** of Exercises 13 and 14.
    What do you observe about the first and third columns? They are the same.
    Complete: The medians of a triangle meet in a point which is __?__ two-thirds
        of the way from any vertex to the opposite side.

C 16. **a.** Draw a large obtuse triangle.
    b. Construct the three altitudes of the triangle.
    c. If you extend the altitudes, will they meet in a point? yes

17. Repeat Exercise 16, using a different-looking obtuse triangle. **c.** yes

18. The goal is to prove that *corresponding altitudes of congruent triangles are equal.* Copy what is shown and write a proof in two-column form.

    Given: $\triangle RST \cong \triangle XYZ$
    $\overline{TV} \perp \overline{RS}$; $\overline{ZW} \perp \overline{XY}$

    Prove: $TV = ZW$

    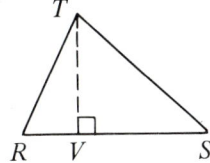 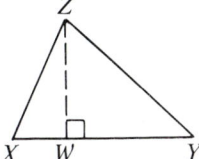

    Strategy for proof:
    A. Use the congruent triangles to prove $RT = XZ$ and $\angle R = \angle X$.
    B. Use perpendicular lines to prove $\angle RVT = \angle XWZ$.
    C. Prove $\triangle RVT \cong \triangle XWZ$.
    D. Use the fact that corr. parts of $\cong \triangle$ are $=$.

Using Congruent Triangles

Experiments

Experiment 1

1. Take a large sheet of paper and fold it down the middle. Cut from the corner to any point on the fold as shown. Unfold.

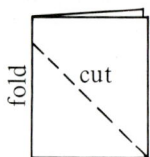

2. What kind of triangle did you form? isosceles

3. Explain why the fold is both an altitude and a median. (See below.)

4. Complete: The median and the altitude to the base of an __?__ triangle are the same segment.   isosceles

   3. The fold is the ⊥ bisector of the base.

Experiment 2

1. Plot points $A(0, 0)$, $B(6, 12)$, and $C(12, 0)$ on graph paper. Draw $\triangle ABC$.

2. Draw dots at the midpoints of the sides of $\triangle ABC$. Draw the three medians. Use P to label the point where the medians meet. Use M to label the midpoint of $\overline{BC}$.

3. M is point (__?__ , __?__) and P is point (__?__ , __?__). $M(9, 6)$
   $P(6, 4)$

4. Measure $\overline{AP}$ and $\overline{AM}$ with your ruler. If you worked carefully, $AP = \frac{2}{3}(AM)$.

Experiment 3

Repeat Experiment 2, using points $A(0, 0)$, $B(0, 6)$, and $C(12, -12)$.

Experiments 2 and 3 suggest this property: *The medians of a triangle intersect in a point which is two-thirds of the way along each median.*

PUZZLE ◆ PROBLEMS

Here is another way to show that the sum of the angles of a triangle equals 180°. Draw a triangle ABC and construct the inscribed circle, as shown on page 164. Cut out the triangle and fold $\angle A$, $\angle B$, and $\angle C$ to the center of the circle. There are six angles at the center, the three original angles and three vertical angles.

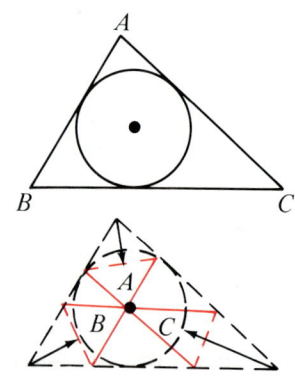

$$2\angle A + 2\angle B + 2\angle C = 360°,$$
$$\text{so } \angle A + \angle B + \angle C = 180°.$$

5 Inscribed and Circumscribed Circles

We know that the medians of a triangle meet in a point. So do the lines containing the altitudes. What about the perpendicular bisectors of the sides?

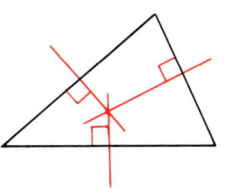

Acute triangle

⊥ bisectors meet inside triangle.

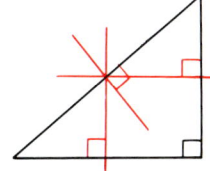

Right triangle

⊥ bisectors meet on triangle.

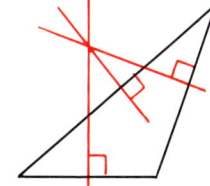

Obtuse triangle

⊥ bisectors meet outside triangle.

This special point in which the perpendicular bisectors meet is used in the following construction.

CONSTRUCTION 7

Given: *A triangle*
Construct: *A circle passing through the vertices of the triangle.*

1.

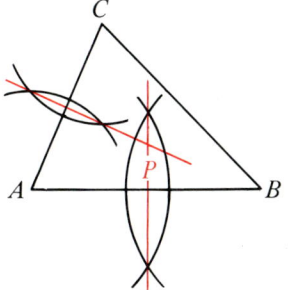

Given $\triangle ABC$, construct the perpendicular bisectors of any two sides. Label the meeting point P.

2.

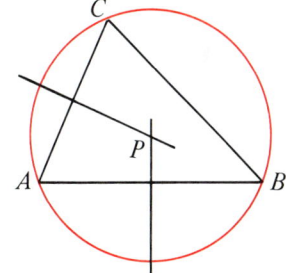

Using P as center and PA as radius, draw a circle. The circle should pass through A, B, and C.

When each vertex of a triangle is a point on a circle, we say that the circle is **circumscribed about** the triangle. In the construction above, for example, the circle shown is circumscribed about $\triangle ABC$. The figure at the right shows another **circumscribed circle**.

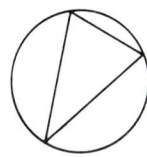

Using Congruent Triangles

A circumscribed circle fits exactly *around* a triangle. On the other hand, an **inscribed circle** (see page 361) fits exactly *inside* a triangle. Here are some inscribed circles.

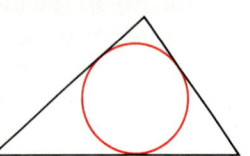

Suppose you are given a triangle, △ DEF. To find the center of the circle which can be inscribed in △ DEF, we use this fact: The angle bisectors of a triangle meet in a point which is also the center of the inscribed circle.

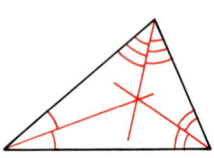   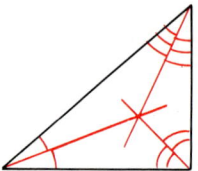   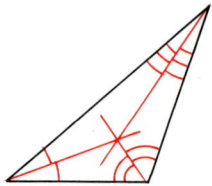

Acute triangle   **Right triangle**   **Obtuse triangle**

CONSTRUCTION 8

Given: *A triangle*
Construct: *An inscribed circle*

1.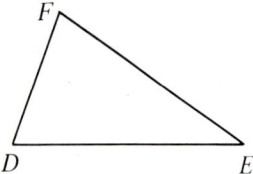

You are given △ DEF.

2.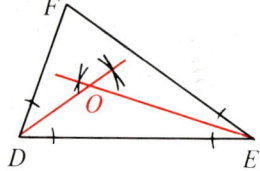

Construct the angle bisectors of any two angles. Label their meeting point O.

3.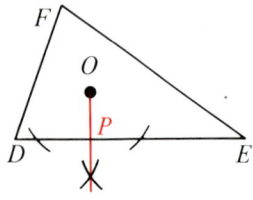

Construct a perpendicular from O to $\overline{DE}$. Label the meeting point P.

4.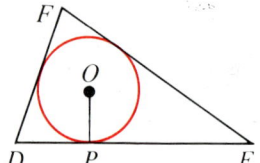

With O as center and OP as radius, draw a circle. The circle should just touch each side of △ DEF.

164   Chapter 4

Classroom Practice

1. Which of the diagrams below show a circle circumscribed about a triangle? b, d

   a. 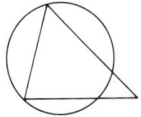 b.  c.

   d. 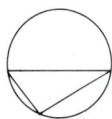 e.  f.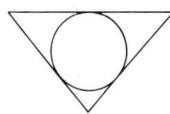

2. Which diagrams above show a circle inscribed in a triangle? e, f

3. The center of a circle inscribed in a triangle is the point at which __?__. the angle bisectors meet

4. The center of a circle circumscribed about a triangle is the point at which __?__. the ⊥ bisectors of the sides meet

5. The perpendicular bisector of $\overline{AC}$ and the perpendicular bisector of $\overline{BC}$ intersect at point X. We wish to show that X lies on the perpendicular bisector of $\overline{AB}$. Supply the required reasons.

   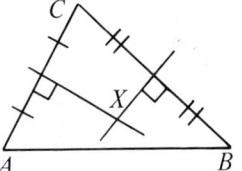

   a. Draw $\overline{AX}$, $\overline{BX}$, and $\overline{CX}$.
   b. $AX = CX$ (Why?) Thm. 1, page 154
   c. $BX = CX$ (Why?) Thm. 1, page 154
   d. Then $AX = BX$. (Why?) Substitution Postulate
   e. X is equidistant from A and B. (Why?) Def. of equidistant
   f. X lies on the perpendicular bisector of $\overline{AB}$. (Why?) Thm. 2, page 155

   Thus the perpendicular bisectors of the three sides of a triangle intersect in one point.

Written Exercises

Sketch each of the following. You need not construct.

 1. A circle inscribed in an isosceles triangle

2. A circle inscribed in a scalene triangle

3. A circle circumscribed about an acute triangle

4. A circle circumscribed about an isosceles right triangle

Using Congruent Triangles **165**

In Exercises 5–7, draw a triangle roughly like the one shown, but much larger. Use Construction 7 to circumscribe a circle about the triangle.

5.    6.    7.

In Exercises 8–10, draw a triangle roughly like the one shown, but much larger. Use Construction 8 to inscribe a circle in the triangle.

8.    9.    10.

B 11. Draw three noncollinear points on your paper and label them X, Y, and Z. Construct a circle that passes through the three points.

12. Repeat Exercise 11, using three different points.

13. Draw a scalene $\triangle ABC$. Circumscribe a circle about $\triangle ABC$ and inscribe a circle in $\triangle ABC$.

14. Repeat Exercise 13, using an isosceles $\triangle ABC$.

C 15. Draw an obtuse $\triangle XYZ$. Construct a circle that passes through Y, Z, and the midpoint of $\overline{XZ}$.

SELF-TEST

Vocabulary

bisector (p. 154)
perpendicular bisector (p. 154)
altitude (p. 158)
median (p. 158)
circumscribed (p. 163)
inscribed (p. 164)

Draw three triangles roughly like, but much larger than, the ones shown.

1. In $\triangle ABC$, construct the altitude from C.   (4-4)

2. In $\triangle ABC$, construct the median from A.

3. Circumscribe a circle about $\triangle DEF$.   (4-5)

4. Find, by construction, the point which would be the center of the circle inscribed in $\triangle GHJ$.

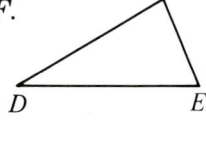
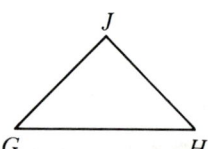

6 Triangles with Two Equal Sides

EXPLORATIONS

1. Draw a large isosceles $\triangle ABC$, with $AC = BC$. Cut out the triangle.
2. Make a fold that bisects $\angle C$.
3. Does vertex A fold onto vertex B? yes
4. Does $\angle A$ have the same measure as $\angle B$? yes

If you worked carefully, your answers to parts 3 and 4 were *yes*. A picture statement of the theorem suggested by the exploration is shown at the right. The exploration suggests a proof involving a "line down the middle."

If 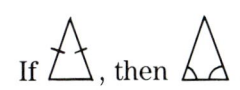.

THEOREM 3

If two sides of a triangle are equal, then the angles opposite those sides are equal.

"Base angles of an isosceles triangle are equal"

Given: $AC = BC$

Prove: $\angle A = \angle B$

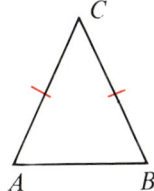 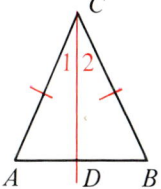

STATEMENTS	REASONS
1. Draw the bisector of $\angle C$.	1. Every angle has a bisector. (Recall Const. 1)
2. $\angle 1 = \angle 2$	2. Definition of angle bisector
3. $AC = BC$	3. Given
4. $CD = CD$	4. From algebra
5. $\triangle ADC \cong \triangle BDC$	5. SAS Postulate
6. $\angle A = \angle B$	6. Corr. parts of $\cong \triangle$ are $=$.

COROLLARY

An equilateral triangle is also equiangular, and each angle has measure 60°.

Using Congruent Triangles

See Exercise 21 on page 171 for an outline of a proof of the corollary on page 167.

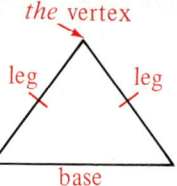
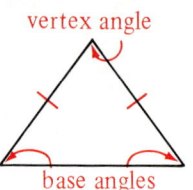

We often use special terms to refer to parts of isosceles triangles.

Some people like to state Theorem 3 in this form:

Base angles of an isosceles triangle are equal.

WAYS TO PROVE TWO ANGLES ARE EQUAL

1. Show that they are corresponding parts of congruent triangles.
2. Show that they are opposite two equal sides of a triangle.
3. Show that they are corresponding angles or alternate interior angles of parallel lines.

Classroom Practice

Exercises 1–6 refer to isosceles △ABC.

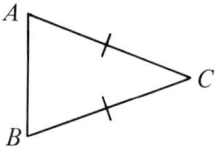

1. Name the base. $\overline{AB}$
2. Name the vertex angle. $\angle C$
3. Name the legs. $\overline{AC}, \overline{BC}$
4. Name the base angles. $\angle A, \angle B$
5. Name the vertices. A, B, C
6. Name *the* vertex. C

In Exercises 7–12, name the angles that must be equal.

7. $\angle A, \angle T$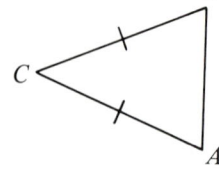
8. $\angle R, \angle D$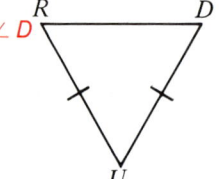
9. $\angle G, \angle S$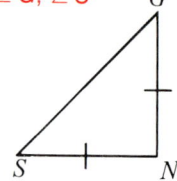

10. $\angle 1, \angle 2$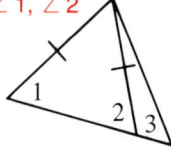
11. $\angle 1, \angle 2$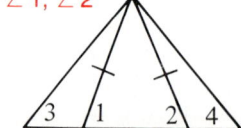
12. $\angle 3, \angle 4$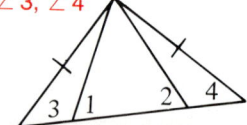

If two sides of a triangle are equal,
then the angles opposite those sides are equal.

13. In the diagram for Exercise 7, you could say $CA = CT$, given.
Then you could conclude: $\angle A = \angle T$.
State the theorem that supports the conclusion.

14. In the diagram for Exercise 9, suppose $\angle N = 90°$.
Then $\angle S = \underline{}°$.   45

15. In the diagram for Exercise 8, suppose that $DU = DR = RU$.
Then $\angle D = \underline{}°$.   60

**In Exercises 16–18, $XZ = YZ$ and $\overline{ZE}$ is drawn in the way described.
State the method you would use to prove that $\triangle XEZ \cong \triangle YEZ$.**

16. Given: $XZ = YZ$
$\overrightarrow{ZE}$ bisects $\angle XZY$.

    Prove: $\triangle XEZ \cong \triangle YEZ$ SAS

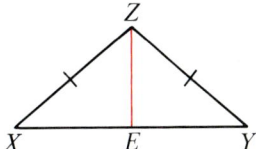

17. Given: $XZ = YZ$
$\overline{ZE}$ is a median.

    Prove: $\triangle XEZ \cong \triangle YEZ$ SSS

18. Given: $XZ = YZ$
$\overline{ZE}$ is an altitude.

    Prove: $\triangle XEZ \cong \triangle YEZ$ HL

Written Exercises

1. 65°; 115°; 115°; 50°
2. 60°; 60°; 120°; 60°
3. 64°; 64°; 116°; 52°
4. 70°; 110°; 110°; 40°
5. 68°; 112°; 112°; 44°
6. 68°; 68°; 112°; 44°
7. 69°; 69°; 111°; 111°
8. 65°; 65°; 115°; 115°

In Exercises 1–10, use the diagram. Complete each row of the table.

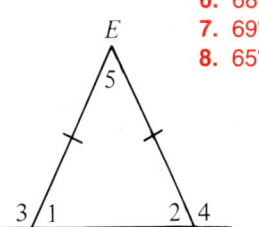

	$\angle 1$	$\angle 2$	$\angle 3$	$\angle 4$	$\angle 5$
1.	65°	?	?	?	?
2.	?	?	120°	?	?
3.	?	?	?	116°	?
4.	?	70°	?	?	?
5.	68°	?	?	?	?
6.	?	?	112°	?	?
7.	?	?	?	?	42°
8.	?	?	?	?	50°
9.	$j°$	?	?	?	?
10.	$k°$	?	?	?	?

Hint: $\angle 1 + \angle 2 = \underline{}°$

Your answers will involve j.

Your answers will involve k.

9. $j°$; $(180 - j)°$; $(180 - j)°$; $(180 - 2j)°$
10. $k°$; $(180 - k)°$; $(180 - k)°$; $(180 - 2k)°$

Using Congruent Triangles   **169**

In Exercises 11–19, equal sides are marked. Find the value of *x*.

SAMPLE

$\angle A + \angle B + \angle C = 180°$
$2x° + 2x° + 92° = 180°$
$4x° + 92° = 180°$
$4x° = 88°$
$x° = 22°$

$\angle A = \angle B = 2x°$

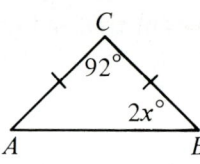

11. 55 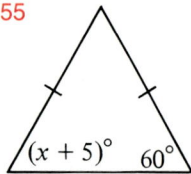 **12.** 45 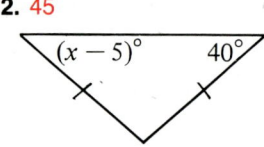 **13.** 9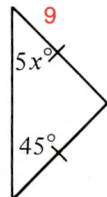

14. 140 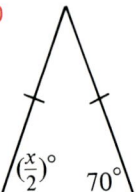 **15.** 6 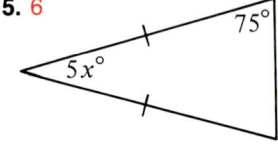 **16.** 10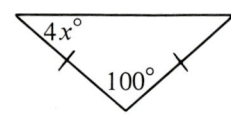

17. 25 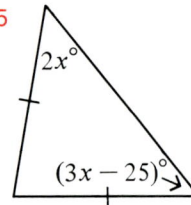 **18.** 40 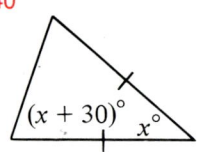 **19.** 180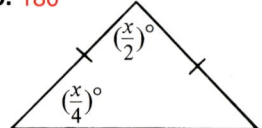

In Exercises 20–22, complete the proofs by supplying the reasons.

20. Given: $AC = BC$
Prove: $\angle 1 = \angle 3$

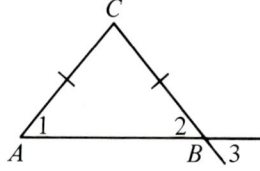

STATEMENTS	REASONS
1. $AC = BC$	1. __?__ Given
2. $\angle 1 = \angle 2$	2. __?__ Base ⚞ of an isos. △ are =.
3. $\angle 3 = \angle 2$	3. __?__ Vertical ⚞ are =.
4. $\angle 1 = \angle 3$	4. __?__ Substitution Postulate

B **21.** Complete this proof of the corollary to Theorem 3:

An equilateral triangle is also equiangular, and each angle has measure 60°.

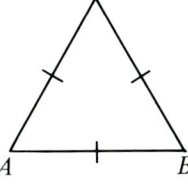

Given: $AB = BC = AC$

Prove: $\angle A = \angle B = \angle C = 60°$

STATEMENTS	REASONS
1. $AC = BC$; $AB = AC$; $AB = BC$	1. __?__ Given
2. $\angle A = \angle B$; $\angle B = \angle C$; $\angle C = \angle A$	2. __?__ Base ⩬ of an isos. △ are =.
3. $\angle A = \angle B = \angle C$	3. __?__ Substitution Postulate
4. $\angle A + \angle B + \angle C = 180°$	4. __?__ The ∠ sum of a △ is 180°.
5. $\angle A + \angle A + \angle A = 180°$, or $3 \cdot \angle A = 180°$	5. __?__ Substitution Postulate
6. $\angle A = 60°$	6. __?__ Division Postulate
7. $\angle A = \angle B = \angle C = 60°$	7. __?__ Substitution Postulate

22. *Given:* $AX = DX$; $AB = DC$

Prove: $\angle 1 = \angle 2$

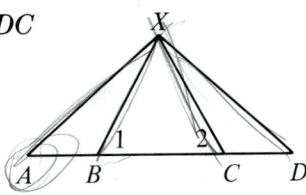

Strategy for proof:
A. Using isosceles $\triangle AXD$, prove that $\angle A = \angle D$.
B. Prove that $\triangle AXB \cong \triangle DXC$, and conclude that $BX = CX$.
C. Using isosceles $\triangle BXC$, prove that $\angle 1 = \angle 2$.

STATEMENTS	REASONS
1. $AX = DX$	1. __?__ Given
2. $\angle A = \angle D$	2. __?__ Base ⩬ of an isos. △ are =.
3. $AB = DC$	3. __?__ Given
4. $\triangle AXB \cong \triangle DXC$	4. __?__ SAS Postulate
5. $BX = CX$	5. __?__ Corr. parts of ≅ ⩬ are =.
6. $\angle 1 = \angle 2$	6. __?__ Base ⩬ of an isos. △ are =.

Using Congruent Triangles

CAREER NOTES

PLUMBER

Have you ever taken something apart just because you were curious about how it worked? Can you figure out why the vacuum cleaner or lawnmower won't run, and then fix it? A plumber must be able to do similar types of problem solving.

Plumbers install and repair fixtures and appliances that are connected to water and sewer lines. They also work with gas pipes that supply energy to ranges, furnaces, and water heaters. Skilled plumbers design complex pipe systems for large commercial and industrial buildings.

It is important for plumbers to be able to visualize three-dimensional objects when looking at building plans and blueprints. Careful measurement is also necessary to avoid wasting materials. For example, a plumber may decide that drilling a hole through a wall or floor will save many lengths of pipe. Distances can also be shortened by joining pipes with 45° elbows.

A plumber uses mathematics often while on the job. He or she may have to calculate water pressure in different locations or determine the weight of water in a tank with a given volume. City or industrial codes may regulate such things as the grade (or slope) of sewer pipes and the rate of absorption of septic systems.

To become a plumber, you must usually complete a 4-year apprenticeship program after high school or vocational school. In many places you must then pass an exam on plumbing skills and local plumbing codes in order to get a license. Most plumbers work for contractors in new construction. Many others are self-employed, specializing in repairs and modernization.

7 Triangles with Two Equal Angles

In Section 6 you saw that when a triangle has two equal sides, it also has two equal angles. Experience suggests that when a triangle has two equal angles, it must also have two equal sides. Expressing this by diagram, we write:

Theorem 3   If 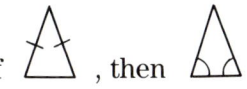 , then .

Converse of Theorem 3   If 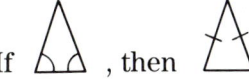 , then .

We shall treat the converse as a theorem.

THEOREM 4

If two angles of a triangle are equal, then the sides opposite those angles are equal.

Given: $\angle A = \angle B$

Prove: $AC = BC$

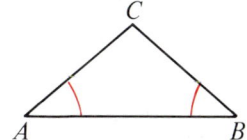 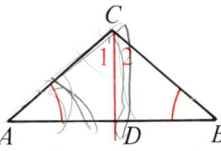

See if you can supply the reasons in the proof below.

STATEMENTS	REASONS
1. Draw the bisector of $\angle C$.	1. Every angle has a bisector. (Recall Construction 1.)
2. $\angle 1 = \angle 2$	2. Def. of bisector
3. $\angle A = \angle B$	3. Given
4. $CD = CD$	4. From algebra
5. $\triangle ACD \cong \triangle BCD$	5. AAS Theorem
6. $AC = BC$	6. Corr. parts of $\cong \triangle$ are =.

COROLLARY

If a triangle is equiangular, it is also equilateral.

Using Congruent Triangles

See Exercise 11 on page 176 for an outline of a proof of the corollary on page 173. The corollary on page 167 states that an equilateral triangle is also equiangular, and that each angle has measure 60°. We can use this fact to construct a 60° angle.

CONSTRUCTION 9

Construct: *A 60° angle*

1.

2.

3.

Draw a segment, $\overline{AB}$.

Set your compass for a radius of *AB*. Draw two arcs, one with *A* as center, and one with *B* as center.

Draw $\overrightarrow{AC}$.
∠*A* is a 60° angle.

Can you explain why ∠*A* must have measure 60°?

Try this:
    1. Draw $\overline{BC}$.
    2. *AB* = *BC* = *AC* by construction.
    3. ∠*A* = ∠*B* = ∠*C* = 60° (Why?) Thm. 3 (twice) and Subst. Post.

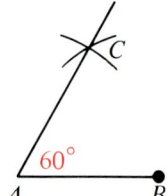

WAYS TO PROVE TWO SEGMENTS ARE EQUAL

1. Show that they are corresponding parts of congruent triangles.

2. Show that they are opposite two equal angles of a triangle.

Classroom Practice

In Exercises 1–6, name the sides that must be equal.

1. OA = OB

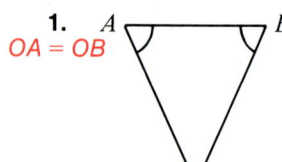

2. JC = JT

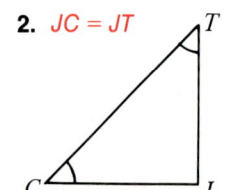

3. RE = RH

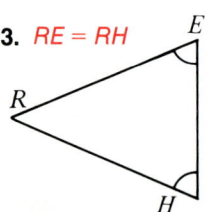

174   Chapter 4

4. ZX = ZY **5.** TA = TB **6.** HD = HG; HE = HF

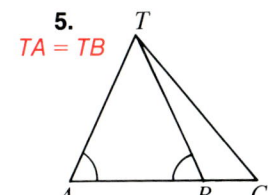

7. If you know that AX = AY, then you can conclude that ∠X = ∠Y. State the theorem that supports the conclusion. Base ⦣ of an isos. △ are =.

8. If you know that ∠X = ∠Y, then you can conclude that AX = AY. State the theorem that supports the conclusion. If 2 ⦣ of a △ are =, then the sides opp. those ⦣ are =.

9. One student decided to try to prove $RT = ST$ as follows:
   a. Bisect ∠RTS with a segment that is perpendicular to $\overline{RS}$.
   b. ∠1 = ∠2
   c. ∠3 = ∠4 (Each has measure 90°.)
   d. TV = TV
   e. △RVT ≅ △SVT
   f. RT = ST
What is wrong with this reasoning? The segment which is ⊥ to $\overline{RS}$ does not necessarily bisect ∠RTS.

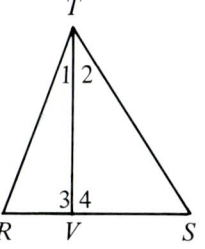

10. a. Construct a 60° angle.
    b. Use your construction from part **a** to construct a 120° angle.

Written Exercises

Equal angles are marked. Find the value of x.

SAMPLE

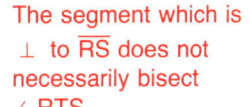

$BC = BA$
$10x = 8x + 4$
$2x = 4$
$x = 2$

A   **1.** 6 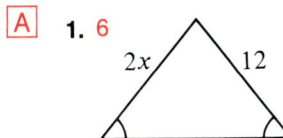   **2.** 11 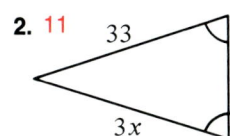   **3.** 6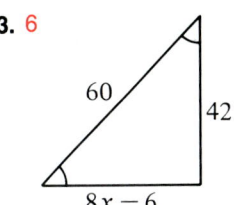

Using Congruent Triangles   **175**

Equal angles are marked. Find the value of x.

4.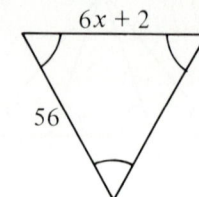

5.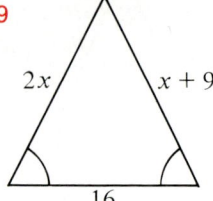

6.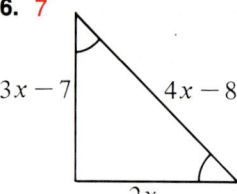

7.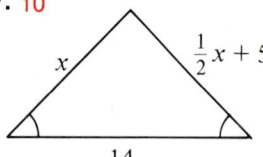

8.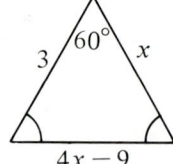

9.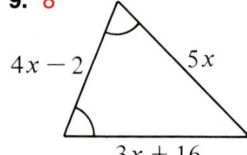

10. Complete the proof by supplying the reasons.

    Given: $\angle 1 = \angle 2$

    Prove: $AC = BC$

    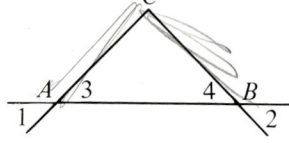

STATEMENTS	REASONS
1. $\angle 3 = \angle 1$; $\angle 4 = \angle 2$	1. __?__ Vertical angles are =.
2. $\angle 1 = \angle 2$	2. __?__ Given
3. $\angle 3 = \angle 4$	3. __?__ Substitution Postulate.
4. $AC = BC$	4. __?__ Thm. 4, page 173

11. Complete this proof of the corollary of Theorem 4:

    If a triangle is equiangular, it is also equilateral.

    Given: $\angle A = \angle B = \angle C$

    Prove: $AB = BC = CA$

STATEMENTS	REASONS
1. $\angle A = \angle C$; $\angle A = \angle B$; $\angle B = \angle C$	1. __?__ Given
2. $AB = BC$; $BC = CA$; $CA = AB$	2. __?__ Thm. 4, page 173
3. $AB = BC = CA$	3. __?__ Substitution Postulate

12. Construct a 60° angle.

13. Use your construction from Exercise 12. Construct a 30° angle.

14. Use your construction from Exercises 12 and 13. Construct a 15° angle.

In Exercises 15–17, refer to the sample on page 80.

B **15.** Construct a 150° angle. Select a strategy.
One choice: Use the fact that 150° = 180° − 30°.
Another choice: Use the fact that 150° = 90° + 60°.

16. Construct a 105° angle.   **17.** Construct a 75° angle.

Equal angles are marked. Find the value of x.

18.
12

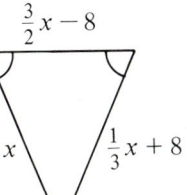

19. 11

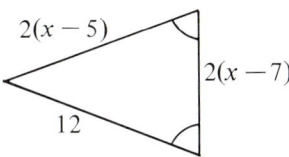

20. 10

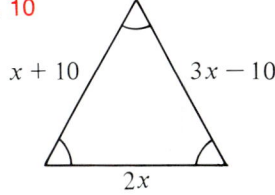

Copy what is shown. Then write a proof in two-column form.

21. *Given:* $JK = NM$; $\angle J = \angle N$

Prove: $\angle 5 = \angle 6$

Strategy for proof:

A. Think of $\angle J$ and $\angle N$ as angles of $\triangle JPN$. Prove that $PJ = PN$.

B. Prove that $\triangle JPK \cong \triangle NPM$ and conclude that $\angle 5 = \angle 6$.

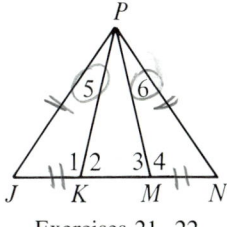

Exercises 21, 22

22. *Given:* $\angle 1 = \angle 4$

Prove: $\triangle PKM$ is an isosceles triangle.

C **23.** The goal of this exercise is to prove the statement: In an isosceles triangle, the bisector of the vertex angle bisects the base and is perpendicular to the base.
  a. Draw and label an appropriate diagram.
  b. List what you are given and what you are to prove in terms of your diagram.
  c. Write a proof in two-column form.

Using Congruent Triangles   **177**

Experiments

Counting Triangles You Cannot Draw

Begin with a very large equilateral triangle.

1. Join the midpoints of the three sides.
   How many equilateral triangles are shown now? 5

2. Using the "inside" triangle, again join the midpoints of the sides. How many equilateral triangles are there now? 9

3. Repeat this process and complete the table.

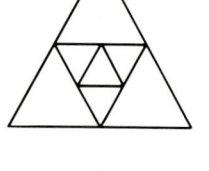

Step		1	2	3	4	5
No. of △	1	5	9	13	17	21

4. Look at your completed table. Each time you join the midpoints of a triangle, how many new triangles are formed? 4

5. How many triangles would be formed if you joined midpoints eight times? 33

SELF-TEST

1. Suppose you know that $TA = TE$.
   Write the theorem that supports the statement: $\angle A = \angle E$.

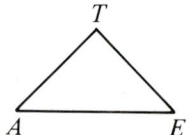

If 2 sides of a △ are =, then the ⩘ opposite those sides are =. *(4-6)*

In each diagram, equal sides are marked. Find the value of y.

2. 60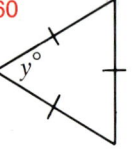

3. 40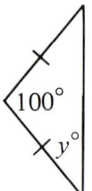

4. 44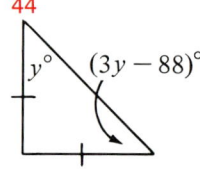

In each diagram, equal angles are marked. Find the value of x.

5. 5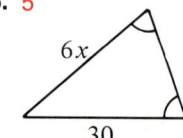

6. 4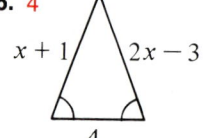

7. 12 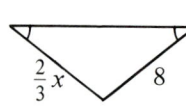 *(4-7)*

APPLICATIONS

THE TRIANGLE INEQUALITY

Suppose we are given the lengths of three line segments. Can we build a triangle with these segments? The SSS Postulate tells us that if we can, there is only one triangle possible.

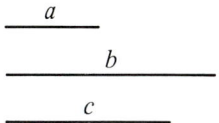
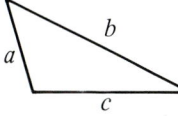

It may not be possible to make any triangle at all. Can you see that there is no triangle with sides of lengths 2, 4, and 10?

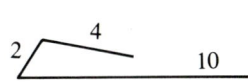
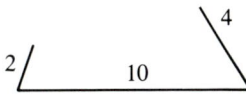

The triangle inequality is a rule which tells us whether three lengths can form the sides of a triangle.

THE TRIANGLE INEQUALITY

In a triangle, the sum of the lengths of any two sides must be greater than the length of the third side.

In the example above, the sum of 2 and 4 is not greater than 10. Therefore, we cannot make a triangle from these lengths.

Some people think of the triangle inequality as a special way of saying "the shortest path between two points is a straight line." You have probably used this idea yourself if you have ever taken a short cut across a corner. Look at △ABC. If the triangle inequality were not true, the "short cut" from A to B would be longer than the "detour" through C!

EXERCISES

Can you draw a triangle using these lengths for the sides?

1. 3, 4, 5   yes
2. 3, 4, 1   no
3. 3, 4, 20   no
4. 7, 7, 7   yes
5. 7, 7, 14   no
6. 7, 7, 0.001   yes

PROBLEM SOLVING STRATEGIES

DIAGRAMS AND CHARTS

If a problem contains a lot of information, it is often helpful to draw a picture or use a chart to organize the data. The following exercises will illustrate the use of a picture and a chart.

EXERCISES

1. Six people are in a room and each one shakes hands with each of the others exactly once. How many handshakes take place?

   A possible picture that represents the conditions given in the problem could be a circle with six points around it. A handshake would then be a line drawn from one point to another.

   The picture shows that A has shaken hands with B, C, D, and E. Complete the diagram to find the number of handshakes that take place. 15 handshakes

2. Digger, Moose, Bigfoot, Goalie, and Fang are all dogs in the same neighborhood. Two are short-haired and three are long-haired. Two have short tails and three have long tails.

   Digger and Moose have the same length hair;
   Fang and Goalie have different length hair.

   Bigfoot and Goalie have the same length tails;
   Moose and Fang have different length tails.

   What is the name of the dog with short hair and a short tail? Fang

Note: Black check marks refer to part (a) only.

	Short hair	Long hair	Short tail	Long tail
Digger	✓	✓	✓✓	
Moose	✓	✓		✓
Bigfoot	✓			✓
Goalie		✓		✓
Fang	✓		✓	

Red check marks indicate the solution to Exercise 2.

Notice the chart shown above. It lists the names of the dogs at the left and their characteristics at the top.

a. Let's assume that the first dog, Digger, is the one with the short hair and a short tail. Check marks have been placed in the appropriate boxes by Digger's name.

One of the clues says that Digger and Moose have the same length hair, so a check mark is placed showing that Moose also has short hair.

Explain why it is impossible to satisfy the condition that Fang and Goalie have different length hair.

Now explain why it is impossible for Digger to have short hair and a short tail.

b. Now assume that the second dog, Moose, is the one with short hair and a short tail. Copy the chart without any check marks. Then fill in the boxes until you come to an impossible situation. You then know that Moose cannot be the dog with the short hair and a short tail.

c. Do the same thing for the rest of the dogs until you find the only dog that can have short hair and a short tail. Fang

d. Find the type of hair and tail each dog has. Start by making check marks for the dog with short hair and a short tail that you found in part (c). Copy the chart without any checkmarks. Then use the given information and parts (a) and (b) to complete the chart.
(See the chart above.)

Using Congruent Triangles

SKILLS REVIEW

DECIMALS AND PERCENTS

Add or subtract. Be sure to keep the decimal points in line.

1. $5.372 + 2.197$ 7.569
2. $91.08 - 16.37$ 74.71
3. $7.65 + 0.829$ 8.479
4. $56.7 - 29.54$ 27.16
5. $26.03 + 62.9$ 88.93
6. $5.003 - 1.315$ 3.688
7. $9.666 - 7.209$ 2.457
8. $1.62 + 4.7$ 6.32

Multiply.

> **SAMPLE**
>
> $39.6 \times 2.04 = 80.784$
>
> 1 decimal place   2 decimal places   3 decimal places

9. 4×39.48 157.92
10. 100×61.2 6120.0
11. 0.2×9.3 1.86
12. 5.9×0.766 4.5194
13. 0.16×365 58.40
14. 35.27×7.75 273.3425
15. 1.3×8.909 11.5817
16. 81×3.14 254.34

Divide. If the division does not "come out even," round the answer to two decimal places.

> **SAMPLE**
>
> $0.51 \overline{)4.7} \rightarrow 0.51 \overline{)4.70\,000} = 9.215$ Round up   *Answer:* 9.22

17. $1.5 \overline{)135}$ 90
18. $0.7 \overline{)2.93}$ 4.19
19. $0.34 \overline{)69.7}$ 205
20. $0.8 \overline{)11}$ 13.75
21. $6.1 \overline{)0.437}$ 0.07
22. $0.03 \overline{)12.8}$ 426.67
23. $0.075 \overline{)5}$ 66.67
24. $9.4 \overline{)6.18}$ 0.66

Write as a percent.

> **SAMPLES** $0.9 = 0.90 = 90\%$ $\frac{9}{25} = \frac{36}{100} = 36\%$
>
> $\frac{3}{8} = 3 \div 8 = 0.375 = 37.5\%$

25. $\frac{3}{4}$ 75%
26. 0.27 27%
27. 0.6 60%
28. $\frac{4}{5}$ 80%
29. $\frac{7}{8}$ 87.5%
30. $0.33\frac{1}{3}$ $33\frac{1}{3}\%$
31. $\frac{1}{2}$ 50%
32. $\frac{7}{10}$ 70%

Complete.

33. 80% of 1125 is __?__.   900
34. 17 is __?__% of 51. $33\frac{1}{3}\%$
35. 25% of __?__ is 96.   384
36. 30% of 15.8 is __?__.   4.74
37. 13 is __?__% of 65.   20%
38. __?__ is 97% of 482.   467.54

CHAPTER REVIEW

CHAPTER SUMMARY

1. One way to prove that two segments or two angles are equal is to show that they are corresponding parts of congruent triangles.

2. Congruent triangles can sometimes be used to show that constructions are correct.

3. Any segment, ray, or line that passes through the midpoint of a line segment is a bisector of the segment. Any point on the perpendicular bisector of $\overline{AB}$ is equidistant from A and B. Any point equidistant from A and B must lie on the perpendicular bisector of $\overline{AB}$.

4. Every triangle has three altitudes. Every triangle has three medians.

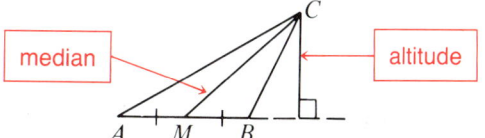

5. A circle can be circumscribed about a triangle (Construction 7). A circle can be inscribed in a triangle (Construction 8).

6. If two sides of a triangle are equal, then the angles opposite those sides are equal. An equilateral triangle is also equiangular, and each angle has measure 60°.

7. If two angles of a triangle are equal, then the sides opposite those angles are equal. If a triangle is equiangular, it is also equilateral.

8. Construction 9 tells how to construct a 60° angle.

REVIEW EXERCISES

Supply the reasons to complete the proof. *(See pp. 144–148.)*

1. Given: $CM = DM$; $EM = FM$
   Prove: $\overline{EC} \parallel \overline{DF}$

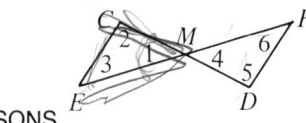

STATEMENTS	REASONS
1. $CM = DM$; $EM = FM$	1. __?__ Given
2. $\angle 1 = \angle 4$	2. __?__ Vertical angles are equal.
3. $\triangle CME \cong \triangle DMF$	3. __?__ SAS Postulate
4. $\angle 3 = \angle 6$	4. __?__ Corr. parts of $\cong \triangle$ are =.
5. $\overline{EC} \parallel \overline{DF}$	5. __?__ If 2 lines and a trans. form = alt. int. $\triangle$, then the lines are $\parallel$.

Using Congruent Triangles

Supply reasons to complete a proof that the following construction is correct. *(See pp. 149–152.)*

2. Angle *A* was given, and angle *B* was constructed to be equal to angle *A*.
   1. $BT = AR$ and $BU = AS$ Construction
   2. $TU = RS$ Construction
   3. $\triangle UBT \cong \triangle SAR$ SSS Postulate
   4. $\angle B = \angle A$ Corr. parts of $\cong \triangle$ are $=$.

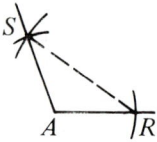 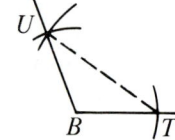

Draw a segment $\overline{CD}$. *(See pp. 154–157.)*

3. Construct the perpendicular bisector of $\overline{CD}$, and label it *k*.

4. Refer to Exercise 3. Suppose you know, for some point *X*, that $CX = DX$. Where must *X* lie?   on line k

5. Suppose *Z* is a point on *k*. What can you say about *Z* with respect to *C* and *D*?   Z is equidistant from C and D.

Refer to the figure shown. *(See pp. 158–162.)*

6. Name an altitude shown. $\overline{RY}$

7. Name a median shown. $\overline{TX}$

8. Draw a large acute triangle, $\triangle MNO$. Construct the altitude from *M*.

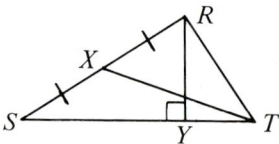

Draw two large figures, an obtuse $\triangle ABC$ and an acute $\triangle DEF$. *(See pp. 163–166.)*

9. Circumscribe a circle about $\triangle ABC$.
10. Inscribe a circle in $\triangle DEF$.

Write a proof of the theorem. *(See pp. 167–171.)*

11. If two sides of a triangle are equal, then the angles opposite those sides are equal.
    Given: $KX = KY$
    Prove: $\angle X = \angle Y$
    (*Hint:* Begin your proof by drawing the bisector of $\angle K$.)

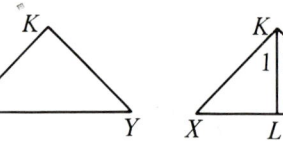 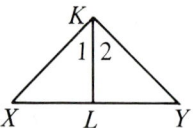

Equal angles are marked. Find *x*. *(See pp. 173–178.)*

12. 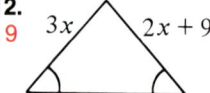   13. 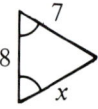   14.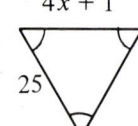

CHAPTER TEST

Given: **M is the midpoint of $\overline{AB}$ and $\overline{CM} \perp \overline{AB}$.**

1. What method could be used to prove $\triangle ACM \cong \triangle BCM$? SAS Post. (4-1)

2. After proving the triangles congruent, you could conclude that $\triangle ABC$ is isosceles. What information and supporting reason could you use? AC = BC; Corr. parts of $\cong$ $\triangle$ are =.

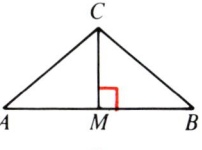

3. Tony was given $\triangle PQR$. By what method did he construct $\triangle XYZ$ congruent to $\triangle PQR$? SAS Post. (4-2)

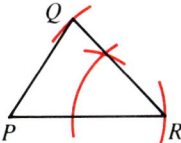 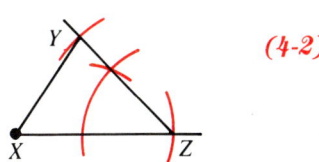

4. Draw a large triangle like $\triangle PQR$. By construction, locate the midpoint M of $\overline{QR}$. (4-3)

6. Any pt. on the $\perp$ bisector of a segment is equidistant from the endpoints of the segment.

Given: **$\overleftrightarrow{RS}$ is the perpendicular bisector of $\overline{AB}$.**

5. Point T is __?__ from A and B. equidistant

6. Why must $RA = RB$ and $SA = SB$? (See above.)

7. Why is $\triangle RAS \cong \triangle RBS$? SSS Postulate

8. Is $\overline{BQ}$ an altitude of $\triangle RBT$? a median? yes; no (4-4)

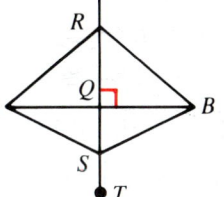

9. Draw a very large obtuse triangle. Draw and label the altitude and the median to the longest side. Which segment is longer?

10. Describe a method you could use to construct a circle that passes through the vertices of a given triangle. (4-5)

11. Draw a large acute triangle. Construct an inscribed circle in it.

Given: **$RS = TS$ and $QS = US$. Write the reason that supports each statement.** 12-13. (See below.)

12. $\angle 1 = \angle 2$ 13. $\angle Q = \angle U$ (4-6)

14. $\angle 3 = \angle 4$ 15. $\triangle QRS \cong \triangle UTS$ AAS Post.

14. If 2 $\triangle$ are supplements of = $\triangle$, then the 2 $\triangle$ are =.

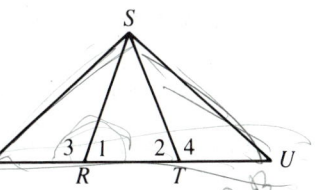

16. In $\triangle ABC$, $\angle A = \angle B$. (4-7)
   If $AB = 4x - 5$, $BC = 3x + 2$, and $AC = 4x - 3$, find x. 5

12–13. If 2 sides of a $\triangle$ are =, then the $\triangle$ opp. these sides are =.

UNIT B CUMULATIVE REVIEW

Complete each statement.

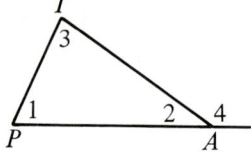

1. In △PAT at the right, ∠4 = ∠ __?__¹ + ∠ __?__³.
2. A triangle with no equal sides is called __?__. scalene
3. A triangle with two equal sides is called __?__. isosceles
4. If a triangle has two 40° angles, the third angle equals __?__°. 100
5. If a triangle has a 110° angle, then the triangle is called __?__. obtuse
6. In a right triangle, the side opposite the right angle is called the __?__. hypotenuse
7. If △AUK ≅ △JIE, then ∠A = ∠ __?__. J
8. If △RMZ ≅ △AKC, then RM = __?__. AK

Using the given information, write SSS, SAS, ASA, AAS, or HL to tell which method you could use to prove △LBU ≅ △LBI.

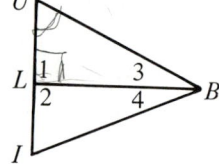

9. $\overline{BL} \perp \overline{UI}$; BU = BI HL
10. BU = BI; UL = IL SSS
11. ∠1 = ∠2; ∠3 = ∠4 ASA
12. ∠U = ∠I; ∠3 = ∠4 AAS
13. ∠1 = 90°; ∠U = ∠I AAS
14. BU = BI; ∠3 = ∠4 SAS

Tell whether each statement is true or false.

15. A segment has exactly one bisector. false
16. If point A is on a perpendicular bisector of $\overline{CD}$, then AC = AD. true
17. The three lines containing the altitudes of a triangle intersect in a point. true
18. Medians and altitudes are always the same segments. false
19. In △ABC, if AB = AC, then ∠B = ∠C. true
20. An equilateral triangle is also equiangular. true
21. Construct an equiangular triangle.
22. Draw an acute △DEF. Circumscribe a circle about △DEF.

UNIT C

Here's what you'll learn in this chapter:

To classify polygons.

To find the interior and exterior angle sums of a convex polygon.

To use the properties of parallelograms, rectangles, rhombuses, and squares.

To prove that a quadrilateral is a parallelogram.

To use properties of trapezoids.

To apply the Midpoints Theorem.

Have you ever noticed how pentagons and hexagons are combined on the surface of a soccer ball? Can you name other polygons that are suggested by the cords of the net shown in the photograph?

Chapter 5

Polygons

1 Introducing Polygons

The word *polygon* comes from the Greek words meaning *many angles*. The polygons shown in the first row below are called **convex polygons.** Do you see how they are different from the polygons in the second row?

One way to tell whether or not a polygon is convex is to imagine fitting a rubber band along the edges of the figure. If the rubber band fits snugly, then the polygon is convex. If it doesn't, the polygon is not convex.

Convex Polygons

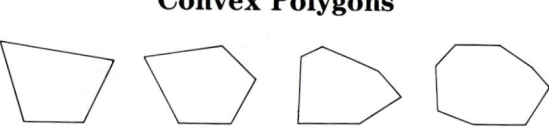

Not Convex Polygons

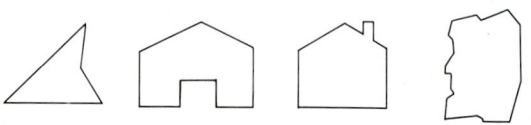

The figures should suggest what a polygon is. Even so, it is difficult to write a definition of *polygon*. Perhaps you and your class would like to try to write one. The figures below are *not* polygons and should not satisfy your definition.

Not Polygons

  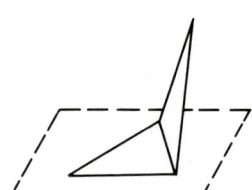

When referring to a polygon, we name its consecutive vertices in order. The *quadrilateral* shown at the top of the next page can be called, for example, *ABCD*, *DCBA*, or *BADC*. A polygon is classified according to the number of sides it has. The table at the right lists some special names.

Number of Sides	Name of Polygon
3	triangle
4	quadrilateral
5	pentagon
6	hexagon
8	octagon
10	decagon

The simplest polygon is a triangle. The terms which we defined for triangles (such as *sides, vertices,* and *angles*) also apply to other polygons.

Two sides which intersect are called **consecutive sides.** The endpoints of a side are called **consecutive vertices.** A segment which joins nonconsecutive vertices is called a **diagonal** of the polygon.

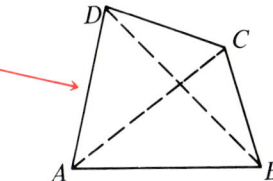

$\overline{AB}$ and $\overline{BC}$ are consecutive sides.
A and B are consecutive vertices.
$\overline{AC}$ and $\overline{BD}$ are diagonals.

Polygons can be equiangular, equilateral, or both as shown below.

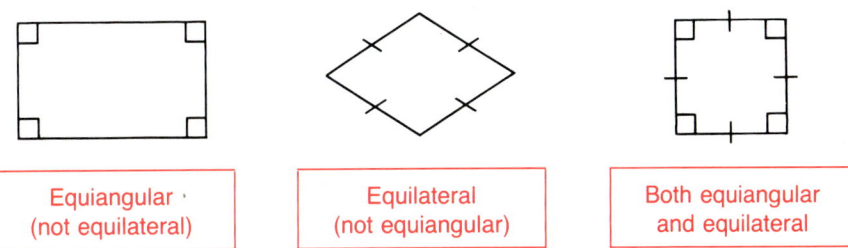

Equiangular (not equilateral)   Equilateral (not equiangular)   Both equiangular and equilateral

If a polygon is both equiangular and equilateral, it is called a **regular polygon.** A regular pentagon is shown at the right. The angle measures for regular polygons are explained in the next section.

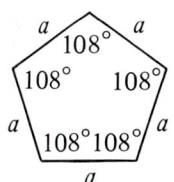

Classroom Practice

For each figure shown, decide (a) if the figure is a polygon and (b) if the figure is a convex polygon.

1. polygon
2. not a polygon
3. convex polygon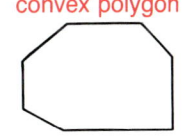
4. polygon

5. not a polygon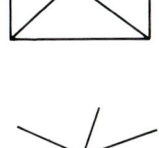
6. not a polygon
7. not a polygon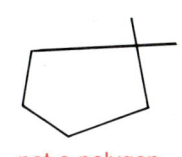
8. convex polygon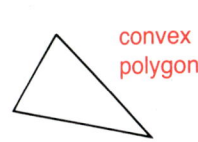

Which of the following are acceptable names for the pentagon shown?

9. *PQRST* acceptable
10. *PTQRS* not accept.
11. *SRQPT* acceptable

12. Name any two consecutive sides of the pentagon.
13. Name any two consecutive vertices of the pentagon.
14. Name as many diagonals as you can. $\overline{QT}, \overline{QS}, \overline{RP}, \overline{RT}, \overline{SP}$
15. A regular polygon with four sides is usually called a __?__. square

Polygons   **191**

Written Exercises

Is the polygon shown a convex polygon?

A 1. yes  2. no  3. yes 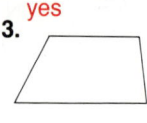 4. yes 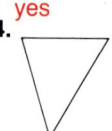 5. no  6. yes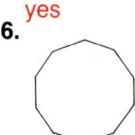

Which of the names below are acceptable for the hexagon shown?

7. ABCDEF — acceptable   8. DCBAFE — acceptable   9. CDFEAB — not acceptable

10. How many diagonals can you draw from vertex A? **3**

11. How many diagonals can you draw from vertex F? **3**

12. $\overline{DE}$ and __?__ are consecutive sides. (Two answers are possible.) $\overline{EF}$ or $\overline{CD}$

Classify each polygon as equiangular, equilateral, or regular.

13. equiangular 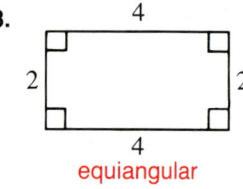   14. regular 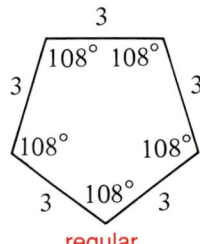   15. equilateral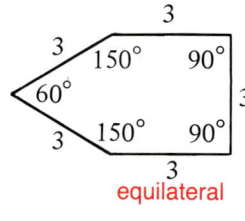

Draw a convex polygon which satisfies the conditions.

16. A quadrilateral which is equiangular but not equilateral

17. A quadrilateral which is equilateral but not equiangular

18. A regular triangle   19. A regular hexagon

B 20. A hexagon which is equilateral but not equiangular

21. A hexagon which is equiangular but not equilateral

In a convex pentagon, you can draw two diagonals from each vertex. This information is given in the table below. Copy and complete the table.

	22.	Sample	23.	24.	25.	26.	27.
Number of sides of a convex polygon	4	5	6	7	8	20	n
Number of diagonals from each vertex	1 ?	2	3 ?	4 ?	5 ?	17 ?	$n-3$?

192   Chapter 5

Copy and complete the table below.
(*Hint:* Draw a diagram for each exercise.)

	28.	29.	30.	31.	32.
Number of sides of a convex polygon	4	5	6	7	8
Total number of diagonals	?	?	?	?	?
	2	5	9	14	20

Experiments

Experiment 1

1. Cut out a strip of paper. A strip two or three centimeters wide cut from the long side of a sheet of notebook paper is convenient. Shade one side of your paper strip.

2. Begin to "tie" the paper as you would to form an ordinary overhand knot (see Figure 1).

Figure 1

3. Carefully press the folds flat. Trim off the excess paper labeled *L* and *R* in Figure 2. The polygon formed is a regular pentagon!

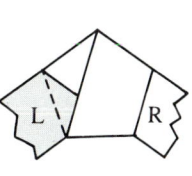

Figure 2

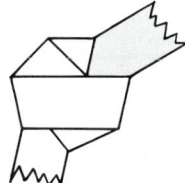

Figure 3

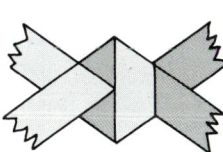

Figure 4

Experiment 2

1. Cut and shade one side of another strip of paper.

2. Begin an overhand knot. Then bring a free end around to the other side and tuck it under itself (see Figure 3).

3. Press the folds flat and trim off the excess paper. You now have a septagon!

Experiment 3

1. Cut two strips from different colored pieces of paper.

2. Make a hexagon by tying the two strips together (see Figure 4).

Polygons   **193**

2 Angle Sums of Polygons

In Section 1 of Chapter 3 we proved that the angle sum of any triangle is 180°. In fact, we can find the angle sum of any convex polygon if we know how many sides the polygon has. The diagrams below suggest a way to find this sum.

Angle Sum of a Quadrilateral

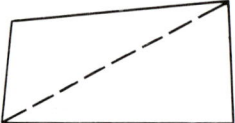

Angle sum of each triangle = 180°
Angle sum of quadrilateral = 2 × 180°
= 360°

Angle Sum of a Pentagon

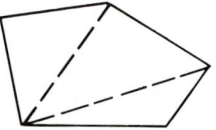

Angle sum of each triangle = 180°
Angle sum of pentagon = 3 × 180°
= 540°

Before reading further, try to find the angle sums of convex polygons with six and seven sides. Check your answers in the table below.

Number of sides of polygon	Number of triangles formed by drawing all the diagonals from one vertex	Angle sum of polygon
4	2	2 × 180° or 360°
5	3	3 × 180° or 540°
6	4	4 × 180° or 720°
7	5	5 × 180° or 900°
8	6	6 × 180° or 1080°

Do you see that if a convex polygon has n sides, you can form $n - 2$ triangles by drawing all the diagonals from one vertex? Therefore, the angle sum of the polygon is the same as the angle sum for $n - 2$ triangles. We state this as our next theorem.

THEOREM 1

If a convex polygon has n sides, then its angle sum is given by the formula

$$S = (n - 2) \times 180°.$$

EXAMPLE A regular polygon has 12 sides.
  a. Find its angle sum.
  b. Find the measure of each angle of the polygon.

  a. $S = (n - 2) \times 180° = (12 - 2) \times 180° = 1800°$
  b. Since the polygon is regular, all 12 angles are equal.
     The measure of each angle is $\frac{1}{12} \times 1800° = 150°$.

The angle sum of a polygon is sometimes called the *interior angle sum* to distinguish it from the *exterior angle sum* of the polygon. The exterior angle sum for the hexagon shown is the sum of the measures of the six indicated angles. (Notice that we consider only one of the two exterior angles at each vertex.) To find this sum, imagine yourself walking along the sides of the polygon. As you reach each vertex, you turn through the number of degrees in an exterior angle. When you return to your starting point, you will have turned through 360°.

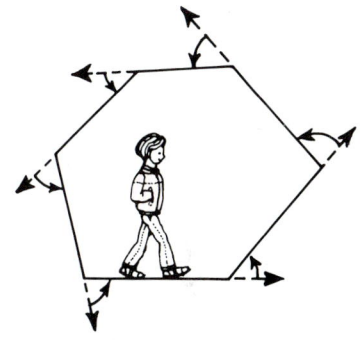

Do you see that given *any* convex polygon with any number of sides, you will turn through 360°? This fact is stated in Theorem 2. A more formal proof is suggested by Exercise 21, page 197.

THEOREM 2

The exterior angle sum of any convex polygon, one angle at each vertex, is 360°.

Classroom Practice

Copy and complete the table for convex polygons.

	Sample	1.	2.	3.	4.	5.	6.
Number of sides	5	6	8	10	12	? 13	? 22
Interior angle sum	3 × 180°	?	?	?	?	11 × 180°	20 × 180°
Exterior angle sum	360°	?	?	?	?	? 360°	? 360°

1. 4 × 180°; 360°
2. 6 × 180°; 360°
3. 8 × 180°; 360°
4. 10 × 180°; 360°

Written Exercises

Copy and complete the table for convex polygons.

Interior angle sum:
1. $2 \times 180°$ 2. $7 \times 180°$
3. $9 \times 180°$ 4. $20 \times 180°$

A

	1.	2.	3.	4.	5.	6.
Number of sides	4	9	11	22	? 12	? 17
Interior angle sum	?	?	?	?	$10 \times 180°$	$15 \times 180°$
Exterior angle sum	? 360°	? 360°	? 360°	? 360°	? 360°	? 360°

7. In a regular triangle each interior angle has measure __?__ 60°, and each exterior angle has measure __?__. 120°

8. **a.** In a regular quadrilateral each interior angle has measure __?__ 90°, and each exterior angle has measure __?__. 90°

   b. A regular polygon with four sides is called a __?__. square

In Exercises 9–11, the polygon shown is a regular polygon.
a. Name the polygon.
b. Find the interior angle sum and the measure of each interior angle.
c. Find the exterior angle sum and the measure of each exterior angle.

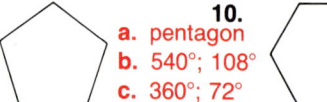

9. **a.** pentagon **b.** 540°; 108° **c.** 360°; 72°
10. **a.** hexagon **b.** 720°; 120° **c.** 360°; 60°
11. **a.** octagon **b.** 1080°; 135° **c.** 360°; 45°

Copy and complete the table for regular convex polygons.

B

	12.	13.	14.	15.	16.
Number of sides	12	? 6	? 9	? 18	? 36
Each interior angle	? 150°	120°	140°	? 160°	? 170°
Each exterior angle	? 30°	? 60°	? 40°	20°	10°

17. The neighboring atoms in a sheet of graphite consist of interlocking regular hexagons. Can you interlock STOP signs as you can hexagons? Use the results of Exercise 11 to explain your answer. No, 135 is not a factor of 360.

18. Could you tile an entrance way with regular pentagons? Use the results of Exercise 9 to explain your answer. No, 108 is not a factor of 360.

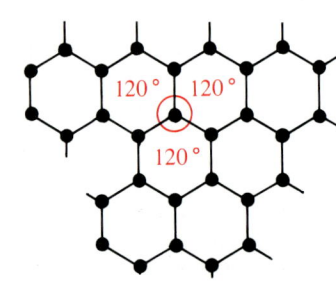

19. Given: Quadrilateral ABCD
    $\angle A = \angle C = x°$
    $\angle B = \angle D = y°$

    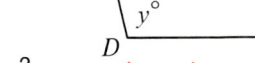

    a. $2x° + 2y° = $ __?__ ° 360   b. $x° + y° = $ __?__ ° 180
    c. $\overline{AD} \parallel \overline{BC}$ because same-side interior angles are __?__. supplementary
    d. $\overline{AB} \parallel \overline{CD}$ because __?__. same-side int. ∠s are supp.

[C] 20. A regular pentagon and two diagonals are shown. Find the measure of ∠1. **36°**

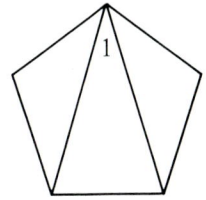

21. The diagram at the right below shows part of a polygon with n sides. Complete each statement.

    a. At vertex A, $\angle 1 + \angle 2 = $ __?__ °. 180
    b. At vertex B, $\angle 3 + \angle 4 = $ __?__ °. 180
    c. At any vertex, the sum of an interior and an exterior angle is __?__°. 180
    d. If the polygon has n sides, then it has __?__ vertices. The sum of the interior and the exterior angles at all the vertices is (__?__ × __?__)°. $n × 180$ $(n - 2) × 180$
    e. By Theorem 1, the sum of the interior angles is __?__°.
    f. By subtracting your answer in part (e) from your answer in (d), show that the sum of the exterior angles, one angle at each vertex, is 360°.
    $(n × 180)° - (n - 2) × 180° = 180° × [n - (n - 2)]$
    $= 180° × 2$
    $= 360°$

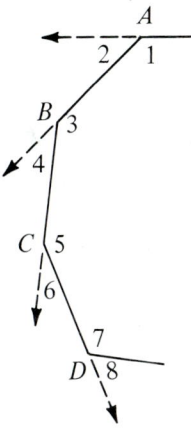

SELF-TEST

Vocabulary

convex polygon (p. 190)
consecutive sides (p. 190)
consecutive vertices (p. 190)
diagonal (p. 190)
regular polygon (p. 191)
interior angle sum (p. 195)
exterior angle sum (p. 195)

1. Draw a quadrilateral which is not convex. *(5-1)*

2. Draw a pentagon which is equilateral but not equiangular.

3. Draw a regular hexagon.

4. Find the measure of each interior angle of the hexagon. **120°** *(5-2)*

5. Find the measure of each exterior angle of the hexagon. **60°**

Polygons   197

3 Special Quadrilaterals

Some special quadrilaterals are defined below.

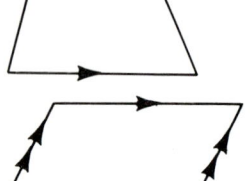

A **trapezoid** is a quadrilateral with just one pair of opposite sides parallel.

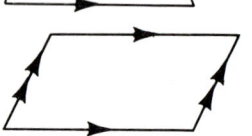

A **parallelogram** is a quadrilateral with both pairs of opposite sides parallel.

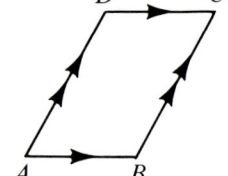

The symbol ▱ stands for parallelogram. In ▱ABCD, $\overline{AD} \parallel \overline{BC}$. Thus ∠A and ∠B must be supplementary angles. Also, ∠B and ∠C are supplementary because $\overline{AB} \parallel \overline{DC}$.

Three other special quadrilaterals are shown below with their definitions.

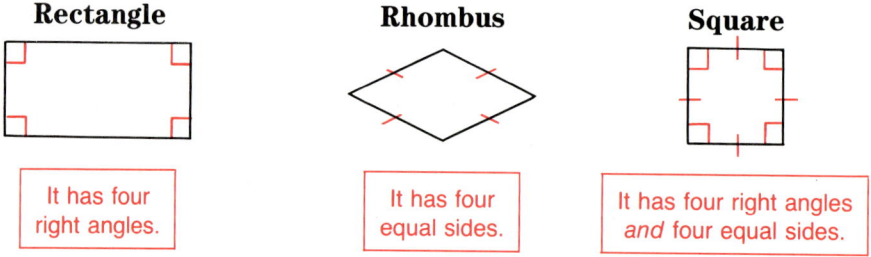

You can show that rectangles, rhombuses, and squares are all types of parallelograms (see Exercises 13–15). The diagram below illustrates how the special quadrilaterals relate to one another. Notice that every square is both a rectangle and a rhombus.

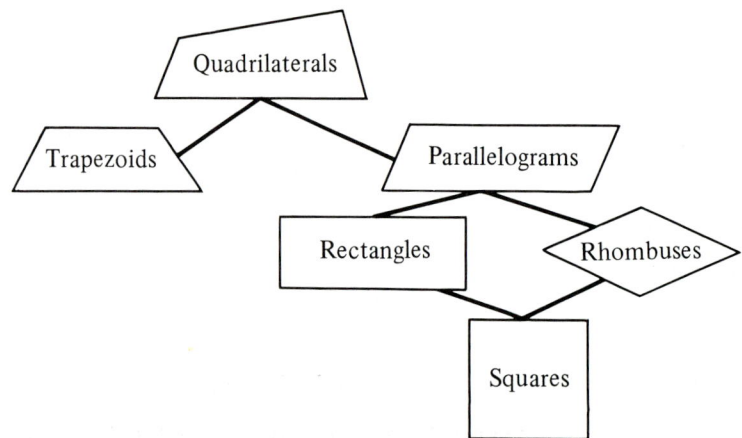

198 Chapter 5

Classroom Practice

Classify each statement as true or false. Refer to the diagram at the bottom of page 198.

1. Every trapezoid is a quadrilateral. true
2. Every rectangle is a quadrilateral. true
3. Every parallelogram is a square. false
4. Every square is a rhombus. true
5. Every trapezoid is a parallelogram. false
6. Every rectangle is a parallelogram. true

7. List each quadrilateral below which appears to be:

   a. a parallelogram. (See below.)
   b. a trapezoid. 1, 6, 9
   c. a rectangle. 3, 5, 12
   d. a rhombus. 3, 7, 11, 12
   e. a square. 3, 12

   a. 2, 3, 5, 7, 8, 11, 12

8. List each figure above which is *both* a rectangle *and* a rhombus. 3, 12
9. List each figure above which is a rectangle but not a square. 5
10. List each figure above which is a rhombus but not a square. 7, 11

In Exercises 11–14, assume that lines that appear to be parallel *are* parallel. Name all quadrilaterals that are parallelograms.

11.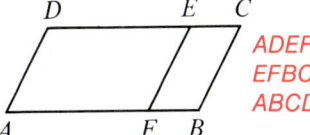
ADEF, EFBC, ABCD

12.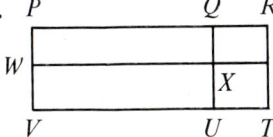
PWXQ, WVTS, WVUX, PVUQ, XUTS, QUTR, QXSR, PVTR, PWSR

13.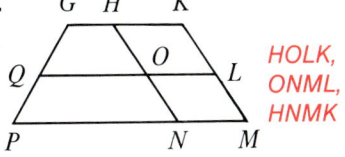
HOLK, ONML, HNMK

14.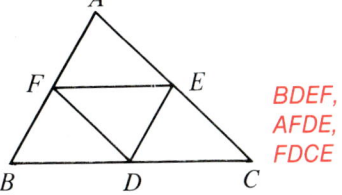
BDEF, AFDE, FDCE

15. Refer to the figure for Exercise 13. Name all trapezoids shown.
GQOH, GPMK, GQLK, GPNH, QONP, QPML
16. Refer to the figure for Exercise 14. Name all trapezoids shown.
FBCE, FDCA, ABDE

Written Exercises

A 1. List each quadrilateral which appears to be:

   a. a parallelogram. 1, 3, 4, 6, 8, 10, 11

   b. a trapezoid. 2, 5, 7, 12

   c. a rectangle. 3, 6, 10

   d. a rhombus. 4, 6, 11

   e. a square. 6

   f. both a rectangle and a rhombus. 6

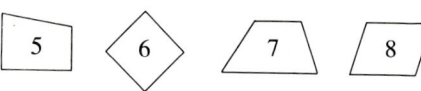
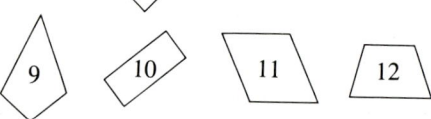

Tell whether each statement is true or false.

2. All squares are rectangles. true
3. All rectangles are squares. false
4. All squares are rhombuses. true
5. All rhombuses are squares. false

6. Imagine pushing against a rectangle with a fixed base to form a parallelogram. If you push against a square with a fixed base, what figure is formed? rhombus

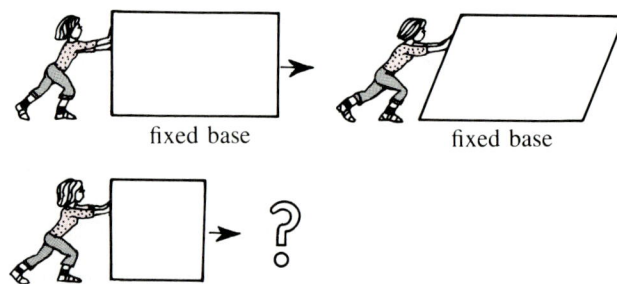

7. There are seven parallelograms in the figure below. Name as many as you can.

APTS, STQB,
PDVU, UVCQ,
APQB, PDCQ,
ADCB

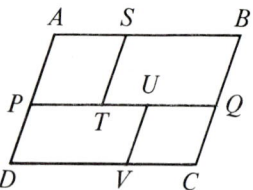

8. Name three parallelograms in the figure below.

WGOE,
EOHX,
WGHX

9. Refer to the figure for Exercise 7. Name two parallelograms which appear to be rhombuses. APTS, UVCQ

10. Refer to the figure for Exercise 8. Name six trapezoids. GZFO, FYHO, GZYH, ZFEW, FYXE, ZYXW

11. In the figure for Exercise 7, ∠D = 70°. Name four other angles with the same measure.
∠B, ∠PUV, ∠UVC, ∠UQC, ∠QTS, ∠AST, ∠APT

12. In the figure for Exercise 8, ∠W = 110°. Find the measure of:
  a. ∠X   b. ∠XHO   c. ∠HOE
   70°   110°   70°

B 13. Our goal is to show that a rhombus must be a parallelogram. Supply the reasons to complete the proof.

Given: ABCD is a rhombus.
Prove: ABCD is a ▱.

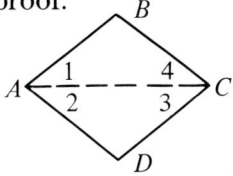

STATEMENTS	REASONS
1. $AB = CD$; $BC = DA$	1. __?__ Given: ABCD is a rhombus.
2. $AC = AC$	2. __?__ From algebra
3. $\triangle ABC \cong \triangle CDA$	3. __?__ SSS Postulate
4. $\angle 1 = \angle 3$; $\angle 2 = \angle 4$	4. __?__ Corr. parts of $\cong \triangle$ are $=$.
5. $\overline{AB} \parallel \overline{DC}$; $\overline{AD} \parallel \overline{BC}$	5. __?__ If 2 lines and a trans. form $=$ alt. int. $\angle$s, then the lines are $\parallel$.
6. ABCD is a ▱.	6. __?__ Def. of parallelogram

14. Our goal is to show that a rectangle must be a parallelogram. Supply the reasons to complete the proof.

Given: ABCD is a rectangle.
Prove: ABCD is a ▱.

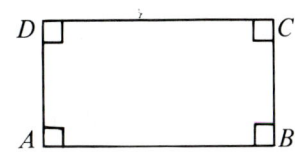

STATEMENTS	REASONS
1. $\angle A = 90°$; $\angle B = 90°$	1. __?__ Given: ABCD is a rectangle.
2. $\overline{AD} \perp \overline{AB}$; $\overline{BC} \perp \overline{AB}$	2. __?__ Def. of $\perp$ lines
3. $\overline{AD} \parallel \overline{BC}$	3. In a plane, if 2 lines are each $\perp$ to a third line, then __?__. they are $\parallel$.
4. $\angle B = 90°$; $\angle C = 90°$	4. __?__ same as 1
5. $\overline{AB} \perp \overline{BC}$; $\overline{DC} \perp \overline{BC}$	5. __?__ same as 2
6. $\overline{AB} \parallel \overline{DC}$	6. __?__ same as 3
7. ABCD is a ▱.	7. __?__ Def. of parallelogram

15. Explain why Exercise 14 lets you conclude that a square must be a parallelogram. Since a square is a rectangle, it has the properties of a rectangle.

16. Construct a rectangle, using straightedge and compass.

17. Construct a trapezoid, using straightedge and compass.

18. Draw a quadrilateral *ABCD* and locate the midpoint of each side. Join the midpoints to form a quadrilateral *EFGH*. What special kind of quadrilateral does *EFGH* appear to be? parallelogram

19. Repeat Exercise 18, beginning with a rectangle *ABCD*. rhombus

20. Repeat Exercise 18, beginning with a rhombus *ABCD*. rectangle

21. Repeat Exercise 18, beginning with a square *ABCD*. square

Using straightedge and compass, construct:

22. A rhombus that has a 60° angle

23. A parallelogram that has a 45° angle

CAREER NOTES

PHYSICAL THERAPIST

Did you ever squeeze a tennis ball, blow up a balloon, or assemble a jigsaw puzzle? There are millions of disabled people of all ages who find these simple tasks next to impossible. Physical therapists assist people who have nerve, joint, bone, and muscle diseases or injuries.

A significant part of the physical therapist's job is to develop and teach programs for treatment. Therapists use a wide range of tests to chart their patients' conditions. They test for muscular strength, muscular development, respiration, and circulation.

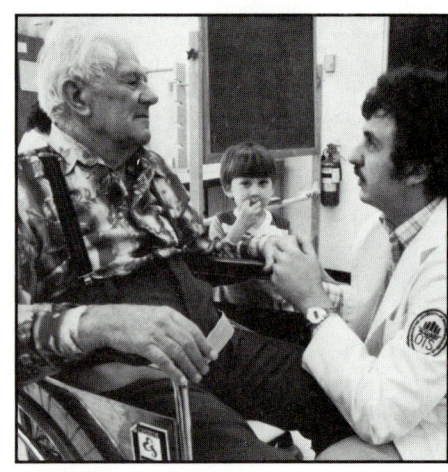

They then develop exercise programs for their patients. For example, a therapist may prescribe a series of breathing exercises for someone with asthma. Often, when a patient's legs are immobile, a therapist will encourage projects that require hand coordination, such as assembling a jigsaw puzzle.

Physical therapy involves a great deal of knowledge and responsibility. High school courses in physical education, science, and mathematics will help you prepare for this career. To obtain a license to practice, applicants must have earned a degree or certificate from a school of physical therapy and have passed an examination.

4 Properties of Parallelograms

Parallelograms have several special properties. Try the explorations below, and see how many properties you can find.

EXPLORATIONS

Carefully draw a very large parallelogram $ABCD$ which is neither a rectangle nor a rhombus. (Side $\overline{AB}$ should be at least 12 cm long.) Use your parallelogram for the exercises that follow.

4. If 2 ∥ lines are cut by a trans. then s.-s. int. ∠s are supp.

A. The Angles of a Parallelogram

1. Carefully measure the four angles of ▱$ABCD$.
2. Compare the measures of consecutive angles: They are supplementary. $\angle A$ and $\angle B$, $\angle B$ and $\angle C$, and so on.
3. Compare the measures of opposite angles: $\angle A$ and $\angle C$, $\angle B$ and $\angle D$. They are equal.
4. How can you tell, without measuring, that $\angle A + \angle B = 180°$? (See How can you tell, without measuring, that $\angle B + \angle C = 180°$? above.)
5. How can you tell that $\angle A = \angle C$, using the two equations from Exercise 4? Subtraction and Substitution Postulates

B. The Sides of a Parallelogram

1. Carefully measure the four sides of ▱$ABCD$.
2. Compare the lengths of opposite sides: $\overline{AB}$ and $\overline{CD}$, $\overline{BC}$ and $\overline{DA}$. They are =.
3. Now draw diagonal $\overline{AC}$. They are ≅.
4. What appears to be true about the two triangles formed?
5. How do you know that $\triangle ABC \cong \triangle CDA$? ASA Post.
6. Use the fact that $\triangle ABC \cong \triangle CDA$ to show that the opposite sides of ▱$ABCD$ are equal.

$AD = CB$ and $DC = BA$ (Corr. parts of ≅ ▵ are =.)

C. The Diagonals of a Parallelogram

1. Draw the other diagonal, $\overline{BD}$.
   Let O be the intersection point of the diagonals.
2. Carefully measure both diagonals. Is $AC = BD$? no
3. Carefully measure $\overline{AO}$ and $\overline{OC}$. Is $AO = OC$? yes
4. Carefully measure $\overline{BO}$ and $\overline{OD}$. Is $BO = OD$? yes
5. Would you say that the diagonals bisect each other? yes

Polygons 203

By now, you have probably discovered many properties of parallelograms. Perhaps you have even proved some of these properties. A summary of them is given below. Proofs of the theorems are left as exercises.

PROPERTIES OF A PARALLELOGRAM

By definition: Opposite sides of a parallelogram are parallel.

THEOREM 3

Opposite sides of a parallelogram are equal.

THEOREM 4

Opposite angles of a parallelogram are equal.

THEOREM 5

Consecutive angles of a parallelogram are supplementary.

THEOREM 6

Diagonals of a parallelogram bisect each other.

Warning: The diagonals of a parallelogram are not usually equal.

Classroom Practice

PQRS is a ▱. State the definition or theorem that justifies each statement.

1. $\overline{PQ} \parallel \overline{SR}$ Definition of ▱
2. $PQ = SR$ Theorem 3
3. $SO = OQ$ Theorem 6
4. $\angle PSR = \angle PQR$ Thm. 4
5. $\angle SPQ + \angle PQR = 180°$ Thm. 5
6. $SP = RQ$ Theorem 3
7. $\overline{SP} \parallel \overline{RQ}$ Definition of ▱
8. $\overline{SQ}$ bisects $\overline{PR}$.   Theorem 6

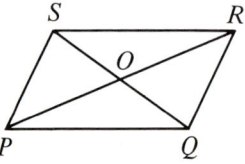

Exercises 1–9

9. Name the four pairs of congruent triangles in the figure above.
△ SPQ ≅ △ QRS; △ PQR ≅ △ RSP; △ POQ ≅ △ ROS; △ SOP ≅ △ QOR

In Exercises 10–17: ACEF is a ▱; BCDG is a ▱. Find the measure of each angle.

10. ∠ C   70°
11. ∠ G   70°
12. ∠ GDC   110°
13. ∠ GBC   110°

Find each length.

14. FE   14
15. AF   7
16. BG   4
17. GD   6

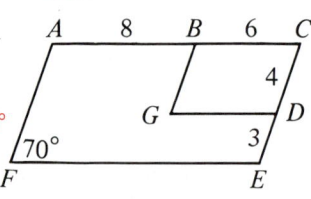

Exercises 10–17

Written Exercises

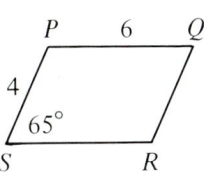

Given: PQRS is a ▱.
Find each length and angle measure.

 1. SR 6   2. RQ 4   3. ∠R 115°   4. ∠Q 65°

Given: ▱MECO and rectangle EXIC lie in the same plane.
Find each length and angle measure.

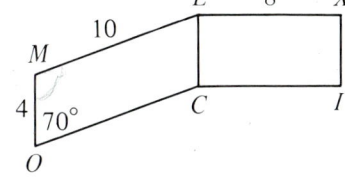

5. OC 10
6. CI 8
7. XI 4
8. ∠I 90°
9. ∠M 110°
10. ∠XEM 160°
11. Is $\overline{MO} \parallel \overline{XI}$? yes Is $\overline{ME} \parallel \overline{CI}$? no

Given: ▱ABCD; ▱AEFG
Find each length and angle measure.

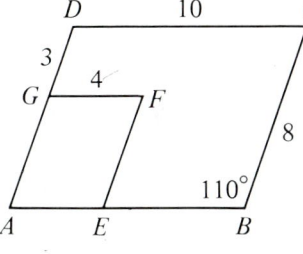

12. AG 5
13. AE 4
14. BE 6
15. ∠C 70°
16. ∠A 70°
17. ∠F 70°
18. Is $\overline{GF} \parallel \overline{DC}$? yes Is $\overline{FE} \parallel \overline{CB}$? yes

Given: ▱TLAS
Find each angle measure.

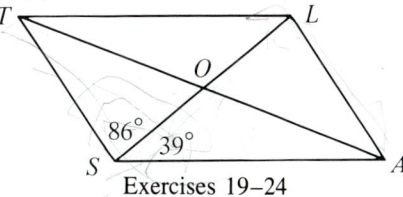

19. ∠ALS 86°
20. ∠TLS 39°
21. ∠LAS 55°

Suppose that LO = 9 and AO = 16.
Find each length.

22. SO 9
23. OT 16
24. SL 18

25. ABCD is a parallelogram. AB = 8 and BC = 10.
    Find the perimeter of the parallelogram.
    (*Hint:* The perimeter is the sum of the lengths of the sides.) 36

26. The length and width of a rectangle are 7 and 5.
    Find the perimeter of the rectangle. 24

Given: ▱QRST

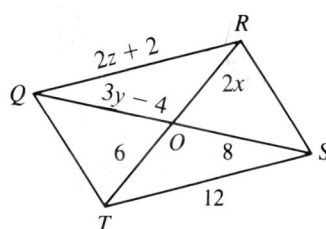

 27. Find the value of x. 3

28. Find the value of y. 4

29. Find the value of z. 5

The purpose of Exercises 30–32 is to prove Theorems 3 and 4.

30. Supply the reasons to complete the proof.

    Given: ▱ABCD

    Prove: △ABC ≅ △CDA

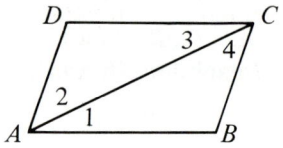

STATEMENTS	REASONS
1. ABCD is a ▱.	1. __?__ Given
2. $\overline{AB} \parallel \overline{DC}$; $\overline{AD} \parallel \overline{BC}$	2. __?__ Def. of parallelogram
3. ∠1 = ∠3; ∠2 = ∠4	3. __?__ If 2 ∥ lines are cut by a trans., then alt int. ⦤ are =.
4. AC = AC	4. __?__ From algebra
5. △ABC ≅ △CDA	5. __?__ ASA Postulate

31. Given: ▱ABCD

    Prove: AB = CD; BC = DA

    (*Hint:* Draw $\overline{AC}$. Then use the result of Exercise 30.)

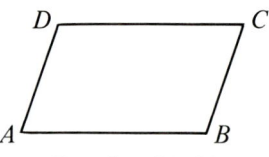

Exercises 31, 32

32. Given: ▱ABCD

    Prove: ∠B = ∠D; ∠A = ∠C

    (*Hint:* To prove that ∠B = ∠D, draw $\overline{AC}$ and use the result of Exercise 30. To prove that ∠A = ∠C, draw $\overline{BD}$.)

33. Complete this proof of Theorem 6 by supplying the reasons.

    Given: ▱EFGH with diagonals meeting at O

    Prove: EO = GO; FO = HO

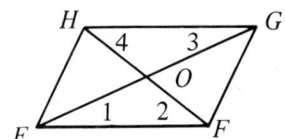

STATEMENTS	REASONS
1. EFGH is a ▱.	1. __?__ Given
2. $\overline{EF} \parallel \overline{GH}$	2. __?__ Def. of parallelogram
3. ∠1 = ∠3; ∠2 = ∠4	3. __?__ If 2 ∥ lines are cut by a trans., then alt. int. ⦤ are =.
4. EF = GH	4. __?__ Opp. sides of a ▱ are =.
5. △EOF ≅ △GOH	5. __?__ ASA Postulate
6. EO = GO; FO = HO	6. __?__ Corr. parts of ≅ △ are =.

34. Write a proof of Theorem 5.

Given: ▱WXYZ

Prove: ∠W and ∠X are supplements.
∠W and ∠Z are supplements.

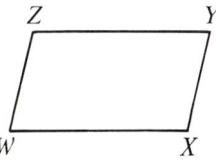

C **35.** Given: ▱ABCD

Prove: $RE = SE$

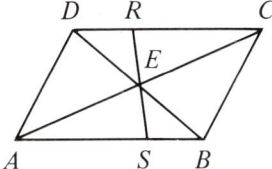

36. Our goal is to show that the distance between two parallel lines is always the same. To do this, we choose any two points on one line and prove that the points are the same distance from the other line. (The distance from a point to a line is defined to be the length of the *perpendicular* segment from the point to the line.)

Given: $l \parallel m$; $\overline{AX} \perp m$; $\overline{BY} \perp m$

Prove: $AX = BY$

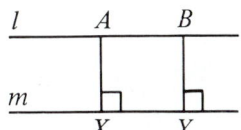

SELF-TEST

Vocabulary

trapezoid (p. 198)   rhombus (p. 198)
parallelogram (p. 198)   square (p. 198)
rectangle (p. 198)

Define each of the following special quadrilaterals. (See Glossary.)

1. Parallelogram
2. Trapezoid
3. Rhombus   *(5-3)*
4. Rectangle
5. Square

ABCD is a parallelogram. Complete each statement.

6. $AB = $ __?__ CD
7. $\overline{BC} \parallel$ __?__ $\overline{AD}$ *(5-4)*
8. $DX = $ __?__ BX
9. $\angle 1 = \angle$ __?__ 5
10. $\angle DAB = \angle$ __?__ BCD
11. $\angle 7 = \angle$ __?__ 3
12. $\angle ABC$ and $\angle$ __?__ are supplementary angles.
     BCD or DAB

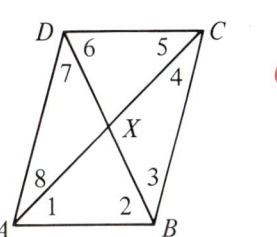

Polygons   **207**

APPLICATIONS

GOLDEN RECTANGLE

Even though all rectangles have four right angles, not all rectangles are similar. Some are much longer than they are wide, while others (squares) are exactly as long as they are wide. In one type of rectangle, the golden rectangle, the length divided by the width will always equal $\frac{1+\sqrt{5}}{2}$, about 1.6. Ancient Greek geometers and architects thought this rectangular shape was the most beautiful.

This photo of the Parthenon shows how closely the dimensions fit within a golden rectangle. The dashed line indicates the original outline of this Greek temple.

You can construct a golden rectangle by following the steps below.

Construct square $ABCD$.

Find the midpoint M of $\overline{AB}$.

Using M as center and MC as radius, draw an arc intersecting $\overleftrightarrow{AB}$ at E.

Construct a perpendicular to $\overleftrightarrow{AB}$ at point E.

Extend $\overline{DC}$ to meet the perpendicular at F.

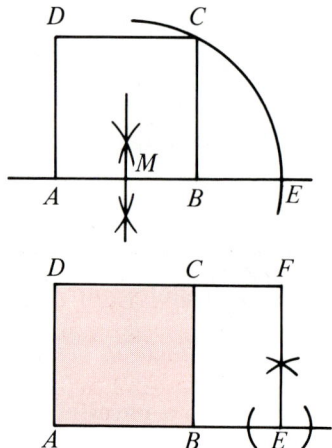

$AEFD$ is a golden rectangle. If we remove the original square $ABCD$, the remaining rectangle $BEFC$ is also a golden rectangle.

Construct a golden rectangle, measure the sides, and divide the length by the width. Is your result about 1.6?

5 Properties of Special Parallelograms

Since rectangles, rhombuses, and squares are parallelograms, they have all the properties listed on page 204. But because they are *special* parallelograms, they have other properties as well. Try the explorations below and see if you can discover what these properties are.

EXPLORATIONS

A. Carefully draw a very large rectangle that is not a square. Draw the two diagonals.
   1. Measure the diagonals. Are they equal? yes
   2. Are the diagonals perpendicular? no
   3. Does each diagonal bisect two angles of the quadrilateral? (For example, in rectangle *ABCD*, does $\overline{BD}$ bisect $\angle ABC$ and $\angle ADC$?) no

   Draw a different-looking rectangle and see whether your answers to questions 1–3 are the same.

B. Carefully draw a very large rhombus that is not a square. Draw the two diagonals.
   Answer questions 1–3 in part A.
   Repeat this experiment with a different-looking rhombus.
   1. no   2. yes   3. yes

C. Carefully draw a large square and its two diagonals.
   Answer questions 1–3 in part A.
   Repeat this experiment with a different square.
   1. yes   2. yes   3. yes

D. Now copy and complete the table.
   In each space write *Y* for *yes* or *N* for *not necessarily*.

	Rectangle	Rhombus	Square
Diagonals are equal.	? Y	? N	? Y
Diagonals are perpendicular.	? N	? Y	? Y
Each diagonal bisects two angles.	? N	? Y	? Y

If you did the exploration exercises carefully, then you will not be surprised by the following theorems.

Polygons   209

THEOREM 7

The diagonals of a rectangle are equal.

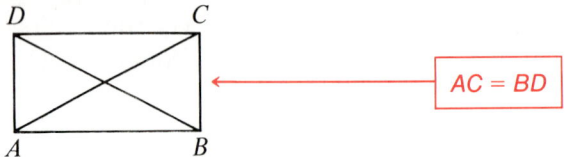

THEOREM 8

The diagonals of a rhombus are perpendicular, and they bisect the angles of the rhombus.

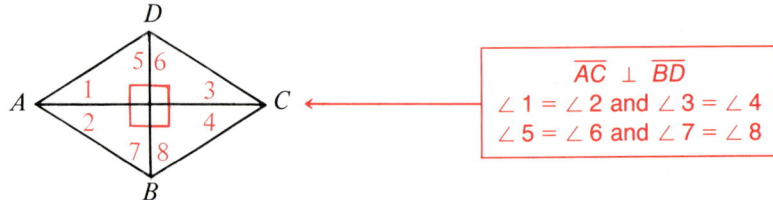

Both of these theorems apply to a square because a square is both a rectangle and a rhombus.

EXAMPLE In rectangle WXYZ, WY = 8.
Find WO, XO, YO, and ZO.

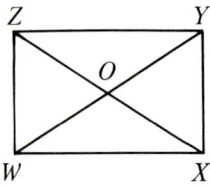

Since the diagonals of a rectangle are equal, WY and XZ both equal 8.

Since the diagonals of a parallelogram bisect each other, $WO = \frac{1}{2} \times 8 = 4$.

Likewise, XO, YO, and ZO are equal to 4.

If half of rectangle WXYZ is removed, we are left with a right triangle. Notice that O is the midpoint of hypotenuse $\overline{WY}$. The fact that WO = XO = YO suggests the following theorem.

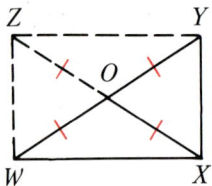

THEOREM 9

The midpoint of the hypotenuse of a right triangle is equidistant from the three vertices.

Classroom Practice

ABCD is a rhombus.

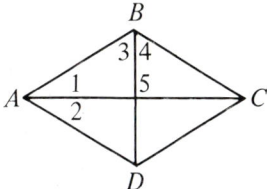

1. Suppose ∠1 = 20°. ∠2 = 20°; ∠3 = ∠4 = 70°; ∠5 = 90°
   Find the measure of each numbered angle.

2. Suppose ∠1 = 30°. ∠2 = 30°; ∠3 = ∠4 = 60°; ∠5 = 90°
   Find the measure of each numbered angle.

PQRS is a rectangle.

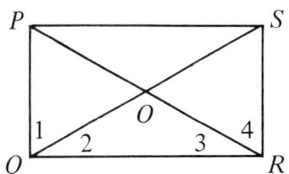

3. Suppose $SQ = 10$.
   Find SO, OQ, PO, and OR. 5; 5; 5; 5

4. Suppose ∠1 = 62°. ∠2 = ∠3 = 28°; ∠4 = 62°
   Find the measure of each numbered angle.

△PQR is a right triangle. O is the midpoint of the hypotenuse.

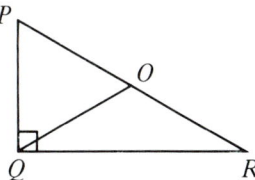

5. $\overline{QO}$ is called a(n) ___median___ ?.
   median/angle bisector/altitude

6. Suppose $PR = 10$. Find PO, OR, and OQ. 5; 5; 5

7. Suppose $QO = 6$. Find PO and PR.
   6   12

Written Exercises

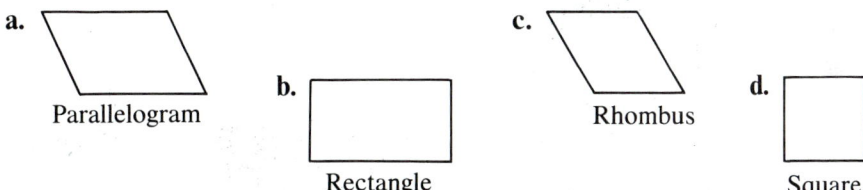

a. Parallelogram   b. Rectangle   c. Rhombus   d. Square

In Exercises 1–10, refer to the quadrilaterals shown above. For each exercise, state which quadrilaterals satisfy the given condition.

1. All angles are right angles. b, d
2. Opposite sides are equal. a, b, c, d
3. Diagonals are equal. b, d
4. Diagonals bisect each other. a, b, c, d
5. Opposite angles are equal. a, b, c, d
6. Diagonals are perpendicular. c, d
7. Each diagonal bisects two angles. c, d
8. All sides are equal. c, d
9. Diagonals are perpendicular bisectors of each other. c, d
10. When a diagonal is drawn, two congruent triangles are formed. a, b, c, d

Polygons   211

STAR is a rhombus.

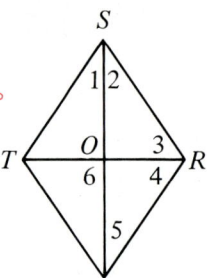

11. Suppose ∠1 = 25°. ∠2 = 25°; ∠3 = ∠4 = 65°; ∠5 = 25°; ∠6 = 90°
    Find the measure of each numbered angle.

12. Suppose ∠5 = 32°. ∠1 = ∠2 = 32°; ∠3 = ∠4 = 58°; ∠6 = 90°
    Find the measure of each numbered angle.

13. Suppose *SA* = 8 and *TR* = 6.
    Find *SO* and *TO*. 4   3

14. Suppose *RA* = 10.
    Find the perimeter of *STAR*. 40

    (*Note:* The perimeter of a polygon is the sum of the lengths of its sides.)

FLAT is a rectangle.

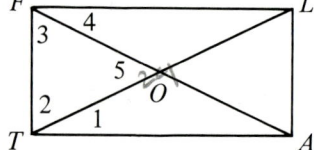

15. Suppose *FL* = 12 and *LA* = 5.
    Find the perimeter of *FLAT*. 34

16. Suppose *FA* = 13. Find *TL*. 13

17. Suppose *FA* = 14. Find *FO* and *TO*. 7; 7

18. Suppose ∠1 = 20°. ∠2 = ∠3 = 70°; ∠4 = 20°; ∠5 = 40°
    Find the measure of each numbered angle.

19. Suppose ∠2 = 75°. ∠1 = ∠4 = 15°; ∠3 = 75°; ∠5 = 30°
    Find the measure of each numbered angle.

20. Suppose ∠5 = 24°. ∠1 = ∠4 = 12°; ∠2 = ∠3 = 78°
    Find the measure of each numbered angle.

DEFG is a square.

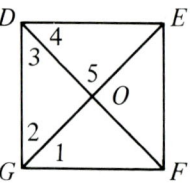

21. Find the measure of each numbered angle. ∠1 = ∠2 = ∠3 = ∠4 = 45°; ∠5 = 90°

22. If *DO* = 8, find *GE*. 16

M is the midpoint of the hypotenuse of right △*SEN*.

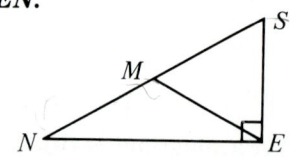

23. If *SN* = 10, find *ME*. 5

24. If *ME* = 7, find *SM* and *SN*.
    7   14

212   Chapter 5

B **25.** The purpose of this exercise is to prove Theorem 7. Supply the reasons needed to complete the proof.

Given: Rectangle ABCD with diagonals $\overline{AC}$ and $\overline{BD}$

Prove: AC = BD

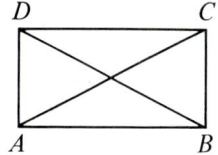

STATEMENTS	REASONS
1. ABCD is a rectangle.	1. __?__ Given
2. AD = BC	2. __?__ Opp. sides of a ▱ are =.
3. ∠DAB = ∠ABC = 90°	3. __?__ Def. of rectangle
4. AB = AB	4. __?__ From algebra
5. △DAB ≅ △CBA	5. __?__ SAS Postulate
6. AC = BD	6. __?__ Corr. parts of ≅ △ are =.

The purpose of Exercises 26 and 27 is to prove Theorem 8.

26. Supply the reasons needed to complete the proof.

Given: Rhombus ABCD with diagonal $\overline{AC}$

Prove: ∠1 = ∠2 and ∠3 = ∠4

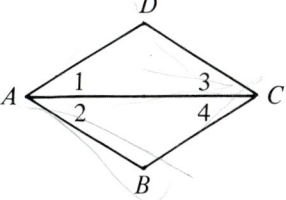

STATEMENTS	REASONS
1. ABCD is a rhombus.	1. __?__ Given
2. AB = AD; BC = DC	2. __?__ Def. of rhombus
3. AC = AC	3. __?__ From algebra
4. △ABC ≅ △ADC	4. __?__ SSS Postulate
5. ∠1 = ∠2 and ∠3 = ∠4	5. __?__ Corr. parts of ≅ △ are =.

C **27.** Copy what is shown and write a two-column proof.

Given: Rhombus ABCD
Diagonals $\overline{AC}$ and $\overline{BD}$ meet at O.

Prove: $\overline{AC} \perp \overline{BD}$

(*Hint:* You know that ∠1 = ∠2 from Exercise 26. Show that △AOD ≅ △AOB.)

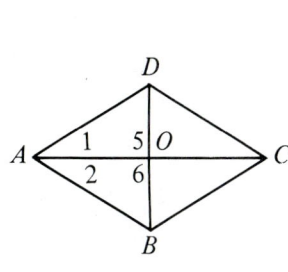

28. Construct a rhombus and draw its diagonals. Then use a protractor to check that the diagonals are perpendicular.

Polygons 213

6 Proving Figures Are Parallelograms

How can you show that quadrilateral *ABCD* is a parallelogram? Of course, one way is to show that its opposite sides are parallel. But there are other ways of showing that *ABCD* is a parallelogram. See if you can decide what these are by trying the explorations below.

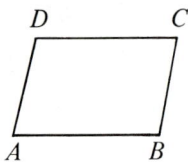

EXPLORATIONS

1. Use lined paper to draw two segments, $\overline{AB}$ and $\overline{DC}$, which are both parallel and equal.

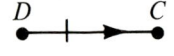

2. Draw $\overline{AD}$ and $\overline{BC}$.
   Do $\overline{AD}$ and $\overline{BC}$ appear to be parallel? yes

3. What kind of quadrilateral is *ABCD*? parallelogram

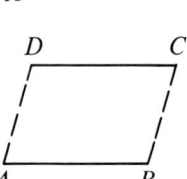

Repeat steps 1–3 using other segments which are both parallel and equal.

THEOREM 10

If a quadrilateral has one pair of opposite sides that are both parallel and equal, then the quadrilateral is a parallelogram.

Given: $\overline{AB} \parallel \overline{DC}$

$AB = DC$

Prove: *ABCD* is a ▱.

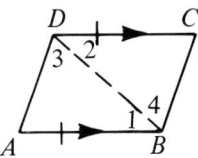

Key steps of proof:

1. Draw $\overline{DB}$. Then $\angle 1 = \angle 2$. (Why?) If 2 ∥ lines are cut by a trans., then alt. int. ▵ are =.

2. △ *ABD* ≅ △ *CDB* (Why?) SAS Postulate

3. $\angle 3 = \angle 4$ (Why?) Corr. parts of ≅ ▵ are =.

4. $\overline{AD} \parallel \overline{BC}$ (Why?) If 2 lines and a trans. form = alt. int. ▵, then the lines are ∥.

5. *ABCD* is a ▱. (Why?) Def. of parallelogram

EXPLORATIONS

If *ABCD* is a parallelogram, we know that *AB* = *DC* and *AD* = *BC*. Now let's look at the converse situation.

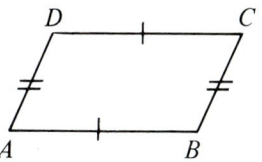

Suppose *AB* = *DC* and *AD* = *BC*.
Can we conclude that *ABCD* is a ▱? yes

Follow the steps below to see how.

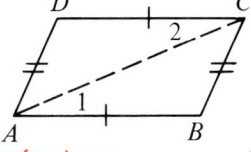

1. Draw $\overline{AC}$.
2. △*ABC* ≅ △*CDA* (Why?) SSS Postulate
3. ∠1 = ∠2 (Why?)
4. $\overline{DC} \parallel \overline{AB}$ (Why?)
5. *ABCD* is a ▱. (Why?)

Hint: The answer is not By definition.

3. Corr. parts of ≅ △ are =.
4. If 2 lines and a trans. form = alt. int. ∠s, then the lines are ∥.

Thm. 10

These five steps are the key ones in the proof of Theorem 11.

THEOREM 11

If a quadrilateral has both pairs of opposite sides equal, then the quadrilateral is a parallelogram.

EXPLORATIONS

We know that when *ABCD* is a parallelogram, diagonals $\overline{AC}$ and $\overline{BD}$ bisect each other. Is the converse true? Follow these steps.
yes

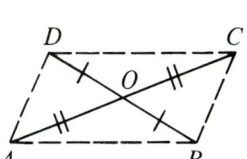

1. Draw two unequal segments, $\overline{AC}$ and $\overline{BD}$, which bisect each other at point *O*.
2. Draw quadrilateral *ABCD*.
3. △*AOB* ≅ △ _?_ and △*AOD* ≅ △ _?_   COD; COB
4. *AB* = *DC* (Why?) *AD* = *BC* (Why?) Corr. parts of ≅ △ are =.
5. *ABCD* is a ▱. (Why?) Thm. 11

These steps are the key ones in the proof of Theorem 12.

THEOREM 12

If a quadrilateral has diagonals that bisect each other, then the quadrilateral is a parallelogram.

Polygons

WAYS TO SHOW THAT A QUADRILATERAL IS A PARALLELOGRAM

1.  2. 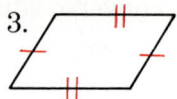 3. 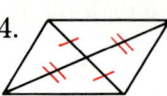 4.

1. Show that both pairs of opposite sides are parallel. (Definition of ▱)
2. Show that one pair of opposite sides are parallel and equal.
3. Show that both pairs of opposite sides are equal.
4. Show that the diagonals bisect each other.

Classroom Practice

State the definition or theorem that allows you to conclude that *CDEF* is a parallelogram.

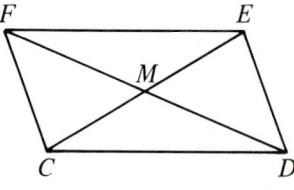

1. $CD = EF$; $DE = CF$ Thm. 11
2. $\overline{CD} \parallel \overline{EF}$; $\overline{DE} \parallel \overline{CF}$ Def. of ▱
3. $CM = ME$; $DM = MF$ Thm. 12
4. $CF = DE$; $\overline{CF} \parallel \overline{DE}$ Thm. 10
5. M is the midpoint of $\overline{CE}$ and $\overline{DF}$.   Thm. 12
6. $CD = EF$; $\angle CDE = 70°$; $\angle DEF = 110°$ Thm. 10

Written Exercises

In Exercises 1–4, write the definition or theorem which supports the statement "*ABCD* is a ▱."

A 1.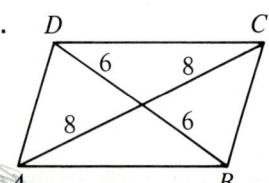
Theorem 12

2.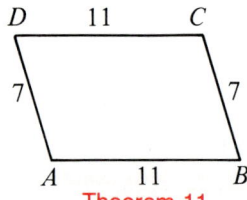
Theorem 11

3.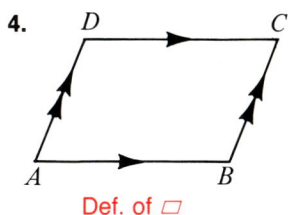
Theorem 10

4. Def. of ▱

In Exercises 5–7, supply the reasons to complete the proofs.

5. Given: $PQ = ST$
$PS = QR$
$\angle 1 = \angle 2$

Prove: $PQTS$ is a ▱.

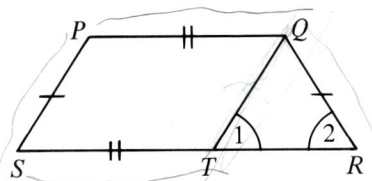

STATEMENTS		REASONS
1. $\angle 1 = \angle 2$	1. __?__	Given
2. $QT = QR$	2. __?__	If 2 ∠s of a △ are =, then the sides opp. those ∠s are =.
3. $PS = QR$	3. __?__	Given
4. $QT = PS$	4. __?__	Substitution Postulate
5. $PQ = ST$	5. __?__	Given
6. $PQTS$ is a ▱.	6. __?__	If a quad. has both pairs of opp. sides =, then the quad. is a ▱.

6. Given: $WXYZ$ is a ▱.
M is the midpoint of $\overline{ZY}$.
N is the midpoint of $\overline{WX}$.

Prove: $WNYM$ is a ▱.

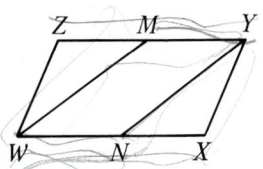

STATEMENTS		REASONS
1. $WX = ZY$	1. __?__	Opp. sides of a ▱ are =.
2. $\frac{1}{2}WX = \frac{1}{2}ZY$	2. __?__	Multiplication Post.
3. $WN = \frac{1}{2}WX$; $MY = \frac{1}{2}ZY$	3. __?__	Given: M is the midpt. of $\overline{ZY}$. N is the midpt. of $\overline{WX}$.
4. $WN = MY$	4. __?__	Substitution Postulate
5. $\overline{WX} \parallel \overline{ZY}$ (and thus $\overline{WN} \parallel \overline{MY}$)	5. __?__	Def. of parallelogram
6. $WNYM$ is a ▱.	6. __?__	If a quad. has one pair of opp. sides both ∥ and =, then the quad. is a ▱.

Polygons 217

10. rhombus; use SAS to show that 4 △, hence 4 sides, are =.
11. square; the sides are = (see Ex. 10) and ⊥.

B 7. Given: ABCD is a ▱.
P is the midpoint of $\overline{AO}$.
Q is the midpoint of $\overline{CO}$.

Prove: PBQD is a ▱.

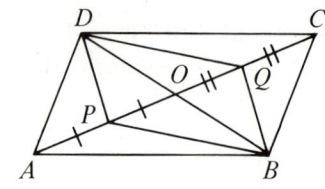

STATEMENTS	REASONS
1. AO = CO	1. __?__ Diags. of a ▱ bisect each other.
2. $\frac{1}{2}AO = \frac{1}{2}CO$	2. __?__ Multiplication Post.
3. $PO = \frac{1}{2}AO$; $QO = \frac{1}{2}CO$	3. __?__ Given: P is the midpt. of $\overline{AO}$. Q is the midpt. of $\overline{CO}$.
4. PO = QO	4. __?__ Substitution Postulate
5. BO = DO	5. __?__ Diags. of a ▱ bisect each other.
6. PBQD is a ▱.	6. __?__ If a quad. has diags. that bisect each other, then the quad. is a ▱.

8. The quadrilaterals numbered 1, 2, 3, and 4 in the diagram are parallelograms. If you wanted to show that quadrilateral 5 is also a parallelogram, which method would you use? (*Hint:* If two lines are parallel to a third line, then they are parallel to each other.) Show that the vertical sides of quad. 5 are both ∥ and =.

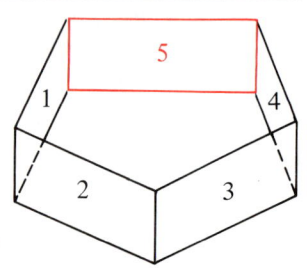

9. Draw a segment $\overline{AB}$ and its midpoint M. Draw a nonperpendicular segment $\overline{XY}$ that has M as its midpoint and is equal in length to $\overline{AB}$. Draw quadrilateral AXBY. What kind of quadrilateral is it? Explain. parallelogram; Thm. 12, page 215

10. Draw a segment $\overline{AB}$ and its midpoint M. Draw a perpendicular segment $\overline{XY}$ that has M as its midpoint but is unequal to $\overline{AB}$. Draw quadrilateral AXBY. What kind of quadrilateral is it? Explain. (See above.)

11. Draw a segment $\overline{AB}$ and its midpoint M. Draw a perpendicular segment $\overline{XY}$ that has M as its midpoint and is equal to $\overline{AB}$. Draw quadrilateral AXBY. What kind of quadrilateral is it? Explain. (See above.)

12. A kitchen utensils rack is built so that PO = QO = RO = SO. No matter how the rack is adjusted, the rows of pegs remain parallel to each other. Explain why this is so. $\overline{RS}$ and $\overline{PQ}$ always bisect each other. Thus PRQS is always a ▱ and so $\overline{RQ}$ ∥ $\overline{PS}$.

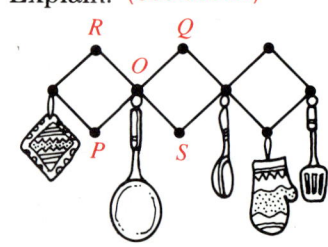

C 13. Explain why the jaws of the pliers shown are always parallel.
The four outer bolts always form a rectangle. The jaws are extensions of opposite, parallel sides.

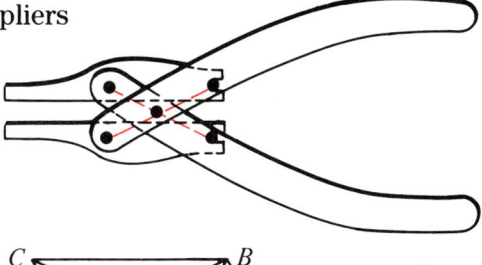

14. Given: ABCD is a ▱.
$\overline{DE} \perp \overline{AC}$
$\overline{BF} \perp \overline{AC}$

Prove: BEDF is a ▱.

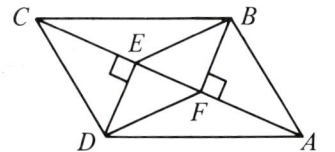

15. Prove that if a quadrilateral has equal opposite angles, then the quadrilateral is a parallelogram. (*Note:* This exercise provides a fifth way to show that a quadrilateral is a parallelogram.)

16. Prove that a parallelogram with perpendicular diagonals must be a rhombus. (*Hint:* Show that all four sides are equal.)

17. Prove that a parallelogram with equal diagonals must be a rectangle. (*Hint:* Show that all four angles are right angles.)

SELF-TEST

1. The diagonals of rectangle ABCD meet at E. If AC = 12, find BD and BE. 12; 6   (5-5)

2. CROK is a rhombus with ∠RCO = 25°. Find the measures of ∠1, ∠2, ∠3, and ∠4. 25°; 90°; 65°; 25°

3. If CR = 7, find the perimeter of CROK. 28

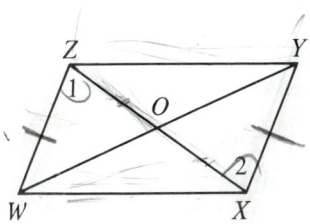

4. Given: WZ = XY
∠1 = ∠2

Prove: WXYZ is a ▱.   (5-6)

Polygons   219

evens

7 Properties of Trapezoids

Recall that a trapezoid is a quadrilateral with just one pair of parallel sides. These parallel sides are called **bases.** The other sides are called **legs.** If the legs are equal, the trapezoid is an **isosceles trapezoid.**

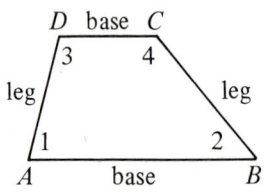
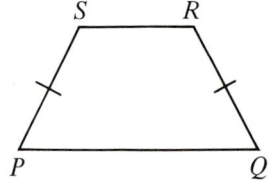

| In trapezoid *ABCD*, ∠ 1 and ∠ 2 are base angles, and so are ∠ 3 and ∠ 4. | Isosceles trapezoid *PQRS* has equal legs, $\overline{PS}$ and $\overline{QR}$. |

The two angles that include a base are called **base angles.** Every trapezoid has two pairs of base angles.

Try the following explorations and see if you can discover the special properties of trapezoids and isosceles trapezoids.

EXPLORATIONS

Draw a large non-isosceles trapezoid *ABCD* with $\overline{AB}$ as one base. Draw a large isosceles trapezoid *PQRS* with $\overline{PQ}$ as one base.

A. The Angles of a Trapezoid

1. Measure base angles, ∠ *A* and ∠ *B*, of trapezoid *ABCD*. Also measure ∠ *C* and ∠ *D*.
2. Measure base angles, ∠ *P* and ∠ *Q*, of trapezoid *PQRS*. Also measure ∠ *R* and ∠ *S*.
3. How are the base angles of an isosceles trapezoid related? They are equal.

B. The Median of a Trapezoid

The segment joining the midpoints of the legs of a trapezoid is called the **median** of the trapezoid.

1. Draw the median, $\overline{EF}$, of trapezoid *ABCD*.
2. Does it appear that the median is parallel to the bases? yes
3. Measure $\overline{AB}$, $\overline{DC}$, and the median $\overline{EF}$. How does *EF* compare with *AB* + *DC* ?

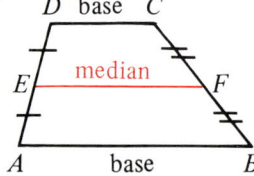

Repeat steps 1–3, using trapezoid *PQRS*.

$EF = \frac{1}{2}(AB + DC)$

220   Chapter 5

By now, you may have discovered the following two theorems.

THEOREM 13

The base angles of an isosceles trapezoid are equal.

Given: PQRS is a trapezoid with SP = RQ.

Prove: ∠P = ∠Q and ∠PSR = ∠QRS

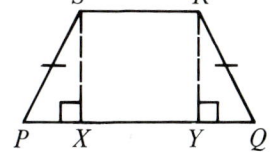

Key ideas of proof:

1. Draw $\overline{SX} \perp \overline{PQ}$ and $\overline{RY} \perp \overline{PQ}$.
2. SX = RY because the distance between two parallel lines is always the same. (See Exercise 36 on page 207.)
3. △SPX ≅ △RQY (Why?) HL Theorem
4. Then ∠P = ∠Q.
5. ∠PSR = ∠QRS (Supplements of equal angles are equal.)

THEOREM 14

The median of a trapezoid has two properties:

(1) It is parallel to the bases.
(2) Its length equals half the sum of the base lengths.

Given: Trapezoid ABCD with median $\overline{MN}$

Prove: (1) $\overline{MN} \parallel \overline{AB}$ and $\overline{MN} \parallel \overline{DC}$

(2) $MN = \frac{1}{2}(AB + DC)$

At this stage of our work, a proof of Theorem 14 would be difficult. A proof using coordinate geometry is included in Chapter 12.

EXAMPLE

A trapezoid and median are shown. Find the value of x.

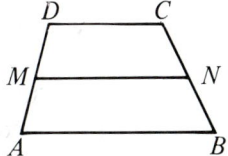

$x = \frac{1}{2}(14 + 6)$
$x = \frac{1}{2}(20)$
$x = 10$

EXAMPLE A trapezoid and its median are shown. Find the value of x.

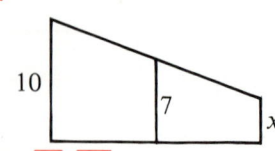

$7 = \frac{1}{2}(10 + x)$
$14 = 10 + x$
$4 = x$

2. a. $\overline{XW}, \overline{YZ}$ b. $\overline{XY}, \overline{WZ}$ c. $\angle X, \angle W; \angle Y, \angle Z$ d. yes e. $\overline{XY}, \overline{WZ}$ f. $\overline{XW}, \overline{YZ}$

Classroom Practice

1. $ABCD$ is a trapezoid.

   a. Name its two bases. $\overline{AB}, \overline{DC}$ b. Name its legs. $\overline{AD}, \overline{BC}$

   c. Name two pairs of base angles. $\angle A, \angle B; \angle C, \angle D$

   d. Does the trapezoid appear to be isosceles? no

   e. The median of the trapezoid joins the midpoint of __?__ and the midpoint of __?__. $\overline{AD}; \overline{BC}$

   f. The median of the trapezoid is parallel to __?__ $\overline{AB}$ and __?__ $\overline{DC}$.

2. $WXYZ$ is a trapezoid. Answer the questions given in Exercise 1. (See above.)

3. Suppose that in Exercise 1, $\angle A = 90°$ and $\angle B = 50°$. Find the measures of $\angle D$ and $\angle C$. $\angle D = 90°; \angle C = 130°$

4. Suppose that in Exercise 2, $WXYZ$ is an isosceles trapezoid. If $\angle Z = 60°$, find the measures of $\angle Y$ and $\angle X$. $\angle Y = 60°; \angle X = 120°$

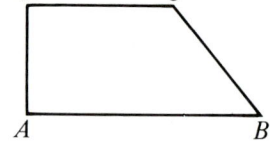

Find the length of the median of each trapezoid.

5. 8

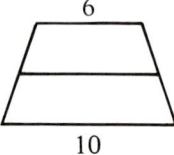

6. 9

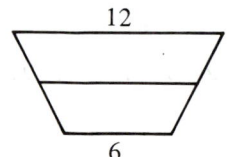

7.

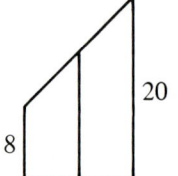

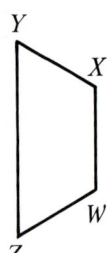

8. You can cut two congruent trapezoids out of cardboard and slide them together as shown. What kind of figure appears to be formed? parallelogram

9. P and R are the midpoints of the opposite sides of the parallelogram shown.

   a. Do $\overline{PQ}$ and $\overline{QR}$ appear to be the medians of the two trapezoids? yes

   b. Does the diagram suggest that $2m = a + b$? yes

   c. Why is $m = \frac{1}{2}(a+b)$? Multiplication Post.

   d. To what theorem is this result related? Theorem 14, page 221

Written Exercises

PQRS is an isosceles trapezoid.

A 1. If ∠P = 50°, find the measures of ∠Q, ∠R, and ∠S.   50° 130° 130°
2. If ∠P = 60°, find the measures of ∠Q, ∠R, and ∠S.
3. If ∠S = 100°, find the measures of ∠P, ∠Q, and ∠R.
4. If ∠R = 110°, find the measures of ∠P, ∠Q, and ∠S.
5. The median of the trapezoid joins the midpoints of __?__ and __?__. $\overline{SP}, \overline{RQ}$
6. If SR = 8 and PQ = 16, find the length of the median. 12

2. 60°, 120°, 120°
3. 80°, 80°, 100°
4. 70°, 70°, 110°

Find the length of the median of each trapezoid.

7. 13   8   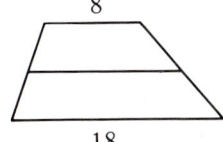   18
8. 18   24   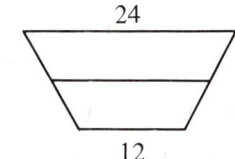   12
9. 12   9   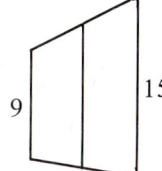   15

10. 16   10   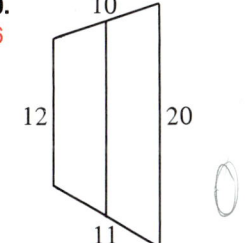   12 20 11
11. 13   8   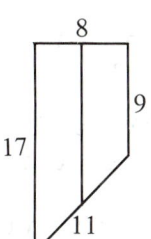   17 9 11
12. 14   9   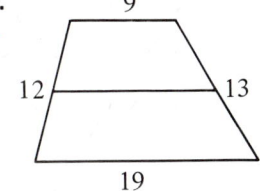   12 13 19

In Exercises 13–21, a trapezoid and its median are shown. Find the value of x.

13. 5   x   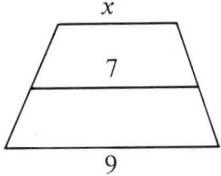   7   9
14. 16   8   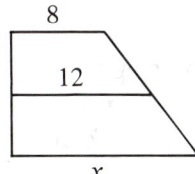   12 x
15. 4   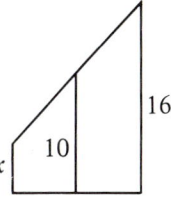   x 10 16

B 16. 2.1   x   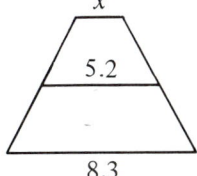   5.2   8.3
17. 4.7   7.9   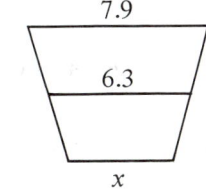   6.3   x
18. 4   x   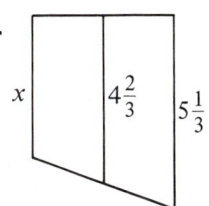 $4\frac{2}{3}$ $5\frac{1}{3}$

Polygons   223

19. 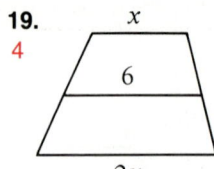   **20.** 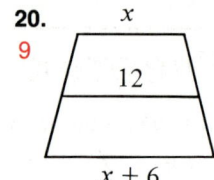   **21.**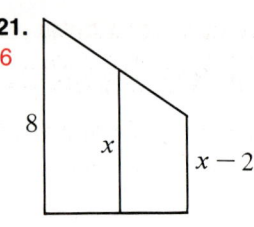

22. a. Theorem 13 can be expressed this way:
   If a trapezoid is isosceles, then __?__. the base angles are equal
  b. What is the converse of this statement? (See below.)
  c. Do you think that the converse is a true statement? yes

  b. If the base ⚞ of a trap. are = , then the trap. is isosceles.

C **23.** Is it possible for the bases of a trapezoid to be equal? Explain.
   No, the figure would then be a ▱ by Thm. 10, page 214.
  24. Is it possible for the diagonals of a trapezoid to bisect each other?
   Explain. No, the figure would then be a ▱ by Thm. 12, page 215.

  25. Construct an isosceles trapezoid.
   Measure the base angles with a protractor.

  26. Our goal is to prove that the diagonals of an
   isosceles trapezoid are equal.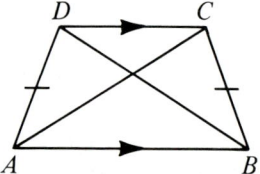

   Given: Trapezoid *ABCD* with $AD = BC$

   Prove: $AC = BD$

   (*Hint:* Use Theorem 13.)

PUZZLE ◆ PROBLEMS

Given information:
(1) Three horses named Steady, Speedy, and Slowpoke ran a race.
(2) Tic, Tac, and Toe each owned one of the horses.
(3) Tac's horse nearly won.
(4) This was Speedy's fifth race.
(5) Toe owned the black horse.
(6) The horse owned by Tic had not
   previously raced.
(7) Steady broke an ankle after the start
   of the race.
(8) The horse that won was brown.

The winning horse was __?__. Slowpoke
The winning horse was owned by __?__.
                                  Tic

8 The Midpoints Theorem

As you study the figures below from left to right, what do you notice? The base $\overline{DC}$ of trapezoid $ABCD$ becomes smaller and smaller. Finally, in figure (d), $\overline{DC}$ has shrunk to a single point and trapezoid $ABCD$ has become $\triangle ABC$.

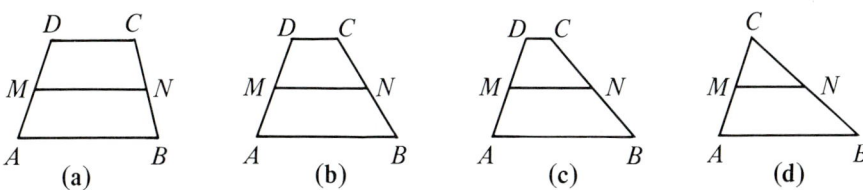

If we apply Theorem 14 to figures (a), (b), and (c) we have the following results for each figure:

(1) $\overline{MN} \parallel \overline{AB}$

(2) $MN = \frac{1}{2}(AB + DC)$

Theorem 14 can also suggest some information about figure (d). If we think of $\triangle ABC$ as a trapezoid $ABCD$ with $DC = 0$, we have

(1) $\overline{MN} \parallel \overline{AB}$

(2) $MN = \frac{1}{2}(AB + 0) = \frac{1}{2}AB$

These results are stated in the following theorem.

> **THEOREM 15 (The Midpoints Theorem)**
>
> *The segment joining the midpoints of two sides of a triangle is parallel to the third side and half as long.*

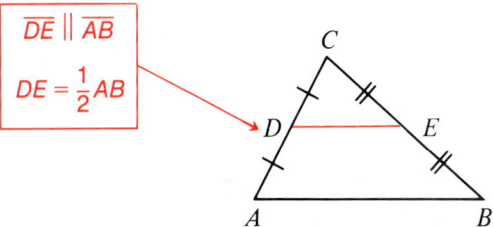

Polygons

EXAMPLE $M, N,$ and O are the midpoints of the sides of $\triangle ABC$. Find $MN, NO,$ and MO.

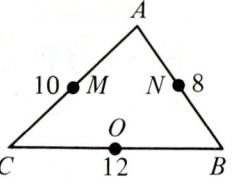

Use the Midpoints Theorem three times.

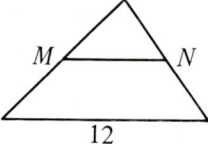 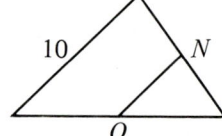 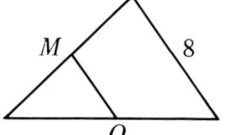

$MN = \frac{1}{2} \times 12 = 6$ $NO = \frac{1}{2} \times 10 = 5$ $MO = \frac{1}{2} \times 8 = 4$

Classroom Practice

$M, N,$ and O are the midpoints of the sides of $\triangle ABC$. Complete each statement.

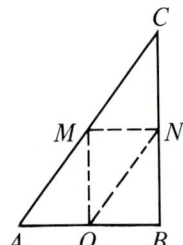

1. If $AB = 8$, then $MN = $ __?__.   4
2. If $AC = 14$, then $NO = $ __?__.   7
3. If $BC = 10$, then $MO = $ __?__.   5
4. If $MN = 11$, then $AB = $ __?__.   22
5. If $NO = 15$, then $AC = $ __?__.   30
6. If $MO = x$, then $BC = $ __?__. $2x$
7. Explain why $AMNO$ must be a parallelogram.   Opposite sides are parallel.
8. Name parallelograms, other than $AMNO$, shown in the figure. $OMNB, OMCN$

Written Exercises

M and N are the midpoints of $\overline{XZ}$ and $\overline{YZ}$. Complete each statement.

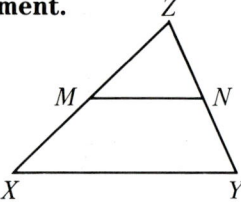

A 1. $\overline{MN} \parallel$ __?__. $\overline{XY}$
2. If $XY = 12$, then $MN = $ __?__.   6
3. If $XY = 20$, then $MN = $ __?__.   10
4. If $MN = 7$, then $XY = $ __?__.   14
5. If $ZM = 10, MN = 12,$ and $ZN = 7$, find the perimeter of $\triangle XYZ$.   58
6. If $XM = 6, MN = 7,$ and $NY = 4$, find the perimeter of $\triangle XYZ$.   34

R, S, and *T* are midpoints of the sides of △*ABC*.
Copy and complete the table.

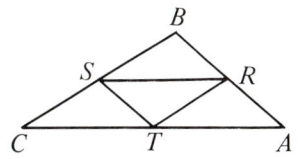

	AB	BC	AC	ST	RT	RS	Perimeter of △ABC	Perimeter of △RST
7.	8	10	12	? 4	? 5	? 6	? 30	? 15
8.	10	10	14	? 5	? 5	? 7	? 34	? 17
9.	? 8	? 12	? 10	4	6	5	? 30	? 15
10.	? 12	? 10	? 14	6	5	7	? 36	? 18
11.	14	? 12	? 8	? 7	6	4	? 34	? 17
12.	? 12	10	16	6	? 5	? 8	? 38	? 19

13. If the perimeter of △*ABC* is 32, the perimeter of △*RST* is ___?___. 16

14. If the perimeter of △*RST* is 25, the perimeter of △*ABC* is ___?___. 50

In Exercises 15–18, exactly one of the lengths represented by *x, y,* and *z* can be found. Find that length.

B

15. *y* = 5

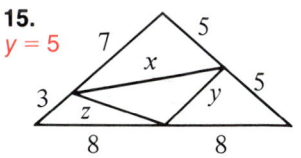

16. *z* = 10

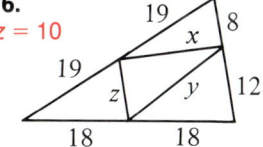

17. *y* = 13

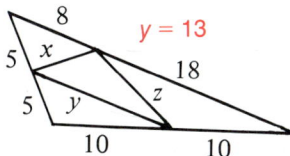

18. *x* = 8

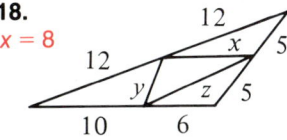

P, Q, and *R* are the midpoints of the sides of △*XYZ*.

19. Find the perimeter of ▱*PQRY*. 44

20. Find the perimeter of ▱*XQRP*. 34

21. a. Name a parallelogram in the figure, other than *PQRY* or *XQRP*. PRZQ

  b. Find the perimeter of this parallelogram. 50

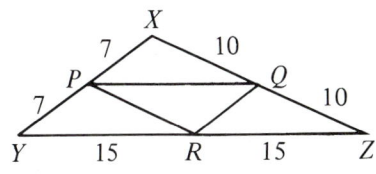

Polygons 227

22. Draw a right △ABC with ∠C = 90°. Let X be the midpoint of $\overline{AC}$. Let Y be the midpoint of $\overline{BC}$. Let Z be the midpoint of $\overline{AB}$.
   a. What kind of quadrilateral is CXZY?   b. If CZ = 10, find XY and AB. 10; 20
   rectangle

C 23. Points M, N, and O are the midpoints of the sides of a triangle, △DEF. Copy the figure shown and construct △DEF.

   M •

   N •

   O •

24. ABCD is a trapezoid with median $\overline{MN}$. Find the values of x, y, and z. x = 4; y = 6; z = 4

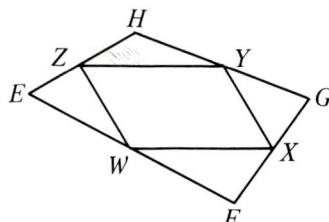

25. Let W, X, Y, and Z be the midpoints of the sides of any quadrilateral EFGH.

   Prove: WXYZ is a ▱.

   (Hint: Draw $\overline{EG}$.)

SELF-TEST

Vocabulary

bases of a trapezoid (p. 220)
legs of a trapezoid (p. 220)
isosceles trapezoid (p. 220)

base angles of a trapezoid (p. 220)
median of a trapezoid (p. 220)

TGOA is an isosceles trapezoid.

1. If GO = 12 and TA = 6, find the length of the median. 9

2. If ∠T = 120°, find the measures of ∠G, ∠O, and ∠A. 60°; 60°; 120°

(5-7)

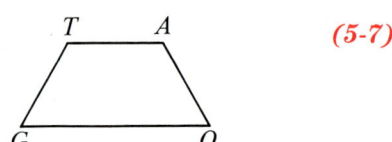

In the diagram, P, K, and T are the midpoints of the sides of △SAL.

3. Name three parallelograms shown. PTKL, SPKT
   PKAT

4. If PT = 5, then LA = __?__. 10

5. If the perimeter of △SAL = 18, then the perimeter of △PKT = __?__. 9

(5-8)

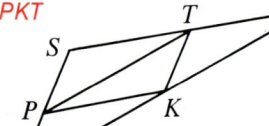

228   Chapter 5

READING GEOMETRY

WORDS AND MEANINGS

Theorem 8 on page 210 says, "The diagonals of a rhombus are perpendicular, and they bisect the angles of the rhombus." This is not a complicated sentence, but it won't make sense to you unless you know the meaning of *diagonals*, *rhombus*, *perpendicular*, *bisect*, and *angle*. Every one of these words stands for a geometric idea. When you read one of these words, you should have a clear picture or mental image to go with it. For example, perpendicular might suggest this:

Just learning words isn't very useful unless you understand the ideas that go with them. You also need to know how the ideas are related. Very often, in geometry, one statement that is known to be true may suggest several other related facts. If you can recall these related facts, you can use them in writing proofs and solving problems. For example, if you know that two given lines are parallel and that they are cut by a transversal, you can safely make statements about the angles that are formed. "Corresponding angles are equal" is one of them.

Sometimes a diagram or chart can help you organize your thoughts around related ideas. A quick look at the diagram at the bottom of page 198 helps you understand how special quadrilaterals are related. The chart on page 346 shows many important relationships that you will need to know when studying circles.

EXERCISES

1. Make a chart to show how triangles can be classified by their angles. Add two columns to indicate where the altitudes and perpendicular bisectors meet. **2.** Yes; If 2 sides of a △ are =, then the ∠s opposite these sides are =.
2. If a triangle is isosceles, do you know something special about its angles? Can you state a theorem that answers this question?
3. In what ways are a square and a rhombus alike? In what ways are they different? Both have 4 equal sides. A square has 4 equal (right) ∠s; a rhombus has 2 pairs of equal ∠s.
4. If you know that two angles are opposite angles of a parallelogram, what other fact can you state about these angles? They are =.
5. If it is given that two sides of a quadrilateral are parallel, can you safely state that the figure is a parallelogram? Explain.
   No; it might be a trapezoid.

Polygons   **229**

SKILLS REVIEW

SQUARES AND SQUARE ROOTS

$5^2 = 25$ ← square of 5 $\sqrt{25} = 5$ ← positive square root of 25

$(-5)^2 = 25$ ← square of -5 $-\sqrt{25} = -5$ ← negative square root of 25

Complete the table.

	1.	2.	3.	4.	5.	6.	7.	8.
Number	64	121	49	81	169	36	100	144
Positive Square Root	8	11	7	9	13	6	10	12
		?	?	?		?	?	?
Negative Square Root	-8	-11	-7	-9	-13	-6	-10	-12
	?		?	?	?	?	?	?

Find the value. Leave your answer in simplest radical form if the number is not a perfect square. Assume that variables represent numbers greater than zero.

SAMPLES $\sqrt{900} = \sqrt{9 \cdot 100} = \sqrt{9} \cdot \sqrt{100} = 3 \cdot 10 = 30$

$\sqrt{72x^2} = \sqrt{36 \cdot 2 \cdot x^2} = \sqrt{36} \cdot \sqrt{2} \cdot \sqrt{x^2} = 6 \cdot \sqrt{2} \cdot x = 6x\sqrt{2}$

9. $\sqrt{225}$ 15   10. $\sqrt{18}$ $3\sqrt{2}$ 11. $\sqrt{49n^2}$ $7n$ 12. $\sqrt{75}$ $5\sqrt{3}$ 13. $\sqrt{63}$ $3\sqrt{7}$
14. $\sqrt{16y^2}$ $4y$ 15. $\sqrt{400}$ 20   16. $\sqrt{125}$ $5\sqrt{5}$ 17. $\sqrt{128}$ $8\sqrt{2}$ 18. $\sqrt{50a^2}$ $5a\sqrt{2}$
19. $\sqrt{98}$ $7\sqrt{2}$ 20. $\sqrt{20k^2}$ 21. $\sqrt{196}$ 14   22. $\sqrt{162}$ $9\sqrt{2}$ 23. $\sqrt{700}$ $10\sqrt{7}$
24. $\sqrt{32b^2}$ 25. $\sqrt{324}$ 18   26. $\sqrt{45}$ $3\sqrt{5}$ 27. $\sqrt{256}$ 16   28. $\sqrt{147}$ $7\sqrt{3}$
20. $2k\sqrt{5}$ 24. $4b\sqrt{2}$

Simplify. Leave no perfect-square factor under the radical sign.

SAMPLES $\sqrt{2} \cdot \sqrt{8} = \sqrt{2 \cdot 8} = \sqrt{16} = 4$ $\dfrac{\sqrt{18}}{\sqrt{2}} = \sqrt{\dfrac{18}{9}} = \sqrt{9} = 3$

29. $\sqrt{12} \cdot \sqrt{3}$ 6   30. $\sqrt{10} \cdot \sqrt{2}$ $2\sqrt{5}$ 31. $\sqrt{5} \cdot \sqrt{15}$ $5\sqrt{3}$ 32. $\sqrt{14} \cdot \sqrt{7}$ $7\sqrt{2}$
33. $\dfrac{\sqrt{48}}{\sqrt{3}}$ 4   34. $\dfrac{\sqrt{42}}{\sqrt{7}}$ $\sqrt{6}$ 35. $\dfrac{\sqrt{32}}{\sqrt{8}}$ 2   36. $\dfrac{\sqrt{54}}{\sqrt{3}}$ $3\sqrt{2}$

CHAPTER REVIEW

CHAPTER SUMMARY

1. A polygon is classified according to the number of sides it has. A regular polygon is both equilateral and equiangular.

2. In a convex polygon with n sides, the interior angle sum is $(n-2) \times 180°$. The exterior angle sum of any convex polygon, one angle at each vertex, is $360°$.

3. Properties of some special quadrilaterals are listed below.

	parallelogram	rectangle	rhombus	square
opposite sides parallel	✓	✓	✓	✓
opposite sides equal	✓	✓	✓	✓
opposite angles equal	✓	✓	✓	✓
consec. angles supplementary	✓	✓	✓	✓
diagonals bisect each other	✓	✓	✓	✓
diagonals equal		✓		✓
diagonals perpendicular			✓	✓
diagonals bisect angles			✓	✓

4. Four ways for showing that a quadrilateral is a parallelogram are given on page 216.

5. A trapezoid is a quadrilateral with just one pair of parallel sides. The median of a trapezoid is parallel to the bases and its length equals half the sum of the base lengths.

6. The Midpoints Theorem states that the segment joining the midpoints of two sides of a triangle is parallel to the third side and half as long.

REVIEW EXERCISES

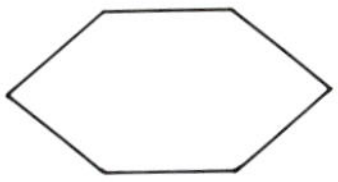

Refer to the figure at the right. *(See pp. 190–193.)*

1. Is the polygon convex? yes
2. Does the polygon appear to be equilateral? yes
3. Does the polygon appear to be equiangular? no
4. Is the polygon regular? no

Solve. *(See pp. 194–197.)*

5. Find the interior angle sum of a pentagon. 540°

6. Find the exterior angle sum of a pentagon. 360°

7. Find the measure of each angle of a regular octagon. 135°

Tell whether each statement is true or false. *(See pp. 198–202.)*

8. Every rhombus is a square. false
9. Every square is a parallelogram. true
10. Every parallelogram is a rectangle. false

Given ▱PQRS. *(See pp. 203–207.)*

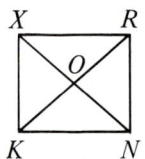

11. If ∠RPQ = 18° and ∠SPR = 30°, find the measures of ∠PRQ and ∠PQR. 30°; 132°

12. If PO = 12 and QO = 7, find the lengths of $\overline{OS}$ and $\overline{PR}$. 7; 24

Given rectangle KNRX and rhombus SLFG. *(See pp. 209–213.)*

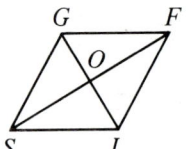

13. If RK = 20, then ON = __?__. 10

14. If ∠RKN = 40°, find the measures of ∠RKX and ∠KXN. 50° 50°

15. If ∠LSG = 60°, find the measures of ∠FSG and ∠LOF. 30° 90°

Write the definition or theorem that supports the statement "ABCD is a parallelogram." *(See pp. 214–219.)*

16.  Thm. 12, p. 215

17. 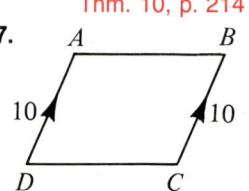 Thm. 10, p. 214

18. 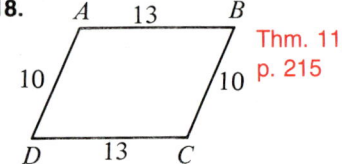 Thm. 11, p. 215

ABXY is an isosceles trapezoid with bases $\overline{AB}$ and $\overline{XY}$. AY = BX = 6, XY = 8, and AB = 14. *(See pp. 220–224.)*

19. Find the length of the median of ABXY. 11

20. If ∠A = 60°, find the measures of ∠B, ∠X, and ∠Y. 60°, 120°, 120°

M, N, and O are the midpoints of the sides of △TRA. RA = 14, AT = 6, and TR = 12. *(See pp. 225–228.)*

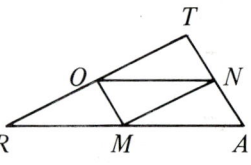

21. MN = __?__ 6

22. The perimeter of △NOM is __?__. 16

CHAPTER TEST

1. Sketch an octagon that is equiangular but not equilateral. *(5-1)*

2. A regular polygon is
   a. equiangular   b. equilateral   c. both (a) and (b) **c**

A regular polygon has ten sides.

3. Find the measure of each exterior angle. **36°** *(5-2)*

4. Find the interior angle sum. **1440°**

5. If *ABCDEF* is a regular hexagon, find the measure of ∠ *AFC*. **60°**

Tell whether each statement is true or false.

6. Some trapezoids are parallelograms. **false** *(5-3)*

7. Some rectangles are squares. **true**   8. Every rhombus is a quadrilateral. **true**

**The diagonals of ▱ WXYZ intersect at point P.
Tell whether the conclusion is true.**

9. *WZ* = *XY* **true**   10. ∠ *WXY* + ∠ *XYZ* = 180° **true**   11. ∠ *ZYW* = ∠ *XWY* **true** *(5-4)*

DAEC is a rectangle.

12. a. ∠ 1 = __?__° **36**   b. ∠ 2 = __?__° **36** *(5-5)*

13. If *AC* = 4.5, then *ED* = __?__. **4.5**

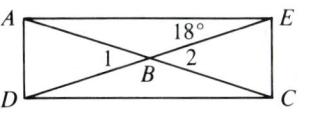

Write the definition or theorem that allows you to conclude that the quadrilateral is a parallelogram.

14. Thm. 10, p. 214   15. Thm. 12, p. 215   16. Thm. 7, p. 72; Def. of ▱ *(5-6)*

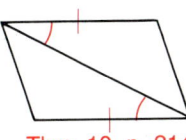 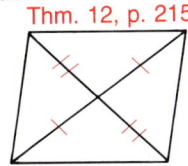 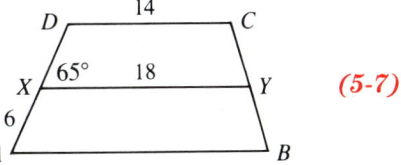

$\overline{XY}$ **is the median of isosceles trapezoid** *ABCD*.

17. Find the measures of ∠ *B* and ∠ *C*. **65°, 115°** *(5-7)*

18. Find the lengths of $\overline{AB}$ and $\overline{BC}$. **22; 12**

In △ *DEF*, *M* is the midpoint of $\overline{DE}$ and *N* is the midpoint of $\overline{EF}$.

19. Explain why ∠ *EMN* = ∠ *D*. Since $\overline{MN}$ ∥ $\overline{DF}$ (Midpoints Thm.) and $\overline{ED}$ is a transversal, corr. ≜ (∠ *EMN* and ∠ *D*) are =. *(5-8)*

20. If *DF* = 5*x* − 1 and *MN* = *x* + 4, *x* = __?__. **3**

Polygons **233**

MIXED REVIEW

1. In $\triangle XYZ$, $\angle X = 53°$, $\angle Y = d°$, and $\angle Z = (2d - 5)°$. Find the value of d. 44

2. Write a two-column proof.

   Given: $\angle 1 = \angle 2$

   Prove: $\angle 3 = \angle 4$

   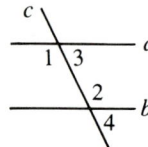

3. In quadrilateral $ABCD$, $\angle A = 90°$, $\angle B = 50°$, and $\angle C = 130°$. Choose the best name for the quadrilateral. b

   a. parallelogram **b.** trapezoid **c.** rectangle **d.** rhombus

In quadrilateral $EFGH$, $\overline{EF} \parallel \overline{GH}$ and $EF = GH$.

4. Which theorem supports the conclusion that $EFGH$ is a parallelogram? If a quad. has one pair of opp. sides both $\parallel$ and $=$, then the quad. is a $\square$.

5. If $EFGH$ is a parallelogram, must $EH = FG$? Explain.
   Yes; opp. sides of a $\square$ are $=$.

Write *yes* or *no* to indicate whether the given method could be used to prove the two triangles congruent.

6. SSS 7. SAS 8. ASA 9. AAS 10. HL
   yes   yes   yes   yes   yes

11. Suppose many isosceles triangles have the same base. Then what must be true of the altitudes of the triangles drawn to the base?
    They all lie on the same line.

12. Write a two-column proof.

    Given: $\angle P = \angle T$; $QR = RS$

    Prove: $PQ = ST$

    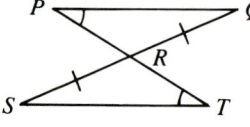

13. Construct a 45° angle.

14. In the diagram, $\angle 1 = 30°$ and $\angle 2 = 20°$. Find the measures of the other numbered angles.

    14. $\angle 3 = \angle 6 = 130°$
         $\angle 4 = 30°$
         $\angle 5 = 20°$
         $\angle 7 = 50°$
         $\angle 8 = 40°$

    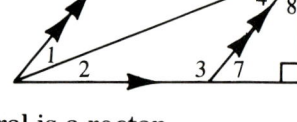

15. Write the converse of the statement "If a quadrilateral is a rectangle, then it has equal diagonals." Is the converse true or false?
    If a quad. has $=$ diags., then it is a rectangle; false.

16. Find the measure of each interior angle of a regular octagon. 135°

17. Use one of the words *always*, *sometimes*, or *never* to complete: An obtuse triangle is __?__ equilateral. never

18. Draw a very large triangle. Construct the inscribed circle. Constr. 8

19. The diagonals of ▱ABCD intersect at O. If $AO = 3x - 4$ and $AC = 4x + 6$, find the numerical length of $\overline{OC}$. 17

20. In $\triangle RST$, $RZ = ZT$, $RW = WS$, $\overline{YW} \perp \overline{RS}$, and $\overline{RX} \perp \overline{TS}$. __?__ is a median, __?__ is a perpendicular bisector, and __?__ is an altitude. $\overline{SZ}$; $\overline{YW}$; $\overline{RX}$

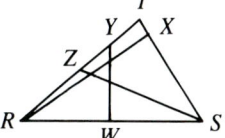

21. Points J and K have coordinates -17 and -5, respectively. Find the coordinate of the midpoint of $\overline{JK}$. -11

22. Can $\triangle ABC$ and $\triangle EDC$ be proved congruent? If so, what postulate or theorem would be simplest to use? yes; HL Theorem

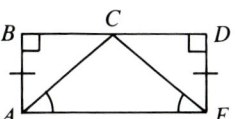

23. Sketch a rhombus that is not a square.

24. Name the four ways to prove that two lines are parallel. See the list on p. 216.

25. In $\triangle JMW$, $JM = 9$, $MW = 7$, and $JW = 10$. Find the perimeter of the triangle formed by joining the midpoints of the sides of $\triangle JMW$. 13

26. *Must* the diagonals of a rhombus have the stated property?
    yes **a.** They are perpendicular.   **b.** They are equal. no
    c. They divide the rhombus into four congruent triangles. yes
    d. They intersect at the midpoint of each diagonal. yes

27. Use a protractor to draw an angle of 125°. Then construct an angle equal to the one you drew.

28. A trapezoid has bases of lengths 9 and 24 and a median of length $3x + \frac{3}{2}$. Find the value of x. 5

29. Supply the reasons to complete the proof.

    Given: A, B, and M are the midpoints of the sides of $\triangle CLK$.

    Prove: $\angle 1 = \angle 2$

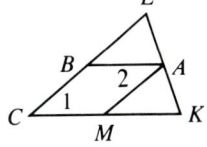

STATEMENTS	REASONS
1. $\overline{AB} \parallel \overline{CM}$; $\overline{AM} \parallel \overline{BC}$	1. __?__ Midpoints Theorem
2. $ABCM$ is a parallelogram.	2. __?__ Def. of ▱
3. $\angle 1 = \angle 2$	3. __?__ Opp. ⚞ of a ▱ are =.

Polygons 235

Here's what you'll learn in this chapter:

To find the area of a rectangle, a square, and a parallelogram.

To find the area of a triangle.

To find the area of a trapezoid.

To use the Pythagorean Theorem and its converse.

To find the circumference and the area of a circle.

The fields on a large farm are often laid out in many different shapes. In order to plan for the best use of the land, a farmer needs to know the area of each field.

Chapter 6

Areas

1 Areas of Rectangles

Which of the rectangles shown at the right covers more surface? Count the squares to find out. *The square covers more surface.*

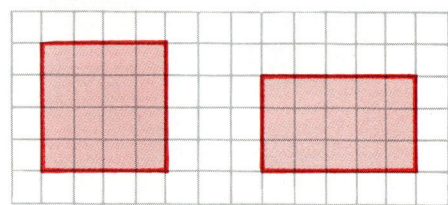

The 4 × 4 (read "4 by 4") rectangle is made up of 16 squares. The 5 × 3 rectangle is made up of 15 squares. We say that the 4 × 4 rectangle has the greater *area*. The amount of surface in a region is its **area.**

A square with sides one unit long has an area of *one square unit.* A common unit of area is the square centimeter.

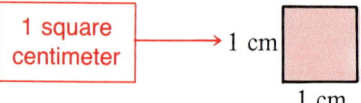

Rectangle *ABCD* has a *base* 5 cm long. Its *height* is 2 cm. Since the rectangle is made up of 10 square centimeters, its area is 10 square centimeters (written 10 cm^2). Notice that the area is the product of the lengths of the sides: 10 = 5 × 2.

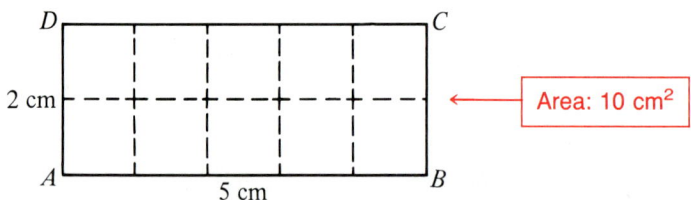

This example suggests the following postulate.

POSTULATE 12

The area of a rectangle is given by the formula:
Area = base × height

EXAMPLE 1   Find the area of a rectangle with base length 12 and height 6.

$A = bh$
$= 12 \times 6$
$= 72$

Answer: 72 square units

EXAMPLE 2   What is the area of a rectangle 2 m long and 40 cm wide?

When the unit of length is 1 centimeter:
Instead of using 2 m, use $2 \times 100 = 200$ cm.

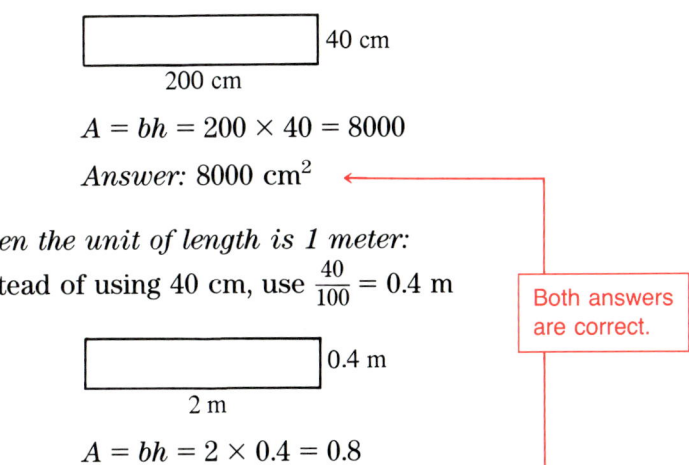

$A = bh = 200 \times 40 = 8000$

Answer: 8000 cm^2

When the unit of length is 1 meter:
Instead of using 40 cm, use $\frac{40}{100} = 0.4$ m

Both answers are correct.

$A = bh = 2 \times 0.4 = 0.8$

Answer: 0.8 m^2

You should see that 8000 cm^2 is the same amount of area as 0.8 m^2.

Recall: Since 1 m = 100 cm,
1 m^2 = 100 cm × 100 cm = 10,000 cm^2.

Every square is a rectangle, so the formula $A = bh$ applies to squares. In a square, b and h are equal. By substituting s for both b and h in the formula of Postulate 12, we obtain Theorem 1.

THEOREM 1

The area of a square is given by the formula:
Area = side squared

EXAMPLE 3   Find the area of a square with sides of length 8.

$A = s^2$
$\quad = 8^2$
$\quad = 8 \times 8 = 64$

Answer: 64 square units

Areas   **239**

Remember: The distance around a region is its **perimeter.**
To find the perimeter of a rectangle, use the
formula $p = 2b + 2h$.

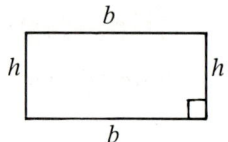

EXAMPLE 4 The perimeter of each rectangle shown below is 100.
Which rectangle has the greatest area?

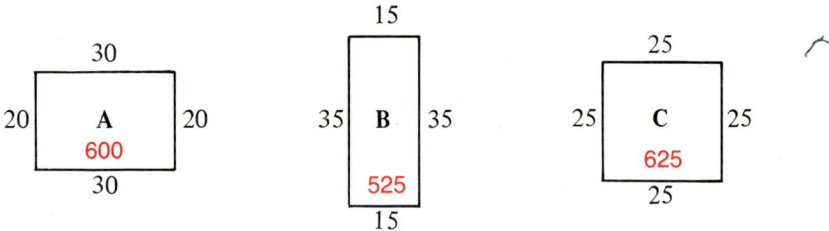

You should be able to show that rectangle C has
the greatest area.

Classroom Practice

1. Find the perimeter of the rectangle shown. 26
2. Find the area of the rectangle shown. 40

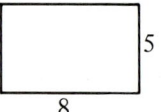

3. Each rectangle in Example 4 has perimeter 100.
   a. Draw two different rectangles, each with perimeter 100.
   b. Find the area of each rectangle you drew. Answers may vary.

4. Do you believe that you could draw a rectangle with perimeter 100 and an area as small as you please? yes

5. Draw a square with sides 1 m long on the chalkboard.
   The area = __?__ m². 1

6. Use the figure of Exercise 5. Each side is __?__ cm long. 100
   The area = __?__ cm². 10,000

7. Large regions are often measured in square kilometers (km²). A large parking lot is in the form of a square, 1.2 km on each side.
   a. The area of the park = __?__ km². 1.44
   b. The area of the park = __?__ m². (1 km = 1000 m) 1,440,000

8. a. Suppose two rectangles have equal areas.
      Must the bases have the same length? no
   b. Suppose two squares have equal areas.
      Must the sides of the squares have the same length? yes

Written Exercises

Find the area of each rectangle.

A 1. 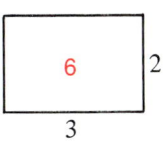   2. 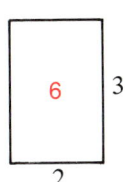   3.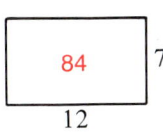

4. 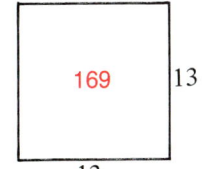   5. 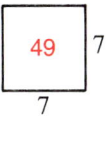   6.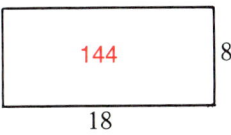

In Exercises 7–12, find the perimeter and the area of a rectangle with base length b and height h. (Perimeter listed first, then area.)

7. $b = 6, h = 2$ 16; 12
8. $b = 10, h = 7$ 34; 70
9. $b = 4, h = 3.2$ 14.4; 12.8
10. $b = 9.6, h = 4$ 27.2; 38.4
11. $b = 7\frac{2}{5}, h = 6$ $26\frac{4}{5}$; $44\frac{2}{5}$
12. $b = j$ cm, $h = k$ cm
     $2j + 2k$ cm; jk cm^2

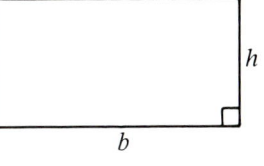

In Exercises 13–18, find the perimeter and the area of a square with sides of length s. (Perimeter listed first, then area.)

13. $s = 1$ 4; 1
14. $s = 2$ 8; 4
15. $s = 3$ 12; 9
16. $s = 4$ 16; 16
17. $s = 5$ 20; 25
18. $s = j$ km
    $4j$ km; j^2 km^2

Copy and complete the table about squares. In Exercises 23 and 24, express your answers in terms of k.

	19.	20.	21.	22.	23.	24.
B side s	? 9	? 5	? 7	? 8	$2k$	$k + 3$
perimeter p	36	20	? 28	? 32	? $8k$	?	$4k + 12$
area A	? 81	? 25	49	64	? $4k^2$	?	$k^2 + 6k + 9$

Find the area (a) in square centimeters and (b) in square meters.

25. A square with sides 60 cm long **a.** 3600 cm^2 **b.** 0.36 m^2
26. A square with sides 0.9 m long **a.** 8100 cm^2 **b.** 0.81 m^2
27. A rectangle 3 m long and 50 cm wide **a.** 15,000 cm^2 **b.** 1.5 m^2
28. A rectangle 5 m long and 70 cm wide **a.** 35,000 cm^2 **b.** 3.5 m^2

Areas

29. An artist plans to design a series of posters. Each poster will have an area of 1855 cm² and a length of 53 cm. How wide will the posters be? 35 cm

30. The area of a rectangular photo album page is 476 cm². The width is 17 cm. Find the length. 28 cm

31. The Nickersons want to plant grass in their new backyard. How many square meters of lawn can they have if they also have a flower garden and a patio? 260 m²

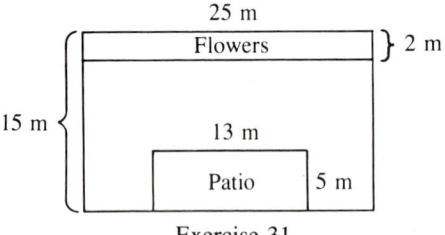

Exercise 31

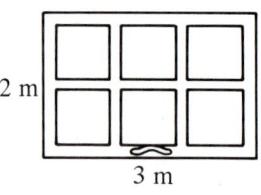

Exercise 32

C 32. Some pleated draperies must be twice as wide as the window that they cover.
   a. How many square meters of material are needed to cover the window shown? 12 m²
   b. The drapery material is sold from a bolt of cloth 1.5 m wide. What length of material should be cut from the bolt? 8 m

33. A piece of sheet metal is cut and bent to form the box shown. The box has no top. Find the area of:
   a. the bottom   b. the two shaded sides   c. the other two sides 108 cm²
      72 cm²   24 cm²

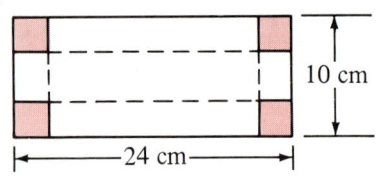

Exercise 33   Exercise 34

34. The metal used for the box in Exercise 33 is 24 cm long and 10 cm wide. The dashed lines indicate folds. The shaded regions indicate wasted sheet metal. How much metal is wasted? 36 cm²

35. The roof of an A-frame cabin is to be shingled at a cost of $90 a square. (A *square*, in shingling, is a region with an area of 100 square feet.) Find the cost of shingling the roof of the cabin shown. $1296

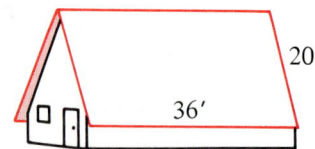

2 Areas of Parallelograms

Suppose you are given the four sticks shown.

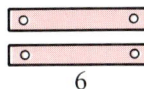 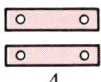
      6   4

You can join the sticks, with equal sides opposite each other, to form a quadrilateral. The quadrilateral has to be a parallelogram. It can be, but does not have to be, a rectangle.

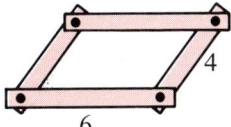 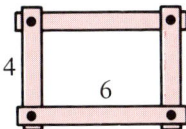

The area of the rectangle is 24, but the area of the parallelogram at the left is less than 24.

*The area of a parallelogram is **not** equal to the product of two consecutive sides, unless the parallelogram happens to be a rectangle.*

The area of the stick parallelogram depends on its *height*. The **height** of a parallelogram is the length of an *altitude*. This is a segment between, and perpendicular to, the lines containing the bases. Two altitudes are drawn to bases $\overline{AB}$ and $\overline{CD}$ in $\square ABCD$. Notice that these altitudes are equal.

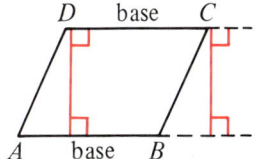

In any parallelogram, either pair of opposite sides may be considered the bases. If $\overline{AD}$ and $\overline{BC}$ are taken as bases, then the red segment is an altitude to those bases.

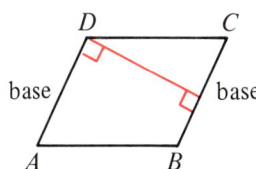

EXPLORATIONS

1. Cut a parallelogram out of cardboard. Cut along the indicated altitude, h.

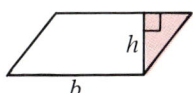

2. Move the triangle as shown to form a rectangle. The area of the rectangle equals bh.

   Therefore the area of the original parallelogram is also equal to bh.

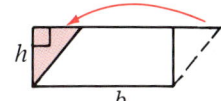

   This exploration suggests the theorem stated on page 244.

THEOREM 2

The area of a parallelogram is given by the formula:
Area = base × height

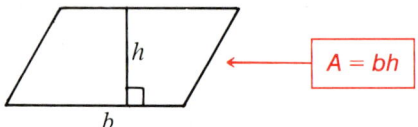

Classroom Practice

Find the area of each parallelogram.

1. 50
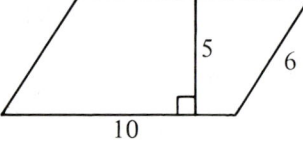

2. 40
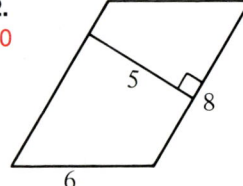

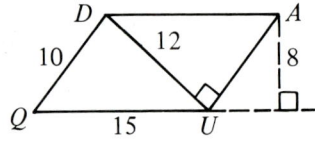

3. Take $\overline{QU}$ as a base of ▱QUAD.
   Area of ▱QUAD = 15 × __?__ 8 = __?__ 120.

4. Take $\overline{UA}$ as a base of ▱QUAD.
   Area of ▱QUAD = 10 × __?__ = __?__.
                        12   120

Exercises 5 and 6 refer to ▱ABCD shown below. The parallelogram is constructed so that its shape may be changed.

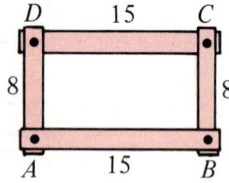 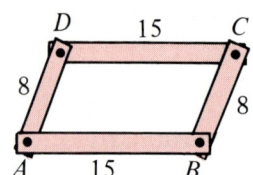 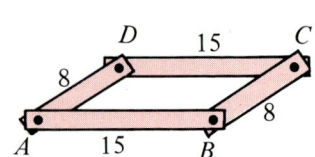

5. Find the greatest possible area that ▱ABCD can have. 120

6. As the height of ▱ABCD decreases, what happens to the area of the parallelogram? The area decreases.

7. For the parallelogram at the right:
   a. Find the perimeter in meters; 8 m
   b. Find the area in square meters. 2.1 m²

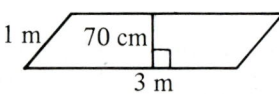

244   Chapter 6

Written Exercises

Find the area of each parallelogram.

A 1. 45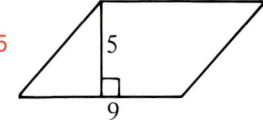
2. 72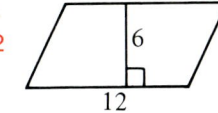
3. 28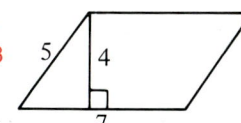

4. 60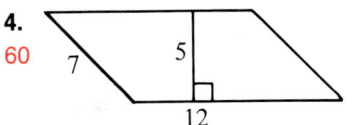
5. 56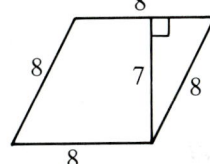
6. 69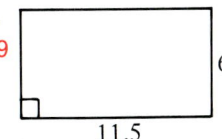

7. 84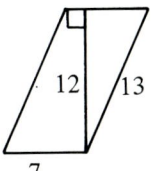
8. 286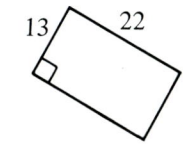
9. 30.4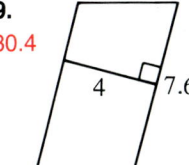

Find the perimeter and the area of each parallelogram. Express your answers in terms of k.

B 10.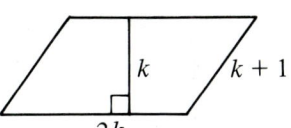
$p = 6k + 2; A = 2k^2$

11.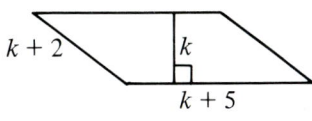
$p = 4k + 14; A = k^2 + 5k$

12.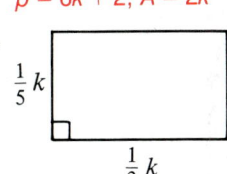
$p = \frac{16}{15}k; A = \frac{1}{15}k^2$

13.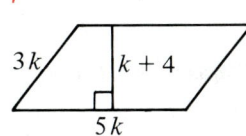
$p = 16k; A = 5k^2 + 20k$

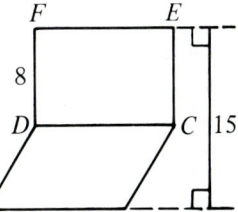

C 14. The area of rectangle $CEFD$ is 96. Find the area of $\square ABCD$. 84

15. Construct a parallelogram $PQRS$. Then construct a rectangle $WXYZ$ so that the area of the rectangle equals the area of the parallelogram.

Areas

SELF-TEST

Vocabulary

area (p. 238) height of a parallelogram (p. 243)
perimeter (p. 240)

1. The sides of a rectangle have lengths 10 and 3. *(6-1)*
   a. The perimeter of the rectangle is __?__. 26
   b. The area of the rectangle is __?__. 30

2. The area of a square is 25 cm^2.
   a. The length of a side of the square is __?__. 5 cm
   b. The perimeter of the square is __?__. 20 cm

Find the area of each parallelogram.

3. 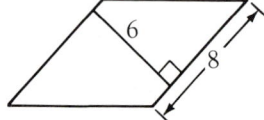 4. 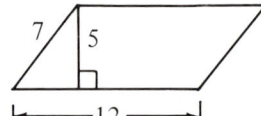 *(6-2)*
48 60

CONSUMER APPLICATIONS

THE COST OF DRIVING

Buying a car is a major investment. It takes money to operate a car, too. The owner must pay for gas, oil, repairs, parking fees, tolls, taxes, and insurance. A car also loses some of its resale value every day due to normal wear and tear. This loss is called *depreciation*. Taking all of these factors into account, we can estimate the average cost of 1 km of driving:

standard-size car: 17.3¢   compact car: 14.5¢   sub-compact car: 14.1¢

For many trips, a car is the most convenient and the least expensive method of transportation. This is particularly true if several people are traveling together. For some short trips, however, public transportation might cost less money. Don't forget that walking is free!

3 Areas of Triangles

Given any parallelogram, you can separate it into two congruent triangles. You simply draw a diagonal. Since congruent triangles have equal areas:

Area of each triangle = $\frac{1}{2}$(Area of parallelogram) = $\frac{1}{2}bh$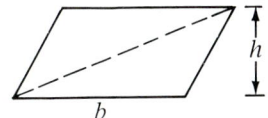

The diagrams below suggest that any triangle is "half" of a parallelogram that can be built on the triangle. The parallelogram has the same base length and height as the triangle.

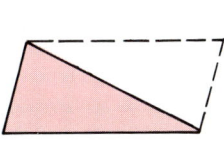 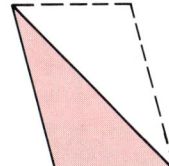

The discussion above contains the key ideas that are used in the proof of Theorem 3.

THEOREM 3

The area of a triangle is given by the formula:

$$\text{Area} = \frac{1}{2} \times base \times height$$

EXAMPLE Find the area of each triangle.

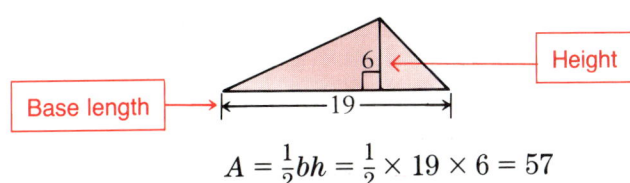

$A = \frac{1}{2}bh = \frac{1}{2} \times 19 \times 6 = 57$

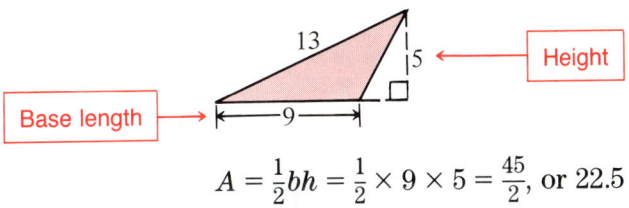

$A = \frac{1}{2}bh = \frac{1}{2} \times 9 \times 5 = \frac{45}{2}$, or 22.5

Areas 247

Classroom Practice

1. $\overline{DE}$ is taken as the base of △ DEF.
   Name the altitude to $\overline{DE}$. *FZ*

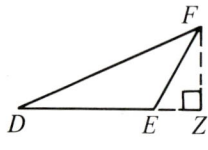

Exercise 1

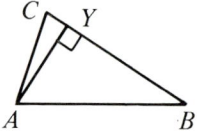

Exercise 2

2. $\overline{BC}$ is taken as the base of △ ABC.
   Name the altitude to $\overline{BC}$. *AY*

Exercises 3–5 refer to right △ JKM.

3. Name the altitude to $\overline{MK}$. *JM*
4. Name the altitude to $\overline{MJ}$. *KM*
5. Name the altitude to $\overline{JK}$. *MN*

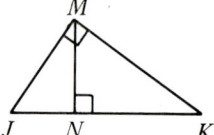

In Exercises 6–10, use the diagram shown.

6. The area of rectangle ABCD is __?__. 24
7. The area of △ AEB is __?__. 4
8. The area of △ ECF is __?__. 4
9. The area of △ FDA is __?__. 6
10. Subtract the areas of these triangles from the area of the rectangle.
    The area of △ AEF is __?__. 10

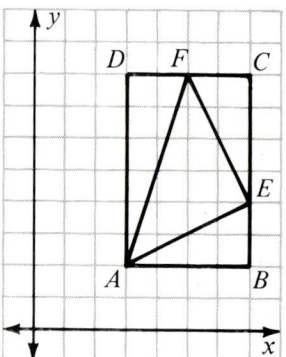

Exercises 11–13 refer to rhombus ABCD with AC = 12 and BD = 20. Recall that the diagonals of a rhombus are perpendicular bisectors of each other.

11. MC = __?__ 6 and MD = __?__ 10
12. Area of △ DMC = __?__ 30
13. Notice that the rhombus is separated into four congruent triangles. The area of rhombus ABCD is __?__. 120

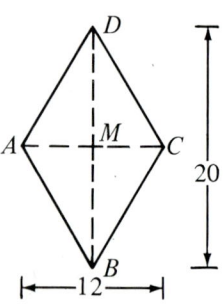

248   Chapter 6

Written Exercises

Find the area of the shaded triangle.

A 1. 14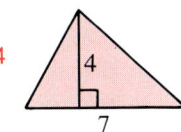

2. 30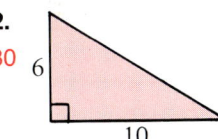

3. 32 m²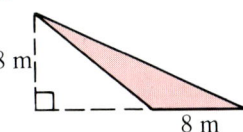

4. 31.5 cm²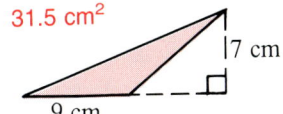

5. 56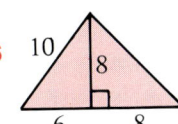

6. 90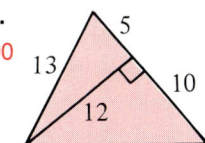

7. 8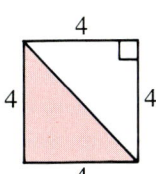

8. 72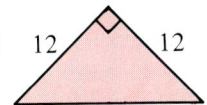

9. 20 m²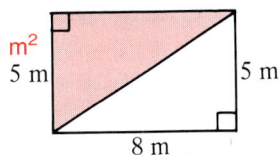

B 10. 43.05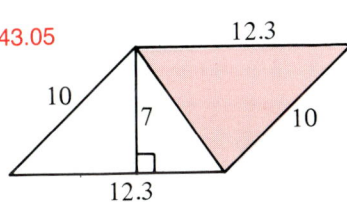

11. 24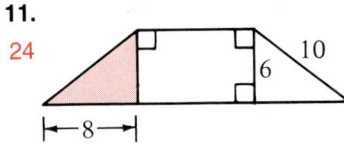

Express the area in terms of t.

12. $20t^2$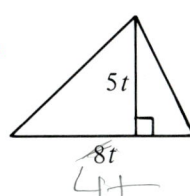

13. $\dfrac{3t^2 - 5t}{2}$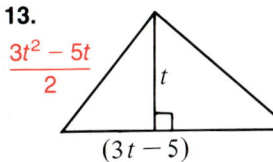

14. $\dfrac{t^2 - 49}{2}$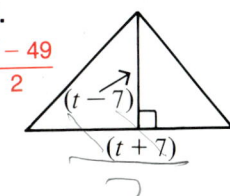

C 15. Draw a triangle ABC. Then construct a rectangle $WXYZ$ so that the area of the rectangle equals the area of the triangle.

   Strategy:

   A. Construct $\overline{WX}$ so that $WX = \frac{1}{2}AB$.
   B. Construct a line perpendicular to $\overline{WX}$ at W.
   C. Mark off Z on the perpendicular so that WZ equals the length of the altitude from C in $\triangle ABC$.
   D. Complete rectangle $WXYZ$.

Areas 249

4 Areas of Trapezoids

These explorations lead to a theorem about the area of a trapezoid.

EXPLORATIONS

Cut out three congruent trapezoids shaped roughly like trapezoid *RSTV*. The bases of your trapezoids should be 11 cm and 21 cm long. The altitudes should be 8 cm long.

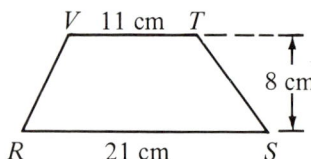

A. Cut along $\overline{VX}$ and $\overline{TY}$.
Since $RS = 21$ and $XY = 11$,
$RX + YS = 21 - 11 = 10$.

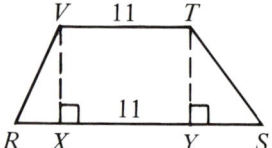

Place the two triangles alongside each other. Notice that rectangle *XYTV* and $\triangle RSZ$ both have altitudes 8 cm long.

Add the area of rectangle *XYTV* and the area of $\triangle RSZ$ to find the area of trapezoid *RSTV*. $88 + 40 = 128$ cm²

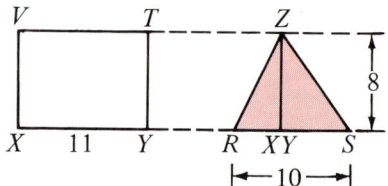

B. Take another copy of trapezoid *RSTV*. Cut along $\overline{VS}$.

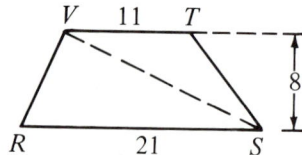
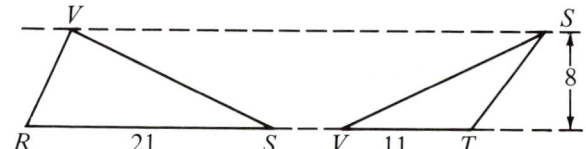

Place the triangles as shown above. Then add the areas of the triangles to find the area of the trapezoid. $84 + 44 = 128$ cm²

C. Take the third copy of the trapezoid.
Draw median $\overline{MN}$.
Using Theorem 14 on page 221:

$$MN = \tfrac{1}{2}(21 + 11) = \tfrac{1}{2}(32) = 16.$$

Cut along $\overline{MX}$ and $\overline{NY}$.

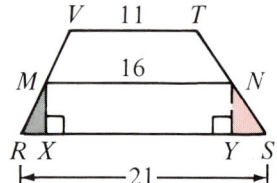

Move the triangles as shown. Find the area of trapezoid *RSTV* by computing the area of the rectangle formed. $16 \times 8 = 128$

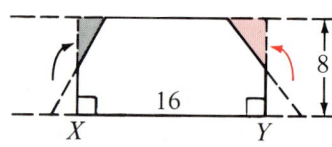

250 Chapter 6

Each of the explorations suggests a way to prove Theorem 4.

THEOREM 4

The area of a trapezoid is given by the formula:

$$\text{Area} = \tfrac{1}{2} \times \text{height} \times \text{sum of the bases}$$

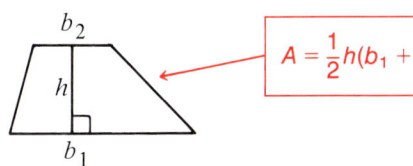

$A = \tfrac{1}{2}h(b_1 + b_2)$

Some students may need an explanation of the use of subscripts.

EXAMPLE Find the area.

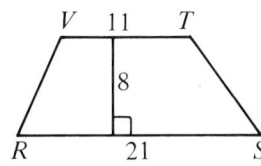

$A = \tfrac{1}{2}h(b_1 + b_2)$
$= \tfrac{1}{2} \times 8 \times (21 + 11)$
$= 4 \times 32$
$= 128$

Did you get this answer in Explorations A, B, and C?

Classroom Practice

Find the area of each trapezoid.

1. 138
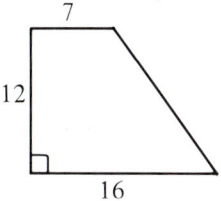

2. 138
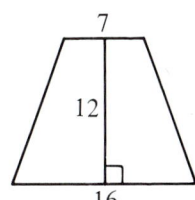

3. 138
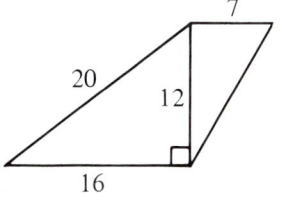

4. 138
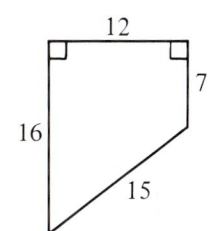

Areas   **251**

The figure shows congruent trapezoids *ABEF* and *DEBC*.

5. Why must quadrilateral *ACDF* be a parallelogram? $\overline{AC} \parallel \overline{FD}$ and AC = FD

6. What is the area of parallelogram *ACDF*? 256

7. What is the area of trapezoid *ABEF*? 128

8. Refer to Explorations A–C. Which suggests that the area of a trapezoid equals the product of the median and the altitude? Expl. C

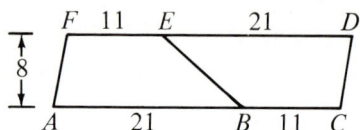

Written Exercises

Select the correct area formula from the column at the right.

A
1. Triangle d
2. Parallelogram c
3. Square a
4. Trapezoid b

a. $A = s^2$
b. $A = \frac{1}{2}h(b_1 + b_2)$
c. $A = bh$
d. $A = \frac{1}{2}bh$

Find the area of each trapezoid.

5. 96
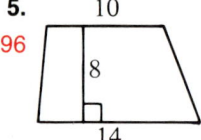

6. 54
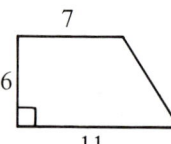

7. 125
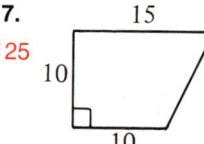

8. 108.5

9. 87.5
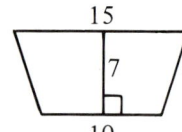

10. 24
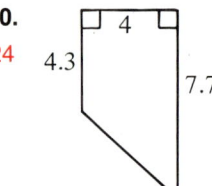

In Exercises 11–16, find the area of each polygon.

11. 28 cm²
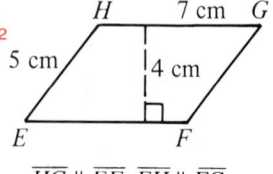
$\overline{HG} \parallel \overline{EF}; \overline{EH} \parallel \overline{FG}$

12. 44 cm²
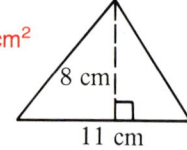

13. 54 cm²
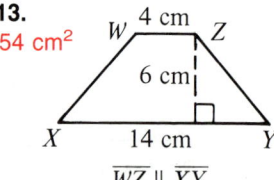
$\overline{WZ} \parallel \overline{XY}$

Find the area of each polygon.

14. 144 cm² — square with sides 12 cm

15. 88 cm²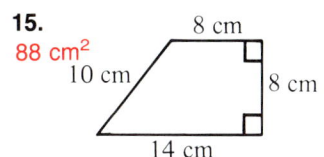

16. 24 cm²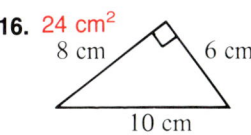

Copy and complete the table about trapezoids.

SAMPLE $A = 30$, $b_1 = 7$, $b_2 = 3$. Find h.

$A = \frac{1}{2}h(b_1 + b_2)$

$30 = \frac{1}{2}h(7 + 3)$

$60 = h(10)$

$6 = h$

	17.	18.	19.	20.	21.	22.	23.	24.
height h	7	8	7	j	? 8	? 10	8	7
base length b_1	10	10	12	k	14	18	5	? 17
base length b_2	4	5	7	n	6	12	? 3	3
area A	? 49	? 60	? 66.5	? $\frac{1}{2}j(k+n)$	80	150	32	70

C 25. A *hip roof* consists of two isosceles trapezoids and two isosceles triangles. How many squares of shingles are needed for the roof shown? (Recall that in talk about roofs, one square means 100 square feet.) 20.4 squares

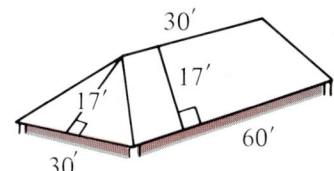

26. The Brock family used to farm one section of land. (A *section* is one square mile.) They sold a strip of land to permit construction of a highway. Then they sold a small triangular part that was awkward to reach to their neighbors.

To the nearest ten thousand square feet, find the area:
a. sold for the highway; 2,120,000 ft²
b. sold to neighbors; 1,330,000 ft²
c. in the Brock's new farm. 24,430,000 ft²

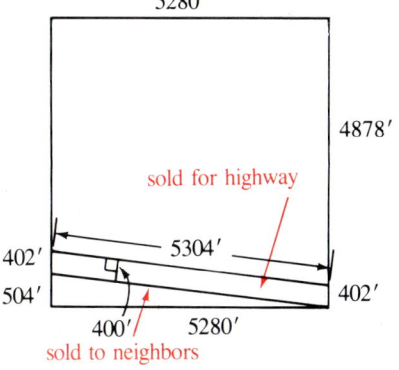

Areas   253

27. Carefully draw a trapezoid *EFGH*. Then construct a rectangle *WXYZ* so that the area of the rectangle equals twice the area of the trapezoid.

Strategy:

A. Construct $\overline{WX}$ so that *WX* equals the sum of the bases of trapezoid *EFGH*.

B. Construct a line perpendicular to $\overline{WX}$ at *W*.

C. Mark off *Z* on the perpendicular so that *WZ* equals the height of trapezoid *ABCD*.

D. Complete rectangle *WXYZ*.

SELF-TEST

Find the area of each triangle.

1. 20 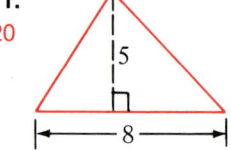 2. 12 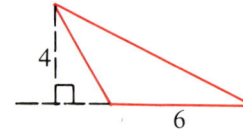 3. 84 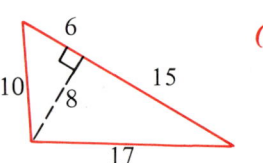 (6-3)

Find the area of each trapezoid.

4. 91 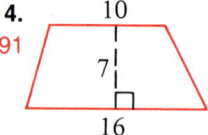 5. 26 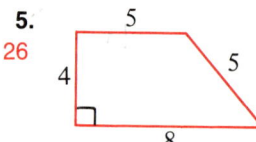 (6-4)

PUZZLE ◆ PROBLEMS

It is easy to join these sticks to form a rectangle with area 24.

How can you join the sticks to form a *nonrectangular* quadrilateral with area 24? Join the longer sticks to each other. Join the shorter sticks to each other. Then join the two portions so that a pair of opp. rt. ≜ are formed, both with sides 4 and 6.

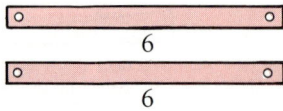
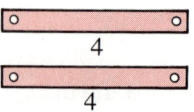

5 The Pythagorean Theorem

EXPLORATIONS

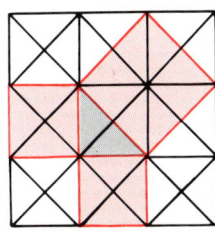

1. Part of a tiled floor is shown.
   Note the right triangle shaded in gray.
   Each side of that triangle is also the side of a square outlined and shaded in red.
   Count tiles to compare the area of the largest square with the sum of the areas of the smaller squares. *equal*

2. Another right triangle and some squares are shown.
   Compare the area of the largest square with the sum of the areas of the smaller squares. *equal*

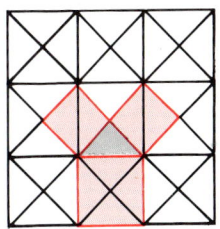

The idea suggested above has often been expressed in this way:
The square on the hypotenuse of a right triangle is equal to the sum of the squares on the legs.

The explorations above deal with special right triangles.
The discussion that follows deals with *all* right triangles.

Begin with *any* right triangle.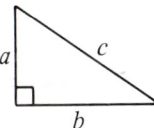

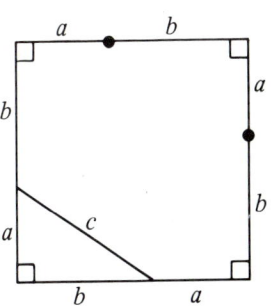

Draw a square, as shown, with sides of length $a + b$.

Locate two new points as shown.

Draw the rest of the inner quadrilateral.

The four right triangles are congruent by the SAS Postulate. Therefore, each hypotenuse has length c.

Each angle of the inner quadrilateral is a right angle (see Exercise 7 at the top of page 257).

The inner quadrilateral is a square with area c^2.

The areas of the two squares are compared on the next page.

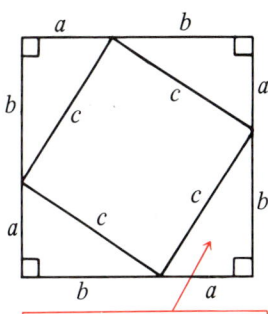

Each right triangle has area $\frac{1}{2} ab$.

$$\begin{array}{ccc}\text{Area of} & \text{Area of} & \text{Areas of}\\ \text{larger square} = \text{smaller square} + \text{four triangles}\\ (a+b)^2 & = \quad c^2 \quad + \quad 4(\tfrac{1}{2}ab)\end{array}$$

$$a^2 + 2ab + b^2 = c^2 + 2ab$$
$$a^2 + b^2 = c^2$$

THEOREM 5 (The Pythagorean Theorem)

In a right triangle, the square of the hypotenuse is equal to the sum of the squares of the legs.

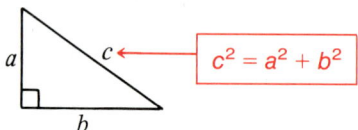

$c^2 = a^2 + b^2$

Theorem 5 is often called the **Pythagorean** (pi-**thag**-uh-**ree**′-an) **Theorem,** in honor of the Greek philosopher and mathematician Pythagoras. He is said to have written a proof of the theorem in the sixth century B.C.

EXAMPLE Find the value of x for each right triangle. Remember that the hypotenuse is the longest side of a right triangle.

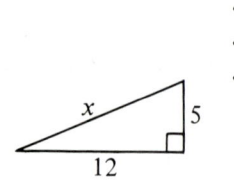

$x^2 = 5^2 + 12^2$
$x^2 = 25 + 144$
$x^2 = 169$
$x = \sqrt{169}$
$x = 13$

$15^2 = x^2 + 9^2$
$225 = x^2 + 81$
$144 = x^2$
$\sqrt{144} = x$
$12 = x$

Classroom Practice

In Exercises 1–6, state whether each equation is written correctly.

1. $a^2 + b^2 = c^2$
yes

2. $c^2 = a^2 + b^2$
no

3. $r^2 = s^2 + t^2$
no

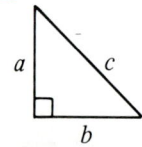
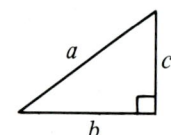
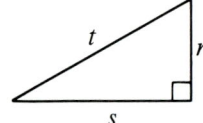

4. $k^2 = l^2 - m^2$
yes
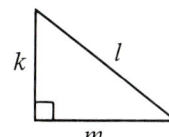

5. $d^2 = e^2 - f^2$
no
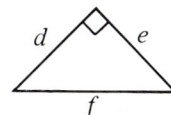

6. $p^2 = r^2 + q^2$
no
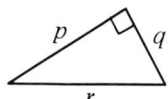

7. *Given:* $\angle 1$ and $\angle 5$ are right angles. Explain why $\angle 3$ must be a right angle. (*Hint:* Use the fact that $\triangle RST \cong \triangle TUV$.)
$\angle 2 = \angle 6$ (Corr. parts of $\cong \triangle$ are $=$.)
$\angle 4 + \angle 6 = 90°$, hence $\angle 4 + \angle 2 = 90°$.
$\angle 3 + \angle 4 + \angle 2 = 180°$, hence $\angle 3 = 90°$.

Written Exercises

In Exercises 1–6, write an equation for finding the value of x. Then solve the equation.

1.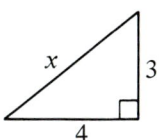
$x^2 = 3^2 + 4^2$; $x = 5$

2.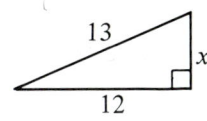
$13^2 = x^2 + 12^2$; $x = 5$

3. $x^2 = 6^2 + 8^2$; $x = 10$

4.
$25^2 = x^2 + 24^2$; $x = 7$

5.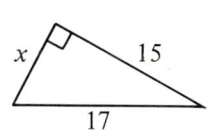
$17^2 = x^2 + 15^2$; $x = 8$

6.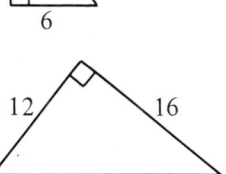
$x^2 = 16^2 + 12^2$; $x = 20$

In Exercises 7–12, the lengths of two sides of a right triangle are given. Find the length of the third side.

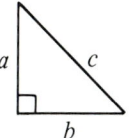

7. $a = 6$, $b = 8$ 10
8. $a = 5$, $b = 12$ 13
9. $a = 12$, $c = 15$ 9
10. $b = 4$, $c = 5$ 3
11. $a = 24$, $c = 26$ 10
12. $b = 16$, $c = 20$ 12

In Exercises 13–21, use the square root table on page 516 to write an approximation for x to the nearest tenth.

SAMPLE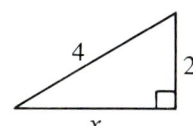
$4^2 = 2^2 + x^2$
$16 = 4 + x^2$
$12 = x^2$
$x = \sqrt{12} \approx 3.5$

Areas   **257**

B

13. 6.4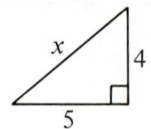

14. 5.3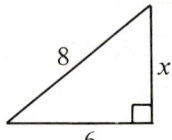

15. 7.8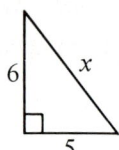

16. 5.7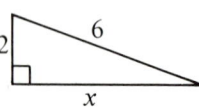

17. 5.2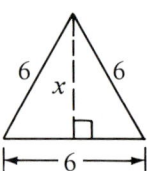

18. 6.9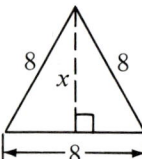

19. 7.1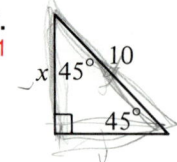

20. 8.5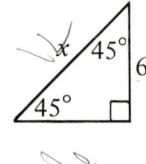

21. 7.2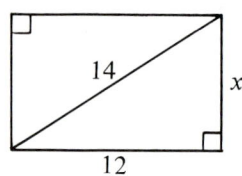

In Exercises 22–27, use the square root table on page 516 to find an approximation for x that is correct to the nearest tenth. Then find the area of each region.

C

22. $x \approx 4.5$; $A \approx 18$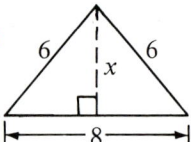

23. $x \approx 7.2$; $A \approx 86.4$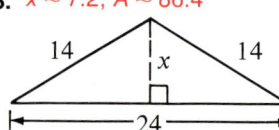

24. $x = 4$; $A = 32$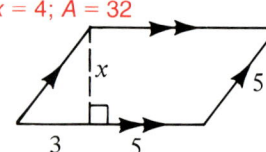

25. 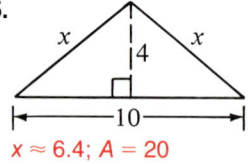 $x \approx 6.4$; $A = 20$

26. 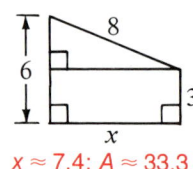 $x \approx 7.4$; $A \approx 33.3$

27. 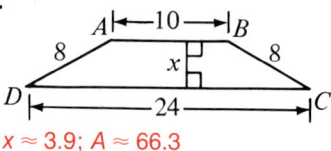 $x \approx 3.9$; $A \approx 66.3$

28. James A. Garfield (1831–1881), twentieth president of the United States, wrote a proof of the Pythagorean Theorem based on the diagram shown.

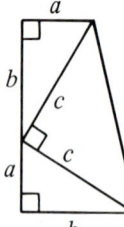

$$\begin{array}{c}\text{Area of}\\\text{the trapezoid}\end{array} = \begin{array}{c}\text{Total area of}\\2 \cong \triangle\end{array} + \begin{array}{c}\text{Area of}\\\text{large rt. } \triangle\end{array}$$

$$\underset{\tfrac{1}{2}(a+b)(a+b)}{?} = \underset{2(\tfrac{1}{2}ab)}{?} + \underset{\tfrac{1}{2}c^2}{?}$$

Supply the missing measures in the equation above. Then simplify to show that $c^2 = a^2 + b^2$.

$$a^2 + 2ab + b^2 = 2ab + c^2$$
$$a^2 + b^2 = c^2$$

6 Converse of Pythagorean Theorem

Suppose you are asked if $\triangle RST$, shown below, is a right triangle. You notice:

$5^2 = 3^2 + 4^2$ because
$25 = 9 + 16$.

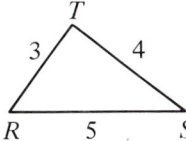

In order to conclude that $\triangle RST$ is a right triangle, you need to know whether the converse of the Pythagorean Theorem is true. Although some theorems have converses that are not true, the Pythagorean Theorem has a converse that *is* true. We omit the proof.

> **THEOREM 6**
>
> *If the square of one side of a triangle is equal to the sum of the squares of the other two sides, then the triangle is a right triangle.*

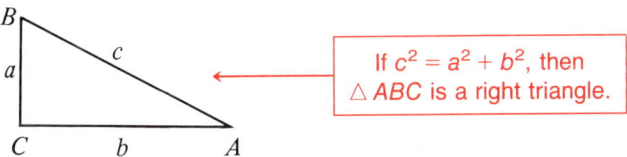

If $c^2 = a^2 + b^2$, then $\triangle ABC$ is a right triangle.

EXAMPLE Decide if each triangle shown is a right triangle.

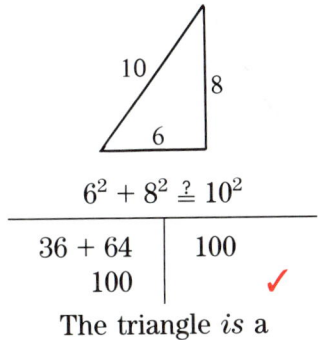

$6^2 + 8^2 \stackrel{?}{=} 10^2$

$36 + 64$	100
100	✓

The triangle *is* a right triangle.

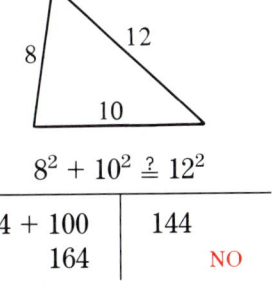

$8^2 + 10^2 \stackrel{?}{=} 12^2$

$64 + 100$	144
164	NO

The triangle *is not* a right triangle.

When you know the lengths of the three sides of a triangle, you can easily decide whether the triangle is acute, right, or obtuse. Look at the strategy outlined on the next page.

Areas 259

> **STRATEGY**
>
> 1. Let c be the longest side. Let a and b be the other two sides.
> 2. Compare c^2 to $a^2 + b^2$.
> 3. If $c^2 < a^2 + b^2$, then the triangle is **acute**.
>   If $c^2 = a^2 + b^2$, then the triangle is a **right** triangle.
>   If $c^2 > a^2 + b^2$, then the triangle is **obtuse**.

Classroom Practice

In each exercise, decide if the triangle is a right triangle. Explain your answers.

1. yes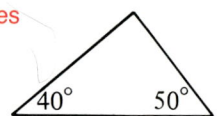

2. no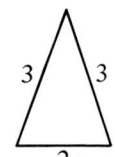

3. no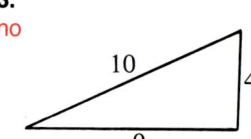

4. yes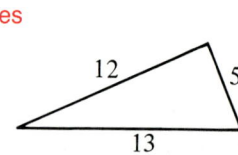

5. no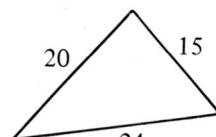

6. yes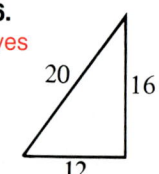

In each exercise, decide if $\triangle ABC$ is a right triangle.

7. $AB = 5$, $BC = 8$, $CA = 9$ no

8. $AB = 2$, $BC = 2$, $CA = 2$ no

9. $AB = 30$ cm, $BC = 40$ cm, $CA = 50$ cm yes

10. To check the accuracy of a corner in a foundation you can measure 7 m out from C to A, and 7 m out from C to B. A string stretched from A to B should be about 10 cm less than 10 m long.

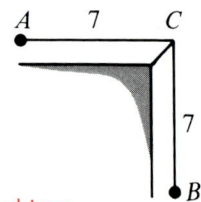

  a. Suppose that $\overline{AB}$ is 10.1 m long.
  Is $\angle ACB$ an acute angle, a right angle, or an obtuse angle? obtuse

  b. Suppose that $\overline{AB}$ is 9.8 m long.
  Is $\angle ACB$ an acute angle, a right angle, or an obtuse angle? acute

Written Exercises

Three sides of a triangle are given. Is the triangle a right triangle?

[A]
1. 3, 4, 5 yes
2. 30, 40, 50 yes
3. 2, 3, 4 no
4. 4, 5, 6 no
5. 8, 15, 17 yes
6. 4, 6, 8 no
7. 10, 15, 20 no
8. 7, 24, 25 yes
9. 0.9, 1.2, 1.5 yes
10. 0.5, 1.2, 1.3 yes
11. 16, 21, 27 no
12. 14, 14, 20 no

Three sides of a triangle are given. Classify the triangle as acute, right, or obtuse. (*Hint:* See the strategy on page 260.)

[B]

13.	14.	15.	16.	17.	18.	19.
6	7	10	7	11	15	12
8	7	10	11	60	16	21
10	10	14	16	61	21	25

13. right
14. obtuse
15. acute
16. obtuse
17. right
18. acute
19. obtuse

[C]
20. Suppose that a, b, and c are the sides of a right triangle. Is a triangle with sides $7a$, $7b$, and $7c$ also a right triangle? yes
Use algebra to show that your answer is correct.

21. Repeat Exercise 20 for a triangle with sides $a + 7$, $b + 7$, and $c + 7$. no

22. Find the area of the quadrilateral shown. 36

23. Sketch □$ABCD$ with $AB = 60$, $BC = 22$, and $AC = 64$. Which diagonal is longer: $\overline{AC}$ or $\overline{BD}$? AC

24. Sketch □$WXYZ$ with $WX = 5$, $WY = 6$, and $XZ = 8$. What special kind of parallelogram is $WXYZ$? rhombus

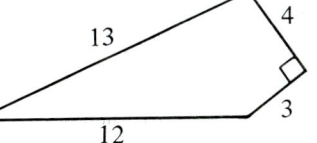

Exercise 22

PUZZLE ♦ PROBLEMS

Estelle, George, Jim, Keiko, Laura, and Luis are running a race.

> Estelle is 20 m ahead of Jim.
> Jim is 50 m ahead of George.
> George is 10 m behind Laura.
> Laura is 50 m behind Keiko.
> Luis is 30 m ahead of Keiko.

Who is winning the race? Luis
Who is second? Third? Fourth? Fifth? Last? Estelle; Keiko; Jim; Laura; George

SELF-TEST

Find the value of x.

1.
2.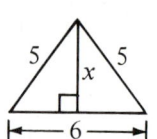
3.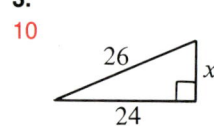
4.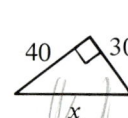

(6-5)

The lengths of three sides of a triangle are given. Is the triangle right, acute, or obtuse?

5. $a = 2, b = 3, c = 4$ obtuse
6. $a = 3, b = 4, c = 5$ right
7. $a = 6, b = 6, c = 6$ acute
8. $a = 2.1, b = 2.8, c = 3.5$ right

(6-6)

CAREER NOTES

AUTO MECHANIC

In 1980, the number of passenger cars in use around the world was about 321,000,000 and in 1981 the number rose to about 331,000,000. Guess how many are in use today. Can you imagine the amount of work involved in keeping these cars running well?

Auto mechanics maintain and repair many different kinds of automobiles. They periodically examine, adjust, and replace car parts. They clean carburetors, balance wheels, replace spark plugs and distributor points, and grind valves.

Auto mechanics may work with sophisticated testing equipment or simple hand tools. Since parts for many cars are measured in metric units, mechanics must be familiar with both metric and nonmetric tools. High school or vocational school courses in machine shop, automobile repair, mathematics, and science help you learn these skills.

It takes three or four years of formal apprenticeship to become a senior auto mechanic. Some senior mechanics work for companies or repair garages. Others operate their own businesses. Mechanics may specialize in repairing particular car parts or particular types of cars.

7 Circumferences of Circles

A **circle** is a figure, in a plane, whose points are all the same distance from a particular point in the plane. That point is called the **center** of the circle. Circle P (written ⊙P) is shown. The distance around ⊙P is called its **circumference.**

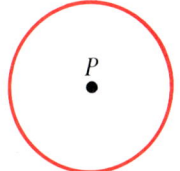

$\overline{PA}$, $\overline{PB}$, and $\overline{PC}$ are *radii* of ⊙P. (The plural of radius is radii.) It follows from the definition of circle that *all radii of a circle are equal.*

$\overline{AC}$ is a *diameter* of ⊙P. Can you see that any diameter of a circle is twice as long as a radius?

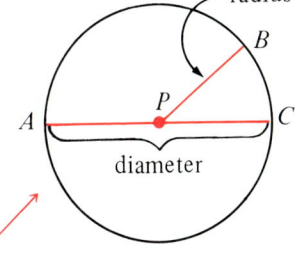

The words *radius* and *diameter* will sometimes refer to lengths of segments. We could say, for instance, that a circle has a radius of 5 and a diameter of 10.

PA = PB = PC;
AC = 2(PA)

A student measured the diameter of a half-dollar coin and reported: The diameter, d, is about 3.0 cm. Then she rolled the coin along a straight line until the coin had gone through one revolution.

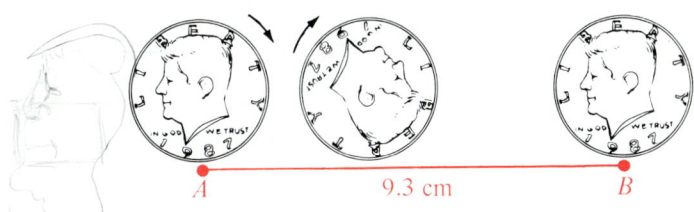

After measuring $\overline{AB}$ she said: The circumference, C, is about 9.3 cm. Next she divided the circumference by the diameter, getting

$$\frac{9.3}{3.0}, \text{ or } 3.1. \quad \leftarrow \quad \boxed{\frac{C}{d} = 3.1}$$

Another student measured the diameter of a tin can. He found that the diameter was about 7.5 cm. The student then measured the distance around the rim of the can and reported: The circumference is about 24.0 cm. In this case,

$$\frac{\text{circumference}}{\text{diameter}} = \frac{24.0}{7.5}, \text{ or } 3.2. \quad \leftarrow \quad \boxed{\frac{C}{d} = 3.2}$$

Areas

Notice that in the two cases on page 263 the values of $\frac{C}{d}$ are almost equal. Try the following experiment yourself.

Select some circular objects such as coins, plates, and wheels. For each object:

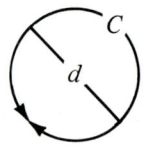

1. measure the diameter and the circumference;
2. divide the circumference by the diameter.

If you work carefully, you should find that $\frac{C}{d}$ is usually a little greater than 3. Mathematicians have shown that the quotient $\frac{C}{d}$ is *exactly* the same for all circles. This quotient is denoted by π, a Greek letter pronounced *pie*. By definition,

$$\pi = \frac{C}{d}$$

The value of π has been computed to thousands of decimal places. Some approximations in common use are the following:

$$\pi \approx 3.14 \qquad \pi \approx 3.1416 \qquad \pi \approx \frac{22}{7}$$

↑ is approximately equal to ↑

None of these is exact. Unless specified otherwise, you may use any approximation that is convenient.

In a circle with radius r, diameter d, and circumference C:

$$\frac{C}{d} = \pi$$
$$C = \pi d$$
$$C = \pi \times 2r \quad \leftarrow \boxed{d = 2r}$$
$$C = 2\pi r$$

The circumference of a circle is given by the formula:

$$C = 2\pi r$$

EXAMPLE Find the circumference of a circle with radius 4 cm. Round your answer to the nearest tenth.

$C = 2\pi r$
$C = 2\pi \times 4 = 8\pi$ ← Exact answer: 8π cm
$C = 8\pi \approx 8 \times 3.14 = 25.12$ ← Approximate answer: 25.1 cm

Note that the exact circumference of a circle is expressed in terms of π. To find an approximate value, use an approximation of π.

Classroom Practice

Exercises 1–8 refer to the diagram below.

1. Name the circle shown. ⊙O
2. Name the center of the circle. O
$\overline{OA}, \overline{OB}, \overline{OC}$
3. Name all the radii shown.
4. Name all the diameters shown. $\overline{AC}$
5. If $OC = 8$, then $OB =$ ___?___. 8
6. If $OC = 3$, then $AC =$ ___?___. 6
7. If $AC = 2x$, then $OB =$ ___?___. x
8. If a radius of the circle is 2 cm long, then the circumference of the circle is ___?___. 4π cm

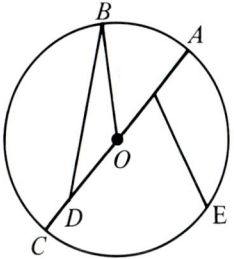

Find the circumference of a circle with the given radius. Express your answers in terms of π.

9.
r	3	6
C	6π	?

12π

10.
r	11	22
C	?	?

22π 44π

11.
r	j	$2j$
C	?	?

$2j\pi$ $4j\pi$

12. Complete the statement suggested by Exercises 9–11:
If the radius of one circle is twice as. . . . great as that of a second circle, the circumference of the first circle is twice that of the second circle.

Find the circumference of a circle with the given radius. Express your answers in terms of π.

13.
r	1	8
C	?	?

2π 16π

14.
r	1	13
C	?	?

2π 26π

15.
r	5	$5k$
C	?	?

10π $10k\pi$

16. Complete the statement suggested by Exercises 13–15:
If the radius of one circle is k times as. . . . great as that of a second circle, the circumference of the first circle is k times that of the second circle.

Areas 265

Written Exercises

In Exercises 1–4, the radius of a circle is given. Find the diameter.

A   **1.** 7   14   **2.** 12 m   24 m   **3.** 5.1 cm   10.2 cm   **4.** x $2x$

In Exercises 5–8, the diameter of a circle is given. Find the radius.

5. 16 cm   8 cm   **6.** 26   13   **7.** 21.6   10.8   **8.** $2y$ y

In Exercises 9–17, find the circumference of a circle with the given radius r or diameter d. Express your answers in terms of π.

SAMPLE 1 $d = 4$ cm $C = \pi d$
$C = \pi \times 4 = 4\pi$

9. $d = 10$ 10π **10.** $d = 13$ 13π **11.** $r = 6$ 12π

12. $d = 9.7$ m 9.7π m   **13.** $d = 33$ m 33π m   **14.** $r = 7.5$ cm 15π cm

15. $r = 0.4$ km 0.8π km   **16.** $r = 16$ cm 32π cm   **17.** $r = 18.4$ km 36.8π km

In Exercises 18–23, find the radius of a circle with the given circumference.

SAMPLE 2 $C = 12\pi$ $C = 2\pi r$
$12\pi = 2\pi r$ ← Divide each side by 2π.
$6 = r$

18. $C = 16\pi$ 8   **19.** $C = 2\pi$ 1   **20.** $C = 61\pi$ cm   30.5 cm

21. $C = 25\pi$ m   12.5 m   **22.** $C = 46\pi$ cm   23 cm   **23.** $C = 53\pi$ m   26.5 m

In Exercises 24–29, find an approximation, correct to the nearest tenth, of the circumference of a circle with the given radius. Use 3.14 for π.

B   **24.** $r = 8$ cm   50.2 cm   **25.** $r = 15$ m   94.2 m   **26.** $r = 8.6$ cm   54.0 cm

27. $r = 2.1$ km   13.2 km   **28.** $r = 37$ m   232.4 m   **29.** $r = 6.5$ m   40.8 m

In Exercises 30–35, find an approximation, correct to the nearest tenth, of the radius of a circle with the given circumference. Use 3.14 for π.

30. $C = 31.4$ m   5 m   **31.** $C = 314$ cm   50 cm   **32.** $C = 1.57$ m   0.3 m

33. $C = 20$ cm   3.2 cm   **34.** $C = 27.5$ cm   4.4 cm   **35.** $C = 9.4$ km   1.5 km

36. Draw two circles. Then construct a circle whose circumference is equal to the sum of the circumferences of the given circles.

Strategy:
A. Let the radii of the given circles be s and t.
B. $2\pi s + 2\pi t = 2\pi(s + t)$
C. Construct a segment with length $s + t$.
D. Construct a circle with radius $s + t$.

In Exercises 37–39, use 3.14 for π. Write each answer correct to the nearest centimeter.

C **37.** The radii of two circles are 10 cm and 11 cm. By how many centimeters is the circumference of the larger circle greater than the circumference of the smaller circle? 6 cm

38. Repeat Exercise 37, using circles with radii 100 cm and 101 cm. 6 cm

39. Wheels revolving in opposite directions have a drive belt as shown. Find the total length of the belt. 1074 cm

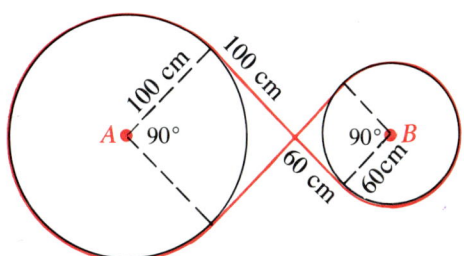

Suggested strategy:
A. $360° - 90° = 270°$
 The belt covers $\frac{270}{360}$, or $\frac{3}{4}$ of $\odot A$.
B. The belt covers $\frac{3}{4}$ of $\odot B$.
C. Add the parts to find the total length of the belt.

40. Look back at Activity 2 on page 4 of Chapter 1. Now you can show that the distance between the earth and the stretched hoop is about 32 cm. Here is one way to do it.

  a. Suppose the radius of the earth is r meters.
  Explain why the stretched hoop has circumference $2\pi r + 2$.

  b. Let R be the radius of the stretched hoop.
  Explain why $2\pi R = 2\pi r + 2$.

  c. Conclude that $2\pi(R - r) = 2$.
  Therefore, $R - r = \frac{2}{2\pi} \approx \frac{2}{2(3.14)} \approx 0.32$.
  The distance between the earth and the hoop is 0.32 m = __?__ cm. 32

  d. Make a sketch showing the stretched hoop around the earth. Show R, r, and $R - r$ on your sketch.

Areas 267

8 Areas of Circles

The diagram represents an apple pie cut into eight equal pieces. Suppose you rearrange these eight pieces as shown below. Notice that the figure formed resembles a parallelogram.

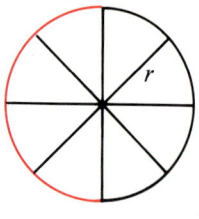
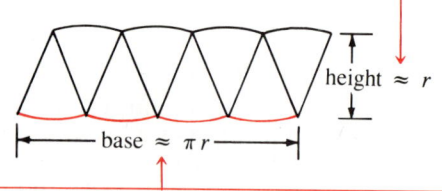

The "height" is approximately equal to r, the radius of the circular pie.

The "base" is approximately equal to half the circumference of the circular pie.

$$\begin{aligned}\text{Area of eight pieces of pie} &\approx \text{Area of parallelogram}\\ &= \text{base} \times \text{height}\\ &\approx \pi r \times r\\ &= \pi r^2\end{aligned}$$

Now suppose you cut the pie into sixteen equal pieces, instead of eight.

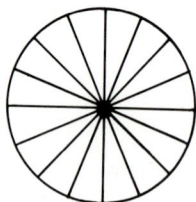
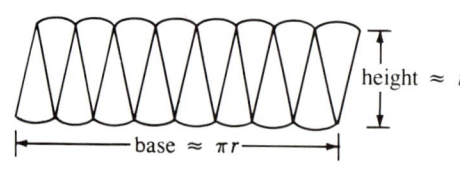

The figure formed looks even more like a parallelogram than the first one does. This means that the base is more nearly equal to πr. The area is more nearly equal to πr^2.

The discussion above suggests that πr^2 is approximately equal to the area of a circle with radius r. It can be proved, in an advanced geometry course, that πr^2 is the *exact* area.

The area of a circle is given by the formula:

$$A = \pi r^2$$

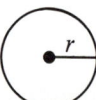

EXAMPLE Find an approximation, correct to tenths, of the area of a circle with diameter 6 m.

Since the diameter is 6 m, the radius is 3 m.

$A = \pi r^2$
$A = \pi \times 3^2 = \pi \times 9 = 9\pi$ ← Exact area: 9π m²
$A = 9\pi \approx 9 \times 3.14 = 28.26$ ← Approximate area: 28.3 m²

Classroom Practice

Exercises 1–8 refer to a circle with radius r, diameter d, circumference C, and area A. Find the missing measures.

	1.	2.	3.	4.	5.	6.	7.	8.
r	5	9	? 4	? 8	? 14	? 6	? 2	? 1
d	? 10	? 18	8	16	? 28	? 12	? 4	? 2
C	? 10π	? 18π	? 8π	? 16π	28π	12π	4π ?	2π ?
A	? 25π	? 81π	? 16π	? 64π	? 196π	? 36π	4π	π

9. Suppose two circles have equal circumferences. Must their radii be equal? Must their areas be equal? yes; yes

Written Exercises

Find the area of a circle with the given radius or diameter. Express your answers in terms of π.

A
1. $r = 5$ 25π
2. $r = 9$ 81π
3. $d = 6$ 9π
4. $d = 8$ cm 16π cm²
5. $r = 1.2$ cm 1.44π cm²
6. $r = 3.5$ m 12.25π m²
7. $d = 7$ cm 12.25π cm²
8. $d = 2.6$ m 1.69π m²
9. $r = 5\frac{5}{6}$ $\frac{1225}{36}\pi$

Find the radius of a circle with the given area.

SAMPLE $A = 49\pi$ cm² $A = \pi r^2$
$49\pi = \pi r^2$
$49 = r^2$
$7 = r$ Answer: 7 cm

Areas   **269**

Find the radius of a circle with the given area.

10. $A = 36\pi$ 6
11. $A = 25\pi$ 5
12. $A = 9\pi$ m² 3 m
13. $A = 81\pi$ cm² 9 cm
14. $A = 100\pi$ cm² 10 cm
15. $A = 64\pi$ cm² 8 cm

Find an approximation, to the nearest tenth, of the area of a circle with the given radius or diameter. Use 3.14 for π.

B 16. $r = 6$ 113.0
17. $r = 7$ 153.9
18. $r = 3$ m 28.3 m²
19. $r = 1.2$ cm 4.5 cm²
20. $d = 1.4$ km 1.5 km²
21. $d = 17$ cm 226.9 cm²

Find the exact area of a circle with the given circumference.

SAMPLE $C = 8\pi$

1. Find r.

$$2\pi r = 8\pi$$

$$r = \frac{8\pi}{2\pi}$$

$$r = 4$$

2. Find the area.

$$A = \pi r^2$$

$$A = \pi(4^2) = 16\pi$$

22. $C = 14\pi$ 49π
23. $C = 48\pi$ 576π
24. $C = 16\pi$ 64π
25. $C = 11\pi$ 30.25π

26. The area of a circle is 4π m². Find the circumference. 4π m
27. The area of a circle is $\frac{49}{4}\pi$ cm². Find the circumference. 7π cm

Find the area in terms of π of the region shaded in red.

C 28. 16π

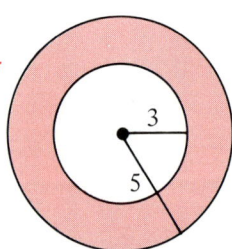

29. 25π − 50

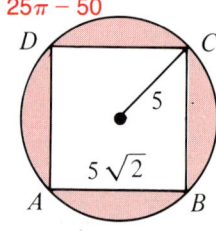

ABCD is a square.

30. 68 − 16π

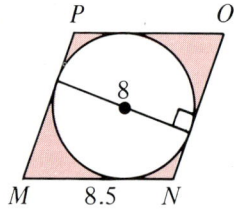

MNOP is a rhombus.

31. 15π

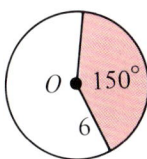

32. 200π

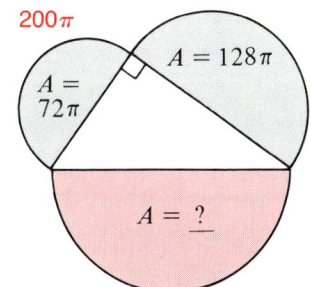

33. Draw two circles. Then construct a circle whose area is equal to the sum of the areas of the given circles.

Strategy:

A. Let the radii of the given circles be s and t.

B. $\pi s^2 + \pi t^2 = \pi(s^2 + t^2)$

C. Construct a right triangle with legs s and t. Let the hypotenuse be r. Then $r^2 = s^2 + t^2$.

D. Construct a circle with radius r.

SELF-TEST

Vocabulary

circle (p. 263)
center of a circle (p. 263)
circumference (p. 263)
radius (p. 263)
diameter (p. 263)

1. Complete: By definition, $\pi = \dfrac{?}{?} \cdot \dfrac{C}{d}$ *(6-7)*

2. Write an approximation for π correct to the nearest hundredth.   3.14

Copy and complete the table about circles.
Some answers may be given in terms of π.

	3.	4.	5.	6.
radius r	5	? 8	? 10	? 6
diameter d	? 10	16	? 20	? 12
circumference C	? 10π	? 16π	20π	? 12π
area A	? 25π	? 64π	? 100π	36π

(6-8)

APPLICATIONS

MIRRORS AND BILLIARDS

When a ray of light strikes a mirror, it is reflected at the same angle at which it arrives. For example, in the diagram, ∠1 = ∠2.

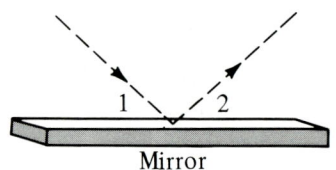

This principle of reflection is applied in the operation of a simple periscope. In the cross section of a periscope shown here, light is reflected twice so that light rays entering the periscope are parallel to light rays leaving the periscope.

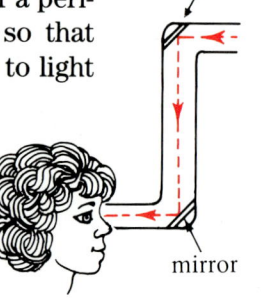

The same principle of reflection also applies to a ball bouncing off the sides of a pocket-billiard table or a miniature-golf course. The ball bounces back at the same angle at which it arrives. A good player uses this principle to sink a shot or to make a hole-in-one.

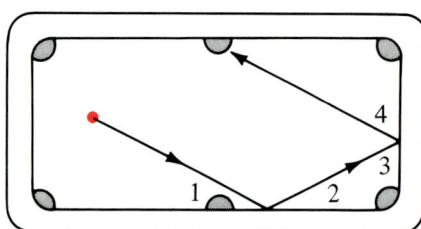

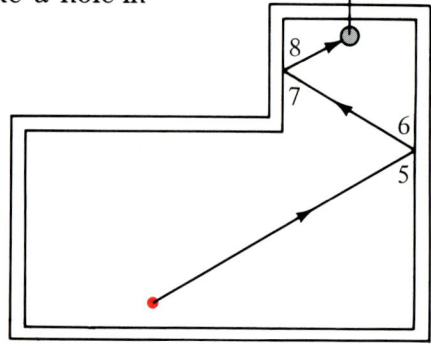

∠1 = ∠2
∠3 = ∠4

∠5 = ∠6
∠7 = ∠8

EXERCISES 1. ∠2 = 60°, ∠3 = 30°, ∠4 = 30°, ∠5 = 120°, ∠6 = 60°

Two mirrors form a 90° angle as shown.

1. If ∠1 = 60° and ∠2 is complementary to ∠3, find the measure of each numbered angle. (See above.)

2. Complete: ∠5 + ∠6 = __?__°. 180

3. Explain why the entering ray is parallel to the reflected ray. If 2 lines and a trans. form supp. s. − s. int. ∠s, then the lines are ∥.

4. ∠2 = 50°, ∠3 = 40°, ∠4 = 40°, ∠5 = 100°, ∠6 = 80°

In the diagram at the right, suppose ∠1 = 50° and ∠2 and ∠3 are complementary angles.

4. Find the measure of each numbered angle. (See above.)

5. Is the entering ray still parallel to the reflected ray? yes

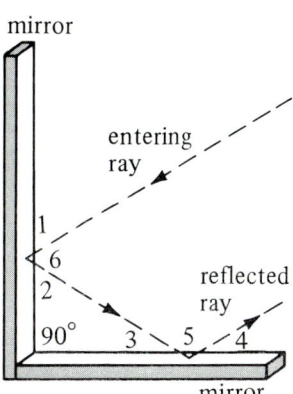

7. Opp. sides of the table are ∥, so alt. int. ∠s are =.

A ball on a billiard table rolls along the path shown. ∠1 = 70°.

6. Find the measure of ∠2. 70°

7. Explain why ∠3 = 70°. (See above.)

8. Find the measures of ∠5 and ∠6. 40°; 40°

9. Explain why $\overrightarrow{DE} \parallel \overrightarrow{BC}$.
If 2 lines and a trans. form = alt. int. ∠s, then the lines are ∥.

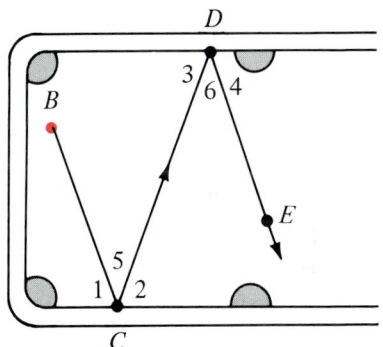

10. A billiard ball rolls from point X to point Y as shown. If the ball is hit hard enough, will it go into a pocket? no

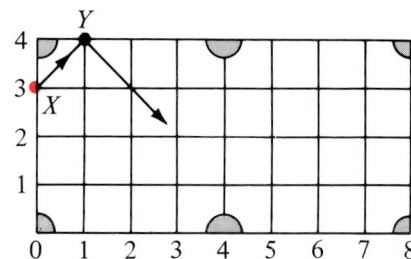

Areas   **273**

PROBLEM SOLVING STRATEGIES

WORKING BACKWARDS

If a problem has a specific goal, and you have to find how to get there, it is sometimes helpful to begin at the goal and then work backwards. You trace your way backwards step by step until you reach the beginning point.

EXERCISES

1. Two people play a game with the understanding that the loser is to double the money of the winner.

    They play the game twice, each person wins once, and they both end up with $6.00.

    How much did each one have at the beginning of the game?

    Let's work backwards. Begin at the end of the game. Here is a diagram showing A and B with $6.00 each at the end of two games.

	Beginning	After Game 1	After Game 2
A	? ←	? ←	$6.00
	Game 1	Game 2	
B	? ←	? ←	$6.00
	Winner has twice as much money as at the beginning.		Winner has twice as much money as after Game 1.

    a. Assume that A lost the last game.

    How much did A have to pay B so that they both end up with $6.00? $3.00

    On a copy of the diagram, fill in the amounts that A and B must have had after Game 1. A: $9.00; B: $3.00

    b. Now assume that B lost the first game.

    How much did B have to pay A in order that A and B end up with the amounts you found in part (a)? $4.50

    Enter the amounts each one must have had at the beginning of the game. A: $4.50; B: $7.50

2. Three people play a game with the understanding that the loser will double the money of the other two.

   At the end of three games each person has lost once, and they each end up with $12.00.

   How much did each person have at the beginning of the game? $19.50
   $10.50
   $6.00

3. A taxi cab driver charged Steve Yamamoto a total of $5.65 for a trip from his home to the local airport. She told Steve that this amount included a flat fee of $2.05 plus $.15 for every $\frac{1}{6}$ of a mile.

   How far is it from Steve's home to the airport?

   a. Subtract the flat fee from the total charge to find the amount Steve paid for mileage. $3.60

   b. Divide your answer to part (a) by $.15 to find out how many sixths of a mile Steve was charged for. 24 sixths of a mile

   c. Multiply your answer to part (b) by $\frac{1}{6}$ to find the distance from Steve's home to the airport. 4 miles

4. The value of a car goes down each year so that it is only worth about $\frac{4}{5}$ of what it was the year before. Marta Ramirez owns a car worth $9600. What was its value last year? You know that last year's value was multiplied by $\frac{4}{5}$ to get $9600. Reverse the operation to find the value of Marta's car last year. $12,000

5. A state engineering department determines that each day they should be able to lay $\frac{3}{8}$ mile of a new highway. How many working days should the construction crew allow for 27 miles of highway? 72 days

6. Given: $AB = ED$
   $AE = BC$
   $\angle 1 = \angle 2$

   Prove: $ABDE$ is a parallelogram.

   a. There are lots of ways to prove that $ABDE$ is a parallelogram. Since you know that $AB = ED$, what additional information would you need to know in order to say that $ABDE$ is a parallelogram? There are two choices: $AE = BD$ or $\overline{AB}$ is parallel to $\overline{DE}$. Which do you think seems the most promising?

   b. Prove that $AE = BD$.

Areas

COMPUTER ACTIVITIES

RANDOM NUMBERS AND π (PI)

Can your computer throw darts? Not really, but it can pick numbers at random. This will simulate dart throwing. The target is a circle on a square board. The computer will use random numbers to show where a "dart" lands. Each dart has an equal chance of landing anywhere on the board. The computer will keep track of whether a dart lands inside or outside the circle.

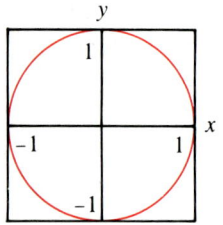

The program below lets you decide how many darts to throw. RND means "random number." Different computers handle random numbers differently. This program is written to work on Apple computers. Modifications are given for TRS-80 and IBM computers. You may need to check on how your computer uses RND. The random numbers in this program are decimals between 0 and 1.

```
10   REM ***SIMULATED DART THROWING
20   LET X = RND(-PEEK(78)-256*PEEK(79))
30   IN = 0
40   T = 0
50   PRINT "HOW MANY DARTS";
60   INPUT T
70   FOR C = 1 TO T
80   X = RND(1)
90   Y = RND(1)
100  REM ***IGNORE DARTS LANDING ON
110  REM ***OR OUTSIDE THE CIRCLE
120  IF X*X + Y*Y > = 1 THEN 150
130  REM ***COUNT DARTS IN CIRCLE
140  IN = IN + 1
150  NEXT C
160  PRINT "I THREW   ";T;"  DARTS"
170  PRINT IN;"  DARTS WENT IN CIRCLE"
180  PRINT "RATIO OF DARTS IN CIRCLE"
190  PRINT "TO TOTAL IS   ";IN/T
200  END
```

On TRS-80 computers, substitute:
20 RANDOM

On IBM computers, substitute:
20 RANDOMIZE VAL(MID$(TIME$,7,2))

On TRS-80 computers, use RND (0) instead of RND(1).

EXERCISES

1. RUN the program. Time the computer. How long does it take your computer to throw Answers will vary; examples are given.
   a. 100 darts? 3 sec.   b. 1000 darts? 25 sec.   c. 10,000 darts? 4 min.

2. a. What is the radius of the circle? 1   b. What is the area of the circle in terms of π? π
   c. How long is a side of the square? 2   d. What is the area of the square? 4

3. RUN the program 6 times. Complete the chart. Answers will vary. Examples are given below.

Number inside circle	74	76	396	378	771	778
Number thrown	100	100	500	500	1000	1000
Ratio: $\frac{\text{darts inside}}{\text{total darts}}$	0.74	0.76	0.79	0.76	0.77	0.78

In Exercise 3, you probably got values between 0.77 and 0.82. The darts are thrown at random, so nobody can be sure what you will get.

The ratio $\frac{\text{darts inside circle}}{\text{total darts thrown}}$ tells what fractional part of the darts landed inside the circle. Suppose the darts land at random on the board. Then the ratio $\frac{\text{darts inside circle}}{\text{total darts thrown}}$ should be about the same as the ratio $\frac{\text{area of circle}}{\text{area of square}}$. Multiply the ratio by 4 (the area of the square). The answer is an approximation of the area of the circle.

4. For each of the RUNs in Exercise 3, copy the value of the ratio. Multiply this by 4 to get the approximate area of the circle. Round areas to the nearest hundredth. Answers will vary. Examples are given below.

Ratio: $\frac{\text{darts inside}}{\text{total darts}}$	0.74	0.76	0.79	0.76	0.77	0.78
Area of circle	2.96	3.04	3.16	3.04	3.08	3.12

5. In Exercise 2b, you found that the area of the circle is exactly π. Give the decimal value of π to the nearest hundredth. 3.14

6. Compare the areas you figured in Exercise 4 to the decimal approximation for π. The calculated area of the circle approximates π.

Areas

SKILLS REVIEW

ALGEBRAIC EXPRESSIONS AND EQUATIONS

Write an expression using the given variable or variables.

1. The quotient when 86 is divided by n $86 \div n$ or $\frac{86}{n}$
2. The third power of k k^3
3. The sum of 13 and three times x $13 + 3x$
4. The product of 4, a, and b $4ab$
5. 18 subtracted from half of c $\frac{1}{2}c - 18$
6. m increased by 41 $m + 41$
7. $(a + b)$ multiplied by 7 $7(a + b)$
8. The ratio of 15 to w $\frac{15}{w}$

Write an equation or an inequality for each statement. Let n be the variable.

9. When 27 is subtracted from some number, the difference is 15. $n - 27 = 15$
10. The product of some number and 7 is not equal to zero. $7n \neq 0$
11. One fifth of some number, minus 13, is equal to 27. $\frac{1}{5}n - 13 = 27$
12. When 7 is subtracted from half of some number, the difference is 18. $\frac{1}{2}n - 7 = 18$
13. Three times some number is greater than 16. $3n > 16$
14. The fifth power of some number, increased by 9, is 41. $n^5 + 9 = 41$
15. The sum of some number and 43 is less than 75. $n + 43 < 75$
16. 16 is the quotient when 480 is divided by some number. $\frac{480}{n} = 16$

Solve.

SAMPLE
$$2x - 3 = 7$$
$$2x - 3 + 3 = 7 + 3 \quad \leftarrow \text{Add 3 to both sides.}$$
$$2x = 10$$
$$x = 5 \quad \leftarrow \text{Divide both sides by 2 to get } x = 5.$$

17. $b + 17 = 41$ 24
18. $x - 9 = -12$ -3
19. $14 = n - 36$ 50
20. $8y = 72$ 9
21. $12m = -48$ -4
22. $-5z = 95$ -19
23. $16 + x = 3$ -13
24. $63 = 7g$ 9
25. $\frac{t}{6} = -4$ -24
26. $\frac{s}{12} = 30$ 360
27. $\frac{a}{51} = 0$ 0
28. $3x + 4 = 19$ 5
29. $38 = 4k - 6$ 11
30. $\frac{a}{6} + 2 = 7$ 30
31. $-13 = \frac{k}{2} - 1$ -24

CHAPTER REVIEW

CHAPTER SUMMARY

1. The following formulas for area can be used.

   Rectangle
   $A = bh$
   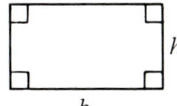

   Square
   $A = s^2$
   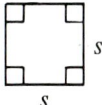

   Parallelogram
   $A = bh$
   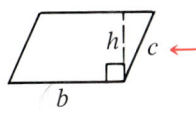

   Unless a parallelogram is also a rectangle, $h \neq c$.

   Triangle
   $A = \frac{1}{2}bh$
   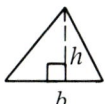

   Trapezoid
   $A = \frac{1}{2}h(b_1 + b_2)$
   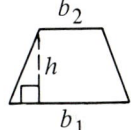

2. The Pythagorean Theorem states:
   In any right triangle, the square of the hypotenuse is equal to the sum of the squares of the legs.
   If $\triangle ABC$ is a right triangle, then $c^2 = a^2 + b^2$.

   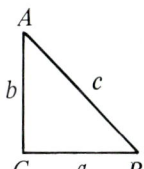

3. The converse of the Pythagorean Theorem is true.
   If $c^2 = a^2 + b^2$, then $\triangle ABC$ is a right triangle.

4. For any circle: $C = 2\pi r$ and $A = \pi r^2$.

REVIEW EXERCISES

Copy and complete the table about rectangles. *(See pp. 238–242.)*

	1.	2.	3.	4.
base b	5	30	7	7.2
height h	2	20	? 7	4
perimeter p	? 14	? 100	? 28	22.4
area A	? 10	? 600	49	28.8

Areas

Find the area of the shaded region. *(See pp. 243–249.)*

5. 50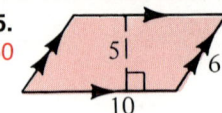
6. 42
7. 45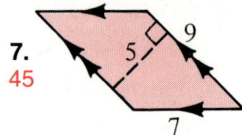

8. 36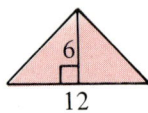
9. 44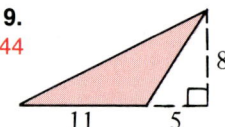
10. 96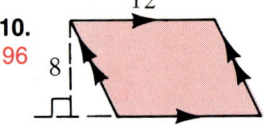

Copy and complete the table about trapezoids. *(See pp. 250–254.)*

	11.	12.	13.	14.
base length b_1	12	11	8.3	7
base length b_2	8	6	4.5	5
height h	5	4	5	? 4
area A	? 50	? 34	? 32	24

In each exercise, find the value of x. *(See pp. 255–258.)*

15. 10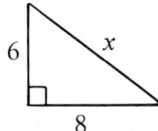
16. 9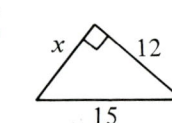
17. 3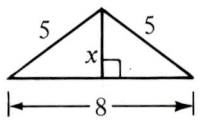

The lengths of the sides of a triangle are given. State whether the triangle is acute, right, or obtuse. *(See pp. 259–261.)*

18. 4, 5, 8   obtuse
19. 11, 60, 61   right
20. 8, 8, 8   acute

Complete the statements about circles. *(See pp. 263–271.)*

21. If $d = 12$, then $C =$ __?__. Answer in terms of π. 12π

22. If $r = 5$, then $A =$ __?__. Answer in terms of π. 25π

23. If $d = 8$, then $A =$ __?__. Use 3.14 for π and round your answer to tenths.   50.2

24. If $r = 12$, then $C =$ __?__. Use 3.14 for π and round your answer to tenths.   75.4

25. If $A = 81\pi$, then $r =$ __?__.   9
26. If $C = 18\pi$, then $r =$ __?__.   9

CHAPTER TEST

1. Find the perimeter and area of a rectangle 10 cm long and 8.5 cm wide. p = 37 cm; A = 85 cm²   (6-1)

2. Find the area of a square with perimeter 12 m. 9 m²

3. Find the perimeter and area of the shaded parallelogram. p = 50; A = 90   (6-2)

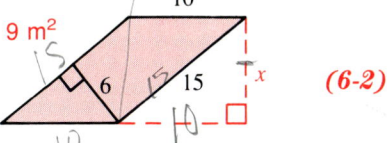

4. Use your answer from Exercise 3 to find the value of x. 9

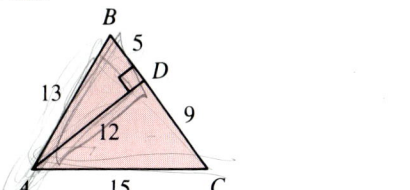

5. Find the area of $\triangle ABC$. 84   (6-3)

6. Find the area of $\triangle ADC$. 54

7. A triangle has base $4x$ and height x. Find its area in terms of x. $2x^2$

8. Find the area of the trapezoid shown. 105   (6-4)

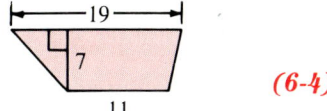

9. A trapezoid with bases 16 and 12 has area 126. Find its height. 9

Find the value of x.

10. 26   11.   12.   (6-5)

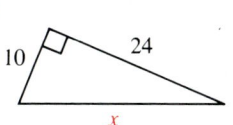 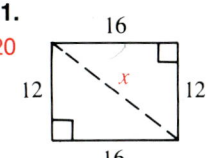 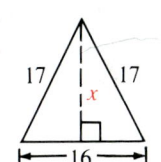

20   15

13. A triangle has sides 5, 6, and 9. Is it acute, right, or obtuse? obtuse   (6-6)

14. A parallelogram has sides of lengths 6 and 8 and a diagonal of length 10. Draw a diagram and explain why the parallelogram must be a rectangle.

15. A circle has diameter 20. Find its circumference to the nearest tenth. (Use 3.14 for π.) 62.8   (6-7)

16. Find the radius of a circle with circumference $\frac{4}{3}\pi$. $\frac{2}{3}$

17. Find the area, in terms of π, of a circle with radius $1\frac{1}{2}$. $\frac{9}{4}\pi$ (6-8)

18. Find the diameter of a circle with an area of 16π m². 8 m

Areas

UNIT C CUMULATIVE REVIEW

Complete using *always*, *sometimes*, or *never*.

1. An equilateral triangle is __?__ a regular polygon. **always**
2. The interior angle sum of a pentagon is __?__ 540°. **always**
3. A parallelogram is __?__ a square. **sometimes**
4. A rhombus is __?__ a parallelogram. **always**
5. Opposite angles of a quadrilateral are __?__ equal. **sometimes**
6. Consecutive angles of a parallelogram are __?__ complementary. **never**
7. The diagonals of a quadrilateral __?__ bisect each other. **sometimes**
8. The diagonals of a rectangle are __?__ equal. **always**

Complete each statement.

9. In rhombus *JKLM*, $\overline{MK}$ __?__ $\overline{LJ}$. **⊥**
10. In △*ABC* below, $MN = $ __?__. **$\frac{1}{2}AB$**
11. In parallelogram *QRST*, $QR = $ __?__ and $QT = $ __?__. **TS; RS**
12. The median of a trapezoid is __?__ to the bases. **parallel**
13. If *M* is the midpoint of hypotenuse $\overline{DE}$ in right △*DEF*, then $MD = $ __?__ $ = $ __?__. **MF; ME**
14. A quadrilateral must be a parallelogram if one pair of opposite sides are both __?__ and __?__. **parallel; equal**

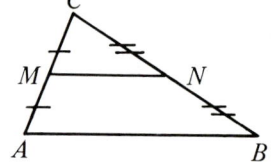

Find the area. Use 3.14 for π and round to the nearest tenth.

15. **78** (rectangle, 13 × 6)
16. **17.6 cm²** (square, 4.2 cm × 4.2 cm)
17. **40** (parallelogram, base 10, height 4, side 5)

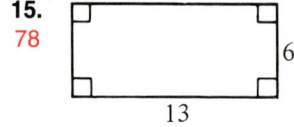 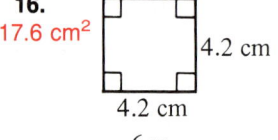 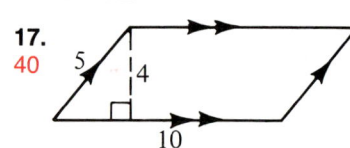

18. **125 m²** (triangle, base 25 m, height 10 m)
19. **48 m²** (trapezoid, bases 6 m and 10 m, height 6 m)
20. **379.9** (circle, radius 11)

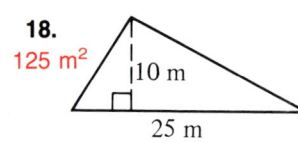 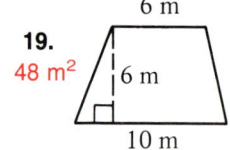 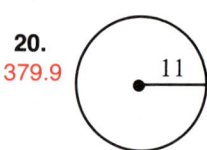

21. Is a triangle with sides 3 m, 4 m, and 6 m a right triangle? **no (obtuse)**
22. Find the exact circumference of a circle with radius 6 m. **12π m**

UNIT D

Here's what you'll learn in this chapter:

To find the ratio of two numbers.

To write proportions in several forms.

To solve problems using proportions.

To make maps and scale drawings.

The very large sand castle, shown above, was built at Pacific Beach in San Diego, California. Although all the measurements in a real castle would be much larger, they would be in proportion to those of the sand castle.

Chapter 7

Ratios and Proportions

1 Ratios

In this chapter and the next, we'll see how ratios and proportions are used by mapmakers, technical artists, and commercial photographers. Before doing so, however, we must review what is meant by *ratio*. In the next section, we'll review proportions.

The **ratio** of two numbers is the quotient of the two numbers.

A ratio is usually expressed in *simplest form*.

$\dfrac{4}{8} = \dfrac{1}{2}$ ← simplest form

$\dfrac{2x^2}{2xy} = \dfrac{2 \cdot x \cdot x}{2 \cdot x \cdot y} = \dfrac{x}{y}$ ← simplest form

Ratio	Written Forms
2 to 3	$\dfrac{2}{3}$ or $2:3$
5 to 1	$\dfrac{5}{1}$ or $5:1$
a to b, $b \neq 0$	$\dfrac{a}{b}$ or $a:b$

EXAMPLE 1 There are 60 streetcars and 72 subway cars in a city transportation system.

    a. Find the ratio of streetcars to subway cars.
    b. Find the ratio of subway cars to streetcars.

a. $\dfrac{60}{72} = \dfrac{5}{6}$

b. $\dfrac{72}{60} = \dfrac{6}{5}$

EXAMPLE 2 A rectangle is 3 m long and 80 cm wide. Find the ratio of its length to its width.

Method 1

Express the length in centimeters. 3 m = 300 cm

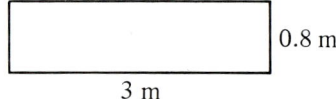

$\dfrac{\text{length}}{\text{width}} = \dfrac{300}{80} = \dfrac{15}{4}$

Method 2

Express the width in meters. 80 cm = 0.8 m

$\dfrac{\text{length}}{\text{width}} = \dfrac{3}{0.8} = \dfrac{30}{8} = \dfrac{15}{4}$

Notice that the ratio is 15:4, no matter which unit is used. Notice also that the ratio contains no units. The ratio is 15:4, and not 15:4 cm or 15:4 m.

In the preceding examples, ratios were used to compare two numbers. Ratios can also be used to compare three or more numbers. For example, in △ABC shown below, *a*, *b*, and *c* are in the ratio 3 to 4 to 5.

This means that

$$a:b = 3:4$$
$$a:c = 3:5$$
$$b:c = 4:5$$

We combine these steps by writing

$$a:b:c = 3:4:5.$$

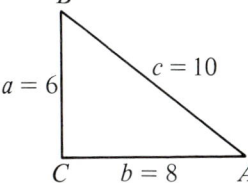

Classroom Practice

Express each ratio in simplest form.

1. $\frac{5}{15}$ $\frac{1}{3}$
2. $\frac{10}{15}$ $\frac{2}{3}$
3. $\frac{9}{12}$ $\frac{3}{4}$
4. $\frac{500}{1000}$ $\frac{1}{2}$
5. 6:8   3:4
6. 20:25   4:5
7. 2:4:6   1:2:3
8. 30:40:50   3:4:5

Given *a* = 3 and *b* = 5, find each ratio.

9. $\frac{a}{b}$ $\frac{3}{5}$
10. $\frac{b}{a}$ $\frac{5}{3}$
11. $\frac{a}{a+b}$ $\frac{3}{8}$
12. $\frac{b}{a+b}$ $\frac{5}{8}$

Use the figure to find each ratio in simplest form.

13. $\frac{AB}{BC}$ $\frac{3}{2}$
14. $\frac{AB}{AC}$ $\frac{3}{5}$
15. $\frac{AC}{BC}$ $\frac{5}{2}$

Written Exercises

Express each ratio in simplest form.

1. $\frac{6}{12}$ $\frac{1}{2}$
2. $\frac{6}{9}$ $\frac{2}{3}$
3. $\frac{18}{24}$ $\frac{3}{4}$
4. $\frac{36}{72}$ $\frac{1}{2}$
5. 21:28   3:4
6. 25:45   5:9
7. 10:20:30   1:2:3
8. 15:25:35   3:5:7

Given *a* = 2 and *b* = 3, find each ratio.

9. $\frac{a}{b}$ $\frac{2}{3}$
10. $\frac{b}{a}$ $\frac{3}{2}$
11. $\frac{a}{a+b}$ $\frac{2}{5}$
12. $\frac{b}{a+b}$ $\frac{3}{5}$

Ratios and Proportions   **287**

Given $c = 4$ and $d = 5$, find each ratio.

13. $\dfrac{c}{d}$ $\dfrac{4}{5}$
14. $\dfrac{d}{c}$ $\dfrac{5}{4}$
15. $\dfrac{d}{c+d}$ $\dfrac{5}{9}$
16. $\dfrac{d-c}{d+c}$ $\dfrac{1}{9}$

Use the figures below to find each ratio in simplest form.

17. $\dfrac{AD}{DB}$ $\dfrac{2}{1}$
18. $\dfrac{AD}{AB}$ $\dfrac{2}{3}$
19. $\dfrac{DB}{AB}$ $\dfrac{1}{3}$

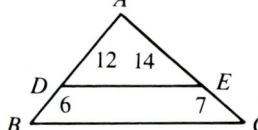

20. $\dfrac{AE}{AC}$ $\dfrac{2}{3}$
21. $\dfrac{AC}{AE}$ $\dfrac{3}{2}$
22. $\dfrac{EC}{AC}$ $\dfrac{1}{3}$

23. $\dfrac{PO}{OS}$ $\dfrac{3}{5}$
24. $\dfrac{PO}{PS}$ $\dfrac{3}{8}$
25. $\dfrac{OS}{PS}$ $\dfrac{5}{8}$

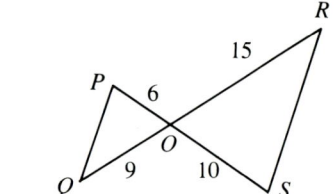

26. $\dfrac{OR}{QO}$ $\dfrac{5}{3}$
27. $\dfrac{OR}{QR}$ $\dfrac{5}{8}$
28. $\dfrac{QR}{PS}$ $\dfrac{3}{2}$

29. $\dfrac{\text{length of } A}{\text{length of } B}$ $\dfrac{3}{2}$
30. $\dfrac{\text{width of } A}{\text{width of } B}$ $\dfrac{2}{3}$

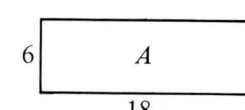 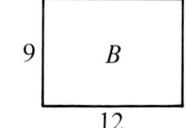

31. $\dfrac{\text{area of } A}{\text{area of } B}$ $\dfrac{1}{1}$
32. $\dfrac{\text{perimeter of } A}{\text{perimeter of } B}$ $\dfrac{8}{7}$

In Exercises 33–36, find, in simplest form:
a. the ratio of the area of the shaded part to the area of the unshaded part;
b. the ratio of the area of the shaded part to the total area.

 33. 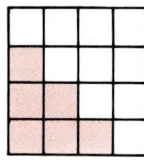 34. 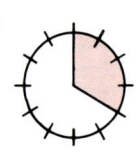 35. 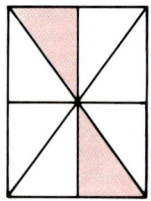 36.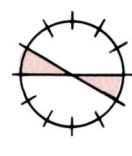

33. a. $\dfrac{3}{5}$ b. $\dfrac{3}{8}$
34. a. $\dfrac{1}{2}$ b. $\dfrac{1}{3}$
35. a. $\dfrac{1}{3}$ b. $\dfrac{1}{4}$
36. a. $\dfrac{1}{5}$ b. $\dfrac{1}{6}$

Express each ratio in simplest form.

37. $\dfrac{30 \text{ cm}}{1 \text{ m}}$ $\dfrac{3}{10}$
38. $\dfrac{4 \text{ mm}}{10 \text{ m}}$ $\dfrac{1}{2500}$
39. $\dfrac{2 \text{ m}}{3 \text{ km}}$ $\dfrac{1}{1500}$

 40. The measures of the angles of a triangle are in the ratio $3:4:5$. Find the measures of the angles.
(*Hint:* Let $3x$ represent the measure of the smallest angle.) $45°, 60°, 75°$

2 Proportions

A **proportion** is an equation which states that two ratios are equal. Here are some examples of proportions.

$$\frac{4}{6} = \frac{2}{3} \qquad 1:5 = 6:30 \qquad \frac{a}{b} = \frac{c}{d} \ (b \neq 0, d \neq 0)$$

The first and last numbers in a proportion are called the *extremes*. The middle numbers are called the *means*.

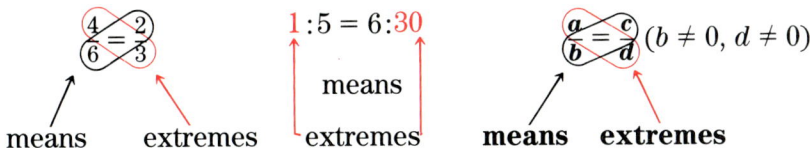

It is sometimes helpful to change a proportion from one form to another. The following properties, based on the postulates of equality, provide ways of changing proportions. The variables used in each property represent nonzero numbers.

PROPERTY 1 (The Cross-Multiplying Property)

In a proportion, the product of the means equals the product of the extremes.

$$10 \times 1 = 5 \times 2 \qquad b \times c = a \times d$$

PROPERTY 2 (The Switching Property)

In a proportion, the means (or the extremes) may be switched.

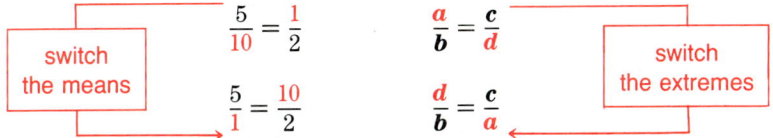

> **PROPERTY 3 (The Overturning Property)**
>
> In a proportion, the ratios may be turned upside-down.

$$\frac{5}{10} = \frac{1}{2} \quad\longrightarrow\quad \frac{10}{5} = \frac{2}{1}$$

$$\frac{a}{b} = \frac{c}{d} \quad\longrightarrow\quad \frac{b}{a} = \frac{d}{c}$$

> **PROPERTY 4 (The Numerator-Changing Property)**
>
> If $\frac{a}{b} = \frac{c}{d}$, then $\frac{a+b}{b} = \frac{c+d}{d}$.
>
> Also, if $\frac{a+b}{b} = \frac{c+d}{d}$, then $\frac{a}{b} = \frac{c}{d}$.

In the following exercises, all variables represent nonzero numbers.

Classroom Practice

Tell whether each of the following is a proportion.

1. $3:6 = 1:2$ yes
2. $\frac{2}{5} = \frac{4}{10}$ yes
3. $\frac{5}{6} = \frac{1}{2} + \frac{1}{3}$ no
4. $\frac{1}{3} = \frac{3}{9}$ yes

Use the Cross-Multiplying Property to complete each statement.

5. If $\frac{6}{8} = \frac{3}{4}$, then $8 \times 3 = \underline{\ ?\ }$. 6×4
6. If $\frac{1}{7} = \frac{2}{14}$, then $\underline{\ ?\ } = 1 \times 14$. 7×2
7. If $\frac{a}{b} = \frac{c}{d}$, then $bc = \underline{\ ?\ }$. ad
8. If $\frac{x}{y} = \frac{p}{q}$, then $py = \underline{\ ?\ }$. qx

Use the Switching Property to complete each statement.

9. If $\frac{4}{6} = \frac{2}{3}$, then $\frac{4}{2} = \underline{\ ?\ }$ and $\frac{3}{6} = \underline{\ ?\ }$. $\frac{6}{3}; \frac{2}{4}$
10. If $\frac{1}{4} = \frac{2}{8}$, then $\frac{1}{2} = \underline{\ ?\ }$ and $\frac{8}{4} = \underline{\ ?\ }$. $\frac{4}{8}; \frac{2}{1}$

Use the Overturning Property to complete each statement.

11. If $\frac{a}{b} = \frac{2}{3}$, then $\frac{b}{a} = \underline{\ ?\ }$. $\frac{3}{2}$
12. If $\frac{2}{9} = \frac{x}{6}$, then $\frac{9}{2} = \underline{\ ?\ }$. $\frac{6}{x}$

Use the Numerator-Changing Property to complete each statement.

13. If $\frac{7}{3} = \frac{x}{6}$, then $\frac{7+3}{3} = \frac{\underline{\quad?\quad}}{6}$. $x+6$

14. If $\frac{w}{x} = \frac{y}{z}$, then $\frac{w+x}{x} = \frac{\underline{\quad?\quad}}{z}$. $y+z$

Find the value of x.

15. $\frac{2}{5} = \frac{x}{20}$
    $5x = \underline{\quad?\quad}$ 40
    $x = \underline{\quad?\quad}$ 8

16. $\frac{5}{6} = \frac{x}{12}$
    $6x = \underline{\quad?\quad}$ 60
    $x = \underline{\quad?\quad}$ 10

17. $\frac{3}{8} = \frac{x}{10}$
    $8x = \underline{\quad?\quad}$ 30
    $x = \underline{\quad?\quad}$ $3\frac{3}{4}$

18. $\frac{6}{x} = \frac{9}{6}$
    $9x = \underline{\quad?\quad}$ 36
    $x = \underline{\quad?\quad}$ 4

Written Exercises

Complete each statement.

[A] 1. If $\frac{a}{2} = \frac{b}{3}$, then $2b = \underline{\quad?\quad}$. $3a$

2. If $\frac{x}{y} = \frac{3}{4}$, then $3y = \underline{\quad?\quad}$. $4x$

3. If $\frac{5}{r} = \frac{3}{5}$, then $\underline{\quad?\quad} = 25$. $3r$

4. If $\frac{7}{c} = \frac{d}{5}$, then $\underline{\quad?\quad} = 35$. cd

5. If $\frac{a}{3} = \frac{b}{4}$, then $\frac{a}{b} = \underline{\quad?\quad}$. $\frac{3}{4}$

6. If $\frac{p}{q} = \frac{5}{6}$, then $\frac{6}{q} = \underline{\quad?\quad}$. $\frac{5}{p}$

7. If $\frac{r}{s} = \frac{2}{3}$, then $\frac{r+s}{s} = \underline{\quad?\quad}$. $\frac{5}{3}$

8. If $\frac{y}{z} = \frac{7}{9}$, then $\underline{\quad?\quad} = \frac{9}{7}$. $\frac{z}{y}$

9. If $\frac{x}{12} = \frac{3}{4}$, then $\frac{x}{3} = \underline{\quad?\quad}$. $\frac{12}{4}$ or $\frac{3}{1}$

10. If $\frac{1}{8} = \frac{9}{x}$, then $\underline{\quad?\quad} = \frac{9+x}{x}$. $\frac{9}{8}$

11. If $\frac{e}{f} = \frac{8}{5}$, then $5e = \underline{\quad?\quad}$. $8f$

12. If $\frac{k}{j} = \frac{9}{4}$, then $4k = \underline{\quad?\quad}$. $9j$

Suppose $\frac{a}{b} = \frac{3}{4}$ **and** $\frac{c}{d} = \frac{5}{7}$. **Find each ratio in simplest form.**

13. $\frac{b}{a}$ $\frac{4}{3}$

14. $\frac{a+b}{b}$ $\frac{7}{4}$

15. $\frac{a+b}{a}$ $\frac{7}{3}$

16. $\frac{2a}{2b}$ $\frac{3}{4}$

17. $\frac{d}{c}$ $\frac{7}{5}$

18. $\frac{c+d}{d}$ $\frac{12}{7}$

19. $\frac{c-d}{c}$ $\frac{-2}{5}$

20. $\frac{3d}{3c}$ $\frac{7}{5}$

Find the value of x.

21. $\frac{x}{6} = \frac{1}{2}$ 3

22. $\frac{4}{3} = \frac{16}{x}$ 12

23. $\frac{1}{3} = \frac{10}{x}$ 30

24. $\frac{9}{x} = \frac{9}{7}$ 7

25. $3:8 = x:16$ 6

26. $9:x = 3:1$ 3

27. $x:10 = 4:5$ 8

28. $3:5 = 24:x$ 40

Ratios and Proportions

Find the value of x.

B 29. $\dfrac{10}{3x} = \dfrac{5}{6}$ *4*

30. $\dfrac{5}{2x} = \dfrac{25}{1}$ *$\dfrac{1}{10}$*

31. $\dfrac{1}{2} = \dfrac{x+1}{8}$ *3*

32. $\dfrac{x-5}{6} = \dfrac{4}{9}$ *$7\dfrac{2}{3}$*

C 33. $\dfrac{3x-2}{4} = \dfrac{x}{3}$ *$\dfrac{6}{5}$*

34. $\dfrac{1}{2} = \dfrac{5x+4}{3}$ *$-\dfrac{1}{2}$*

35. $\dfrac{x}{3} = \dfrac{x+2}{5}$ *3*

36. $\dfrac{x-5}{x} = \dfrac{3}{4}$ *20*

37. $\dfrac{x+2}{x-1} = \dfrac{x+3}{x-3}$ *-1*

38. $\dfrac{(2x-3)^2}{(2x+1)(x-3)} = \dfrac{2}{1}$ *$7\dfrac{1}{2}$*

CALCULATOR ACTIVITIES

Suppose the sides of a triangle have lengths *a*, *b*, and *c*. The area of the triangle is given by the formula

$$A = \sqrt{s(s-a)(s-b)(s-c)} \text{ where } s = \dfrac{a+b+c}{2}.$$

To use this formula:

1. First compute *s*. Then compute $s - a$, $s - b$, and $s - c$.
2. Substitute in the formula $\sqrt{s(s-a)(s-b)(s-c)}$.

EXAMPLE   Find the area.

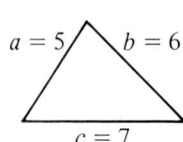
$a = 5$, $b = 6$, $c = 7$

1. $s = \dfrac{5+6+7}{2}$; $s = 9$.
   $s - a = 4$, $s - b = 3$, and $s - c = 2$.
2. $\sqrt{s(s-a)(s-b)(s-c)} = \sqrt{9 \times 4 \times 3 \times 2} \approx 14.696938$

To the nearest tenth, the area is 14.7.

Find the area of the triangle specified.
Round your answers to the nearest tenth.

	1.	2.	3.	4.	5.	6.	7.	8.
a	6	7	6	8	11	21	16	13.1
b	8	5	6	6	15	17	19	12.0
c	10	10	6	6	17	10	12	14.8

1. $A = 24$
2. $A \approx 16.2$
3. $A \approx 15.6$
4. $A \approx 17.9$
5. $A \approx 81.3$
6. $A = 84$
7. $A \approx 95.5$
8. $A \approx 74.8$

3 Using Proportions

Proportions can be used to solve many kinds of problems. You probably remember, from studying algebra, how to solve Example 1.

EXAMPLE 1

Four cans of paint cost $30.
How much will six cans of paint cost?

1. Summarize your data.

Number of cans	4	6
Cost	$30	$x

2. Write a proportion.

$$\frac{4}{30} = \frac{6}{x}$$

$$4x = 180$$
$$x = 45$$ *Answer:* $45

EXAMPLE 2

John Kaylen's summer job is painting houses and barns. He has just used three cans of paint to paint the front of the barn shown. How much more paint will he need for the remaining sides?

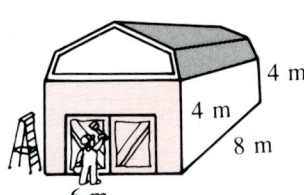

1. Estimate the area of the painted and unpainted sides. (Ignore the areas of doors.)

   Area of painted side = $6 \times 4 = 24$ m^2

   Area of unpainted sides = $\underbrace{8 \times 4}_{\text{side}} + \underbrace{6 \times 4}_{\text{back}} + \underbrace{8 \times 4}_{\text{side}} = 88$ m^2

2. Summarize your data.

	Area	Number of cans
Painted	24 m^2	3
Unpainted	88 m^2	x

3. Write a proportion.

$$\frac{24}{88} = \frac{3}{x}$$

$$24x = 264$$
$$x = 11$$

Answer: He will need about 11 more cans of paint.

Ratios and Proportions   293

Classroom Practice

**Make up a problem to go with each table below.
Then write a proportion which can be solved for x.** Answers will vary.

1.
Number of cans	4	6
Cost	$7	$x

$\frac{4}{7} = \frac{6}{x}$

2.
Number of boxes	8	20
Cost	$3	$x

$\frac{8}{3} = \frac{20}{x}$

3.
Painted area	10 m²	65 m²
Cost	$4	$x

$\frac{10}{4} = \frac{65}{x}$

4.
Cost of gasoline	$6	$9
Distance traveled	280 km	x km

$\frac{6}{280} = \frac{9}{x}$

5. Six grapefruits cost $1.92. How many can you buy for $4.80? 15 grapefruits

6. Three tickets to a track meet cost $15.90. How many can you buy for $26.50? 5 tickets

Written Exercises

A 1. Two concert tickets cost $34. How much would five tickets cost? $85

2. Three cans of cat food cost 93¢. How much would ten cans cost? $3.10

3. Three copies of *Sports Today* cost $5.85. How much would five copies cost? $9.75

4. Four containers of yogurt cost $2.52. How much would six containers cost? $3.78

5. Two books of stamps cost $8.80. How many books can you buy with $22.00? 5 books

6. Two packages of mints cost 69¢. How many packages can you buy for $2.25? 6 packages

7. It costs $11 to paint the front of the barn shown. How much will it cost to paint the right side? $16.50

8. Two cans of paint are needed to paint the front of the barn shown. How much paint is needed to paint the other three sides? 8 cans

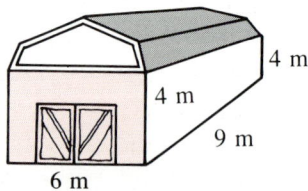

B 9. The Fongs would like to redo their kitchen floor. They find that linoleum will cost $16 per square meter. Carpeting will cost $20 per square meter. The new kitchen floor will cost $192 if linoleum is used. How much will it cost if carpeting is used? $240

10. It has taken Alexis 30 minutes to mow a 6 m by 32 m strip of lawn. How much more time will she need to finish the job? (Assume that she mows at a constant rate.) 65 min, or 1 h 5 min

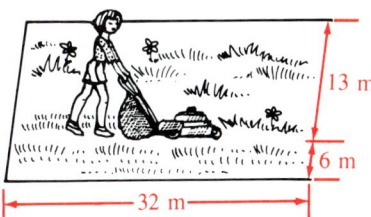

11. The Carmona Construction Company is laying a brick patio. Thirty rows of bricks have been installed. Twenty more rows are needed to finish the job. If 600 bricks have already been used, how many will be used for the entire job? 1000 bricks

SELF-TEST

Vocabulary

ratio (p. 286) means (p. 289)
proportion (p. 289) extremes (p. 289)

Express each ratio in simplest form.

1. $\frac{8}{12}$ $\frac{2}{3}$
2. $\frac{25}{15}$ $\frac{5}{3}$
3. $25:40$ $\frac{5}{8}$
4. $9:18:27$ $1:2:3$ *(7-1)*

Find the value of x.

5. $\frac{x}{6} = \frac{7}{3}$ 14
6. $\frac{10}{x} = \frac{8}{12}$ 15
7. $\frac{4}{6} = \frac{6}{x}$ 9
8. $x:6 = 3:9$ 2   *(7-2)*

Solve.

9. Three loaves of bread cost $2.67. How much will 5 loaves cost? $4.45   *(7-3)*

10. A rug 4 m wide and 5 m long costs $132. How much would a 3 m by 4 m rug of the same material cost? $79.20

PUZZLE ◆ PROBLEMS

Ron, Jo, and Bobbie each won a contest during class day at their high school. One student won the photography exhibit, another, the math contest, and the third, the track race. The photographer, who is an only child, does not know how to play tennis. Bobbie is a friend of the math whiz. Jo often plays tennis with Bobbie's sister. Which student won each event? photography: Ron, track: Bobbie, math: Jo

4 Maps and Scale Drawings

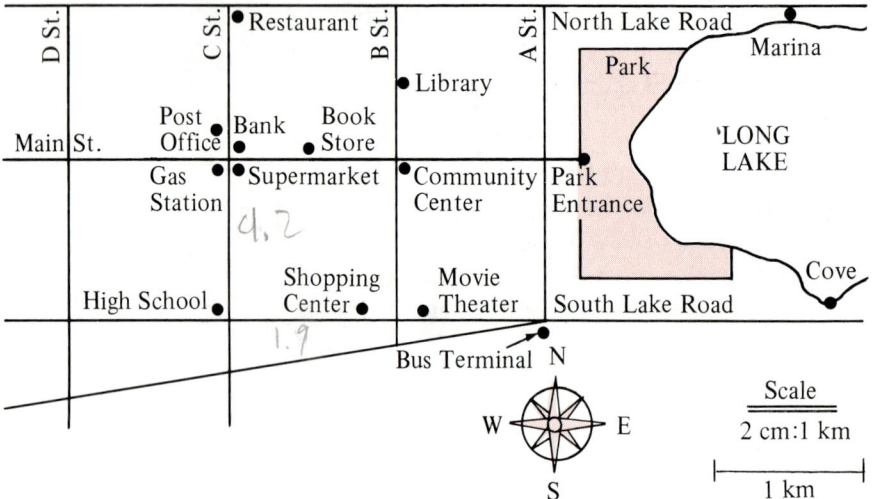

The map above shows part of the town of Lakeville. Notice that the **scale** of the map is

$$2 \text{ cm} : 1 \text{ km}.$$

This means that a distance of 2 cm *on the map* represents an *actual* distance of 1 km in Lakeville. For example, the distance between the high school and the post office is about 2.4 cm. To find the actual distance, we write a proportion:

$$\frac{2 \text{ cm}}{1 \text{ km}} = \frac{2.4 \text{ cm}}{x \text{ km}}$$

$$\frac{2}{1} = \frac{2.4}{x}$$

$$2x = 2.4$$

$$x = 1.2 \quad \longleftarrow \boxed{\text{The actual distance is about 1.2 km.}}$$

A *scale drawing* of a plot of land is very much like a map. So is the floor plan of a house. In the floor plan on the next page, the scale is listed as $1:200$. This means that

$$\frac{\text{distance in drawing}}{\text{corresponding distance in house}} = \frac{1}{200}.$$

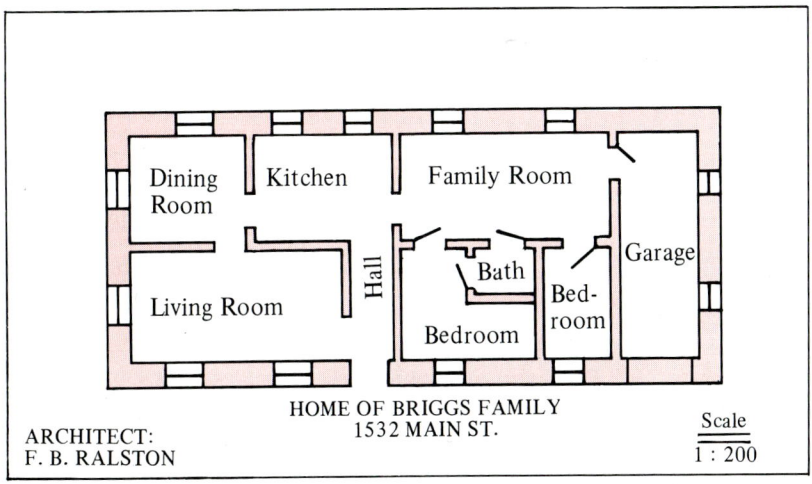

Classroom Practice

Exercises 1–4 refer to the floor plan above.

1. A scale of 1:200 means that
   1 cm in the floor plan = 200 cm in the house.
   Then 1 cm in the floor plan = __?__ m in the house. 2

Use a ruler to find each distance to the nearest centimeter. Then find the actual distance in the house.

2. The length, the width and the perimeter of the living room   3 cm, 2 cm, 10 cm; 6 m, 4 m, 20 m

3. The length and the width of the family room   3 cm, 1 cm; 6 m, 2 m

4. The perimeter of the entire house   24 cm; 48 m

Refer to the map of Lakeville on page 296. Use a ruler to find the distances referred to in the first row of the table. Measure to the nearest centimeter. Then copy and complete the table.

		Post Office to Park Entrance	Gas Station to Marina	Community Center to High School
5.	Distance on the map in cm	? 5	? 10	? 5
6.	Actual distance when scale is 2 cm : 1 km	? 2.5	? 5	? 2.5

Ratios and Proportions

Written Exercises

Use the map of Lakeville on page 296, with scale 1 cm : 800 m. Find each actual distance in the town.

A
1. Gas station to community center 2080 m
2. Bus terminal to movie theater 1360 m
3. Shopping center to library 2960 m
4. Park entrance to supermarket 3760 m
5. Restaurant to shopping center 4880 m
6. High school to bookstore 2480 m
7. Marina to cove by boat 3120 m
8. Marina to cove by car 9280 m

Use the floor plan on page 297. Find the actual length, width, perimeter, and area of each room. Use the scale 1 cm : 250 cm.

9. Dining room
   $l = 4$ m; $w = 3.75$ m
   $p = 15.5$ m; $A = 15$ m²
10. The smaller bedroom
    $l = 2.25$ m; $w = 3.75$ m
    $p = 12$ m; $A = 8.4$ m²
11. Kitchen
    $l = 4.75$ m
    $w = 3.5$ m
    $p = 16.5$ m
    $A = 16.6$ m²
12. Garage
    $l = 2.5$ m
    $w = 7.75$ m
    $p = 20.5$ m
    $A = 19.4$ m²

Use the map of Sequoia National Park to find the distance between the following places to the nearest kilometer.

B
13. Crystal Cave and Silver City 21 km
14. Mt. Whitney and Triple Divide Peak 18 km
15. Park Headquarters and Kern Canyon Ranger Station 38 km
16. Giant Forest and Garfield Grove of Big Trees 24 km

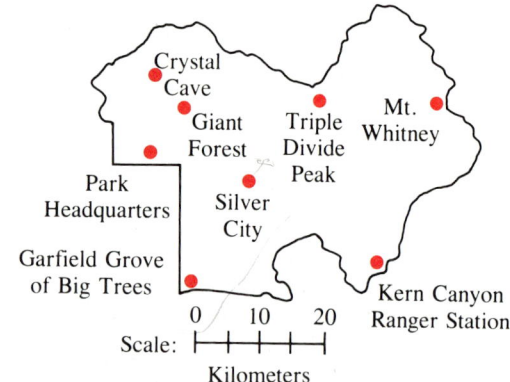

17. The diagram shows three cities in New Hampshire.
    a. Measure the map distances between the cities to the nearest millimeter. (See below.)
    b. Write a proportion to find the actual distance between Portsmouth and Concord. $\frac{48}{60} = \frac{56}{x}$; $x = 70$; 70 km
    c. Find the distance between Manchester and Concord. 25 km

5 Making Maps and Scale Drawings

When making a map or a scale drawing, you must first decide on the scale you are going to use. For example, suppose you are making a scale drawing of your classroom.

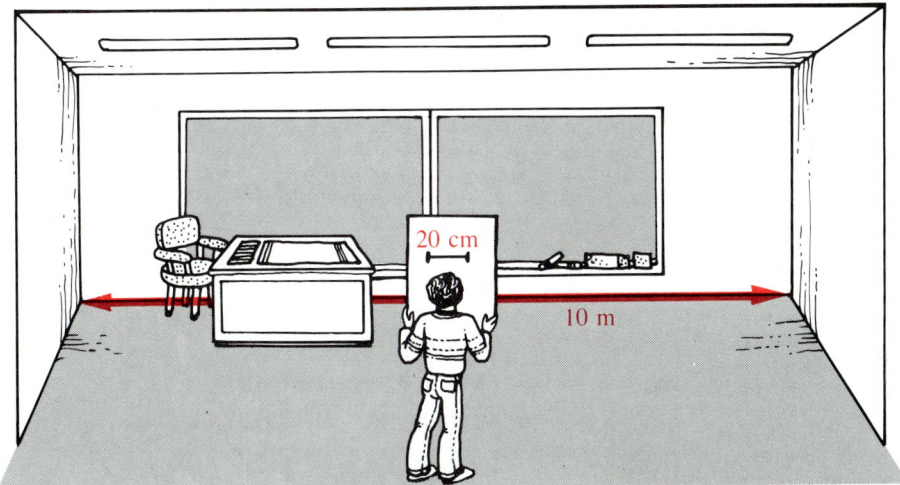

Suppose your classroom is 10 m long and you choose to represent this distance by a line segment 20 cm long. Then your scale is:

$$\frac{\text{length of classroom in drawing}}{\text{actual length of classroom}} = \frac{20 \text{ cm}}{10 \text{ m}} = \frac{20 \text{ cm}}{1000 \text{ cm}} = \frac{1}{50}$$

Measure the width and other distances in meters. Express these measurements in centimeters and multiply each by the *scale ratio*, $\frac{1}{50}$, to find the corresponding distance in your drawing.

Classroom Practice

Make a scale drawing of your classroom. Follow the instructions above. Suggested scale: 1 cm : 1 m or $\frac{1}{100}$

Written Exercises

A 1. Make a scale drawing of an Olympic-sized swimming pool which is 50 m long and 21 m wide. Suggested scale: 1 cm : 5 m or $\frac{1}{500}$

2. Draw the cover of this book, using the scale 1:2. Drawing should be 9.6 cm × 12 cm.

3. Repeat Exercise 2, using the scale 1:4.
   Drawing should be 4.8 cm × 6 cm.

4. In Frazier Park, the high school is 3 km east of the junior high school.

   The shopping mall is 4 km south of the junior high school and is 5 km from the high school.

   Draw a map showing the high school, junior high, and mall.

   Be sure to include the scale on your map.

5. The recreation center pool measures 42 m by 15 m.

   The pool is surrounded by a deck 3 m wide.

   Make a scale drawing of the pool and deck with scale 1 cm : 6 m.

6. Town A is 6 km south of city B.

   City C is 3 km northwest (NW) of A.

   Find the distance, to the nearest kilometer, from B to C. 4 km

7. Boothbay is 12 km west of New Harbor.

   Monhegan Island is 20 km southeast (SE) of New Harbor.

   Make a scale drawing and find the distance, to the nearest kilometer, from Boothbay to Monhegan Island. 30 km

B 8. Dallas, Texas, is about 300 km west of Shreveport, Louisiana.

   San Antonio, Texas, is located about 620 km southwest (SW) of Shreveport.

   To the nearest 50 km, how far is it from Dallas to San Antonio? 500 km

9. Campsite *A* is 6 km north of campsite *B*.

   From *A*, a fire tower is sighted on a bearing of 75°. From *B*, the same fire tower is sighted on a bearing of 60°.

   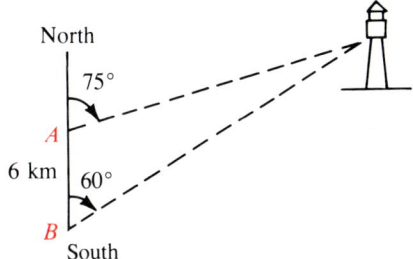

   a. To the nearest kilometer, how far is it from *A* to the fire tower? 20 km
   b. To the nearest kilometer, how far is it from *B* to the fire tower? 22 km

10. Campsite *C* is 2 km north of campsite *D*.

    From *C*, a waterfall is sighted at a point *W*, directly east of *C*. From *D*, the same waterfall is sighted on a bearing of 70°.

    a. Make a scale drawing showing *C*, *D*, and *W*.
    b. Find the distances to the nearest kilometer from *C* and *D* to the waterfall. from C: 6 km   from D: 6 km

C 11. Feingold's Furniture wants to put a new display in its front window. Design a furnished living room by making a scale drawing.
    Scale drawings will vary.

SELF-TEST

Vocabulary
scale (p. 296)   scale drawing (p. 296)

Use the diagram at the right.

1. Measure the map distance between Mesa and Tempe to the nearest centimeter. Then give the scale of the map. (7-4)
   3 cm; 1 cm : 3 km or $\frac{1}{300,000}$
2. Find the actual distance between Mesa and Scottsdale. 12 km
3. Find the actual distance between Scottsdale and Tempe. 8.4 km
4. Make a scale drawing of an ice hockey rink that is 61 m long and 30.5 m wide. Suggested scale: 1 cm : 5 m or $\frac{1}{5000}$ (7-5)

Ratios and Proportions

Applications

USING A COMPASS

A magnetic compass is a basic navigational tool for finding direction of travel with respect to magnetic north. A magnetic compass measures direction in degrees, like the protractors you have used to measure angles. Although many protractors only indicate measures from 0° to 180°, a compass includes measures from 0° to 360°, that is, a compass is a complete circle.

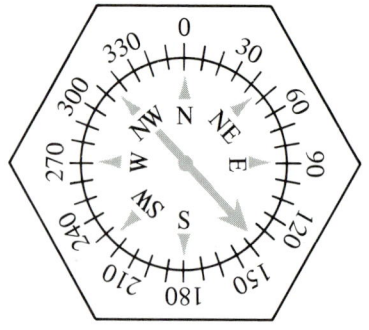

Look at the compass shown here. Notice that 0° indicates a northerly reading while 180° indicates a southerly direction. If you head toward 135°, your direction of travel is southeast. See if you can determine the reading for a southwesterly direction of travel. 225°

Because magnetic north is not located at the north pole, navigators make corrections to accommodate for the difference between the actual direction of travel they wish to take and the magnetic compass bearing they will use. The difference between magnetic north and actual north in any particular area of the earth is called *magnetic variation*. As this variation is not the same everywhere, it is important to know the magnetic variation for your particular location before you use a compass to find your direction.

Because a magnetic compass is such a basic navigational tool, navigational charts are printed with one compass showing actual or true north, and another showing magnetic north. The example shown below is taken from a navigational chart. It shows a magnetic variation of 10° east. This means that the magnetic compass needle will point 10° east of true north. In this location, a magnetic compass reading due west or 270° indicates a true direction of 280°.

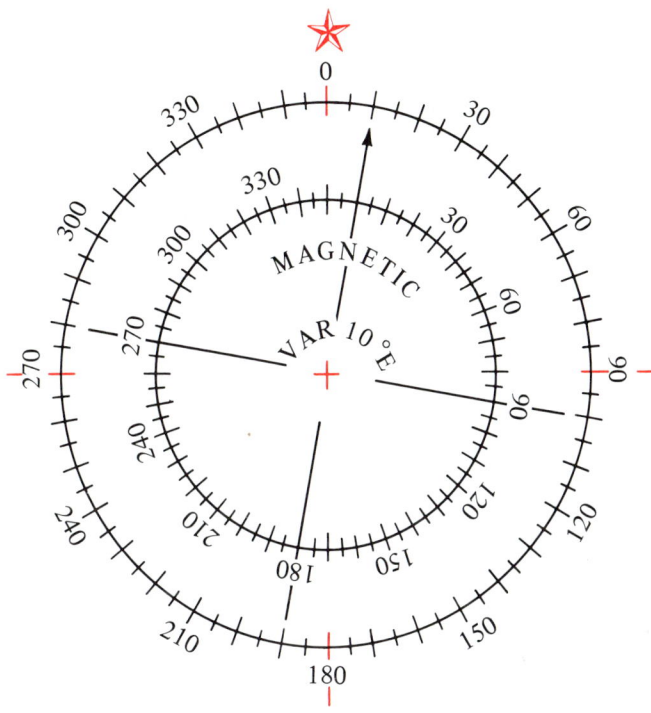

Suppose you want to head in the true direction of 45°. Find the compass heading by subtracting the magnetic variation, 10°, from the true direction. The magnetic compass heading that you should use is 45° − 10° = 35°.

EXERCISES

If there is no magnetic variation, find the compass heading for:

1. An easterly course 90°
2. A northwesterly course 315°

Find the proper magnetic compass heading for the following:

3. A southwesterly course; magnetic variation 10°E 215°
4. A westerly course; magnetic variation 7°W 277°

Ratios and Proportions

SKILLS REVIEW

FRACTIONAL AND DECIMAL EQUATIONS

Solve.

SAMPLES

$$\frac{2x}{5} + 1 = 9$$
$$\frac{2x}{5} + 1 - 1 = 9 - 1$$
$$\frac{2x}{5} \cdot \frac{5}{1} = 8 \cdot \frac{5}{1}$$
$$2x = 40$$
$$x = 20$$

$$0.12x = 3$$
$$100 \times 0.12x = 100 \times 3$$
$$12x = 300$$
$$x = 25$$

$$\frac{n}{2} + \frac{n}{5} = 7$$
$$\frac{5n}{10} + \frac{2n}{10} = 7$$
$$\frac{5n + 2n}{10} = 7$$
$$\frac{7n}{10} \cdot \frac{10}{1} = 7 \cdot \frac{10}{1}$$
$$7n = 70$$
$$n = 10$$

Solve.

1. $\frac{a}{9} + 3 = 13$ 90
2. $30 + \frac{d}{4} = 51$ 84
3. $\frac{b}{2} + 17 = 0$ -34
4. $0.6x = 42$ 70
5. $0.25y = 3.5$ 14
6. $78 = 1.3m$ 60
7. $\frac{2n}{3} = \frac{4}{7}$ $\frac{6}{7}$
8. $\frac{3x}{5} + 2 = 11$ 15
9. $\frac{5b}{2} - 12 = -7$ 2
10. $\frac{k}{2} + \frac{k}{4} = 9$ 12
11. $\frac{c}{3} - \frac{2c}{9} = \frac{1}{6}$ $\frac{3}{2}$
12. $\frac{3m}{10} - \frac{m}{2} = 4$ -20
13. $\frac{1}{4} + \frac{2}{3} = \frac{a}{6}$ $5\frac{1}{2}$
14. $\frac{2m}{3} + \frac{m}{2} = \frac{7}{2}$ 3
15. $\frac{4k}{5} - \frac{2k}{3} = -4$ -30

Write an equation. Then solve.

16. When half a number is subtracted from 4 times the number, the difference is 35. What is the number?   10

17. Regina is 105 cm tall. She is 5 cm more than $\frac{5}{6}$ as tall as her brother Ramon. How tall is Ramon?   120 cm

18. One week Wally's Wheel City sold 27 ten-speed bicycles. This was $\frac{3}{5}$ of all the bicycles sold that week. How many bicycles did the shop sell in all?   45 bicycles

CHAPTER REVIEW

CHAPTER SUMMARY

1. A ratio is a quotient of two numbers.
   The ratio of 2 to 3 is written $\frac{2}{3}$ or $2:3$.

2. To find the ratio of two measures, first express both measures in terms of the same unit. Note that there are *no* units in the final answer.

3. A proportion is an equation which states that two ratios are equal. The first and the last numbers in a proportion are called the extremes. The middle numbers are called the means.

4. Four helpful properties for changing the form of a proportion:
   a. The Cross-Multiplying Property
   b. The Switching Property
   c. The Overturning Property
   d. The Numerator-Changing Property

5. Proportions are used in problem solving.

6. Proportions are used in making and reading maps and scale drawings. The scale of a map is the ratio of the map distance to the actual distance.

REVIEW EXERCISES

Express each ratio in simplest form. *(See pp. 286–288.)*

1. $\frac{9}{15}$ $\frac{3}{5}$
2. $\frac{56}{16}$ $\frac{7}{2}$
3. $\frac{24x}{9x}$ $\frac{8}{3}$
4. $21:14$ $3:2$
5. $ab:ac$ $b:c$
6. $48:30:24$ $8:5:4$
7. $8\text{ m}:40\text{ cm}$ $20:1$
8. $20\text{ m}:2\text{ km}$ $1:100$
9. $20\text{ min to }3\text{ h}$ $1:9$

Use the figure to find each ratio in simplest form. *(See pp. 286–288.)*

10. $\frac{AB}{DC}$ $\frac{9}{5}$
11. $\frac{AE}{ED}$ $\frac{1}{2}$
12. $\frac{ED}{FC}$ $\frac{3}{4}$
13. $\frac{AD}{BC}$ $\frac{3}{4}$

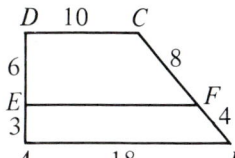

Ratios and Proportions

Complete each statement. All variables represent nonzero numbers. *(See pp. 289–292.)*

14. If $\frac{a}{4} = \frac{b}{5}$, then $\frac{a}{b} = $ __?__. $\frac{4}{5}$

15. If $\frac{m}{6} = \frac{3}{2}$, then __?__ $= 18$. $2m$

16. If $\frac{p}{4} = \frac{2}{5}$, then $\frac{p+4}{4} = $ __?__. $\frac{7}{5}$

17. If $\frac{x}{9} = \frac{4}{y}$, then $\frac{9}{x} = $ __?__. $\frac{y}{4}$

Find the value of x. *(See pp. 289–292.)*

18. $\frac{x}{8} = \frac{3}{4}$ 6

19. $\frac{5}{6} = \frac{x}{18}$ 15

20. $x:8 = 5:2$ 20

Solve. *(See pp. 293–295.)*

21. Two granola bars cost 90¢. What is the cost of five bars? $2.25

22. It costs $27.50 to paint a fence 3 m high and 25 m long. How much would it cost to paint another fence of the same type if it is 2 m high and 15 m long? $11.00

Use the diagram at the right. *(See pp. 296–298.)*

23. Measure the map distance between Bartlesville and Stillwater to the nearest centimeter. Then give the scale of the map. (See below.)

24. Find the actual distance between Stillwater and Tulsa. 68 km

25. Find the actual distance between Bartlesville and Tulsa. 51 km

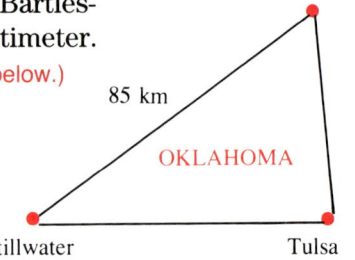

23. 5 cm; $\frac{1}{1,700,000}$ or 1 cm : 17 km

Draw a map. Be sure to include the scale. *(See pp. 299–301.)*

26. The Masconomet Regional High School is 1 km south of the county courthouse, and the local business district is 3 km west of the school. Draw a map showing the school, courthouse, and business district.

CHAPTER TEST

If $a = 6$ and $b = 4$, find each ratio in simplest form.

1. $\dfrac{a}{a+b}$ $\quad \tfrac{3}{5}$
2. $a:b:(a-b)$ $\quad 3:2:1$

(7-1)

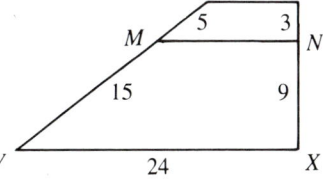

Find each ratio in simplest form.

3. $\dfrac{XY}{WZ}$ $\quad \tfrac{3}{5}$
4. $\dfrac{YZ}{WX}$ $\quad \tfrac{1}{3}$
5. $\dfrac{WM}{XY}$ $\quad \tfrac{5}{4}$

6. If $\dfrac{e}{f} = \dfrac{j}{k}$, then $\dfrac{k}{f} = \underline{}$ and $fj = \underline{}$. $\quad \tfrac{j}{e};\ ek$ (7-2)

7. If $\dfrac{r}{s} = \dfrac{2}{5}$, then $\dfrac{s}{r} = \underline{}$. $\quad \tfrac{5}{2}$
8. If $\dfrac{r}{s} = \dfrac{2}{5}$, then $\dfrac{r+s}{s} = \underline{}$. $\quad \tfrac{7}{5}$

9. If $\dfrac{x}{28} = \dfrac{3}{4}$, find the value of x. $\quad 21$

10. One week Carlos earned \$85.50 for working 18 hours at Yamamoto's Furniture Store. How much did Carlos earn the following week when he worked 25 hours? $\quad \$118.75$ (7-3)

11. Elaine found that a car trip of 136 km required 16 L of gas. How much gas would a trip of 61.2 km at the same speed require? $\quad 7.2$ L

Use the diagram at the right.

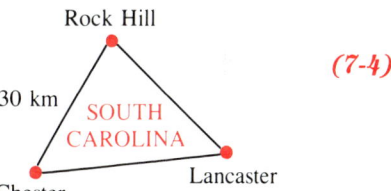

12. Measure the distance between Chester and Rock Hill to the nearest centimeter. The scale is 1 cm: $\underline{}$ km. $\quad 15$ (7-4)

13. Find the actual distance between Chester and Lancaster. $\quad 40.5$ km

14. Find the map distance in centimeters that would represent an actual distance of 80 km. $\quad 5.3$ cm

Brenda and Larry attend a school that is 3 km northwest of their house. The post office is 1 km west of the school.

15. Draw a map showing the house, the school, and the post office. Include the scale that you use. (7-5)

16. Find the distance, to the nearest kilometer, from the post office to the house. $\quad 4$ km

Ratios and Proportions 307

MIXED REVIEW

1. Express in "If... then" form and tell if the statement is true: "Every triangle with a 60° angle is equilateral."

   If a triangle has a 60° angle, then it is equilateral. false

2. Supply the reasons to complete the proof.
   Given: $\overline{CX}$ is the perpendicular bisector of $\overline{AB}$.
   Prove: $\overline{CX}$ is the angle bisector of $\angle ACB$.

   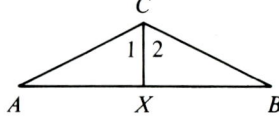

STATEMENTS	REASONS
1. $\overline{CX}$ is the ⊥ bisector of $\overline{AB}$.	1. __?__ Given
2. $\overline{CX} \perp \overline{AB}$; $AX = BX$	2. __?__ Def. of ⊥ bisector
3. $\angle CXA = 90° = \angle CXB$	3. __?__ Def. of ⊥ lines
4. $CX = CX$	4. __?__ From algebra
5. $\triangle CXA \cong \triangle CXB$	5. __?__ SAS Postulate
6. $\angle 1 = \angle 2$	6. __?__ Corr. parts of ≅ △ are =.
7. $\overline{CX}$ is the bisector of $\angle ACB$.	7. __?__ Def. of ∠ bisector

3. An isosceles triangle has a base of 10 cm and a height of 12 cm. Find the length of each leg. 13 cm

4. Find the diameter of a circle with circumference 17π cm. 17 cm

5. $\angle 1$ and $\angle 2$ are same-side interior angles formed by two parallel lines. If $\angle 1 = 75°$, find the measure of $\angle 2$. 105°

6. In quadrilateral $ABCD$, the diagonals intersect at point O. If $AO = BO = CO = DO$, what is the best name for $ABCD$: parallelogram, rectangle, rhombus, or square? rectangle

7. An isosceles triangle has a vertex angle of 28°. Find the measures of the exterior angles of the triangle. 104°, 104°, 152°

8. Find the perimeter of $\triangle JKL$. 36

9. Find the value of x: $\frac{x-5}{6} = \frac{3}{4}$. 9.5

   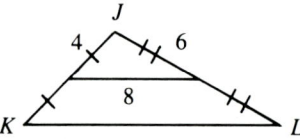

10. Find the area of a triangle with sides of lengths 15, 17, and 8. 60 sq. units

11. In $\triangle GWC$, $\angle G = \angle W$. If $GW = 5t + 1$, $WC = 4t + 3$, and $CG = 6t - 3$, find the value of t. 3

Write SSS, SAS, ASA, AAS, or HL to indicate the method you would use to prove two triangles congruent. If no method applies, write *none*.

12.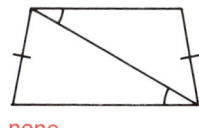
none

13.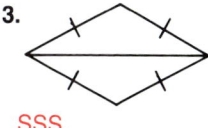
SSS

14.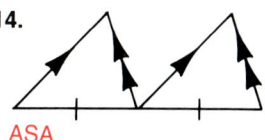
ASA

15. Construct a 60° angle. Construct the angle bisector.

16. Find the area, to the nearest tenth, of a circle with radius 2.4. Use 3.14 for π. 18.1 square units

17. Find the measure of each numbered angle shown.

18. Draw a large $\triangle PQR$ with obtuse $\angle P$. Construct altitude $\overline{PX}$ of the triangle.

$\angle 1 = 75°$
$\angle 2 = 60°$
$\angle 3 = 60°$
$\angle 4 = 135°$

Find the area of the parallelogram, rectangle, and trapezoid.

19.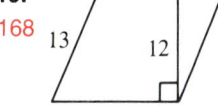
168

20.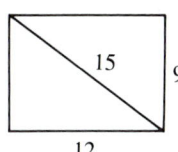
108

21.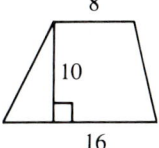
120

22. Write a two-column proof.
    Given: $\angle P = \angle R$; $\angle Q = \angle S$.
    Prove: $PS = QR$.
    (*Hint:* Use the result of Exercise 15 on page 219.)

    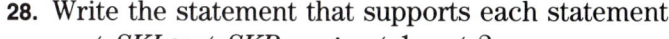

23. Eight oranges cost $1.92. How many can you buy for $1.20? 5 oranges

24. In a regular polygon with 15 sides, find the interior angle sum and the measure of *each* exterior angle. S = 2340°; 24°

25. A distance on a map is 5.6 cm. If the scale is 1 cm : 400 km, find the actual distance in kilometers. 2240 km

26. Find the area of a square with perimeter $8x$. $4x^2$

27. Find the ratio 80¢ : $2 in simplest form. 2:5

Given: $\overline{SI} \perp \overline{KI}$; $\overline{SP} \perp \overline{KP}$; $SI = SP$

28. Write the statement that supports each statement.
    a. $\angle SKI \cong \angle SKP$ b. $\angle 1 = \angle 2$ (See below.)

29. What method could be used to prove that $\triangle SKI \cong \triangle SKP$? HL Thm.

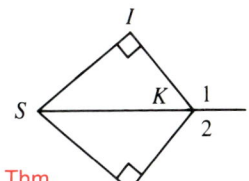

28. a. Corr. parts of $\cong \triangle$ are =.   b. Supplements of = $\angle$s are =.

Here's what you'll learn in this chapter:

To tell whether two polygons are similar.

To find missing parts of similar triangles.

To use properties related to parallel lines.

To find the perimeters and areas of two similar polygons.

Nuts, bolts, pipes, and many other objects used in building remind us of geometric figures. The hexagonal nuts shown in the photograph are of different sizes, but they are similar because they have the same shape.

Chapter 8

Similar Polygons

1 Defining Similar Polygons

Two polygons with the same shape are called *similar* polygons. Quadrilaterals *ABCD* and *WXYZ* are similar polygons.

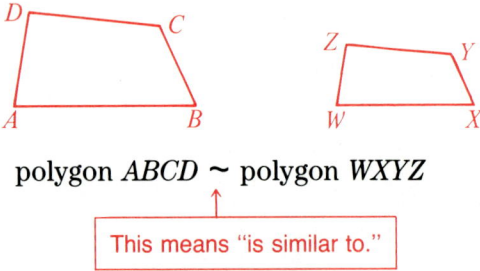

polygon *ABCD* ~ polygon *WXYZ*

This means "is similar to."

Recall that we name congruent polygons by listing their vertices in corresponding order. We name similar polygons the same way:

polygon *ABCD* ~ polygon *WXYZ*

When two polygons are similar, the following are true:

1. Corresponding angles are equal.

2. Corresponding sides are in proportion (have equal ratios).

Since quadrilateral *ABCD* ~ quadrilateral *WXYZ*, we can conclude:

1. $\angle A = \angle W$, $\angle B = \angle X$, $\angle C = \angle Y$, $\angle D = \angle Z$
2. $\frac{AB}{WX} = \frac{BC}{XY} = \frac{CD}{YZ} = \frac{DA}{ZW}$

Suppose that you wish to show that two polygons are similar. To do this, you must show that *both* conditions of the definition are true.

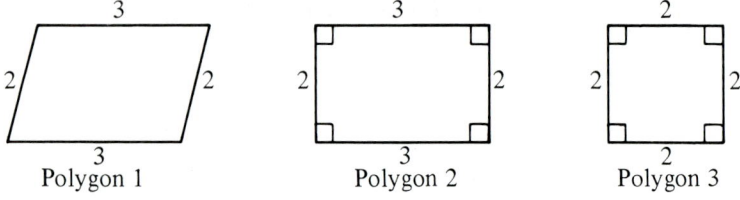

Polygon 1   Polygon 2   Polygon 3

The corresponding sides of Polygons 1 and 2 are in proportion. However, corresponding angles are not equal. Therefore the polygons are not similar.

The corresponding angles of Polygons 2 and 3 are equal. However, their corresponding sides are not in proportion. Therefore Polygons 2 and 3 are not similar.

EXAMPLE △PQR ~ △STU.
Find the values of x and y.

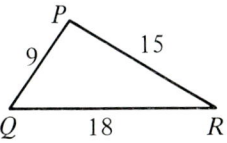

Since corresponding sides are in proportion:

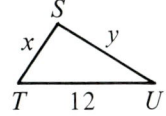

$$\frac{PQ}{ST} = \frac{QR}{TU} = \frac{RP}{US}$$

$$\frac{9}{x} = \frac{18}{12} = \frac{15}{y}$$

$$\downarrow$$

$$\frac{9}{x} = \frac{3}{2} = \frac{15}{y}$$

$\frac{9}{x} = \frac{3}{2}$ $\qquad$ $\frac{3}{2} = \frac{15}{y}$

$18 = 3x$ $\qquad\qquad$ $3y = 30$

$6 = x$ $\qquad\qquad\quad$ $y = 10$

If two polygons are similar, the ratio of two corresponding sides is called the *scale factor*. The scale factor of △ABC to △DEF is $\frac{5}{10}$, or $\frac{1}{2}$. Likewise, the scale factor of △DEF to △ABC is $\frac{2}{1}$. Notice that the ratio of the perimeters of these triangles is equal to the scale factor:

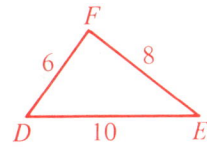

$$\frac{\text{perimeter of } \triangle DEF}{\text{perimeter of } \triangle ABC} = \frac{10 + 8 + 6}{5 + 4 + 3} = \frac{24}{12} = \frac{2}{1}.$$

Classroom Practice

In Exercises 1–3, state why the two quadrilaterals are, or are not, similar.

1.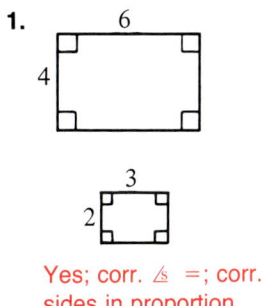

Yes; corr. ⦞ =; corr. sides in proportion

2.

10
2
4
1

No; corr. sides not in proportion

3.

4
4   4
4

No; corr. ⦞ not equal

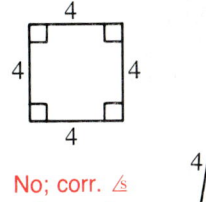

Similar Polygons

△ABC ~ △DEF. Complete each statement.

4. The scale factor of △ABC to △DEF is __?__. $\frac{2}{3}$

5. $\frac{x}{9} = \frac{10}{?}$ $\frac{10}{15}$

6. $\frac{8}{y} = \frac{x}{?}$ $\frac{x}{9}$

7. ∠C = ∠ __?__ F

8. ∠ __?__ = ∠D A

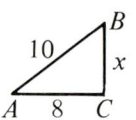
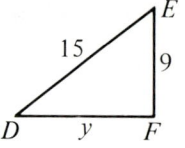

Tell whether each statement is true or false.

9. All squares are similar. true

10. All rectangles are similar. false

11. All equilateral triangles are similar. true

12. All isosceles triangles are similar. false

13. All rhombuses are similar. false

14. Every polygon is similar to itself. true

Written Exercises

Match the vertices of rectangles ABCD and WXYZ like this:

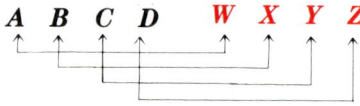

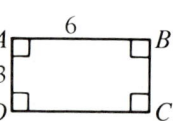

A 1. Are corresponding angles equal? yes

2. Are corresponding sides in proportion? no

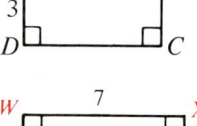

3. Are the rectangles similar? no

Match the vertices of quadrilaterals EFGH and PQRS like this:

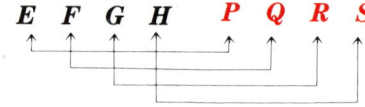

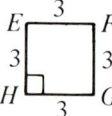

4. Are corresponding angles equal? no

5. Are corresponding sides in proportion? yes

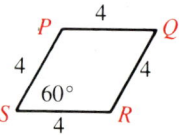

6. Are the quadrilaterals similar? no

In Exercises 7–9, state why the two polygons are, or are not, similar.

7.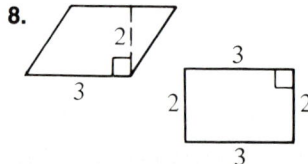
Yes; polygons ≅

8.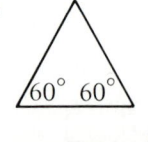
No; corr. ⦞ not =

9.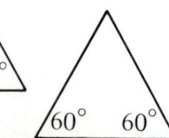
Yes; both equiangular and equilateral

In Exercises 10–12, the two polygons are similar. Complete each statement.

10.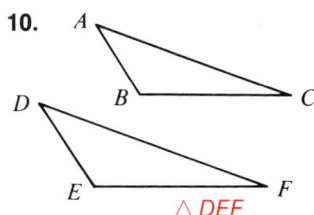

△ABC ~ __△DEF__

$\dfrac{AB}{DE} = \dfrac{BC}{?} = \dfrac{CA}{?}$

EF   FD

11.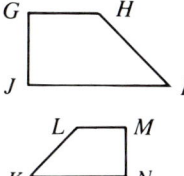

trapezoid GHIJ ~ __trapezoid MLKN__

$\dfrac{GH}{ML} = \dfrac{HI}{?} = \dfrac{IJ}{?}$

LK   KN

12.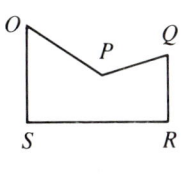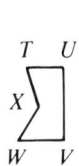

pentagon OPQRS ~ __pentagon WXTUV__

$\angle Q = \angle \underline{\ ?\ }$; $\dfrac{RS}{?} = \dfrac{SO}{?}$

T   UV   VW

13. Are rectangles 1 and 2 similar? **yes**

14. What is the scale factor of rectangle 1 to rectangle 2? $\frac{2}{3}$

15. a. $\dfrac{\text{perimeter of rectangle 1}}{\text{perimeter of rectangle 2}} = \dfrac{2+4+2+4}{?+?+?+?} = \dfrac{12}{?}$ 3 + 6 + 3 + 6; 18

   b. Is the ratio of the perimeters equal to the scale factor? **yes**

16. a. $\dfrac{\text{area of rectangle 1}}{\text{area of rectangle 2}} = \dfrac{8}{?}$ 18

   b. Is the ratio of the areas equal to the scale factor? **no**

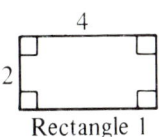

Rectangle 1

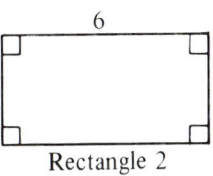

Rectangle 2

In Exercises 17–19, the two polygons shown are similar. Find the values of x, y, and z.

B 17.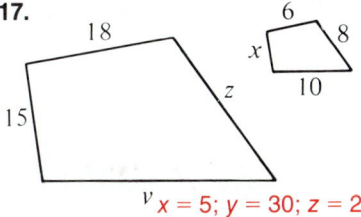

x = 5; y = 30; z = 24

18.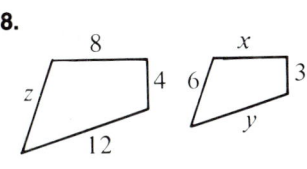

x = 6; y = 9; z = 8

19.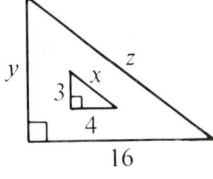

x = 5; y = 12; z = 20

20. The length and width of a rectangle are 18 and 12. A similar rectangle has length 27. What is its width? **18**

21. The sides of a triangle are 6, 8, and 9. The shortest side of a similar triangle is 15. How long are its other sides? **20, 22.5**

22. a. Plot points $A(-4, 8)$, $B(-4, 2)$, and $C(5, 2)$ on graph paper. Draw △ABC.
    b. Plot points $D(5, -3)$ and $E(5, -5)$ on the same graph.
    c. Locate a point F so that △ABC ~ △DEF. F_1 (8, −5); F_2 (2, −5)

23. Repeat Exercise 22, using points $A(4, 10)$, $B(4, 4)$, $C(0, 4)$, $D(6, 1)$, and $E(6, 13)$.
    c. F_1 (14, 13); F_2 (−2, 13)

Similar Polygons

C 24. The length and width of a rectangle are represented by x and y ($x \neq y$). The length and width of another rectangle are represented by $x + 2$ and $y + 2$. Are the rectangles similar? no

25. The length and width of a rectangle are represented by x and y ($x \neq y$). The length and width of another rectangle are represented by $2x$ and $2y$. Are the rectangles similar? yes

26. When a card 8 cm long is cut in half, each piece has the same shape as the original card. Find the *exact* value of x. (The answer involves a square root.) $x = \sqrt{32} = 4\sqrt{2}$

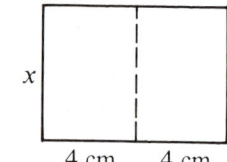

CAREER NOTES

ACTUARY

Did you know that in a recent year, a residential fire occurred in the United States about once every 49 seconds? Or that there were 2379 structural fires per day? This means that there were 868,335 structural fires per year, 643,592 (or about 74%) of which were residential fires!

Actuaries use information like that given above to establish the policy rates charged by insurance companies. First they collect all kinds of information about the circumstances in which certain types of losses occur. Then they organize this data into tables and charts.

Graphs often help actuaries to show the relationship between different factors. For instance, people who smoke often die younger than non-smokers. On the other hand, those who exercise regularly often live longer than those who don't. To find out more about life insurance, read the Consumer Application on page 349.

Often actuaries use computers to analyze data on the frequency of actual losses. This helps them to estimate the probability of other losses that might occur in the future. Policy-rate scales are based on such probabilities.

2 The AA Postulate

You can prove that two polygons are similar by using the definition of *similar*. That is, by showing that

      (1) corresponding angles are equal, and
      (2) corresponding sides are in proportion.

There is, however, a more direct method for proving two *triangles* similar. The exploration below will help you discover this method.

EXPLORATIONS

1. Draw any scalene △ABC.

2. Construct $\overline{DE}$ with length 2(AB).

3. Copy ∠A at D. Copy ∠B at E. Let F be the point shown.

4. Measure $\overline{AC}$ and $\overline{DF}$. Then measure $\overline{BC}$ and $\overline{EF}$.
   Remember that $\frac{AB}{DE} = \frac{1}{2}$. Is $\frac{AC}{DF} \approx \frac{1}{2}$? Is $\frac{BC}{EF} \approx \frac{1}{2}$? yes; yes
   Do you think that corresponding sides are in proportion? yes

5. Remember that ∠A = ∠D and ∠B = ∠E. Is ∠C = ∠F? yes
   Do you think that corresponding angles are equal? yes

6. Does it seem that △ABC ~ △DEF? yes

If you repeat the exploration using different-looking triangles, you will obtain the same result.

POSTULATE 13 (The AA Postulate)

If two angles of one triangle are equal to two angles of another triangle, then the triangles are similar.

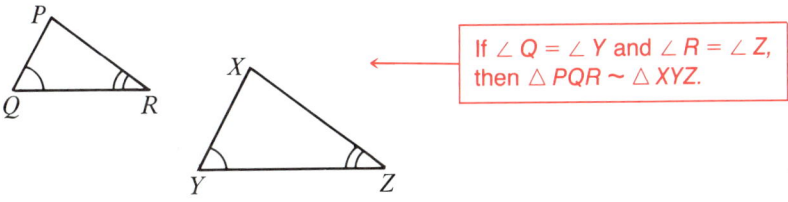

If ∠Q = ∠Y and ∠R = ∠Z, then △PQR ~ △XYZ.

Similar Polygons   **317**

EXAMPLE

Given: $\angle B = \angle E$

Prove: $\triangle ABC \sim \triangle DEC$

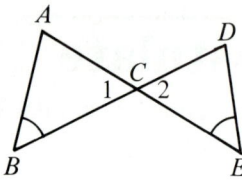

STATEMENTS	REASONS
1. $\angle B = \angle E$	1. Given
2. $\angle 1 = \angle 2$	2. Vertical angles are equal.
3. $\triangle ABC \sim \triangle DEC$	3. AA Postulate

Suppose, in the proof above, we were asked to prove $\frac{AB}{DE} = \frac{BC}{EC}$. We would need just one additional step:

4. $\frac{AB}{DE} = \frac{BC}{EC}$	4. Corresponding sides of $\sim$ $\triangle$ are in proportion.

The definition of similar triangles may be used in a proof in much the same way that the definition of congruent triangles was used.

> The definition of **congruent triangles** tells us:
> 1. Corresponding angles of congruent triangles are equal.
> 2. Corresponding sides of congruent triangles are equal.
>
> The definition of **similar triangles** tells us:
> 1. Corresponding angles of similar triangles are equal.
> 2. Corresponding sides of similar triangles are in proportion.

Classroom Practice

In each exercise, state whether the two triangles are similar.

1. yes
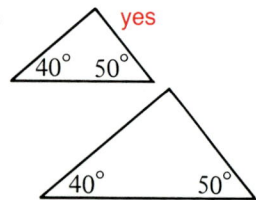

2. yes
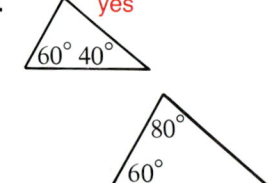

3. no
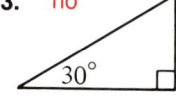
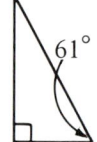

4. yes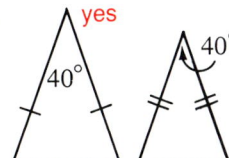
5. no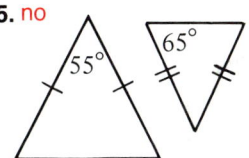
6. yes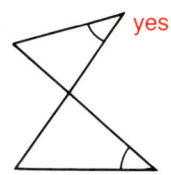

7. Exercise 4 suggests: If the vertex angles of two isosceles △ __?__.
   are equal, the triangles are similar.

Complete each proportion.

8. △ABC ~ △DEF
$$\frac{AB}{\underset{DE}{?}} = \frac{BC}{\underset{EF}{?}} = \frac{AC}{\underset{DF}{?}}$$

9. △RST ~ △XYZ
$$\frac{XY}{\underset{RS}{?}} = \frac{XZ}{\underset{RT}{?}} = \frac{YZ}{\underset{ST}{?}}$$

10. △KLM ~ △RST
$$\frac{?}{KL}_{RS} = \frac{?}{LM}_{ST} = \frac{?}{KM}_{RT}$$

11. Complete the proof.
    Given: ∠1 = ∠2
    Prove: $\frac{AB}{AD} = \frac{AC}{AE}$

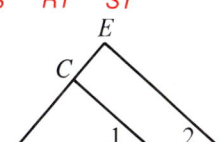

STATEMENTS	REASONS
1. ∠1 = ∠2	1. __?__ Given
2. ∠A = ∠A	2. __?__ From algebra
3. △ABC ~ △ADE	3. __?__ AA Postulate
4. $\frac{AB}{AD} = \frac{AC}{AE}$	4. __?__ Corr. sides of ~ △ are in proportion.

Use the figure from Exercise 11. Assume that △ABC ~ △ADE.

12. If AB = 8, AD = 12, and BC = 6, find DE. 9
13. If AB = 15, AC = 9, and AE = 15, find AD. 25

Written Exercises

In each exercise, state whether the two triangles are similar.

A

1. no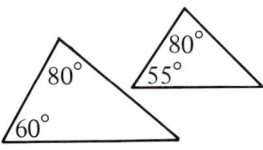
2. no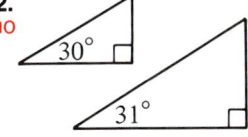
3. yes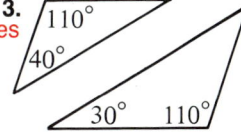
4. yes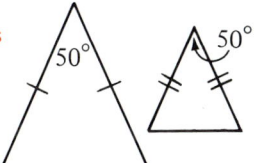
5. no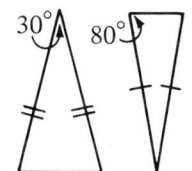
6. yes

Similar Polygons   319

State whether the two triangles are similar.

7. yes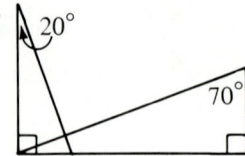

8. yes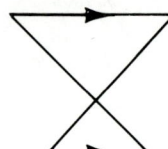

9. yes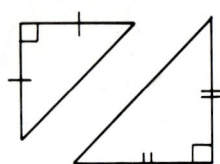

Complete each statement.

10.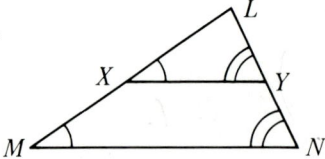

a. △LMN ~ △ __?__   LXY
b. $\dfrac{LX}{LM} = \dfrac{LY}{?} = \dfrac{XY}{?}$ LN   MN

11.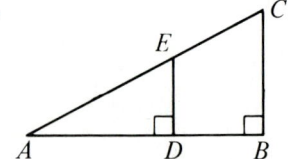

a. △ABC ~ △ __?__   ADE
b. $\dfrac{AB}{AD} = \dfrac{BC}{?} = \dfrac{CA}{?}$ DE   EA

12.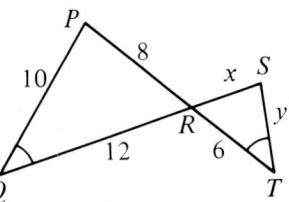

a. △PQR ~ △ __?__   STR
b. $\dfrac{12}{6} = \dfrac{?}{x}$ and $\dfrac{12}{6} = \dfrac{?}{y}$ 8; 10
c. $x = $ __?__ and $y = $ __?__   4   5

13.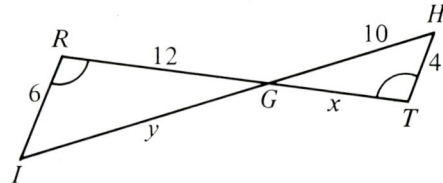

a. △RIG ~ △ __?__   THG
b. $\dfrac{4}{6} = \dfrac{x}{?}$ and $\dfrac{4}{6} = \dfrac{?}{y}$ 12; 10
c. $x = $ __?__ and $y = $ __?__   8   15

14.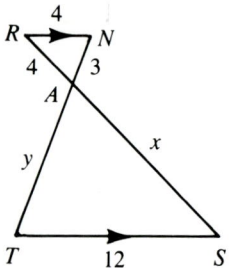

a. △SAT ~ △ __?__   RAN
b. $x = $ __?__   12
c. $y = $ __?__   9

15.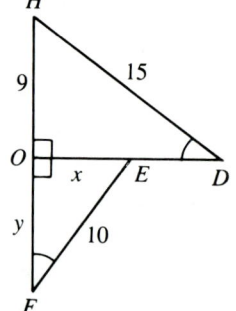

a. △FOE ~ △ __?__   DOH
b. $x = $ __?__   6
c. $y = $ __?__   8

320   Chapter 8

Supply the reasons to complete each proof.

B 16.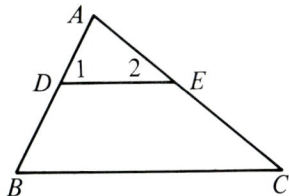

Given: $\overline{DE} \parallel \overline{BC}$

Prove: $\dfrac{AD}{AB} = \dfrac{DE}{BC}$

1. $\overline{DE} \parallel \overline{BC}$ Given
2. $\angle B = \angle 1$ Corr. angles
3. $\angle C = \angle 2$ Corr. angles
4. $\triangle ADE \sim \triangle ABC$ AA Post.
5. $\dfrac{AD}{AB} = \dfrac{DE}{BC}$ Corr. sides of ~ △ are in proportion.

17.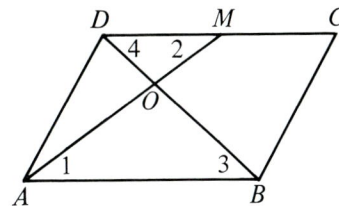

Given: $ABCD$ is a ▱.

Prove: $\dfrac{AO}{MO} = \dfrac{AB}{MD}$

1. $ABCD$ is a ▱. Given
2. $\overline{DM} \parallel \overline{AB}$ Opp. sides of a ▱ are ∥.
3. $\angle 1 = \angle 2$; $\angle 3 = \angle 4$ Alt. int. angles
4. $\triangle AOB \sim \triangle MOD$ AA Postulate
5. $\dfrac{AO}{MO} = \dfrac{AB}{MD}$ Corr. sides of ~ △ are in proportion.

In the figure shown, $ABCD$ is a parallelogram.

18. $\triangle ABO \sim \triangle \underline{\ ?\ }$ EDO   19. $\triangle DOA \sim \triangle \underline{\ ?\ }$ BOF
20. $\triangle AFB \sim \triangle \underline{\ ?\ }$ EAD; also $\triangle AFB \sim \triangle \underline{\ ?\ }$ EFC
21. $\triangle AED \sim \triangle \underline{\ ?\ }$ FAB; also $\triangle AED \sim \triangle \underline{\ ?\ }$ FEC

22. Given: $\angle X = \angle Y$

    Prove: $\dfrac{XS}{YS} = \dfrac{ST}{SR}$

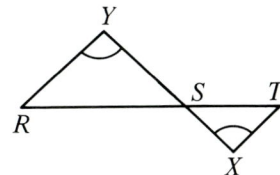

C 23. Given: $\overline{EA} \parallel \overline{PL}$

    Prove: $\dfrac{SE}{SA} = \dfrac{SP}{SL}$

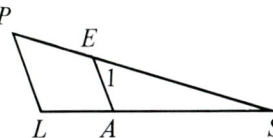

24. Prove the following: If two similar triangles have scale factor k, then their corresponding altitudes have scale factor k.

    Given: $\triangle ABC \sim \triangle DEF$; $\dfrac{AC}{DF} = k$

    $\overline{CX}$ and $\overline{FY}$ are corresponding altitudes.

    Prove: $\dfrac{CX}{FY} = k$

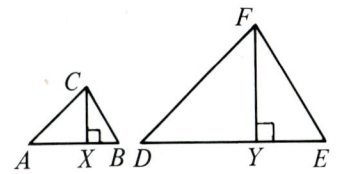

Similar Polygons

SELF-TEST

The two polygons given are similar. Complete each statement.

1. $\triangle UNO \sim \triangle \underline{\ ?\ }$ JKP (8-1)

2. $\dfrac{UN}{JK} = \dfrac{NO}{?} = \dfrac{OU}{?}$
    KP   PJ

3. The scale factor is $\underline{\ ?\ } : \underline{\ ?\ }$.
    4:8 or 1:2

4. $\angle U = \angle \underline{\ ?\ }$ J

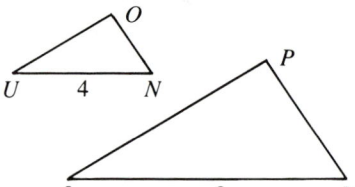

State whether the two polygons are, or are not, similar.

5.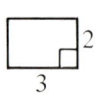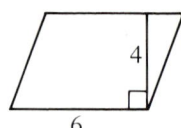
   not similar

6. 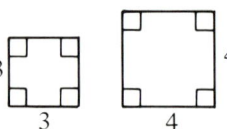 (8-2)
   similar

7.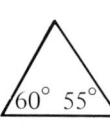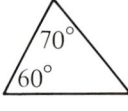
   not similar

8.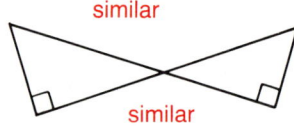
   similar

CALCULATOR ACTIVITIES

Can you see that any two equilateral triangles are similar polygons?

  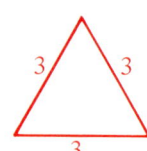

The area of an equilateral triangle is given by the formula $A = \dfrac{s^2\sqrt{3}}{4}$, where s is the length of a side.

Use a calculator to compute the perimeter and area of each triangle described. ($\sqrt{3} \approx 1.732$) Round each area to the nearest tenth.

	s	1	2	3	4	5	6	7
1.	perimeter	3	? 6	? 9	? 12	? 15	? 18	? 21
2.	area	0.4	? 1.7	? 3.9	? 6.9	? 10.8	? 15.6	? 21.2

3 A Special Case of Similar Triangles

In the figure at the right, $\overline{DE} \parallel \overline{BC}$. Do you see that $\triangle ADE \sim \triangle ABC$? (For a proof, see Exercise 16 on page 321.) Since corresponding sides of these triangles are in proportion, we can find the values of x and y.

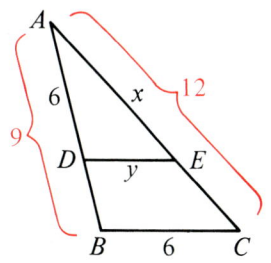

$\triangle ADE \sim \triangle ABC$

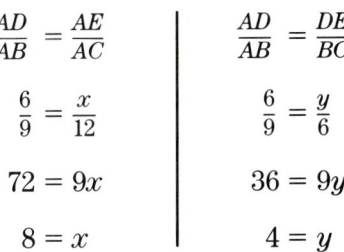

$$\frac{AD}{AB} = \frac{AE}{AC} \qquad \frac{AD}{AB} = \frac{DE}{BC}$$

$$\frac{6}{9} = \frac{x}{12} \qquad \frac{6}{9} = \frac{y}{6}$$

$$72 = 9x \qquad 36 = 9y$$

$$8 = x \qquad 4 = y$$

Classroom Practice

Complete each statement. Then find the values of x and y.

1.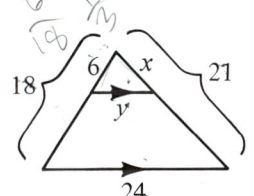

    $\frac{6}{18} = \frac{x}{?} = \frac{y}{?}$
    21   24
    $x = 7$; $y = 8$

2.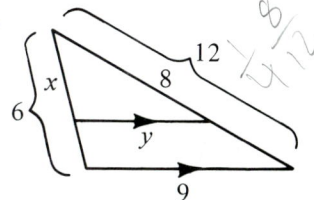

    $\frac{x}{6} = \frac{8}{?} = \frac{y}{?}$
    12   9
    $x = 4$; $y = 6$

3.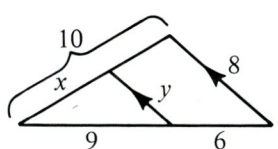

    $\frac{10}{x} = \frac{15}{?} = \frac{8}{y}$
    9
    $x = 6$; $y = 4.8$

Written Exercises

Complete each statement. Then find the values of x and y.

A

1.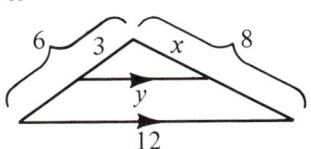

    $\frac{3}{6} = \frac{x}{?} = \frac{y}{?}$
    8   12
    $x = 4$; $y = 6$

2.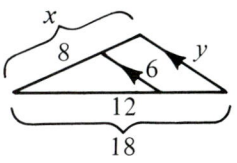

    $\frac{18}{12} = \frac{x}{?} = \frac{y}{?}$
    8   6
    $x = 12$; $y = 9$

3.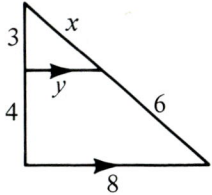

    $\frac{x}{x+6} = \frac{3}{?} = \frac{y}{?}$ 7; 8
    $x = 4\frac{1}{2}$; $y = 3\frac{3}{7}$

Similar Polygons

Find the values of x and y.

4.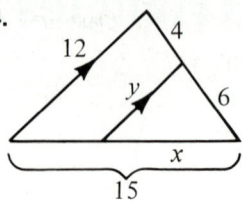
$x = 9;\ y = 7.2$

5.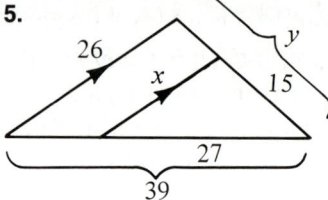
$x = 18;\ y = 21\frac{2}{3}$

6.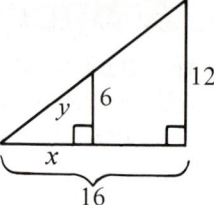
$x = 8;\ y = 10$

7.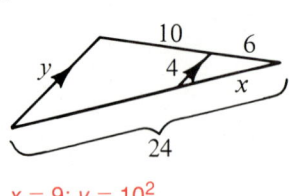
$x = 9;\ y = 10\frac{2}{3}$

8.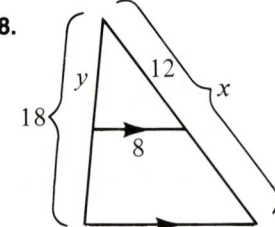
$x = 22.5;\ y = 9.6$

9.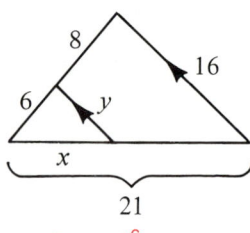
$x = 9;\ y = 6\frac{6}{7}$

Find the value of x.

B **10.**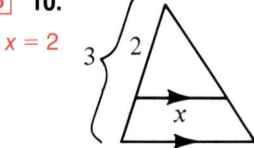
$x = 2$

11. $x = 7$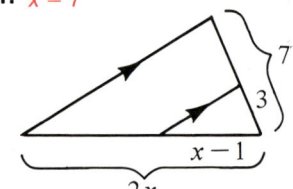

12. $x = 2\frac{2}{3}$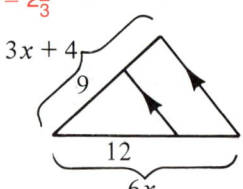

PUZZLE ◆ PROBLEMS

Try to draw each figure below without lifting your pencil from the paper and without retracing any line. It is impossible to do this for one of the figures shown. Which one? The center figure

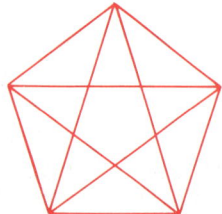 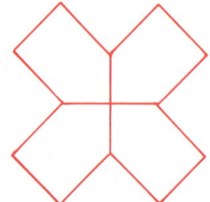 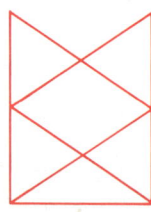

4 The Triangle Proportionality Theorem

In the figure at the right, $\overline{PQ} \parallel \overline{YZ}$. We have seen that this means that $\triangle XYZ \sim \triangle XPQ$. Since corresponding sides of these similar triangles are in proportion,

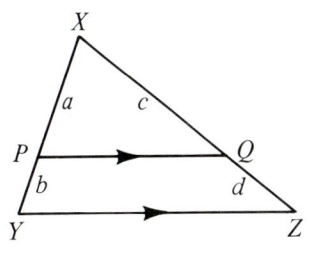

$$\frac{XY}{XP} = \frac{XZ}{XQ}$$

or $\quad \dfrac{a+b}{a} = \dfrac{c+d}{c}$

$$\frac{b}{a} = \frac{d}{c} \quad \text{Use properties of proportions.}$$

$$\frac{a}{b} = \frac{c}{d}$$

The discussion above proves the following useful theorem.

> **THEOREM 1 (The Triangle Proportionality Theorem)**
>
> *If a line intersects a triangle and is parallel to one side, then it divides the other two sides proportionally.*

If , then $\dfrac{a}{b} = \dfrac{c}{d}$.

EXAMPLE Find the values of x and y.

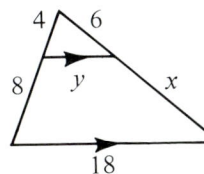

Caution! Notice that $\frac{4}{8} = \frac{6}{x}$, but $\frac{4}{8} \neq \frac{y}{18}$. (Why?)

Use the Triangle Proportionality Theorem to find x and use corresponding sides of similar triangles to find y.

$$\frac{4}{8} = \frac{6}{x} \qquad\qquad \frac{4}{4+8} = \frac{y}{18}$$

$$4x = 48 \qquad\qquad 72 = 12y$$

$$x = 12 \qquad\qquad 6 = y$$

Using properties of proportions, it can be shown that the following proportions are true when $\overline{PQ} \parallel \overline{YZ}$:

$$\frac{XP}{XY} = \frac{XQ}{XZ} \qquad \frac{XP}{PY} = \frac{XQ}{QZ} \qquad \frac{PY}{XY} = \frac{QZ}{XZ}$$

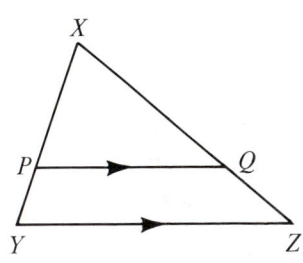

We shall agree that Theorem 1 may be used to justify each of these proportions.

Similar Polygons

Theorem 1 has a corollary. For a proof, see Exercise 19.

> **COROLLARY**
>
> If three parallel lines intersect two transversals, they divide the transversals proportionally.

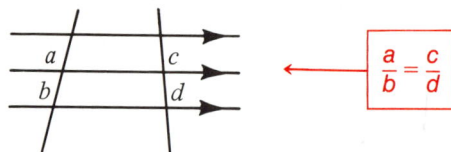

Classroom Practice

Complete each statement.

1. $\dfrac{PS}{SQ} = \dfrac{PT}{?}$
   TR

2. $\dfrac{PQ}{PS} = \dfrac{PR}{?}$
   PT

3. $\dfrac{ST}{QR} = \dfrac{PS}{?}$
   PQ

4. $\dfrac{SQ}{PQ} = \dfrac{TR}{?}$
   PR

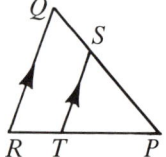

Use the figure below to find the value of each of the following.

5. $\dfrac{BD}{BA}$ $\dfrac{2}{5}$

6. $\dfrac{BE}{EC}$ $\dfrac{2}{3}$

7. $\dfrac{BE}{BC}$ $\dfrac{2}{5}$

8. $\dfrac{EC}{BC}$ $\dfrac{3}{5}$

9. $\dfrac{DE}{AC}$ $\dfrac{2}{5}$

10. $\dfrac{BA}{DA}$ $\dfrac{5}{3}$

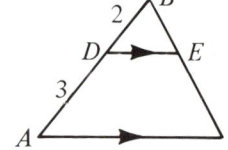

Complete each statement. Then find the values of x and y.

11. $\dfrac{x}{4} = \dfrac{9}{?}$ and $\dfrac{y}{16} = \dfrac{9}{?}$
    3 12
    $x = 12$
    $y = 12$

12. $\dfrac{x}{3} = \dfrac{9}{?}$ and $\dfrac{y}{20} = \dfrac{9}{?}$
    6 15
    $x = 4\tfrac{1}{2}$
    $y = 12$

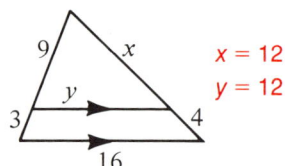
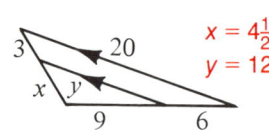

13. If $\dfrac{AB}{BC} = \dfrac{5}{7}$, then $\dfrac{DE}{EF} = $ ___?___ . $\dfrac{5}{7}$

14. If $\dfrac{AB}{BC} = \dfrac{5}{7}$, then $\dfrac{DF}{DE} = $ ___?___ . $\dfrac{12}{5}$

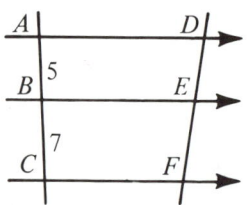

Written Exercises

For each figure below, complete the statement: $\frac{a}{b} = \frac{?}{?}$.

A

1. 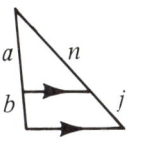 $\frac{n}{j}$

2. 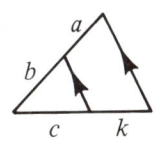 $\frac{k}{c}$

3. 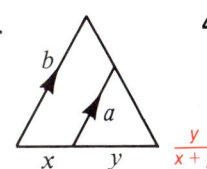 $\frac{y}{x+y}$

4. 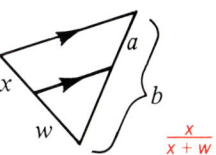 $\frac{x}{x+w}$

Complete each statement.

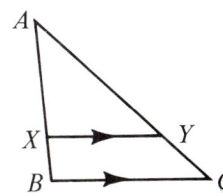

5. $\frac{AX}{XB} = \frac{AY}{?}$ YC

6. $\frac{AX}{AB} = \frac{AY}{?}$ AC

7. $\frac{XY}{BC} = \frac{AY}{?}$ AC

8. $\frac{XB}{AB} = \frac{YC}{?}$ AC

Find the values of x and y.

9.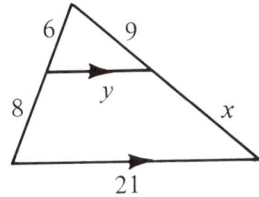
$x = 12$; $y = 9$

10.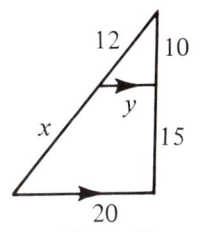
$x = 18$; $y = 8$

11.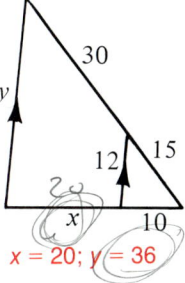
$x = 20$; $y = 36$

12.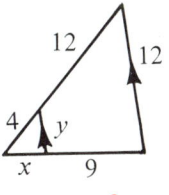
$x = 3$; $y = 3$

13.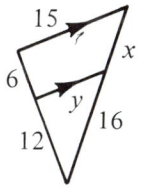
$x = 8$; $y = 10$

14.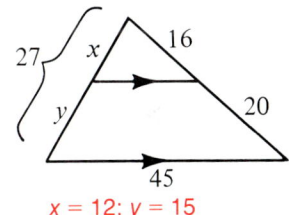
$x = 12$; $y = 15$

B

15.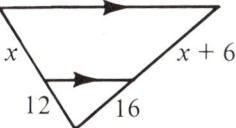

16. $18\frac{2}{3}$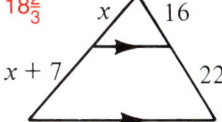

17. $13\frac{1}{3}$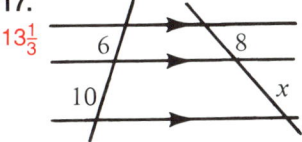

18. 18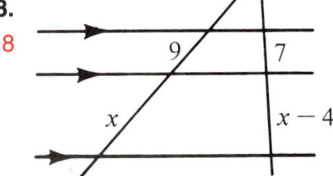

Similar Polygons   **327**

19. Supply the reasons for the following proof of the Corollary to Theorem 1.

Given: $k \parallel l \parallel m$

Prove: $\dfrac{a}{b} = \dfrac{c}{d}$

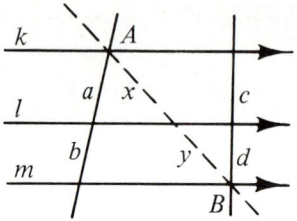

STATEMENTS	REASONS
1. $k \parallel l \parallel m$	1. __?__ Given
2. Draw $\overleftrightarrow{AB}$.	2. Through any two points there is __?__. exactly one line
3. $\dfrac{a}{b} = \dfrac{x}{y}$; $\dfrac{x}{y} = \dfrac{c}{d}$	3. __?__ If a line intersects a △ and is ∥ to one side, then it divides the other 2 sides proportionally.
4. $\dfrac{a}{b} = \dfrac{c}{d}$	4. __?__ Substitution Postulate

20. As shown, lots A and B are bounded on either side by Hill Street and River Road. Use the diagram to find x and y, the distance each lot has along River Road. $x = 150;\ y = 100$

21. Which lot has the greater area? How much greater? Lot A has the greater area by 1800 m².

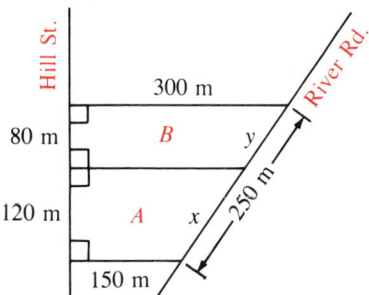

In Exercises 22 and 23, use the statement: The bisector of an angle of a triangle divides the opposite side in the same ratio as the other two sides of the triangle.

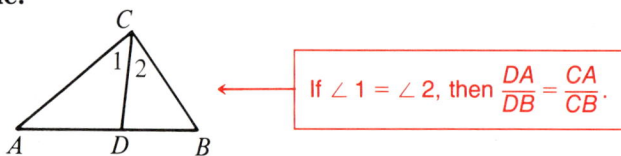

If ∠1 = ∠2, then $\dfrac{DA}{DB} = \dfrac{CA}{CB}$.

Complete each statement. Then find the value of x.

 22.

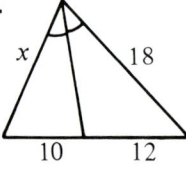

$\dfrac{10}{12} = \dfrac{x}{?}$ 18

$x = 15$

23.

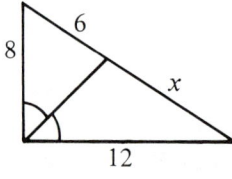

$\dfrac{x}{6} = \dfrac{?}{?}$ $\dfrac{12}{8}$

$x = 9$

5 Perimeters and Areas

Have you ever seen a billboard like this one?

The increased view is represented by pictures of two similar windows. The dimensions of the larger window are *twice* the corresponding dimensions of the smaller one. Yet the larger window seems huge in comparison with the smaller one. This is because its area is *four times as great*. Do you see why?

$$\begin{aligned}
\text{Area of a rectangle} &= \text{base} \times \text{height} \\
\text{Area of smaller rectangle} &= xy \\
\text{Area of larger rectangle} &= (2x)(2y) \\
&= 4xy \\
&= 4(\text{area of smaller rectangle})
\end{aligned}$$

Suppose that the dimensions of the larger rectangle were three times those of the smaller rectangle. Then the larger area would be *nine* times the smaller one:

$$\begin{aligned}
\text{Area of smaller rectangle} &= xy \\
\text{Area of larger rectangle} &= (3x)(3y) \\
&= 9xy \\
&= 9(\text{area of smaller rectangle})
\end{aligned}$$

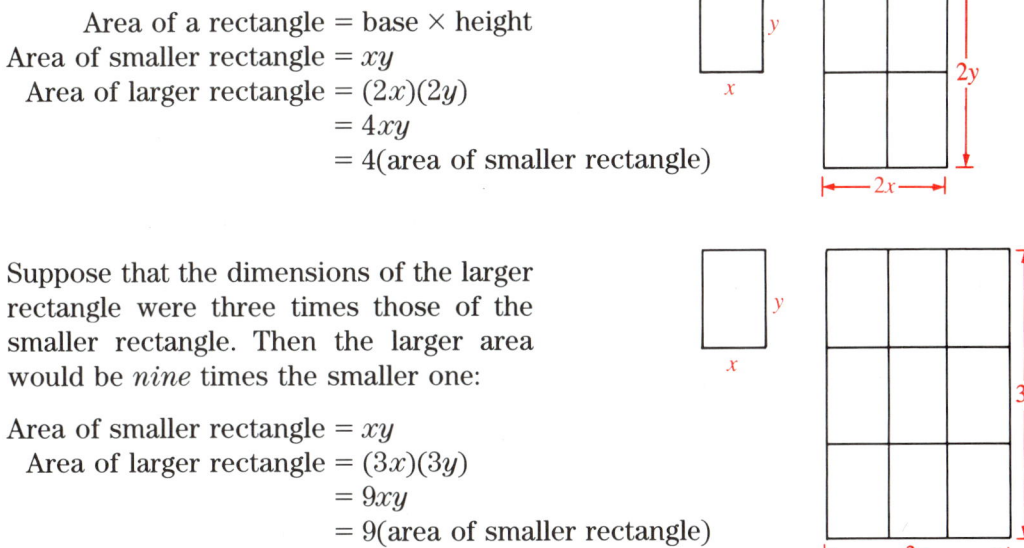

Our work in Section 1 suggested that the perimeters of similar polygons are related in a special way. The discussion above suggests that the areas of similar polygons are also related.

Similar Polygons **329**

THEOREM 2

If two similar polygons have a scale factor $a:b$, then
(1) the ratio of their perimeters is $a:b$;
(2) the ratio of their areas is $a^2:b^2$.

Theorem 2 is true, not only for polygons, but for figures such as circles, which are not polygons.

EXAMPLE The similar figures below have the scale factor $3:2$. Find
a. the ratio of their perimeters;
b. the ratio of their areas.

Using Theorem 2:
a. the ratio of their perimeters is $3:2$;
b. the ratio of their areas is $3^2:2^2$, or $9:4$.

Classroom Practice

Each exercise of the table refers to two similar figures. Copy and complete the table.

	1.	2.	3.	4.	5.	6.	7.	8.
Scale factor	1:4	1:5	2:5	3:7	?	?	?	?
Ratio of perimeters	?	?	?	?	2:3	3:8	?	?
Ratio of areas	?	?	?	?	?	?	1:4	36:25

Answers above the table: 1:2, 6:5 (scale factor for 7, 8); 1:4, 1:5, 2:5, 3:7, 2:3, 3:8, 1:2, 6:5 (ratio of perimeters); 1:16, 1:25, 4:25, 9:49, 4:9, 9:64 (ratio of areas).

The purpose of Exercises 9 and 10 is to prove Theorem 2 for similar triangles. The triangles shown have the scale factor $a:b$.

9. $\dfrac{\text{perimeter of triangle 1}}{\text{perimeter of triangle 2}} = \dfrac{at + ak + an}{bt + bk + bn}$

Show that this ratio is equal to $a:b$.

$\dfrac{at + ak + an}{bt + bk + bn} = \dfrac{a(t + k + n)}{b(t + k + n)} = \dfrac{a}{b}$; $a:b$

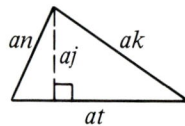 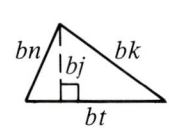

Triangle 1   Triangle 2

10. Recall from Exercise 24, page 321, that corresponding altitudes of similar triangles have the same ratio as corresponding sides. Write an expression for the ratio of the areas, and show that this ratio is equal to $a^2:b^2$.

Written Exercises

Each exercise refers to two similar figures with the given scale factor. Find the ratio of the perimeters and the ratio of the areas.

A 1. 1:2 1:2 1:4
2. 1:3 1:3 1:9
3. 2:3 2:3 4:9
4. 5:4 5:4 25:16

Each exercise refers to two similar figures. Copy and complete the table.

5. 3:5
6. 7:4
7. 2:3
8. 9:5
9. 3:7
10. 10:1

	5.	6.	7.	8.	9.	10.
Scale factor	?	?	?	?	?	?
Ratio of perimeters	3:5	7:4	?	?	?	?
Ratio of areas	?	?	4:9	81:25	9:49	100:1

Answers: 9:25, 49:16, 2:3, 9:5, 3:7, 10:1

By measuring each pair of similar figures, find: a. their scale factor; b. the ratio of their perimeters; c. the ratio of their areas.

11.
a. 1:2
b. 1:2
c. 1:4

12.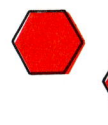
a. 5:11
b. 5:11
c. 25:121

13.
a. 3:5
b. 3:5
c. 9:25

B 14. Refer to the floor plan of a house on page 297. Find the ratio of the floor area in the actual house to the floor area in the drawing. 40,000:1

15. The widths of two similar rectangles are 15 and 25. Find the ratio of their perimeters and the ratio of their areas. 3:5; 9:25

16. The bases of two similar triangles are 18 and 27. Find the ratio of their perimeters and the ratio of their areas. 2:3; 4:9

17. Given: $\triangle ABC \sim \triangle DEF$ with scale factor 3:4.
   a. If $\triangle ABC$ has perimeter 15, then $\triangle DEF$ has perimeter __?__. 20
   b. If $\triangle ABC$ has area 18, then $\triangle DEF$ has area __?__. 32

18. Given: $\triangle PQR \sim \triangle XYZ$ with scale factor 3:5.
   a. If $\triangle PQR$ has perimeter 30, then $\triangle XYZ$ has perimeter __?__. 50
   b. If $\triangle PQR$ has area 36, then $\triangle XYZ$ has area __?__. 100

19. Given: $\triangle LMN$; O is the midpoint of $\overline{LN}$; P is the midpoint of $\overline{MN}$.

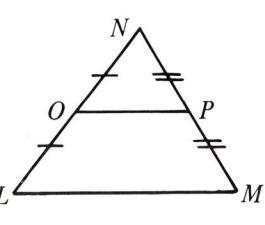

a. 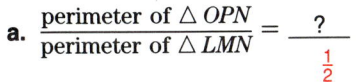 $= \dfrac{?}{\frac{1}{2}}$

b. 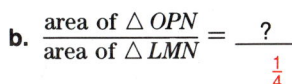 $= \dfrac{?}{\frac{1}{4}}$

Similar Polygons 331

20. The ceilings of two rooms are similar rectangles with scale factor 2:5. It costs $9 to paint the ceiling of the smaller room. How much does it cost to paint the ceiling of the larger room? $56.25

C 21. The perimeters of two similar hexagons are in the ratio 3:4. The sum of the areas of the hexagons is 70 cm². Find the area of each hexagon. 25.2 cm², 44.8 cm²

22. The areas of two similar pentagons are 45 cm² and 80 cm². The sum of the perimeters of the pentagons is 35 cm. Find the perimeter of each pentagon. 15 cm, 20 cm

CALCULATOR ACTIVITIES

A landscape architect designs a park pathway with round stepping stones. The stones come in four different sizes.

Recall the formulas for the circumference and the area of a circle:

$$C = 2\pi r$$
$$C = \pi d$$
$$A = \pi r^2$$

Use a calculator to find the missing measures. ($\pi \approx 3.14$) Round your answers to one decimal place.

		Radius	Diameter	Circumference	Area
1.	Small stones	6.2 cm	?	?	?
2.	Medium stones	?	23 cm	?	?
3.	Large stones	?	?	100 cm	?
4.	Extra large stones	?	?	?	1200 cm²
		11.5 cm	12.4 cm	38.9 cm	120.7 cm²
		15.9 cm	31.8 cm	72.2 cm	415.3 cm²
		19.5 cm	39 cm	122.5 cm	793.8 cm²

Experiments

1. Draw a small pentagon *ABCDE* near the center of a large sheet of paper.

2. Choose a point *O* inside the pentagon. Draw dotted rays from *O* through each vertex.

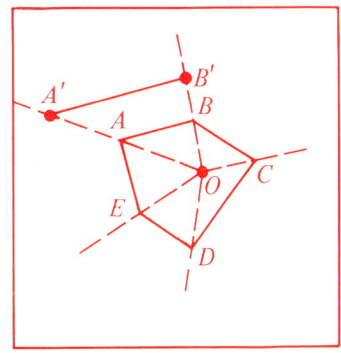

3. As shown, locate point A' on $\overrightarrow{OA}$ so that
$$OA' = 2(OA).$$
Locate point B' on $\overrightarrow{OB}$ so that $OB' = 2(OB)$.
Locate points C', D', and E' in the same way.

4. Draw pentagon $A'B'C'D'E'$.

5. Using a protractor, compare the measures of corresponding angles of the pentagons.

6. Using a ruler, compare the lengths of corresponding sides of the pentagons.

7. What can you conclude about pentagons *ABCDE* and $A'B'C'D'E'$? Can you prove your conclusion?   *ABCDE* ~ $A'B'C'D'E'$

SELF-TEST

Find the values of *x* and *y*.

1. (8-4)

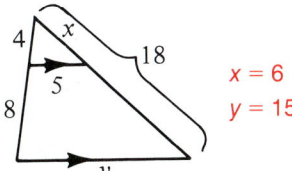

$x = 6$
$y = 15$

2. (8-5)

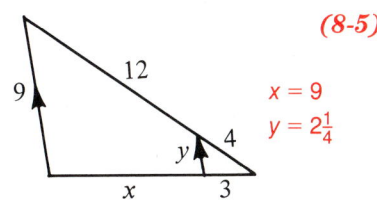

$x = 9$
$y = 2\frac{1}{4}$

The table refers to two similar figures. Copy and complete the table.

	3.	4.	5.
Scale factor	1:3	? 3:7	? 5:6
Ratio of perimeters	? 1:3	3:7	? 5:6
Ratio of areas	? 1:9	? 9:49	25:36

APPLICATIONS

INDIRECT PROOF

The method we have been using to prove statements is called *direct proof*. Sometimes a direct proof is very difficult or impossible to use, so we use an indirect method of reasoning instead. We call this method of reasoning *indirect proof*. We often use it in everyday situations.

Suppose you are watching the basketball playoffs on television. The Pivots lead the Chargers by one point in the final seconds of the game. One of the Chargers takes a shot and misses. A teammate tips up the rebound for a basket just as the final buzzer goes off. No fouls are called, but suddenly the television goes blank. Did the final basket count?

Later you learn that the Pivots won the game. You conclude that the final basket did not count. The next paragraph outlines the reasoning used to arrive at this conclusion.

First, assume that the final basket counts. That basket puts the Chargers ahead by one point, and they win the game. This result contradicts the fact that the Pivots won the game. Our assumption must be incorrect. We conclude that the final basket did not count.

To write an indirect proof:

1. Begin by assuming that what you wish to prove is not true.
2. Reason logically until you reach a contradiction of a known fact.
3. Conclude that your assumption is *false* and that what you wish to prove is *true*.

| EXAMPLE 1 | *Given:* Lines r and s with transversal t, $\angle 1 \neq \angle 2$.
Prove: r is not parallel to s.

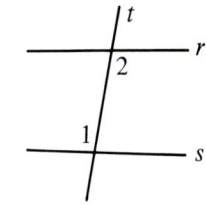

Proof: Assume that r is parallel to s. Then, since $\angle 1$ and $\angle 2$ are alternate interior angles, $\angle 1 = \angle 2$. This contradicts the given information, $\angle 1 \neq \angle 2$. Our assumption, r is parallel to s, must be false. We conclude that r is not parallel to s.

| EXAMPLE 2 | *Given:* line l and point P.
Prove: There is only one line through P perpendicular to l.

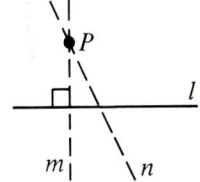

Proof: We know from Construction 3, page 26, that we can draw a line through P perpendicular to l. Assume that there is another line n through P perpendicular to l.

If two lines are perpendicular to a third line, they are parallel to each other, so line m is parallel to line n. But m and n intersect at P. This contradicts the fact that m is parallel to n. We conclude that there is only one line through P perpendicular to l.

Notice that an indirect proof depends on our ability to reason until we have found a contradiction. Indirect proof is sometimes called proof by contradiction.

1. Assume $a \| b$. Then since $\angle 1$ and $\angle 2$ are corr. $\angle s$, $\angle 1 = \angle 2$.
This contradicts the given information that $\angle 1 \neq \angle 2$.
Thus we can conclude that $a \not\| b$.

EXERCISES

Write an indirect proof for the following.

1. *Given:* Lines a and b with transversal t; $\angle 1 \neq \angle 2$.
   Prove: a is not parallel to b.

2. *Given:* Lines l and m; l is not perpendicular to m.
   Prove: $\angle 1 \neq \angle 2$ Assume $\angle 1 = \angle 2$. Since $\angle 1 + \angle 2 = 180°$, $\angle 1 = 90°$. Hence $\ell \perp m$. This contradicts the given information that ℓ is not $\perp$ to m. Thus $\angle 1 \neq \angle 2$.

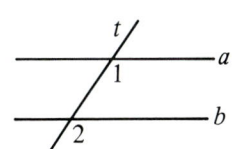

Similar Polygons

PROBLEM SOLVING STRATEGIES

AUXILIARY LINES

In solving a geometry problem, you often first draw a clear diagram. Then you mark in the given information on your diagram. Next you focus your attention on what you have to prove. Finally you try to figure out a way to get from the known to the unknown. To do this, it is often necessary to draw in auxiliary, or helping lines.

You have already seen auxiliary lines used in your study of geometry. For example, the proof of Theorem 1 on page 92 depended on drawing a line through C, parallel to $\overline{AB}$. The proof of Theorem 3 on page 167 depended on drawing the bisector of $\angle C$.

Auxiliary lines were also used often in Chapter 5. They helped us to find the interior and exterior angle sums of a polygon on pages 194–195. The key steps used to prove Theorems 10, 11, and 12 on pages 214–215 also use auxiliary lines.

EXERCISES

1. In the figure at the right, $AB = 5$, $BC = 6$, and $CD = 7$.
   The angles at B and C are both 90°.
   Find the distance AD.

   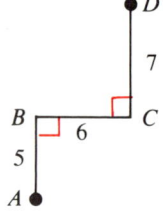

   One way to solve the problem is to see $\overline{AD}$ as the hypotenuse of a right triangle.
   Draw a line through A, parallel to $\overline{BC}$.
   Extend $\overline{DC}$ until it meets the first line you drew.
   Use the right triangle ADX to find the length of $\overline{AD}$. $6\sqrt{5}$

2. Prove Theorem 3 on page 167 by drawing the median from C to $\overline{AB}$ (instead of the bisector of $\angle C$).

   Given: $AC = BC$

   Prove: $\angle A = \angle B$

336   Chapter 8

3. Prove Theorem 1 on page 92 by drawing the auxiliary lines shown in the diagram below.

Given: △ABC
Prove: ∠1 + ∠2 + ∠3 = 180°

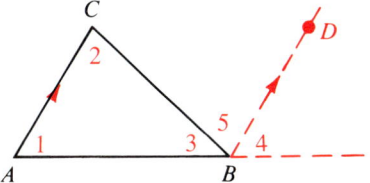

4. The figure below shows △ABC with medians $\overline{BE}$ and $\overline{CD}$ intersecting at T.

If AB = 6 and AC = 8, find the ratio DT:TC.

Let's draw an auxiliary line.

Since we need the ratio DT:TC, draw a line through D, parallel to TE, and intersecting $\overline{AE}$ at X.

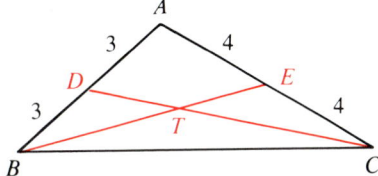

We now have $\overline{TE} \parallel \overline{DX}$ and intersecting two sides of △CDX.
 a. Find the length of $\overline{AX}$ and of $\overline{XE}$. AX = XE = 2
 b. Find the ratio CE:EX. 2:1
 c. Find the ratio CT:TD. 2:1

Look ahead to Chapter 9 where you will work with circles. Name the auxiliary lines shown in the diagrams that go with each of the exercises below.

5. Exercise 30 on page 363 $\overline{OT}$ and $\overline{OV}$

6. Exercise 31 on page 367 $\overline{OA}$

7. Exercise 32 on page 367 $\overline{CX}$ (or $\overline{OC}$ and $\overline{OX}$)

8. Exercise 33 on page 367 $\overline{CX}$ (or $\overline{OC}$ and $\overline{OX}$)

9. Name the auxiliary line used in the example on page 437 of Chapter 11.
Altitude $\overline{DE}$

Similar Polygons **337**

SKILLS REVIEW

THE DISTRIBUTIVE PROPERTY

Simplify. Use the distributive property.

SAMPLE $3(2 + 4x) - 8x + 1$
$6 + 12x - 8x + 1$
$4x + 7$

1. $5(x + 8) + 3x - 7$ $8x + 33$
2. $2(3 - 6k^2) + 9k^2$ $6 - 3k^2$
3. $-(a + 9 - 6a) - 4a$ $a - 9$
4. $8b^3 + 4(6 - 2b^3)$ 24
5. $5 - 3(n + 1) - 6n$ $2 - 9n$
6. $x^2 - (12 - x^2) + 16$ $2x^2 + 4$
7. $0.2(5 - 3m) + m$ $1 + 0.4m$
8. $4(9 - 1.75x) + 7x$ 36
9. $10(1.5a + 6.13) - 7a$ $8a + 61.3$

Solve.

SAMPLE $5(m + 4) = 35$
$5m + 20 = 35$
$5m = 15$
$m = 3$

10. $2(x - 10) = 6$ 13
11. $5(w + 3) = 45$ 6
12. $3(2n + 1) = 21$ 3
13. $11(a + 7) = 0$ -7
14. $7x = 4(x - 9)$ -12
15. $6(3b - 4) = 12$ 2
16. $39 = 13(x + 4)$ -1
17. $3(4 + 2m) = 3m$ -4
18. $6(2k + 1) - 9k = 0$ -2
19. $9y = 4(2y - 12)$ -48
20. $7(9 + 2c) = 5c$ -7
21. $13 + 10d = 9(d + 5)$ 32
22. $5(32 - x) = 3x$ 20
23. $a + 7 = 2(5a - 1)$ 1
24. $7x - 3(5x + 1) = 13$ -2

Write an equation. Then solve.

25. At a garage sale, Tanya bought 5 records that cost $(6x - 5)$ cents each. She spent $7.25 in all. Find the value of x. Then find the cost of each record. $x = 25$; $1.45

26. A rectangular window has length and width as shown. The perimeter is 5.1 m. How many meters wide is the window? How many meters long?   1.1 m; 1.45 m

27. The perimeter of a square is 128. The length of a side is $3n + 5$. Find the value of n.   9

$5k$ cm

$(4k - 6)$ cm

CHAPTER REVIEW

CHAPTER SUMMARY

1. If two figures have the same shape, they are similar. Two things are true of similar polygons:
   (a) Corresponding angles are equal.
   (b) Corresponding sides are in proportion (have equal ratios).

2. To show that two polygons are similar, you must show that both (a) and (b) above are true. However, to show that two *triangles* are similar, you only need to show that two angles of one triangle are equal to two angles of the other triangle.

3. The Triangle Proportionality Theorem states:

   If [figure], then $\frac{a}{b} = \frac{c}{d}$.

4. If three parallel lines intersect two transversals, they divide the transversals proportionally.

5. The scale factor of the similar figures shown is $3:2$. This is the ratio of a pair of corresponding sides. The ratio of the perimeters is also $3:2$. The ratio of the areas is $3^2:2^2$, or $9:4$.

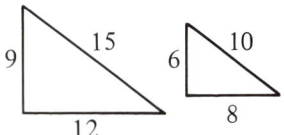

REVIEW EXERCISES

In Exercises 1–6, the two polygons given are similar. Complete each statement. *(See pp. 312–316.)*

1. $\triangle LAF \sim \triangle \underline{\quad?\quad}$ JIG

2. $\frac{LA}{JI} = \frac{LF}{?} = \frac{AF}{?}$ JG; IG

3. $\angle A = \angle \underline{\quad?\quad}$ I

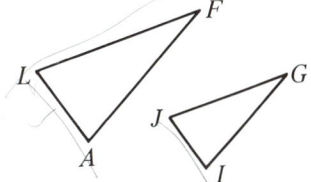

4. Trapezoid $SKON \sim$ trapezoid $\underline{\quad?\quad}$ FLAP

5. The scale factor is $\frac{?}{2} : \frac{?}{3}$.

6. $\frac{14}{21} = \frac{x}{?}$ 27

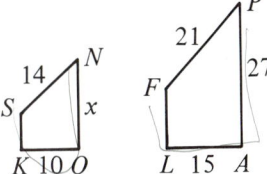

Similar Polygons   **339**

In the diagram at the right, $\overline{UV} \parallel \overline{RS}$. Complete each statement. *(See pp. 317–322.)*

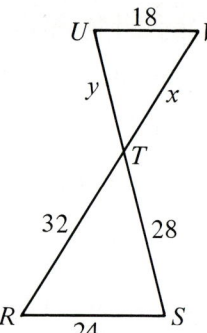

7. $\angle R = \angle \underline{\ ?\ }$ and $\angle S = \angle \underline{\ ?\ }$ V; U

8. $\triangle RST \sim \triangle \underline{\ ?\ }$ VUT

9. $\frac{18}{24} = \frac{x}{?}$ 32

10. $x = \underline{\ ?\ }$ 24

11. $y = \underline{\ ?\ }$ 21

For each figure, complete the statement $\frac{a}{b} = \frac{?}{?}$. *(See pp. 323–324.)*

12. 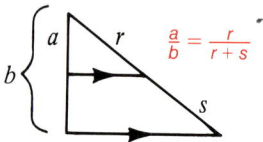 $\frac{a}{b} = \frac{r}{r+s}$

13. 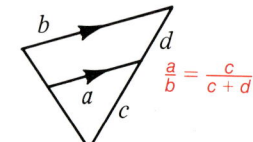 $\frac{a}{b} = \frac{c}{c+d}$

Find the value of x. *(See pp. 325–328.)*

14. 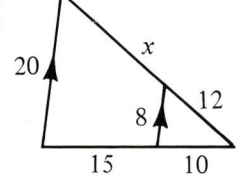   18

15. 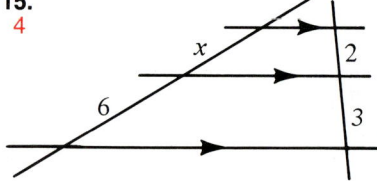   4

For the pair of similar rectangles shown, complete each statement. *(See pp. 329–332.)*

16. The scale factor is $\underline{\ ?\ } : \underline{\ ?\ }$.   6:5

17. The ratio of their perimeters is $\underline{\ ?\ } : \underline{\ ?\ }$.   6:5

18. The ratio of their areas is $\underline{\ ?\ } : \underline{\ ?\ }$.   36:25

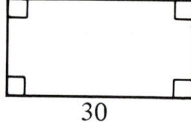

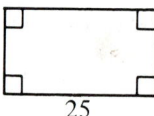

CHAPTER TEST

The two triangles are similar.

1. $\triangle BCD \sim \triangle \underline{\ ?\ }$ ADB   2. $\angle CBD = \angle \underline{\ ?\ }$ DAB   (8-1)

3. Find the value of x. 8   4. Find the value of y. 16

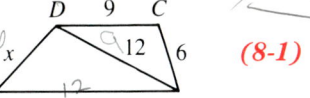

5. Complete the proof.   (8-2)

Given: $\angle 1 = \angle 2$

Prove: $\dfrac{RU}{RS} = \dfrac{UV}{ST}$

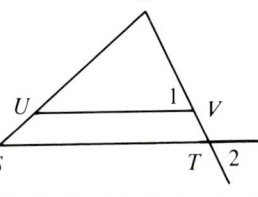

STATEMENTS	REASONS
1. $\angle 1 = \angle 2$	1. __?__ Given
2. $\angle 2 = \angle VTS$	2. __?__ Vert. $\angle$ are =.
3. $\angle 1 = \angle VTS$	3. __?__ Substitution Postulate
4. $\angle R = \angle R$	4. __?__ From algebra
5. $\triangle RUV \sim \triangle RST$	5. __?__ AA Postulate
6. $\dfrac{RU}{RS} = \dfrac{UV}{ST}$	6. __?__ Corr. sides of $\sim \triangle$ are in proportion.

In Questions 6–8, refer to the diagram.

6. $\dfrac{8}{12} = \dfrac{4}{?} = \dfrac{6}{?}$ 7. $x = \underline{\ ?\ }$ 2   8. $y = \underline{\ ?\ }$ (8-3)

  4 + x; y

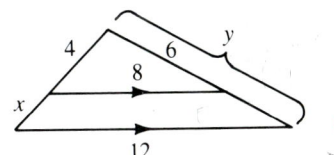

For each figure, complete the statement $\dfrac{a}{b} = \dfrac{?}{?}$.   (8-4)

9.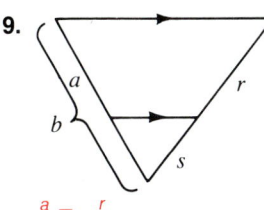

$\dfrac{a}{b} = \dfrac{r}{r+s}$

10.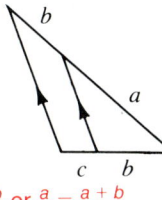

$\dfrac{a}{b} = \dfrac{b}{c}$ or $\dfrac{a}{b} = \dfrac{a+b}{b+c}$

11.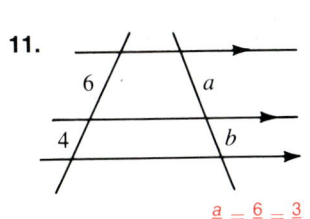

$\dfrac{a}{b} = \dfrac{6}{4} = \dfrac{3}{2}$

The bases of two similar parallelograms are 9 cm and 12 cm.   (8-5)

12. Find their scale factor. 3:4   13. Find the ratio of their areas. 9:16

14. If the perimeter of the smaller parallelogram is 30 cm, what is the perimeter of the larger parallelogram? 40 cm

Similar Polygons   **341**

UNIT D CUMULATIVE REVIEW

Express each ratio in simplest form.

1. $\frac{11}{77}$ $\frac{1}{7}$
2. $\frac{45}{100}$ $\frac{9}{20}$
3. $\frac{20 \text{ cm}}{4 \text{ m}}$ $\frac{1}{20}$
4. $8:16:24$ $1:2:3$

Find the value of x.

5. $\frac{x}{2} = \frac{6}{3}$ 4
6. $\frac{1}{7} = \frac{2}{x}$ 14
7. $\frac{2}{x} = \frac{4}{22}$ 11
8. $10:15 = x:3$ 2

9. Three cassette tapes cost $24. How much will five tapes cost? $40

10. At a shopping center, Hal's Hardware is 40 m south of Sound City. The supermarket is 60 m east of Sound City. Draw a map showing the hardware store, music store, and supermarket. Be sure to include the scale.

State whether the two polygons are, or are not, similar.

11. not similar
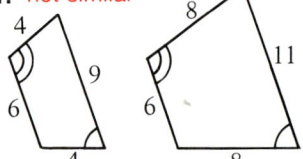

12.
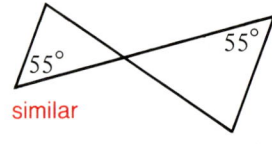
similar

13. similar
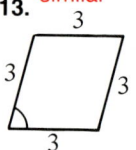

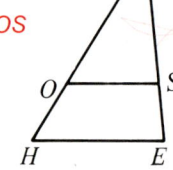

In the diagram, $\overline{OS} \parallel \overline{HE}$. Complete each statement.

14. $\triangle ROS \sim \triangle \underline{\ ?\ }$ RHE
15. $\angle RHE = \angle \underline{\ ?\ }$ ROS
16. $\frac{?}{RH} = \frac{RS}{RE}$ RO
17. $\frac{RO}{OH} = \frac{RS}{?}$ SE
18. If $RS = 8$, $SE = 4$, and $OS = 6$, then $HE = \underline{\ ?\ }$. 9

The quadrilaterals shown are similar. Complete each statement.

19. $\angle E = \angle \underline{\ ?\ }$ A
20. The scale factor is $\underline{\ ?\ } : \underline{\ ?\ }$. 2:1
21. $x = \underline{\ ?\ }$ 2
22. The ratio of the perimeters is $\underline{\ ?\ } : \underline{\ ?\ }$. 2:1
23. The ratio of the areas is $\underline{\ ?\ } : \underline{\ ?\ }$. 4:1

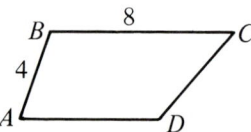

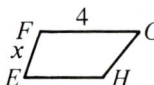

UNIT
E

Here's what you'll learn in this chapter:

To define a circle and the points, lines, and segments related to circles.

To apply theorems relating tangents and radii.

To construct a tangent to a circle at a given point on the circle.

To classify and measure arcs.

To use theorems involving the chords of a circle.

To draw a circle inscribed in a polygon and a circle circumscribed about a polygon.

To use theorems relating angle measure and arc measure.

Notice the circular pattern in which the reflecting mirrors shown above are arranged. These movable mirrors focus sunlight on a boiler, in order to make steam for generating electricity.

Chapter 9

Circles

1 Basic Terms

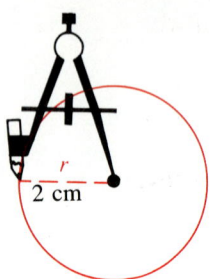

Set your compass for a radius of 2 cm.
Now draw a complete curve.
You know, of course, that the curve is a *circle*.
The pin point is the *center* of the circle.

This experiment should remind you that a **circle** is a figure, in a plane, whose points are the same distance from a particular point in the plane. The table below shows some important definitions relating to circles.

radius	1. a segment that joins the center and a point on the circle 2. the length of such a segment
chord	a segment that joins two points on the circle
diameter	1. a chord that passes through the center 2. the length of such a chord
secant	a line that contains a chord
tangent	a line, in the plane of the circle, that intersects the circle in exactly one point
point of tangency	the point in which a tangent intersects the circle

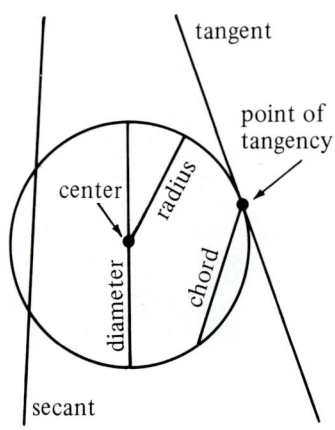

From the definition of a circle, you can see that:

(1) All radii of a circle are equal.

(2) All diameters of a circle are equal.

Many of the terms used for circles are also used for *spheres*. A **sphere** is a figure in space whose points are the same distance from a particular point. That point is called the *center* of the sphere.

$\overline{PA}$, $\overline{PB}$, and $\overline{PC}$ are radii of the sphere.
$\overline{BC}$ is a diameter.
$\overleftrightarrow{AD}$ is a tangent.
$\overleftrightarrow{BC}$ is a secant.

Classroom Practice

Exercises 1–10 refer to ⊙O (circle with center O) in the diagram.

5. *F is not on the circle.*

1. Name all the radii shown. $\overline{OD}, \overline{OA}, \overline{OB}$
2. Name a secant shown. $\overleftrightarrow{CD}$
3. Is $\overleftrightarrow{CD}$ a chord? no (See above.)
4. Is $\overline{CD}$ a chord? yes
5. Explain why $\overline{AF}$ is not a chord.
6. Name a tangent to ⊙O. $\overleftrightarrow{BE}$
7. Is $\overline{BD}$ a chord? a diameter? yes; yes
8. Name a point of tangency. B
9. How many diameters that contain point C can be drawn? that contain point O? one; an infinite number
10. How many secants containing both A and C can be drawn? one 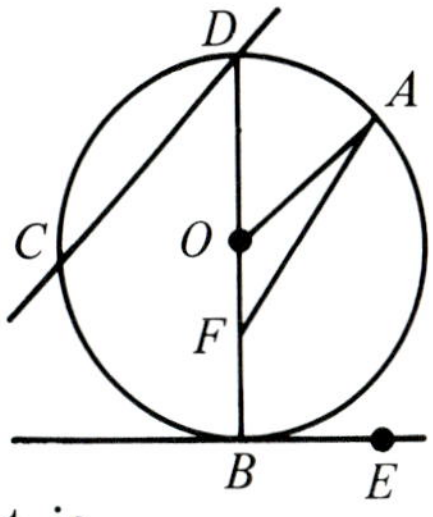
11. State a definition of a radius of a sphere.
    A segment joining the center and a point on the sphere.
12. State a definition of a tangent to a sphere.
    A line or plane that intersects the sphere in exactly one point.
13. Set a basketball on top of a desk and you see a sphere that is tangent to a plane. Hold a piece of cardboard to illustrate another plane tangent to the sphere. In how many positions can you hold the cardboard to show a tangent plane? An infinite number
14. Refer to Exercise 13. Hold the piece of cardboard so that it is both tangent to the ball and parallel to the desk top. Consider the point where the cardboard touches the ball and the point where the desk top touches the ball. What can you say about the segment that joins these two points? It is a diameter of the sphere.

Written Exercises

Exercises 1–10 refer to ⊙A. Name each of the following.

A
1. A circle ⊙A
2. Four radii $\overline{AR}, \overline{AC}, \overline{AB}, \overline{AD}$
3. Two diameters $\overline{RB}, \overline{CD}$
4. Two tangents $\overleftrightarrow{GE}, \overleftrightarrow{BF}$
5. A chord that is not a diameter $\overline{BC}$
6. Two secants, each containing a diameter $\overleftrightarrow{RB}, \overleftrightarrow{CD}$
7. A secant that does not contain a diameter $\overleftrightarrow{BC}$
8. Two points of tangency R, B
9. How many tangents containing both point A and point D can be drawn? none
10. How many chords containing both point R and point C can be drawn? one

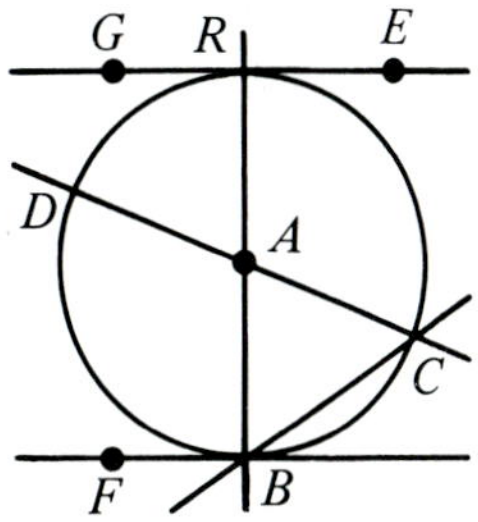

Circles

The length of a radius is given. Find the length of a diameter.

11. 19 38
12. 5.3 10.6
13. $7\frac{1}{2}$ 15
14. $3k$ $6k$

The length of a diameter is given. Find the length of a radius.

15. 26 13
16. 2.4 1.2
17. $8\frac{1}{2}$ $4\frac{1}{4}$
18. $5k$ $\frac{5k}{2}$

Refer to the diagram of the sphere with center O. Name each of the following.

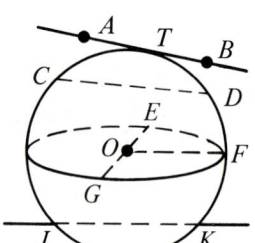

B **19.** Eight points on the sphere J, K, E, F, G, C, D, T

20. Three radii of the sphere $\overline{OE}, \overline{OG}, \overline{OF}$

21. A diameter of the sphere $\overline{EG}$

22. Three chords of the sphere $\overline{JK}, \overline{EG}, \overline{CD}$

23. A secant of the sphere $\overleftrightarrow{JK}$

24. A tangent to the sphere $\overrightarrow{AB}$

25. Write a definition of a diameter of a sphere. A chord that passes through the center of the sphere

26. Write a definition of a chord of a sphere.
A segment that joins two points on the sphere

For each exercise draw a circle with radius 8. Draw two points R and S that satisfy the conditions stated. If the conditions cannot be satisfied, write *not possible*. (Drawings omitted)

C **27.** $\overline{RS}$ is a chord and $RS = 10$.

28. R and S both lie inside the circle and $RS = 17$. not possible

29. R and S both lie outside the circle and $RS = 7$.

30. R lies inside the circle.
S lies outside the circle.
$\overleftrightarrow{RS}$ intersects the circle in exactly one point. not possible

31. $\overline{OA}$ and $\overline{OB}$ are radii.
$\angle A = 5x - 3; \angle B = 3x + 17$.
Find the value of x. 10

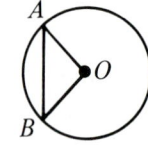

32. Point P is the center of the circle.
Find the measure of $\angle E$. 25°

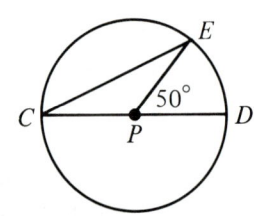

Experiments

1. Copy the diagram shown.
2. Construct a line, through O, perpendicular to chord $\overline{AB}$. Let C be the point of intersection with $\overline{AB}$.
3. What appears to be true about AC and BC? AC = BC
4. Prove that your conclusion is true.

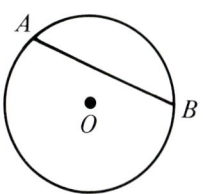

CONSUMER APPLICATIONS

LIFE INSURANCE

A life insurance policy is an agreement that you and your insurance company make with each other. You agree to pay the insurance company a certain fee (or *premium*) each year. In return, the company agrees to pay your family a certain amount when you die.

In order to make money, life insurance companies must take in more money by selling premiums than they pay out. Statisticians who compute insurance risks and premiums are called actuaries. To find out more about what they do, read the Career Notes on page 316.

Many different life insurance policies are offered by many different companies. There are two general types of life insurance with which you should be familiar.

Permanent life insurance is the kind most people know. This type of insurance is part financial protection in case of death and part savings plan. The face amount of the policy is paid out by the insurance company either when the insured person dies or when the insured person reaches an age (often 65 or 70) specified in the policy.

Term life insurance does not provide a savings feature. The principal is paid out by the insurance company only if the insured person dies within the term of years specified in the policy.

Which type of insurance is better? Term insurance costs less, but there is no guarantee of ever collecting any money from the insurance company. Permanent life insurance costs more, but provides a savings plan in addition to financial protection against sudden death. Before you buy insurance, you should consult an insurance broker or a banker.

2 Tangents

EXPLORATIONS

1. Draw circles and tangents like those shown.

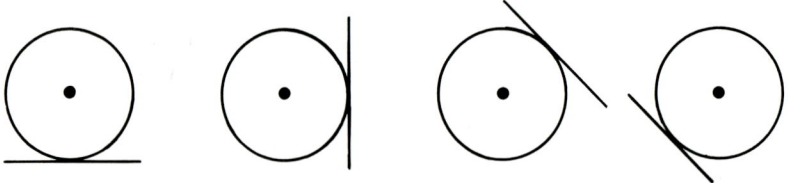

   In each case draw the radius to the point of tangency. What do you observe? The radii are ⊥ to the tangents.

2. Draw circles and radii like those shown.

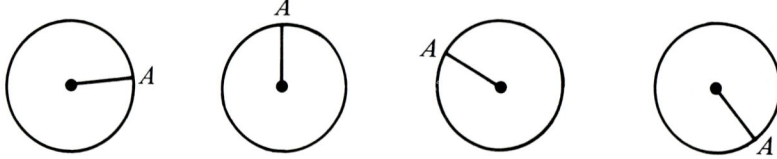

   In each case draw a line that passes through A and is perpendicular to the radius. What do you observe? These lines are tangents.

These explorations suggest two theorems that we state but do not prove.

THEOREM 1

A radius drawn to a point of tangency is perpendicular to the tangent.

THEOREM 2

If a line lies in the plane of a circle and is perpendicular to a radius at its outer endpoint, the line is tangent to the circle.

If line t is tangent to ⊙O, then $\overline{OA} \perp t$.

If line $t \perp \overline{OA}$ at point A, then t is tangent to ⊙O.

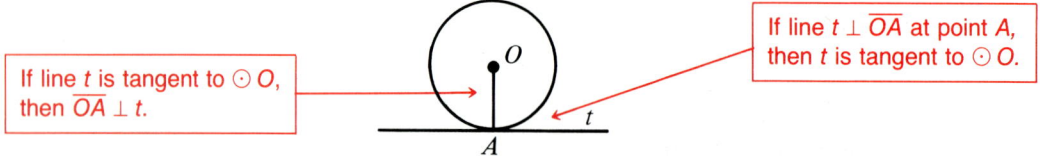

A tangent to a circle has been defined as a line. However, we sometimes say that a *segment* is tangent to a circle. For example, since $\overleftrightarrow{PX}$ is tangent to the circle at X, $\overline{PX}$ may also be called a tangent to the circle.

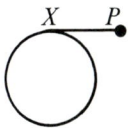

A line that is tangent to each of two coplanar circles is called a **common tangent.**

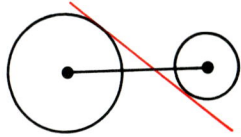

A common *internal* tangent intersects the segment joining the centers.

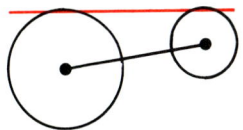

A common *external* tangent does not intersect the segment joining the centers.

When two coplanar circles are tangent to a line at one point, we say that the *circles are tangent to each other*.

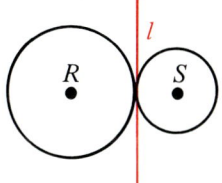

Circles R and S are *externally* tangent. Line l is a common internal tangent.

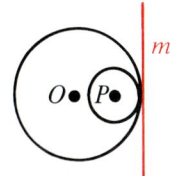

Circles O and P are *internally* tangent. Line m is a common external tangent.

CONSTRUCTION 10

Given: *Point A on $\odot O$*
Construct: *A tangent to $\odot O$ at point A*

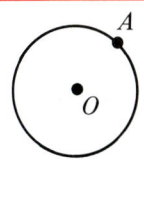

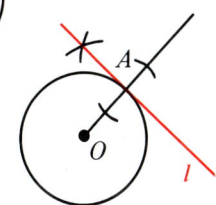

1. You are given $\odot O$ with point A.
2. Draw $\overrightarrow{OA}$.
   Using Construction 2, page 26, construct a line perpendicular to $\overrightarrow{OA}$ at point A. Call the line l.
3. Line l is tangent to $\odot O$.

Classroom Practice

1. If line t is tangent to $\odot O$ at point A, what must be true about $\overline{OA}$ and line t? State the theorem that supports your conclusion. *$\overline{OA} \perp t$; a radius drawn to a point of tangency is $\perp$ to the tangent.*

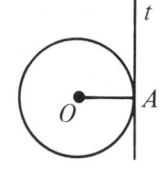

2. If line t and $\odot O$ lie in the same plane, and if line t is perpendicular to $\overline{OA}$, what must be true about line t and $\odot O$? State the theorem that supports your conclusion. *t is a tangent to $\odot O$; if a line lies in the plane of a $\odot$ and is $\perp$ to a radius at its outer endpoint, the line is tangent to the $\odot$.*

In each diagram, which lines are common external tangents and which are common internal tangents?

3.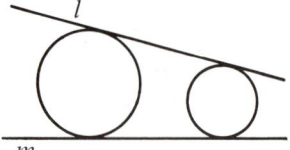

4.

3. *l and m both are common external tangents.*

4. *s is a common internal tangent. r is a common external tangent.*

5.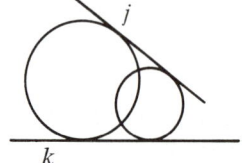

6.

5. *j and k both are common external tangents.*

6. *t is a common external tangent.*

7. In the figures for Exercises 3–6, state whether the two circles are externally tangent, internally tangent, or are not tangent.
not tangent; externally tangent; not tangent; internally tangent

Draw a circle O on the chalkboard. Choose a point P outside the circle. Draw tangents, $\overline{PX}$ and $\overline{PY}$, to the circle.

8. What appears to be true about PX and PY? *$PX = PY$*

9. Complete the listing of what is given.
   Given: $\odot O$; $\overline{PX}$ and $\overline{PY}$ are . . . *tangents to $\odot O$ at X and Y respectively.*
   Prove: $PX = PY$

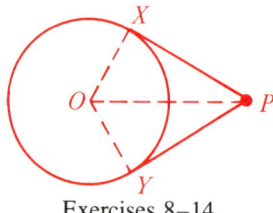

Exercises 8–14

By drawing $\overline{OX}$, $\overline{OY}$, and $\overline{OP}$, we form two triangles that appear to be congruent.

10. State a theorem that supports the statements: $\overline{OX} \perp \overline{PX}$; $\overline{OY} \perp \overline{PY}$. *A radius drawn to a pt. of tangency is $\perp$ to the tangent.*

11. State a reason that supports the statement: $OX = OY$. *Radii of a circle are =.*

12. Explain why right $\triangle PXO \cong$ right $\triangle PYO$. *HL Theorem*

13. State a reason that supports the statement: $PX = PY$. *Corr. parts of $\cong \triangle$ are =.*

14. Complete this statement of the theorem that has just been proved: If two tangent segments are drawn to . . . *a circle from an exterior point, then the segments are equal.*

Written Exercises

In each exercise, state the number of common internal tangents and the number of common external tangents that can be drawn to the two circles.

A 1.

   internal: 2; external: 2

2.

   internal: 1; external: 2

3.

   internal: none; external: 2

4.

   internal: none; external: 2

5.

   internal: none
   external: none

6.

   internal: none; external: 1

7. Which of the six diagrams above shows two circles that are externally tangent? two circles that are internally tangent?
   Ex. 2   Ex. 6

8. Draw a figure showing two circles that are tangent to a third circle but are not tangent to each other.

In Exercises 9–12, line t lies in the plane of the three circles and intersects each of the circles in exactly one point, X. Classify each statement as true or false.

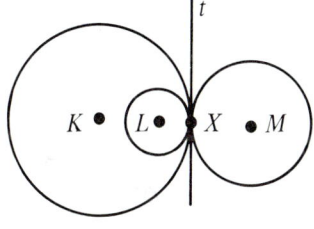

9. Line t is a common tangent of $\odot L$ and $\odot M$. true

10. Line t is a common external tangent of $\odot L$ and $\odot M$. false

11. Line t is a common external tangent of $\odot K$ and $\odot L$. true

12. $\odot M$ and $\odot K$ are internally tangent circles. false

In Exercises 13 and 14, draw a figure roughly like the one shown, but larger.

13. Construct a line tangent to $\odot P$ at point X.

    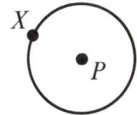

14. Construct lines tangent to $\odot O$ at points D and E.

    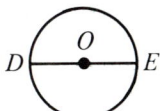

If possible, draw a diagram to illustrate the conditions. (Drawings omitted)

B 15. Two circles with five common tangents not possible

16. Two tangent circles with three common tangents

Circles 353

If possible, draw a diagram to illustrate the conditions.

17. Two externally tangent circles with two common internal tangents not possible

18. Two internally tangent circles with one common external tangent

Find the required length.

19.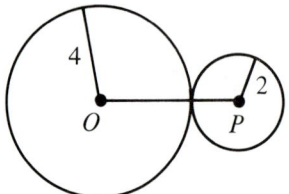

    $OP = $ __?__ 6

20.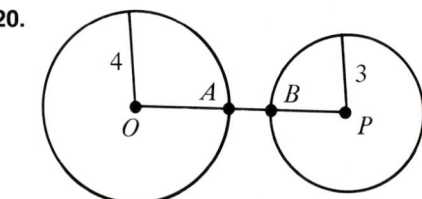

    $OP = 9$, $AB = $ __?__ 2

21.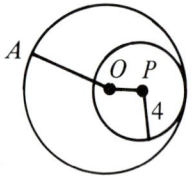

    $OA = 7$ $OP = $ __?__ 3

22.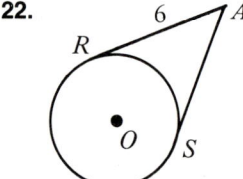

    $AS = $ __?__ 6
    (*Hint:* See Exercise 14, page 353.)

[C] 23.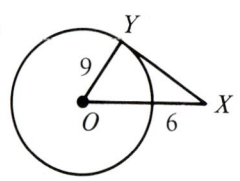

    $XY = $ __?__ 12

24.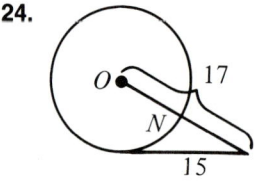

    $ON = $ __?__ 8

25. Refer to the diagram at the bottom of page 351.
    Explain why line l must be tangent to $\odot O$. l is a line in the plane of the $\odot$ that is $\perp$ to a radius ($\overline{OA}$) at its outer endpoint (A). By Thm. 2, l is tangent to $\odot O$.

PUZZLE ◆ PROBLEMS

Using ordinary household waxed paper, cut out a circle with as large a radius as convenient. Mark the center. Fold the paper so that some point of the circle falls on the center O. Make a sharp crease, then unfold. Make about twenty different creases in the same way. Hold the paper at arm's length. The creases suggest a circle!

3 Arcs and Central Angles

An *arc* is part of a circle.

Classifying Arcs:

Semicircle   Minor Arc   Major Arc

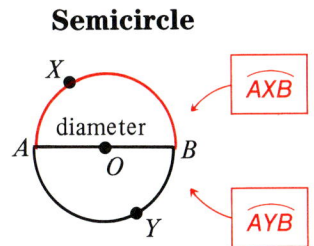 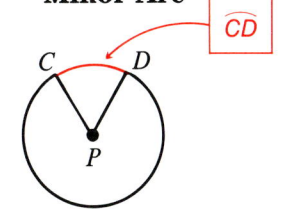 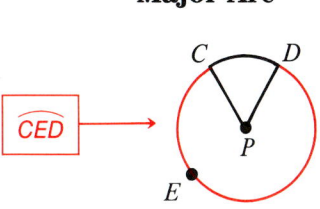

Exactly half a circle   Less than a semicircle   Greater than a semicircle

Notice that you need three letters to name a semicircle or a major arc.
$\overarc{AXB}$ is read "arc *AXB*."

A **central angle** of a circle is an angle whose vertex is the center of the circle.

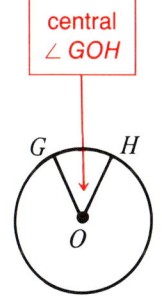

> *The degree measure of a minor arc is defined to be the measure of its central angle.*

By definition, $\overarc{GH} = \angle GOH$. You may read this as "arc *GH* equals angle *GOH*." Just remember that this means that the *measure* of the angle equals the *measure* of the arc.

Finding the Measure of an Arc:

Semicircle   Minor Arc   Major Arc

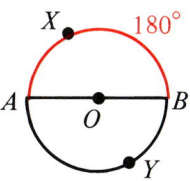 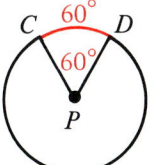 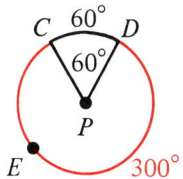

$\overarc{AXB} = \overarc{AYB} = 180°$ $\overarc{CD} = \angle CPD = 60°$ $\overarc{CED} = 360° - \overarc{CD} = 300°$

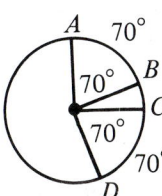

If two arcs of a circle have equal measures, they are called **equal arcs.**

Thus, in the diagram, $\overarc{AB} = \overarc{CD}$ because each arc has measure 70°.

Circles   **355**

THEOREM 3
In a circle, equal central angles have equal minor arcs.

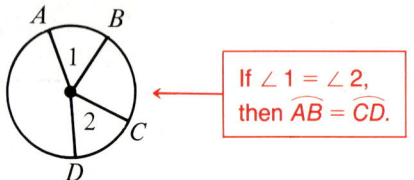

Strategy for proof:

$\widehat{AB} = \angle 1$ (Why?)
$\angle 1 = \angle 2$ (Why?)
$\angle 2 = \widehat{CD}$ (Why?)
$\widehat{AB} = \widehat{CD}$

THEOREM 4
In a circle, equal minor arcs have equal central angles.

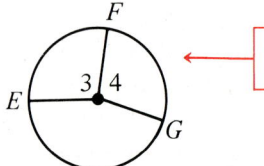

If $\widehat{EF} = \widehat{FG}$, then $\angle 3 = \angle 4$.

The proof is left as Written Exercise 26 on page 358.

Classroom Practice

Exercises 1–6 refer to the diagram shown.

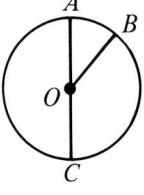

1. Name a minor arc. $\widehat{AB}, \widehat{BC}$
2. Name a major arc. $\widehat{ACB}, \widehat{BAC}$
3. Name a semicircle. $\widehat{ABC}$
4. Name two central angles.
   $\angle AOB, \angle BOC, \angle AOC$
5. If $\angle AOB = 40°$, then $\widehat{AB} = \underline{\quad?\quad}°$. 40
6. If $\angle AOB = 40°$, then $\widehat{ACB} = \underline{\quad?\quad}°$. 320

$\overline{AD}$ and $\overline{EC}$ are diameters of $\odot O$. $\overline{OB}$ is a radius.
Classify each statement as true or false.

7. $\angle AOC = 100°$ true
8. $\angle AOE = 50°$ false
9. $\angle AOE$ is a central angle. true
10. $\widehat{ABC}$ is a minor arc. true
11. $\widehat{AED}$ is a semicircle. true
12. $\widehat{BCD}$ is a major arc. false
13. $\widehat{AB} = 50°$ true
14. $\widehat{ABD} = 180°$ true
15. $\widehat{AB} = \widehat{BC}$ true
16. $\widehat{BD} = 100°$ false
17. $\widehat{CDA}$ is a major arc. true
18. $\widehat{DC} = \widehat{EA}$ true

Written Exercises

In Exercises 1–9, radii, diameters, and chords of ⊙O are shown. State the measure of ∠1.

A 1. 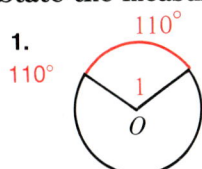   2. 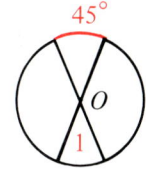   3.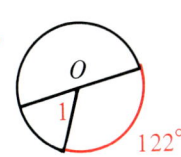

4. 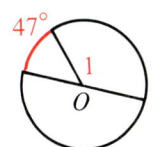   5. 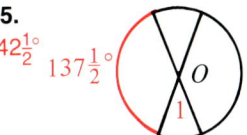   6.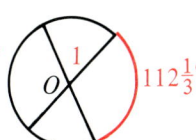

7. 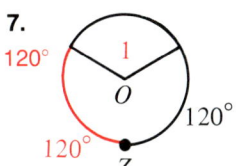   8. 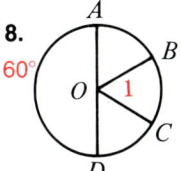   9.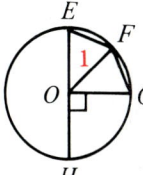

In Exercises 10–12, find the measure of the indicated arc.

10. $\widehat{AB}$ = ___?___°   72
11. $\widehat{JL}$ = ___?___°   49
12. $\widehat{RSQ}$ = ___?___°   220

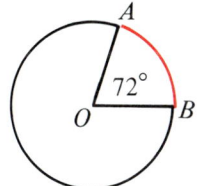   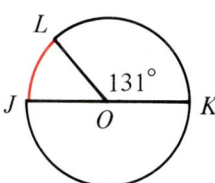   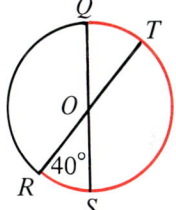

In the figure, diameter $\overline{XY}$ bisects acute ∠POQ.   (See below.)

13. Name five minor arcs.   14. Name five major arcs.   (See below.)

15. Name two semicircles.   16. Name three *pairs* of equal arcs.
$\widehat{XPY}$, $\widehat{XQY}$

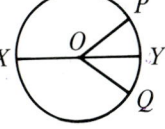

$\widehat{PY} = \widehat{YQ}$; $\widehat{XP} = \widehat{XQ}$; $\widehat{XPY} = \widehat{XQY}$; $\widehat{XPQ} = \widehat{XQP}$; $\widehat{YXQ} = \widehat{YXP}$

In the figure, $\overline{BC}$ is a diameter of ⊙O.

B 17. If ∠1 = 40°, then $\widehat{AB}$ = ___?___°.   140

18. If ∠1 = 36°, then ∠3 = ___?___°. (Hint: AO = BO)   18

19. If ∠1 = 42°, then ∠2 = ___?___°.   21

20. If ∠2 = 25°, then ∠1 = ___?___°.   50

21. If ∠3 = 28°, then ∠1 = ___?___°.   56   22. If ∠2 = 31°, then $\widehat{ACB}$ = ___?___°.   242

13. $\widehat{XP}$, $\widehat{PY}$, $\widehat{PQ}$, $\widehat{YQ}$, $\widehat{QX}$ 14. $\widehat{XPQ}$, $\widehat{PQX}$, $\widehat{YQP}$, $\widehat{QXP}$, $\widehat{QXY}$

Circles

In Exercises 23–25, $\overline{AB}$ is a diameter of $\odot O$.

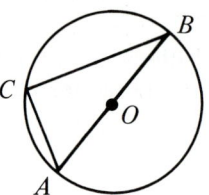

C 23. If $\widehat{AC} = 50°$, then $\angle ACB = \underline{\ ?\ }°$. 90
(Hint: Draw $\overline{OC}$. Find $\angle ACO$ and $\angle BCO$.)

24. If $\widehat{BC} = 120°$, then $\angle ACB = \underline{\ ?\ }°$. 90

25. If $\widehat{AC} = j°$, then $\angle ACB = \underline{\ ?\ }°$. 90

26. Write a proof of Theorem 4.

SELF-TEST

Vocabulary

circle (p. 346)
chord (p. 346)
diameter (p. 346)
secant (p. 346)
tangent (p. 346)
point of tangency (p. 346)
sphere (p. 346)

common tangent (p. 351)
internal tangent (p. 351)
external tangent (p. 351)
semicircle (p. 355)
minor arc (p. 355)
major arc (p. 355)
central angle (p. 355)

Name each of the following.

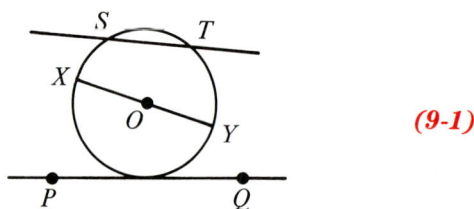

1. A radius $\overline{OX}$; $\overline{OY}$

2. A diameter $\overline{XY}$

3. A secant $\overleftrightarrow{ST}$ (9-1)

4. A tangent $\overrightarrow{PQ}$

5. A chord that is not a diameter $\overline{ST}$

6. Construct a circle O. Use X to label a point on the circle.   (9-2)
Construct a tangent to circle O at X.

7. How many common tangents can be drawn to the two circles? 2

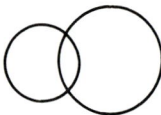

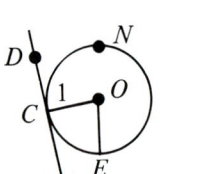

8. If $\angle 1 = 90°$, then $\overleftrightarrow{CD}$ is $\underline{\ ?\ }$ to $\odot O$. tangent   (9-3)

9. If $\widehat{CE} = 80°$, then $\angle COE = \underline{\ ?\ }°$. 80

10. If $\widehat{CE} = 80°$, then $\widehat{CNE} = \underline{\ ?\ }°$. 280

4 Chords

Chord $\overline{JK}$ separates $\odot O$ into two arcs: minor arc JK and major arc JZK. We call $\widehat{JK}$ *the* arc of chord $\overline{JK}$.

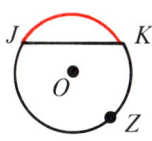

If $\widehat{CM} = \widehat{MD}$, then M is the **midpoint** of $\widehat{CD}$.
A line, ray, or segment that contains M **bisects** $\widehat{CD}$.
In the diagram, line j bisects $\widehat{CD}$.

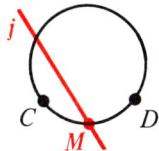

The *distance from center O to chord* $\overline{AB}$ is the length of the shortest segment that can be drawn from O to a point on $\overline{AB}$. It can be shown that the shortest segment is always the *perpendicular* segment from the point to the line.

OX is the distance from O to $\overline{AB}$.

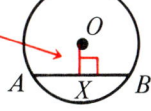

EXPLORATIONS

1. Draw a circle with center O. Then draw equal chords $\overline{WX}$ and $\overline{YZ}$ as shown. Draw four additional segments and prove that $\triangle WOX \cong \triangle YOZ$. Then use corresponding angles of the congruent triangles to prove that $\widehat{WX} = \widehat{YZ}$.

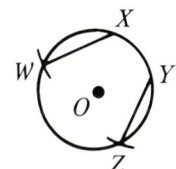

2. Draw a circle with center P. On $\odot P$ mark off an arc, $\widehat{EF}$. Choose a point G on the circle, and then a point H so that it looks as if $\widehat{GH} = \widehat{EF}$. Assume that $\widehat{GH}$ really does equal $\widehat{EF}$. Draw $\overline{EF}$ and $\overline{GH}$. Try to prove that $EF = GH$. In a $\odot$, equal minor arcs have equal central angles; SAS Postulate; corr. parts of $\cong \triangle$ are $=$.

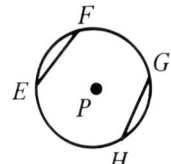

3. Draw a circle with center Q. Then draw equal chords $\overline{RS}$ and $\overline{TV}$. Draw a segment that will help you think of the distance from Q to $\overline{RS}$. Also draw a segment that will remind you of the distance from Q to $\overline{TV}$. What appears to be true about the segments that you have drawn? They have the same length.

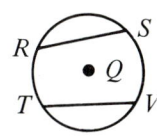

4. Draw a large circle. Then draw a chord and the diameter that is perpendicular to the chord. Do you see any equal segments? any equal arcs? The chord is divided into equal segments. There are pairs of equal minor arcs and equal major arcs.

These explorations lead to the following theorems.

THEOREM 5

In a circle, equal chords have equal arcs and equal arcs have equal chords.

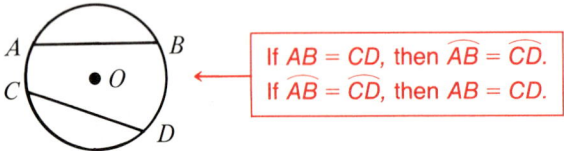

If $AB = CD$, then $\overarc{AB} = \overarc{CD}$.
If $\overarc{AB} = \overarc{CD}$, then $AB = CD$.

Do you see that Theorem 5 is a "double" theorem? It states both a theorem and its converse. Theorem 6 is also a "double" theorem.

THEOREM 6

In a circle, equal chords are equidistant from the center. Chords that are equidistant from the center are equal.

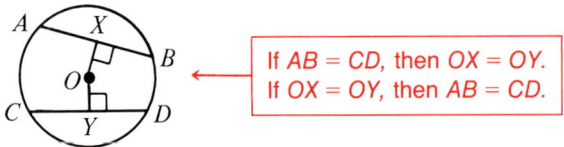

If $AB = CD$, then $OX = OY$.
If $OX = OY$, then $AB = CD$.

We omit the proof.

THEOREM 7

A diameter that is perpendicular to a chord bisects the chord and its arc.

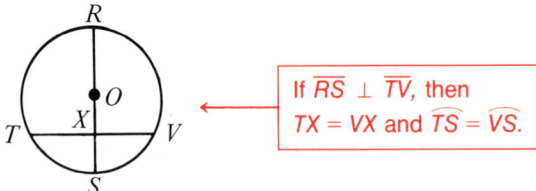

If $\overline{RS} \perp \overline{TV}$, then
$TX = VX$ and $\overarc{TS} = \overarc{VS}$.

The strategy for a proof is left as Exercise 30 on page 363.

We have used the terms *inscribed* and *circumscribed* with respect to triangles and circles. As the diagrams show, the terms are applied to other polygons, too.

A polygon is *inscribed in* a circle when each vertex lies on the circle. A polygon is *circumscribed about* a circle when each side is tangent to the circle.

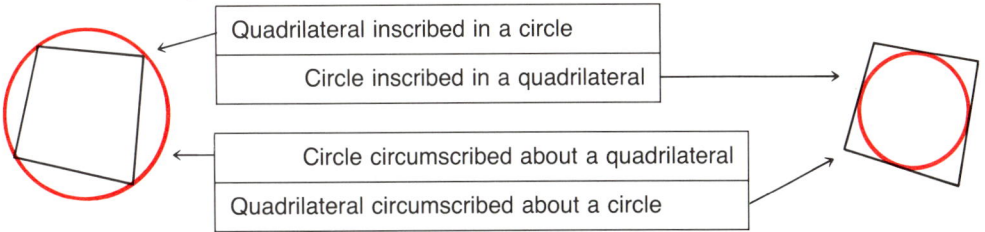

Quadrilateral inscribed in a circle
Circle inscribed in a quadrilateral
Circle circumscribed about a quadrilateral
Quadrilateral circumscribed about a circle

In Chapter 4, you learned how to circumscribe a circle about a triangle (Construction 7). You also learned how to inscribe a circle in a triangle (Construction 8).

Classroom Practice

$\overline{AB}$ **is a diameter of** $\odot O$. $\overline{AB} \perp \overline{XY}$.

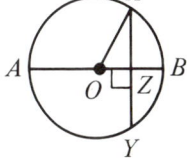

1. $XZ =$ __?__ ZY
2. $\widehat{XB} =$ __?__ $\widehat{BY}$
3. The midpoint of $\widehat{XY}$ is __?__. B
4. $\widehat{AX} = 180° -$ __?__ and $\widehat{AY} = 180° -$ __?__ $\widehat{XB}$; $\widehat{BY}$
5. From Exercises 3 and 4, we conclude: __?__ = __?__. $\widehat{AX}$; $\widehat{AY}$

Exercises 6–10 refer to the figure.

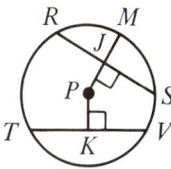

6. $\overline{PM}$ is a __?__ of $\widehat{RS}$. bisector
7. If $PJ = PK$, then $RS =$ __?__. TV
8. Which is longer: $\overline{PS}$ (not drawn) or $\overline{JS}$? PS
9. Compare the length of $\overline{TK}$ with the length of $\overline{TV}$. $TK = \frac{1}{2} TV$
10. If $RS = TV$, then $\widehat{RS} =$ __?__. $\widehat{TV}$

Tell whether each statement is true or false.

11. If a polygon is inscribed in a circle, the sides of the polygon are chords of the circle. true
12. If a rectangle is circumscribed about a circle, then the rectangle must be a square. true

Tell whether each statement is true or false.

13. A triangle inscribed in a circle must be isosceles. false

14. It is possible to draw an inscribed quadrilateral in such a way that one of its sides is a diameter of the circle. true

Written Exercises

State whether the polygon is inscribed in the circle, is circumscribed about the circle, or neither.

A 1.
inscribed

2.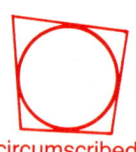
circumscribed

3.
circumscribed

4.
neither

5.
neither

6.
inscribed

Exercises 7–12 refer to $\odot O$ with $\overline{OR} \perp \overline{AB}$.

7. If $RB = 5$, $AB = $ __?__ . 10

8. If $AB = 14$, $AR = $ __?__ . 7

9. If $RB = 4$ and $OR = 3$, $OB = $ __?__ . 5

10. If $OB = 10$ and $RB = 8$, $OR = $ __?__ . 6

11. If $OB = 10$ and $AR = 6$, $OR = $ __?__ . 8

12. If $OB = 17$ and $AB = 30$, $OR = $ __?__ . 8

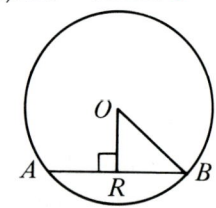

In each exercise, a circle O is shown. Can you conclude that $\overset{\frown}{AB} = \overset{\frown}{BC}$? Write *yes* or *no*.

13. yes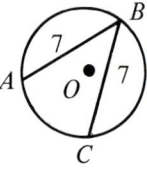

14. yes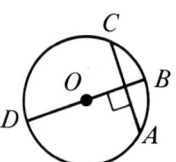

15. no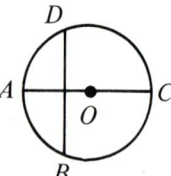

16. yes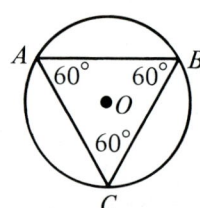

B 17. Here is a way to construct an inscribed regular hexagon:
   a. Construct a circle.
   b. Without changing your compass, mark off points around the circle. You should end up at the starting point.
   c. Draw the hexagon.

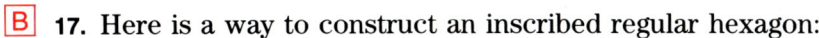

18. Construct a circle. Then inscribe an equilateral triangle in the circle. (*Hint:* See Exercise 17.)

19. Construct a circle. Inscribe a square in the circle. (*Hint:* Draw any diameter. Use Construction 2, page 26, to get a diameter that is perpendicular to the first one. Connect endpoints.)

20. Construct a circle. Inscribe a regular octagon in the circle. (*Hint:* Construct perpendicular diameters as in Exercise 19. Then construct two additional diameters that bisect central angles.)

In each exercise, a circle O is shown. Find ST.

21. 8
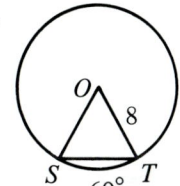

22. 4
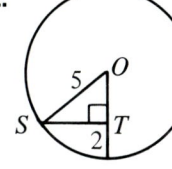

23. 10
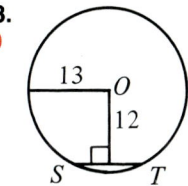

Draw the figure described. If the figure cannot be drawn, write *not possible*. (You need not construct.) (Drawings omitted)

24. A rectangle inscribed in a circle

25. A trapezoid circumscribed about a circle

26. A parallelogram, not a rectangle, inscribed in a circle not possible

C 27. Explain why the construction described in Exercise 17 works.

28. Construct a circle. Then construct a square that is circumscribed about the circle.

29. One student insisted: A diameter that bisects a chord must be perpendicular to the chord. Draw a figure that would show that the student is mistaken.

30. *Given:* $\odot O$ with $\overline{RS} \perp \overline{TV}$.
    State a strategy you could use to show that
    $TX = VX$ and $\widehat{TS} = \widehat{VS}$.
    You need not write out the proof.

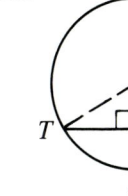

Circles 363

5 Inscribed Angles

An angle is *inscribed in* a circle and is an **inscribed angle** if its vertex lies on the circle and its sides are chords of the circle. Each diagram below shows an ∠C inscribed in ⊙O.

Inscribed Angles

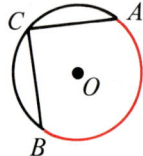 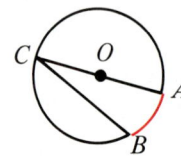 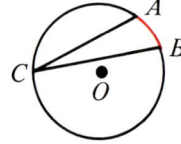

In each case, ∠C **intercepts** an arc, $\widehat{AB}$.

EXPLORATIONS

1. Draw a circle O with a radius at least 5 cm long.
2. Choose three points A, B, and C on the circle so that $\widehat{ACB}$ is a major arc.
3. Draw $\overline{CA}$ and $\overline{CB}$. Use a protractor to measure ∠ACB.
4. Draw $\overline{OA}$ and $\overline{OB}$. Use a protractor to measure ∠AOB.

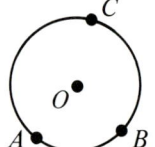

How does the measure of ∠ACB compare with that of ∠AOB? Recall that ∠AOB = $\widehat{AB}$.

∠ACB = $\tfrac{1}{2}$∠AOB

How does the measure of ∠ACB compare with that of $\widehat{AB}$?

∠ACB = $\tfrac{1}{2}\widehat{AB}$

Even if your angle measures are not exact, your results should suggest the following theorem.

THEOREM 8

An inscribed angle is equal to half its intercepted arc.

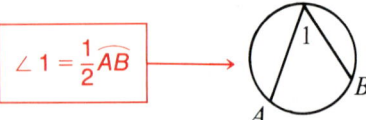

See Written Exercises 31–33.

364   Chapter 9

EXAMPLE Find the measures of ∠1, ∠2, and ∠3.

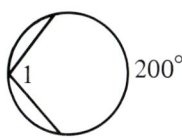 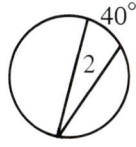 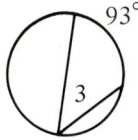

$\angle 1 = \frac{1}{2}(200°) = 100°$ $\angle 2 = \frac{1}{2}(40°) = 20°$ $\angle 3 = \frac{1}{2}(93°) = 46.5°$

In the first three figures below, ∠A is an inscribed angle formed by chords $\overline{AX}$ and $\overline{AY}$. As point Y moves along $\overset{\frown}{XA}$ toward point A, $\overline{AY}$ becomes more and more like a tangent to the circle.

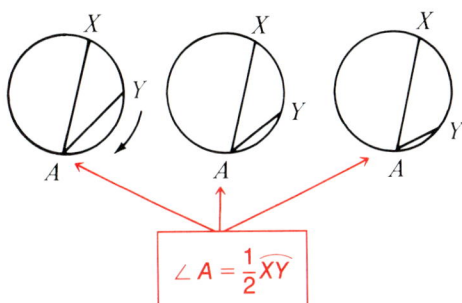 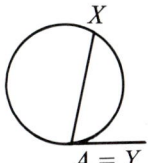

$\angle A = \frac{1}{2}\overset{\frown}{XY}$

What seems to be true about the measure of ∠A?

THEOREM 9

An angle formed by a chord and a tangent is equal to half its intercepted arc.

$\angle 1 = \frac{1}{2}\overset{\frown}{AB}$ →

Classroom Practice

State whether or not ∠1 is an inscribed angle.

1. yes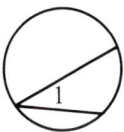
2. yes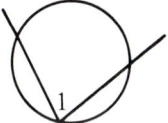
3. no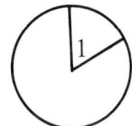

Circles **365**

State whether or not ∠1 is an inscribed angle.

4.
no

5.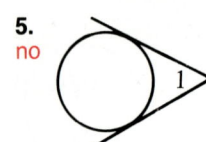
no

6. yes

In the diagram, $\overrightarrow{CD}$ is a tangent. Complete each statement.

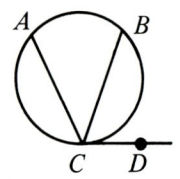

7. The inscribed angle pictured is ∠ _?_ . ACB
8. ∠ACB intercepts _?_ . $\widehat{AB}$ 9. ∠BCD intercepts _?_ . $\widehat{BC}$
10. ∠ACB = $\frac{1}{2}$ _?_ 　$\widehat{AB}$ 11. ∠BCD = $\frac{1}{2}$ _?_ 　$\widehat{BC}$ 12. ∠ACD = $\frac{1}{2}$ _?_ 　$\widehat{ABC}$

Written Exercises

In Exercises 1–8, use the diagram.

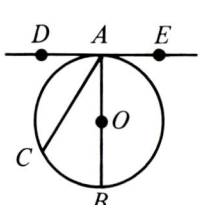

A 1. If $\widehat{KN}$ = 50°, then ∠1 = _?_ °. 25
2. If $\widehat{NM}$ = 110°, then ∠2 = _?_ °. 55
3. If $\widehat{KN}$ = 70° and $\widehat{NM}$ = 115°, then $\widehat{KNM}$ = _?_ °. 185
4. If $\widehat{KPM}$ = 170° and $\widehat{KP}$ = 100°, then $\widehat{PM}$ = _?_ °. 70
5. If ∠1 = 28°, then $\widehat{KN}$ = _?_ °. 56
6. If ∠2 = 76°, then $\widehat{NM}$ = _?_ °. 152
7. If ∠1 = 24° and ∠2 = 68°, then $\widehat{KNM}$ = _?_ °. 184
8. If ∠1 = 28° and ∠2 = 69°, then $\widehat{KPM}$ = _?_ °. 166

In the diagram, $\overleftrightarrow{DE}$ is tangent to ⊙O at point A.

9. ∠BAE = _?_ ° 90   10. ∠DAB = _?_ ° 90
11. If ∠CAB = 30°, then ∠DAC = _?_ °. 60
12. If ∠DAC = 55°, then ∠CAB = _?_ °. 35
13. If ∠CAB = 32°, then $\widehat{BC}$ = _?_ °. 64
14. If $\widehat{AC}$ = 100°, then ∠CAB = _?_ °. 40

Exercises 15–20 refer to the diagram.

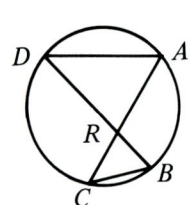

B 15. If $\widehat{AB}$ = 96°, then ∠D = _?_ ° and ∠C = _?_ °. 48; 48
16. If $\widehat{DC}$ = 118°, then ∠A = _?_ ° and ∠B = _?_ °. 59; 59

17. If ∠D = 44°, then $\widehat{AB}$ = __?__° and ∠C = __?__°. 88; 44

18. If ∠D = 47°, then $\widehat{AB}$ = __?__° and ∠C = __?__°. 94; 47

19. Is ∠D = ∠C? Is ∠A = ∠B? yes
yes

20. Is △ADR ~ △BCR? yes

In each diagram, find the measure of ∠1.

21. 51°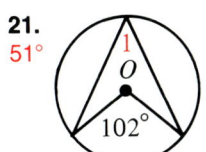

22. 40°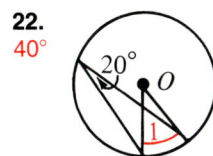

23. 37°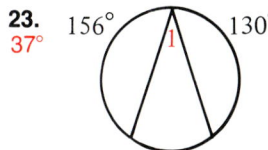

24. 43°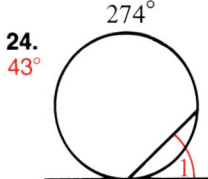

25. 94°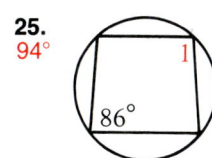

26. 70°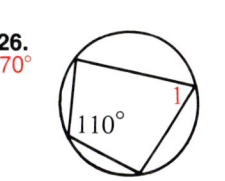

△*CAB* is isosceles with *AC* = *AB*.

27. If ∠A = 55°, then $\widehat{AB}$ = __?__°. 125

28. If $\widehat{CB}$ = k°, then ∠B = __?__°. $90 - \frac{k}{4}$
(Answer in terms of k.)

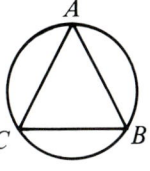

29. *DEFGH* is a regular pentagon.
  a. $\widehat{DE}$ = __?__° and $\widehat{EF}$ = __?__° 72; 72
  b. Find the measure of each numbered angle.

29. b. ∠1 = 36°
∠2 = 36°
∠3 = 36°
∠4 = 108°

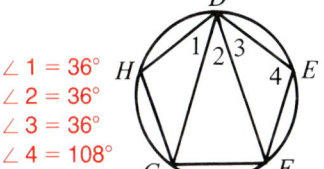

C **30.** Explain why the opposite angles of a quadrilateral inscribed in a circle must be supplementary. (See below.)

In Exercises 31–33, write a strategy for proving the three cases of Theorem 8. In each case, $\overline{CA}$ and $\overline{CB}$ are chords of ⊙*O*. Show that ∠ACB = $\frac{1}{2}\widehat{AB}$. You need not write out the proof.

∠6 = $\frac{1}{2}\widehat{XB}$
∠5 = $\frac{1}{2}\widehat{XA}$
∠6 − ∠5 = $\frac{1}{2}(\widehat{XB} - \widehat{XA})$
∠ACB = $\frac{1}{2}\widehat{AB}$

31. *Case 1*   ∠1 = ∠ACB + ∠2
∠2 = ∠ACB
2(∠ACB) = ∠1
= $\widehat{AB}$
∠ACB = $\frac{1}{2}\widehat{AB}$

32. *Case 2*   ∠3 = $\frac{1}{2}\widehat{AX}$
∠4 = $\frac{1}{2}\widehat{XB}$
∠3 + ∠4 = $\frac{1}{2}(\widehat{AX} + \widehat{XB})$
∠ACB = $\frac{1}{2}\widehat{AB}$

33. *Case 3*

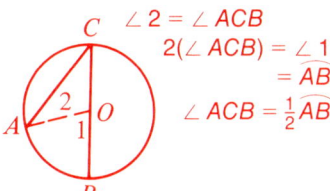 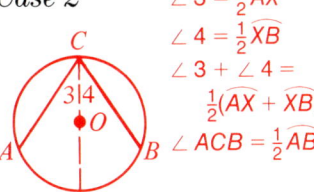 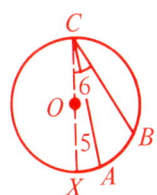

Hint:
∠ACB + ∠2 = ∠1

Hint: Use Case 1.
$\widehat{AB} = \widehat{AX} + \widehat{XB}$

Hint: Use Case 1.
$\widehat{AB} = \widehat{XB} - \widehat{XA}$

30. Their sum is $\frac{1}{2}$ the number of degrees in the circle, or $\frac{1}{2} \cdot 360° = 180°$; hence they are supplementary.

SELF-TEST

Vocabulary

midpoint of an arc (p. 359) inscribed angle (p. 364)
bisector of an arc (p. 359) intercepted arc (p. 364)

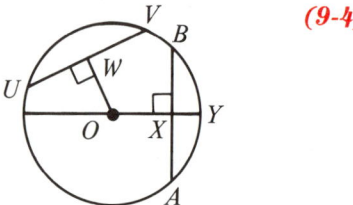

1. If $AB = 12$, then $AX = $ __?__. 6 *(9-4)*
2. If $\overset{\frown}{BY} = 32°$, then $\overset{\frown}{AY} = $ __?__ °. 32
3. If $AB = 12$ and $OW = OX$, then $UV = $ __?__. 12

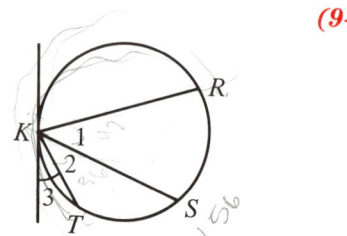

4. If $\overset{\frown}{RS} = 80°$, then $\angle 1 = $ __?__ °. 40 *(9-5)*
5. If $\angle 2 = 38°$, then $\overset{\frown}{ST} = $ __?__ °. 76
6. If $\angle 1 = 42°$ and $\angle 2 = 36°$, then $\overset{\frown}{RST} = $ __?__ ° and $\overset{\frown}{RKT} = $ __?__ °. 156, 204
7. If $\overset{\frown}{TK} = 60°$, then $\angle 3 = $ __?__ °. 30

8. Draw a rhombus inscribed in a circle.

PUZZLE ◆ PROBLEMS

The project on page 354 involved folding waxed paper to form a circle. You can fold paper to form other curves. For this project, use a rectangular sheet of waxed paper.

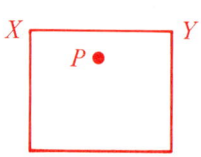

Select a side $\overline{XY}$ and a point P.

Fold the paper so that some point of $\overline{XY}$ falls on point P.

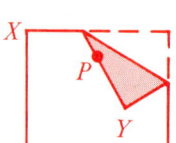

Make a sharp crease, then unfold.

Make about twenty different creases in the same way.

Hold the paper at arm's length and the creases will suggest a *parabola*.

6 Other Angles

EXPLORATIONS

A. We wish to find the measure of ∠1 without using a protractor.

1. Draw $\overline{AD}$. You can easily find the measures of ∠ADC and ∠DAB.

2. Notice that ∠1 is an exterior angle of △AXD.

   $\angle 1 = \frac{1}{2}(\underset{70}{\underline{}}°) + \frac{1}{2}(\underset{40}{\underline{}}°)$

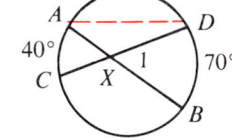

B. We wish to find the measure of ∠P without using a protractor.

1. Draw $\overline{ST}$. You can easily find the measures of ∠RST and ∠STV.

2. Notice that ∠RST is an exterior angle of △TSP.

   $\angle P = \frac{1}{2}(\underset{100}{\underline{}}°) - \frac{1}{2}(\underset{50}{\underline{}}°)$

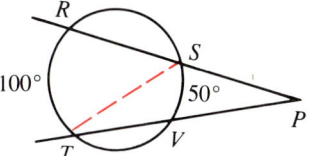

These explorations suggest two theorems—and their proofs.

THEOREM 10

An angle formed by two chords is equal to half the sum of the intercepted arcs.

$\angle 1 = \frac{1}{2}(\widehat{AC} + \widehat{BD})$

THEOREM 11

An angle formed by two secants is equal to half the difference of the intercepted arcs.

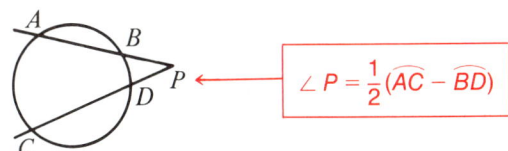

$\angle P = \frac{1}{2}(\widehat{AC} - \widehat{BD})$

EXAMPLE In each diagram, find the measure of ∠1.

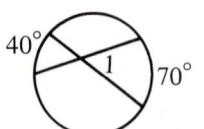
$\angle 1 = \frac{1}{2}(40° + 70°)$
$= \frac{1}{2}(110°)$
$= 55°$

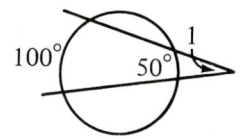
$\angle 1 = \frac{1}{2}(100° - 50°)$
$= \frac{1}{2}(50°)$
$= 25°$

Did you get these answers when you tried the explorations on the previous page?

SUMMARY OF ANGLE AND ARC RELATIONSHIPS

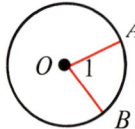
∠1 is a central angle.
$\angle 1 = \widehat{AB}$

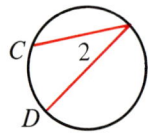
∠2 is an inscribed angle.
$\angle 2 = \frac{1}{2}\widehat{CD}$

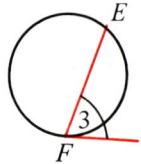
∠3 is formed by a chord and a tangent.
$\angle 3 = \frac{1}{2}\widehat{EF}$

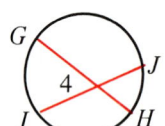
∠4 is formed by two chords.
$\angle 4 = \frac{1}{2}(\widehat{GI} + \widehat{JH})$

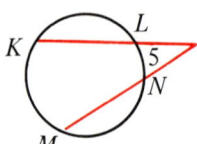
∠5 is formed by two secants.
$\angle 5 = \frac{1}{2}(\widehat{KM} - \widehat{LN})$

Classroom Practice

Find the measure of each numbered angle.

1.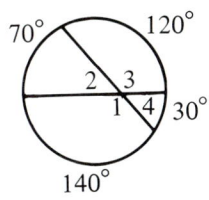
∠1 = 130°; ∠2 = 50°;
∠3 = 130°; ∠4 = 50°

2.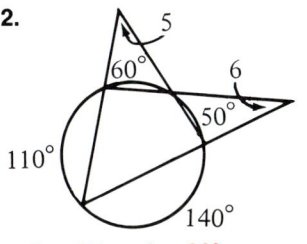
∠5 = 40°; ∠6 = 30°

3.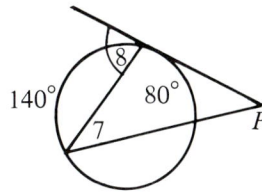
∠7 = 40°; ∠8 = 70°

4. Using the diagram for Exercise 3, find ∠P.
(*Hint:* ∠8 is an exterior angle of the triangle.) ∠P = 30°

Written Exercises

Find the measure of ∠1.

A

1. 40°
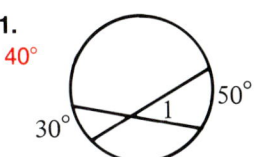

2. 44°
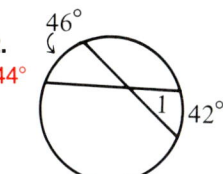

3. 38°
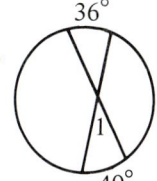

4. 110°
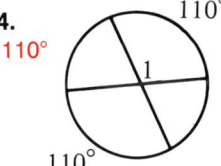

5. 80°
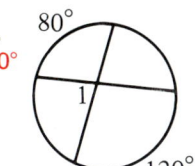

6. 70°
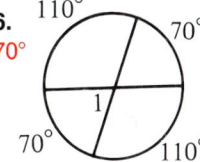

7. 35°
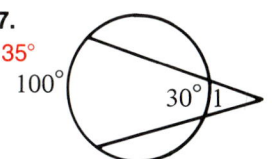

8. 50°
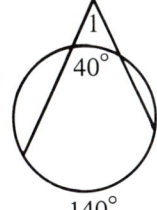

9. 59°
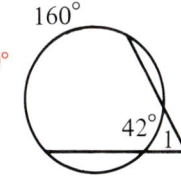

10. 40°
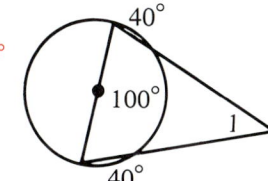

11. 45°
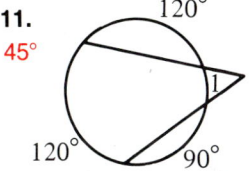

12. 45°
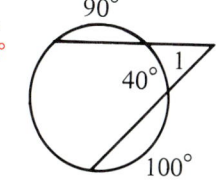

Circles   371

Find the value of x.

SAMPLE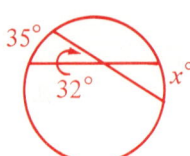

$32 = \frac{1}{2}(35 + x)$

$64 = 35 + x$

$29 = x$

B 13. 39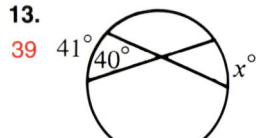

14. 90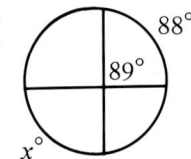

15. 180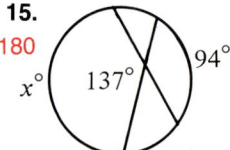

16. 50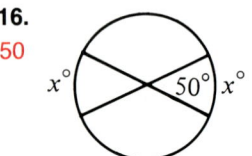

17. 84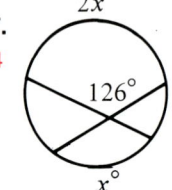

18. 143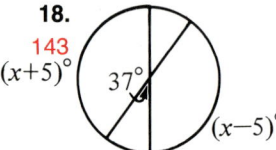

19. 58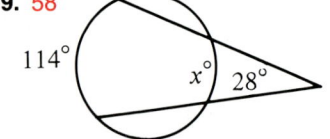

20. 100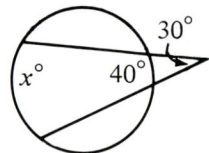

C 21. 27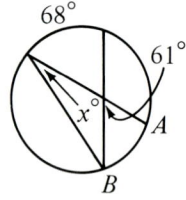

22. 89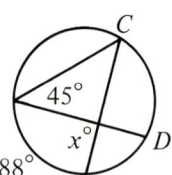

23. 40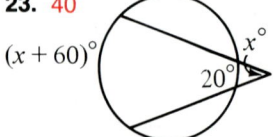

24. 32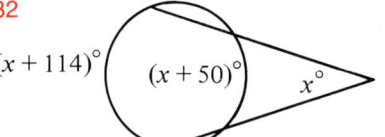

7 Segments of Chords

Do you believe that the equation

$$AX \cdot XB = CX \cdot XD$$

can possibly be correct? Some people do not believe it until they see a proof.

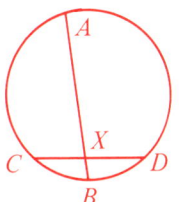

THEOREM 12

If two chords intersect inside a circle, the product of the lengths of the segments of one chord equals the product of the lengths of the segments of the other.

Given: Chords $\overline{AB}$ and $\overline{CD}$ intersect at point X inside the circle.

Prove: $AX \cdot XB = CX \cdot XD$

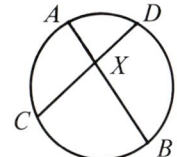 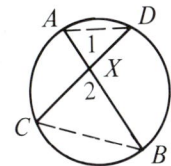

STATEMENTS	REASONS
1. Draw $\overline{AD}$ and $\overline{BC}$.	1. Through any two points there is exactly one line.
2. $\angle A = \frac{1}{2}\widehat{BD}$; $\angle C = \frac{1}{2}\widehat{BD}$	2. An inscribed angle is equal to __?__. half its intercepted arc
3. $\angle A = \angle C$	3. Substitution Postulate
4. $\angle 1 = \angle 2$	4. __?__ Vertical angles are equal.
5. $\triangle AXD \sim \triangle CXB$	5. AA Postulate
6. $\dfrac{AX}{CX} = \dfrac{XD}{XB}$	6. Corresponding sides of $\sim \triangle$ are in proportion.
7. $AX \cdot XB = CX \cdot XD$	7. A property of proportions

Sometimes it is simpler to apply Theorem 12 by thinking of a figure like the one shown.

$p \times q = r \times s$

Circles 373

Classroom Practice

In each exercise, find the length x.

1.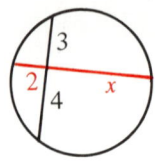

   $2x =$ __?__ 12
   $x =$ __?__ 6

2.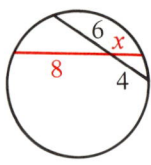

   $8x =$ __?__ 24
   $x =$ __?__ 3

3.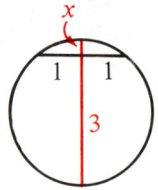

   $3x =$ __?__ 1
   $x =$ __?__ $\frac{1}{3}$

Exercises 4–7 lead to a formula about the lengths associated with secants to a circle from an outside point.

4. Name two angles of $\triangle PBC$ that are equal to two angles of $\triangle PDA$. $\angle P = \angle P, \angle B = \angle D, \angle BCP = \angle DAP$

5. $\triangle PBC \sim \triangle$ __?__ . PDA

6. Write a proportion that involves PB, PA, PD, and PC. $\frac{PB}{PD} = \frac{PC}{PA}$

7. $PB \cdot$ __?__ $= PD \cdot$ __?__ . PA; PC

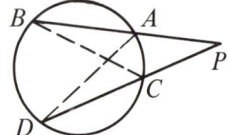

8. Look at the figures below. As point B moves along $\overset{\frown}{BA}$ toward point A, $\overrightarrow{PB}$ becomes more and more like a tangent to the circle.

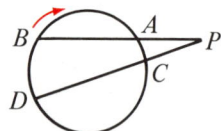 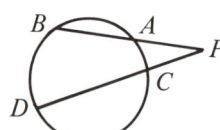 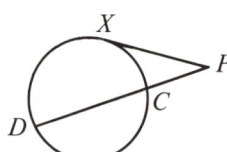

In the third diagram, A and B have become the same point, X. Look at the equation in Exercise 7. Suggest an equation involving PX, PC, and PD. $(PX)^2 = PC \cdot PD$

Written Exercises

In each exercise, find the length x.

SAMPLE	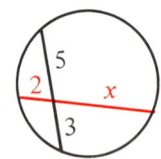	$2x = 3 \cdot 5$ $2x = 15$ $x = \dfrac{15}{2} = 7\dfrac{1}{2}$

374   Chapter 9

A

1. 2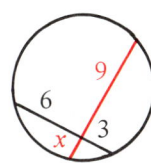

2. 2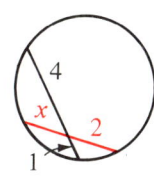

3. 4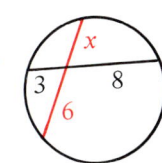

4. $5\frac{1}{4}$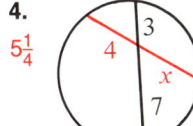

5. $8\frac{3}{4}$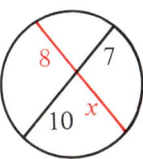

6. $5\frac{5}{11}$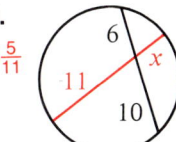

7. $2\frac{3}{4}$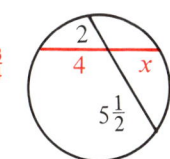

8. 24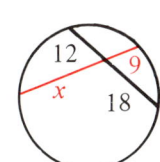

9. 6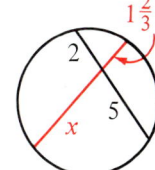

B

10. 11.88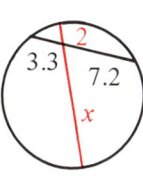

11. 0.9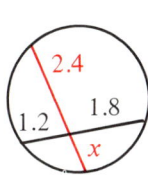

12. $1\frac{1}{2}$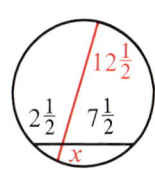

Exercises 13–16 lead to the same equation that was developed in Exercise 8, page 374. In the diagram, $\overline{PX}$ is a tangent segment and $\overleftrightarrow{PD}$ is a secant.

13. $\angle PXC = \frac{1}{2}\underline{\ ?\ }^{\overset{\frown}{XC}}$; $\angle PDX = \frac{1}{2}\underline{\ ?\ }^{\overset{\frown}{XC}}$
    $\angle PXC = \angle PDX$; $\angle P = \angle P$

14. $\triangle PXC \sim \triangle \underline{\ ?\ } PDX$

15. $\frac{PX}{?} = \frac{PC}{?}$ PD; PX

16. $PX \times \underline{\ ?\ }_{PX} = PC \times \underline{\ ?\ }_{PD}$

Secants of circles are shown. Find the value of x.

C

17. 6

Hint: $4 \times 12 = 8x$

18. 12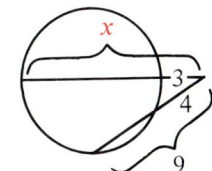

Secants and tangents of circles are shown. Find the value of x.

19.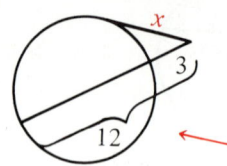

20.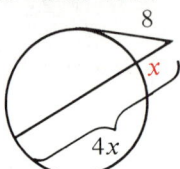

Hint: See Exercise 16.

SELF-TEST

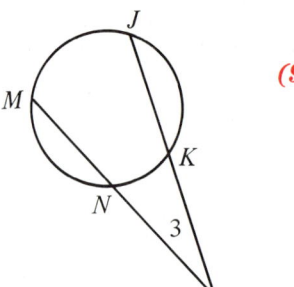

1. If $\widehat{JM} = 100°$ and $\widehat{KN} = 50°$, then $\angle 3 = \underline{\quad?\quad}°$. 25   (9-6)
2. If $\widehat{JK} = 108°$, $\widehat{JM} = 96°$, and $\widehat{MN} = 94°$, then $\angle 3 = \underline{\quad?\quad}°$. 17
3. If $\angle 3 = 40°$ and $\widehat{KN} = 30°$, then $\widehat{JM} = \underline{\quad?\quad}°$. 110

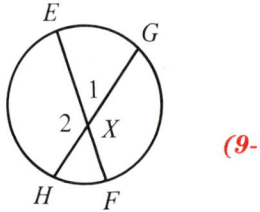

4. If $\widehat{GE} = 60°$ and $\widehat{HF} = 40°$, then $\angle 1 = \underline{\quad?\quad}°$. 50
5. If $\widehat{GE} = 67°$ and $\widehat{HF} = 41°$, then $\angle 2 = \underline{\quad?\quad}°$. 126
6. If $\angle 2 = 130°$ and $\widehat{EH} = 140°$, then $\widehat{GF} = \underline{\quad?\quad}°$. 120
7. If $EX = 15$, $XF = 8$, and $GX = 12$, then $XH = \underline{\quad?\quad}$. 10
8. If $GX = 9$, $XH = 8$, and $EX = 12$, then $FX = \underline{\quad?\quad}$. 6

(9-7)

PUZZLE ◆ PROBLEMS

Suppose you have a circle, but have lost track of its center. How can you find the center?

Draw two chords, $\overline{AB}$ and $\overline{CD}$. Construct the perpendicular bisector of each chord. The two perpendicular bisectors intersect at the center of the circle.

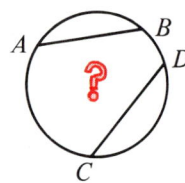

Try the construction when $\overline{AB}$ and $\overline{CD}$ are parallel. Why doesn't the construction work? The ⊥ bisectors are the same line.

Return to the original problem. Can you discover a different way to construct the center?

Applications

USING CIRCLES TO SET UP SCHEDULES

Five teams enter a tournament. Each team is to play each other team once. Begin arranging a schedule by dividing a circle into five equal arcs. Call the points, and the teams, A, B, C, D, and E.

FIRST ROUND
Draw the diameter containing A.
Connect other points as shown.
Games: A-bye B-E C-D
Bye means that the team does not play.

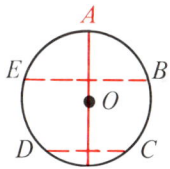

SECOND ROUND
Draw the diameter containing B.
Games: B-bye C-A D-E

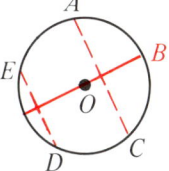

THIRD, FOURTH, AND FIFTH ROUNDS
Draw diameters containing C, D, and E.
In the five rounds there will be a total
of ten games, not counting byes.

For a tournament of six teams, use five equal arcs as above. But when you draw the diameter containing A, let that mean that A will play the sixth team, O. There won't be any byes. In the five rounds there will be a total of fifteen games.

For a tournament of seven teams—or of eight teams—use seven equal arcs. You should get 21 games and 28 games, respectively.

Circles   **377**

COMPUTER ACTIVITIES

TEN-SPEED BICYCLES

A ten-speed bike doesn't have ten gears! It has two sprockets (gears) in front and five gears in back. Each front sprocket can be connected to each back gear with the chain. The pedals turn the front sprocket. When the pedals turn once, the front sprocket also turns once.

The back gears have fewer teeth than the front sprockets. Therefore, they turn faster. For example, a bike might have 40 front sprocket teeth and 20 back gear teeth. Every time the pedals turned once, the gear would turn twice. The back wheel is connected to a rear gear. So for one pedal turn, you would have two turns of the wheels.

Different bikes have different gear setups. Here are the numbers of teeth on one typical bike.

FRONT: 52,40   REAR: 28,24,20,17,14

"First gear" is easiest to pedal. The chain is on the smaller front sprocket (40 teeth) and the biggest gear (28 teeth). This is the gear for climbing hills. "Tenth gear" is the hardest to pedal. The chain is on the larger front sprocket (52 teeth) and the smallest gear (14 teeth). This gear is used for fast cruising on the level or downhill.

The gear ratio indicates the degree of effort needed to use each gear. The program below will figure the gear ratio for each of the ten gear combinations. This is found by dividing the number of front sprocket teeth by the number of back gear teeth. The lower the gear ratio, the easier it is to pedal.

```
10   REM ***FINDING GEAR RATIOS
20   DATA 28,24,20,17,14
30   DATA 28,24,20,17,14
40   FOR G = 1 TO 10
50   READ B
60   REM ***FIGURE GEAR RATIOS
70   REM ***FOR 40 FRONT TEETH
80   IF G <= 5 THEN R = 40/B
90   REM ***FOR 52 FRONT TEETH
100  IF G > 5 THEN R = 52/B
110  PRINT "GEAR ";G;" HAS RATIO ";R
120  NEXT G
200  END
```

Notice the DATA and READ statements in lines 20, 30, and 50. This is a way to give the computer information it needs when the values are known ahead of time. Then the computer doesn't have to stop to get the values, as it does for INPUT statements.

EXERCISES

1. RUN the program. Complete the chart. Round gear ratios to the nearest hundredth.

Gear		1	2	3	4	5	6	7	8	9	10
Teeth	Front	40	40	40	40	40	52	52	52	52	52
	Rear	28	24	20	17	14	28	24	20	17	14
Gear ratio		?	?	?	?	?	?	?	?	?	?
		1.43	1.67	2.00	2.35	2.86	1.86	2.17	2.60	3.06	3.71

2. List the gears in order. Go from easiest (1st) to hardest (10th).   1, 2, 6, 3, 7, 4, 8, 5, 9, 10

3. Suppose your bike had 36 and 50 teeth on the two front sprockets. What would you have to change in the program? (Ignore REM statements.)
   80 IF G < = 5 THEN R = 36/B
   100 IF G > 5 THEN R = 50/B

4. Your tire is 27 inches in diameter.
   a. What is the diameter in feet?   2.25 ft
   b. What is the circumference? Use $\pi \approx 3.14$. (Round to the nearest tenth of a foot.) This is how far the bike goes when the wheel turns once.   7.1 ft

5. How many times does the wheel turn in a mile (5280 feet)? Divide 5280 by the answer to Exercise 4(b). (Round to the nearest whole number.)   744 turns per mile

Make the following changes in the program. Put the answer to Exercise 5 in the blank. The revised program finds the number of times you must turn the pedals to go a mile in each gear.

   105 T = __?__ / R
   115 PRINT "GEAR: ";G;"TURNS/MILE: ";INT(T)

6. RUN the program. Complete the chart.

Gear	1	2	3	4	5	6	7	8	9	10
Pedal turns per mile	?	?	?	?	?	?	?	?	?	?
	520	446	372	316	260	400	343	286	243	200

7. How many times do you turn the pedals to go
   a. five miles in 1st gear?   b. five miles in 10th gear?   1000 pedal turns
   c. ten miles in 5th gear?   d. ten miles in 8th gear?   2860 pedal turns
      a., c. 2600 pedal turns

SKILLS REVIEW

QUOTIENTS OF SQUARE ROOTS

Do you remember this rule? $\sqrt{\dfrac{a}{b}} = \dfrac{\sqrt{a}}{\sqrt{b}}$ and $\dfrac{\sqrt{a}}{\sqrt{b}} = \sqrt{\dfrac{a}{b}}$

Find the square root. Express the result in simplest form. Assume that any variable represents a number greater than zero.

SAMPLES

$\sqrt{\dfrac{100}{49}} = \dfrac{\sqrt{100}}{\sqrt{49}}$
$= \dfrac{10}{7}$ or $1\dfrac{3}{7}$

$\sqrt{\dfrac{5}{64}} = \dfrac{\sqrt{5}}{\sqrt{64}}$
$= \dfrac{\sqrt{5}}{8}$

$\sqrt{\dfrac{8n^2}{9}} = \dfrac{\sqrt{4 \cdot 2 \cdot n^2}}{\sqrt{9}}$
$= \dfrac{2n\sqrt{2}}{3}$

1. $\sqrt{\dfrac{1}{81}}$ $\dfrac{1}{9}$
2. $\sqrt{\dfrac{4}{49}}$ $\dfrac{2}{7}$
3. $\sqrt{\dfrac{121}{36}}$ $\dfrac{11}{6}$
4. $\sqrt{\dfrac{9}{25}}$ $\dfrac{3}{5}$
5. $\sqrt{\dfrac{169}{16}}$ $\dfrac{13}{4}$

6. $\sqrt{\dfrac{2}{49}}$ $\dfrac{\sqrt{2}}{7}$
7. $\sqrt{\dfrac{3}{100}}$ $\dfrac{\sqrt{3}}{10}$
8. $\sqrt{\dfrac{7}{36}}$ $\dfrac{\sqrt{7}}{6}$
9. $\sqrt{\dfrac{5}{16}}$ $\dfrac{\sqrt{5}}{4}$
10. $\sqrt{\dfrac{11}{64}}$ $\dfrac{\sqrt{11}}{8}$

11. $\sqrt{\dfrac{36x^2}{25}}$ $\dfrac{6x}{5}$
12. $\sqrt{\dfrac{4b^2}{81}}$ $\dfrac{2b}{9}$
13. $\sqrt{\dfrac{2c^2}{100}}$ $\dfrac{c\sqrt{2}}{10}$
14. $\sqrt{\dfrac{7a^2}{16}}$ $\dfrac{a\sqrt{7}}{4}$
15. $\sqrt{\dfrac{18n^2}{49}}$

16. $\sqrt{\dfrac{20k^2}{81}}$ $\dfrac{2k\sqrt{5}}{9}$
17. $\sqrt{\dfrac{75x^2}{36}}$ $\dfrac{5x\sqrt{3}}{6}$
18. $\sqrt{\dfrac{28a^2}{25}}$ $\dfrac{2a\sqrt{7}}{5}$
19. $\sqrt{\dfrac{45d^2}{144}}$ $\dfrac{d\sqrt{5}}{4}$
20. $\sqrt{\dfrac{32b^2}{49}}$

15. $\dfrac{3n\sqrt{2}}{7}$ $\dfrac{4b\sqrt{2}}{7}$

Express in simplest radical form.

SAMPLE

$\dfrac{\sqrt{2}}{\sqrt{5}} = \dfrac{\sqrt{2} \cdot \sqrt{5}}{\sqrt{5} \cdot \sqrt{5}}$
$= \dfrac{\sqrt{10}}{5}$ or $\dfrac{1}{5}\sqrt{10}$

← Multiply the numerator and denominator by a number that will make the denominator a whole number.

21. $\dfrac{\sqrt{3}}{\sqrt{7}}$ $\dfrac{\sqrt{21}}{7}$
22. $\dfrac{\sqrt{6}}{\sqrt{3}}$ $\sqrt{2}$
23. $\dfrac{\sqrt{5}}{\sqrt{8}}$ $\dfrac{\sqrt{10}}{4}$
24. $\dfrac{\sqrt{11}}{\sqrt{2}}$ $\dfrac{\sqrt{22}}{2}$
25. $\dfrac{\sqrt{7}}{\sqrt{5}}$ $\dfrac{\sqrt{35}}{5}$

26. $\sqrt{\dfrac{1}{8}}$ $\dfrac{\sqrt{2}}{4}$
27. $\sqrt{\dfrac{4}{11}}$ $\dfrac{2\sqrt{11}}{11}$
28. $\sqrt{\dfrac{16}{7}}$ $\dfrac{4\sqrt{7}}{7}$
29. $\sqrt{\dfrac{3}{2}}$ $\dfrac{\sqrt{6}}{2}$
30. $\sqrt{\dfrac{2}{5}}$ $\dfrac{\sqrt{10}}{5}$

31. $\dfrac{5\sqrt{6}}{\sqrt{3}}$ $5\sqrt{2}$
32. $\dfrac{3\sqrt{8}}{\sqrt{2}}$ 6
33. $\dfrac{6\sqrt{5}}{\sqrt{3}}$ $2\sqrt{15}$
34. $\dfrac{10\sqrt{2}}{\sqrt{5}}$ $2\sqrt{10}$
35. $\dfrac{4\sqrt{7}}{\sqrt{2}}$ $2\sqrt{14}$

CHAPTER REVIEW

CHAPTER SUMMARY

1. Many of the terms used with circles are discussed on page 346.

2. If line t is tangent to $\odot O$ at point A, then $\overline{OA} \perp t$.
   If line $t \perp \overline{OA}$ at point A, then t is tangent to $\odot O$.
   Construction 10 shows how to construct a tangent to a circle at a given point.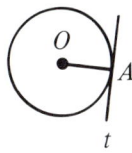

3. Given two circles, a common internal tangent intersects the segment joining the centers, and a common external tangent does not. Two circles may be externally tangent or internally tangent.

4. Arcs are classified on page 355.

5. In a circle, equal central angles have equal arcs, and equal minor arcs have equal central angles.

6. Given the figure at the right. When any one of these three statements is true, both of the others are also true.

   $QR = ST$
   $\overset{\frown}{QR} = \overset{\frown}{ST}$
   $OJ = OK$

   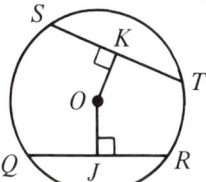

7. A diameter that is perpendicular to a chord bisects the chord and its arc.

8. A polygon is inscribed in a circle when each vertex lies on the circle. A polygon is circumscribed about a circle when each side is tangent to the circle.

9. A summary of angle and arc relationships is given on page 370.

10. If two chords intersect inside a circle, then the product of the lengths of the segments of one chord equals the product of the lengths of the segments of the other.

REVIEW EXERCISES

Refer to the figure and use letters to name the following.
(See pp. 346–349)

1. A radius $\overline{OB}$ or $\overline{OC}$
2. A diameter $\overline{CB}$
3. A secant $\overleftrightarrow{BC}$
4. A tangent $\overrightarrow{BE}$

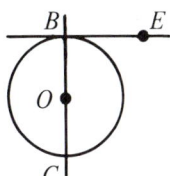

Refer to the diagram to complete each statement. *(See pp. 350–354.)*

5. The number of common external tangents that can be drawn to $\odot R$ and $\odot S$ is __?__. 2

6. The number of common internal tangents that can be drawn to $\odot R$ and $\odot S$ is __?__. 2

7. $\odot T$ and $\odot$ __?__ are externally tangent. R

8. $\odot T$ and $\odot$ __?__ are internally tangent. S

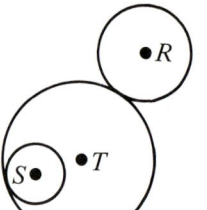

Write numerical values to complete the statements.
(See pp. 355–358.)

9. $\overarc{DAB} = $ __?__ ° 180

10. If $\angle 1 = 50°$, $\overarc{AB} = $ __?__ °. 50

11. If $\overarc{BC} = 48°$, $\angle 2 = $ __?__ °. 48

12. If $\overarc{AD} = 126°$, $\angle 1 = $ __?__ °. 54

13. If $\overarc{AB} = \overarc{BC}$ and $\angle 1 = 47°$, then $\angle 2 = $ __?__ °. 47

Write numerical values to complete the statements.
(See pp. 359–363.)

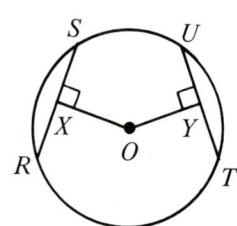

14. If $TU = 7$ and $\overarc{RS} = \overarc{TU}$, then $RS = $ __?__. 7

15. If $\overarc{RS} = 70°$ and $TU = RS$, then $\overarc{TU} = $ __?__ °. 70

16. If $RS = TU$ and $OX = 5$, then $OY = $ __?__. 5

17. If $OX = OY$ and $TU = 4$, then $RS = $ __?__. 4

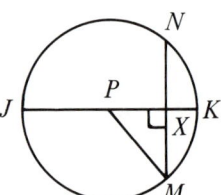

18. If $\overarc{MN} = 100°$, $\overarc{MK} = $ __?__ °. 50

19. If $\overarc{MK} = 48°$, $\overarc{JM} = $ __?__ °. 132

20. If $MN = 14$, $MX = $ __?__. 7

21. If $MN = 16$ and $PX = 6$, $MP = $ __?__. 10

Find the value of x. *(See pp. 364–376.)*

22.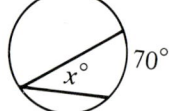
35

23.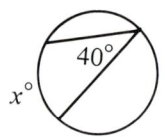
80

24.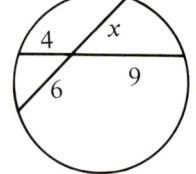
6

25.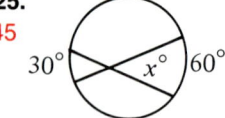
45

26.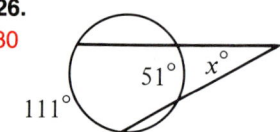
30

27.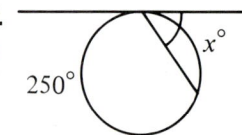
55

Chapter Test

1. Sketch a circle with center *P* and a diameter $\overline{AB}$.   (9-1)

2. Sketch a tangent to $\odot P$ at point *A*.

3. Sketch two circles so that the circles have exactly one common tangent. Are the circles internally or externally tangent?   **internally tangent**   (9-2)

4. Construct a tangent *t* to a $\odot O$ at any point *X* on the circle.

In $\odot O$, $\angle 1 = \angle 2 = \angle 3$. $\widehat{AB} = 120°$ or $\widehat{BC} = 60°$

5. Name a minor arc and its measure.   (9-3)

6. Name a major arc and its measure.   6. $\widehat{ACB} = 240°$ or $\widehat{BAC} = 300°$

7. If you draw $\overrightarrow{OD}$, which intersects $\odot O$ at point *X*, then $\underline{\ ?\ } = \underline{\ ?\ } = \underline{\ ?\ }$.
   $\widehat{AX} \quad \widehat{XB} \quad \widehat{BC}$

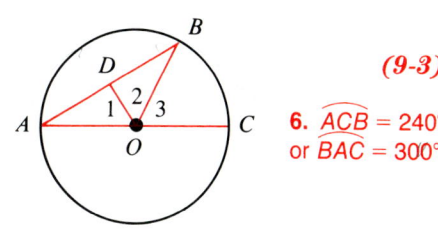

Find the value of *x*.

8. 15, 30, 8, 8, *x*   9. 55, *x*°, *r*°, *r*°, 70°   10. 8, 10, 4, *x*   (9-4)

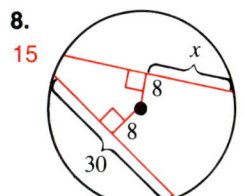 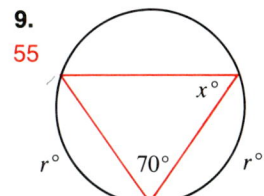 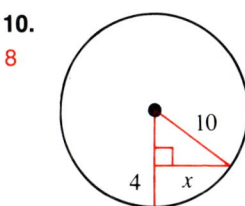

11. $\angle STR = \underline{\ ?\ }°$ 62   (9-5)

12. $\angle STQ = \underline{\ ?\ }°$ 47

13. Name an angle equal to $\angle RTP$. $\angle RST$

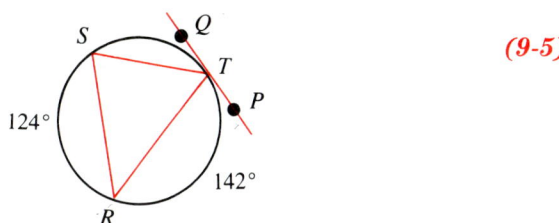

Find the value of *x*.

14. 122   15. 31   16. 122   (9-6)

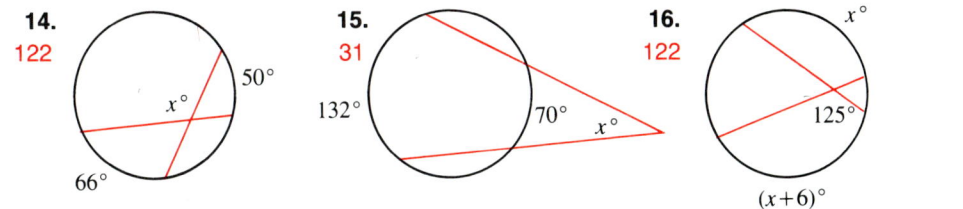

$\widehat{AB}$ and $\widehat{CD}$ are arcs of $\odot O$. $\overline{AB}$ and $\overline{CD}$ intersect at point *J*.

17. If $AJ = 9$, $BJ = 8$, and $CJ = 6$, then $CD = \underline{\ ?\ }$ 18   (9-7)

18. If $AJ = x$, $BJ = x + 5$, and $CJ = DJ = x + 2$, find the value of *x*.   4

Circles   **383**

MIXED REVIEW

1. An angle inscribed in a circle intercepts a semicircle. Is the angle acute, right, or obtuse? right

2. Supply the reason to support each statement about ⊙O.
   a. ∠AOB = ∠COD   **b.** $\widehat{AB} = \widehat{CD}$ **c.** AB = CD

   a. Vert. ⩞ are =.   **b.** In a ⊙, = central ⩞ have = minor arcs.   **c.** In a ⊙, = arcs have = chords.

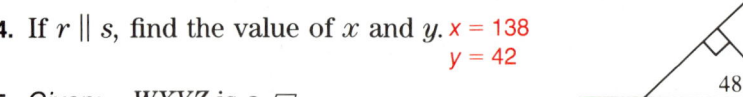

3. The measure of each interior angle of a regular polygon is 144°. How many sides does the polygon have? 10 sides

4. If r ∥ s, find the value of x and y. x = 138, y = 42

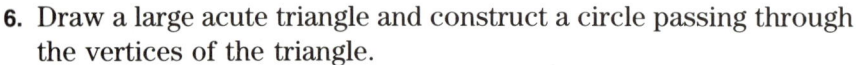

5. Given: WXYZ is a ▱.
   $\overline{WY}$ and $\overline{XZ}$ intersect at P.

   Prove: △WPX ≅ △YPZ

6. Draw a large acute triangle and construct a circle passing through the vertices of the triangle.

7. A triangle has sides of lengths 7, 12, and 14. Is it acute, right, or obtuse? obtuse

In the figure, FM = MG and FN = NH.

8. Why must $\overline{MN}$ and $\overline{GH}$ be parallel?

9. △MNF ~ △ ? GHF

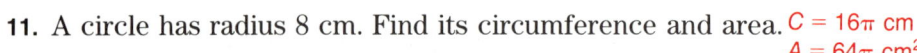

10. If △MNF has area 12, find the area of △FGH. 48

11. A circle has radius 8 cm. Find its circumference and area. C = 16π cm, A = 64π cm²

12. A rhombus has diagonals of 6 and 8. Find its perimeter. 20

Find the value of x.

13. 9   14. 13   15. 13½

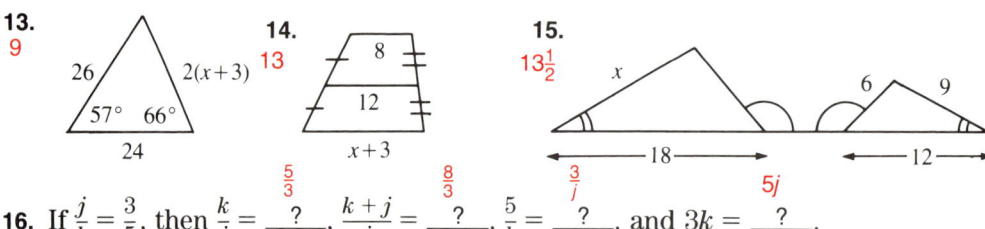

16. If $\frac{j}{k} = \frac{3}{5}$, then $\frac{k}{j} = $? $\frac{5}{3}$, $\frac{k+j}{j} = $? $\frac{8}{3}$, $\frac{5}{k} = $? $\frac{3}{j}$, and $3k = $? $5j$.

17. Construct a circle. Then construct a circumscribed triangle by constructing tangents to the circle at three appropriately chosen points on the circle.

18. The three __?__ of any triangle always meet inside the triangle.   angle bisectors, medians

Find the area of the triangle, parallelogram, and trapezoid shown.

19. 68.4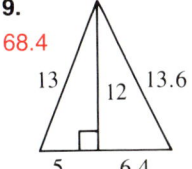

20. 180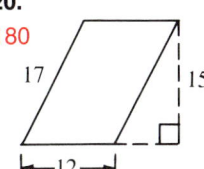

21. 119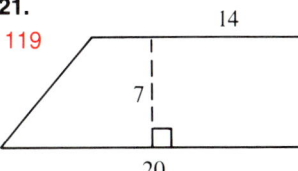

22. Find the measure of each numbered angle shown.   ∠1 = 103°; ∠2 = 63°; ∠3 = 108.5°

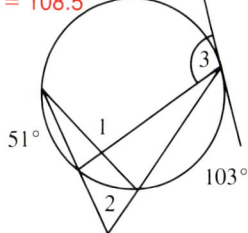

23. A bank account of $600 earned $3.50 for one month's interest. How much interest did an account of $420 earn? $2.45

24. Consider the statement "If a polygon is regular, then it is equilateral." Write the converse and tell if the converse is true.   If a polygon is equilateral, then it is regular; false.

25. Given: $\overline{RT}$ bisects ∠QRS.
    $\overline{TR}$ bisects ∠QTS.

    Prove: $QT = ST$

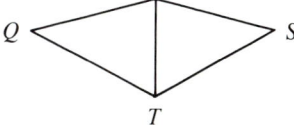

26. In △RST, RS = ST. If $RS = 3x - 1$, $RT = 4x - 6$, and $ST = 2x + 4$, find the numerical lengths of the sides.   RS = RT = ST = 14

27. How many tangent lines contain a given point inside a circle?   none

28. Could the information be used to prove that quadrilateral GRIN is a parallelogram?
    a. $RI = GN$ and $\overline{GR} \parallel \overline{NI}$ no   b. ∠G = ∠I and ∠R = ∠N.   yes
    c. $GR = NI$ and $RI = GN$ yes   d. Consecutive angles are supplementary   yes

Find the value of x.

29. 10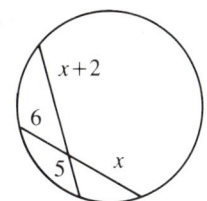

30. $9\frac{1}{3}$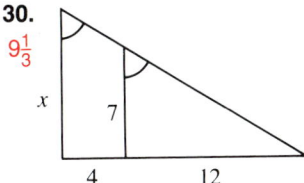

31. 5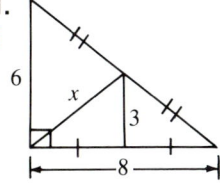

32. Draw a large scalene, obtuse triangle. Sketch and label the median and the altitude to the longest side.

Circles

Here's what you'll learn in this chapter:

To state the positions of points and lines with respect to each other.

To find the lateral area, total area, and volume of a right prism.

To find the lateral area, total area, and volume of a right circular cylinder.

To find the lateral area, total area, and volume of a regular pyramid.

To find the lateral area, total area, and volume of a right circular cone.

To find the area and volume of a sphere.

Both surface areas and volumes are important in the design and construction of buildings. Architects and contractors must make many accurate measurements. Surface area is also important to the window washers shown above.

Chapter 10

Areas and Volumes of Solids

1 Lines and Planes in Space

Look at lines *l* and *m* in the diagram. The lines do not meet, but they are not parallel. They are *skew* lines. **Skew lines** are two lines that do not lie in any one plane.

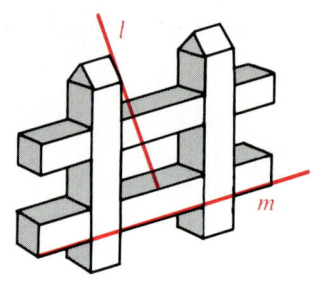

Possible positions of two lines:

  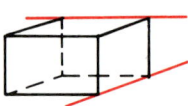

| Intersecting lines are coplanar. | Parallel lines are coplanar. | Skew lines are not coplanar. |

Look at line *t* and plane *P* at the right. The line and the plane are *parallel*. A line and a plane are **parallel** when they do not have any point in common. Any segment of the line is also said to be parallel to the plane.

Possible positions of a line and a plane:

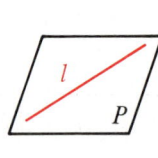 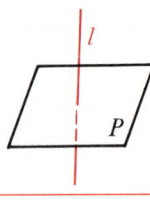 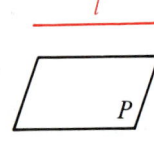

| *l* lies in *P*. *P* contains *l*. | *l* intersects *P*. *P* intersects *l*. | *l* and *P* are parallel. *l* ∥ *P* |

A line and a plane can intersect in a special way. Notice in the diagram that an axle is perpendicular to each spoke of a wagon wheel. The diagram suggests a line *perpendicular* to a plane.

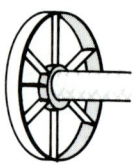

Line *l* is **perpendicular** to plane *P* when *l* is perpendicular to *every* line in *P* that passes through the point of intersection.

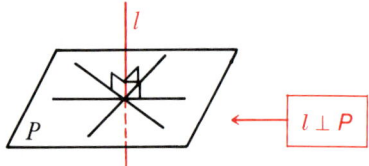

Suppose that $l \perp P$ and that plane M contains line l. Then plane M is said to be **perpendicular** to plane P. Perpendicular planes are one special kind of intersecting planes.

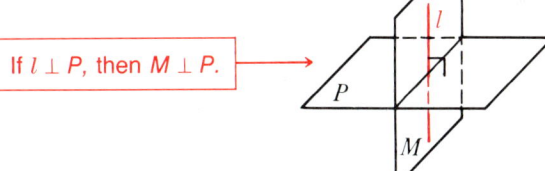

If $l \perp P$, then $M \perp P$.

Possible positions of two planes:

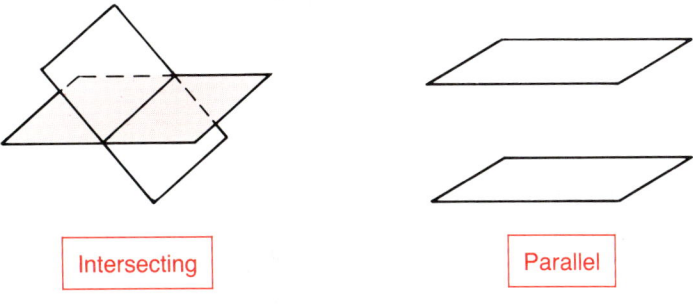

Intersecting   Parallel

10. a. the lines of intersection of the floor and ceiling with a wall compared with the intersection of that wall and an adjacent wall

Classroom Practice

Use your classroom and its furnishings to find examples that suggest each figure named. Examples:

1. Two lines that intersect *the corner of a room*
2. Two lines that are parallel *top and bottom edges of a wall*
3. Two lines that are skew *side edge of floor and top edge of front wall*
4. Two planes that intersect *wall and ceiling*
5. Two planes that are parallel *ceiling and floor*
6. Two planes that are perpendicular *floor and front wall*

7. A line and a plane such that the line
   a. lies in the plane
   b. intersects the plane and is perpendicular to the plane
   c. intersects the plane but is not perpendicular to the plane
   d. is parallel to the plane

   a. *a pencil on a desk*
   b. *the line of intersection of 2 walls and the floor*
   c. *a nonvertical chair leg and the floor*
   d. *the intersection of 2 walls and a third wall*

8. a. Four points that all lie in one plane *the bottoms of 4 chair legs*
   b. Four points that do not all lie in any one plane
      3 noncollinear points on a desk and a point on the wall
9. Two parallel planes that are intersected by a third plane
   the floor and ceiling and a wall
10. Two lines that are both perpendicular to a third line and
    a. are parallel to each other   b. are perpendicular to each other
       (See above.)   *the 3 lines of intersection in a corner of the room*
11. Alonzo said: "If two planes are perpendicular, every line in one of the planes must be perpendicular to the other plane."
    Explain why Alonzo was wrong. *Any 2 points determine a line, so some lines in one plane will be parallel to the second plane, while others will intersect the plane but not be perpendicular to it.*

Areas and Volumes of Solids

12. Lee wanted to set a flagpole upright, and said to a friend: "Stand over there and tell me when it's straight up." Could the friend tell for sure that the pole was upright? no

Written Exercises

Use a straightedge to draw each figure.

[A] 1. A line in a plane
2. A line intersecting a plane
3. A line parallel to a plane
4. A line perpendicular to a plane

A rectangular solid is shown. Name the figures described. In some cases, more than one answer is possible.

5. Three lines parallel to $\overleftrightarrow{AB}$ $\overleftrightarrow{PQ}, \overleftrightarrow{SR}, \overleftrightarrow{DC}$

6. Four lines perpendicular to $\overleftrightarrow{BQ}$ $\overleftrightarrow{PQ}, \overleftrightarrow{AB}, \overleftrightarrow{RQ}, \overleftrightarrow{BC}$

7. Four lines skew to $\overleftrightarrow{AD}$ $\overleftrightarrow{PQ}, \overleftrightarrow{SR}, \overleftrightarrow{QB}, \overleftrightarrow{RC}$

8. A plane parallel to plane $ADSP$ $BCRQ$

9. Four planes, each one perpendicular to plane $ADSP$ $PQRS, SDCR, ABCD, ABQP$

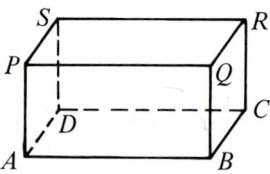

Exercises 5–14

10. Two lines that are perpendicular to $\overleftrightarrow{PS}$ and are parallel to each other $\overleftrightarrow{PA}$ and $\overleftrightarrow{SD}$ (or $\overleftrightarrow{SR}$ and $\overleftrightarrow{PQ}$)

11. Two lines that are perpendicular to $\overleftrightarrow{PS}$ and are skew to each other $\overleftrightarrow{PQ}$ and $\overleftrightarrow{SD}$ (or $\overleftrightarrow{SR}$ and $\overleftrightarrow{PA}$)

12. Two lines that are perpendicular to $\overleftrightarrow{PS}$ and are perpendicular to each other $\overleftrightarrow{PQ}$ and $\overleftrightarrow{PA}$ (or $\overleftrightarrow{SR}$ and $\overleftrightarrow{SD}$)

13. $\overleftrightarrow{BD}$ and $\overleftrightarrow{QS}$ (not drawn) are __?__ lines. parallel

14. $\overleftrightarrow{PA}$ and $\overleftrightarrow{AC}$ are __?__ lines. perpendicular

Quadrilateral *JKLM* is a square. The lengths of $\overline{VJ}, \overline{VK}, \overline{VL}$, and $\overline{VM}$ are equal. Tell whether each statement is true or false.

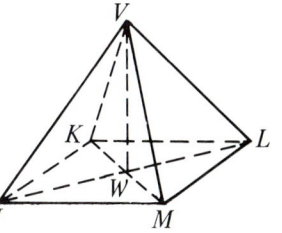

15. $JW = WL$ true
16. $JW = KW$ true
17. $\angle JKL > 90°$ false
18. $\angle VJM < \angle VMJ$ false
19. $\overleftrightarrow{JL}$ and $\overleftrightarrow{VK}$ are skew.   true
20. $\triangle MVL \cong \triangle WVL$ false
21. $\angle JWM > 90°$ false
22. $\triangle VKJ$ is a right $\triangle$.   false
23. $VM = VW$ false
24. $\overline{VW} \perp$ plane $JKLM$ true

Draw a figure to illustrate each statement. Some statements have more than one solution.

B 25. Two lines that are perpendicular to the same line *l* do not have to be parallel to each other.

26. Two lines that are skew to the same line *l* do not have to be skew to each other.

27. Line *j* lies in a plane *P* and line *k* lies in a plane *Q*, but lines *j* and *k* do not have to be skew lines.

28. Two planes both perpendicular to plane *X* do not have to be parallel to each other.

29. A line that is perpendicular to a line in a plane does not have to be perpendicular to the plane.

Line *b* and plane *Q* have the positions described. Tell how many planes can be drawn that contain line *b* and are, at the same time, perpendicular to plane *Q*.

30. *b* lies in *Q*.   one
31. $b \parallel Q$ one
32. $b \perp Q$ an infinite number
33. *b* intersects *Q*, but *b* is not $\perp$ to *Q*.   one

C 34. In the figure:
$j \parallel X$; *V* contains *j*;
V intersects *X* in *k*.
Explain why *j* and *k* must be parallel lines.

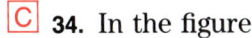

Since $j \parallel X$, *j* does not intersect any line in *X*.
j and *k* are coplanar and do not intersect since *k* is in *X*.
Hence, by definition, $j \parallel k$.

35. In the figure:
$P \parallel Q$; *R* intersects *P* in *l*, and *Q* in *m*.
Explain why *l* and *m* must be parallel lines.

No line in *P* intersects a line in *Q*, so *l* and *m* do not intersect. Since they are coplanar, $l \parallel m$.

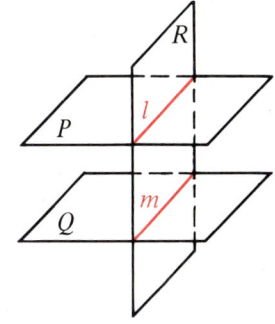

Experiments

To make an interesting surface like the one pictured at the right, take a long strip of paper as shown below. Give the strip a half-twist.

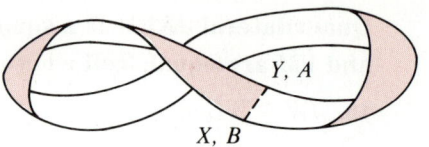

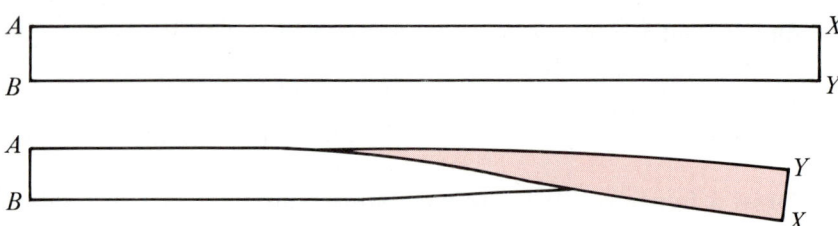

Tape the ends together, joining *X* and *B*, *Y* and *A*. The resulting figure is called a *Möbius band*.

1. Draw a pencil line down the middle. Do you return to your starting point? How many sides does the band have? **yes; one**
2. Cut along your pencil line. What happens to the band? **It is still a single strip.**
3. Take another long strip of paper. Give the paper a full twist and tape the ends together. Then cut the band down the middle, as before. What happens to the band? **It becomes 2 interlocked strips.**

CONSUMER APPLICATIONS

CALLING LONG DISTANCE

It's fun to make long distance calls to friends or family members, but a long distance call costs more than one across town. You can save on telephone bills by knowing how long distance rates vary at different times of the day. The chart below shows a sample rate structure for a long distance direct-dial call.

BUSINESS HOURS (8 A.M.–5 P.M.)		EVENINGS (5 P.M.–11 P.M.)		NIGHTS (11 P.M.–8 A.M.)	
First min	Each add. min	First min	Each add. min	First min	Each add. min
$0.56	$0.23	$0.36	$0.14	$0.22	$0.09

2 Right Prisms

The figure shows a *rectangular solid*. You see many rectangular solids every day, for example, bricks and cereal boxes.

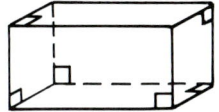

Some people have trouble drawing rectangular solids. One way to proceed is this:

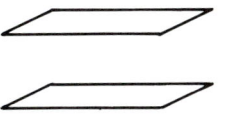

Draw a parallelogram.
Draw a second parallelogram directly below the first one.

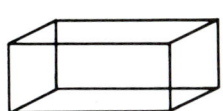

Draw four vertical segments.

When you look at a box, you cannot see some of the edges. Because of this fact, some people draw dashed lines to show hidden segments. Other people don't show the hidden edges at all.

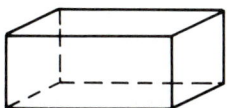   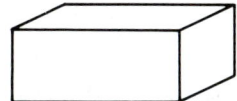

A rectangular solid is the most common kind of **right prism.** Two other right prisms are pictured below. The shaded *faces*, called *bases*, are congruent and parallel. The other faces of each prism are called *lateral faces*. Other parts of a prism are named as shown.

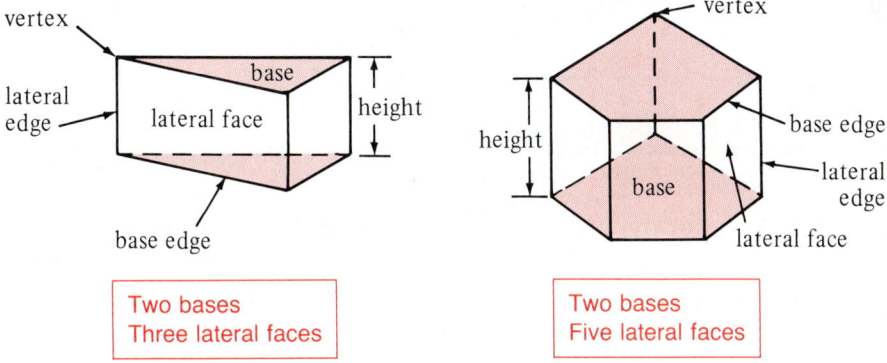

Two bases
Three lateral faces

Two bases
Five lateral faces

In a right prism, each lateral edge is perpendicular to the bases. Any segment, such as a lateral edge, that is perpendicular to the bases and has an endpoint on each base is called an *altitude* of the prism. The length of an altitude is the *height*, h, of the prism.

The **lateral area** (L.A.) of a prism is the sum of the areas of the lateral faces. Can you see that the lateral area of the prism shown is the sum of the areas of six rectangles?

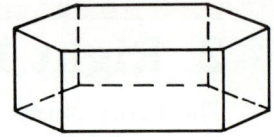

The **total area** (T.A.) of a prism is the sum of the areas of all the faces. Using B to represent the area of a base,

> *Total Area = Lateral Area + 2 × Area of a Base*
> T.A. = L.A. + 2B

EXAMPLE 1 The base of a right prism is a 5 by 3 rectangle. The height is 2. Find:
a. the lateral area;   **b.** the total area.

a. The area of the closest face is $5 \times 2 = 10$.
The area of an end face is $3 \times 2 = 6$.
L.A. = 10 + 6 + 10 + 6 = 32

b. T.A. = L.A. + 2B
T.A. = $32 + 2(5 \times 3) = 32 + 30 = 62$

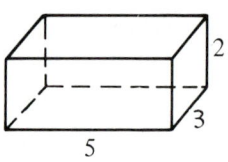

A prism has *volume* as well as area. When each edge of a rectangular solid is one unit long, the volume of the solid is one cubic unit. Can you see that the rectangular solid shown below can be filled with blocks that are 1 cm cubes?

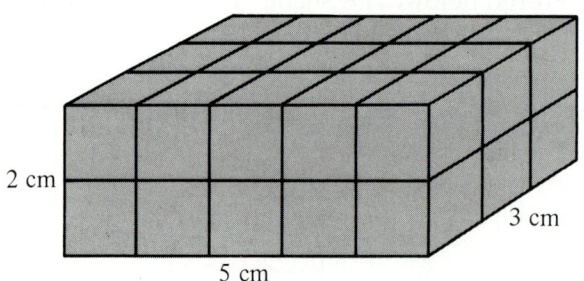

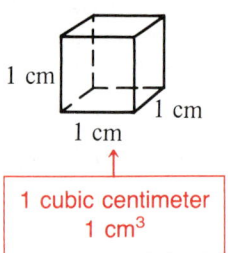

1 cubic centimeter
1 cm³

Fifteen blocks form a bottom layer, fifteen more a top layer. The box can be filled with 30 blocks, and the volume, V, is 30 cm³. Notice that the volume is the product of the base area, B, and the height, h:

$$V = 15 \times 2$$

> *In any right prism:*   *Volume = Base × height*
> $V = Bh$

394   Chapter 10

EXAMPLE 2 The base of a rectangular solid is a 6 cm square. The height of the solid is 10 cm. Find the volume.

$B = s^2 = 6^2 = 36$
$V = Bh$
$V = 36 \times 10 = 360$
Answer: 360 cm³

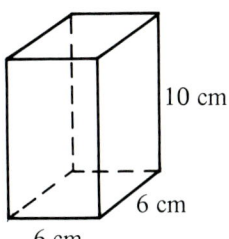

Classroom Practice

For each right prism, state the number of:

a. lateral edges   b. lateral faces   c. bases
d. base edges   e. faces   f. vertices

1.
a. 3
b. 3
c. 2
d. 6 e. 5 f. 6

2. a. 4 b. 4 c. 2
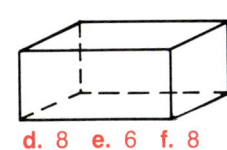
d. 8 e. 6 f. 8

3. a. 6 b. 6 c. 2
d. 12 e. 8 f. 12
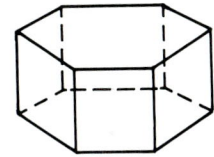

Tell why the solid pictured is not a right prism.

4.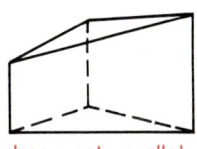
base not parallel

5.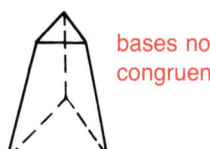
bases not congruent

6.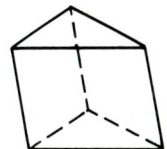
lateral edge not ⊥ to base

7. One student said that the diagram at the left shows a right prism, but the one at the right does not. Was the student correct? Explain. No; the prism is still a right prism when tipped on its side.

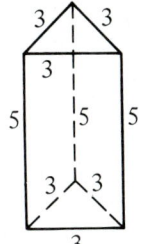
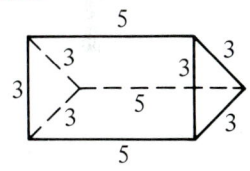

8. The bases of the right prism shown are isosceles trapezoids. Find the lateral area, the total area, and the volume of the prism. L.A. = 60; T.A. = 100; V = 60

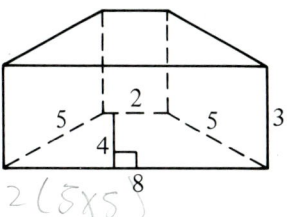

Tell whether each statement is true or false.

9. Every right prism has two bases. true

10. Every right prism has a rectangular base. false

11. Each base of a right prism must have at least four edges. false

Areas and Volumes of Solids

Tell whether each statement is true or false.

12. Every lateral face of a right prism must be a rectangle. true

13. The length of an altitude of a prism is called its height. true

14. In any right prism, T.A. = 2(L.A.) + B. false

Written Exercises

Exercises 1–10 refer to the rectangular solid.

1. $BE =$ __?__   4 cm **2.** $DG =$ __?__   4 cm

3. $DB =$ __?__   8 cm **4.** $AB =$ __?__   5 cm

5. The perimeter of quadrilateral $AFHC =$ __?__.   24 cm

6. The perimeter of a base = __?__.   18 cm

7. The area of a base = __?__.   20 cm^2

8. The lateral area of the prism = __?__.   144 cm^2

9. The total area of the prism = __?__.   184 cm^2 **10.** The volume of the prism = __?__.   160 cm^3

Exercises 11–14 deal with right prisms. The number of sides in the base is not specified. Complete the table.

	B	h	V	
11.	30 cm^2	4 cm	?	120 cm^3
12.	?	5 cm	100 cm^3	20 cm^2
13.	?	2 cm	100 cm^3	50 cm^2
14.	64 cm^2	?	256 cm^3	4 cm

Exercises 15–18 deal with a rectangular solid having dimensions l, w, and h. Draw a figure if you need one. Complete the table.

	l	w	h	L.A.	T.A.	V	
15.	6	2	3	?	?	?	48; 72; 36
16.	8	3	4	?	?	?	88; 136; 96
17.	10	4	?	?	?	120	3; 84; 164
18.	?	6	6	?	?	540	15; 252; 432

Exercises 19–22 deal with *cubes*. A cube is a rectangular solid with all edges equal. Complete the table.

	Edge	B	L.A.	T.A.	V
19.	2	? 4	? 16	? 24	? 8
20.	? 4	16	? 64	? 96	? 64
21.	5	? 25	? 100	? 150	? 125
22.	? 10	? 100	? 400	? 600	1000

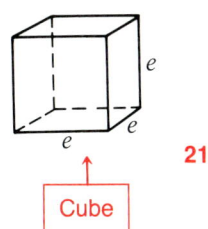

Cube

21. L.A. = 100;
T.A. = 150;
V = 125

23. Examine your answers for Exercises 19 and 20 as a pair. Then look at those for Exercises 21 and 22 as a pair. These exercises suggest that when each edge of a cube is doubled:
   a. the area is multiplied by __?__; 4
   b. the volume is multiplied by __?__. 8

C 24. One cube has edges 10 cm long. Another cube has edges 30 cm long. Find the ratio of the:
   a. total area of the smaller cube to that of the larger cube; $\frac{1}{9}$
   b. volume of the smaller cube to that of the larger cube. $\frac{1}{27}$

25. A rough-cut piece of lumber has a 2-inch by 4-inch end. Before being sold, the end is planed down to $1\frac{1}{2}$ inches by $3\frac{1}{2}$ inches. What percent of the wood is lost? 34.4%

26. a. How many square centimeters of aluminum foil are needed to cover the block of cheese pictured? (Ignore overlap.) 2560 cm²
   b. Suppose the block is cut into cubes with 2 cm edges. How many square centimeters of foil are needed to wrap all the individual cubes? 24,576 cm²

27. How many liters of water are needed to fill the swimming pool shown to within 0.2 m of the top edge? (*Hints:* Think of PQRST as the base of a right prism. 1 m³ contains 1000 L.) 88,750 L

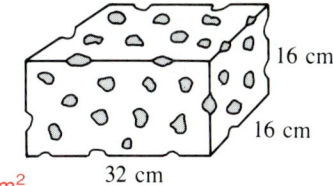

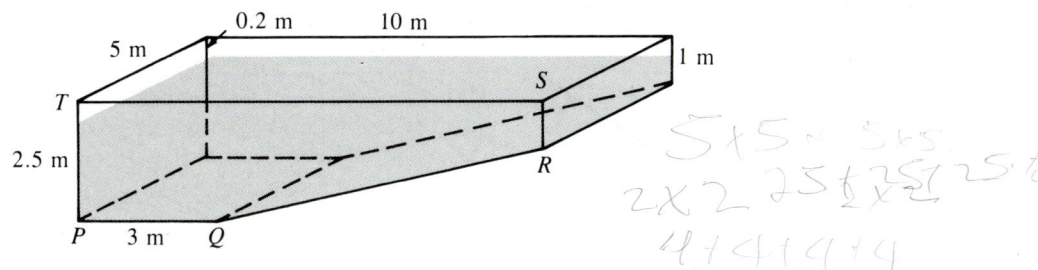

Areas and Volumes of Solids   397

3 Right Circular Cylinders

A **right circular cylinder** is like a right prism except that its bases are congruent circles instead of congruent polygons. The *radius* of a base is also called the radius of the cylinder.

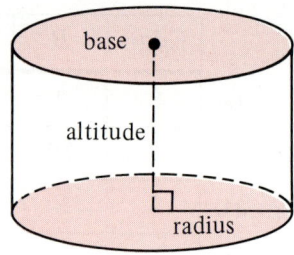

Right Circular Cylinder

As in right prisms, an *altitude* of a right circular cylinder is a segment that is perpendicular to the bases and has an endpoint in each base. The length of an altitude is the *height*, h, of the cylinder. We'll use the word *cylinder* to mean a right circular cylinder.

To find the lateral area of a cylinder, imagine that the diagram shows a tin can with its ends removed. If you cut along $\overline{AB}$, you can unroll the metal and lay it out flat.

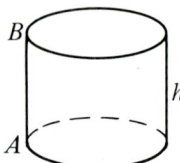 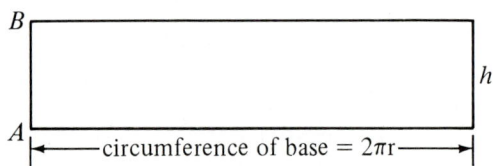

The lateral surface of the cylinder becomes a rectangular region.

$$\begin{aligned} \text{L.A. of cylinder} &= \text{area of rectangle} \\ &= 2\pi r \times h \\ &= 2\pi rh \end{aligned}$$

The formula for the total area of a cylinder is much like the formula for the total area of a right prism:

$$\text{T.A.} = \text{L.A.} + 2B$$
$$\text{T.A.} = 2\pi rh + 2 \times \pi r^2$$

You can find the volume of a cylinder in a similar way:

$$V = Bh$$
$$V = \pi r^2 \times h$$

In any (right circular) cylinder:

$$\text{L.A.} = 2\pi rh \qquad \begin{aligned}\text{T.A.} &= \text{L.A.} + 2B \\ &= 2\pi rh + 2\pi r^2\end{aligned} \qquad \begin{aligned}V &= Bh \\ &= \pi r^2 h\end{aligned}$$

EXAMPLE A cylinder has a radius of 3 cm and a height of 4 cm. Find:
a. the lateral area   b. the total area   c. the volume

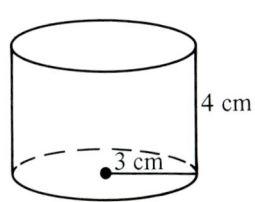

a. L.A. $= 2\pi rh = 2\pi \times 3 \times 4 = 24\pi$ cm^2
b. T.A. $=$ L.A. $+ 2B$
   $= 24\pi + 2\pi r^2$
   $= 24\pi + (2\pi \times 3^2)$
   $= 24\pi + 18\pi = 42\pi$ cm^2
c. $V = \pi r^2 h = \pi \times 3^2 \times 4 = 36\pi$ cm^3

Classroom Practice

1. When you use the formula T.A. = L.A. + 2B, what should you substitute for L.A.? (Use the figure shown.) 2πrh

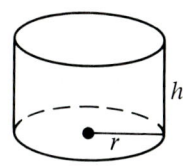

2. When you use the formula $V = Bh$, what formula do you use to find B? (Use the figure shown.) Area of a ⊙ = πr²

3. The lateral area of a new piece of chalk is much greater than the area of a base. Name an object, shaped like a cylinder, in which:
   a. the area of a base is much greater than the lateral area; example: a coin
   b. the area of a base is roughly equal to the lateral area.
      any object where r = h, such as a can of tuna

4. Why is it incorrect to call the solid shown a cylinder?
   The bases are not congruent.

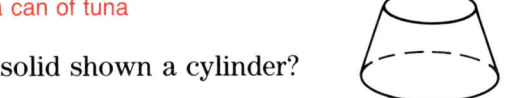

5. Why is it incorrect to call the solid shown a right cylinder?
   The sides are not ⊥ to the bases.

6. For the cylinder shown, find:
   a. the lateral area; 12π
   b. the total area; 30π
   c. the volume. 18π

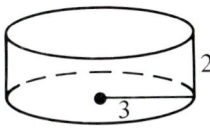

Written Exercises

Find the indicated values for the cylinder shown.

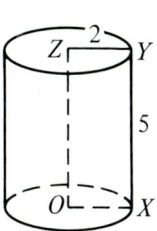

 A 1. OX 2
2. OZ 5
3. Height of cylinder 5
4. Circumference of base 4π
5. Area of base 4π
6. Lateral area 20π
7. Total area 28π
8. Volume 20π

Areas and Volumes of Solids

Complete the table, which refers to cylinders.
C represents the circumference of a base.

	r	h	C	B	L.A.	T.A.	V	
9.	5	2	?10π	?25π	?20π	?70π	?50π	
10.	8	3	?16π	?64π	?48π	?	?	176π; 192π
11.	1	6	?2π	?π	?12π	?14π	?6π	
12.	6	1	?12π	?36π	?12π	?84π	?36π	
B 13.	?3	2	6π	?9π	?12π	?30π	?18π	
14.	?6	4	?12π	36π	?48π	?	?	120π; 144π
15.	3	?4	?6π	?9π	?24π	?42π	36π	
16.	?2	5	?4π	?4π	?20π	?28π	20π	

Use 3.14 for π. Write your answers correct to the nearest integer.

17. In a cylinder: $d = 6$; $h = 8$. $V = $ __?__   226

18. In a cylinder: $r = 10$; $h = 3$. T.A. = __?__   816

19. One can of nuts is twice as tall as another, but only half as wide. Which can holds more nuts? the shorter can

20. If the can on the left costs $1.09 and the one on the right costs $2.19, which is the better buy? the can on the left

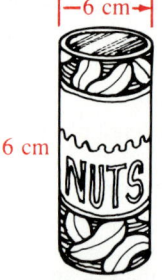

C 21. A chemical company has developed an additive that cuts down on friction, permitting water to flow through a hose twice as fast. Using this additive, a fire department can replace 7 cm hose with 5 cm hose, handle a hose considerably lighter, yet deliver the same amount of water.

Show that the amount of water in a 7 cm hose is about twice the amount in a 5 cm hose. (Use the same length for each hose.)
The ratio is 49 : 25, or approximately 2.

SELF-TEST

Vocabulary

skew lines (p. 388)
parallel planes (p. 388)
perpendicular planes (p. 388)
rectangular solid (p. 393)

right prism (p. 393)
lateral area (p. 394)
right circular cylinder (p. 398)

Refer to the rectangular solid shown.

1. Name two planes that are parallel to $\overline{CR}$. planes *TUFE, TEDS*   (10-1)

2. Name two planes that are perpendicular to $\overline{DE}$. planes *CDSR, FETU*

3. Name any line shown that is skew to $\overleftrightarrow{UT}$.
   $\overleftrightarrow{CF}, \overleftrightarrow{DE}, \overleftrightarrow{CR}, \overleftrightarrow{DS}$

In the rectangular solid shown, let $CD = 8$, $DE = 5$, and $DS = 4$.

4. Total area = __?__ 184   (10-2)

5. Volume = __?__ 160

**In a certain cylinder, $r = 8$ and $h = 4$.
Find the indicated values in terms of π.**

6. L.A. = __?__ 64π (10-3)

7. T.A. = __?__ 192π

8. $V =$ __?__ 256π

PUZZLE ◆ PROBLEMS

The faces of a cube are marked with these six symbols:

Three views of the cube are shown below.
Which symbols are on opposite sides of the cube?

Areas and Volumes of Solids

4 Regular Pyramids

The ancient Egyptians built *pyramids* that still stand. Because the bases are squares, the pyramids are called *square* pyramids. We say that a pyramid is *regular* if its base is a regular polygon and the top lies directly over the center of the base.

The table and diagram below explain some of the terms used for regular pyramids.

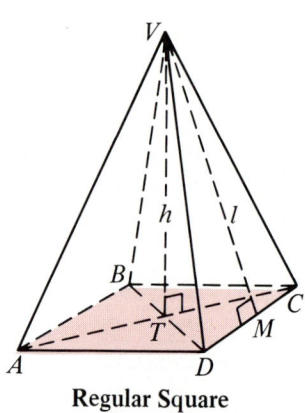

Regular Square Pyramid

vertices	points A, B, C, D, and V
the vertex	point V
the base	square $ABCD$
lateral edges	$\overline{VA}$, $\overline{VB}$, $\overline{VC}$, and $\overline{VD}$
base edges	$\overline{AB}$, $\overline{BC}$, $\overline{CD}$, and $\overline{DA}$
faces	$\triangle VDC$ is one of four congruent lateral faces. Base $ABCD$ is also a face.
center of base	point T
altitude	$\overline{VT}$ (Note: $\overline{VT} \perp$ plane $ABCD$)
height	VT, also h, the length of $\overline{VT}$
slant height	$\overline{VM}$, also l, the length of $\overline{VM}$ (Note: $\overline{VM} \perp \overline{CD}$; M is the midpoint of $\overline{CD}$.)

The lateral faces of a pyramid are always triangular. To find the lateral area of a pyramid, use the formula for the area of a triangle. To find the total area, you simply add the lateral area and the area of the base. To find the volume, use the formula that we state without proof.

In any regular pyramid: T.A. = L.A. + B $V = \frac{1}{3}Bh$

EXAMPLE The base of a regular square pyramid has edges of length 6. The height of the pyramid equals 4 and the slant height equals 5. Find: **a.** the lateral area; **b.** the total area; **c.** the volume.

Draw a diagram.

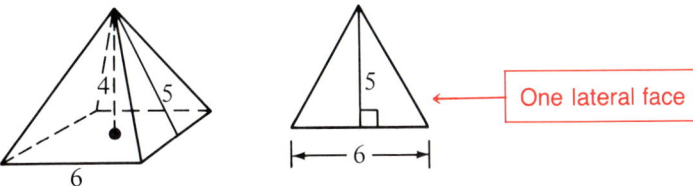

One lateral face

a. The area of one lateral face $= \frac{1}{2} \times 6 \times 5 = 15$
L.A. $= 4 \times 15 = 60$

b. $B = 6^2 = 36$
T.A. $=$ L.A. $+ B = 60 + 36 = 96$

c. $V = \frac{1}{3} Bh = \frac{1}{3} \times 36 \times 4 = 48$

Classroom Practice

Use the diagram of a regular triangular pyramid to name:

1. A lateral edge   2. The vertex V
3. The altitude $\overline{VW}$ 4. The base $\triangle XYZ$
5. A slant height $\overline{VM}$ 6. A lateral face
7. Name all the edges that equal $\overline{VY}$.
8. Which is longer: $\overline{VW}$ or $\overline{VM}$? VM
9. Which is longer: $\overline{VM}$ or $\overline{VX}$? VX
10. Name the hypotenuse of right $\triangle YVM$. $\overline{VY}$
11. Does $\triangle VYZ$ have to be isosceles? equilateral? no yes
12. $\angle XYZ = \underline{\ ?\ }^\circ$ $\angle XWY = \underline{\ ?\ }^\circ$ $\angle WXY = \underline{\ ?\ }^\circ$
      60   120   30

1. $\overline{VX}, \overline{VY},$ or $\overline{VZ}$
6. $\triangle VXY, \triangle VYZ,$ or $\triangle VXZ$
7. $\overline{VX}, \overline{VZ}$

Exercises 1–12

Base of pyramid

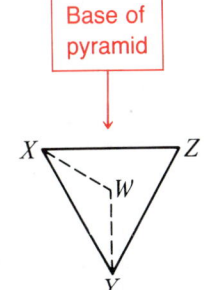

Exercise 12

The diagram shows a regular square pyramid.

13. Find the lateral area. 320 square units
14. Find the total area. 576 square units
15. Find the volume. 512 cubic units

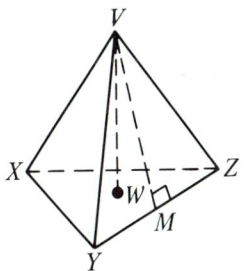

Areas and Volumes of Solids   **403**

Written Exercises

**Refer to the regular square pyramid shown.
Tell whether each statement is true or false.**

  1. The pyramid has five faces. true

2. All eight edges of the pyramid must be equal. false

3. VM is the height of the pyramid. false

4. $KM = \frac{1}{2} CF$ true   5. $KM = ME$ true

6. If $\overline{KC}$ were drawn, $\angle VKC$ would be a right angle. true

7. L.A. = 4 × area of quadrilateral $CDEF$ false   8. T.A. = L.A. + 2B false

9. VM must be greater than VK. true   10. VM must be greater than KM. true

Draw a regular pyramid whose base is the polygon named.

11. Triangle   12. Square   13. Pentagon   14. Hexagon

Exercises 15–20 refer to a regular pyramid whose base is an 8 by 8 square. The height equals 3 and the slant height equals 5.

15. Draw a figure.   16. Perimeter of the base = __?__ 32

17. Area of the base = __?__ 64   18. Lateral area of the pyramid = __?__ 80

19. Total area of the pyramid = __?__ 144   20. Volume of the pyramid = __?__ 64

**In Exercises 21 and 22, a regular pyramid is described.
Find the lateral area, the total area, and the volume.**

  21. The base is a 12 cm by 12 cm square. The height equals 8 cm.
The slant height equals 10 cm. L.A. = 240 cm²; T.A. = 384 cm²; V = 384 cm³

22. The base is a 16 cm by 16 cm square. The height equals 15 cm.
The slant height equals 17 cm. L.A. = 544 cm²; T.A. = 800 cm²; V = 1280 cm³

23. The base of a regular pyramid is a hexagon with sides 4 m long.
The slant height equals 10 m. Find the lateral area. 120 m²

24. In a certain regular pyramid: $B = 52$; $h = 6$; $l = 7$. Decide which you
can find, the lateral area or the volume. Then find it. V = 104

25. A cube has a 6 cm edge. A pyramid is formed by joining the vertices of one face to X, the center of the opposite face. Draw a figure. Find the volume of the pyramid. 72 cm³

26. Repeat Exercise 25, using Y, the center of the cube, instead of X. 36 cm³

Experiments

The purpose of this experiment is to find a relation between the number of vertices, the number of faces, and the number of edges that a solid has. Divide the class into ten committees. Each committee can work with one of the solids below.

1. prism with triangular base
2. pyramid with triangular base
3. prism with square base
4. pyramid with square base
5. prism with hexagonal base
6. pyramid with hexagonal base

7. 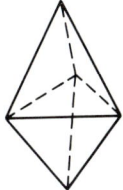 8. 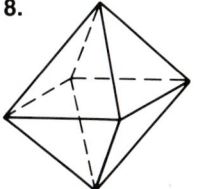 9. 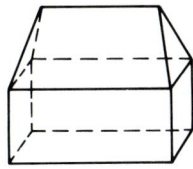 10.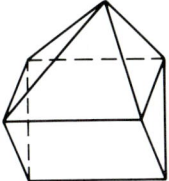

For each figure, count:

V, the number of vertices;
F, the number of faces;
E, the number of edges.

Copy and complete the table at the chalkboard.

	1.	2.	3.	4.	5. $V = 12$	6.	7.	8.	9. $V = 10$	10.
V	$?_6$	$?_4$	$?_8$	$?_5$	?	$?_7$	$?_5$	$?_6$	?	$?_7$
F	$?_5$	$?_4$	$?_6$	$?_5$	$?_8$	$?_7$	$?_6$	$?_8$	$?_9$	$?_8$
E	$?_9$	$?_6$	?	$?_8$	?	?	$?_9$	?	?	?
			12		18	12		12	17	13

Study the columns.
Find a formula that tells how V, F, and E are related.

$E = V + F - 2$

PUZZLE ◆ PROBLEMS

Sam stared at a cactus in the Painted Desert until he lost his memory. The cactus is 0.5 m high and its shadow is 1 m long. Sam's shadow is 3.5 m long. Can you tell him how tall he is? 1.75 m

5 Right Circular Cones

A **right circular cone** is very much like a regular pyramid. Notice that the base of a cone is circular. As the diagram shows, we use many of the same terms to describe a cone as we use for other solids.

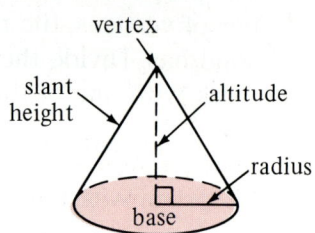

Right Circular Cone

The formula for the volume of a pyramid suggests the formula for the volume of a cone:

$$V = \tfrac{1}{3} Bh; \quad V = \tfrac{1}{3} \times \pi r^2 \times h$$

The formula, stated below, for the lateral area of a cone will be verified in the exercises.

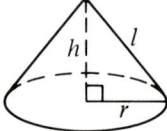

In any right circular cone: L.A. $= \pi r l$

$$\text{T.A.} = \text{L.A.} + B \qquad V = \tfrac{1}{3} Bh$$
$$= \pi r l + \pi r^2 \qquad = \tfrac{1}{3}\pi r^2 h$$

In the remainder of the course, the word *cone* will mean right circular cone.

EXAMPLE   For the cone pictured, find:
a. the lateral area   **b.** the total area   **c.** the volume

a. L.A. $= \pi r l$
$= \pi \times 6 \times 10 = 60\pi$

b. T.A. $= $ L.A. $+ B$
$= 60\pi + \pi r^2$
$= 60\pi + (\pi \times 6^2)$
$= 60\pi + 36\pi = 96\pi$

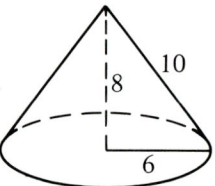

c. $V = \tfrac{1}{3} Bh$
$= \tfrac{1}{3} \times \pi r^2 \times h$
$= \tfrac{1}{3} \times 36\pi \times 8 = 96\pi$

Classroom Practice

1. In the example on page 406, the number of cubic units in the volume is equal to the number of square units in the total area. Do you think that this is true for all cones? Find the lateral area, the total area, and the volume for a cone with $r = 8$, $h = 6$, and $l = 10$.
   No; L.A. = 80π; T.A. = 144π; V = 128π

Draw a circle whose radius is actually 6 cm.
Draw two radii that form a 120° angle.

2. The area of the shaded region is what fractional part of the area of the circle? $\frac{1}{3}$

3. Find the area of the shaded region. 12π cm²

4. The length of $\overarc{AB}$ is what fractional part of the circumference of the circle? $\frac{1}{3}$

5. Find the length of $\overarc{AB}$. 4π cm

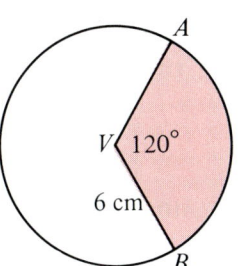

Cut out the shaded region. Curve the paper to form a cone, letting $\overline{VA}$ and $\overline{VB}$ come together.

6. The slant height of the cone equals __?__. 6 cm

7. The lateral area of the cone equals __?__. 12π cm²
   (Answer this the easy way. See Exercise 3.)

8. The circumference of the base of the cone equals __?__. 4π cm
   (Again, the answer is easy. See Exercise 5.)

9. Use your answer for Exercise 8 to find the radius of the base of the cone. 2

10. Show that your answers for Exercises 6, 7, and 9 satisfy the formula L.A. = $\pi r l$. L.A. = $\pi r \ell = \pi \cdot 2 \cdot 6 = 12\pi$, which agrees with the answer for Exercise 7.

Written Exercises

Find the lateral area of each cone.

A 1. 54π 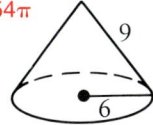 2. 16π 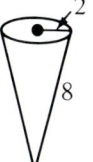 3. 30π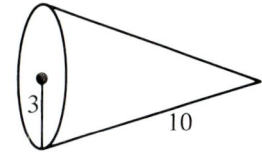

4–6. Find the total areas of the cones shown in Exercises 1–3.
4. 90π 5. 20π 6. 39π

Areas and Volumes of Solids

Find the volume of each cone.

7. 180π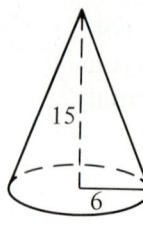

8. 196π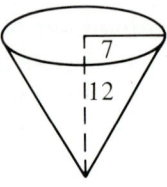

9. 48π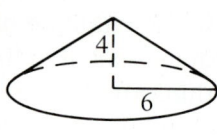

Copy and complete the table, which deals with cones. (*C* represents the circumference of the base.)

	r	*C*	*B*	*h*	*l*	L.A.	T.A.	*V*
10.	3	? 6π	? 9π	4	5	? 15π	? 24π	? 12π
11.	12	? 24π	? 144π	9	15	? 180π	? 324π	? 432π
12.	8	? 16π	? 64π	15	17	? 136π	? 200π	? 320π
13.	? 6	12π	? 36π	8	10	? 60π	? 96π	? 96π
14.	? 4	? 8π	16π	3	5	? 20π	? 36π	? 16π
15.	5	? 10π	? 25π	? 12	? 13	? 65π	? 90π	100π

16. A piece of sheet metal is cut and formed into the lateral surface of the cone pictured. There are 0.24 g in 1 cm² of the metal. Find the total number of grams of metal. (Use 3.14 for π.) 37.68 g

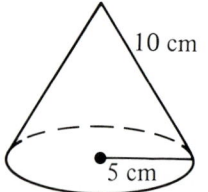

17. The solid hardwood cone pictured has a 16 cm radius and a 35 cm altitude. There are 0.62 g of wood in 1 cm³ of the hardwood used. Find the total number of grams of wood, to the nearest 10 g. (Use 3.14 for π.) 5810 g

Refer to the diagram below of a cone and a regular square pyramid.

18. Find the ratio of the lateral areas of the solids. $\frac{\pi}{4}$

19. Find the ratio of the total areas of the solids. $\frac{\pi}{4}$

20. Find the ratio of the volumes of the solids. $\frac{\pi}{4}$

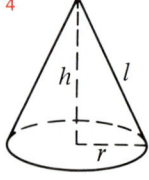 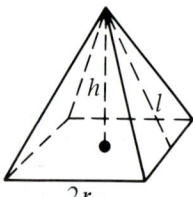

Experiments

For this experiment you need a small piece of cardboard with a bright, solid-colored surface. Close your left eye. Gaze straight ahead. Hold the card in your right hand with your right arm stretched out to the side. The color of the card cannot be accurately judged. Slowly move your unbent arm forward. At a certain stage you begin to see the color clearly. Why?

The back of the retina of your eye is lined with *rods* and *cones*. The rods—long, thin cylinders—register only in black and white. An object that you view from the side is seen primarily by the rods. The cones, which are able to distinguish color, are concentrated near the center of your eye.

CALCULATOR ACTIVITIES

Solid metal cones are to be made of various metals. How many cones of the size specified can be made from 10 kg (10,000 g) of the metal named? Round your answer to the nearest 10 cones.

EXAMPLE

Bronze cones: $r = 1.5$ cm; $h = 4$ cm.
There are 8.8 g in 1 cm³ of bronze.

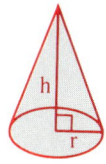

$$\text{Number of cones} = \left(\frac{\text{total number of}}{\text{grams of metal}}\right) \div \left(\frac{\text{total number of}}{\text{grams in one cone}}\right)$$

$$= 10{,}000 \div \left(\frac{1}{3}\pi r^2 h \times 8.8\right)$$

$$= 10{,}000 \div \left(\frac{1}{3}\pi \times 1.5^2 \times 4 \times 8.8\right) \approx 120$$

Copy and complete the table.

		r (in cm)	h (in cm)	Grams in 1 cm³	No. of cones
1.	Aluminum	2	3	2.7	? 290
2.	Brass	2.6	4	8.6	? 40
3.	Iron	5.2	7.6	7.9	? 10
4.	Magnesium	3.2	4.1	1.7	? 130
5.	Nickel	3	8	8.9	? 10

6 Spheres

Look at the circle and the sphere. Each has radius r.

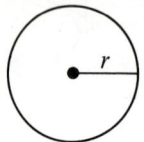 

Recall that the area of the circle equals πr^2. It happens that the surface area of the sphere is four times the area of the circle, or $4\pi r^2$.

The formula for the volume of a sphere is stated without proof.

In any sphere: $\quad A = 4\pi r^2 \quad\quad V = \frac{4}{3}\pi r^3$

EXAMPLE The diameter of a sphere is 8.
a. Find the area and the volume in terms of π.
b. Find approximations for the area and the volume. Round your answers to the nearest integer.

Since $d = 8$, $r = 4$.

a. $A = 4\pi r^2 \quad\quad\quad V = \frac{4}{3}\pi r^3$

$ = 4\pi \times 4^2 \quad\quad = \frac{4}{3}\pi \times 4^3$

$ = 64\pi \quad\quad\quad\quad = \frac{256\pi}{3}$

The area is 64π, and the volume is $\frac{256\pi}{3}$.

b. $A = 64\pi \quad\quad\quad\quad\quad V = \frac{356\pi}{3}$

$A \approx 64 \times 3.14 = 200.96 \quad\quad V \approx \frac{256 \times 3.14}{3} = 267.95$

$A \approx 201 \quad\quad\quad\quad\quad\quad\quad V \approx 268$

To the nearest integer, the area is 201, and the volume is 268.

Classroom Practice

Find the area and the volume of a sphere with the given radius. Express your answers in terms of π. (See below.)

1. $r = 2$ cm
2. $r = 5$ cm
3. $r = 0.1$ cm
4. $r = \frac{1}{4}$

5. If the area of a sphere is 36π, what is its radius? 3

6. "Half" of a sphere is called a *hemisphere*. Find the area and the volume of the hemisphere shown. $A = 128\pi$; $V = 341\frac{1}{3}\pi$

1. $A = 16\pi$; $V = \frac{32}{3}\pi$
2. $A = 100\pi$; $V = \frac{500}{3}\pi$
3. $A = 0.04\pi$; $V = \frac{0.004}{3}\pi$
4. $A = \frac{\pi}{4}$; $V = \frac{1}{48}\pi$

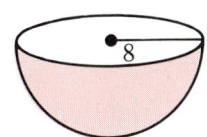

Written Exercises

Find the area and the volume of a sphere with the given radius. Express your answers in terms of π.

[A]
1. $r = 6$
   $A = 144\pi$
   $V = 288\pi$
2. $r = 12$
   $A = 576\pi$
   $V = 2304\pi$
3. $r = 8$ $A = 256\pi$
   $V = \frac{2048}{3}\pi$
4. $r = \frac{1}{2}$ $A = \pi$
   $V = \frac{\pi}{6}$

Find the area and the volume of a sphere with the given diameter. Use 3.14 for π. Round your answers to the nearest integer.

5. $d = 18$
   $A \approx 1017$
   $V \approx 3052$
6. $d = 2$ m
   $A \approx 13$ m²
   $V \approx 4$ m³
7. $d = 12$ cm
   $A \approx 452$ cm²
   $V \approx 904$ cm³
8. $d = 0.6$
   $A \approx 1$
   $V \approx 0.1$
   (Accept $V = 0$, to the nearest integer.)

Copy and complete the table about spheres. Write your answers in terms of π.

[B]

	9.	10.	11.	12.	13.	14.	15.	16.
r	$\frac{1}{8}$	?$\frac{3}{8}$	?5	?6	$\frac{3}{2}$	?6	?7	?$\sqrt{3}$
d	?$\frac{1}{4}$	$\frac{3}{4}$	?10	?12	?3	?12	?14	$2\sqrt{3}$
A	?$\frac{1}{16}\pi$	?$\frac{9}{16}\pi$	100π	?144π	?9π	144π	?196π	?12π
V	?	?	?	288π	?	?	$\frac{1372\pi}{3}$	?
	$\frac{1}{384}\pi$	$\frac{9}{128}\pi$	$\frac{500}{3}\pi$		$\frac{9}{2}\pi$	288π		$4\sqrt{3}\pi$

Areas and Volumes of Solids

A sphere fits inside a cylinder as shown.

C 17. Show that the area of the sphere equals the lateral area of the cylinder. $A = 4\pi n^2 = 2\pi n \cdot 2n = $ L.A.

18. Find the ratio of the volume of the sphere to the volume of the cylinder. $\frac{2}{3}$

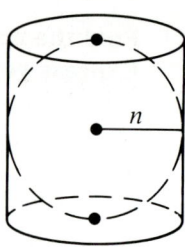

SELF-TEST

Vocabulary
regular pyramid (p. 402):
  vertices   faces
  the vertex   center of base
  the base   altitude
  lateral edges   height
  base edges   slant height

right circular cone (p. 406)
sphere (p. 410)
hemisphere (p. 411)

Exercises 1–4 refer to a regular pyramid whose base is an 18 by 18 square. The height equals 12 and the slant height equals 15.

1. Draw a figure.   *(10-4)*
2. L.A. = __?__   540
3. T.A. = __?__   864
4. V = __?__   1296

Refer to the cone shown. Answer in terms of π.

5. Slant height = __?__   5   *(10-5)*
6. L.A. = __?__ 15π
7. T.A. = __?__ 24π
8. V = __?__ 12π

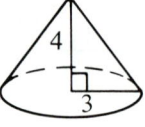

The radius of a certain sphere is 4. Use 3.14 for π. Round to the nearest integer.

9. A = __?__   201   *(10-6)*
10. V = __?__   268

APPLICATIONS

SURFACE AREA AND VOLUME

Geometry tells us that, in general, a small object has more surface area *per unit of volume* than a large object has. Let us compare a baseball with radius 3.5 cm and a basketball with radius 12.4 cm.

Baseball

$$\frac{\text{Surface area}}{\text{Volume}} \approx \frac{153.9}{179.5} \approx 0.86$$

Basketball

$$\frac{\text{Surface area}}{\text{Volume}} \approx \frac{1931.2}{7982.4} \approx 0.24$$

Now let us see how this principle affects a relatively small animal, a mouse, and a larger animal, a human being. Having more skin area per unit of volume, a mouse loses body heat faster than a person does. Just to keep warm, a mouse must eat and burn more food relative to its size. A mouse needs to eat about one half of its weight in food each day. How much food would you have to consume if you ate at that rate?

The relationship between surface area and volume also helps to explain the difference between a mouse's lung and a human's lung. In order to do its job well, the lung must have enough surface area in its air sacs. A small animal like a mouse needs only a simple lung to provide enough air-sac surface. A larger animal like a human being needs a more complex lung to provide enough surface area for breathing. A giant mouse with a simple lung could not breathe enough oxygen to keep it alive.

EXERCISE

Compute and compare the ratio of total surface area to volume for a small cube with edges 2 cm long and a larger cube with edges 60 cm long. small cube: $\frac{T.A.}{V} = \frac{3}{1}$; larger cube: $\frac{T.A.}{V} = \frac{1}{10}$

7 Similar Solids (Optional)

Two solids are called *similar solids* if they have the same shape. The triangular prisms shown are similar.

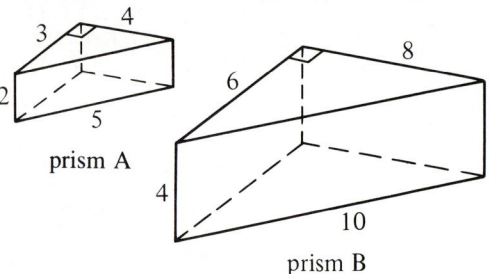

prism A

prism B

Notice that in similar solids:
(1) corresponding angles are equal;
(2) corresponding segments are in proportion.

EXPLORATIONS

Recall, from Chapter 8, that the perimeters and areas of similar figures are related. Let's see whether the surface areas and volumes of similar solids are related. Use the similar triangular prisms shown above. Copy and complete the table.

	prism A	prism B	ratio in simplest form
hypotenuse of base	5	10	1:2
height	2	4	1:2
base area	6	24	1:4
lateral area	24	96	1:4
total area	36	144	1:4
volume	12	96	1:8

You should have found that corresponding lengths are in the ratio $1:2$. Now notice: the base areas, lateral areas, and total areas are all in the ratio $1:4$, or $1^2:2^2$. The volumes are in the ratio $1:8$, or $1^3:2^3$. This exploration suggests the following theorem.

THEOREM 1

Suppose two similar solids have scale factor $a:b$. Then:
(1) the ratio of corresponding segments is $a:b$;
(2) the ratio of the areas is $a^2:b^2$;
(3) the ratio of the volumes is $a^3:b^3$.

EXAMPLE Suppose the two right cylinders shown are similar. Find:

a. the scale factor
b. the height of the larger cylinder
c. the ratio of the base areas
d. the ratio of the lateral areas
e. the ratio of the volumes

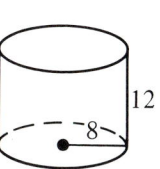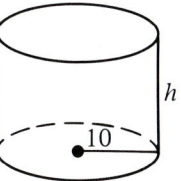

a. The scale factor is $8:10$, or $4:5$.
b. Write a proportion and solve for h. Then $h = 15$.
c. ratio of base areas $= 4^2:5^2 = 16:25$
d. ratio of lateral areas $= 16:25$
e. ratio of volumes $= 4^3:5^3 = 64:125$

$$\frac{4}{5} = \frac{12}{h}$$
$$4h = 60$$
$$h = 15$$

Classroom Practice

Refer to the similar pyramids shown.

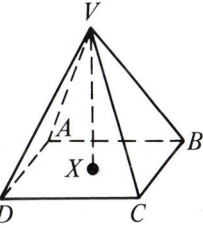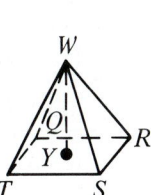

1. $\angle DVC$ corresponds to $\angle \underline{\ ?\ }$. TWS
2. $\overline{BC}$ and $\underline{\ ?\ }$ are corresponding segments. RS
3. $\dfrac{VC}{WS} = \dfrac{DA}{?}$ TQ
4. $\dfrac{AB}{QR} = \dfrac{?}{WY}$ VX

5. Suppose the pyramids have scale factor $10:7$. Find the ratio of:
   a. the base areas   b. the total areas   c. the volumes
      $\frac{100}{49}$ $\frac{100}{49}$ $\frac{1000}{343}$

Written Exercises

Copy and complete the table about similar solids.

5. volumes: $8:125$
6. areas: $9:4$
7. areas: $1:16$

	1.	2.	3.	4.	5.	6.	7.	8.
ratio of corresponding segments	3:4	4:3	1:6	7:9	2:5	3:2	1:4	10:1
					?	?	?	?
ratio of areas	?	?	?	?	4:25	?	?	100:1
ratio of volumes	?	?	?	?	?	27:8	1:64	?

areas: 1. 9:16   2. 16:9   3. 1:36   4. 49:81   1000:1
volumes: 27:64   64:27   1:216   343:729

Areas and Volumes of Solids **415**

Exercises 9–11 refer to two similar prisms with scale factor 3:5.

B 9. The shortest edge of the smaller prism is 6 cm long. How long is the shortest edge of the larger prism? 10 cm

10. The base area of the larger prism is 50 cm². Find the base area of the smaller prism. 18 cm²

11. The volume of the smaller prism is 216 cm³. Find the volume of the larger prism. 1000 cm³

12. The total areas of two similar cones have the ratio 9:4. Find the ratio of their volumes. 27:8

13. Any two spheres are similar solids. The volumes of two spheres are 288π and 7776π, respectively. Find the ratios of their areas. 1:9

A town has two water towers that are similar solids. Two corresponding lengths are labeled.

C 14. Twenty liters of paint were needed to paint the smaller tower. How much paint is needed for the larger tower? 45 L

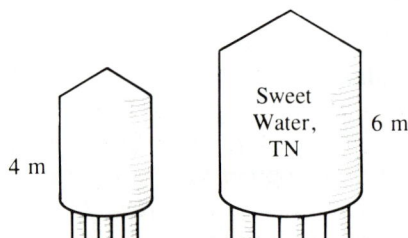

15. The capacity of the larger tank is 180,000 liters. What is the capacity of the smaller tank? 53,333 L

16. For two spheres with scale factor $a:b$, prove:
    a. the ratio of their areas is $a^2:b^2$;
    b. the ratio of their volumes is $a^3:b^3$.

17. A leg bone of a certain animal has a cross-section area of 15 cm². The bone supports about 50 kg. Imagine a similar, but larger, animal with scale factor 10:1.
    a. In the smaller animal, the bone supports __?__ kg/cm². $3\frac{1}{3}$
    b. In the larger animal, the corresponding bone would have a cross-section area of __?__, and the bone would have to support __?__ kg. Thus it would have to support __?__ kg/cm². 1500 cm²; 50,000; $33\frac{1}{3}$
    c. Comparing (a) and (b), comment on the statement "Bigger is better." In this case, bigger is not better. The bones of the larger animal would be under 10 times as much strain.

READING GEOMETRY

THREE-DIMENSIONAL FIGURES

Most of your work in this book has involved plane figures. Since plane figures have only two dimensions, they can easily be pictured by diagrams on a flat piece of paper. Figures in space have three dimensions, so it isn't as easy to represent them in this way. For example, you already know that a plane can only be suggested by a drawing. You have to imagine it as extending endlessly.

When studying figures in space, models can help you see relationships more clearly. The walls, floor, and ceiling of your classroom can be used as models of parallel planes and intersecting planes. Large pieces of cardboard can be held to represent planes in different positions, and pencils or pieces of stiff wire can be used for lines. Boxes and cans make good models of rectangular solids, cubes, and cylinders. Balls are good models of spheres. Some sets of building blocks contain cones and pyramids as well as more common shapes. Test your powers of observation by looking for models of solids around you.

Making models of solid figures can be fun. To do this, you need patterns. A pattern for a square pyramid is shown at the right. Make a large copy of it on stiff paper or cardboard, cut it out, fold on the dashed lines, and tape the edges together. Your school library may have a book of patterns for other models, or try making your own.

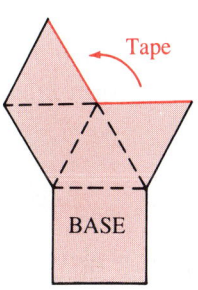

Page 393 explains an easy way to draw a rectangular solid. How could you use this method to draw a cylinder? Try copying some of the diagrams in Chapter 10.

EXERCISES

1. Use a milk carton to make a pattern for a rectangular solid. Look carefully at the carton in order to cut it so that you can spread it out flat in one piece.

2. Look at the diagram of the triangular pyramid on page 403. Try making a model of this figure. (What kind of triangle forms the base of this pyramid?) equilateral

3. Try the Mechanical Drawing Application on page 431. Go to it now and do Exercises 1–3.

PROBLEM SOLVING STRATEGIES

COUNTEREXAMPLES

One way you can disprove a statement is to find an example for which the statement is not true. We call this a *counterexample*.

If someone says to you that all triangles are acute, you could draw an obtuse triangle. If someone says that the diagonals of a parallelogram bisect the angles of the parallelogram, you could draw a picture similar to the one below.

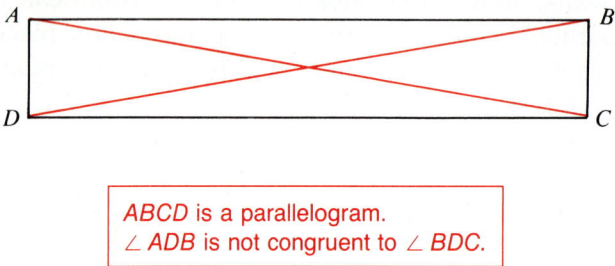

ABCD is a parallelogram.
∠ADB is not congruent to ∠BDC.

To disprove a statement, you need to find an example that shows the conclusion false when the hypothesis is true.

EXERCISES

Draw a picture that shows that each of the following statements is FALSE.

1. If a triangle is isosceles, then it can't be a right triangle.

2. The diagonals of a parallelogram must always be perpendicular to each other.

3. If a quadrilateral is equilateral, then it is equiangular.

4. If a quadrilateral is equiangular, then it is equilateral.

5. If the diagonals of a trapezoid are perpendicular, the trapezoid is a rhombus.

6. All isosceles triangles are acute.

7. If a figure is a polygon, then it is a triangle.

8. If △XYZ is obtuse, then ∠XYZ is obtuse.

9. Two circles which do not intersect have two common external tangents.

10. If two chords are perpendicular, then one of them must pass through the center of the circle.

Give a counterexample to disprove each statement.

> **SAMPLE** All even integers are divisible by 4.
>
> The even integers are made up of 2, 4, 6, 8, 10, ... and −2, −4, −6, −8, −10, ...
>
> Many of these numbers are not divisible by 4. You need only come up with one of them to have a counterexample. For instance:
>
> 10 is an even integer.
> 10 is not divisible by 4.
>
> Therefore, not all even integers are divisible by 4.

The altitudes of any obtuse △ meet outside the △.

11. The altitudes of a triangle meet inside the triangle. **12.–20.** Answers may vary.

12. $(a + b)^2$ is always equal to $a^2 + b^2$. $(3 + 4)^2 = 49$; $3^2 + 4^2 = 25$

13. If the ratio of the sides of two similar polygons is 1:2, then the ratio of their areas is 1:2.

14. The square of any integer is odd. $6^2 = 36$ which is even.

15. If $x > 5$, then $x = 6$. $7 > 5$ and $7 \neq 6$

16. If $AM = MB$, then M is the midpoint of $\overline{AB}$.

17. If $x^2 = y^2$, then $x = y$. $3^2 = 9 = (-3)^2$ and $3 \neq -3$

18. If two angles are congruent, they must be vertical angles.

19. If two angles are complementary, their measures are each 45°.

20. Equal fractions have the same denominator. $\frac{2}{3} = \frac{4}{6}$

21. If $m^2 > n^2$, then $m > n$. $(-5)^2 = 25 > 16 = 4^2$ and $-5 < 4$

22. If $(x - 4)(x + 6) = 0$, then x must equal 4. $x = -6$ will also make $(x - 4)(x + 6) = 0$.

23. If $\angle 1 = \angle 2$, then $\angle 1$ and $\angle 2$ are vertical angles.
$\angle 1$ and $\angle 2$ could be the two equal angles of an isosceles triangle.

SKILLS REVIEW

WORKING WITH POLYNOMIALS

Multiply. The samples will help you remember some patterns.

SAMPLES

1. $2x^2(x - 3y) = (2x^2)(x) - (2x^2)(3y) = 2x^3 - 6x^2y$

2. $(a - 4b)(2a + 9b) = 2a^2 + 9ab - 8ab - 36b^2 = 2a^2 + ab - 36b^2$

3. $(c + 7d)(c + 7d) = c^2 + 7cd + 7cd + 49d^2 = c^2 + 14cd + 49d^2$

4. $(12 + 2k)(12 - 2k) = 144 - 24k + 24k - 4k^2 = 144 - 4k^2$

1. $a^3(3a^2 - ab - b^2)$ $3a^5 - a^4b - a^3b^2$
2. $(a + 5)(a - 3)$ $a^2 + 2a - 15$
3. $(r + 4)(r + 4)$ $r^2 + 8r + 16$
4. $(m + 8)(m - 8)$ $m^2 - 64$
5. $(10 - 3x)(10 - 3x)$ $100 - 60x + 9x^2$
6. $c^2d^3(cd^2 - 7cd)$ $c^3d^5 - 7c^3d^4$
7. $(b - 13)(b - 13)$ $b^2 - 26b + 169$
8. $(m + 2n)(m - n)$ $m^2 + mn - 2n^2$
9. $(x - 6)(x - 9)$ $x^2 - 15x + 54$
10. $(6w + 2x)(6w - 2x)$ $36w^2 - 4x^2$
11. $-6k(5h + 4hk - k^4)$ $-30hk - 24hk^2 + 6k^5$
12. $(2d + 7c)(2d + 7c)$ $4d^2 + 28cd + 49c^2$
13. $(a + 12b)(a - 12b)$ $a^2 - 144b^2$
14. $(3y + 4z)(5y + 7z)$ $15y^2 + 41yz + 28z^2$
15. $(8k - h)(3k + 2h)$ $24k^2 + 13hk - 2h^2$
16. $(2b + 6)(b - 5)$ $2b^2 - 4b - 30$
17. $(3m + 2n)(3m - 2n)$ $9m^2 - 4n^2$
18. $-ab^4(2a^2b^2 - 5b^3)$ $-2a^3b^6 + 5ab^7$

Factor.

SAMPLES

1. $4d^6 + 14cd^8 = 2d^6(2 + 7cd^2)$
2. $n^2 - 11n - 26 = (n - 13)(n + 2)$
3. $x^2 - 6x + 9 = (x - 3)(x - 3)$, or $(x - 3)^2$
4. $100c^2 - 9d^2 = (10c + 3d)(10c - 3d)$

19. $15a^2x - 10ax^2$ $5ax(3a - 2x)$
20. $x^2 + 6x + 8$ $(x + 4)(x + 2)$
21. $y^2 - 20y + 100$ $(y - 10)(y - 10)$, or $(y - 10)^2$
22. $4m^2 - 9$ $(2m - 3)(2m + 3)$
23. $a^2 + 2a - 35$ $(a + 7)(a - 5)$
24. $14b^7 - 7b^6 + 28b^5$ $7b^5(2b^2 - b + 4)$
25. $k^2 + 14k + 49$ $(k + 7)(k + 7)$, or $(k + 7)^2$
26. $m^2 + 17m + 16$ $(m + 16)(m + 1)$
27. $c^2 - 13c + 30$ $(c - 10)(c - 3)$
28. $25x^2 - 121y^2$ $(5x - 11y)(5x + 11y)$
29. $27a^3b^3 + 6ab^2c$ $3ab^2(9a^2b + 2c)$
30. $b^2 - 8b + 12$ $(b - 6)(b - 2)$
31. $144 + 24n + n^2$ $(12 + n)(12 + n)$, or $(12 + n)^2$
32. $1 - 36x^2y^2$ $(1 - 6xy)(1 + 6xy)$
33. $h^2 + 6h - 16$ $(h + 8)(h - 2)$
34. $r^2 - 16r + 64$ $(r - 8)(r - 8)$, or $(r - 8)^2$
35. $n^2 - 12n + 27$ $(n - 9)(n - 3)$
36. $a^2 - a - 72$ $(a - 9)(a + 8)$
37. $k^2 - 9k - 22$ $(k - 11)(k + 2)$
38. $c^2 + 18c + 81$ $(c + 9)(c + 9)$, or $(c + 9)^2$
39. $r^2 + 11r + 28$ $(r + 7)(r + 4)$
40. $144 - 25y^2$ $(12 - 5y)(12 + 5y)$
41. $d^2 - 20d + 19$ $(d - 19)(d - 1)$
42. $m^2 - 2m - 63$ $(m - 9)(m + 7)$

CHAPTER REVIEW

CHAPTER SUMMARY

1. Two lines can intersect, be parallel, or be skew. A line can lie in a plane, intersect the plane, or be parallel to the plane. Two planes intersect, or else they are parallel.

    If line $j \perp$ plane M, and if plane N contains line j, then $N \perp M$.

    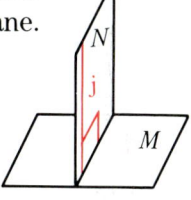

2. In any **right prism**:
    L.A. = sum of the areas of the lateral faces
    T.A. = L.A. + 2B
    V = Bh

    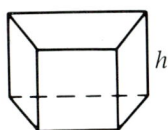

3. In any **right circular cylinder**:
    L.A. = $2\pi rh$
    T.A. = L.A. + 2B = $2\pi rh + 2\pi r^2$
    V = Bh = $\pi r^2 h$

    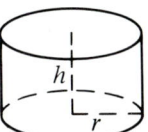

4. In any **regular pyramid**:
    L.A. = sum of the areas of the lateral faces
    T.A. = L.A. + B
    $V = \frac{1}{3}Bh$

    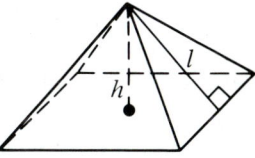

5. In any **right circular cone**:
    L.A. = πrl
    T.A. = L.A. + B = $\pi rl + \pi r^2$
    $V = \frac{1}{3}Bh = \frac{1}{3}\pi r^2 h$

    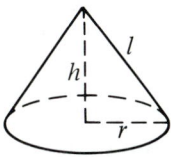

6. In any **sphere**:
    $A = 4\pi r^2$
    $V = \frac{4}{3}\pi r^3$

7. If two similar solids have scale factor $a:b$, then:
    a. the ratio of corresponding segments is $a:b$;
    b. the ratio of the areas is $a^2:b^2$;
    c. the ratio of the volumes is $a^3:b^3$.

REVIEW EXERCISES

Tell whether each statement is true or false. (*See pp. 388–391.*)

1. If two planes do not intersect, they must be parallel. true
2. If two lines do not intersect, they must be parallel. false
3. If a line doesn't lie in a plane, the line must be parallel to the plane. false
4. If line $\overleftrightarrow{AB}$ is perpendicular to plane R, then $\overleftrightarrow{AB}$ is perpendicular to every line in R. false

Refer to the right prism shown. (*See pp. 393–397.*)

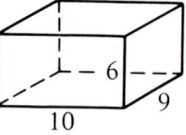

5. $B = $ __?__ 90
6. L.A. = __?__ 228
7. T.A. = __?__ 408
8. $V = $ __?__ 540

Refer to the cylinder shown. Answer in terms of π. (*See pp. 398–400.*)

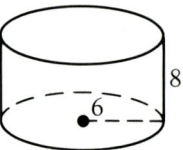

9. $C = $ __?__ 12π
10. L.A. = __?__ 96π
11. T.A. = __?__ 168π
12. $V = $ __?__ 288π

Refer to the regular square pyramid shown. (*See pp. 402–404.*)

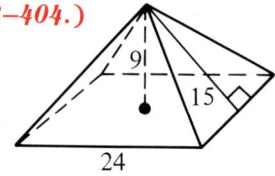

13. $B = $ __?__ 576
14. L.A. = __?__ 720
15. T.A. = __?__ 1296
16. $V = $ __?__ 1728

Refer to the right circular cone shown. Answer in terms of π. (*See pp. 406–408.*)

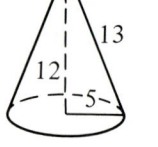

17. $C = $ __?__ 10π
18. L.A. = __?__ 65π
19. T.A. = __?__ 90π
20. $V = $ __?__ 100π

Answer the questions about spheres in terms of π. (*See pp. 410–412.*)

21. When $r = 5$, $A = $ __?__. 100π
22. When $d = 8$, $V = $ __?__. $\frac{256}{3}\pi$

(Optional) Copy and complete the table which deals with similar solids. (*See pp. 414–416.*)

	23.	24.	25.
ratio of corresponding segments	2:5	? 7:9	? 4:3
ratio of areas	? 4:25	49:81	? 16:9
ratio of volumes	8: ? 125	343: ? 729	64:27

CHAPTER TEST

Complete each statement with *always*, *sometimes*, or *never*.

1. Two planes that do not intersect are __?__ parallel. always *(10-1)*
2. Two lines that do not intersect are __?__ parallel. sometimes
3. If line j is perpendicular to plane P, then line j is __?__ perpendicular to every line in P. never

A rectangular solid has length 10, width 8.5, and height 6.

4. L.A. = __?__ 222   5. T.A. = __?__ 392   6. V = __?__ 510   *(10-2)*

7. If a cube has sides of length x, find its total area and volume in terms of x. T.A. = $6x^2$; $V = x^3$

A cylinder has a radius of 7 and a height of 5. Find the indicated values in terms of π.

8. The circumference of a base 14π
9. The area of a base 49π *(10-3)*
10. The lateral area and the total area L.A. = 70π; T.A. = 168π
11. The volume 245π

Refer to the regular square pyramid shown.

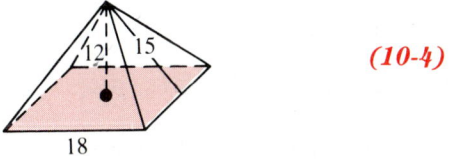

12. B = __?__ 324   13. L.A. = __?__ 540   *(10-4)*
14. T.A. = __?__ 864   15. V = __?__ 1296

Refer to the right circular cone shown. Give answers in terms of π.

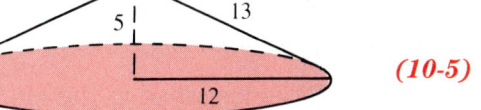

16. B = __?__ 144π 17. L.A. = __?__ 156π *(10-5)*
18. T.A. = __?__ 300π 19. V = __?__ 240π

A sphere has a diameter of 9 cm.

20. Find the area in terms of π. 81π cm^2 *(10-6)*
21. Find the volume to the nearest integer. Use 3.14 for π. 382 cm^3
22. Find the exact volume of a sphere with area 16π. $10\frac{2}{3}\pi$

(Optional) Two similar cones have heights of 6 cm and 8 cm.

23. Find the ratio of their total areas. 9:16   *(10-7)*
24. If the volume of the smaller cone is 96π cm^3, find the volume of the larger cone. 227.6π cm^3

UNIT E CUMULATIVE REVIEW

Complete each statement.

1. Line *l* is a common internal tangent of ⊙S and __?__. ⊙R
2. Line *l* is a common external tangent of __?__ and ⊙S. ⊙O

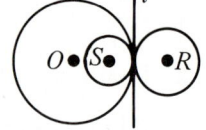

3. If $\overleftrightarrow{AX}$ is a tangent of ⊙O, then ∠1 = __?__°. 90
4. If ∠AOD = 150°, then $\widehat{AD}$ = __?__°. 150
5. If ∠BOA = ∠COD and $\widehat{AB}$ = 47°, then $\widehat{CD}$ = __?__°. 47
6. If $\widehat{PQ} = \widehat{RS}$ and $\widehat{RS}$ = 100°, then $\widehat{PQ}$ = __?__°. 100
7. If OY = OZ, then PQ = __?__. RS
8. If RS = 10, then RZ = __?__. 5
9. If $\widehat{PQ}$ = 98°, then $\widehat{PM}$ = __?__°. 49

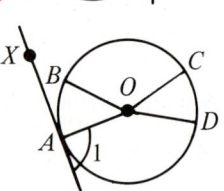

Tell whether each statement is true or false.

10. If $\widehat{CF}$ = 60°, then ∠CEF = 30°. true
11. If ∠CEA = 70°, then $\widehat{EBC}$ = 140°. true
12. ∠BDE = $\frac{1}{2}(\widehat{BE} + \widehat{EF})$ false
13. ∠EAF = $\frac{1}{2}(\widehat{EF} - \widehat{BE})$ true
14. BD · DF = ED · DC true
15. △DEF is inscribed in the circle. false

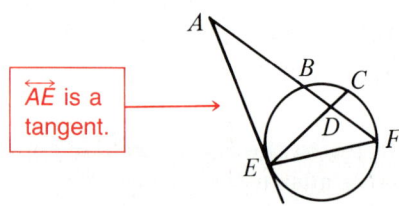

Exercises 10–15

16. Skew lines sometimes intersect. false
17. Two intersecting lines must be coplanar. true
18. Find the volume of a right prism with base area 25 and height 7. 175
19. Find the lateral area of a cylinder with radius 4 and height 7. 56π
20. A regular square pyramid has base edges that are 4 m long. The height is 6 m. Find the volume of the pyramid. 32 m³
21. Find the total area of a cone with radius 3 and slant height 5. 24π
22. Find the area and volume of a sphere with radius 3 cm. 36π; 36π

UNIT F

Here's what you'll learn in this chapter:

To solve problems involving some common right triangle lengths.

To apply theorems about 45°-45°-90° triangles and 30°-60°-90° triangles.

To find the length of the diagonal of a rectangular solid.

To use right triangles to solve problems involving pyramids and cones.

To use the tangent, sine, and cosine ratios of an acute angle of a right triangle.

You can see right angles almost everywhere you look in the photograph above. Most of the triangles shown are right triangles.

Chapter 11

Right Triangles

1 Reviewing Right Triangles

In this section, we shall review right triangles. The Pythagorean Theorem and its converse are the most important statements to remember when working with right triangles. Certain right triangles appear often in geometry and are listed at the right.

Some Common Right Triangle Lengths	
3-4-5	5-12-13
6-8-10	
9-12-15	8-15-17

The first entry in the table, 3-4-5, represents the triangle shown. Remember: The triangle must be a right triangle because

$$3^2 + 4^2 = 5^2.$$

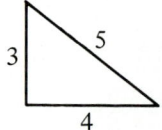

EXAMPLE Find the exact values of x and y in the diagram.

1. Since the four sides are equal, the quadrilateral is a rhombus.

2. The diagonals of a rhombus bisect each other. Therefore $x = 5$.

3. The diagonals of a rhombus are perpendicular. Therefore:

$$5^2 + y^2 = 8^2$$
$$25 + y^2 = 64$$
$$y^2 = 39$$
$$y = \sqrt{39}$$

Take the square root of both sides.

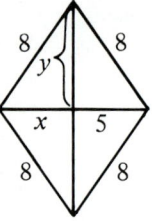

Classroom Practice

Show that each equation is true.

1. $6^2 + 8^2 = 10^2$
   $36 + 64 = 100$

2. $5^2 + 12^2 = 13^2$
   $25 + 144 = 169$

3. $8^2 + 15^2 = 17^2$
   $64 + 225 = 289$

Find the exact value of x.

4.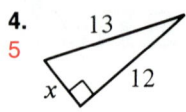

5.

6.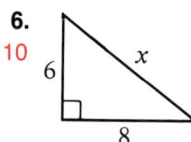

428 Chapter 11

7.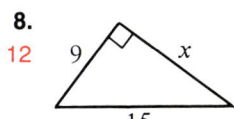

8.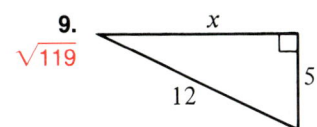

9.

Find the exact values of x and y.

10. 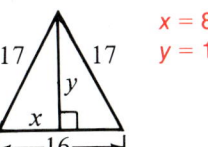 $x = 13,$ $y = 15$

11. 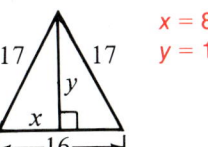 $x = 8,$ $y = 15$

12. 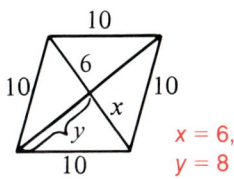 $x = 6,$ $y = 8$

Written Exercises

Find the exact value of x.

A 1.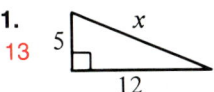

2.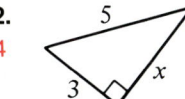

3.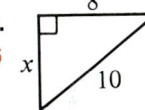

4.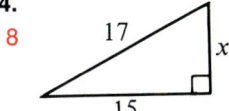

5.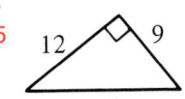

6.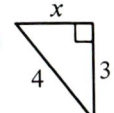

Find the exact values of x and y.

7. $x = 8,$ $y = \sqrt{37}$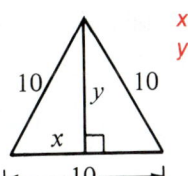

8. $x = 5,$ $y = 5\sqrt{3}$

9. 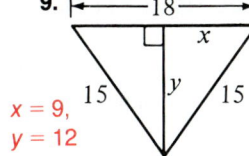 $x = 9,$ $y = 12$

10. 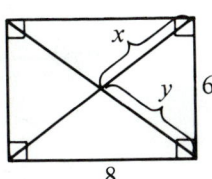 $x = 5,$ $y = 5$

11. 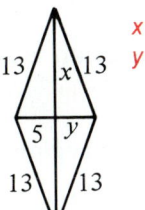 $x = 12,$ $y = 5$

12. $x = \sqrt{51},$ $y = 7$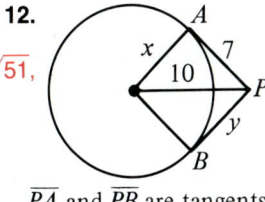

$\overline{PA}$ and $\overline{PB}$ are tangents.

13. 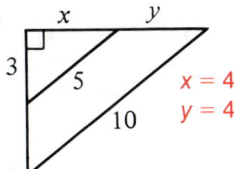 $x = 4,$ $y = 4$

14. 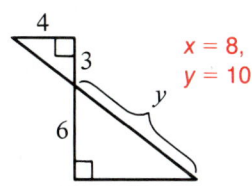 $x = 8,$ $y = 10$

15. 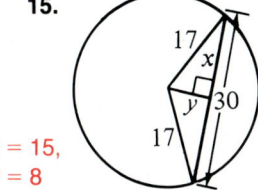 $x = 15,$ $y = 8$

Right Triangles   **429**

Find the area. (*Hint:* First find *x*.)

16.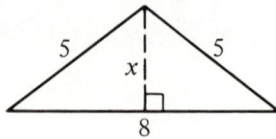
$x = 3$; $A = 12$ sq. units

17.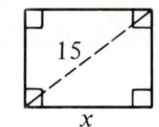
$x = 12$; $A = 108$ sq. units

18.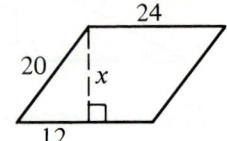
$x = 16$; $A = 384$ sq. units

[B] 19. Is $\angle C$ a right angle? Explain. yes; $6^2 + 8^2 = 10^2$

20. Why must $\overline{AB}$ be a diameter of the circle? (*Hint:* What is the measure of $\widehat{ADB}$?) 180°

21. What is the area of the circle? 25π

22. What is the circumference of the circle? 10π

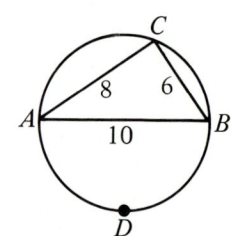

23. Find the values of p and q. $p = 20$, $q = 15$

24. Is $p^2 + q^2 = r^2$? yes

25. Is $\triangle JKL$ a right triangle? yes

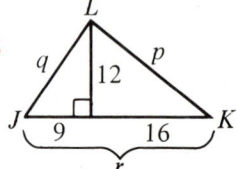

26. Find the values of x and y. $x = 10$, $y = 17$

27. Is $x^2 + y^2 = z^2$? no

28. Is $\triangle STU$ a right triangle? no

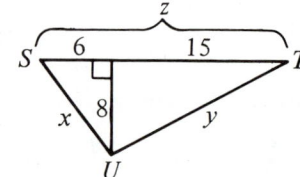

$ABCD$ is a parallelogram. $AB = 15$, $AC = 24$, and $BD = 18$.

29. $AO = \underline{\ ?\ }$ 12   30. $BO = \underline{\ ?\ }$ 9

31. Is $\triangle AOB$ a right triangle? Explain. yes; $12^2 + 9^2 = 15^2$

32. What special kind of parallelogram is $ABCD$? Explain.
rhombus; the diagonals are perpendicular

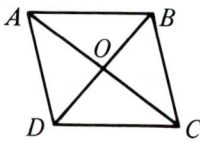

Find the exact area of each shaded region.

[C] 33.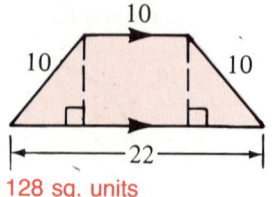
128 sq. units

34.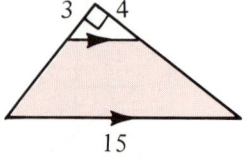
48 sq. units

35.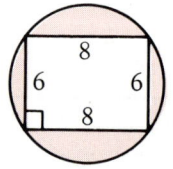
$25\pi - 48$ sq. units

36. A right triangle has legs of length 30 and 40. How long is the median drawn to the hypotenuse? 25

APPLICATIONS

MECHANICAL DRAWING

Technical artists prepare mechanical drawings for architects and machinists. Many mechanical drawings show a building, a machine, or a tool, as seen from different views. For example, a calculator might be drawn like this.

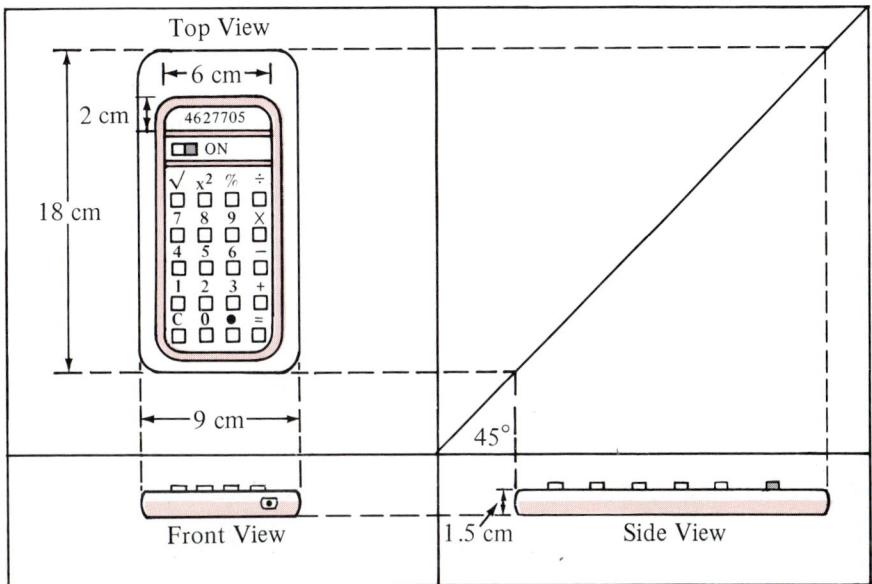

Each length in a mechanical drawing is drawn to scale. Suppose the actual calculator is 9 cm wide. If this width is represented by a segment 3 cm long, then the scale is 3:9, or 1:3.

EXERCISES

Sketch each object as seen from the top, front, and side.

1.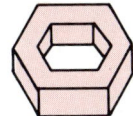

2.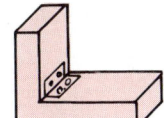

3.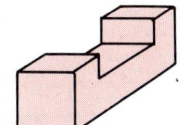

4. Make a mechanical drawing of a milk container, showing the top, the front, and the side.

Right Triangles

2 Special Right Triangles

The two triangles shown are special right triangles. The first one is an isosceles right triangle. It is often called a 45°-45°-90° triangle. The other special triangle is a 30°-60°-90° triangle.

The Pythagorean Theorem can be used to prove two theorems about these special triangles.

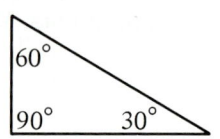

THEOREM 1

In a 45°-45°-90° triangle, the hypotenuse is $\sqrt{2}$ times as long as a leg.

Given: A 45°-45°-90° triangle

Prove: $h = l\sqrt{2}$

Proof: $h^2 = l^2 + l^2$
$h^2 = 2l^2$
$h = \sqrt{2l^2}$
$h = l\sqrt{2}$ ← $\sqrt{2} \approx 1.4$

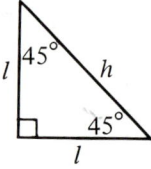

EXAMPLE 1   For each diagram below, find the value of *t*.

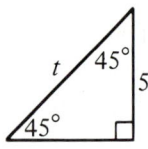 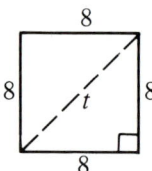 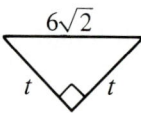

hypotenuse = leg × $\sqrt{2}$ hypotenuse = leg × $\sqrt{2}$ hypotenuse = leg × $\sqrt{2}$
$t = 5\sqrt{2}$ $t = 8\sqrt{2}$ $6\sqrt{2} = t\sqrt{2}$
                                                                                $6 = t$

THEOREM 2

In a 30°-60°-90° triangle,
(1) the hypotenuse is twice as long as the shorter leg;
(2) the longer leg is $\sqrt{3}$ times as long as the shorter leg.

Given: A 30°-60°-90° triangle

Prove: **a.** $h = 2s$ **b.** $l = s\sqrt{3}$

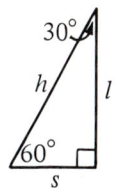

Proof: Notice that a 30°-60°-90° triangle is "half" of an equilateral triangle.

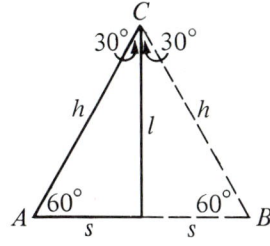

  a. Since $\triangle ABC$ is equilateral, $h = 2s$.

  b. To find l, use the Pythagorean Theorem.

$$l^2 + s^2 = h^2$$
$$l^2 + s^2 = (2s)^2$$
$$l^2 + s^2 = 4s^2$$
$$l^2 = 3s^2$$
$$l = \sqrt{3s^2}$$
$$l = s\sqrt{3} \quad \longleftarrow \boxed{\sqrt{3} \approx 1.7}$$

EXAMPLE 2 In each triangle below, the length of one side is given. Find the lengths of the other two sides.

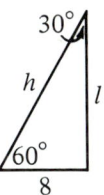

hypotenuse = 2 × shorter leg
$h = 2 \times 8$
$h = 16$

longer leg = shorter leg × $\sqrt{3}$
$l = 8\sqrt{3}$

hypotenuse = 2 × shorter leg
$12 = 2s$
$6 = s$

longer leg = shorter leg × $\sqrt{3}$
$l = 6\sqrt{3}$

Classroom Practice

$\triangle ABC$ is an isosceles right triangle.

1. $\angle A = \underline{\;?\;}°$ 45
2. $\angle B = \underline{\;?\;}°$ 45
3. If $AC = 3$, then $BC = \underline{\;?\;}$ and $AB = \underline{\;?\;}$. 3; $3\sqrt{2}$
4. If $AC = 4$, then $BC = \underline{\;?\;}$ and $AB = \underline{\;?\;}$. 4; $4\sqrt{2}$
5. If $BC = 9$, then $AB = \underline{\;?\;}$. $9\sqrt{2}$
6. If $AB = 7\sqrt{2}$, then $BC = \underline{\;?\;}$. 7
7. If $AB = 2\sqrt{2}$, then $AC = \underline{\;?\;}$. 2

Right Triangles

△ XYZ, shown below, is a 30°-60°-90° triangle.

8. Which leg is the shorter leg? $\overline{YZ}$
9. In a 30°-60°-90° triangle, the shorter leg is *always* opposite the __?__° angle. 30
10. Which leg is the longer leg? $\overline{XZ}$
11. In a 30°-60°-90° triangle, the longer leg is *always* opposite the __?__° angle. 60
12. If $YZ = 5$, then $XY =$ __?__ and $XZ =$ __?__. 10; $5\sqrt{3}$
13. If $YZ = 6$, then $XY =$ __?__ and $XZ =$ __?__. 12; $6\sqrt{3}$
14. If $XY = 8$, then $YZ =$ __?__ and $XZ =$ __?__. 4; $4\sqrt{3}$
15. If $XY = 6$, then $YZ =$ __?__ and $XZ =$ __?__. 3; $3\sqrt{3}$

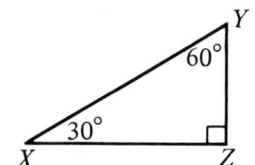

Written Exercises

In Exercises 1–9, find the exact values of x and y.

[A] **1.** 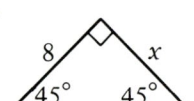 $x = 8$, $y = 8\sqrt{2}$
2. 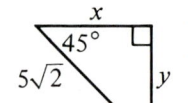 $x = 5$, $y = 5$
3. 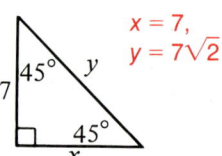 $x = 7$, $y = 7\sqrt{2}$

4. 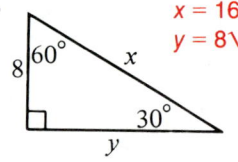 $x = 16$, $y = 8\sqrt{3}$
5. 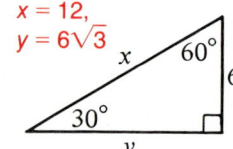 $x = 12$, $y = 6\sqrt{3}$
6. 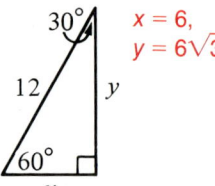 $x = 6$, $y = 6\sqrt{3}$

7. 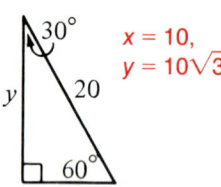 $x = 10$, $y = 10\sqrt{3}$
8. 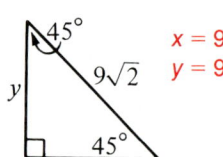 $x = 9$, $y = 9$
9. 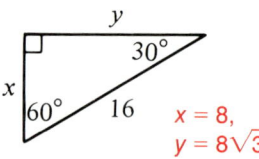 $x = 8$, $y = 8\sqrt{3}$

In the diagram shown, △ ABC is equilateral. $\overline{AD}$ is an altitude.

10. ∠ ABD = __?__° 60
11. ∠ ADB = __?__° 90
12. ∠ BAD = __?__° 30
13. ∠ CAD = __?__° 30

Suppose $AB = 6$. Find:

14. BD 3
15. AD $3\sqrt{3}$
16. Area of △ ABC $9\sqrt{3}$

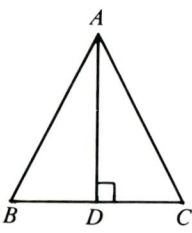

Exercises 10–16

In the diagram shown, *ABCD* is a square. Suppose *AB* = 6. Find:

17. *AC* 6√2   18. *AO* 3√2   19. *BO* 3√2

Suppose *AO* = 4. Find:

20. *BO* 4   21. Area of △*AOB* 8   22. Area of *ABCD* 32

Suppose *DO* = 5. Find:

23. *DC* 5√2   24. Area of △*DOC* 12½   25. Area of *ABCD* 50

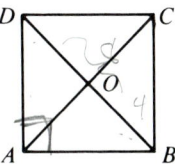

Exercises 17–25

In Exercises 26–28, find the exact value of *t*.

SAMPLE

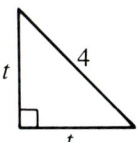

Method 1

Hypotenuse = leg × √2 (Thm. 1)
$$4 = t\sqrt{2}$$
$$\frac{4}{\sqrt{2}} = t$$

Simplify.

$$t = \frac{4}{\sqrt{2}} = \frac{4 \cdot \sqrt{2}}{\sqrt{2} \cdot \sqrt{2}} = \frac{4\sqrt{2}}{2} = 2\sqrt{2}$$

Method 2

$$t^2 + t^2 = 4^2 \text{ (Pythagorean Thm.)}$$
$$2t^2 = 16$$
$$t^2 = 8$$
$$t = \sqrt{8}$$

Simplify.

$$t = \sqrt{8} = \sqrt{4 \times 2} = 2\sqrt{2}$$

B 26. 3√2

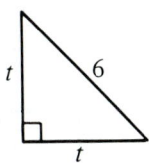

27.

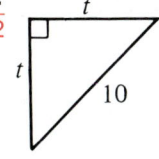

5√2

28. 4√2

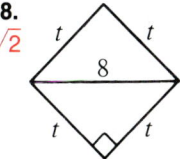

29. Each edge of the cube shown has length 6. Find the length of each side of the red triangle. 6√2, 6√2, 6√2

30. What is the measure of each angle of the red triangle? 60°

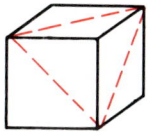

△ABC is a 30°-60°-90° triangle. $\overline{CD}$ is an altitude.
Copy and complete the table.

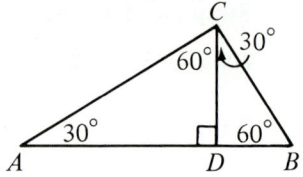

	AB	BC	AC	CD	AD	DB
Sample	4	2	$2\sqrt{3}$	$\sqrt{3}$	3	1
31.	12	? 6	? $6\sqrt{3}$	? $3\sqrt{3}$	? 9	? 3
32.	? 20	10	? $10\sqrt{3}$	? $5\sqrt{3}$	? 15	? 5
33.	? 8	? 4	$4\sqrt{3}$	? $2\sqrt{3}$	? 6	? 2
34.	? 12	? 6	? $6\sqrt{3}$	? $3\sqrt{3}$	? 9	3
C 35.	? 28	? 14	? $14\sqrt{3}$	$7\sqrt{3}$	? 21	? 7
36.	? 12	? 6	? $6\sqrt{3}$	? $3\sqrt{3}$	9	? 3

CALCULATOR ACTIVITIES

A jewelry designer plans to make earrings that are hollow spheres.
Find the total number of grams in each earring described.
Round your answers to the nearest tenth of a gram.

EXAMPLE   The outer diameter is 1 cm, and the inner diameter is 0.9 cm.
The alloy to be used has 8.6 g/cm³.

Total number of grams =
$$\left[\frac{4}{3}\pi(0.5)^3 - \frac{4}{3}\pi(0.45)^3\right]8.6 \approx 1.2$$

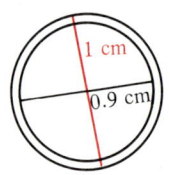

Copy and complete the table.

		Outer diam. (in cm)	Inner diam. (in cm)	Grams in 1 cm³	Total no. of grams
1.	Bronze	0.8	0.6	8.8	? 1.4
2.	German Silver	0.8	0.66	8.3	? 1
3.	Silver	0.8	0.65	10.5	? 1.3
4.	Gold	0.7	0.54	19.3	? 1.9

3 Using Special Right Triangles

In this section, we will apply the theorems about 45°-45°-90° triangles and 30°-60°-90° triangles to many figures you have studied in this course. Here is an example.

EXAMPLE Find the area of ▱ABCD if ∠A = 60°, AD = 8, and AB = 10.

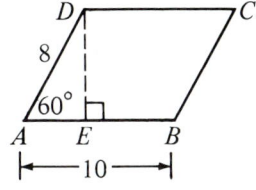

1. Draw altitude $\overline{DE}$.
2. Notice that △ADE is a 30°-60°-90° triangle.
3. $AE = \frac{1}{2} \times 8 = 4$ $DE = 4\sqrt{3}$
4. Area = base × height = $10 \times 4\sqrt{3} = 40\sqrt{3}$

Classroom Practice

Exercises 1–6 refer to the diagram.

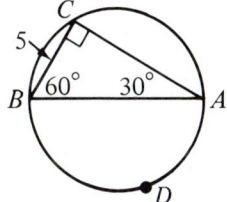

1. △ABC is a __?__°-__?__°-__?__° triangle. 30; 60; 90
2. Since BC = 5, AB = __?__. 10
3. Since ∠C = 90°, $\stackrel{\frown}{BDA}$ = __?__°. 180
4. $\overline{AB}$ is a __?__ of the circle. diameter
5. Find the radius of the circle. 5
6. Find the area of the circle. 25π

WXYZ is a rectangle. XZ = 12 and ∠ZXW = 30°. Find:

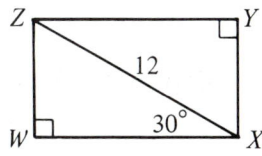

7. ZW 6
8. WX 6√3
9. The area of rectangle WXYZ 36√3

Written Exercises

Find the area of each triangle.

A
1. 2

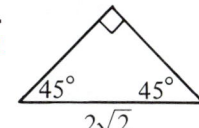

2. 8√3

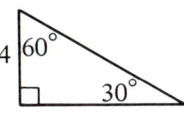

3. 50√3

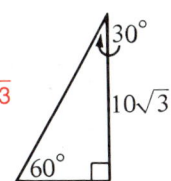

Right Triangles   **437**

Find the area of each parallelogram. Refer to the example, page 437.

4. 60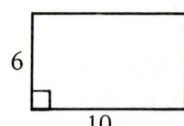
5. 30√3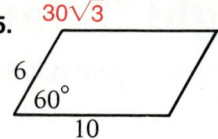
6. 30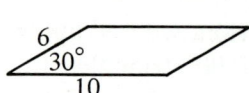

For each triangle shown, find the value of *x*. Then find the area of the triangle.

7.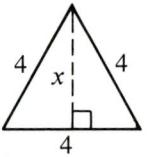
x = 2√3; A = 4√3

8.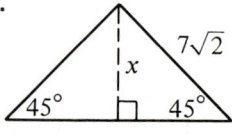
x = 7; A = 49

9.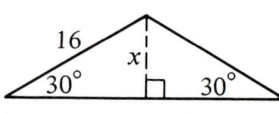
x = 8; A = 64√3

$\overline{TA}$ and $\overline{TB}$ are tangents to ⊙O.
OA = 4 and ∠ATB = 60°.
Find each of the following.

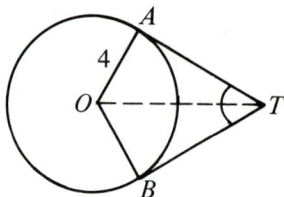

10. ∠OAT 90°   11. ∠OTA 30°
12. AT 4√3   13. BT 4√3

Study the diagram. Then find each of the following.

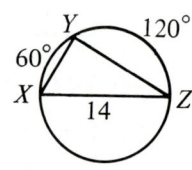

14. ∠Z 30°   15. ∠X 60°   16. ∠Y 90°
17. XY 7   18. YZ 7√3   19. The area of △XYZ $\frac{49}{2}$√3

PQRS is a rhombus. ∠SPQ = 120° and SP = 8.
Find each of the following.

 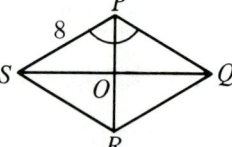

20. ∠SPO 60°   21. ∠POS 90°   22. ∠PSO 30°
23. PO 4   24. SO 4√3   25. The area of PQRS 32√3

26. Is △EFG ~ △HIJ? yes
27. What is the scale factor of △EFG to △HIJ? 5:3
28. What is the ratio of their perimeters? 5:3
29. What is the ratio of their areas? 25:9

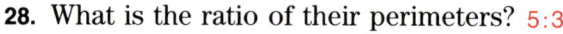 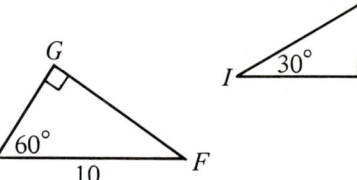

QRST is a square inscribed in ⊙O. The radius of ⊙O is 4.

30. Find the area of the circle. 16π
31. Find the area of the square. 32
32. Find the area of the shaded region. 18.24

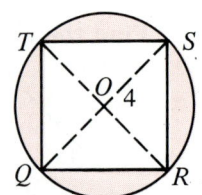

C **33.** Find the area of trapezoid ABCD if ∠A = ∠B = 45°, AB = 16, and DC = 6. 55

34. Find the area of a regular hexagon with sides 2 cm long. 6√3 cm²

35. The wrench shown just fits the nut. Find the value of x. √3 cm

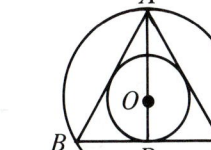

△ABC is an equilateral triangle with sides 6 cm long. OD is the radius of the inscribed circle. OA is the radius of the circumscribed circle.

36. AD = __?__ 3√3   **37.** OD = __?__ √3   **38.** OA = __?__ 2√3

Hint: See the Experiments on page 162.

SELF-TEST

Find the exact values of x and y.

1. 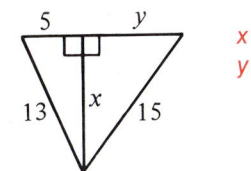   x = 12, y = 9   (11-1)

2. x = 2, y = 2√2

3. 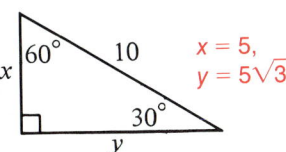 x = 5, y = 5√3   (11-2)

Find the area of each polygon.

4. 18 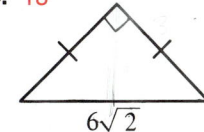   **5.** 27√3 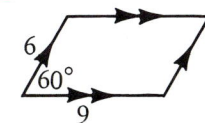   **6.** 27√3 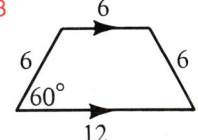   (11-3)

Right Triangles

4 Diagonals of Rectangular Solids

The rectangular solid shown is 3 cm long, 1 cm wide, and 2 cm high. $\overline{HB}$ is called a *diagonal* of the solid. A rectangular solid has four diagonals, which are all equal in length. The other diagonals, not drawn in the figure, are $\overline{EC}$, $\overline{GA}$, and $\overline{FD}$.

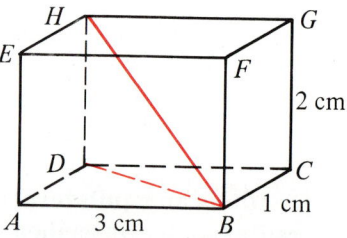

We can use the Pythagorean Theorem to find the length of $\overline{HB}$.

1. First study rectangle *ABCD*.

   $(DB)^2 = 1^2 + 3^2$
   $(DB)^2 = 1 + 9 = 10$
   $DB = \sqrt{10}$

   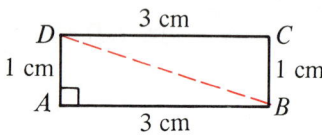

2. Now study right $\triangle HDB$. Notice

   $(HB)^2 = (HD)^2 + (DB)^2$

   $(HB)^2 = \;\;2^2\;\; + \;\;10$
   $(HB)^2 = 14$

   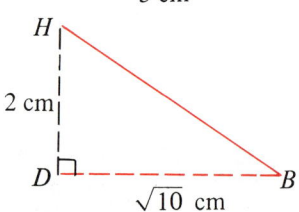

3. $\overline{HB}$ is exactly $\sqrt{14}$ cm long. If we use the square root table on page 516, we find that $\overline{HB}$ is approximately 3.742 cm long.

The method illustrated above can be used to find the length of a diagonal of any rectangular solid.

1. $x^2 = a^2 + b^2$

2. $d^2 = x^2 + c^2$
   $d^2 = (a^2 + b^2) + c^2$ ← Substitute using Step 1.

3. $d = \sqrt{a^2 + b^2 + c^2}$

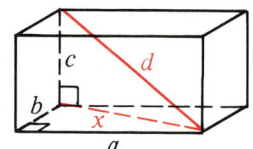

If a rectangular solid has dimensions *a*, *b*, and *c*, then a diagonal has length

$$d = \sqrt{a^2 + b^2 + c^2}.$$

EXAMPLE A rectangular solid is shown. Find the length of a diagonal of the solid.

$$d = \sqrt{5^2 + 4^2 + 7^2}$$
$$= \sqrt{25 + 16 + 49}$$
$$= \sqrt{90}$$
$$= \sqrt{9 \cdot 10}$$
$$= 3\sqrt{10}$$

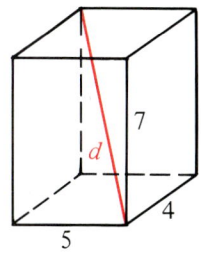

Classroom Practice

The rectangular solid shown has dimensions 4, 2, and 3.

1. $x^2 = 2^2 + 4^2$
   $x^2 = \underline{\ ?\ }$ 20

2. $d^2 = 3^2 + x^2$
   $d^2 = \underline{\ ?\ } + \underline{\ ?\ }$ 9; 20

3. $d = \sqrt{\underline{\ ?\ }}$ 29

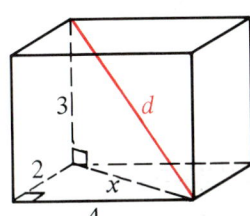

4. Using the formula for a diagonal of a rectangular solid:
   $d = \sqrt{\underline{\ ?\ }^2 + \underline{\ ?\ }^2 + \underline{\ ?\ }^2}$ 4; 2; 3
   $d = \sqrt{\underline{\ ?\ } + \underline{\ ?\ } + \underline{\ ?\ }}$ 16; 4; 9
   $d = \sqrt{\underline{\ ?\ }}$ 29

The rectangular solid shown has dimensions 6, 2, and 3.

5. $x^2 = \underline{\ ?\ }^2 + \underline{\ ?\ }^2$ 2; 6
   $x^2 = \underline{\ ?\ }$ 40

6. $d^2 = x^2 + \underline{\ ?\ }^2$ 3
   $d^2 = \underline{\ ?\ } + \underline{\ ?\ }$ 40; 9

7. $d = \sqrt{\underline{\ ?\ }}$ 49

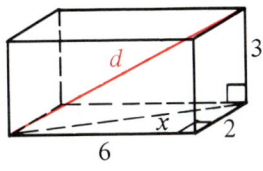

8. Use the formula for a diagonal of a rectangular solid to show that your answer for Exercise 7 is correct. $d = \sqrt{6^2 + 2^2 + 3^2} = 7$

Exercises 9–11 refer to the rectangular solid shown.

9. Find the value of x.   5

10. Find the height c.   12

11. Find the volume of the solid.   144

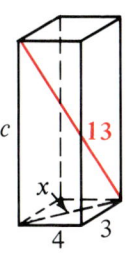

Right Triangles   **441**

Written Exercises

Copy and complete the table about the rectangular solid shown.

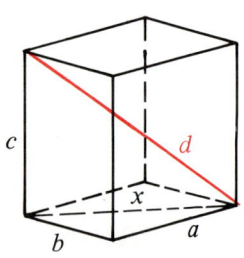

	a	b	c	x	d
1.	2	4	5	? 2√5	? 3√5
2.	3	4	5	? 5	? 5√2
3.	2	1	3	? √5	? √14
4.	3	2	6	? √13	? 7
5.	5	2	6	? √29	? √65
6.	6	2	5	? 2√10	? √65
7.	9	12	? 8	? 15	17
8.	3	? 4	? 12	5	13

The figure shows a cube with edges 1 cm long.

9. Show that $x = \sqrt{2}$. $x^2 = 1^2 + 1^2$, so $x = \sqrt{2}$

10. Show that $d = \sqrt{3}$. $d = \sqrt{1^2 + 1^2 + 1^2} = \sqrt{3}$

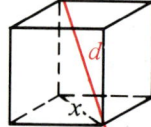

11. A rectangular solid is 21 cm long, 20 cm wide, and 8 cm high. Find the length of a diagonal to the nearest tenth. 30.1 cm

12. A rectangular solid is 20 cm long, 15 cm wide, and 10 cm high. Find the length of a diagonal to the nearest tenth. 26.9 cm

13. Will a very thin metal rod, 70 cm long, fit inside a box that is 51 cm long, 43 cm wide, and 21 cm high? Explain. No; the diagonal of the box is d, with $d = \sqrt{51^2 + 43^2 + 21^2} \approx 69.936$ cm

PUZZLE ◆ PROBLEMS

Make a 3-4-5 triangle with 12 straws of equal length.

a. By moving only 3 straws, make a polygon with area 4.

b. By moving only 2 straws, make a polygon with area 5.

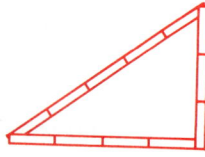

5 Right Triangles in Pyramids and Cones

In the previous section, we used right triangles to solve problems involving rectangular solids. In this section, we'll use right triangles to work with pyramids and cones.

EXAMPLE 1 For the regular pyramid shown, find:
   a. the lateral area;
   b. the total area;
   c. the volume.

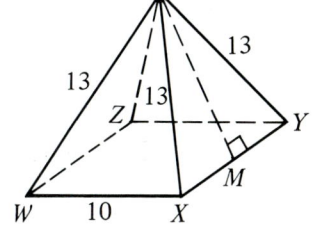

a. Notice that $\triangle VXM$ is a 5-12-13 right triangle with $VM = 12$.

The area of one lateral face $= \frac{1}{2} \times 10 \times 12 = 60$

L.A. $= 4 \times 60 = 240$

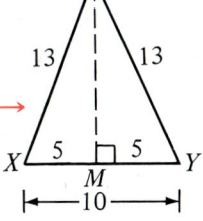

one lateral face

b. T.A. = L.A. + B
      ↓   ↓
    = 240 + 10^2
    = 340

c. To find the volume, we need to find the height, h.
To do this, we use right $\triangle VOM$.
We know from part (a) that $VM = 12$.
We also know that $OM = 5$. (Why?) $OM = \frac{1}{2} WX$

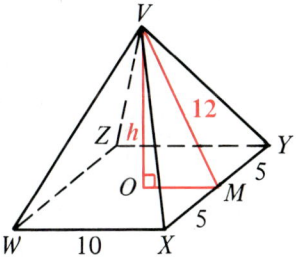

Now we can use the Pythagorean Theorem to find h.

$h^2 + 5^2 = 12^2$
$h^2 + 25 = 144$
$h^2 = 119$
$h = \sqrt{119}$

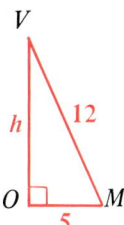

Recall the formula for the volume of a pyramid:

$V = \frac{1}{3}Bh$

$= \frac{1}{3} \times 10^2 \times \sqrt{119}$

$= \frac{100}{3}\sqrt{119}$

Right Triangles   **443**

EXAMPLE 2 For the cone shown, find:
a. the lateral area;
b. the total area;
c. the volume.

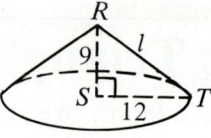

a. To find the lateral area, we must find l, the slant height of the cone. Notice that $\triangle RST$ is a 9-12-15 triangle with $l = 15$.

$$\begin{aligned} \text{L.A.} &= \pi r l \\ &= \pi \times 12 \times 15 \\ &= 180\pi \end{aligned}$$

b. $\begin{aligned} \text{T.A.} &= \text{L.A.} + B \\ &= 180\pi + \pi r^2 \\ &= 180\pi + (\pi \times 12^2) \\ &= 324\pi \end{aligned}$

c. $V = \frac{1}{3}Bh = \frac{1}{3}\pi r^2 h = \frac{1}{3}\pi \times 12^2 \times 9 = 432\pi$

Classroom Practice

Exercises 1–7 refer to the regular square pyramid shown.

1. The area of lateral face $VRS =$ __?__. 15
2. The lateral area of the pyramid = __?__. 60
3. The area of the base = __?__. 36
4. The total area of the pyramid = __?__. 96
5. $OM =$ __?__ 3
6. The height, VO, of the pyramid = __?__. 4
7. The volume of the pyramid = __?__. 48

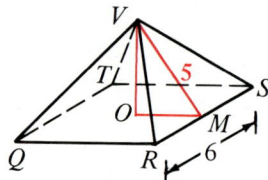

Exercises 8–11 refer to the cone pictured below.

8. The lateral area of the cone = __?__. 255π
9. The total area of the cone = __?__. 480π
10. The height, h, of the cone = __?__. 8
11. The volume of the cone = __?__. 600π

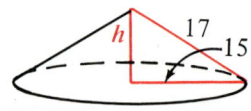

Written Exercises

Copy and complete the table about regular square pyramids. Draw an accurate diagram for each exercise if you wish.

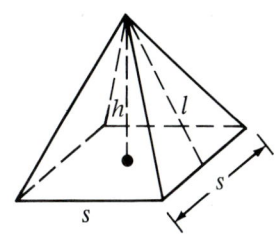

	s	h	l	L.A.	T.A.	V
A 1.	8	3	? 5	? 80	? 144	? 64
2.	10	12	? 13	? 260	? 360	? 400
3.	12	8	? 10	? 240	? 384	? 384
4.	24	? 5	13	? 624	? 1200	? 960
5.	12	? 8	10	? 240	? 384	? 384
6.	? 12√3	6	12	? 288√3	? T.A. = 288√3 + 432	? 864

Copy and complete the table about cones. Draw an accurate diagram for each exercise if you wish.

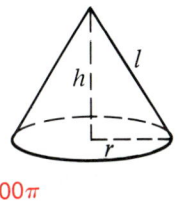

	r	h	l	L.A.	T.A.	V
7.	8	6	? 10	? 80π	? 144π	? 128π
8.	5	12	? 13	? 65π	? 90π	? 100π
9.	15	8	? 17	? 255π	? 480π	? 600π
10.	9	12	? 15	? 135π	? 216π	? 324π
11.	? 15	8	17	? 255π	? 480π	? 600π
12.	1	? √3	2	? 2π	? 3π	? $\frac{\sqrt{3}}{3}\pi$

The four faces of the pyramid shown are equilateral triangles.

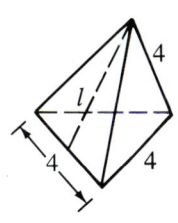

B 13. Find the slant height l. $2\sqrt{3}$

14. Find the lateral area of the pyramid. $12\sqrt{3}$

15. Find the total area of the pyramid.
    Remember: the four faces are congruent. $16\sqrt{3}$

A regular square pyramid has base edges that are 16 cm long and lateral edges that are 17 cm long. Find each of the following.

16. The slant height of the pyramid 15 cm

17. The lateral area of the pyramid 480 cm²

18. The total area of the pyramid 736 cm²

19. The volume of the pyramid $\frac{256\sqrt{161}}{3}$ cm³

Right Triangles 445

20. A regular square pyramid has base edges that are 18 cm long and lateral edges that are 15 cm long. Find the lateral area, the total area, and the volume of the pyramid. L.A. = 432 cm²; T.A. = 756 cm²; V = 324√7 cm³

A cone with radius 6 is inscribed in a sphere with radius 10.

C 21. Find the height of the cone. 18

22. Find the volume of the cone. 216π

23. Find the volume of the sphere. $\frac{4000}{3}\pi$

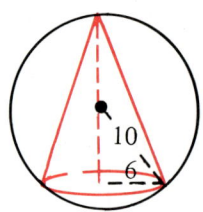

SELF-TEST

1. A rectangular solid is 8 cm long, 8 cm wide, and 4 cm high. Find the length of a diagonal. 12 cm *(11-4)*

2. A cone has height 3 and slant height 5. What is its radius? 4

3. Find the volume of the cone described in Exercise 2. 16π

4. A regular square pyramid has height 12 and base edges 10. Find its lateral area. 260 *(11-5)*

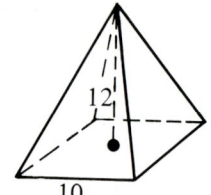

PUZZLE ◆ PROBLEMS

An ant sitting at the corner of an open box smells some sugar at the center of the bottom. The dotted path shows the shortest route to the sugar. How far does the ant have to crawl? 10

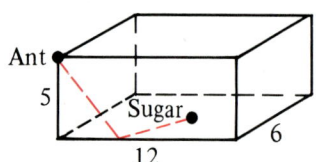
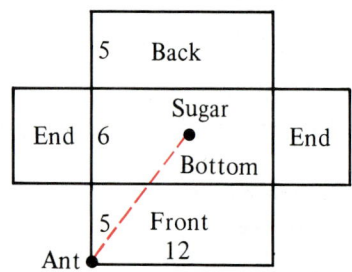

Hint: Imagine unfolding the box as shown.

6 The Tangent Ratio

The word *trigonometry* comes from Greek words which mean "triangle measurement." In this course, the study of trigonometry will be limited to *right* triangles.

In each right triangle below, an angle is marked in color. The diagram indicates the leg *adjacent to* (next to) this angle and the leg *opposite* this angle.

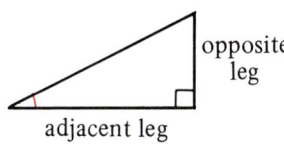
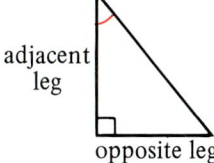

We are now ready to define the first important trigonometric ratio.

tangent of $\angle A = \dfrac{\text{leg opposite } \angle A}{\text{leg adjacent to } \angle A}$

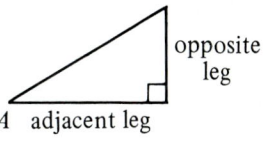

We can abbreviate this by writing:

$$\tan A = \frac{\text{opposite}}{\text{adjacent}}$$

EXAMPLE 1 In the right triangle shown, find tan A and tan B.

$\tan A = \dfrac{\text{leg opposite } \angle A}{\text{leg adjacent to } \angle A} = \dfrac{3}{4}$

$\tan B = \dfrac{\text{leg opposite } \angle B}{\text{leg adjacent to } \angle B} = \dfrac{4}{3}$

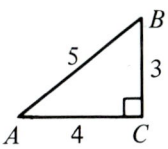

In the right triangles shown, $\angle R = \angle X$.
Do you see that $\triangle RST \sim \triangle XYZ$? (Why?)
Therefore,

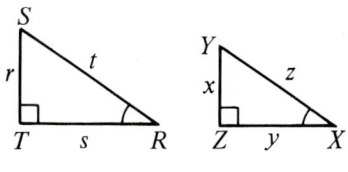

AA Postulate

$\dfrac{r}{x} = \dfrac{s}{y}$ ← Corr. sides of $\sim \triangle$ are proportional.

$\dfrac{r}{s} = \dfrac{x}{y}$ ← Use the Switching Property, page 289.

$\tan R = \tan X$

The discussion above shows that if $\angle R = \angle X$, then tan R = tan X. This means that the value of the tangent ratio does not depend on the size of a right triangle, but only on the size of an angle.

Right Triangles   **447**

On page 517, there is a table that lists values of the tangent ratio for selected acute angles. Most of the values in the table are not exact. They are rounded to four decimal places.

Part of the table is shown at the right. Notice that:

tan 15° = 0.2679.

Angle	Tangent
5°	0.0875
10°	0.1763
15°	0.2679
20°	0.3640

For convenience, we'll use = rather than ≈.

EXAMPLE 2 Find the value of y to the nearest tenth.

$\tan 20° = \frac{y}{10}$

From the table → $0.3640 = \frac{y}{10}$

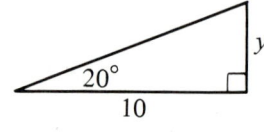

$10 \times 0.3640 = y$
$3.64 = y$

To the nearest tenth, $y = 3.6$.

Classroom Practice

Use the table on page 517 to complete the following.

1. tan 15° = __?__ 0.2679
2. tan 30° = __?__ 0.5774
3. tan 65° = __?__ 2.1445
4. tan __?__ = 0.3640 20°
5. tan __?__ = 1.7321 60°
6. tan __?__ = 0.0875 5°

Use the definition of tangent to find the value of tan A for each right triangle shown. Express your answers in simplest form.

7. $\frac{1}{2}$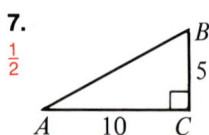

8. $\frac{5}{8}$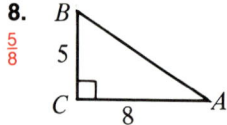

9. $\frac{4}{3}$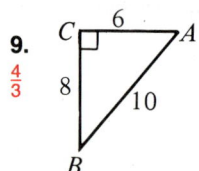

10–12. Find tan B for each right triangle above. Express your answers in simplest form. 10. 2   11. $\frac{8}{5}$ 12. $\frac{3}{4}$

13. What do you know about the legs of a 45°-45°-90° triangle? They are equal.

14. Without using the table, state the value of tan 45°. 1

15. The diagram shows a 30°-60°-90° triangle. The length of the longer leg is exactly $\sqrt{3}$. This is approximately 1.732.
 a. Using the diagram, find tan 60°. 1.732
 b. Using the table, find tan 60°. 1.7321

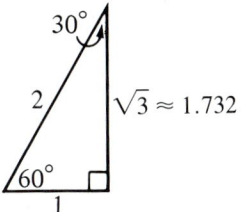

There are two ways to find the value of t.

16. a. From the figure, $\tan 55° = \frac{?}{?} \cdot \frac{4}{t}$
 b. Use the table: tan 55° = __?__. 1.4281
 c. From parts (a) and (b) we have $1.4281 = \frac{4}{t}$.
 Then $1.4281t = 4$, and $t = \frac{4}{1.4281}$.
 Correct to tenths, $t =$ __?__. 2.8

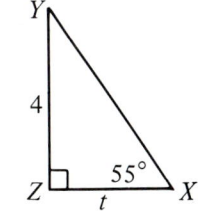

17. Here is an easier way to find the value of t.
 a. Find the measure of $\angle Y$. 35°
 b. From the figure, $\tan Y = \frac{?}{?} \cdot \frac{t}{4}$
 c. Use the table to find the value of tan Y. 0.7002
 d. From parts (b) and (c) we have $0.7002 = \frac{t}{4}$.
 Correct to tenths, $t =$ __?__. 2.8
 e. Compare your answer for part (d) with your answer for Exercise 16, part (c). They are the same.

Written Exercises

Use the table on page 517 to complete the following.

[A] **1.** tan 5° = __?__ 0.0875   **2.** tan 25° = __?__ 0.4663   **3.** tan 70° = __?__ 2.7475

4. tan __?__ = 3.7321 75°   **5.** tan __?__ = 0.5774 30°   **6.** tan __?__ = 0.1763 10°

Find tan A and tan B for each right triangle shown. Express your answers in simplest form.

7. 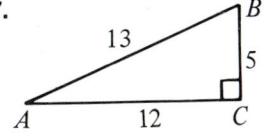 tan $A = \frac{5}{12}$; tan $B = \frac{12}{5}$

8. 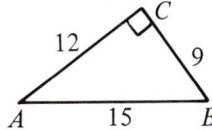 tan $A = \frac{3}{4}$; tan $B = \frac{4}{3}$

9. 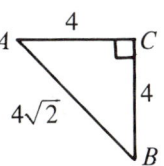 tan $A = 1$; tan $B = 1$

10. 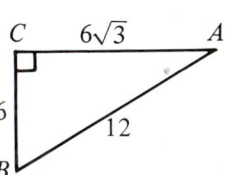 tan $A = \frac{\sqrt{3}}{3}$; tan $B = \sqrt{3}$

Find the value of y to the nearest tenth. Use the table on page 517.

11. 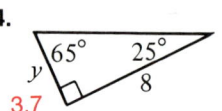 15°, 10, 2.7, y

12. y, 1.4, 35°, 2

13. 3, 40°, y, 2.5

B 14. 65°, 25°, y, 8, 3.7

15. 70°, 5, 20°, y, 13.7

16. 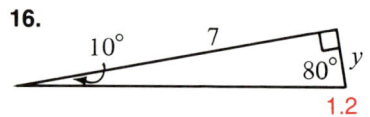 10°, 7, 80°, y, 1.2

17. The diagram shows a loading ramp. Find x, the distance the floor of the truck is from the ground, to the nearest tenth of a meter. **0.8 m**

18. The shadow of a lighthouse is 30 m long. The angle between the ground and the line to the sun is 35°. Find x, the height of the lighthouse to the nearest meter. **21 m**

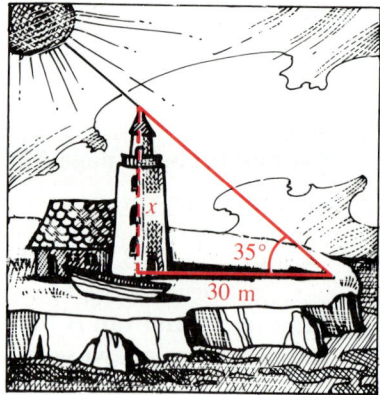

For each right triangle shown, do three things:
a. find tan A as a fraction;
b. find tan A as a decimal;
c. find the measure of $\angle A$ to the nearest degree by using the table on page 517.

C 19. 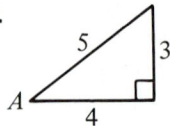 5, 3, 4, A
   a. $\frac{3}{4}$
   b. 0.75
   c. 37°

20. 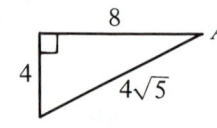 8, A, 4, $4\sqrt{5}$
   a. $\frac{1}{2}$ b. 0.5   c. 27°

21. 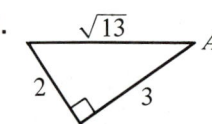 $\sqrt{13}$, A, 2, 3
   a. $\frac{2}{3}$ b. $0.\overline{6}$ c. 34°

Exercises 22-24 refer to the rectangular solid shown.

22. $x = $ __?__ 5

23. $\tan \angle ABC = $ __?__ 0.4000

24. To the nearest degree, $\angle ABC = $ __?__ °. 22

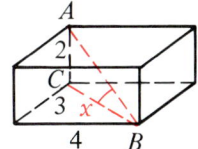

CAREER NOTES

AIRCRAFT ASSEMBLY TECHNICIAN

Suppose someone gave you a model airplane kit with forty parts inside. How long do you think it would take you to put it together? Four hours? Eight hours? Two days? Can you imagine how long it would take you to assemble an actual airplane with its thousands or even millions of parts?

Teams of aircraft assembly technicians divide the job of airplane construction into a large variety of highly specialized tasks. Some assembly technicians join complete sections of the aircraft. For example, they may use bolts, rivets, drills, and solder to join the landing gear to the fuselage. Other technicians work on a smaller scale, assembling engine or auxiliary component parts. Still others specialize in electronic equipment including tiny circuits and modules.

All aircraft assembly technicians must interpret engineering specifications on blueprints for both mechanical and electronic assemblies. Skilled assemblers need high school or vocational training with courses in mathematics, electronics, and mechanical drawing. Two to four years of plant experience are necessary to fully master aircraft assembly skills.

7 The Sine and Cosine Ratios

The tangent ratio involves the two legs of a right triangle. Two other important trigonometric ratios involve the hypotenuse and a leg of a right triangle. These ratios are called the *sine* and the *cosine*.

TRIGONOMETRIC RATIOS

tangent of $\angle A = \dfrac{\text{leg opposite } \angle A}{\text{leg adjacent to } \angle A}$

sine of $\angle A = \dfrac{\text{leg opposite } \angle A}{\text{hypotenuse}}$

cosine of $\angle A = \dfrac{\text{leg adjacent to } \angle A}{\text{hypotenuse}}$

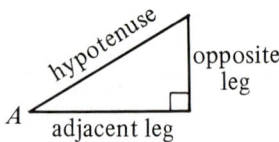

We write:

$$\tan A = \frac{\text{opposite}}{\text{adjacent}} \qquad \sin A = \frac{\text{opposite}}{\text{hypotenuse}} \qquad \cos A = \frac{\text{adjacent}}{\text{hypotenuse}}$$

In the previous section, we used the table on page 517 in our work with tangents. Notice that the table lists values of the sine ratio and the cosine ratio, in addition to those of the tangent ratio. Remember that most of the values are approximate. We shall use the trigonometric table in the examples which follow.

EXAMPLE 1   Find the value of y to the nearest tenth.

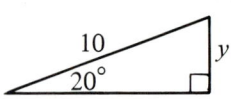

$$\sin 20° = \frac{\text{opposite}}{\text{hypotenuse}}$$

$$\sin 20° = \frac{y}{10}$$

From the table ⟶ $0.3420 = \dfrac{y}{10}$

$$10 \times 0.3420 = y$$

$$3.42 = y$$

To the nearest tenth, $y = 3.4$.

452   Chapter 11

EXAMPLE 2   A guy wire is stretched from the top of a tower to a point 10 m from the base of the tower. The wire makes a 65° angle with the ground. Find z, the length of the wire, to the nearest meter.

$$\cos 65° = \frac{\text{adjacent}}{\text{hypotenuse}}$$

$$0.4226 = \frac{10}{z}$$

$$0.4226z = 10$$

$$z = \frac{10}{0.4226}$$

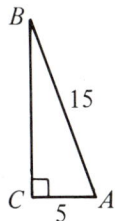

To the nearest meter, $z = 24$ m.

EXAMPLE 3   Find the measure of $\angle A$ to the nearest five degrees.

The given sides of the triangle are the leg adjacent to $\angle A$ and the hypotenuse. Therefore, we use the cosine ratio.

1. Find $\cos A$ as a fraction: $\cos A = \frac{\text{adjacent}}{\text{hypotenuse}}$

$$\cos A = \frac{5}{15} = \frac{1}{3}$$

2. Express $\cos A$ as a decimal:

$$\cos A = 0.3333\ldots$$

$$\begin{array}{r} 0.3333\ldots \\ 3\overline{)1.000} \end{array}$$

3. Look in the table under *Cosine* to find the entry closest to 0.3333.

To the nearest five degrees, $\angle A = 70°$.

Classroom Practice

Match each name with the correct expression.

1. sine **b.**
2. cosine **c.**
3. tangent **a.**

a. $\frac{\text{opposite}}{\text{adjacent}}$

b. $\frac{\text{opposite}}{\text{hypotenuse}}$

c. $\frac{\text{adjacent}}{\text{hypotenuse}}$

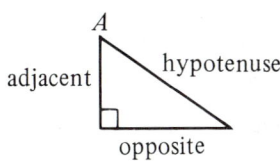

Right Triangles

Refer to △PQR. Find each ratio.

4. sin P $\frac{3}{5}$
5. cos P $\frac{4}{5}$
6. tan P $\frac{3}{4}$
7. sin Q $\frac{4}{5}$
8. cos Q $\frac{3}{5}$
9. tan Q $\frac{4}{3}$

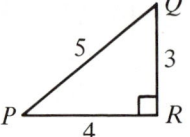

Use the table on page 517 to find the following.

10. sin 25° 0.4226
11. cos 40° 0.7660
12. tan 75° 3.7321
13. sin 50° 0.7660

Use the table on page 517 to find the measure of each angle to the nearest degree.

14. sin A = 0.2588   15°
15. cos P = 0.6428   50°
16. tan R = 0.3750   21°
17. sin S = 0.3500   20°

In each exercise, state an equation you could use to find the value of x.

18.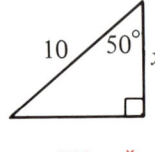
cos 50° = $\frac{x}{10}$

19.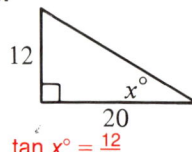
tan x° = $\frac{12}{20}$

20.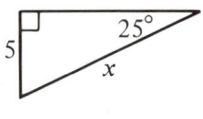
sin 25° = $\frac{5}{x}$

Written Exercises

Use the table on page 517 to find the following.

A 1. sin 5° 0.0872
2. cos 15° 0.9659
3. tan 45° 1.0000
4. cos 80° 0.1736

Use the table on page 517 to find the measure of ∠A to the nearest degree.

5. sin A = 0.9659   75°
6. cos A = 0.5736   55°
7. tan A = 2.1400   65°
8. cos A = 0.4900   61°

Refer to △ABC. Find each ratio.

9. sin A $\frac{5}{13}$
10. cos A $\frac{12}{13}$
11. tan A $\frac{5}{12}$
12. sin B $\frac{12}{13}$
13. cos B $\frac{5}{13}$
14. tan B $\frac{12}{5}$

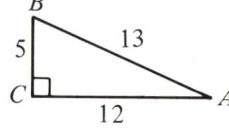

15. In △ABC above, tan B = $\frac{12}{5}$ = 2.4000. Use the trigonometric table to find the measure of ∠B to the nearest degree. 67°

Refer to △XYZ. Find each ratio.

16. sin X $\frac{7}{25}$
17. cos X $\frac{24}{25}$
18. tan X $\frac{7}{24}$
19. sin Y $\frac{24}{25}$
20. cos Y $\frac{7}{25}$
21. tan Y $\frac{24}{7}$

22. In △XYZ above, sin X = $\frac{7}{25}$ = 0.2800. Use the trigonometric table to find the measure of ∠X to the nearest degree. 16°

23. Find sin A as a fraction. $\frac{1}{2}$ 24. Find sin A as a decimal. 0.5

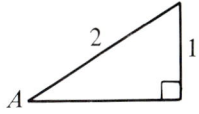

25. Use the table on page 517 to find the measure of $\angle A$ to the nearest degree. 30°

26. Find cos B as a fraction. $\frac{9}{10}$ 27. Find cos B as a decimal. 0.9

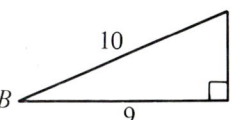

28. Use the table on page 517 to find the measure of $\angle B$ to the nearest degree. 26°

Find the value of y to the nearest tenth. Use the table on page 517.

29. 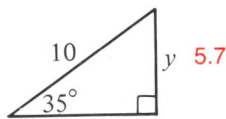 5.7

30. 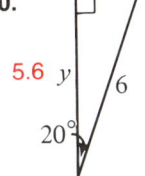 5.6

31. 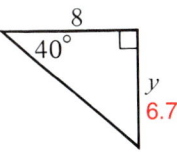 6.7

B 32. 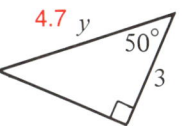 4.7

33. 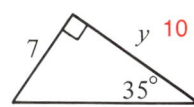 10

34. 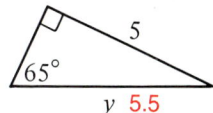 5.5

35. A tree trimmer has a ladder 4 m long. It is leaning against a palm tree at a 70° angle with the ground. How far up the tree does the ladder reach? 3.8 m

36. A golfer stands 80 m from the Number 5 hole. The golfer's aim is poor, and the ball goes 15° off course and lands at the left of the hole.
   a. How far is the ball from the hole? 21.4 m
   b. How far is the ball from the golfer? 82.8 m

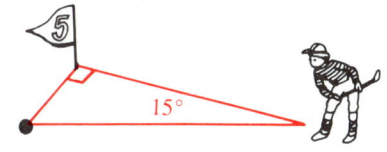

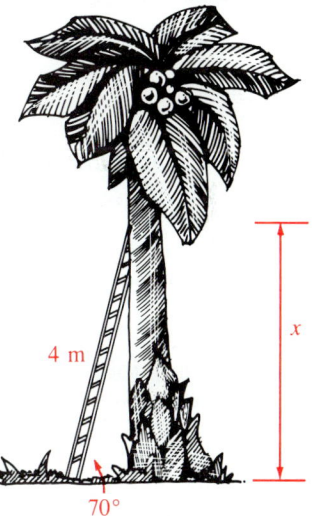

Find the measure of $\angle A$ to the nearest degree.

C 37. 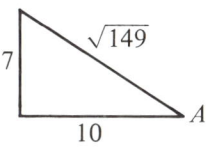 35°

38. 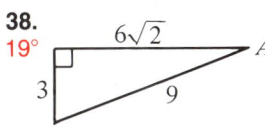 19°

39. 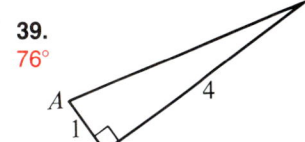 76°

Right Triangles   455

SELF-TEST

Vocabulary

tangent (p. 447)   sine (p. 452)   cosine (p. 452)

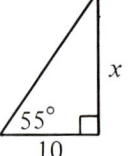

1. Find the value of x to the nearest tenth. Use the table on page 517. 14.3   (11-6)

Complete each statement.

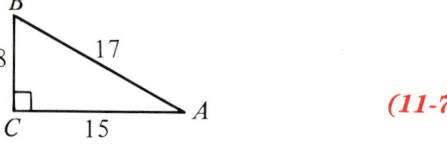

2. $\tan A = $ __?__ $\frac{8}{15}$
3. $\sin A = $ __?__ $\frac{8}{17}$ (11-7)
4. $\cos B = $ __?__ $\frac{8}{17}$
5. $\sin B = $ __?__ $\frac{15}{17}$

6. Find the values of y and z to the nearest tenth. Use the table on page 517. $y = 16.4$; $z = 11.5$

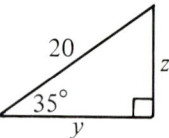

PUZZLE ◆ PROBLEMS

1. ENCOUNTER is a game that is played on graph paper. The players choose any two starting points, A and B, on the grid.

2. Player A draws a horizontal segment from point A to any other point of the grid. (The segment should not be very long.)

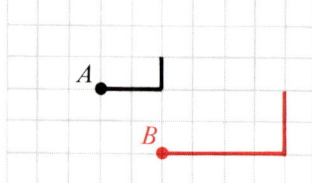

3. Player B starts at point B and draws a segment in the same direction as player A's segment, but twice as long.

4. Player A joins a vertical segment to the end of the first segment drawn.

5. Player B draws a segment in the same direction as player A's second segment, but twice as long.

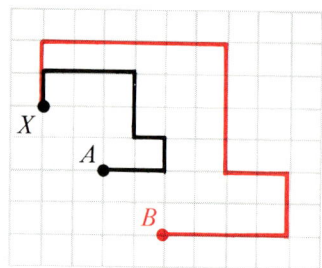

6. Taking turns, the players draw horizontal segments. Then they both draw vertical segments. Each time, player B draws a segment in the same direction as player A's, but twice as long.

The object of the game is for player A to move, in as few moves as possible, to a point X that will force B to move to the same point.

APPLICATIONS

ANGLES OF ELEVATION AND DEPRESSION

Sometimes you can not make measurements using a meter stick, yard stick, or tape measure. Airplanes and ships often have sophisticated electronic devices or computers that figure out altitudes, ocean depth, and other distances. Trigonometry can also be used to find distances that are hard to measure physically.

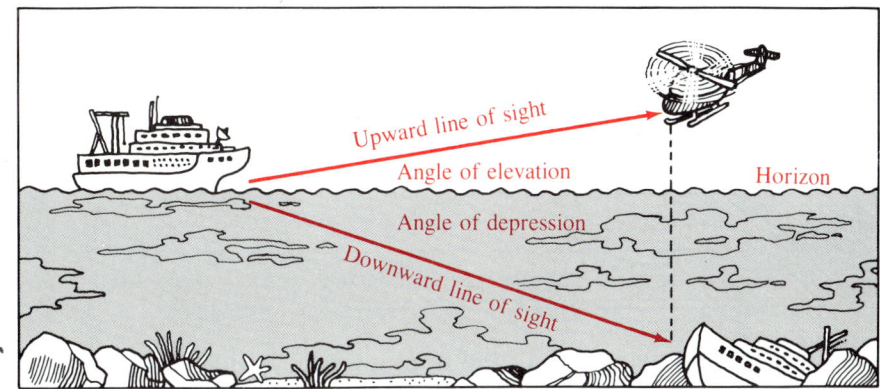

The ship in the diagram is taking scientists and historians on a search for sunken treasure. Their helicopter has spotted what appears to be a shipwreck on the ocean floor. While it hovers above the wreck, the ship's crew determines the relative positions of the ship, helicopter, and wreck. They look out over the ocean into the horizon. Then they adjust their line of sight upward and downward.

If the crew looks *up*, they see the helicopter. Their line of sight forms an **angle of elevation** with the horizon. Looking *downward*, their line of sight forms an **angle of depression** with the horizon.

Now consider the perspectives looking *from* the helicopter and wreck *to* the ship. Notice the angles that are formed by these lines of sight and the vertical line from *H* to *W*. These are also called angles of elevation and depression. The helicopter pilot looks *down* to the ship. The angle indicated at *H* is called an *angle of depression* of *S*. A scuba diver exploring the wreck would have to look *up* to the ship. The angle indicated at *W* is called an *angle of elevation* of *S*.

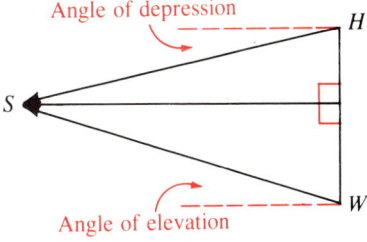

Right Triangles **457**

Notice that the line of sight is always the hypotenuse of a right triangle. The vertical and horizontal distances are the legs of the triangle. Therefore, given an angle of elevation or depression and one of these distances, we can use one of the trigonometric ratios to find a missing distance.

EXAMPLE An Olympic swimming pool is 50 m long. From the opposite end of the pool, the angle of elevation of the highboard diving platform is about 11°. How high is the platform?

Step 1 Draw a diagram to illustrate the problem. Include all the given information in relation to a right triangle.

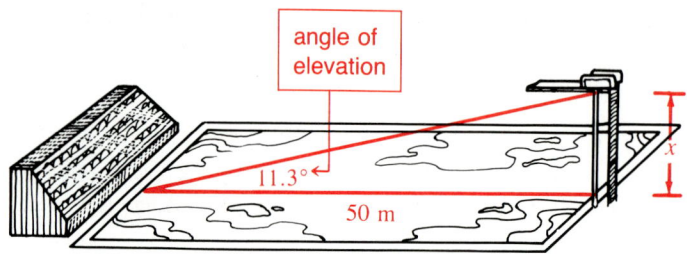

Step 2 Write an equation using one of the trigonometric ratios. The platform ladder is opposite the angle of elevation. The length of the pool is adjacent to the angle of elevation.

$$\tan 11° = \frac{\text{opposite}}{\text{adjacent}} = \frac{x}{50}$$

Step 3 Solve the equation.
$$\tan 11° = \frac{x}{50}$$
$$0.1944 = \frac{x}{50}$$
$$50(0.1944) = x$$
$$9.72 = x$$

To the nearest whole number, the height of the diving platform is 10 m.

EXERCISES

1. $\overleftrightarrow{PR}$ is horizontal
   a. For a person at Q, what is the angle of elevation of S? 15°
   b. From S, what is the angle of depression of Q? 15°
   c. Someone at P measures the angle of elevation of S. Is the measure greater than or less than 15°? less than 15°

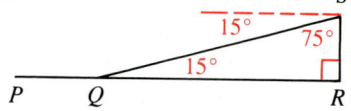

2. The lines shown are horizontal and vertical except for $\overleftrightarrow{AD}$ and $\overleftrightarrow{CD}$. Give the number of the angle and its special name when:
   a. An observer at D sights A. ∠2; ∠ of elevation
   b. An observer at D sights C. ∠3; ∠ of depression
   c. An observer at A sights D. ∠5; ∠ of depression
   d. An observer at C sights D. ∠8; ∠ of elevation

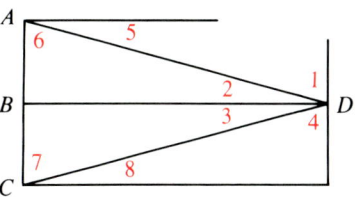

Draw a diagram to illustrate each situation below.

3. A kite flies at an angle of elevation of 50°. All 120 m of string have been let out. You want to find the kite's height.

4. The height of a lighthouse is 16 m. From the top of the lighthouse, the angle of depression of a nearby boat is 40°. You want to find the distance from the base of the lighthouse to the boat.

5. A sloping theater aisle is 80 m long. From the lowest point next to the stage, the angle of elevation of the highest point is 10°. You want to find the *horizontal* distance between the stage and the rear of the theater.

6–8. Write an equation to find the missing distances in Exercises 3–5. (*Hint:* Each exercise requires a different trigonometric ratio.) (See below.)

9–11. Solve the equations you wrote for Exercises 6–8. Round your answers to the nearest whole number. (See below.)

6. $\sin 50° = \frac{x}{120}$
7. $\tan 40° = \frac{16}{x}$
8. $\cos 10° = \frac{x}{80}$
9. 92 m
10. 19 m
11. 79 m

PUZZLE ◆ PROBLEMS

Begin with a right triangle.
Draw squares, as shown, and cut them out.

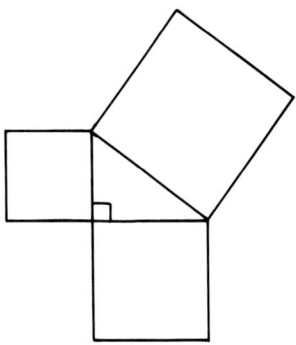

Place the two smaller squares next to one another.
Locate point X so that $VX = AB$.
Cut along $\overline{AX}$ and $\overline{WX}$.
Try to rearrange the five pieces formed to cover the largest square.

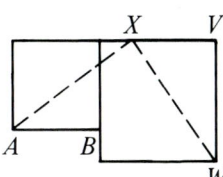

Right Triangles **459**

SKILLS REVIEW

ALGEBRAIC FRACTIONS: MULTIPLICATION AND DIVISION

Simplify.

SAMPLE 1 $\quad \dfrac{-5x^2}{20xy} = \dfrac{-1 \cdot 5 \cdot x \cdot x}{4 \cdot 5 \cdot x \cdot y} = \dfrac{-x}{4y} \ or \ -\dfrac{x}{4y}$

SAMPLE 2 $\quad \dfrac{7x - 21}{x^2 - 9} = \dfrac{7(x-3)}{(x+3)(x-3)} = \dfrac{7}{x+3}$

1. $\dfrac{-3a^2b^2}{12ab^2} \quad -\dfrac{a}{4}$
2. $\dfrac{2h^4k^5}{10h^3k} \quad \dfrac{hk^4}{5}$
3. $\dfrac{11b^2c^4}{-33bc^2} \quad -\dfrac{bc^2}{3}$
4. $\dfrac{-4r^3s}{-16rt} \quad \dfrac{r^2s}{4t}$

5. $\dfrac{2m}{am+bm} \quad \dfrac{2}{a+b}$
6. $\dfrac{6x^2}{3xy-9x^2} \quad \dfrac{2x}{y-3x}$
7. $\dfrac{2a+2b}{a^2-b^2} \quad \dfrac{2}{a-b}$
8. $\dfrac{x+4}{x^2+6x+8} \quad \dfrac{1}{x+2}$

9. $\dfrac{a^2-64}{a+8} \quad a-8$
10. $\dfrac{6r-6s}{r^2-s^2} \quad \dfrac{6}{r+s}$
11. $\dfrac{d^2-2d-15}{13d+39} \quad \dfrac{d-5}{13}$
12. $\dfrac{3a-21}{a^2-14a+49} \quad \dfrac{3}{a-7}$

Multiply.

SAMPLE $\quad \dfrac{a^2-2ab+b^2}{10} \cdot \dfrac{5}{a-b} = \dfrac{5(a^2-2ab+b^2)}{10(a-b)} = \dfrac{5(a-b)(a-b)}{2 \cdot 5(a-b)} = \dfrac{a-b}{2}$

13. $\dfrac{x}{y} \cdot \dfrac{5y}{2x} \quad \dfrac{5}{2}, \text{ or } 2\tfrac{1}{2}$
14. $\dfrac{8a^2}{b} \cdot \dfrac{b^2}{2a} \quad 4ab$
15. $\dfrac{a+b}{3} \cdot \dfrac{9}{ab+b^2} \quad \dfrac{3}{b}$

16. $\dfrac{7x}{9y} \cdot \dfrac{18}{35x^2} \quad \dfrac{2}{5xy}$
17. $\dfrac{4}{x+2} \cdot \dfrac{x^2-4}{12} \quad \dfrac{x-2}{3}$
18. $\dfrac{7x-21}{3} \cdot \dfrac{15}{x-3} \quad 35$

19. $\dfrac{x-5}{y} \cdot \dfrac{y^3}{x^2-25} \quad \dfrac{y^2}{x+5}$
20. $\dfrac{m-3}{m+3} \cdot \dfrac{3+m}{3-m} \quad -1$
21. $\dfrac{x^2+4x+4}{28} \cdot \dfrac{7}{x+2} \quad \dfrac{x+2}{4}$

Divide.

SAMPLE $\quad \dfrac{a^3}{3} \div \dfrac{a^4}{3a^2+3} = \dfrac{a^3}{3} \cdot \dfrac{3a^2+3}{a^4} = \dfrac{a \cdot a \cdot a \cdot 3(a^2+1)}{3 \cdot a \cdot a \cdot a \cdot a} = \dfrac{a^2+1}{a}$

22. $\dfrac{b}{3} \div \dfrac{1}{9} \quad 3b$
23. $\dfrac{m^2}{n^2} \div 2m \quad \dfrac{m}{2n^2}$
24. $3r^2s^2 \div \dfrac{12s^3}{5t^2} \quad \dfrac{5r^2t^2}{4s}$

25. $\dfrac{a+b}{12} \div \dfrac{a+b}{6b} \quad \dfrac{b}{2}$
26. $\dfrac{2\pi r - 2\pi s}{9} \div \dfrac{2\pi}{3} \quad \dfrac{r-s}{3}$
27. $\dfrac{4c-12}{7} \div (c-3) \quad \dfrac{4}{7}$

28. $\dfrac{n^2-64}{5} \div (n+8) \quad \dfrac{n-8}{5}$
29. $\dfrac{d^2+d^3}{2c+4} \div \dfrac{d^2}{6} \quad \dfrac{3(1+d)}{c+2}$
30. $\dfrac{3y}{y^2-49} \div \dfrac{1}{y+7} \quad \dfrac{3y}{y-7}$

CHAPTER REVIEW

CHAPTER SUMMARY

1. Right triangle lengths you should know are included in the table on page 428.

2. In a 45°-45°-90° triangle, the hypotenuse is $\sqrt{2}$ times as long as a leg.

3. In a 30°-60°-90° triangle, the hypotenuse is twice as long as the shorter leg, and the longer leg is $\sqrt{3}$ times as long as the shorter leg.

4. If a rectangular solid has dimensions a, b, and c, then a diagonal has length $d = \sqrt{a^2 + b^2 + c^2}$.

5. You can often use the Pythagorean Theorem to find the slant height or the height of a regular pyramid or cone.

6. Three trigonometric ratios are defined for any acute angle of a right triangle as follows:

$\tan A = \dfrac{\text{opposite}}{\text{adjacent}}$

$\sin A = \dfrac{\text{opposite}}{\text{hypotenuse}}$

$\cos A = \dfrac{\text{adjacent}}{\text{hypotenuse}}$

REVIEW EXERCISES

Find the exact values of x and y. *(See pp. 428–436.)*

1.

$x = 10, y = 15$

2.

$x = 12, y = 15$

3.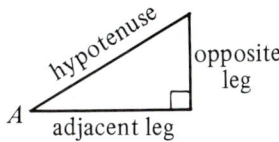

$x = 2, y = 4\sqrt{2}$

4.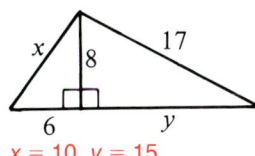

$x = 18, y = 9\sqrt{3}$

5.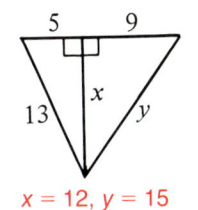

$x = 4, y = 4\sqrt{2}$

6.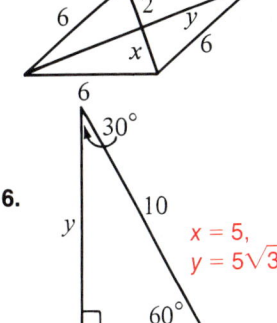

$x = 5, y = 5\sqrt{3}$

Right Triangles   **461**

Find the exact area of each shaded figure. *(See pp. 437–439.)*

7.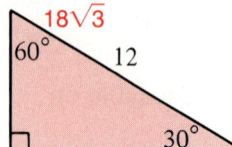
8.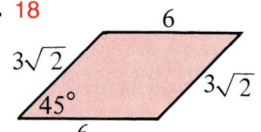
9.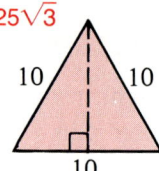

Refer to the rectangular solid shown. *(See pp. 440–442.)*

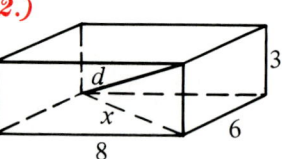

10. Find the value of x. 10
11. Find the length of the diagonal d. $\sqrt{109}$

Refer to the regular square pyramid and cone shown at the right. *(See pp. 443–446.)*

12. Find the lateral area of the pyramid. 240
13. Find the total area of the pyramid. 384
14. Find the height of the pyramid. 8
15. Find the volume of the pyramid. 384

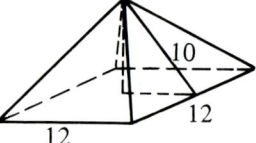

16. Find the slant height of the cone. 15
17. Find the lateral area of the cone. 135π
18. Find the total area of the cone. 216π
19. Find the volume of the cone. 324π

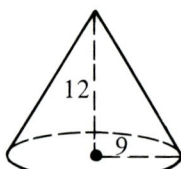

Use the diagram to find the following. *(See pp. 447–455.)*

20. $\tan A$ $\frac{3}{4}$ 21. $\cos A$ $\frac{4}{5}$
22. $\sin B$ $\frac{4}{5}$ 23. $\tan B$ $\frac{4}{3}$

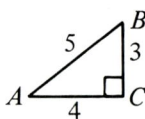

Find the value of x to the nearest tenth. Use the table on page 517. *(See pages 447–455.)*

24.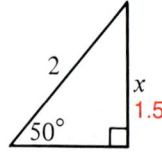
25.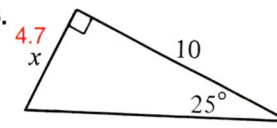
26.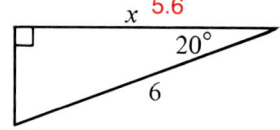

CHAPTER TEST

Find the exact values of x and y.

1. $x = 8, y = 10$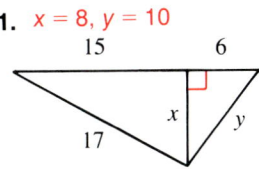

2. $x = 2\sqrt{7}, y = 9$

3. 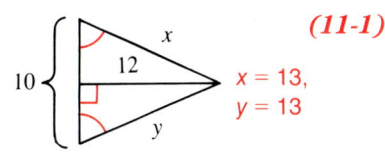 $x = 13, y = 13$

4. $x = 3, y = 3\sqrt{2}$

5. 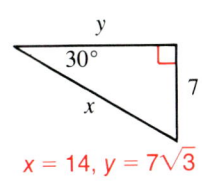 $x = 14, y = 7\sqrt{3}$

6. 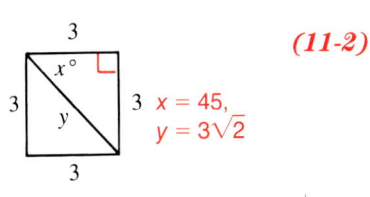 $x = 45, y = 3\sqrt{2}$

(11-1)

(11-2)

Find the area of each figure described.

7. The parallelogram shown $12\sqrt{2}$

8. An isosceles right triangle with a hypotenuse of $9\sqrt{2}$ $\quad \frac{81}{2}$

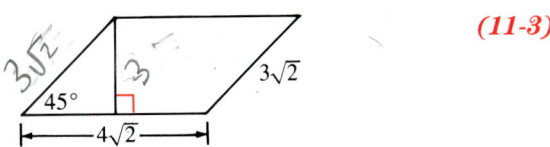

(11-3)

A rectangular solid has length 4, width 3, and height 12.

9. Sketch the figure.
10. Find the length of a diagonal of a base. 5
11. Find the length of a diagonal of the rectangular solid. 13

(11-4)

A regular square pyramid has height 16 and base edges 24.

12. $l = \underline{\ ?\ }$ 20
13. L.A. $= \underline{\ ?\ }$ 960
14. T.A. $= \underline{\ ?\ }$ 1536
15. $V = \underline{\ ?\ }$ 3072

(11-5)

A cone has height 21 and slant height 29.

16. $r = \underline{\ ?\ }$ 20
17. L.A. $= \underline{\ ?\ }$ 580π
18. T.A. $= \underline{\ ?\ }$ 980π
19. $V = \underline{\ ?\ }$ 2800π

20. In $\triangle ABC$, $AC = 3$, $BC = 4$, and $AB = 5$. Find $\tan A$ and $\tan B$. $\frac{4}{3}; \frac{3}{4}$

(11-6)

21. Write an equation that can be used to find the value of x.

22. If $\tan 50° = 1.1918$, find the value of x to the nearest tenth. 23.8

21. $\tan 50° = \frac{x}{20}$

Find the value of x to the nearest tenth. Use the following as needed: $\sin 15° = 0.2588$, $\cos 15° = 0.9659$, and $\tan 15° = 0.2679$.

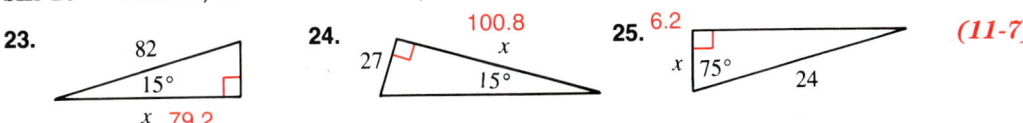

23. x 79.2
24. x 100.8
25. x 6.2

(11-7)

MIXED REVIEW

Tell whether each statement is true or false.

1. If a triangle has two equal angles, then it must be isosceles. true
2. If the diagonals of a quadrilateral are perpendicular, then the quadrilateral is a rhombus. false
3. The converse of a true "If . . . then" statement is not always true. true
4. If two lines do not intersect, then they must be parallel. false
5. The ratio of the areas of two similar polygons is equal to the ratio of two corresponding sides. false
6. An exterior angle of a regular polygon is always obtuse. false
7. Complete the following proof.

    Given: $ABCD$ is an isosceles trapezoid
    $\overline{DE} \perp \overline{AB}$; $\overline{CF} \perp \overline{AB}$

    Prove: $\angle ADE = \angle BCF$

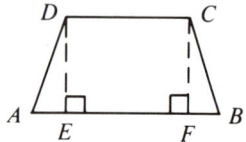

STATEMENTS	REASONS
1. $ABCD$ is an isos. trapezoid	1. __?__ Given
2. $AD = BC$	2. __?__ Def. of isos. trapezoid
3. $\angle A = \angle B$	3. __?__ Base $\triangle$ of an isos. trap. are =.
4. $\overline{DE} \perp \overline{AB}$; $\overline{CF} \perp \overline{AB}$	4. __?__ Given
5. $\angle DEA = 90° = \angle CFB$	5. __?__ Def. of $\perp$ lines
6. $\triangle AED \cong \triangle$ __?__ BFC	6. __?__ AAS Theorem
7. __?__ $\angle ADE = \angle BCF$	7. __?__ Corr. parts of $\cong \triangle$ are =.

8. In $\triangle MLS$, $\angle M = 90°$, $LM = 15$, and $SM = 8$. Find the values of $\sin S$, $\cos S$, and $\tan S$. $\sin S = \frac{15}{17}$; $\cos S = \frac{8}{17}$; $\tan S = \frac{15}{8}$
9. Explain why the median of a triangle divides it into two triangles with equal areas. (*Hint:* The triangles have the same height.)
   The $\triangle$ have the same height and equal bases.
10. Draw a line l and a point P not on l. Construct a line through P and parallel to l. Label each pair of alternate interior angles.

11. ∠1 is a central angle of ⊙O and ∠2 is an inscribed angle of ⊙O. If both ∠1 and ∠2 intercept $\widehat{AB}$ and $\widehat{AB}$ = 50°, ∠1 = __?__° and ∠2 = __?__°. 50 ; 25

12. The bases of a right prism are 3-4-5 right triangles. The height of the prism is 10. Find the lateral area, the total area, and the volume of the prism. L.A. = 120; T.A. = 132; V = 60

Find the values of x and y.

13.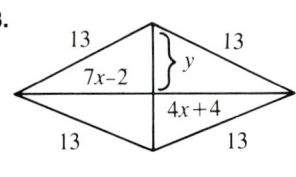
x = 2, y = 5

14.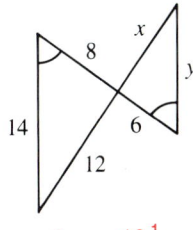
x = 9, y = 10½

15.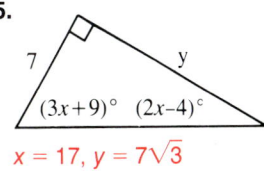
x = 17, y = 7√3

16. Which *one* of the following is *not* true? **b.**
    a. ∠XMN = ∠Y   b. $\dfrac{XM}{MY} = \dfrac{MN}{YZ}$
    c. YZ = 2(MN)

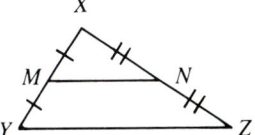

17. A cone has slant height 29 and radius 21. Find its volume. 2940π

18. Find the area of a trapezoid with bases of 11 and 20, and height 8. 124

19. Draw a large obtuse triangle. Construct and label the altitude to the shortest side.

20. Suppose R and S lie on the perpendicular bisector of $\overline{FG}$. What postulate or theorem supports the statement?
    a. RF = RG and SF = SG   Any point on the ⊥ bisector of a segment is equidistant from the endpoints of the segment.
    b. △RFS ≅ △RGS (Hint: RS = RS)
       SSS Postulate

Find the value of t.

21. 24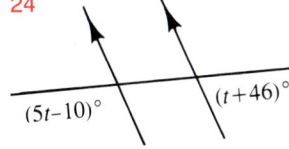

22. 16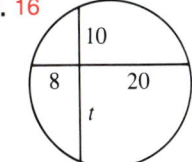

23.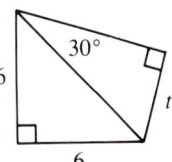
3√2

24. Find the sum of the measures of the angles of a decagon. 1440°

25. Find the diameter of a sphere with surface area 400π. 20

Here's what you'll learn in this chapter:

To specify points by their coordinates in the coordinate plane.

To find the distance between two points in the coordinate plane.

To apply the midpoint formula.

To find the slope of a line.

To find the slopes of parallel and perpendicular lines.

To graph the line specified by a given equation.

To prove theorems using coordinate geometry.

The intersecting lines drawn to help you locate points on the coordinate plane are called a grid. Such a grid is suggested by the construction of the power switching yard in Vancouver, Washington, shown above.

Chapter 12

Coordinate Geometry

1 Points and Coordinates

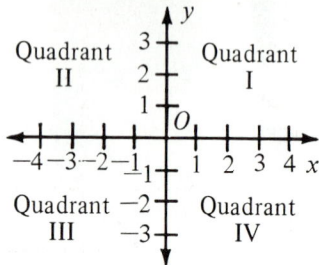

The *coordinate plane* has an *x*-axis and a *y*-axis, as shown in the diagram. These **coordinate axes** (**ax**-eez) are number lines that intersect in a point called the origin, point *O*. The axes separate the plane into four quadrants.

Every point in the coordinate plane can be named by two numbers, called the **coordinates** of the point. In the diagram, the ***x*-coordinate** of point *A* is 2. The ***y*-coordinate** is 4. We can refer to point *A* as the point (2, 4). Notice that the origin is the point (0, 0).

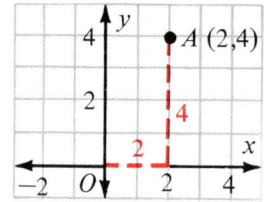

Point *B*, with coordinates -2 and 3, has been *plotted* (drawn) in this diagram. We can name point *B* as the point $(-2, 3)$. Can you name the coordinates of points *C* and *D*? (Remember that the *x*-coordinate is always named first.)

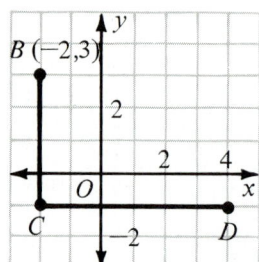

From the diagram you can conclude:

$BC = 4$ $CD = 6$ $\angle BCD = 90°$

Classroom Practice

Exercises 1–5 refer to points that are labeled in Figure 1.

1. Point *B* lies on an axis, not in a quadrant. Name two other points that do not lie in quadrants. **E, O**

2. Name the points that lie in:
   a. quadrant I. **D, A**   b. quadrant II. **F, G**
   c. quadrant III. **H, J** d. quadrant IV. **C**

3. Name the points that have:
   a. *x*-coordinate 0. **E, O**
   b. a positive *x*-coordinate. **A, B, C, D**
   c. a negative *x*-coordinate. **F, G, H, J**

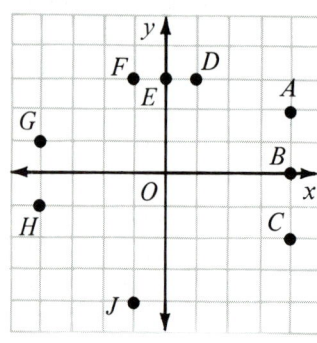

Figure 1

468   Chapter 12

4. Name the points in Figure 1 that have:
   a. y-coordinate 0. *B, O*
   b. a positive y-coordinate. *A, D, E, F, G*
   c. a negative y-coordinate. *C, H, J*

   $A(4, 2); B(4, 0); C(4, -2);$
   $D(1, 3); E(0, 3); F(-1, 3);$
   $G(-4, 1); H(-4, -1); J(-1, -4);$

5. State the coordinates of each point in Figure 1. *O(0, 0)*

6. Think of all the points that lie on the x-axis. What can you say about the y-coordinate of each of those points? *The y-coordinate of each is zero.*

7. Think of all the points that have an x-coordinate equal to 0. What can you say about each of those points? *They are on the y-axis.*

8. From the diagram, what can you conclude about ST? about $\angle RTS$? *ST = 5; $\angle RTS = 45°$*

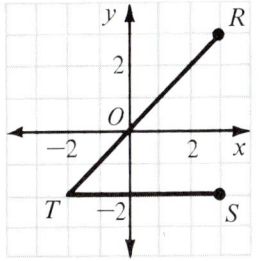

Exercise 8

The legs of the right triangle are parallel to the coordinate axes. State the coordinates of point C.

9.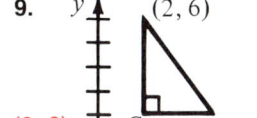
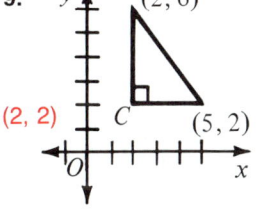

10.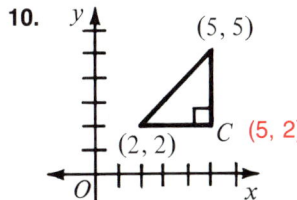

11.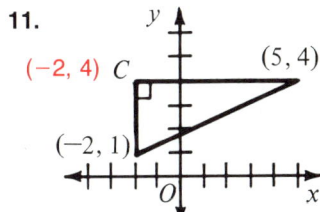

12.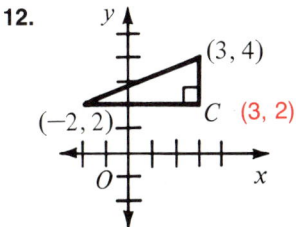

13.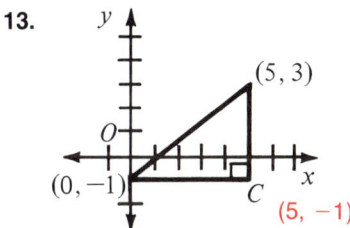

14.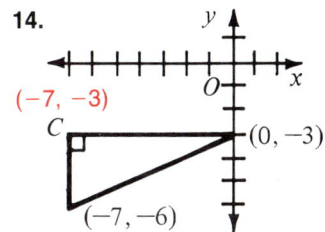

Coordinate Geometry   **469**

Written Exercises

A 1. Point $(3, -1)$ lies in quadrant __?__ IV, and point $(-3, 1)$ lies in quadrant __?__. II

2. Are $(-5, -4)$ and $(-4, -5)$ the same point? no Do the points lie in the same quadrant? yes

3. If the x-coordinate and the y-coordinate of a point are both negative numbers, the point must lie in quadrant __?__. III

4. Point $(-4, 0)$ lies on the __?__-axis. x

5. If the x-coordinate of a point is 0, the point must lie on the __?__-axis. y

6. Use the diagram to find the coordinates of each of the labeled points.
   A(0, 3); B(2, 4); SAMPLE Z(3, −2)
   C(3, 1); D(1, 0); E(1, −3); F(3, −4); G(0, −4);
   H(−3, −2); J(−3, 0); K(−3, 3); L(−2, 2); M(−1, 4)

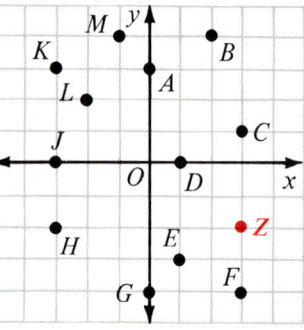

Exercise 6

For each exercise, draw a pair of coordinate axes.
a. Plot points R and S. Draw $\overleftrightarrow{RS}$.   b. State which point (A, B, or C) lies on $\overleftrightarrow{RS}$.

	R	S	A	B	C	
7.	(−2, −3)	(−2, 1)	(2, 2)	(4, −2)	(−2, −2)	b. C
8.	(0, 0)	(−3, 2)	(3, 2)	(−6, 4)	(1, 1)	b. B
9.	(1, 1)	(2, 2)	(5, 0)	(6, 7)	(−3, −3)	b. C
10.	(5, 5)	(−5, 0)	(−7, −1)	(−2, 1)	(7, 7)	b. A

In Exercises 11–16, a base of the rectangle is parallel to the x-axis. State the coordinates of points J and K.

11. J(2, 2); K(5, 5)

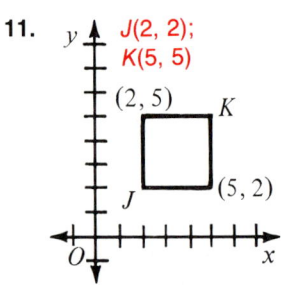

12. J(3, −2); K(7, −8)

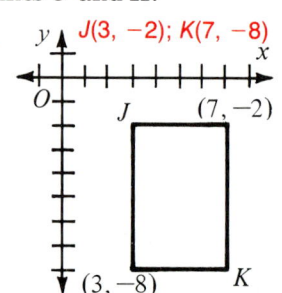

13. J(−5, 5); K(5, 2)

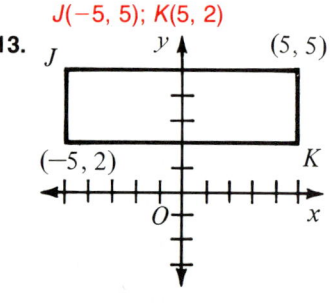

470   Chapter 12

14. J(−6, −5); K(0, −2)

15. J(−3, −1); K(5, −4)

16. J(−3, −2); K(2, 3)

In Exercises 17–22, find the distance between the points named.

17. (4, 0) and (9, 0) 5

18. (0, 7) and (0, 2) 5

19. (5, 2) and (2, 2) 3

20. (4, 1) and (4, 6) 5

21. (1, 3) and (−5, 3) 6

22. (−2, 4) and (−2, −3) 7

Plot the four points and draw quadrilateral ABCD. State whether the quadrilateral is a square, rectangle, parallelogram, or trapezoid. (Use the name that gives the best description of the figure.)

	A	B	C	D	
B 23.	(−3, −2)	(6, −2)	(5, 1)	(−1, 1)	trapezoid
24.	(2, 1)	(6, 3)	(6, 0)	(2, −2)	parallelogram
25.	(−1, −1)	(1, 1)	(5, −3)	(3, −5)	rectangle
26.	(3, 2)	(5, −1)	(8, 1)	(6, 4)	square

Plot points R and S and draw $\overleftrightarrow{RS}$. Point T is to lie on $\overleftrightarrow{RS}$, but only one coordinate of T is given. Find the other coordinate of T.

	R	S	T	
27.	(0, −2)	(−3, 0)	(−6, ?)	2
28.	(9, 9)	(6, 4)	(0, ?)	−6
29.	(6, −5)	(10, 5)	(?, 0)	8
30.	(−6, −1)	(9, −7)	(?, −3)	−1

C **31.** Points (0, 0), (0, 5), and (3, 4) are three of the vertices of a parallelogram. Find three possible positions for the fourth vertex. (−3, 1); (3, −1); (3, 9)

32. Repeat Exercise 31 but use points (−1, 2), (5, 0), and (7, 4).
(1, 6); (13, 2); (−3, −2)

Coordinate Geometry **471**

2 Distance between Two Points

You can count squares to check the lengths of $\overline{RT}$ and $\overline{ST}$ marked in the diagram. You cannot find the length of $\overline{RS}$ by counting squares. However, you can think this way:

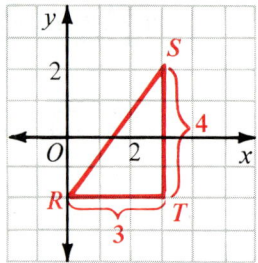

$\triangle RST$ is a right triangle.
The lengths of the legs are 3 and 4.
Thus, the length RS of the hypotenuse must be 5.

EXAMPLE Find the distance between $A(-2, 2)$ and $B(5, -1)$.

Plot points A and B. Then draw horizontal and vertical segments to form right $\triangle ACB$. Find AC and BC by counting.

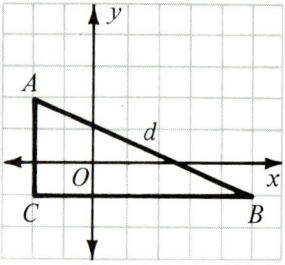

Use the Pythagorean Theorem.

$AC = 3 \qquad BC = 7$
$d^2 = 3^2 + 7^2$
$d^2 = 9 + 49 = 58$
$d = \sqrt{58}$

By using the square root table on page 516, you can approximate d to the nearest tenth.

$d = \sqrt{58} \approx 7.6$

Classroom Practice

State the lengths of the two legs of the right triangle.

1.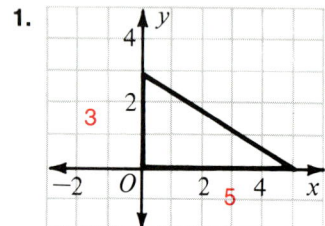

2.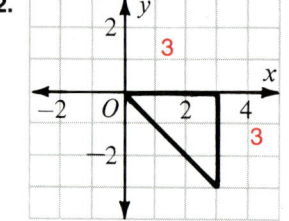

3.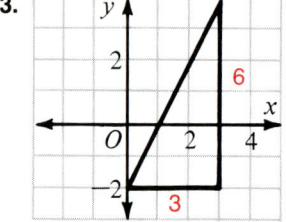

4.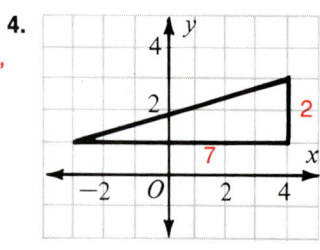

5.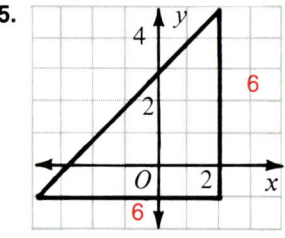

6.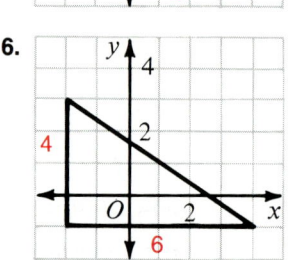

7. Leg $\overline{RT}$ of the right triangle is parallel to the x-axis. Answer in terms of a, b, and c.

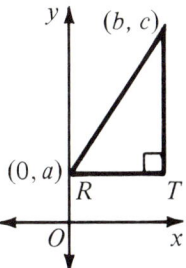

   a. The x-coordinate of point T is __?__. b
   b. The y-coordinate of point T is __?__. a
   c. The length of $\overline{RT}$ is __?__. b

8. Leg $\overline{VW}$ of the right triangle is parallel to the x-axis. Answer in terms of j, k, p, and q.

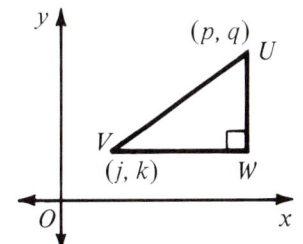

   a. The x-coordinate of point W is __?__. p
   b. The length of $\overline{VW}$ is __?__. p − j
   c. The y-coordinate of point W is __?__. k
   d. The length of $\overline{UW}$ is __?__. q − k

Written Exercises

In Exercises 1–12, draw axes and plot points A and B. Then draw a right triangle you can use to find AB. Then find AB. Leave your answer in radical form if it isn't a whole number.

SAMPLE $A(-3, 1)$ and $B(5, 4)$

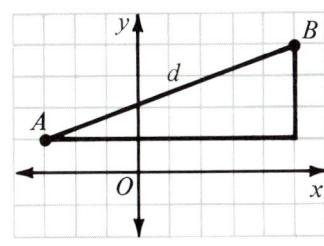

$d^2 = 8^2 + 3^2$
$d^2 = 64 + 9$
$d^2 = 73$
$d = \sqrt{73}$

A
1. $A(0, 4)$ and $B(5, 2)$ $d = \sqrt{29}$
2. $A(2, 2)$ and $B(6, 7)$ $d = \sqrt{41}$
3. $A(-2, -1)$ and $B(3, 5)$ $d = \sqrt{61}$
4. $A(-5, 3)$ and $B(4, -4)$ $d = \sqrt{130}$
5. $A(2, 1)$ and $B(4, 2)$ $d = \sqrt{5}$
6. $A(3, 5)$ and $B(5, 8)$ $d = \sqrt{13}$
7. $A(-2, 1)$ and $B(-5, 4)$ $d = 3\sqrt{2}$
8. $A(-6, 7)$ and $B(-4, -2)$ $d = \sqrt{85}$
9. $A(2, 4)$ and $B(-2, 1)$ $d = 5$
10. $A(-3, 3)$ and $B(1, 6)$ $d = 5$
11. $A(4, -2)$ and $B(-2, 3)$ $d = \sqrt{61}$
12. $A(-2, -3)$ and $B(1, 5)$ $d = \sqrt{73}$

Find the distance between points A and B correct to tenths. Use the square root table on page 516.

B 13. $A(-2, -2)$ and $B(3, 3)$ 7.1   14. $A(2, -2)$ and $B(-4, 4)$ 8.5

15. $A(-4, -3)$ and $B(-6, -7)$ 4.5   16. $A(0, 6)$ and $B(-4, -1)$ 8.1

Find the perimeter and the area of the triangle shown.

17. 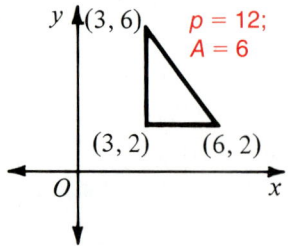   18.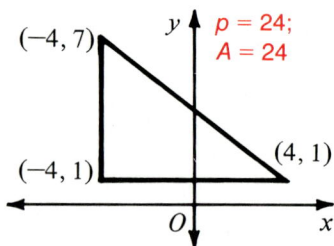

19. 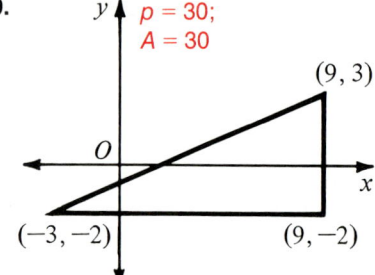   20.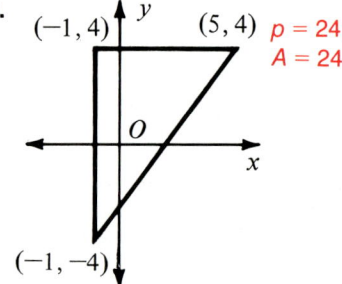

Find the perimeter and area of the square with the given vertices.

C 21. Vertices: $(0, 0)$; $(4, 3)$; $(1, 7)$; $(-3, 4)$ $p = 20$; $A = 25$

22. Vertices: $(2, -1)$; $(4, 1)$; $(6, -1)$; $(4, -3)$ $p = 8\sqrt{2}$; $A = 8$

Find the perimeter and the area of the quadrilateral shown.

23. 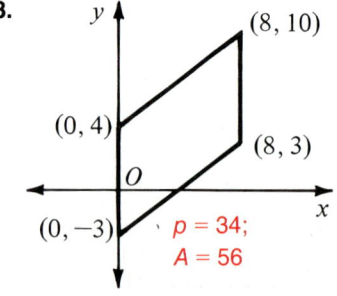   24.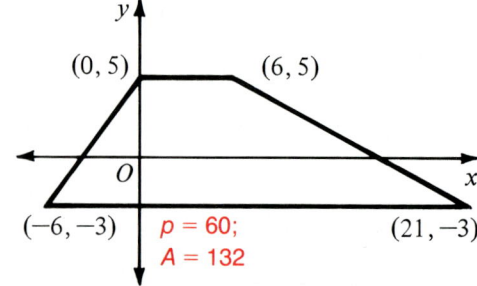

CALCULATOR ACTIVITIES

The rise, run, and length of a ramp are dependent upon each other. When one changes, the others do also.

The tangent ratio can be used to describe the slope of the ramp.

$$\text{Slope} = \frac{\text{rise}}{\text{run}}$$
$$= \frac{\text{leg opposite } \angle A}{\text{leg adjacent to } \angle A}$$
$$= \text{tangent of } \angle A$$

Similarly, $\frac{\text{rise}}{\text{ramp length}} = \text{sine of } \angle A$

and $\frac{\text{run}}{\text{ramp length}} = \text{cosine of } \angle A$.

Assume that $\angle A$ is known in each case. State which trigonometric ratio you would use to find the unknown length.

1. Known: rise   Unknown: **a.** run   **b.** ramp length   **a.** tangent   **b.** sine
2. Known: run   Unknown: **a.** rise   **b.** ramp length   **a.** tangent   **b.** cosine
3. Known: ramp length   Unknown: **a.** rise   **b.** run   **a.** sine   **b.** cosine

Use a calculator to find the trigonometric ratios for each measure of $\angle A$ to four decimal places. Then calculate the missing lengths, rounding to one decimal place.

	$\angle A$	sin $\angle A$	cos $\angle A$	tan $\angle A$	Rise	Run	Ramp
						2.1 m	16.3 m	16.4 m
4.	7°	?	?	?	2 m	?	?
5.	12°	?	?	?	?	10 m	?
6.	20°	?	?	?	?	?	5 m
		0.1219	0.9925	0.1228	1.7 m	4.7 m	10.2 m
		0.2079	0.9781	0.2126			
		0.3420	0.9397	0.3640			

Coordinate Geometry

3 Midpoint of a Segment

You should remember that the **average** of two numbers is equal to half the sum of the numbers.

EXAMPLE 1 The average of 4 and 10 is $\frac{4+10}{2} = \frac{14}{2} = 7$.

EXAMPLE 2 The average of 4 and -10 is $\frac{4+(-10)}{2} = \frac{-6}{2} = -3$.

EXPLORATIONS

Part A

1. Points R and S lie on a number line. Find the coordinate of the midpoint of $\overline{RS}$.

   a. 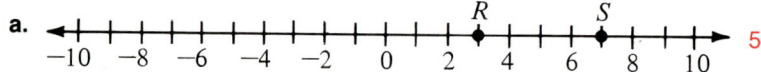 5

   b. 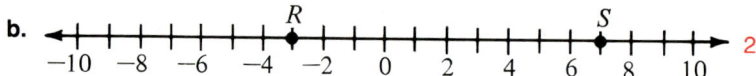 2

   c. 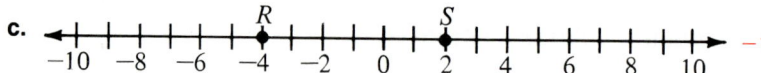 -1

   d. 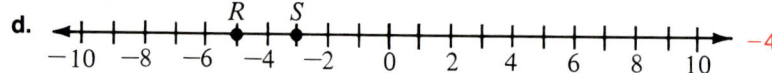 -4

2. Suppose, in Exercise 1, that each number line were vertical rather than horizontal. Would the answers be any different? If necessary, draw vertical number lines to check your answer. no

3. Find the average of each pair of numbers.
   a. 3 and 7 5   b. -3 and 7 2   c. -4 and 2 -1 d. -5 and -3 -4

4. Compare your answers to Exercises 1 and 3. Complete the following statement about points R and S on a number line:

   The coordinate of the midpoint of $\overline{RS}$ is equal to the __?__ of the coordinates of R and S.   average

Part B

Given: $\overline{AB}$ with midpoint M; the horizontal and vertical segments shown.

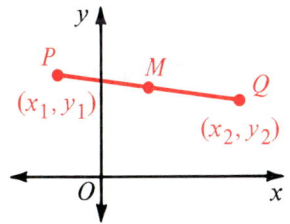

1. The x-coordinate of point C is __?__. 5

2. Because S is the midpoint of $\overline{AC}$, the x-coordinate of S is __?__. 3

3. The x-coordinate of M is __?__. 3

4. You know that the x-coordinate of A is 1, and the x-coordinate of B is 5. Then what is a simple way of finding the x-coordinate of midpoint M? The x-coordinate of M is the average of 1 and 5.

5. The y-coordinate of A is 2 and of B is 8. The y-coordinate of M is __?__. (*Hint:* Use point T.) 5

The explorations above suggest this strategy:

> To find the coordinates of the midpoint of a segment when you know the coordinates of the endpoints:
> Find the average of the x-coordinates.
> Find the average of the y-coordinates.

A convenient way to name points is suggested by the diagram at the right.

Point P has coordinates (x_1, y_1).

This is read "x-sub-1, y-sub-1."

THE MIDPOINT FORMULA

Let $P(x_1, y_1)$ and $Q(x_2, y_2)$ be any two points. Then the midpoint of $\overline{PQ}$ is the point $\left(\dfrac{x_1 + x_2}{2}, \dfrac{y_1 + y_2}{2}\right)$.

EXAMPLE 1 Find the coordinates of the midpoint of the segment that joins points (1, 2) and (5, 8).

The coordinates are $\left(\dfrac{1+5}{2}, \dfrac{2+8}{2}\right)$, or (3, 5).

EXAMPLE 2 Find the coordinates of the midpoint of the segment that joins points $(-2, 4)$ and $(-6, 7)$.

The coordinates are $\left(\dfrac{-2 + (-6)}{2}, \dfrac{4 + 7}{2}\right)$, or $\left(\dfrac{-8}{2}, \dfrac{11}{2}\right)$, or $\left(-4, 5\dfrac{1}{2}\right)$.

Classroom Practice

1. State the x-coordinate of the midpoint of each segment.

a.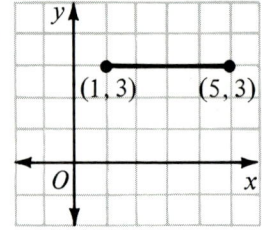

b.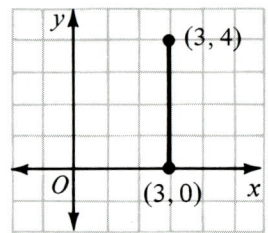

c.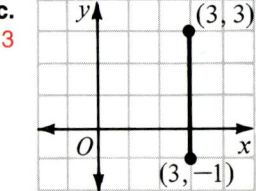

d.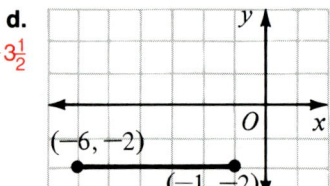

2. State the y-coordinate of the midpoint of each segment shown in Exercise 1. **a.** 3 **b.** 2 **c.** 1 **d.** -2

3. State the coordinates of the midpoint of each segment.

a.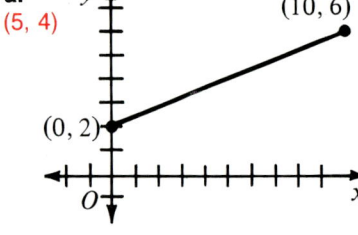

b.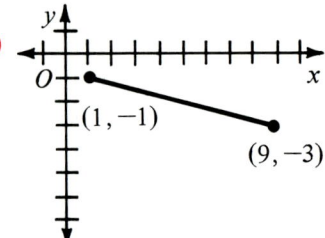

c.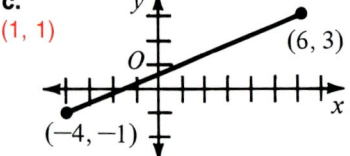

d.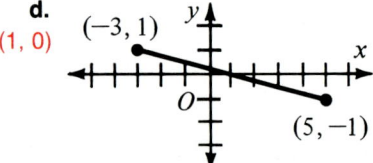

478 Chapter 12

Written Exercises

Plot points C and D. Draw $\overline{CD}$. Mark the midpoint of $\overline{CD}$ and label it M. State the coordinates of M.

[A]
1. $C(2, 0)$ and $D(8, 6)$ (5, 3)
2. $C(3, 1)$ and $D(-7, 1)$ (-2, 1)
3. $C(-5, 2)$ and $D(3, 6)$ (-1, 4)
4. $C(0, 1)$ and $D(6, 6)$ $(3, 3\frac{1}{2})$

State the coordinates of the midpoint of the segment.

5. (4, -2)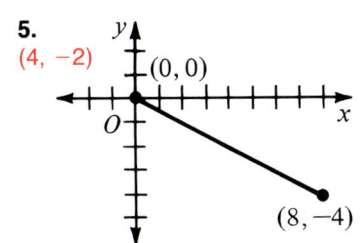

6. (-1, 0)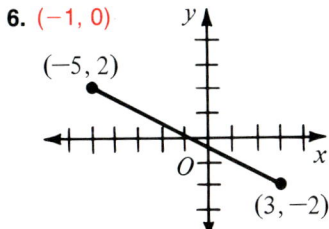

7. (0, -3)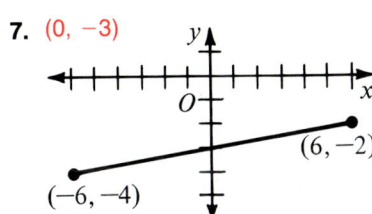

8. $(-1, -\frac{1}{2})$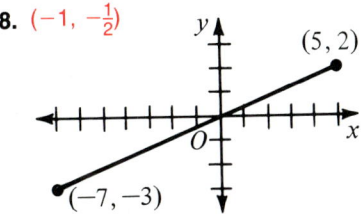

Plot points E and M and draw $\overleftrightarrow{EM}$. Mark point F in such a position that M is the midpoint of $\overline{EF}$. From your drawing state the coordinates of F.

9. $E(0, 3)$ and $M(4, 3)$ (8, 3)
10. $E(5, -1)$ and $M(5, 2)$ (5, 5)
11. $E(2, 2)$ and $M(5, 4)$ (8, 6)
12. $E(-2, 1)$ and $M(1, 0)$ (4, -1)

Use the midpoint formula to find the coordinates of the midpoint of the segment that joins the points named.

13. $(2, 5)$ and $(8, 1)$ (5, 3)
14. $(0, -6)$ and $(4, -2)$ (2, -4)
15. $(-3, -5)$ and $(7, -5)$ (2, -5)
16. $(11, 5)$ and $(-1, 7)$ (5, 6)
17. $(2, 1)$ and $(8, 8)$ $(5, 4\frac{1}{2})$
18. $(0, 5)$ and $(-6, -8)$ $(-3, -1\frac{1}{2})$

[B]
19. $(1, 2.5)$ and $(5, 3.2)$ (3, 2.85)
20. $(-1, 1.6)$ and $(2, 6.4)$ (0.5, 4)
21. $(0, -3)$ and $\left(1\frac{1}{2}, -7\right)$ $(\frac{3}{4}, -5)$
22. $\left(-\frac{1}{3}, 0\right)$ and $\left(4\frac{1}{3}, 5\right)$ $(2, 2\frac{1}{2})$

Coordinate Geometry 479

M is the midpoint of $\overline{AB}$. The coordinates of A and M are given. Use the midpoint formula to find the coordinates of B.

SAMPLE $A(4, -3)$ and $M(6, 1)$
Let (t, u) be the coordinates of B.

$$\frac{4+t}{2} = 6 \qquad \frac{-3+u}{2} = 1$$
$$4 + t = 12 \qquad -3 + u = 2$$
$$t = 8 \qquad u = 5$$

The coordinates of point B are $(8, 5)$.

23. $A(0, 3)$ and $M(5, 4)$ (10, 5)
24. $A(-1, 1)$ and $M(2, 5)$ (5, 9)
25. $A(1, 5)$ and $M(2, 2)$ (3, −1)
26. $A(4, 1)$ and $M(2, 0)$ (0, −1)

SELF-TEST

Vocabulary

coordinate axes (p. 468) y-coordinate (p. 468)
coordinate(s) (p. 468)   average (p. 476)
x-coordinate (p. 468)   midpoint of a segment (p. 477)

The sides of rectangle *RSTU* are parallel to the coordinate axes.

1. Point T has coordinates (__?__, __?__). (5, −4)   *(12-1)*

2. $RU = $ __?__ 6

3. $SU = $ __?__ 10   Exercises 1–3, 7–8

In Exercises 4–6, use points $A(-9, -5)$ and $B(3, 0)$.

4. The distance between the origin and point B is __?__. 3

5. The distance between A and B is __?__. 13

6. The midpoint of $\overline{AB}$ has coordinates (__?__, __?__).    *(12-3)*

In Exercises 7–8, refer to the diagram above.

7. The midpoint of $\overline{RS}$ has coordinates (__?__, __?__). (1, 2)

8. The midpoint of $\overline{SU}$ has coordinates (__?__, __?__). (1, −1)

CALCULATOR ACTIVITIES

If you know two sides and the included angle of a triangle, you can find its area using this formula:

Area = $\frac{1}{2}$ (product of two sides) × (sine of included angle)

   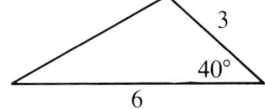   Area = $\frac{1}{2}$ (6 × 3) sin 40° = 5.787

This formula can be used to find the area of any regular polygon. For example, the area of a regular octagon may be found by circumscribing a circle about it. Then the center of the circle is joined to the vertices of the polygon. The octagon is made up of 8 isosceles triangles.

Each central angle measures $\frac{360°}{8}$ = 45°.

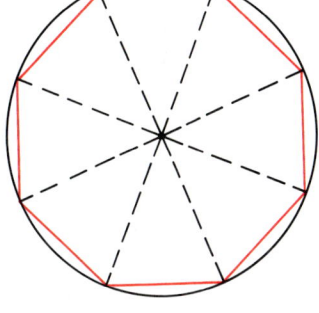

Suppose the radius of the circle is 10.
Then,

Area of octagon = 8 × Area of one triangle
= 8 × $\left[\frac{1}{2}(10 \times 10) \sin 45°\right]$
= 35.35

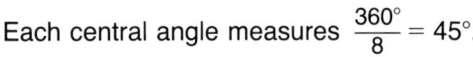

Use a calculator to find the area of each polygon. Round your answers to two decimal places.

1. Two sides of a triangle are 6 and 10 and the included angle is 70°. **28.19**

2. Two sides of a triangle are 17 and 23 and the included angle is 41°. **128.26**

3. A regular hexagon is inscribed in a circle with radius 10. **259.81**

4. A regular polygon with 9 sides is inscribed in a circle with radius 1. **2.89**

5. A regular polygon with 100 sides is inscribed in a circle with radius 4. **50.23**

Applications

LOCUS

Sometimes we need to identify the points that meet one or more requirements. For example, we might want to specify the service contour of a radio station, that is, the locations at which the radio signal can be received.

The mathematical name for specifying points is *locus*, the Latin word for "place." A **locus** is the group made up of all the points that satisfy some condition.

EXPLORATIONS

1. Draw any point P near the middle of a sheet of paper.
2. Locate any point Q that is 5 cm from P.
3. Locate several other points, each 5 cm from P.
4. Look at all the points that are 5 cm from P.
   What kind of figure do the points seem to form? circle
   Complete: The locus of points that are 5 cm from P is
   a __?__ with __?__ = 5 cm. circle; radius

EXAMPLE 1   Specify the divider of State Highway 35 as a locus of points.

The divider line is the locus of points equidistant from the parallel edges of the road.

Example 1 shows that the locus of points equidistant from two parallel lines, and contained in the plane of the lines, is the line coplanar with the given lines that is halfway between them. The locus described in the Explorations, above, also consists of coplanar points.

Example 2 illustrates a locus of points that are not coplanar.

EXAMPLE 2 Find the locus of points *in space* that are 5 cm from a given point P.

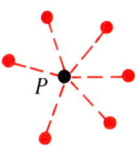

The locus is a sphere with center P and radius 5 cm.

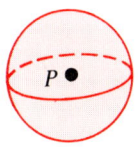

When you specify a locus you should include a clear description and a diagram of the locus.

Some loci (plural of *locus*) are determined by more than one condition, as illustrated in Example 3.

EXAMPLE 3 Mayberry Mall is shaped like a trapezoid, as shown. The architect decides to install a fountain in the mall. The fountain is to be equidistant from Store A and Store B. The fountain should also be the same distance from the specialty shops along the north and south sides of the mall. Find the location of the fountain.

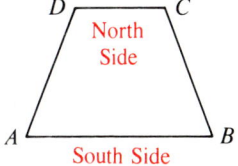

Step 1 The locus of points equidistant from A and B and in the plane of the mall is the perpendicular bisector of $\overline{AB}$.

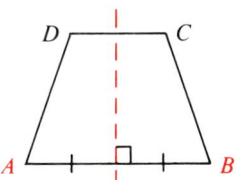

Step 2 The locus of points equidistant from the north ($\overline{CD}$) and south ($\overline{AB}$) sides of the mall is __?__. (Hint: $\overline{AB} \parallel \overline{CD}$) a line parallel to $\overline{AB}$ and $\overline{CD}$ and equidistant from $\overline{AB}$ and $\overline{CD}$

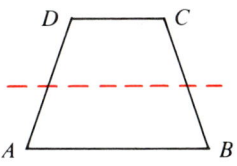

Step 3 The required locus must satisfy both conditions. The required locus is the intersection of the loci in Steps 1 and 2. The fountain should be located at point F.

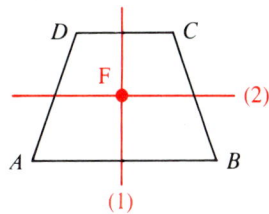

EXERCISES

1. **a.** Draw a line t on your paper.
   Locate many points that are 3 cm from t.
   b. Draw and label the locus of points that are in the plane of the paper and are 3 cm from t.
   c. Describe the locus. The locus is lines m and n, ∥ to t, each 3 cm from t.
2. Refer to Exercise 1. Draw and describe the locus of points in space that are 3 cm from t. The locus is a cylindrical surface with axis t and radius 3 cm.
3. Sketch and describe the locus of points in space equidistant from two parallel planes R and S. (*Hint:* It may be helpful to use two pieces of paper to represent two parallel planes.) The locus is a plane ∥ to planes R and S, midway between them.
4. **a.** Draw a large acute angle, $\angle ANG$.
   b. Find a point that is the same distance from $\overrightarrow{NA}$ as from $\overrightarrow{NG}$.
   c. Find many points that satisfy this condition.
   d. Describe the locus of points that are equidistant from the sides of an angle. The locus is $\overrightarrow{NX}$, the ∠ bisector of $\angle ANG$.
5. Refer to Step 1 of Example 3. State the theorem that allows you to conclude that the perpendicular bisector of $\overline{AB}$ contains the points that are equidistant from A and B. (See below.)
6. Specify the plane of the net in a tennis court as a locus of points. (*Hint:* See Example 1.) The plane is the locus of points that lie on the ⊥ bisecting plane of the court, above the ground.

5. Any pt. on the ⊥ bisector of a segment is equidistant from the end pts. of the segment.

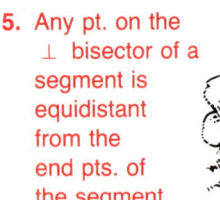
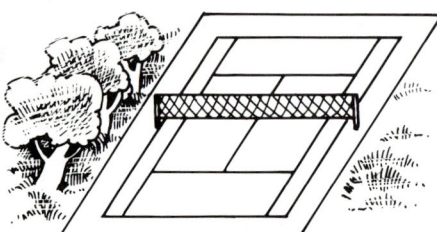

7. The chandelier should be installed at the intersection of the diagonals of the rectangular ceiling.

7. A chandelier is to be installed in a rectangular room. The light must be positioned at an equal distance from each of the four corners of the ceiling.
   Draw a diagram and write a description of the required position. (See above.)
8. What is the locus of points in space that are equidistant from points J and K? Include a diagram and a description. The locus is the ⊥ bisecting plane of $\overline{JK}$.

9. What is the locus in space of points that are 1 m from plane *P*?
The locus is 2 planes ∥ to plane *P*, each 1 m from it.

10. Draw two points, *R* and *S*. Find the locus of points that are at a distance of *RS* from point *R* and lie in the plane of the paper.
The locus is a circle with center *R* and radius of length *RS*.

11. A 7 m hose is attached to a 15 m by 12 m building at point *X*. Show the region that can be reached by the hose.

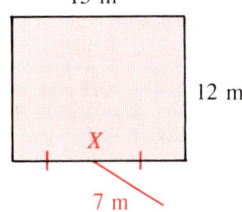

12. Repeat Exercise 11 if the hose is attached at point *Y*, shown at the right.

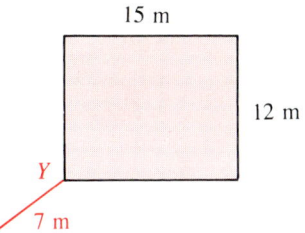

13. A regional recreation center is planned. The mayors of La Palma, Eastbury, and Azalia want the center to be an equal distance from the three towns. Find this location. The location is the intersection of the ⊥ bisectors of the sides of the △ formed by the 3 towns.

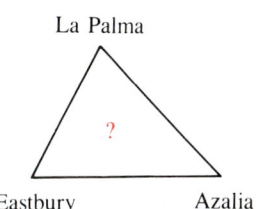

14. The Fongs have a triangular garden. They want to plant a rosebush in the garden. The rosebush should be an equal distance from the edges of the garden. Sketch the garden and determine the location of the rosebush. (*Hint:* The center of an inscribed circle is equidistant from the sides of a triangle.) The location is the intersection of the ∠ bisectors of the △.

Cora and Octavio live 5 km from each other. How many possible meeting points do they have if they meet at a point that is:

15. equidistant from their houses? an infinite number

16. 4 km from Cora's house and 3 km from Octavio's house? two

17. 2 km from Cora's house and 1 km from Octavio's house? none

18. 2.4 km from Cora's house and 2.6 km from Octavio's house? one

19. Draw two points, *X* and *Y*. Construct (not draw) the locus of points in the plane of the paper that are equidistant from *X* and *Y* and at a distance of 2(*XY*) from point *Y*.

Coordinate Geometry 485

4 Slope of a Line

A car can make the climb at the left, but not the climb at the right.

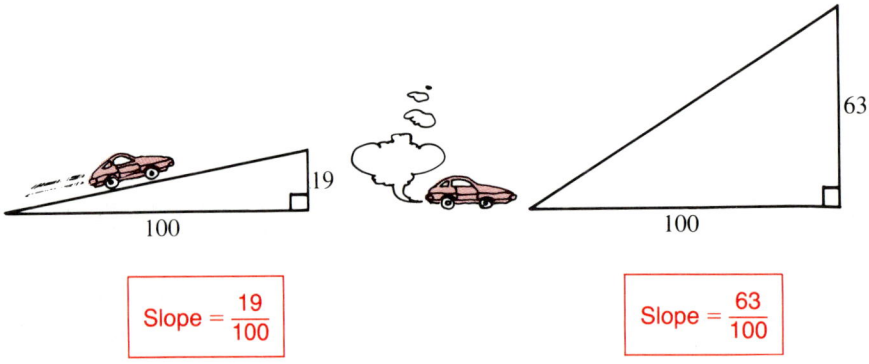

The *slope* of a road is the ratio of *rise* to *run*.

$$\text{Slope} = \frac{\text{rise}}{\text{run}}$$

Lines have slopes, too.

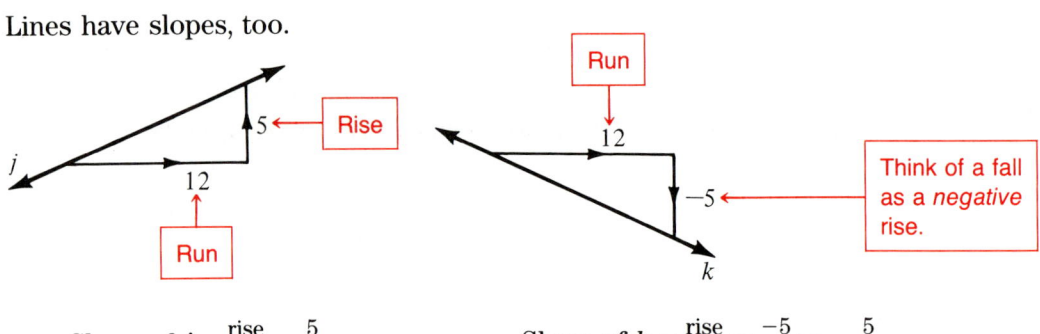

$$\text{Slope of } j = \frac{\text{rise}}{\text{run}} = \frac{5}{12} \qquad \text{Slope of } k = \frac{\text{rise}}{\text{run}} = \frac{-5}{12} = -\frac{5}{12}$$

Count squares in this figure and you find that the rise is 4, the run is 3, and the slope is $\frac{4}{3}$. Instead of counting squares you can subtract coordinates, as shown in the figure.

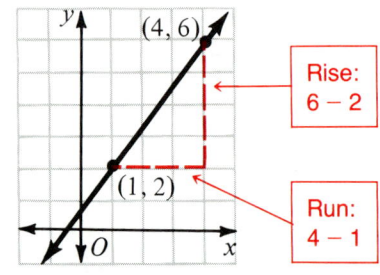

$$\text{Slope} = \frac{\text{rise}}{\text{run}} = \frac{6-2}{4-1} = \frac{4}{3}$$

The slope of the line that joins points (x_1, y_1) and (x_2, y_2) is defined by

$$\text{Slope} = \frac{y_2 - y_1}{x_2 - x_1}$$

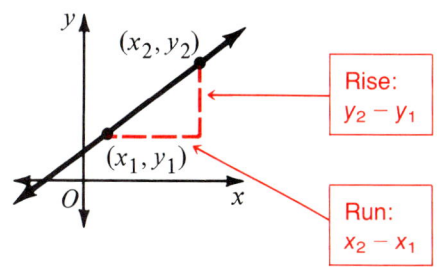

Here are some more facts that will help you understand the meaning of "slope."

1. When a line *rises from left to right*, the slope is a *positive* number.

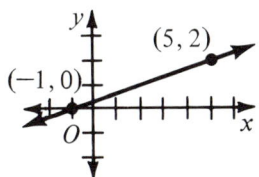

$$\frac{2 - 0}{5 - (-1)} = \frac{2}{6} = \frac{1}{3}$$

2. When a line *falls from left to right*, the slope is a *negative* number.

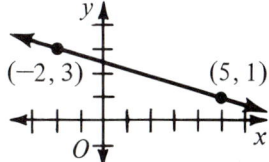

$$\frac{1 - 3}{5 - (-2)} = \frac{-2}{7} = -\frac{2}{7}$$

3. A *horizontal* line does not rise or fall. The slope of a horizontal line is zero.

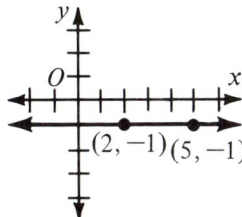

$$\frac{-1 - (-1)}{5 - 2} = \frac{0}{3} = 0$$

4. The word "slope" does not apply to a *vertical* line. We say that the slope of a vertical line is *not defined*.

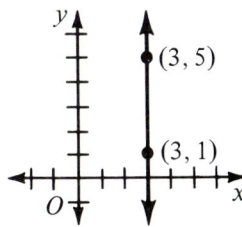

Note that the expression $\frac{5-1}{3-3}$ does not represent a number.

Coordinate Geometry

EXPLORATIONS

Part A

1. Think of going from A to B.

   Run = __?__ 5   Rise = __?__ 4

   Slope = $\frac{\text{rise}}{\text{run}}$ = __?__ $\frac{4}{5}$

2. Think of going from B to A.

   Run = -5 Rise = -4

   Slope = $\frac{\text{rise}}{\text{run}}$ = __?__ $\frac{4}{5}$

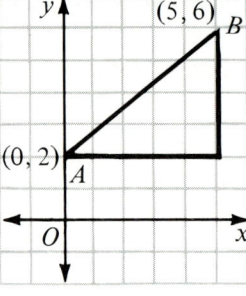

Part A suggests that when you compute the slope by using the ratio $\frac{\text{rise}}{\text{run}}$, you may start at *either* point. Test this with other points.

Part B

Copy and complete the table.

	(x_1, y_1)	(x_2, y_2)	$y_2 - y_1$	$x_2 - x_1$	Slope
1.	(0, 2)	(5, 6)	? 4	? 5	? $\frac{4}{5}$
2.	(5, 6)	(0, 2)	? -4	? -5	? $\frac{4}{5}$

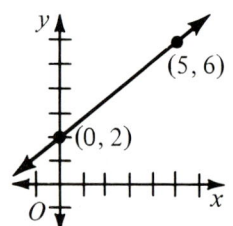

Part B suggests that when you compute the slope by using $\frac{y_2 - y_1}{x_2 - x_1}$, it doesn't matter which point you choose as (x_2, y_2).

Part C

Use the points named to compute the slope of the line.

all slopes: $-\frac{1}{2}$

Points	Slope
1. R and S	__?__
2. R and T	__?__
3. R and U	__?__
4. R and V	__?__
5. S and U	__?__
6. S and V	__?__

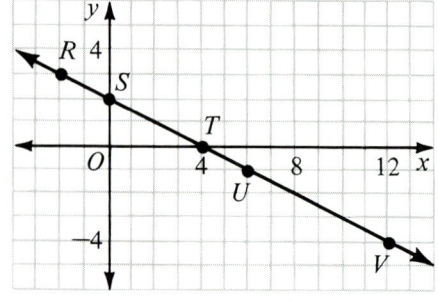

Part C suggests that when you compute the slope of a line, you may use *any* two points on the line.

Classroom Practice

For each diagram, think of going from point *R* to point *S*.
a. Tell whether the rise is positive, negative, or zero.
b. Tell whether the run is positive, negative, or zero.
c. Tell whether the slope is positive, negative, zero, or not defined.

1.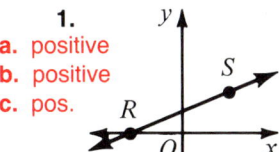
a. positive
b. positive
c. pos.

2.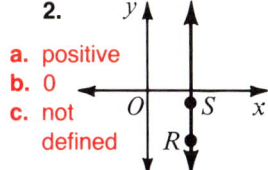
a. positive
b. 0
c. not defined

3.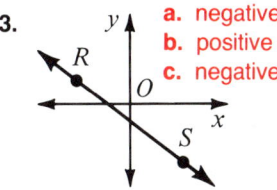
a. negative
b. positive
c. negative

4.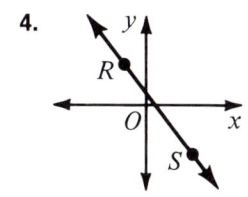
a. negative
b. positive
c. negative

5.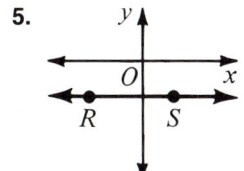
a. 0
b. positive
c. 0

6.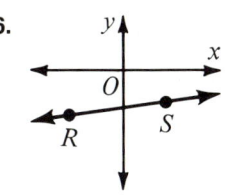
a. positive
b. positive
c. positive

Written Exercises

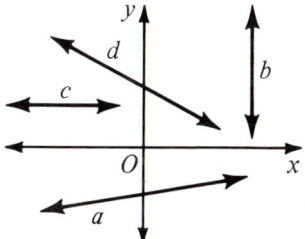

A 1. Name the line that fits the description.
   a. The slope is positive. *a*
   b. The slope is zero. *c*
   c. The slope is negative. *d*
   d. Slope is not defined for the line. *b*

State the slope of the line shown. If the slope is not defined for the line, write *not defined*.

2. $\frac{3}{5}$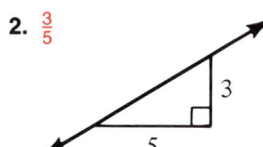

3. $\frac{1}{3}$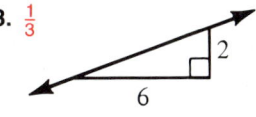

4. 0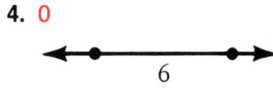

5. $-\frac{7}{3}$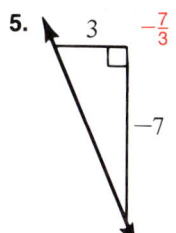

6. -1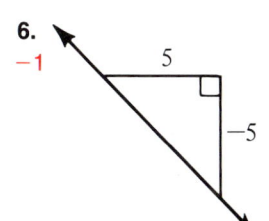

7. 6

Coordinate Geometry

Take P as point (x_1, y_1) and Q as point (x_2, y_2). Copy and complete the table. When $\overleftrightarrow{PQ}$ does not have a slope, write *not defined*.

	P	Q	$y_2 - y_1$	$x_2 - x_1$	Slope
8.	$(0, 0)$	$(6, 5)$	? 5	? 6	? $\frac{5}{6}$
9.	$(0, 0)$	$(-3, -7)$	? -7	? -3	? $\frac{7}{3}$
10.	$(2, 1)$	$(6, -2)$	? -3	? 4	? $-\frac{3}{4}$
11.	$(6, -2)$	$(2, 1)$	? 3	? -4	? $-\frac{3}{4}$
12.	$(-3, -1)$	$(-5, 4)$	? 5	? -2	? $-\frac{5}{2}$
13.	$(-5, 4)$	$(-3, -1)$	? -5	? 2	? $-\frac{5}{2}$

The coordinates of the vertices of quadrilateral *ABCD* are given. State which sides of the quadrilateral have positive slopes.

B 14. $A(-5, 2)$; $B(4, 6)$; $C(2, -3)$; $D(-1, -2)$ $\overline{AB}, \overline{BC}$

15. $A(3, 4)$; $B(6, -5)$; $C(-7, -1)$; $D(-2, 6)$ $\overline{CD}$

16. $A(8, -10)$; $B(-6, -2)$; $C(-4, 6)$; $D(10, -2)$ $\overline{BC}, \overline{AD}$

17. $A(-5, 0)$; $B(1, 1)$; $C(1, 6)$; $D(-4, 5)$ $\overline{AB}, \overline{CD}, \overline{AD}$

In Exercises 18–21, a line is described. State three points (any three points) that the line passes through. You may use graph paper, but write your answers in the form: Points (__?__, __?__), (__?__, __?__), and (__?__, __?__). *(Answers will vary.)*

For example:
18. A line through the origin with slope equal to 1 (1, 1), (2, 2), and (−3, −3)

19. A line through point $(0, 4)$ with slope equal to 1 (1, 5), (2, 6), and (−1, 3)

20. A line containing points $(2, 0)$ and $(0, 3)$ (4, −3), (−2, 6), and (−4, 9)

21. A line through point $(0, -4)$ with slope equal to 0 (−1, −4), (2, −4), and (6, −4)

C 22. Rosalie says: "No matter how great a number you choose, I can find a line whose slope is equal to that number." Isabella chooses the number *one billion*. Rosalie could respond by saying: "Take the line that contains points (__?__, __?__) and (__?__, __?__)." (Answers will vary.) (1, 1 billion), (2, 2 billion)

23. You are given points (a, b) and (c, d) on a line. Show that you get the same value for the slope whether you choose (a, b) to be (x_1, y_1) or (x_2, y_2). *Suggested strategy:* Write two expressions for the slope. Then use the algebraic property: $d - b = -(b - d)$.

5 Parallel and Perpendicular Lines

The purpose of this section is to show how you can use the slopes of two lines to decide if the lines are parallel or perpendicular. Since vertical and horizontal lines are special, they will be considered in the exercises, rather than the lesson itself.

Suppose lines p and q are parallel. Draw two vertical segments and label lengths as shown.

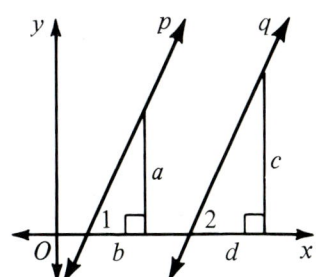

1. $\angle 1 = \angle 2$ (Why?) If 2 ∥ lines are cut by a trans., then corr. ∠s are =.
2. The triangles are similar. (Why?) AA Postulate
3. $\frac{a}{c} = \frac{b}{d}$. (Why?) Def. of similar polygons
4. $\frac{a}{b} = \frac{c}{d}$. (Why?) Property of proportions
5. Slope of line p = slope of line q. (Def. of slope)

We have just shown that if two lines are parallel, then their slopes are equal. The converse is also true. (See Exercise 21.)

> *If two lines are parallel, then they have equal slopes.*
>
> *If two lines have equal slopes, then they are parallel.*

EXPLORATIONS

Draw a line through the origin and point (2, 3). If your line is carefully drawn, you will see that it also contains points $(-2, -3)$, (4, 6), and (6, 9).

1. Use a compass to construct, or use a protractor to draw, lines perpendicular to the given line at points $(-2, -3)$, (0, 0), (2, 3), (4, 6), and (6, 9).

2. On each of the perpendiculars choose a convenient point and label it with its coordinates.

3. Find the slope of each perpendicular.

4. If you worked carefully, you found that each perpendicular has slope $-\frac{2}{3}$. Notice that the given line has slope $\frac{3}{2}$. Describe a relation between the numbers $\frac{3}{2}$ and $-\frac{2}{3}$. $\frac{3}{2} \cdot (-\frac{2}{3}) = -1$

Coordinate Geometry

Suppose a line p has slope $\frac{4}{5}$. From the Explorations on page 491, what would you guess to be the slope of a line perpendicular to p? The Explorations suggest that there is a simple relationship between the slopes of two perpendicular lines.

> *If two lines are perpendicular, then the product of their slopes is -1.*
>
> *If the product of the slopes of two lines is -1, then the lines are perpendicular.*

Classroom Practice

1. Line l has slope $\frac{2}{7}$.
   a. The slope of any line parallel to l equals __?__. $\frac{2}{7}$
   b. The slope of any line perpendicular to l equals __?__. $-\frac{7}{2}$

2. Line j has slope -5.
   a. The slope of any line parallel to j equals __?__. -5
   b. The slope of any line perpendicular to j equals __?__. $\frac{1}{5}$

3. Suppose one line has slope $\frac{3}{11}$ and another line has slope $\frac{4}{11}$. Can the lines be parallel? Explain. No; $\frac{3}{11} \neq \frac{4}{11}$

4. Suppose one line has slope $\frac{2}{5}$ and another line has slope $\frac{5}{2}$. Are the lines perpendicular? Explain. No; $\frac{2}{5} \cdot \frac{5}{2} \neq -1$

5. *Given:* The slopes of two lines are 1 and -1.
   Are the lines parallel? Are they perpendicular? Explain.
   No; $1 \neq -1$ Yes; $1(-1) = -1$

6. *Given:* Points $A(0, 1)$, $B(3, 3)$, $C(6, 5)$, and $D(12, 9)$.
   a. The slope of $\overleftrightarrow{AB}$ equals __?__. $\frac{2}{3}$
   b. The slope of $\overleftrightarrow{CD}$ equals __?__. $\frac{2}{3}$
   c. Do you think that $\overleftrightarrow{AB}$ and $\overleftrightarrow{CD}$ are parallel lines? yes
   d. Plot points A, B, C, and D. Do you need to change your answer to part (c)? Explain. They are the same line.

Written Exercises

1. A given line has slope $-\frac{5}{7}$.
   a. Any line parallel to the given line has slope __?__. $-\frac{5}{7}$
   b. Any line perpendicular to the given line has slope __?__. $\frac{7}{5}$

2. A given line has slope 2.
   a. Any line parallel to the given line has slope __?__. 2
   b. Any line perpendicular to the given line has slope __?__. $-\frac{1}{2}$

3. A given line is vertical.
   a. The slope of the given line is __?__. not defined
      0/not defined
   b. The slope of a line perpendicular to the given line is __?__. 0
      0/not defined

4. A given line rises to the right.
   a. The slope of the line is __?__. positive
      positive/negative
   b. The slope of a line perpendicular to the given line is __?__. negative
      positive/negative

The slopes of two lines are given. Are the lines parallel, perpendicular, or neither?

5. $\frac{4}{8}$ and $\frac{1}{2}$ parallel

6. $\frac{7}{2}$ and $\frac{2}{7}$ neither

7. $-\frac{4}{3}$ and $\frac{3}{4}$ perpendicular

8. 3 and -3 neither

Copy and complete the table.

	A	B	C	D	Slope of $\overleftrightarrow{AB}$	Slope of $\overleftrightarrow{CD}$	Is $\overleftrightarrow{AB} \parallel \overleftrightarrow{CD}$?	Is $\overleftrightarrow{AB} \perp \overleftrightarrow{CD}$?
Sample	$(-1, 1)$	$(3, 5)$	$(0, -2)$	$(3, 2)$	1	$\frac{4}{3}$	No	No
9.	$(2, 2)$	$(5, 2)$	$(-4, 4)$	$(0, 4)$	? 0	? 0	? Yes	? No
10.	$(0, 0)$	$(3, 4)$	$(-1, -2)$	$(2, 2)$	? $\frac{4}{3}$	? $\frac{4}{3}$	? Yes	? No
11.	$(3, -1)$	$(2, 1)$	$(5, 0)$	$(3, 4)$	? -2	? -2	? Yes	? No
12.	$(0, 3)$	$(2, 0)$	$(0, 0)$	$(6, -4)$	? $-\frac{3}{2}$	? $-\frac{2}{3}$	? No	? No

Coordinate Geometry

∠F is a right angle. Find the slope of $\overleftrightarrow{EF}$ and the slope of $\overleftrightarrow{FG}$.

B 13.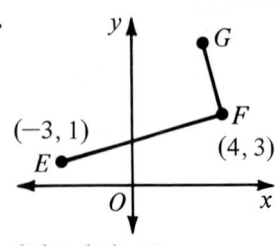

14.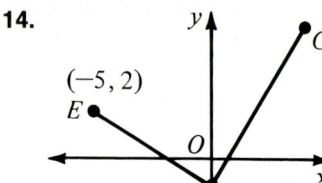

$\overleftrightarrow{EF}: \frac{2}{7}$; $\overleftrightarrow{FG}: -\frac{7}{2}$

$\overleftrightarrow{EF}: -\frac{3}{5}$; $\overleftrightarrow{FG}: \frac{5}{3}$

For Exercises 15–18, a strategy is:
Draw a diagram. Then find and use slopes as necessary.
Determine if the diagonals of quadrilateral *RSTV* are perpendicular.

15. $R(0,0)$; $S(4, 3)$; $T(1, 7)$; $V(-3, 4)$ yes

16. $R(-2, 6)$; $S(5, 5)$; $T(7, -3)$; $V(-3, -4)$ no

What special kind of figure is quadrilateral *ABCD*?

17. $A(-1, 0)$; $B(0, 3)$; $C(6, 3)$; $D(5, 0)$ parallelogram

18. $A(0, 0)$; $B(4, 2)$; $C(2, 3)$; $D(0, 2)$ trapezoid

C 19. You are given the figure shown. *M* and *N* are the midpoints of $\overline{RV}$ and $\overline{ST}$.
   a. *M* has coordinates (? , ?). (1, 3)
   b. *N* has coordinates (? , ?). (6, 3)
   c. Slope of $\overline{MN}$ = ? ; slope of $\overline{RS}$ = ? ; 0; 0
      slope of $\overline{VT}$ = ? . 0
   d. $\overline{MN} \parallel \overline{RS} \parallel \overline{VT}$. Why? Their slopes are equal.
   e. RS = ? ; VT = ? ; MN = ?   8; 2; 5
      Does $MN = \frac{1}{2}(RS + VT)$? yes

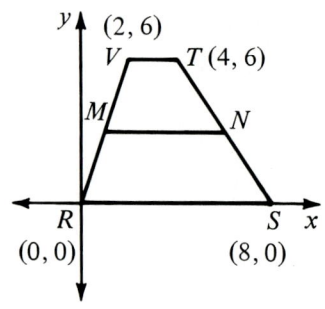

20. Given: $\triangle ABC$; *M* and *N* are the midpoints of $\overline{AC}$ and $\overline{BC}$.
   a. *M* has coordinates (? , ?). (−1, 3)
   b. *N* has coordinates (? , ?). (3, 6)
   c. Slope of $\overline{MN}$ = ? ; $\frac{3}{4}$
      slope of $\overline{AB}$ = ? . $\frac{3}{4}$
   d. Is $\overline{MN} \parallel \overline{AB}$? yes
   e. MN = ? and AB = ?   5; 10
   f. Does $MN = \frac{1}{2}AB$? yes

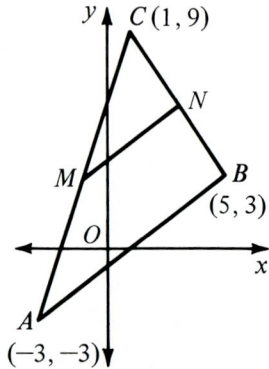

21. Our goal is to show that:
If two lines have equal slopes, then the lines are parallel.

Given: Lines p and q
Slope of p = slope of q
Prove: $p \parallel q$

Suggested strategy:
Take points B and E so that $AB = DE$.
Draw vertical segments $\overline{BC}$ and $\overline{EF}$.
Show that $\frac{BC}{AB} = \frac{EF}{DE}$ and conclude that $BC = EF$.
Prove that $\triangle ABC \cong \triangle DEF$.
Since $\angle 1 = \angle 2$, you can conclude that $p \parallel q$.

22. The statements on pages 491 and 492 do not apply to the special case of vertical lines. Complete the sentence that follows to show that more complicated statements could cover this special case: When two lines are perpendicular, either the product of their slopes is -1, or .... one line has undefined slope and the other has slope 0.

SELF-TEST

Vocabulary

slope of a line (p. 486)   rise (p. 486)   run (p. 486)

1. The slope of a ___?___ line equals 0. horizontal   (12-4)
horizontal/vertical

2. The slope of a ___?___ line is not defined. vertical
horizontal/vertical

3. You are given point $A(2, 0)$ and point $B(5, 4)$.
Find the slope of $\overleftrightarrow{AB}$. $\frac{4}{3}$

4. You are given point $B(5, 4)$ and point $C(8, -2)$.
Find the slope of $\overleftrightarrow{BC}$. -2

5. If the slope of line j equals $\frac{3}{7}$, then the slope of any line parallel   (12-5)
to j equals ___?___. $\frac{3}{7}$

6. If the slope of line k equals $-\frac{5}{2}$, then the slope of any line perpendicular to k equals ___?___. $\frac{2}{5}$

6 Equations and Lines

Let's see how the line shown in the figure is related to the equation $2x + 3y = 12$. Do the coordinates of the four labeled points make the equation a true statement when they are substituted for x and y?

Point (0, 4)
$2x + 3y = 12$
$2(0) + 3(4) \stackrel{?}{=} 12$
$0 + 12 \stackrel{?}{=} 12$ ✓
Yes

Point (3, 2)
$2x + 3y = 12$
$2(3) + 3(2) \stackrel{?}{=} 12$
$6 + 6 \stackrel{?}{=} 12$ ✓
Yes

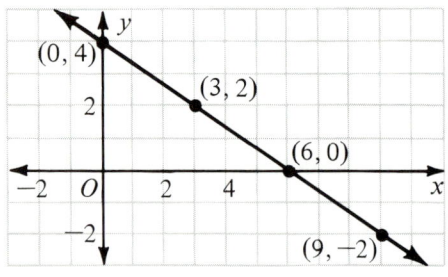

Point (6, 0)
$2x + 3y = 12$
$2(6) + 3(0) \stackrel{?}{=} 12$
$12 + 0 \stackrel{?}{=} 12$ ✓
Yes

Point (9, −2)
$2x + 3y = 12$
$2(9) + 3(-2) \stackrel{?}{=} 12$
$18 - 6 \stackrel{?}{=} 12$ ✓
Yes

Suppose you kept on testing the coordinates of different points. You would find that the coordinates of any point on the line make the equation a true statement. The coordinates of any other point do not. The line is called the **graph** of the equation $2x + 3y = 12$.

In general, it is true that:

> *If a point lies on the graph of an equation, then its coordinates make the equation a true statement.*

> *If the coordinates of a point make an equation a true statement, then the point lies on the graph of the equation.*

The facts you discovered about the equation $2x + 3y = 12$ are true for every equation in the form $ax + by = c$.

> *If an equation can be written in the form $ax + by = c$, then the graph of the equation is a line. (Assume that a and b are not both equal to zero.)*

The examples that follow show how to graph an equation that can be written in the form $ax + by = c$.

EXAMPLE 1 Draw the graph of $2x + y = 6$.

Pick any three numbers to use for x or for y. Find the coordinates of three points.

Let $x = 0$:
$2x + y = 6$
$2(0) + y = 6$
$0 + y = 6$
$y = 6$

Let $y = 0$:
$2x + y = 6$
$2x + 0 = 6$
$2x = 6$
$x = 3$

Let $x = 1$:
$2x + y = 6$
$2(1) + y = 6$
$2 + y = 6$
$y = 4$

Now plot the points:
$(0, 6)$; $(3, 0)$; $(1, 4)$.
Then draw the line.

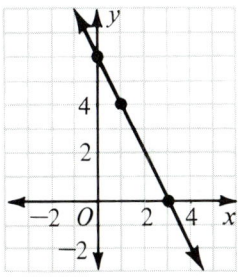

EXAMPLE 2 Draw the graph of $x - 3y = 3$.

Let $x = 0$:
$x - 3y = 3$
$0 - 3y = 3$
$-3y = 3$
$y = -1$

Let $y = 0$:
$x - 3y = 3$
$x - 3(0) = 3$
$x - 0 = 3$
$x = 3$

Let $x = 7$:
$x - 3y = 3$
$7 - 3y = 3$
$-3y = -4$
$y = 1\frac{1}{3}$

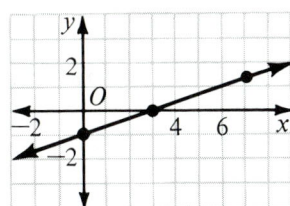

Notice in Example 2, that when you let $x = 7$, you get a fractional value for y. You have two choices:

(1) estimate the position of $\left(7, 1\frac{1}{3}\right)$; or

(2) forget that point and try other values for x and y until you get a pair of coordinates that doesn't involve fractions.

Coordinate Geometry

EXAMPLE 3 Draw the graph of $x = 2$.

The equation can be rewritten in the form $x + 0y = 2$.

If you try to let x be any number other than 2, you cannot find a value for y.

On the other hand, no matter what value you assign to y, the value of x is always 2.

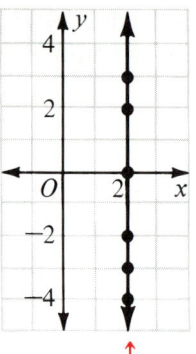

The graph of $x = 2$ is a vertical line.

Classroom Practice

1. The graph of the equation $y = x - 2$ is shown. Does the point lie on the line?

   a. $(2, 0)$ yes
   b. $(0, 2)$ no
   c. $(3, 1)$ yes
   d. $(1, 3)$ no
   e. $(4, 4)$ no
   f. $(1, -1)$ yes
   g. $(-1, -3)$ yes
   h. $\left(2\frac{1}{2}, \frac{1}{2}\right)$ yes

   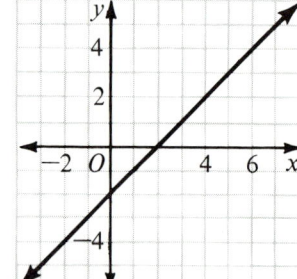

2. Do the coordinates of the point make the equation $2x - y = 4$ a true statement?

   a. $(2, 2)$ no   b. $(4, 0)$ no   c. $(2, 0)$ yes   d. $(1, 1)$ no
   e. $(10, 16)$ yes   f. $(0, -4)$ yes   g. $(-3, -10)$ yes   h. $(-2, 0)$ no

3. Name the coordinates of five points (any five) that lie on the graph of the equation $y = 2x$. For example: $(-3, -6), (0, 0), (1, 2), (2, 4), (6, 12)$

4. The equation $y = 5$ is given. Note that the equation can be written in the form $0x + y = 5$.

   a. When $x = 0$, $y =$ __?__. 5
   b. When $x = -13$, $y =$ __?__. 5
   c. No matter what value you assign to x, $y =$ __?__. 5
   d. The graph of $y = 5$ is a _____?_____ line. horizontal
      horizontal/vertical

Written Exercises

A 1. State the coordinates of four points (any four) that lie on the line at the left, below. For example: (0, −1), (3, −3), (−3, 1), (−6, 3)

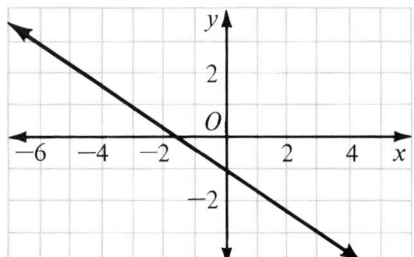
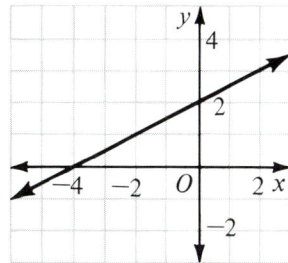

2. One of the points named does not lie on the line at the right, above. Which point? (1, 3)

$(-4, 0)$ $(1, 3)$ $\left(1, 2\tfrac{1}{2}\right)$ $(0, 2)$

Copy and complete the table of values.

3. $x + y = 6$

x	y
0	? 6
6 ?	0
2	? 4

4. $x - y = 6$

x	y
0	? −6
6 ?	0
4	? −2

5. $y = 3x$

x	y
0	? 0
0 ?	0
3	? 9

6. $y = x + 4$

x	y
0	? 4
−4 ?	0
3	? 7

7. $2x - 3y = 12$

x	y
0	? −4
6 ?	0
12 ?	4

8. $3x - y = -6$

x	y
0	? 6
−2 ?	0
−1 ?	3

9–14. Draw the graphs of the equations in Exercises 3–8. If your points do not lie on a line, check the work you did earlier.

15. Draw the graph of $x = 5$. (*Hint:* See Example 3, page 498.)

16. Draw the graph of $y = -4$.

Make a table of values for plotting three points. Plot the points and draw the graph of the equation.

B 17. $3x + 2y = 9$ 18. $x - 4y = 6$

19. $4x + 3y = 6$ 20. $2x - 3y = -12$

Coordinate Geometry

For each exercise, graph the two equations on the same pair of axes. Inspect your drawing and state the coordinates of the point where the graphs intersect.

21. $x + y = 6$
    $x - y = 2$
    (4, 2)

22. $y = x + 2$
    $y = -x$
    (−1, 1)

23. $2x + y = 6$
    $3x - y = 4$
    (2, 2)

24. $2x + y = -1$
    $x + 2y = 4$
    (−2, 3)

Graph the three equations on the same pair of axes. Find the area of the triangle formed by the three lines.

C 25. $x = 1$
    $y = -1$
    $y = x + 2$
    A = 8

26. $x = 2$
    $y = 2$
    $x + y = 6$
    A = 2

27. $y = -1$
    $y = 3x + 2$
    $3x + y = 2$
    A = 3

CAREER NOTES

CARPENTER

Have you ever put together a picture frame or bookcase? Do you enjoy helping build props for your school play? If so, you have been using carpentry skills.

Carpenters build, remodel, and repair structures made of wood. Rough carpenters work on a building's framework. They put up scaffolding and make the forms into which concrete is poured. Finish carpenters do more detailed work. They put in paneling, cabinets, window sashes, doors, and molding.

Being able to make careful, accurate measurements is an important part of a carpenter's job. Carpenters often apply their knowledge of arithmetic, geometry, and informal algebra. For instance, you have learned in geometry that the diagonals of a rectangle are equal. The converse of this theorem is also true. Knowing this, carpenters can measure the diagonals of a house's foundation. If they are the same length, the foundation is a rectangle. This then tells the carpenters that the corners of the foundation are all right angles.

A four-year apprenticeship program with classes in drafting, blueprint reading, mathematics, and wood shop is good preparation for a career in carpentry.

7 Coordinate Geometry Proofs

Recall Theorem 14 in Chapter 5:

> The median of a trapezoid has two properties.
> (1) It is parallel to the bases.
> (2) Its length equals half the sum of the base lengths.

This theorem was stated without proof. We are now able to prove the theorem by using coordinates.

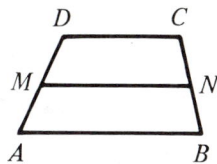

Given: Trapezoid $ABCD$ with median $\overline{MN}$

Prove: 1. $\overline{MN} \parallel \overline{AB} \parallel \overline{DC}$
         2. $MN = \frac{1}{2}(AB + DC)$

Proof:
Draw coordinate axes as shown.
Then A is point $(0, 0)$, and the y-coordinate of B is 0.
Assign coordinates to B, C, and D as shown.
(Because $\overline{DC}$ is horizontal, D and C must have equal y-coordinates.)

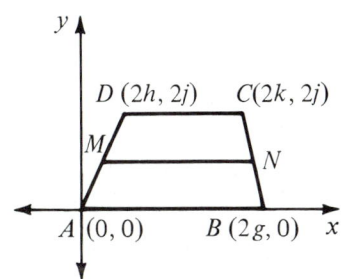

The coordinates of M are $\left(\frac{2h + 0}{2}, \frac{2j + 0}{2}\right)$, or (h, j).

The coordinates of N are $\left(\frac{2g + 2k}{2}, \frac{0 + 2j}{2}\right)$, or $(g + k, j)$.

1. Slope of $\overline{MN} = \dfrac{j - j}{(g + k) - h} = \dfrac{0}{g + k - h} = 0$.
   Slope of $\overline{AB}$ = slope of $\overline{DC} = 0$.
   Therefore, $\overline{MN} \parallel \overline{AB} \parallel \overline{DC}$.

2. $MN = g + k - h$
   $AB = 2g - 0 = 2g$
   $DC = 2k - 2h$

   $AB + DC = 2g + 2k - 2h$, and $\frac{1}{2}(AB + DC) = g + k - h$.

   Because both MN and $\frac{1}{2}(AB + DC)$ equal $g + k - h$,
   $MN = \frac{1}{2}(AB + DC)$.

Coordinate Geometry

Classroom Practice

The purpose of these exercises is to use coordinates to prove Theorem 15 from Chapter 5: The segment joining the midpoints of two sides of a triangle is parallel to the third side and half as long.

Given: $\triangle ABC$;
midpoints D and E as shown

Prove: 1. $\overline{DE} \parallel \overline{AB}$
2. $DE = \frac{1}{2}AB$

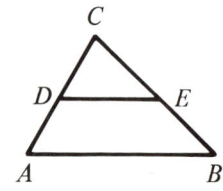

Proof:
Draw axes and assign coordinates as shown.

1. a. The coordinates of D are (__?__, __?__). (h, j)
   b. The coordinates of E are (__?__, __?__). $(h + g, j)$
   c. Slope of $\overline{AB} = \dfrac{0}{2g} = $ __?__. 0
   d. Slope of $\overline{DE} = \dfrac{j - j}{(g + h) - h} = $ __?__. 0
   e. $\overline{DE} \parallel \overline{AB}$ because __?__. The slopes are equal.

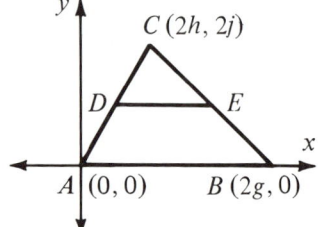

2. a. $DE = (g + h) - h = $ __?__. g
   b. Because $AB = 2g$, $\frac{1}{2}AB = $ __?__. g
   c. Because both DE and $\frac{1}{2}AB$ equal __?__ g, $DE = \frac{1}{2}AB$.

Written Exercises

1. Axes have been drawn for a particular rectangle. Complete a proof that the diagonals are equal.

   a. Because $\overline{VT}$ is horizontal, the y-coordinate of T is __?__. 5
   b. Because $\overline{ST}$ is vertical, the x-coordinate of T is __?__. 4
   c. In right $\triangle VRS$, $(VS)^2 = 5^2 + $ __?__^{2}, and $VS = \sqrt{\text{\underline{?}}}$. 41
   d. In right $\triangle TSR$, $(RT)^2 = $ __?__$^2 + $ __?__^{2}, and $RT = \sqrt{\text{\underline{?}}}$. 41
   e. Because both VS and RT equal __?__ $\sqrt{41}$, $VS = RT$.

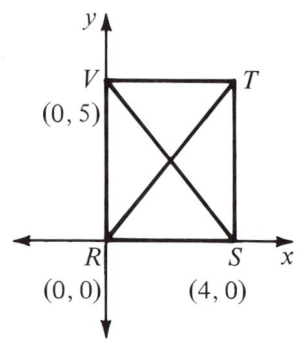

2. Axes have been drawn for a particular square. Complete a proof that the diagonals are perpendicular.

   a. The coordinates of T are ($\underline{\ ?\ }^5$, $\underline{\ ?\ }^5$).
   b. Slope of $\overline{RT} = \dfrac{5-0}{5-0} = \underline{\ ?\ }^1$
   c. Slope of $\overline{VS} = \dfrac{0-5}{5-0} = \underline{\ ?\ }^{-1}$
   d. (Slope of $\overline{RT}$) · (Slope of $\overline{VS}$) = $\underline{\ ?\ }_1 \cdot \underline{\ ?\ }_{-1} = \underline{\ ?\ }_{-1}$, so $\overline{RT} \perp \overline{VS}$.

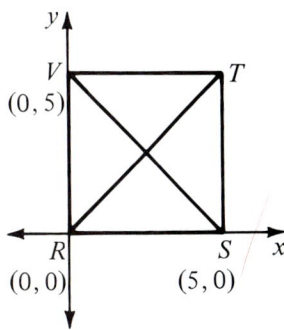

3. Axes have been drawn for a particular right triangle. Complete a proof that the midpoint M of the hypotenuse is equidistant from the three vertices.

   a. The coordinates of M are ($\underline{\ ?\ }^2$, $\underline{\ ?\ }^3$).
   b. Draw horizontal and vertical segments from M as shown.
   c. The coordinates of J are ($\underline{\ ?\ }^0$, $\underline{\ ?\ }^3$), and the coordinates of K are ($\underline{\ ?\ }^2$, $\underline{\ ?\ }^0$).
   d. In right $\triangle TJM$, $(MT)^2 = 3^2 + \underline{\ ?\ }^2$, and $MT = \sqrt{\underline{\ ?\ }}$. 2; 13
   e. In right $\triangle MKS$, $(MS)^2 = \underline{\ ?\ }^2 + \underline{\ ?\ }^2$, and $MS = \sqrt{\underline{\ ?\ }}$. 2; 3; 13
   f. In right $\triangle MKR$, $(MR)^2 = \underline{\ ?\ }^2 + \underline{\ ?\ }^2$, and $MR = \sqrt{\underline{\ ?\ }}$. 3; 2; 13
   g. Because MT, MS, and MR all equal $\sqrt{\underline{\ ?\ }_{13}}$, $MT = MS = MR$.

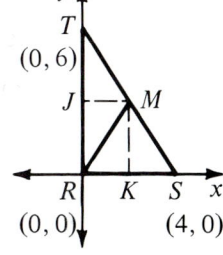

In Exercises 4–6 copy the diagram shown. See Exercises 1–3 for ideas about proofs.

B

4. *Prove:* The diagonals of any rectangle are equal.
5. *Prove:* The diagonals of any square are perpendicular.
6. *Prove:* The midpoint of the hypotenuse of any right triangle is equidistant from the three vertices.

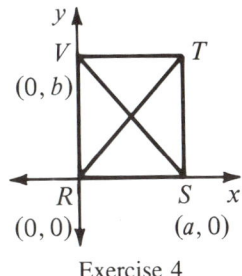
Exercise 4

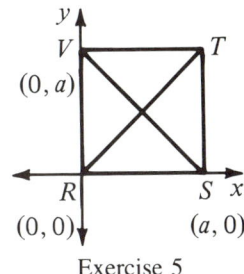
Exercise 5

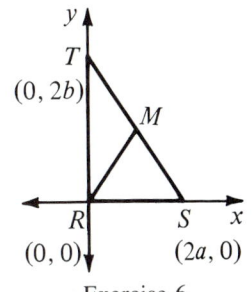
Exercise 6

Coordinate Geometry   503

In Exercises 7 and 8, copy the diagrams shown.

C **7. Prove:** The medians drawn to the legs of an isosceles triangle are equal. (*Hint:* Notice the x-coordinate of T. What must the x-coordinate of S be?) 4a

8. Prove: The segments joining, in order, the midpoints of the sides of any quadrilateral form a parallelogram.

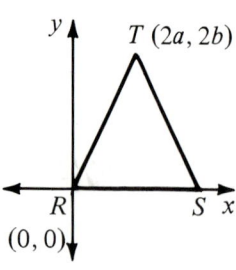

Exercise 7

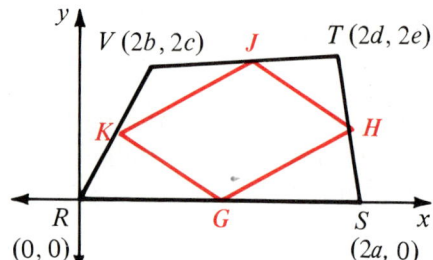

Exercise 8

SELF-TEST

Vocabulary

graph of an equation (p. 496)

1. Do the coordinates of the point $(5, -3)$ make the equation $2x - 3y = 1$ a true statement? no   *(12-6)*

Copy and complete each table of values.

2. $x + y = 8$

x	y
0	? 8
8 ?	0
3	? 5

3. $4x - 3y = 12$

x	y
0	? -4
3 ?	0
5	? $\frac{8}{3}$

4. Given the equation $2x + y = 10$, make a table of values, plot the points, and draw the graph of the equation.

5. You are given points $A(0, 0)$, $B(5, 0)$, $C(6, 4)$, and $D(1, 4)$. Prove that quadrilateral $ABCD$ is a parallelogram.   *(12-7)*

Applications

SOME SHAPES IN NATURE

Many of the shapes you have studied in this course occur in nature. Here are a few examples.

1.

   Did you guess that this is a much-enlarged picture of a snowflake? A snowflake is sometimes described as *hexagonal*.

2.

   A crystal of Iceland spar has the form of a *parallelepiped* (a solid with six faces, each a parallelogram). The crystal has an interesting property. Set a piece of Iceland spar over one segment and you see two segments.

   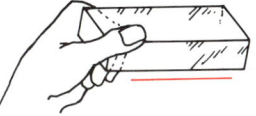
   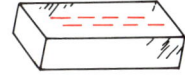

3.

   Quartz crystals suggest plane surfaces bounded by various kinds of polygons.

Coordinate Geometry

PROBLEM SOLVING STRATEGIES

PATTERNS

What is the sum of the interior angles of a polygon? On page 194 you looked at the sum of the interior angles of simple polygons and observed a pattern: The sum of the interior angles of a polygon with n sides is given by the formula

$$S = (n - 2) \times 180°.$$

EXERCISES

1. How many squares are there in this picture? 204

   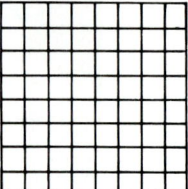

   Fill in the following chart. Use the pattern formed by the answers to predict the number of squares in question.

Number of Rows/Columns		Number of Squares
1	□	1
2		$5 = 4 + 1$
3		$14 = 9 + 4 + 1$
4		$\dfrac{?}{30} = \dfrac{?}{16} + \dfrac{?}{9} + \dfrac{?}{4} + \dfrac{?}{1}$

2. A square 12 units on each side has 100 congruent circles inscribed in it. What is the sum of the areas of the circles? 36π

   To answer this question, let's look at some data. Remember that the area of a circle is πr^2.

Number of Circles		Radius of Each Circle	Sum of Areas
1	(square with 1 circle, side 12)	$\frac{1}{2}(12) = 6$	$1 \times 36\pi$
4	(square with 4 circles, side 12)	$\frac{1}{2}\left(\frac{12}{2}\right) = 3$	$4 \times 9\pi$
9	(square with 9 circles, side 12)	$\frac{1}{2}\left(\frac{12}{3}\right) = 2$	$\underline{}_{9} \times \underline{}_{4} \pi$

3. If ten points are arranged around a circle and all the possible lines connecting the points in pairs are drawn, how many such lines will there be? Extend the table for five points and then predict the answer to the question. number of pts.: 5; number of lines: 10 = 4 + 3 + 2 + 1
number of pts.: 10; number of lines: 45

Number of Points		Number of Lines
2	(circle with 2 points, 1 line)	1
3	(circle with 3 points, triangle)	3 = 2 + 1
4	(circle with 4 points, all lines)	6 = 3 + 2 + 1

Coordinate Geometry

COMPUTER ACTIVITIES

THE BEST BOX

You are designing the package for a new product. The package is a cardboard box. You decide that the box will have a square base. It has to hold 1000 cm^3 of the new product. There are many possible choices for the dimensions of the box. For instance, it could be 5 cm by 5 cm by 40 cm. Or it could be 8 cm by 8 cm by about 15.6 cm. Either box would have a square base and hold 1000 cm^3 of the product.

However, you also have to think about the cost of the cardboard. The less cardboard you use, the less it will cost to make the box. You want the lowest cost box that will satisfy your requirements.

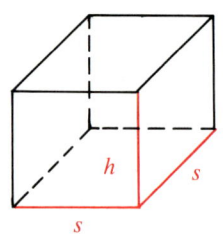

The volume of the box is the amount of product it can hold. The surface area of the box is the area of the cardboard. Your problem is to design the box with: (1) a square base, (2) a volume of 1000 cm^3, and (3) the smallest possible area.

In the program below, you choose the length of each side of the square base of the box. The computer figures the height and total area. *Note:* The complicated expression in line 60 is just the computer's way of rounding a decimal to thousandths.

```
10  REM ***BOX VOLUME AND TOTAL AREA
20  PRINT "LENGTH OF A SIDE OF"
30  PRINT "THE SQUARE BASE = ";
40  INPUT S
50  REM ***FIGURE HEIGHT IF VOLUME = 1000
60  H = INT(1000*(1000/S/S + .0005)) / 1000
70  REM ***FIGURE THE TOTAL AREA
80  TA = 2*S*(S + 2*H)
90  REM ***GIVE RESULTS
100 PRINT "HEIGHT: ";H
110 PRINT "TOTAL AREA: ";TA
120 PRINT
130 PRINT "DO YOU WANT TO TRY AGAIN";
140 INPUT Q$
150 IF Q$ = "YES" THEN 10
200 END
```

EXERCISES

1. RUN the program. Complete the chart.

Side of base	4	5	6	7	8	9	10	11	12	13	14
Height	?	?	?	?	?	?	?	?	?	?	?
Total area	?	?	?	?	?	?	?	?	?	?	?

2. a. What seems to be the smallest possible total area? 600 cm^2
   b. What are the dimensions of the box with this area?
   Give length, width, and height. 10 cm × 10 cm × 10 cm

3. a. RUN the program.
   Use a base side 0.5 cm more than in Exercise 2b. H = 9.070; T.A. = 601.440
   b. RUN the program.
   Use a base side of 0.5 cm less than in Exercise 2b. H = 11.080; T.A. = 601.540
   c. Do either of these boxes have a smaller total area
   than the box in Exercise 2? no

Next you work on a box that has to hold 2000 cm^3. It will also have a square base. Change the program as shown.

```
50  REM *** FIGURE HEIGHT IF VOLUME IS 2000
60  H = INT(1000 * (2000/S/S + .0005)) / 1000
```

4. RUN the changed program. Experiment until you think you have found the box with the smallest possible total area.
   a. What is this area? T.A. = 952.4592
   b. What are the dimensions of the box? Give length, width, and height.
   Round to the nearest tenth of a centimeter. 12.6 cm × 12.6 cm × 12.6 cm

5. a. What is special about the dimensions of the boxes in Exercises 2 and 4? They are all equal.
   b. What is the special name for this kind of box? cube

6. A box has to hold 8000 cm^3. You are to use the least cardboard possible. What should the dimensions be? Try to answer without using the computer. (*Hint:* If you decide to use the computer, you will need to change lines 50 and 60 again.) 20 cm × 20 cm × 20 cm

Coordinate Geometry

SKILLS REVIEW

ALGEBRAIC FRACTIONS: ADDITION AND SUBTRACTION

Simplify.

SAMPLE 1

$$\frac{6ab}{cd} - \frac{3ab+2}{cd} = \frac{6ab-(3ab+2)}{cd}$$

The denominators are the same. — Be careful with signs!

$$= \frac{6ab-3ab-2}{cd}$$

$$= \frac{3ab-2}{cd}$$

1. $\frac{12x}{7} - \frac{8x}{7}$ $\frac{4x}{7}$
2. $\frac{2ab}{3z} + \frac{4ab}{3z}$ $\frac{2ab}{z}$
3. $\frac{4xy}{5} - \frac{9xy}{5}$ $-xy$

4. $\frac{11d^2}{2c} - \frac{9d^2}{2c}$ $\frac{d^2}{c}$
5. $\frac{6b^3}{a} + \frac{2b}{a}$ $\frac{6b^3+2b}{a}$
6. $\frac{-8k}{7} + \frac{3k}{7}$ $\frac{-5k}{7}$

7. $\frac{11}{x+y} - \frac{10}{x+y}$ $\frac{1}{x+y}$
8. $\frac{4x+2}{k} - \frac{2+4x}{k}$ 0
9. $\frac{4m-n}{5n} + \frac{n+m}{5n}$ $\frac{m}{n}$

10. $\frac{5h+2k}{h+k} + \frac{h+4k}{h+k}$ 6
11. $\frac{b^2}{b+1} - \frac{1}{b+1}$ $b-1$
12. $\frac{a^2-6a}{a-3} + \frac{9}{a-3}$ $a-3$

SAMPLE 2

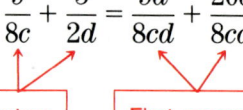

The denominators are different.   *First rename the fractions.*   *The new denominator can be 8cd.*

13. $\frac{a}{5} + \frac{a}{2}$ $\frac{7a}{10}$
14. $\frac{x}{3} - \frac{x}{8}$ $\frac{5x}{24}$
15. $\frac{2a}{3} + \frac{a}{6}$ $\frac{5a}{6}$
16. $\frac{n}{10} + \frac{4n}{5}$ $\frac{9n}{10}$

17. $\frac{5}{2a} + \frac{1}{a}$ $\frac{7}{2a}$
18. $\frac{3}{h} - \frac{1}{3h}$ $\frac{8}{3h}$
19. $\frac{7}{ab} - \frac{5a}{b}$ $\frac{7-5a^2}{ab}$
20. $\frac{2}{d} + \frac{3}{d^2}$ $\frac{2d+3}{d^2}$

21. $\frac{3m}{4} - \frac{2m}{3}$ $\frac{m}{12}$
22. $\frac{2a}{5} + \frac{a}{4}$ $\frac{13a}{20}$
23. $\frac{7}{x} + \frac{1}{2}$ $\frac{14+x}{2x}$
24. $\frac{1}{y} - \frac{1}{z}$ $\frac{z-y}{zy}$

25. $\frac{4}{3n} - \frac{2}{n}$ $\frac{-2}{3n}$
26. $\frac{a^2c}{b} + \frac{a^2b}{c}$ $\frac{a^2c^2+a^2b^2}{bc}$
27. $\frac{5x}{6} + \frac{x}{2}$ $\frac{4x}{3}$
28. $\frac{3m}{5} - \frac{m}{10}$ $\frac{m}{2}$

CHAPTER REVIEW

CHAPTER SUMMARY

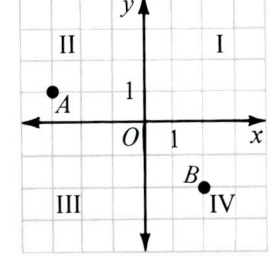

1. In the coordinate plane, the x-axis and the y-axis intersect at the origin, point $(0, 0)$.

   Point $A(-3, 1)$ lies in quadrant II.

   Point $B(2, -2)$ lies in quadrant IV.

2. The Pythagorean Theorem can be used to find the distance between two points in the coordinate plane.

3. Given $P(x_1, y_1)$ and $Q(x_2, y_2)$.
   The midpoint M of $\overline{PQ}$ has coordinates $\left(\dfrac{x_1 + x_2}{2}, \dfrac{y_1 + y_2}{2}\right)$.

4. The line that joins points (x_1, y_1) and (x_2, y_2) has slope $\dfrac{y_2 - y_1}{x_2 - x_1}$.
   The slope of a vertical line is not defined.

5. The following statements apply to nonvertical lines:

   If two lines are parallel, then their slopes are equal. If two lines have equal slopes, then they are parallel.

   If two lines are perpendicular, then the product of their slopes is -1. If the product of the slopes of two lines is -1, then the lines are perpendicular.

6. If an equation can be written in the form $ax + by = c$ (a and b not both 0), then its graph is a straight line. We find the graph by first preparing a table of values.

7. Coordinate geometry can be used to prove theorems.

REVIEW EXERCISES

Refer to the diagram at the right. *(See pp. 468–471.)*

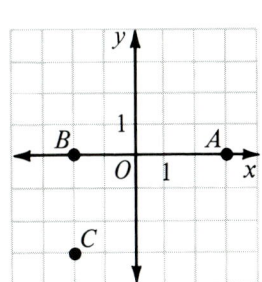

1. The coordinates of point A are __?__. $(3, 0)$

2. The coordinates of point B are __?__. $(-2, 0)$

3. Point B lies on the __?__-axis. x

4. Suppose D is located so that quadrilateral $ABCD$ is a rectangle. Point D has coordinates $(\underset{3}{\underline{\ ?\ }}, \underset{-3}{\underline{\ ?\ }})$.

5. Point C lies in quadrant __?__. III

Coordinate Geometry

Find the distance between the points named. Use graph paper if you wish. *(See pp. 472–474.)*

6. $(0, 0)$ and $(6, 8)$ 10
7. $(3, 0)$ and $(3, 7)$ 7
8. $(0, 2)$ and $(4, 5)$ 5
9. $(-1, -1)$ and $(7, -7)$ 10

State the coordinates of the midpoint of the segment that joins the points named. *(See pp. 476–480.)*

10. $(2, 0)$ and $(6, 0)$ $(4, 0)$
11. $(3, -8)$ and $(3, 0)$ $(3, -4)$
12. $(3, 4)$ and $(5, -4)$ $(4, 0)$
13. $(3, 4)$ and $(-5, 1)$ $(-1, \frac{5}{2})$

State the slope of the line. If the slope is not defined for a particular line, write *not defined*. *(See pp. 486–490.)*

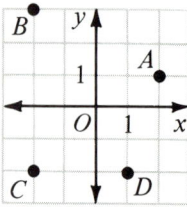

14. $\overleftrightarrow{OA}$ $\frac{1}{2}$
15. $\overleftrightarrow{OB}$ $-\frac{3}{2}$
16. $\overleftrightarrow{BA}$ $-\frac{1}{2}$
17. $\overleftrightarrow{CD}$ 0
18. $\overleftrightarrow{CA}$ $\frac{3}{4}$
19. $\overleftrightarrow{CB}$ not defined

A certain line t has slope $\frac{2}{3}$. Tell whether each statement is true or false. *(See pp. 491–495.)*

20. Line t rises to the right. true
21. Every line that is perpendicular to t has slope $\frac{3}{2}$. false
22. The line that contains points $(-1, 0)$ and $(3, 6)$ is parallel to t. false
23. Every line that is parallel to t has slope $\frac{2}{3}$. true

Line j has the equation $2x - y = 10$. *(See pp. 496–500.)*

24. Does j pass through the origin? no
25. Does j pass through the point $(5, 1)$? no
26. On line j, when $y = 4$, $x = $ ___?___. 7

Points A, B, C, and D have the coordinates shown. *(See pp. 501–504.)*

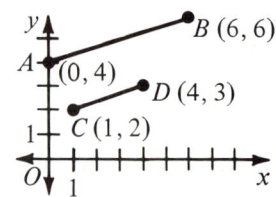

27. Prove: $\overleftrightarrow{AB} \parallel \overleftrightarrow{CD}$
28. Prove: $CD = \frac{1}{2}AB$

Chapter Test

Consider points $A(-3, 0)$, $B(-2, -1)$, and $C(-2, 3)$.

1. Plot points A, B, and C.
2. Which point lies in quadrant III? B   *(12-1)*
3. Find the distance between points B and C. 4

Find the distance between the points named.

4. $A(0, 0)$ and $B(-9, -12)$ 15
5. $A(2, 5)$ and $B(9, 4)$ $5\sqrt{2}$ *(12-2)*
6. Find the perimeter and area of the triangle with vertices $A(-3, 0)$, $B(3, 0)$, and $C(0, 4)$. $p = 16$; $A = 12$

Find the coordinates of the midpoint of $\overline{AB}$.

7. $A(-7, -1)$ and $B(-3, 5)$ $(-5, 2)$
8. $A(0, 6)$ and $B(3, -10)$ $(\frac{3}{2}, -2)$ *(12-3)*
9. M is the midpoint of $\overline{XY}$, X is the point $(2, 3)$, and M is point $(5, 1)$. Find the coordinates of point Y. $(8, -1)$

You are given points $R(2, -5)$, $S(-3, -5)$ and $T(2, -10)$. Find the slope of the line, if possible, or write *not defined*.

10. $\overleftrightarrow{RS}$ 0
11. $\overleftrightarrow{ST}$ -1
12. $\overleftrightarrow{RT}$ not defined   *(12-4)*

Suppose line t has slope $-\frac{1}{4}$. Complete each statement.

13. Line t __?__ from left to right.
    a. rises   **b. falls**   c. neither
14. If $\overleftrightarrow{AB} \perp t$, then $\overleftrightarrow{AB}$ has slope __?__.   *(12-5)*
    a. -4 b. $\frac{1}{4}$ **c. 4**
15. The line through points $(2, 5)$ and $(10, 3)$ __?__ line t.
    a. is parallel to   b. is perpendicular to   c. neither

Line j has equation $2x - 3y = 6$.

16. Find the missing coordinates of the following points on line j.   *(12-6)*
    a. $P(0, \_?\_)$ -2 b. $Q(\_?\_, 0)$ 3   c. $R(\_?\_, 2)$ 6
17. Show that $(2, -1)$ does not lie on line j. $2(2) - 3(-1) = 4 + 3 = 7 \neq 6$
18. Draw the graph of line j.

You are given points $A(0, 0)$, $B(4, 2)$, $C(1, 8)$, and $D(-3, 6)$.

19. Prove: $AC = BD$
20. Prove: $ABCD$ is a rectangle.   *(12-7)*

UNIT F CUMULATIVE REVIEW

2. $3\sqrt{3}$ **3.** $3\sqrt{2}$

Refer to the figure shown. Find each of the following.

1. SR 3   **2.** SE **3.** TS **4.** TA 5

5. The perimeter of $\triangle TAR$ 12

6. The area of $\triangle STA$ $10\frac{1}{2}$

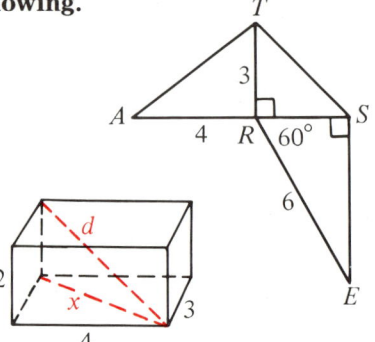

Refer to the rectangular solid shown.

7. Find the value of x. 5

8. Find the length d of the diagonal. $\sqrt{29}$

9. A regular square pyramid has height 15 and slant height 17.
  a. How long is an edge of the base?   **b.** What is the volume?   **a.** 16   **b.** 1280

10. A cone has radius 5 and height 12. Find:
  a. the slant height   **b.** the total area   **c.** the volume
       13   90π 100π

Refer to $\triangle GHI$. Express each ratio in simplest form.

11. $\sin G$ $\frac{4}{5}$ **12.** $\cos I$ $\frac{4}{5}$ **13.** $\tan G$ $\frac{4}{3}$ **14.** $\sin I$ $\frac{3}{5}$

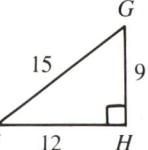

Use the table on page 517 to complete each statement.

15. $\sin B =$ __?__   **16.** $\tan A =$ __?__   0.4663
         0.9063

17. To the nearest tenth, $y =$ __?__.   4.7

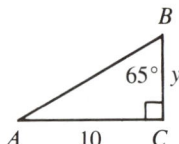

Complete each statement.

18. Point D lies in quadrant __?__. III

19. The coordinates of point A are (__?__, __?__). -2, 3

20. The length of $\overline{AD}$ is __?__. $\sqrt{17}$

21. The midpoint of $\overline{DC}$ is __?__. $(0, -2)$

22. The slope of $\overleftrightarrow{AD}$ is __?__. 4

23. The slope of $\overleftrightarrow{DC}$ is __?__. $-\frac{1}{3}$

24. Any line perpendicular to $\overleftrightarrow{AB}$ has slope __?__. 3

25. Prove that quadrilateral $ABCD$ is a parallelogram.

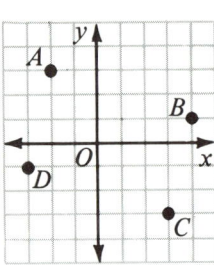

Table of Squares

Number	Square
n	n^2
1	1
2	4
3	9
4	16
5	25
6	36
7	49
8	64
9	81
10	100
11	121
12	144
13	169
14	196
15	225
16	256
17	289
18	324
19	361
20	400
21	441
22	484
23	529
24	576
25	625

Number	Square
n	n^2
26	676
27	729
28	784
29	841
30	900
31	961
32	1024
33	1089
34	1156
35	1225
36	1296
37	1369
38	1444
39	1521
40	1600
41	1681
42	1764
43	1849
44	1936
45	2025
46	2116
47	2209
48	2304
49	2401
50	2500

Number	Square
n	n^2
51	2601
52	2704
53	2809
54	2916
55	3025
56	3136
57	3249
58	3364
59	3481
60	3600
61	3721
62	3844
63	3969
64	4096
65	4225
66	4356
67	4489
68	4624
69	4761
70	4900
71	5041
72	5184
73	5329
74	5476
75	5625

Number	Square
n	n^2
76	5776
77	5929
78	6084
79	6241
80	6400
81	6561
82	6724
83	6889
84	7056
85	7225
86	7396
87	7569
88	7744
89	7921
90	8100
91	8281
92	8464
93	8649
94	8836
95	9025
96	9216
97	9409
98	9604
99	9801
100	10,000

Table of Square Roots

The square roots are given correct to three decimal places.

Number	Positive Square Root	Number	Positive Square Root	Number	Positive Square Root	Number	Positive Square Root
n	$\sqrt{n}$	n	$\sqrt{n}$	n	$\sqrt{n}$	n	$\sqrt{n}$
1	1	26	5.099	51	7.141	76	8.718
2	1.414	27	5.196	52	7.211	77	8.775
3	1.732	28	5.292	53	7.280	78	8.832
4	2	29	5.385	54	7.348	79	8.888
5	2.236	30	5.477	55	7.416	80	8.944
6	2.449	31	5.568	56	7.483	81	9
7	2.646	32	5.657	57	7.550	82	9.055
8	2.828	33	5.745	58	7.616	83	9.110
9	3	34	5.831	59	7.681	84	9.165
10	3.162	35	5.916	60	7.746	85	9.220
11	3.317	36	6	61	7.810	86	9.274
12	3.464	37	6.083	62	7.874	87	9.327
13	3.606	38	6.164	63	7.937	88	9.381
14	3.742	39	6.245	64	8	89	9.434
15	3.873	40	6.325	65	8.062	90	9.487
16	4	41	6.403	66	8.124	91	9.539
17	4.123	42	6.481	67	8.185	92	9.592
18	4.243	43	6.557	68	8.246	93	9.644
19	4.359	44	6.633	69	8.307	94	9.695
20	4.472	45	6.708	70	8.367	95	9.747
21	4.583	46	6.782	71	8.426	96	9.798
22	4.690	47	6.856	72	8.485	97	9.849
23	4.796	48	6.928	73	8.544	98	9.899
24	4.899	49	7	74	8.602	99	9.950
25	5	50	7.071	75	8.660	100	10

Table of Trigonometric Ratios

Angle	Sine	Cosine	Tangent	Angle	Sine	Cosine	Tangent
1°	.0175	.9998	.0175	46°	.7193	.6947	1.0355
2°	.0349	.9994	.0349	47°	.7314	.6820	1.0724
3°	.0523	.9986	.0524	48°	.7431	.6691	1.1106
4°	.0698	.9976	.0699	49°	.7547	.6561	1.1504
5°	.0872	.9962	.0875	50°	.7660	.6428	1.1918
6°	.1045	.9945	.1051	51°	.7771	.6293	1.2349
7°	.1219	.9925	.1228	52°	.7880	.6157	1.2799
8°	.1392	.9903	.1405	53°	.7986	.6018	1.3270
9°	.1564	.9877	.1584	54°	.8090	.5878	1.3764
10°	.1736	.9848	.1763	55°	.8192	.5736	1.4281
11°	.1908	.9816	.1944	56°	.8290	.5592	1.4826
12°	.2079	.9781	.2126	57°	.8387	.5446	1.5399
13°	.2250	.9744	.2309	58°	.8480	.5299	1.6003
14°	.2419	.9703	.2493	59°	.8572	.5150	1.6643
15°	.2588	.9659	.2679	60°	.8660	.5000	1.7321
16°	.2756	.9613	.2867	61°	.8746	.4848	1.8040
17°	.2924	.9563	.3057	62°	.8829	.4695	1.8807
18°	.3090	.9511	.3249	63°	.8910	.4540	1.9626
19°	.3256	.9455	.3443	64°	.8988	.4384	2.0503
20°	.3420	.9397	.3640	65°	.9063	.4226	2.1445
21°	.3584	.9336	.3839	66°	.9135	.4067	2.2460
22°	.3746	.9272	.4040	67°	.9205	.3907	2.3559
23°	.3907	.9205	.4245	68°	.9272	.3746	2.4751
24°	.4067	.9135	.4452	69°	.9336	.3584	2.6051
25°	.4226	.9063	.4663	70°	.9397	.3420	2.7475
26°	.4384	.8988	.4877	71°	.9455	.3256	2.9042
27°	.4540	.8910	.5095	72°	.9511	.3090	3.0777
28°	.4695	.8829	.5317	73°	.9563	.2924	3.2709
29°	.4848	.8746	.5543	74°	.9613	.2756	3.4874
30°	.5000	.8660	.5774	75°	.9659	.2588	3.7321
31°	.5150	.8572	.6009	76°	.9703	.2419	4.0108
32°	.5299	.8480	.6249	77°	.9744	.2250	4.3315
33°	.5446	.8387	.6494	78°	.9781	.2079	4.7046
34°	.5592	.8290	.6745	79°	.9816	.1908	5.1446
35°	.5736	.8192	.7002	80°	.9848	.1736	5.6713
36°	.5878	.8090	.7265	81°	.9877	.1564	6.3138
37°	.6018	.7986	.7536	82°	.9903	.1392	7.1154
38°	.6157	.7880	.7813	83°	.9925	.1219	8.1443
39°	.6293	.7771	.8098	84°	.9945	.1045	9.5144
40°	.6428	.7660	.8391	85°	.9962	.0872	11.4301
41°	.6561	.7547	.8693	86°	.9976	.0698	14.3007
42°	.6691	.7431	.9004	87°	.9986	.0523	19.0811
43°	.6820	.7314	.9325	88°	.9994	.0349	28.6363
44°	.6947	.7193	.9657	89°	.9998	.0175	57.2900
45°	.7071	.7071	1.0000	90°	1.0000	0.0000	Undefined

Postulates

THE ADDITION POSTULATE
page 29

If $a = b$ and $c = d$, then $a + c = b + d$.

THE SUBTRACTION POSTULATE
page 29

If $a = b$ and $c = d$, then $a - c = b - d$.

THE MULTIPLICATION POSTULATE
page 29

If $a = b$, then $ac = bc$.

THE DIVISION POSTULATE
page 30

If $a = b$ and $c \neq 0$, then $\dfrac{a}{c} = \dfrac{b}{c}$.

THE SUBSTITUTION POSTULATE
page 30

If $a = b$, then a can be substituted for b in any equation or inequality.

POSTULATE 1
page 34

Through any two points there is exactly one line.

POSTULATE 2
page 34

Through any three noncollinear points there is exactly one plane.

POSTULATE 3
page 34

If two points lie in a plane, then the line joining them lies in that plane.

POSTULATE 4
page 34

If two planes intersect, then their intersection is a line.

POSTULATE 5 (The Ruler Postulate)
page 34

Each point on a line can be paired with exactly one real number called its coordinate. The distance between two points is the positive difference of their coordinates.

POSTULATE 6 (The Protractor Postulate)
page 35

Suppose O is a point of $\overleftrightarrow{XY}$. Consider all rays with endpoint O which lie on one side of $\overleftrightarrow{XY}$. Each ray can be paired with exactly one real number between 0 and 180.

POSTULATE 7
page 67

If two parallel lines are cut by a transversal, then corresponding angles are equal.

POSTULATE 8
page 72

If two lines and a transversal form equal corresponding angles, then the lines are parallel.

POSTULATE 9 (The SSS Postulate)
page 107

If three sides of one triangle are equal to the corresponding parts of another triangle, the triangles are congruent.

POSTULATE 10 (The SAS Postulate)
page 116

If two sides and the included angle of one triangle are equal to the corresponding parts of another triangle, the triangles are congruent.

POSTULATE 11 (The ASA Postulate)
page 121

If two angles and the included side of one triangle are equal to the corresponding parts of another triangle, the triangles are congruent.

POSTULATE 12
page 238

The area of a rectangle is given by the formula:
Area = base × height

POSTULATE 13 (The AA Postulate)
page 317

If two angles of one triangle are equal to two angles of another triangle, then the triangles are similar.

Theorems

CHAPTER 2 Introducing Proof

THEOREM 1
page 46
If two angles are complements of equal angles (or of the same angle), then the two angles are equal.

THEOREM 2
page 47
If two angles are supplements of equal angles (or of the same angle), then the two angles are equal.

THEOREM 3
page 47
Vertical angles are equal.

THEOREM 4
page 69
If two parallel lines are cut by a transversal, then alternate interior angles are equal.

THEOREM 5
page 69
If two parallel lines are cut by a transversal, then same-side interior angles are supplementary.

THEOREM 6
page 72
If two lines and a transversal form equal alternate interior angles, then the lines are parallel.

THEOREM 7
page 72
If two lines and a transversal form supplementary same-side interior angles, then the lines are parallel.

THEOREM 8
page 72
In a plane, if two lines are each perpendicular to a third line, then the two lines are parallel.

CHAPTER 3 Triangles

THEOREM 1
page 92
The angle sum of a triangle is 180°.

COROLLARY
page 93
An exterior angle of a triangle is equal to the sum of the two opposite angles of the triangle.

THEOREM 2
page 126
(AAS Theorem) If two angles and a non-included side of one triangle are equal to the corresponding parts of another triangle, the triangles are congruent.

THEOREM 3
page 126
(HL Theorem) If the hypotenuse and a leg of one right triangle are equal to the corresponding parts of another right triangle, the triangles are congruent.

CHAPTER 4 Using Congruent Triangles

THEOREM 1
page 154
Any point on the perpendicular bisector of a segment is equidistant from the endpoints of the segment.

THEOREM 2
page 155
Any point that is equidistant from the endpoints of a segment is on the perpendicular bisector of the segment.

THEOREM 3
page 167
If two sides of a triangle are equal, then the angles opposite those sides are equal.

COROLLARY
page 167
An equilateral triangle is also equiangular, and each angle has measure 60°.

THEOREM 4
page 173
If two angles of a triangle are equal, then the sides opposite those angles are equal.

COROLLARY
page 173
If a triangle is equiangular, it is also equilateral.

CHAPTER 5 Polygons

THEOREM 1 If a convex polygon has n sides, then its angle sum is given by the formula:
page 194 $S = (n - 2) \times 180°$

THEOREM 2 The exterior angle sum of any convex polygon, one angle at each vertex, is 360°.
page 195

THEOREM 3 Opposite sides of a parallelogram are equal.
page 204

THEOREM 4 Opposite angles of a parallelogram are equal.
page 204

THEOREM 5 Consecutive angles of a parallelogram are supplementary.
page 204

THEOREM 6 Diagonals of a parallelogram bisect each other.
page 204

THEOREM 7 The diagonals of a rectangle are equal.
page 210

THEOREM 8 The diagonals of a rhombus are perpendicular, and they bisect the angles of the
page 210 rhombus.

THEOREM 9 The midpoint of the hypotenuse of a right triangle is equidistant from the three
page 210 vertices.

THEOREM 10 If a quadrilateral has one pair of opposite sides that are both parallel and equal, then
page 214 the quadrilateral is a parallelogram.

THEOREM 11 If a quadrilateral has both pairs of opposite sides equal, then the quadrilateral is a
page 215 parallelogram.

THEOREM 12 If a quadrilateral has diagonals that bisect each other, then the quadrilateral is a
page 215 parallelogram.

THEOREM 13 The base angles of an isosceles trapezoid are equal.
page 221

THEOREM 14 The median of a trapezoid has two properties:
page 221 **(1)** It is parallel to the bases.
  (2) Its length equals half the sum of the base lengths.

THEOREM 15 (The Midpoints Theorem) The segment joining the midpoints of two sides of a triangle
page 225 is parallel to the third side and half as long.

CHAPTER 6 Areas

THEOREM 1 The area of a square is given by the formula:
page 239 Area = side squared

THEOREM 2 The area of a parallelogram is given by the formula:
page 244 Area = base × height

THEOREM 3 The area of a triangle is given by the formula:
page 247 Area = $\frac{1}{2}$ × base × height

THEOREM 4 The area of a trapezoid is given by the formula:
page 251 Area = $\frac{1}{2}$ × height × sum of the bases

THEOREM 5 *page 256*	(The Pythagorean Theorem) In a right triangle, the square of the hypotenuse is equal to the sum of the squares of the legs.
THEOREM 6 *page 259*	If the square of one side of a triangle is equal to the sum of the squares of the other two sides, then the triangle is a right triangle.

CHAPTER 8 Similar Polygons

THEOREM 1 *page 325*	(The Triangle Proportionality Theorem) If a line intersects a triangle and is parallel to one side, then it divides the other two sides proportionally.
COROLLARY *page 326*	If three parallel lines intersect two transversals, they divide the transversals proportionally.
THEOREM 2 *page 330*	If two similar polygons have a scale factor $a{:}b$, then: (1) the ratio of their perimeters is $a{:}b$; (2) the ratio of their areas is $a^2{:}b^2$.

CHAPTER 9 Circles

THEOREM 1 *page 350*	A radius drawn to a point of tangency is perpendicular to the tangent.
THEOREM 2 *page 350*	If a line lies in the plane of a circle and is perpendicular to a radius at its outer endpoint, the line is tangent to the circle.
THEOREM 3 *page 356*	In a circle, equal central angles have equal minor arcs.
THEOREM 4 *page 356*	In a circle, equal minor arcs have equal central angles.
THEOREM 5 *page 360*	In a circle, equal chords have equal arcs and equal arcs have equal chords.
THEOREM 6 *page 360*	In a circle, equal chords are equidistant from the center. Chords that are equidistant from the center are equal.
THEOREM 7 *page 360*	A diameter that is perpendicular to a chord bisects the chord and its arc.
THEOREM 8 *page 364*	An inscribed angle is equal to half its intercepted arc.
THEOREM 9 *page 365*	An angle formed by a chord and a tangent is equal to half its intercepted arc.
THEOREM 10 *page 369*	An angle formed by two chords is equal to half the sum of the intercepted arcs.
THEOREM 11 *page 369*	An angle formed by two secants is equal to half the difference of the intercepted arcs.
THEOREM 12 *page 373*	If two chords intersect inside a circle, the product of the lengths of the segments of one chord equals the product of the lengths of the segments of the other.

CHAPTER 10 Areas and Volumes of Solids

THEOREM 1 *page 414*	Suppose two similar solids have scale factor $a{:}b$. Then: (1) the ratio of corresponding segments is $a{:}b$; (2) the ratio of the areas is $a^2{:}b^2$; (3) the ratio of the volumes is $a^3{:}b^3$.

Theorems

CHAPTER 11 Right Triangles

THEOREM 1 In a 45°-45°-90° triangle, the hypotenuse is $\sqrt{2}$ times as long as a leg.
page 432

THEOREM 2 In a 30°-60°-90° triangle:
page 432
(1) the hypotenuse is twice as long as the shorter leg;
(2) the longer leg is $\sqrt{3}$ times as long as the shorter leg.

Constructions

CONSTRUCTION 1 *Given:* An angle
page 25 *Construct:* A bisector of the angle

CONSTRUCTION 2 *Given:* A line and a point on the line
page 26 *Construct:* A perpendicular to the line through the point

CONSTRUCTION 3 *Given:* A line and a point not on the line
page 26 *Construct:* A perpendicular to the line through the point

CONSTRUCTION 4 *Given:* An angle
page 77 *Construct:* An angle equal to the given angle

CONSTRUCTION 5 *Given:* A line and a point not on the line
page 78 *Construct:* A line that passes through the point and is parallel to the given line

CONSTRUCTION 6 *Given:* A segment
page 154 *Construct:* A perpendicular bisector of the segment

CONSTRUCTION 7 *Given:* A triangle
page 163 *Construct:* A circle passing through the vertices of the triangle

CONSTRUCTION 8 *Given:* A triangle
page 164 *Construct:* An inscribed circle

CONSTRUCTION 9 *Construct:* A 60° angle
page 174

CONSTRUCTION 10 *Given:* Point A on $\odot O$
page 351 *Construct:* A tangent to $\odot O$ at point A

Review Exercises

CHAPTER 1

In the diagram, $\overline{AD} \perp \overline{FC}$ and $\angle DOE = 30°$.
a. Find the measures of the following angles.
b. Then classify each as acute, right, or obtuse.

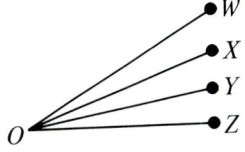

1. $\angle BOA$ acute
2. $\angle BOF$ obtuse
3. $\angle COD$ right
4. $\angle BOD$ obtuse
5. $\angle BOA$ and $\angle DOE$ are called __?__ angles. vertical
6. Name an angle equal to $\angle BOD$. $\angle EOA$

In the diagram, $\overrightarrow{OX}$ bisects $\angle WOY$ and $\overrightarrow{OY}$ bisects $\angle XOZ$.

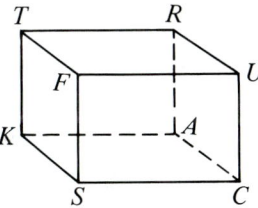

7. $\angle WOX = \angle$ __?__   XOY
8. $\angle YOZ = \angle$ __?__   XOY
9. Which postulate allows you to conclude that $\angle WOX = \angle YOZ$? Substitution

Refer to the rectangular solid shown.

10. $K, S, F,$ and __?__ are coplanar points. T
11. Plane $TFUR \parallel$ plane __?__ KSCA
12. Name two planes that intersect in $\overleftrightarrow{FS}$. TFSK; UFSC
13. How many planes contain points $R, U, F,$ and S? none
14. Construct a 90° angle.

CHAPTER 2

Given: $\overline{AB} \perp \overline{BC}$ and $\overline{AD} \perp \overline{DC}$.

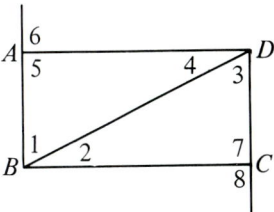

1. $\angle 1$ and $\angle$ __?__ are called complementary angles. 2
2. $\angle 5$ and $\angle 6$ are called __?__ angles. supplementary
3. If $\angle 1 = \angle 3$, state the theorem that allows you to conclude that $\angle 2 = \angle 4$. If 2 ▵ are comp. of = ▵, then the 2 ▵ are =.
4. If $\angle 5 = \angle 7$, state the theorem that allows you to conclude that $\angle 6 = \angle 8$. If 2 ▵ are supp. of = ▵, then the 2 ▵ are =.

Exercises 5–11 refer to the figure at the right.

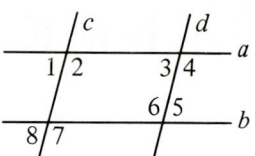

5. ∠1 and ∠8 are called __?__ angles. corresponding
6. ∠2 and ∠ __?__ are called same-side interior angles. 3
7. ∠6 and ∠7 are called __?__ angles. alternate interior

Use the given information to tell which lines must be parallel.

8. ∠3 = ∠5   a ∥ b   9. ∠1 = ∠3   c ∥ d   10. ∠8 = ∠5   c ∥ d   11. ∠4 + ∠5 = 180°   a ∥ b

12. Write a complete proof in two-column form.
    Given: ∠2 = ∠3
    $\overrightarrow{QS}$ bisects ∠PQR.
    Prove: $\overleftrightarrow{PQ} \parallel \overleftrightarrow{SR}$

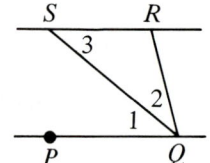

13. Draw an acute angle, ∠A.
    Then construct an angle equal to ∠A.

CHAPTER 3

In each exercise, find the value of x.

1. 75
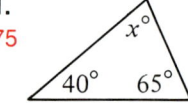

2. 30
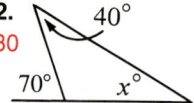

3. 30
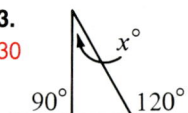

4. In △ABC, AB = AC.
   a. △ABC must be a(n) __?__ triangle. isosceles
   b. Is it possible that △ABC is an equilateral triangle? yes
   c. Is it possible that △ABC is an obtuse triangle? yes
   d. Is it possible that △ABC is an acute triangle? yes

5. Suppose △ANT ≅ △LER.
   a. AT = __?__   LR   b. __?__ = ER   NT   c. ∠N = ∠ __?__   E

Write SSS, SAS, ASA, AAS, or HL to indicate a method you could use to prove two triangles congruent.

6. SAS
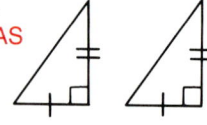

7. ASA
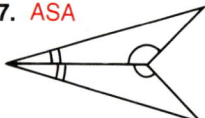

8. AAS
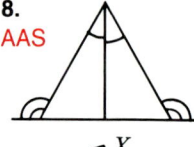

9. Write a complete proof in two-column form.
   Given: $\overline{AB} \perp \overline{XY}$
   $AX = AY$
   Prove: △ABX ≅ △ABY

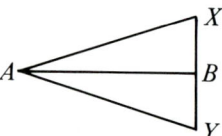

CHAPTER 4

1. Suppose that △ RUN ≅ △ TEX. What reason can you write to support the statements: UN = EX and ∠ N = ∠ X? Corresponding parts of ≅ △ are =.

2. Draw a large △ ABC with $\overline{AB}$ clearly longer than $\overline{AC}$. Draw the altitude $\overline{AH}$ from point A and the median $\overline{AM}$ from point A.

3. Draw a figure like the one for Exercise 2, but much larger. Construct the perpendicular bisectors of any two sides of the triangle. Then construct a circle circumscribed about △ ABC.

In each diagram, equal angles are marked. Find the value of x.

4. 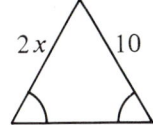 5 , 2x , 10

5. 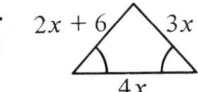 2x + 6 , 3x , 4x , 6

6. 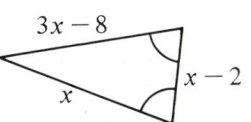 3x − 8 , x − 2 , 4 , x

Write complete proofs in two-column form.

7. Given: ∠ A = ∠ X
         AB = XY
         AC = XZ
   Prove: BC = YZ

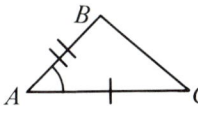

8. Given: RS = TS
   Prove: ∠ 3 = ∠ 2

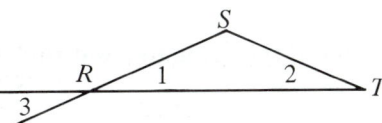

CHAPTER 5

Tell whether each statement is true or false.

1. An octagon has ten sides. false
2. A regular polygon is equilateral. true
3. Every square is a rectangle. true
4. Every rectangle is a square. false
5. The interior angle sum of a hexagon is 720°. true
6. In a regular pentagon, each interior angle has measure 110°. false
7. The diagonals of a rhombus are perpendicular. true
8. The diagonals of a trapezoid bisect each other. false
9. The diagonals of a parallelogram bisect each other. true

Review Exercises   525

Tell whether each statement is true or false.

10. The legs of a trapezoid are parallel. false

11. If the opposite angles of a quadrilateral are equal, then it must be a parallelogram. true

12. If a quadrilateral has two 80° angles and two 100° angles, then it must be a parallelogram. false

13. Base angles of an isosceles trapezoid are supplementary. false

14. The diagonals of a rectangle are equal. true

15. Write a complete proof in two-column form.
    Given: X, Y, and Z are the midpoints of the sides of △ABC.
    Prove: AXYZ is a ▱.

CHAPTER 6

1. A rectangle has sides 7 m and 3 m. Find the perimeter and the area. $p = 20$ m; $A = 21$ m²

2. Find the area of a square whose perimeter is 12 cm. 9 cm²

Find the area of the parallelogram, the triangle, and the trapezoid.

3. 40
4. 30
5. 63

In each diagram, find the value of x.

6.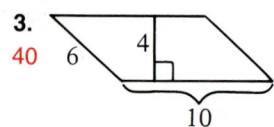
7.
8.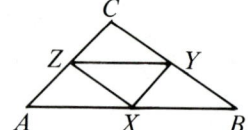

Three sides of a triangle are given. Classify the triangle as acute, right, or obtuse.

9. 3, 4, 5 right
10. 3, 4, 6 obtuse
11. 7, 7, 7 acute

Find the circumference of each circle, in terms of π, given:

12. $r = 6$ 12π
13. $r = 1.5$ 3π
14. $d = 8$ 8π

15–17. Use 3.14 for π. Find the area of each circle described above, correct to tenths. **15.** 113.0 **16.** 7.1 **17.** 50.2

CHAPTER 7

Express each ratio in simplest form.

1. $\dfrac{15}{9}$ $\dfrac{5}{3}$
2. $\dfrac{24}{36}$ $\dfrac{2}{3}$
3. $\dfrac{15}{20}$ $\dfrac{3}{4}$
4. $\dfrac{3ac}{9ab}(a \neq 0, b \neq 0)$ $\dfrac{c}{3b}$
5. $18:20$ $9:10$
6. $9:12:15$ $3:4:5$
7. $2\text{ m}:40\text{ cm}$ $5:1$
8. $3\text{ km}:500\text{ m}$ $6:1$

9. $\dfrac{SA}{AL}$ $\dfrac{3}{2}$
10. $\dfrac{SR}{ST}$ $\dfrac{3}{5}$
11. $\dfrac{AR}{LT}$ $\dfrac{3}{5}$
12. $\dfrac{TR}{RS}$ $\dfrac{2}{3}$

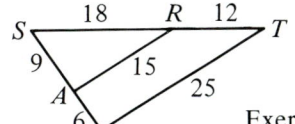

Exercises 9–12

Find the value of x.

13. $\dfrac{4}{x} = \dfrac{6}{9}$ 6
14. $\dfrac{18}{x} = \dfrac{9}{4}$ 8
15. $x:9 = 6:27$ 2
16. $\dfrac{x-3}{4} = \dfrac{x}{8}$ 6

17. Two bags of cookies cost $1.78. How much will 5 bags cost? $4.45

Exercises 18 and 19 refer to the diagram.

18. Measure the distance on the map between Livonia and Geneva to the nearest centimeter. Then give the scale of the map. 4 cm; 1 cm : 25 km

19. Find the actual distance between Raleigh and Geneva. 62.5 km

20. Make a scale drawing of a soccer field 100 m by 60 m.

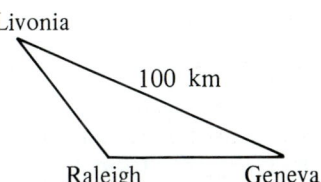

CHAPTER 8

State whether the following polygons must be similar.

1. Any two right triangles no
2. Any two regular pentagons yes
3. Any two congruent polygons yes
4. Any two isosceles trapezoids no

Find the values of x and y in each diagram.

5.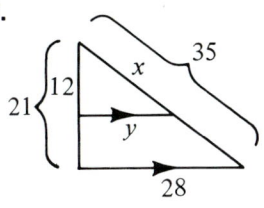

   $x = 20;\ y = 16$

6.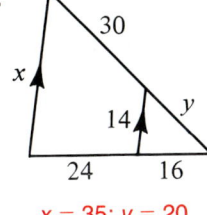

   $x = 35;\ y = 20$

7.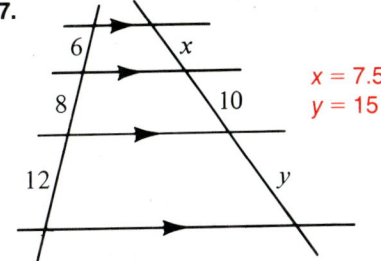

   $x = 7.5$
   $y = 15$

Exercises 8–10 refer to the figure shown.

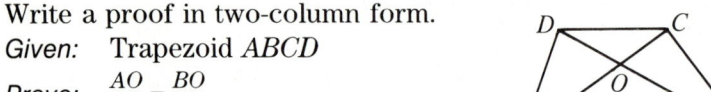

8. Which postulate supports the statement:

    △ RCA ~ △ SBU? AA Postulate

9. Find the ratio of the perimeters of the triangles. $\frac{3}{4}$

10. Find the ratio of the areas of the triangles. $\frac{9}{16}$

11. Write a proof in two-column form.
    Given: Trapezoid ABCD
    Prove: $\frac{AO}{CO} = \frac{BO}{DO}$
    (Hint: Find two similar triangles.)

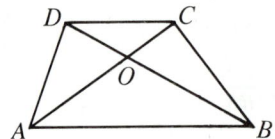

CHAPTER 9

In the diagram, $\overrightarrow{UV}$ is tangent to $\odot O$. $\widehat{AB} = 70°$ and $\widehat{BC} = 80°$. Find the measure of the arc or angle.

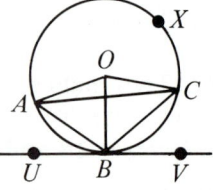

1. $\widehat{ABC}$ 150°
2. $\widehat{AXC}$ 210°
3. ∠ OBU 90°
4. ∠ AOB 70°
5. ∠ ACB 35°
6. ∠ CBV 40°

$\overline{YZ}$ is a diameter of $\odot P$. $\overline{YZ} \perp \overline{RT}$.

7. If $\widehat{RT} = 100°$, then $\widehat{RZ} = \underline{\ 50\ }°$ and $\widehat{RY} = \underline{\ 130\ }°$.
8. If $RS = 7$, then $ST = \underline{\ 7\ }$ and $RT = \underline{\ 14\ }$.
9. If $PR = 10$ and $PS = 8$, then $RS = \underline{\ 6\ }$.
10. If $YZ = 30$ and $RT = 18$, then $PS = \underline{\ 12\ }$.

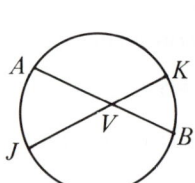

11. If $\widehat{AJ} = 60°$ and $\widehat{KB} = 40°$, then ∠ AVJ = $\underline{\ 50\ }°$.
12. If $\widehat{JB} = 170°$ and ∠ AVK = 155°, then $\widehat{AK} = \underline{\ 140\ }°$.
13. Suppose $\overrightarrow{AK}$ and $\overrightarrow{JB}$ intersect at P (not shown). If $\widehat{AJ} = 70°$ and $\widehat{KB} = 40°$, then ∠ APJ = $\underline{\ 15\ }°$.
14. If $AV = 12$, $VB = 9$, and $JV = 18$, then $VK = \underline{\ 6\ }$.

15. Construct a circle. Then construct a tangent to the circle.

CHAPTER 10

1. Name three possible positions of two different lines, j and k. intersecting, parallel, skew
2. Name three possible positions of a line l and a plane X. intersecting, parallel, line contained in the plane

3. Can two planes, P and Q, be called *skew* planes? no

A rectangular solid is shown. Find:

4. The area of a base 120
5. The lateral area 220
6. The total area 460
7. The volume 600

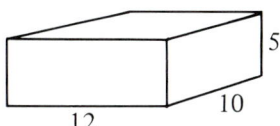

In each exercise, find:
a. the lateral area   b. the total area   c. the volume.
In Exercises 8 and 9, express your answers in terms of π.

8.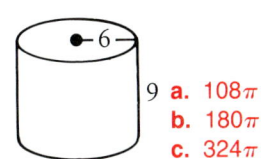
Cylinder
a. 108π
b. 180π
c. 324π

9.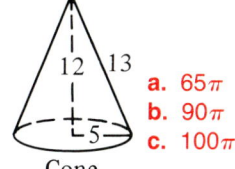
Cone
a. 65π
b. 90π
c. 100π

10.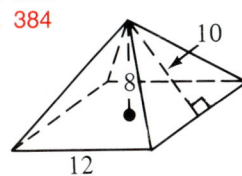
Regular square pyramid
a. 240
b. 384
c. 384

11. In a certain cylinder, the radius is 3 and the height is 2. Find the volume correct to tenths. (Use 3.14 for π.) 56.5

12. A sphere has radius 6. Find the area and volume in terms of π.   A = 144π   V = 288π

13. The area of a sphere is 36π. Find the volume in terms of π. 36π

14. The slant height of a cone is 8, and the lateral area is 24π. Find the diameter. 6

CHAPTER 11

Find the exact values of x and y.

1.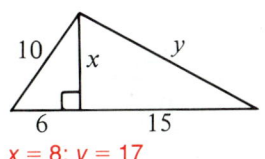
x = 8; y = 17

2.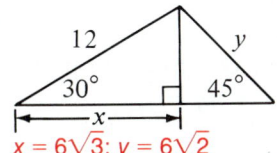
x = 6√3; y = 6√2

3.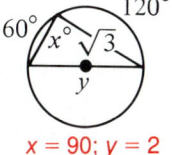
x = 90; y = 2

Find the area of each polygon.

4. A square with diagonals of length 8√2   64
5. A rectangle with width 5 and diagonals of length 13   60
6. A 30°-60°-90° triangle with hypotenuse of length 4   2√3
7. An equilateral triangle with sides 6 cm long   9√3 cm²
8. A parallelogram with sides 4 m and 8 m, and with a 60° angle   16√3
9. An isosceles triangle with sides of length 13, 13, and 10   60

Review Exercises   **529**

10. A box is 12 cm by 9 cm by 7 cm. Find the length of a diagonal. $\sqrt{274}$

11. Suppose a cone has diameter 12 and height 8. Find:
    a. the radius   b. the slant height   c. the total area   d. the volume.
       6   10 96π 96π

Find x, correct to tenths. Use the table on page 517.

12.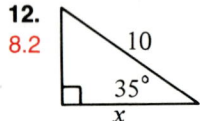
8.2

13.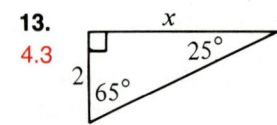
4.3

14.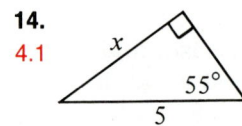
4.1

CHAPTER 12

1. Point B lies in quadrant __?__. II

2. The coordinates of point C are (__?__, __?__). (1, 3)

3. $AC =$ __?__ 3, and $OB =$ __?__. $\sqrt{5}$

4. The midpoint of $\overline{OB}$ has coordinates (__?__, __?__). $(-1, \frac{1}{2})$

5. The slope of $\overline{AC}$ is __?__, and the slope of $\overline{DE}$ is __?__. $\frac{4}{3}$
   0

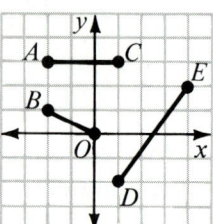

6. Does point (3, 1) lie on the graph of the equation $5x - y = 13$? no

7. The slope of line j is $\frac{3}{8}$. What is the slope of:
   a. any line parallel to j? $\frac{3}{8}$ b. any line perpendicular to j? $-\frac{8}{3}$

8. The vertices of a triangle are $K(-2, 3)$, $E(3, 5)$, and $L(7, -5)$. What angle of $\triangle KEL$ is a right angle? $\angle KEL$

9. Use the equation $2x + 3y = 12$.
   a. Copy and complete the table of values.
   b. Using a pair of coordinate axes, plot the points. Then draw the graph of $2x + 3y = 12$.

10. Draw the graph of the equation $x = 3$.

11. The vertices of a triangle are $A(-1, -1)$, $B(3, -3)$, and $C(3, 2)$. Prove that $\triangle ABC$ is an isosceles triangle. $AC = BC = 5$

x	y
0	? 4
6 ?	0
3	? 2
-3	? 6

Appendix: Transformational Geometry

SYMMETRY

Perhaps you have made a design by placing some ink on a folded piece of paper and then pressing the paper together. The figure is said to be **symmetric** with respect to the line of the fold. The two figures on opposite sides of the line of symmetry are identical in shape and size. However, they are reversed in direction.

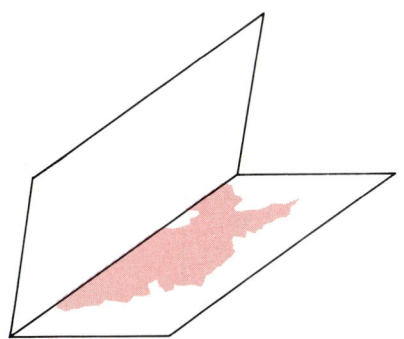
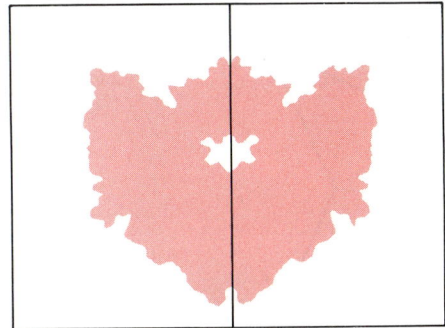

Line symmetry occurs frequently in art and nature as illustrated below. The Mexican art and the butterfly each have one line of symmetry.

The daffodil shows that it is possible to have more than one line of symmetry for the same figure.

Transformational Geometry

Two points are symmetric to one another if the line of symmetry is the perpendicular bisector of the segment joining the two points.

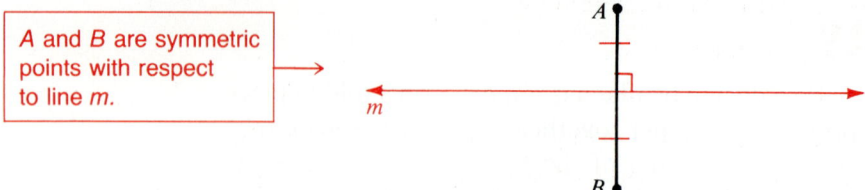

A and B are symmetric points with respect to line *m*.

A figure is said to have line symmetry in a geometric sense if *every* point of the figure is symmetric to another point of the figure.

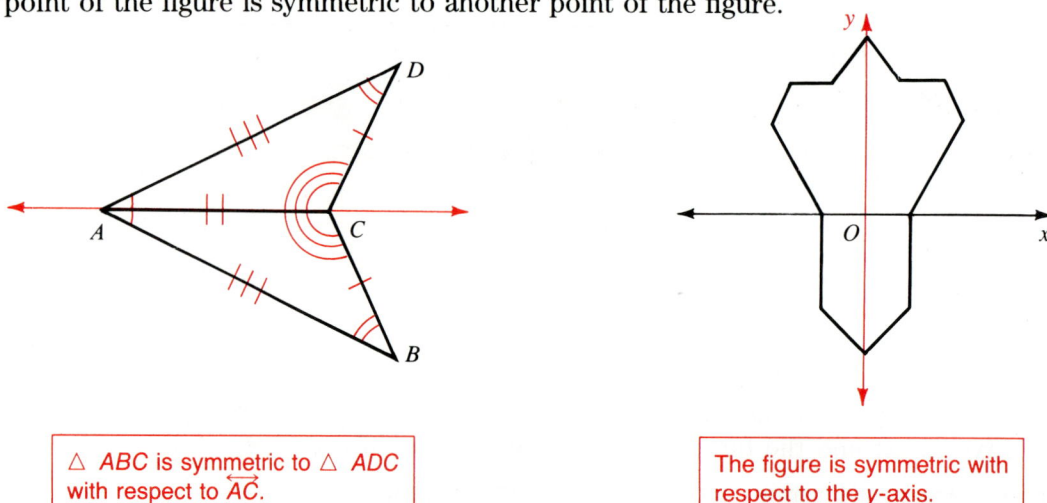

△ ABC is symmetric to △ ADC with respect to $\overleftrightarrow{AC}$.

The figure is symmetric with respect to the *y*-axis.

EXPLORATIONS

A. Fold a sheet of paper and draw a figure as shown. With scissors cut along the drawn line. Unfold the paper to reveal the symmetric figure.

Create an original piece of symmetric art in a similar manner. Trade your artwork with a classmate and then find each other's symmetries.

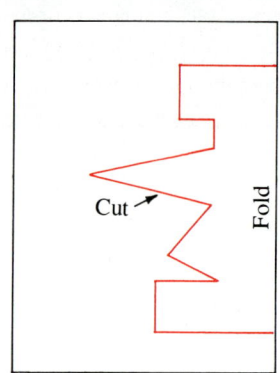

B. On a sheet of graph paper, draw axes.

Locate the following points:
 $(0, -8)$, $(3, -4)$, $(8, -4)$, $(5, 0)$, $(8, 4)$, $(3, 4)$, and $(0, 8)$.

Use a ruler to connect the points in the given order. Then locate the points needed to complete a figure symmetric to the *y*-axis. Draw the figure.

C. Draw another set of axes on your graph paper.

Draw a dashed line through the points $(0, 0)$, $(3, 3)$, and $(-3, -3)$. It will bisect the first and third quadrants.

Locate the following points:
$(7, 7)$, $(-1, 7)$, $(-4, 4)$, and $(-4, -4)$.

Use a ruler to connect the points in the given order. Then locate the points needed to complete a figure that is symmetric with respect to the dashed line. Draw the figure.

EXERCISES

For each figure, tell whether the dashed line is a line of symmetry.

1. yes

2. no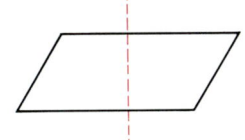

3. yes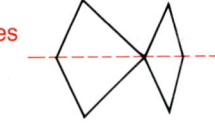

4. no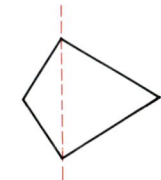

5. no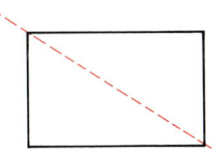

6. yes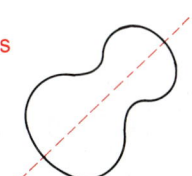

These symbols were used by alchemists during the Middle Ages to represent copper, lead, and tin, respectively. How many lines of symmetry can be drawn for each figure?

7. 2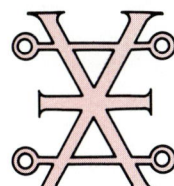

8. 2

9. 2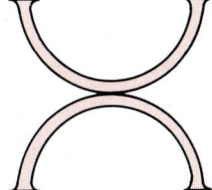

For each figure, state the number of lines of symmetry that can be drawn.

10. An equilateral triangle 3   **11.** A circle an infinite number   **12.** A square 4

13. The figure shown at the right 8

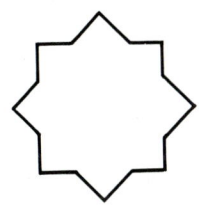

Transformational Geometry   **533**

TRANSLATIONS

When the position, size, or shape of a figure is changed according to a given rule, it is called a **geometric transformation.** We will study three types of transformations in which the new figure is congruent to the original figure. Since congruent figures have the same size and shape, we will only be discussing changes in position.

In the game of chess, the knight must move two spaces parallel to one side of the board and one space parallel to the other side. Such a move is a special kind of transformation that is called a **translation.** By the rules of chess, each of the other pieces must move or slide to another square in a special manner which also represents a translation.

A translation may be thought of as a slide by a figure in any direction without turning. If $\triangle ABC$ slides to a new position, $\triangle A'B'C'$ (read A prime, B prime, C prime), the new triangle, is said to be the **image** of the original triangle.

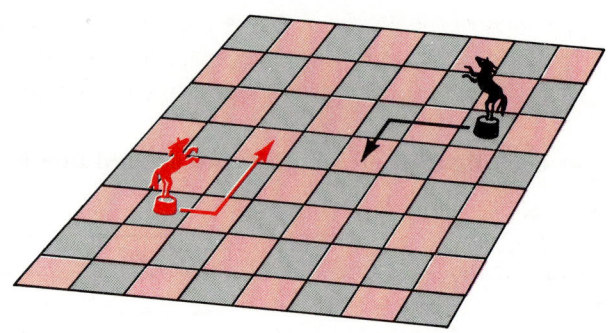

$\triangle RST$ has vertices with coordinates as shown. If the triangle slides 6 units to the right and 2 units up, the coordinates of the vertices of the image triangle are $R'(4, -1)$, $S'(4, 3)$, and $T'(0, 3)$. Notice that these coordinates can be obtained by adding 6 to the x-coordinate and 2 to the y-coordinate of each vertex of $\triangle RST$. This suggests an algebraic representation of a translation as follows.

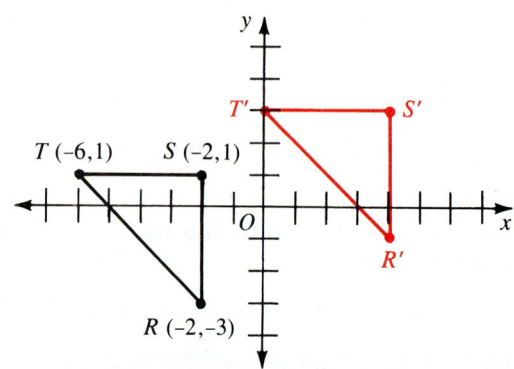

534   Transformational Geometry

> If the coordinates of some point in a given figure are (x, y), then the coordinates of its image point by a translation of a units horizontally and b units vertically are $(x + a, y + b)$.

If a and b are: *Then the translation is:*
  positive   to the right and up
  negative   to the left and down

EXAMPLE   If the coordinates of one point of a figure are $(3, 8)$ and its image point under a translation is at $(3 + 4, 8 - 5)$, or $(7, 3)$, the entire figure is translated 4 units to the right and 5 units down.

EXPLORATIONS

From a sheet of graph paper, carefully cut out a triangle as shown and label the vertices A, B, and C. (Save this triangle for use in the next two sections.)

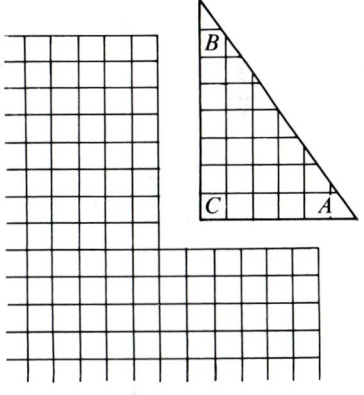

On another sheet of graph paper, draw and label the x and y axes.

Part A
  Step 1   Place the cutout triangle on the graph paper with two vertices at the given points.

  Step 2   Write the coordinates of the third vertex.

  Step 3   Physically move the triangle as directed without turning.

  Step 4   Write the coordinates of the three vertices of the image triangle.

1. $A(2, 2)$, $C(-4, 2)$; 4 units right, 2 units down $B(-4, 10)$; $A'(6, 0)$, $B'(0, 8)$, $C'(0, 0)$

2. $B(1, -3)$, $C(9, -3)$; 3 units left, 2 units up $A(9, 3)$; $A'(6, 5)$, $B'(-2, -1)$, $C'(6, -1)$

3. $A(-6, 0)$, $B(0, -8)$; 6 units right, 8 units up $C(0, 0)$; $A'(0, 8)$, $B'(6, 0)$, $C'(6, 8)$

4. $C(-6, -1)$, $A(-6, -7)$; 2 units left, 1 unit down $B(2, -1)$; $A'(-8, -8)$, $B'(0, -2)$, $C'(-8, -2)$

5. $A(-2, 1)$, $B(-8, 9)$; 10 units right $C(-8, 1)$; $A'(8, 1)$, $B'(2, 9)$, $C'(2, 1)$

6. $B(10, 5)$, $C(2, 5)$; 6 units down $A(2, -1)$; $A'(2, -7)$, $B'(10, -1)$, $C'(2, -1)$

Transformational Geometry   **535**

Part B

Step 1 Given the coordinates of three points, use your cutout triangle to find the coordinates of the other three points.

Step 2 Describe the translation.

(See below.)

	A	B	C	A'	B'	C'	Translation
7.	(7, 2)	(1, ?10)	(1, 2)	(8, 4)	(2, ?12)	(2, ?4)	?
8.	(−5, ?−4)	(3, 2)	(−5, 2)	(−6, ?−7)	(2, −1)	(−6, ?−1)	?
9.	(−6, 3)	(0, −5)	(0, ?3)	(0, ?3)	(6, −5)	(6, ?3)	?
10.	(4, 3)	(−4, ?−3)	(4, −3)	(0, ?7)	(−8, ?1)	(0, 1)	?
11.	(1, ?−2)	(−5, 6)	(−5, −2)	(1, ?−3)	(−5, 5)	(−5, −3)	?
12.	(3, 6)	(−5, 0)	(3, ?0)	(7, 1)	(−1, ?−5)	(7, ?−5)	?

7. 1 right, 2 up
8. 1 left, 3 down
9. 6 right
10. 4 left, 4 up
11. 1 down
12. 4 right, 5 down

EXERCISES

State whether each of the following represents a translation.

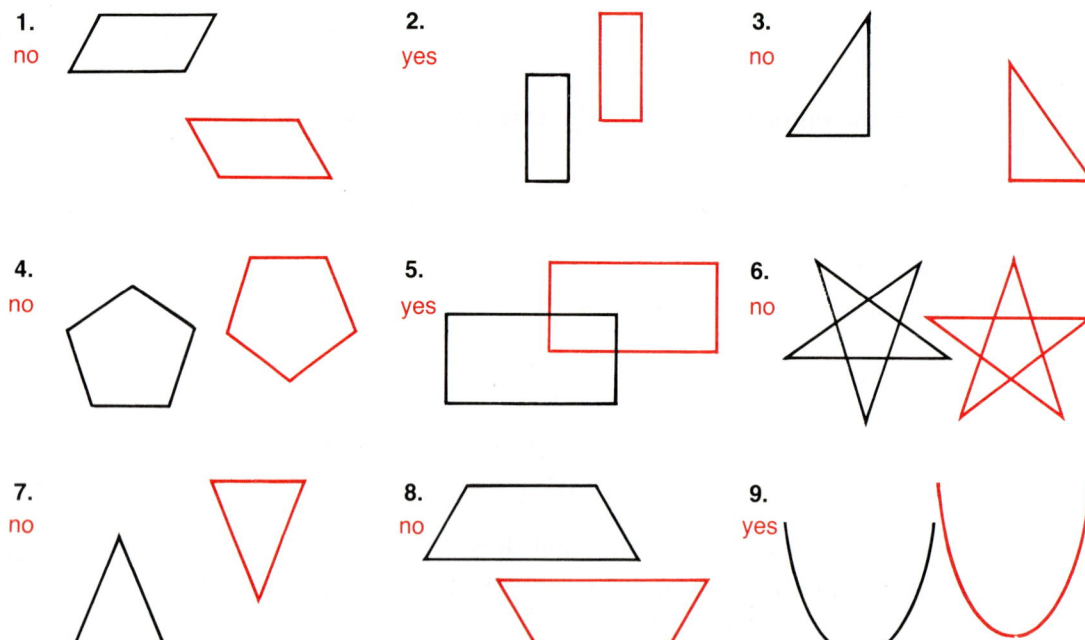

1. no
2. yes
3. no
4. no
5. yes
6. no
7. no
8. no
9. yes

REFLECTIONS

A reflection is another type of geometric transformation in which the image is congruent to the original figure. You are already familiar with reflections in a mirror or still water.

Reflection transformations are closely related to the symmetric figures discussed previously. Symmetry involves a single figure. Each point is symmetric to another point of the figure. Reflection involves a figure and its image. They are symmetric to one another with respect to a line of reflection. This line is the perpendicular bisector of the segments joining the points in the figure below with the corresponding points in the image.

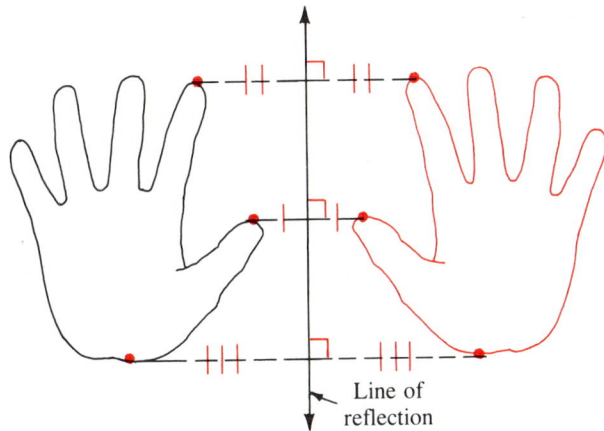

Line of reflection

In the coordinate plane, three lines have special significance as lines of reflection: the y-axis, the x-axis, and the line whose equation is $y = x$. The last one bisects the first and third quadrants.

Transformational Geometry   **537**

In each case below, observe the relationship between the coordinates of a point and its image.

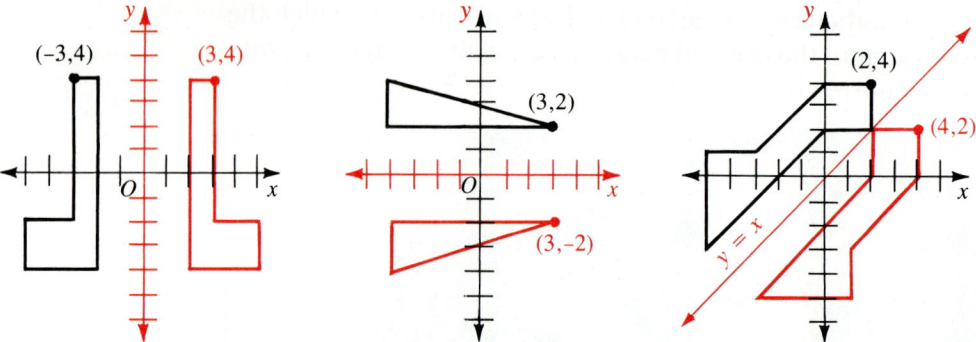

EXPLORATIONS

Use the cutout triangle from the previous section for these exercises. Turn the triangle over and write A', B', and C' on the back of A, B, and C, respectively. A reflection will require you to flip the triangle to the other side of the line of reflection.

On a sheet of graph paper, draw and label the x-axis, the y-axis, and the line whose equation is $y = x$.

Part A

Step 1   Place the cutout triangle on the graph paper, with two vertices at the given points.

Step 2   Write the coordinates of the third vertex.

Step 3   Physically move the triangle as directed.

Step 4   Write the coordinates of the vertices of the image triangle.

For Exercises 1–3, (a) reflect in the y-axis and then (b) reflect in the x-axis.

1. (a) $A'(-8, 1)$ $B'(-2, 9)$ $C'(-2, 1)$
   (b) $A'(8, -1)$ $B'(2, -9)$ $C'(2, -1)$
2. (a) $A'(-6, 4)$ $B'(2, -2)$ $C'(-6, -2)$
   (b) $A'(6, -4)$ $B'(-2, 2)$ $C'(6, 2)$

1. $A(8, 1)$, $C(2, 1)$ $B(2, 9)$ 2. $B(-2, -2)$, $C(6, -2)$ $A(6, 4)$ 3. $A(-6, 0)$, $B(0, -8)$ $C(0, 0)$
   (a) $A'(6, 0)$ $B'(0, -8)$ $C'(0, 0)$
   (b) $A'(-6, 0)$ $B'(0, 8)$ $C'(0, 0)$

For Exercises 4–9, reflect in the line $y = x$.

4. $B(2, 1)$, $C(10, 1)$ $A(10, 7)$ 5. $A(3, 1)$, $B(11, 7)$ $C(3, 7)$ 6. $A(1, -8)$, $C(1, -2)$ $B(9, -2)$
   $A'(7, 10)$, $B'(1, 2)$, $C'(1, 10)$ $A'(1, 3)$, $B'(7, 11)$, $C'(7, 3)$ $A'(-8, 1)$, $B'(-2, 9)$, $C'(-2, 1)$
7. $B(4, 4)$, $C(-4, 4)$ 8. $A(5, -4)$, $B(-1, 4)$ 9. $A(-3, -3)$, $C(-3, 3)$ $B(5, 3)$
   $A(-4, -2)$, $A'(-2, -4)$, $C(-1, -4)$, $A'(-4, 5)$, $B'(4, -1)$, $A'(-3, -3)$, $B'(3, 5)$, $C'(3, -3)$
   $B'(4, 4)$, $C'(4, -4)$ $C'(-4, -1)$

Part B

Given the coordinates of three points and a line of reflection, use your cutout triangle to find the coordinates of the other three points.

	A	B	C	Reflection	A'	B'	C'
10.	(9, 1)	(3, ?9)	(3, 1)	y-axis	(−9, 1)	(−3, 9)	(−3, 1)
11.	(5, ?−2)	(−1, 6)	(−1, −2)	y-axis	(−5, −2)	(1, 6)	(1, −2)
12.	(−2, 0)	(6, ?6)	(−2, 6)	x-axis	(−2, 0)	(6, −6)	(−2, −6)
13.	(2, 5)	(8, −3)	(8, ?5)	x-axis	(2, −5)	(8, 3)	(8, −5)
14.	(2, ?1)	(8, −7)	(8, 1)	line $y = x$	(1, 2)	(−7, 8)	(1, 8)
15.	(4, −4)	(−2, 4)	(−2, ?−4)	line $y = x$	(−4, 4)	(4, −2)	(−4, −2)

Your observations should lead you to the following conclusions.

1. If a figure is reflected in the y-axis and the coordinates of a point of the figure are (a, b), the coordinates of its image are $(-a, b)$.

2. If a figure is reflected in the x-axis and the coordinates of a point of the figure are (a, b), the coordinates of its image are $(a, -b)$.

3. If a figure is reflected in the line whose equation is $y = x$ and the coordinates of a point of the figure are (a, b), the coordinates of its image are (b, a).

EXAMPLES The image of $(3, 5)$ reflected in the y-axis is $(-3, 5)$.
The image of $(3, 5)$ reflected in the x-axis is $(3, -5)$.
The image of $(3, 5)$ reflected in the line $y = x$ is $(5, 3)$.

EXERCISES

Write the coordinates of the image of each point when the given reflection line is used.

1. $(7, 1)$; y-axis $(-7, 1)$
2. $(-6, 2)$; y-axis $(6, 2)$
3. $(4, 6)$; x-axis $(4, -6)$
4. $(5, -8)$; x-axis $(5, 8)$
5. $(2, 9)$; $y = x$ $(9, 2)$
6. $(-7, 4)$; $y = x$ $(4, -7)$

Consider only capital letters for the following exercises.

7. Find all letters of the alphabet that look like the same letter when reflected in the y-axis. A, H, I, M, O, T, U, V, W, X, Y

8. Find all letters of the alphabet that look like the same letter when reflected in the x-axis. B, C, D, E, H, I, K, O, X

9. Which letters look the same under both reflections? H, I, O, X

Transformational Geometry

ROTATIONS

The burro in the picture helps make molasses by turning the machinery that squeezes cane. It rotates about a central point, always staying at the same distance from that point.

The last geometric transformation we will consider is the rotation. In a rotation, each point of the given figure moves about a center point as if it were attached to the end of a spoke on a wheel. Each point rotates through the same angle and the image figure is congruent to the original figure. We describe a counterclockwise rotation as being positive and a clockwise rotation as being negative.

To do the rotations that follow, you will need a ruler, a protractor, and a compass.

EXAMPLE Rotate $\overline{AB}$ by 60° about center O.

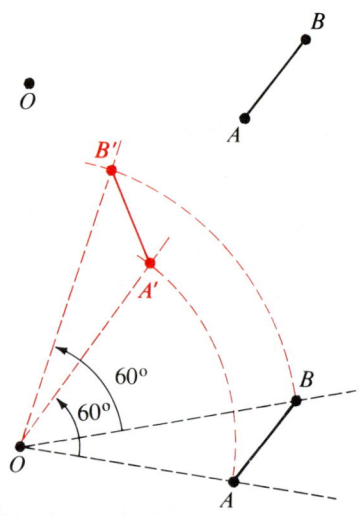

1. Draw $\overrightarrow{OA}$.

2. With a protractor, measure an angle of 60° counterclockwise from $\overrightarrow{OA}$, with O as vertex. Draw the second side of the angle.

3. With compass point at O, strike an arc through A to intersect the other side of the angle. Label the point of intersection A'.

4. Repeat Steps 1–3, starting with $\overrightarrow{OB}$ and locating B.

5. Draw $\overline{A'B'}$. This is the image of $\overline{AB}$ under the rotation.

In the coordinate plane we frequently use the origin as the center of rotation.

EXAMPLE Rotate $\triangle DEF$ by 90° about the origin.

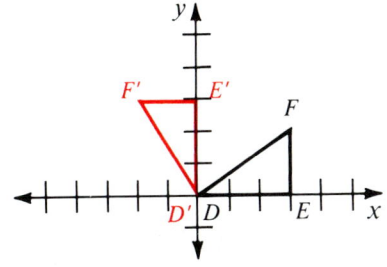

In this case, D remains where it is since it is the center of rotation. E is on the x-axis and a 90° rotation places its image on the y-axis. Since $\overline{EF}$ is vertical, a 90° rotation requires its image to be horizontal. Since $\triangle DEF \cong \triangle D'E'F'$, we can locate F' and draw the sides of $\triangle D'E'F'$.

EXAMPLE Rotate $\triangle RST$ by 90° about the origin.

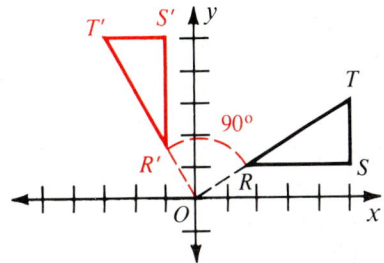

In this case, we must first locate one image point. Since $\angle ROR' = 90°$, $\overline{OR} \perp \overline{O'R'}$. We also know that $OR = O'R'$. Recall your earlier studies of slopes and distances in the coordinate plane. This should lead you to discover that R' must be at $(-1, 2)$ if R is at $(2, 1)$. We may then use the ideas from the first example to complete $\triangle R'S'T'$.

EXPLORATIONS

A. Draw a figure similar to the one shown.

Using a ruler, protractor, and compass, find the image of $\triangle ABC$ under a rotation of 70° about O.

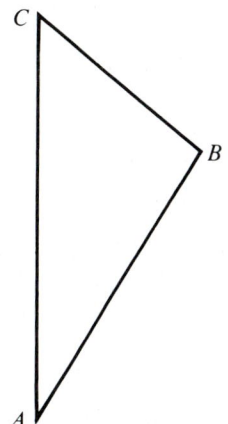

Transformational Geometry

B. Use the cutout triangle from the previous two sections.

Step 1 On a sheet of graph paper, draw and label axes.

Step 2 Place the cutout triangle on the graph paper with two vertices at the given points.

Step 3 Write the coordinates of the third vertex.

Step 4 Rotate the triangle 90° counterclockwise about the indicated point as the center of rotation.

Step 5 Write the coordinates of the three vertices of the image triangle.

1. $A(6, 0), C(0, 0)$; center at the origin $B(0, 8); A'(0, 6), B'(-8, 0), C'(0, 0)$
2. $B(2, 0), C(10, 0)$; center at the origin $A(10, 6); A'(-6, 10), B'(0, 2), C'(0, 10)$
3. $C(-1, 0), A(-7, 0)$; center at the origin $B(-1, -8); A'(0, -7), B'(8, -1), C'(0, -1)$
4. $B(-8, -7), A(0, -1)$; center at point $(0, -1)$ $C(0, -7); A'(0, -1), B'(6, -9), C'(6, -1)$
5. $A(8, 1), C(2, 1)$; center at point $(2, 1)$ $B(2, 9); A'(2, 7), B'(-6, 1), C'(2, 1)$

EXERCISES

Identify each of the following transformations as a translation, reflection, or rotation.

1. reflection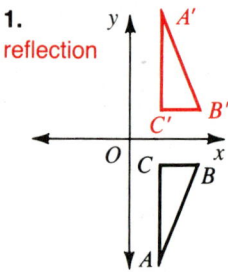
2. rotation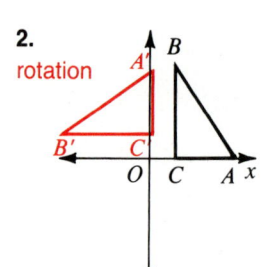
3. translation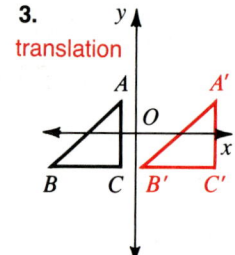
4. rotation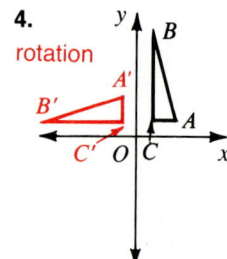
5. translation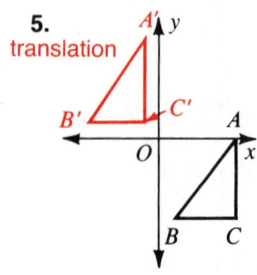
6. reflection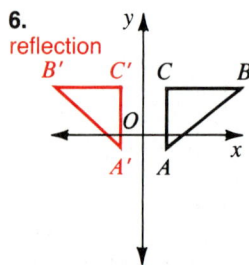
7. reflection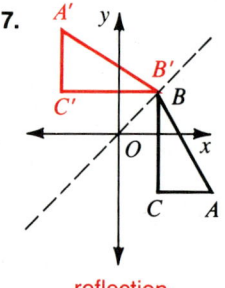
8. translation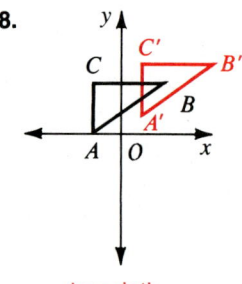

TESSELLATIONS

The design shown at the right is called a *tessellation* (from the Latin word meaning *tile*). Notice that congruent copies of pink and white arrows completely fill the plane without overlapping or leaving gaps. Although you see just a small part of the total tessellation, the pattern continues repeating left and right, up and down.

If you slide the entire pattern so that one arrow moves onto another arrow of the same color, all the arrows will move exactly onto arrows of matching color. Thus we say that the tessellation has *translational symmetry*.

In addition to translational symmetry, this tessellation has *line symmetry* because reflecting the design in various lines (in this case, the lines are horizontal) will cause the design to reflect exactly onto itself.

The tessellation at the right is an abstract geometrical design found in the Casa Real of the Alhambra in Granada, Spain. This tiling has translational symmetry (along diagonals) and line symmetry (in various horizontal and vertical lines).

This tessellation also has *rotational symmetry* because a rotation of 180° about any of the points where two grey (or two pink) tiles touch, or about any of the centers of the large white tiles, will move the pattern exactly onto itself.

Transformational Geometry   543

A tessellation of the letter F is shown in the diagram at the right. It does not have line symmetry. It has more than one kind of rotational symmetry, though, because you can rotate any F-shape 90°, 180°, or 270° onto another F-shape.

If the F-shapes are colored as shown, the new tessellation will not have rotational symmetry because a 90°, 180°, or 270° clockwise rotation will map a black F-shape onto a red, grey, or pink F-shape. Adding colors reduces the number of symmetries of a tessellation. Do you see how removing colors from the Alhambra tessellation on the previous page would increase the number of symmetries? Find a point that would then be the center of 90°, 180°, and 270° rotational symmetry.

Centers of large white tiles

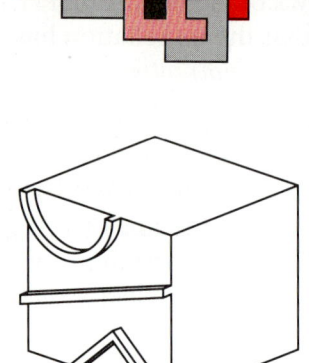

EXPLORATIONS

A. Cut a potato in half and cut a small cube (3 cm on a side) from one half. Use a felt tip marker to draw an abstract design like the one shown and then cut around this design so it is raised above the rest of the potato. Press the design into a flat cloth soaked with food coloring and then stamp the potato cube onto a piece of paper. Make 3 other stampings on the paper, each time rotating the potato cube 90°.

(1)   (2)   (3)   (4)

 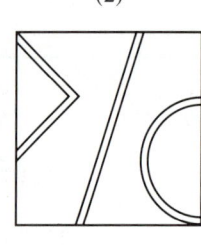  

Make a tessellation by stamping a sheet of paper repeatedly. Some possibilities are shown below.

2	2	2	2
4	4	4	4
2	2	2	2
4	4	4	4

1	2	1	2
4	3	4	3
1	2	1	2
4	3	4	3

544   Transformational Geometry

B. Cut another 3 cm cube from a potato. Onto this cube stamp the potato cube used in Part A. Cut around this design to form a mirror image stamp of the first design. Then stamp it and 3 rotations of it on a piece of paper.

(5) (6) (7) (8)

 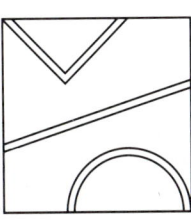 

Make a tessellation using both stamps. Some possibilities are shown below.

 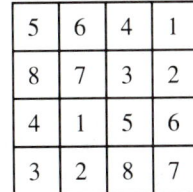 

EXERCISES

State whether each tessellation has (a) translational symmetry, (b) line symmetry, (c) rotational symmetry (if so, list what kinds; for example, 90°).

1. 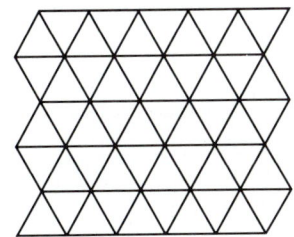 a. yes
   b. yes
   c. yes
      (60°, 120°, 180,° 240°, 300°, 360°)

2.  a. yes
   b. no
   c. yes
      (180°, 360°)

Draw copies of each figure on graph paper so that they form a tessellation.

3. 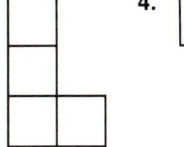 4.  5.  6.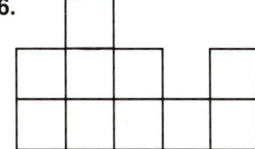

Transformational Geometry **545**

Tell whether or not a tessellation can be made with the given figure.

7. An isosceles triangle **yes**
8. A scalene triangle **yes**
9. A circle **no**
10. A regular pentagon **no**
11. A regular hexagon **yes**
12. A regular octagon **no**
13. A rectangle **yes**
14. A rhombus **yes**
15. A trapezoid **yes**

16. You can make a tessellation by tracing around *any* quadrilateral, placing copies of the quadrilateral systematically as shown.

    a. The tessellation shown has many symmetry points but none of these are at vertices of the quadrilateral. Where are they? **at edge midpoints**
    b. What other kind of symmetry does this tessellation have? **translational**

17. The basic unit of the tessellation in Exercise 2 is shown at the right. It is formed by altering two opposite sides of a square in exactly the same way. Use this idea to make an original tessellation design.

    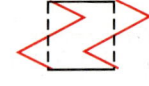

    Check students' drawings.

18. Repeat Exercise 17, starting with a parallelogram instead of a square. Alter both pairs of opposite sides in different ways to create a tessellation.

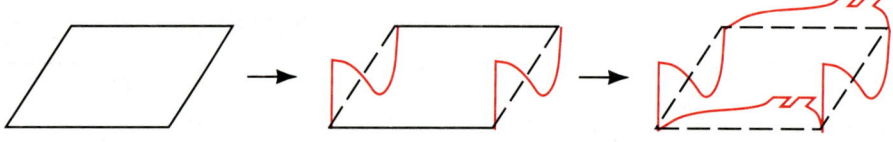

Glossary

A

acute angle (p. 15) An angle whose measure is between 0° and 90°.

acute triangle (p. 97) A triangle with three acute angles.

altitude of a triangle (p. 158) A segment, drawn from any vertex, perpendicular to the line that contains the opposite side. The opposite side is considered to be the *base* of the triangle.

angle (p. 10) A figure formed by two rays or two segments with a common endpoint. *B* is the *vertex* of ∠ *ABC*. $\overrightarrow{BA}$ and $\overrightarrow{BC}$ are the *sides* of ∠ *ABC*.

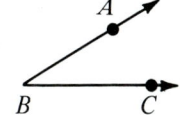

arc (p. 355) Part of a circle. A *semicircle*, such as $\overarc{ABC}$, is exactly half a circle. A *minor arc*, such as $\overarc{AB}$ or $\overarc{BC}$, is less than a semicircle. A *major arc*, such as $\overarc{BAC}$ or $\overarc{ACB}$, is greater than a semicircle.

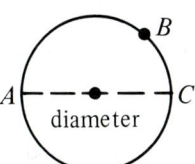

area (p. 238) The amount of surface in a region. Area is measured in square units.

B

bisector of an angle (p. 20) A ray or a line which divides an angle into two equal angles. If ∠ *AOX* = ∠ *BOX*, $\overrightarrow{OX}$ bisects ∠ *AOB*.

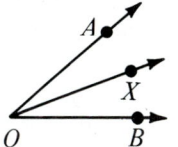

bisector of an arc (p. 359) A line, ray, or segment that contains the midpoint of an arc.

bisector of a segment (p. 154) A segment, ray, or line that contains the midpoint of a segment.

C

central angle of a circle (p. 355) An angle whose vertex is the center of a circle.

chord of a circle (p. 346) A segment that joins two points on a circle.

circle (p. 263) A figure, in a plane, whose points are all the same distance from a particular point in the plane. This point is the *center* of the circle.

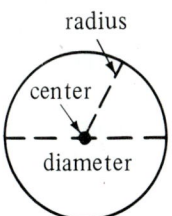

circumference (p. 263) The distance around a circle.

circumscribed circle (pp. 163, 361) A circle is circumscribed about a polygon when each vertex of the polygon lies on the circle.

circumscribed polygon (p. 361) A polygon is circumscribed about a circle when each side of the polygon is tangent to the circle.

collinear points (p. 6) Points that lie on one line.

common tangent (p. 351) A line that is tangent to each of two coplanar circles. A common *internal* tangent intersects the segment joining the centers of the circles. A common *external* tangent does not intersect the segment joining the centers of the circles.

complementary angles (p. 46) Two angles whose measures total 90°.

conclusion (p. 50) A result reached by reasoning. In a statement of the form "If A, then B," B is the conclusion.

cone (p. 406) The *base* of a right circular cone is circular. Any segment that joins the vertex to a point on the circle is a *slant height* of the cone.

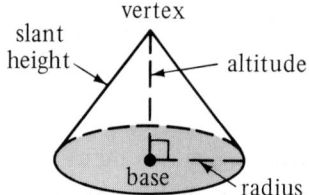

congruent triangles (p. 102) $\triangle ABC \cong \triangle DEF$ if the following statements are true:
$\angle A = \angle D \quad \angle B = \angle E \quad \angle C = \angle F$
$AB = DE \quad BC = EF \quad AC = DF$

consecutive sides of a polygon (p. 190) Two sides which intersect.

consecutive vertices of a polygon (p. 190) The endpoints of a side.

constructing a geometric figure (p. 25) Drawing a figure using only a straightedge and a compass.

converse of a statement (p. 50) A statement formed by exchanging the hypothesis and the conclusion of the given statement.

convex polygon (p. 190) Imagine fitting a rubber band along the edges of a polygon. If the rubber band fits snugly, the polygon is convex.

coordinate(s) (pp. 7, 468) On a number line, the number paired with a point. In the coordinate plane, the numbers which are paired with a point. Point (2, 4) has *x-coordinate* 2 and *y-coordinate* 4.

coordinate axes (p. 468) The horizontal and vertical number lines in the coordinate plane.

coordinate plane (p. 468) A plane which contains a horizontal number line (the *x-axis*) and a vertical number line (the *y-axis*). Every point in the coordinate plane can be named by a pair of numbers.

coplanar lines (p. 66) Lines that lie on one plane.

coplanar points (p. 33) Points that lie on one plane.

corollary of a theorem (p. 93) A statement which can be proved easily by applying a theorem.

corresponding parts of two polygons (p. 102) In $\triangle ABC$ and $\triangle DEF$, if point A is matched with point D, B with E, and C with F, then (1) the *corresponding sides* are $\overline{AB}$ and $\overline{DE}$, $\overline{BC}$ and $\overline{EF}$, and $\overline{AC}$ and $\overline{DF}$; (2) the *corresponding angles* are $\angle A$ and $\angle D$, $\angle B$ and $\angle E$, and $\angle C$ and $\angle F$.

cylinder (p. 398) In a right circular cylinder, the *bases* are congruent circles which are parallel. Any segment that is perpendicular to the bases and has an endpoint in each base is an *altitude* of the right circular cylinder.

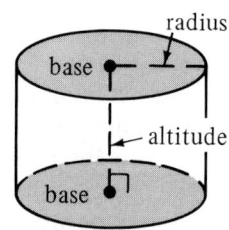

D

diagonal of a polygon (p. 190) A segment which joins two nonconsecutive vertices of the polygon.
diameter of a circle (p. 346) A chord that passes through the center of a circle. Also, the length of such a chord.
distance from a point to a line (p. 359) The length of the perpendicular segment from the point to the line.

E

equal angles (p. 15) Two angles whose measures are equal.
equal arcs of a circle (p. 355) Two arcs whose measures are equal.
equal segments (p. 144) Two segments whose lengths are equal.
equal sides (p. 97) Two sides whose lengths are equal.
equiangular polygon (p. 191) A polygon with all angles equal.
equiangular triangle (p. 97) A triangle with all angles equal.

equidistant points (p. 154) If $AB = AC$, then A is equidistant from B and C.

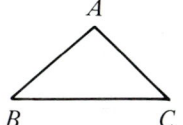

equilateral polygon (p. 191) A polygon with all sides equal.
equilateral triangle (p. 97) A triangle with all sides equal.

G

graph of an equation (p. 496) The geometric figure which contains all the points whose coordinates make the equation a true statement, and no other points.

H

hypotenuse of a right triangle (p. 97) The side opposite the right angle. The longest side of a right triangle.
hypothesis (p. 50) Information that is given. In a statement of the form "If A, then B," A is the hypothesis.

I

if . . . then statement (p. 50) A statement of the form "If A, then B."
inscribed angle (p. 364) An angle whose vertex lies on a circle and whose sides are chords of the circle.
inscribed circle (pp. 164, 361) A circle is inscribed in a polygon when each side of the polygon is tangent to the circle.
inscribed polygon (p. 361) A polygon is inscribed in a circle when each vertex of the polygon lies on the circle.

intercepted arc(s) (pp. 364, 365, 369) In each figure, the angle intercepts the arc or arcs shown in color.

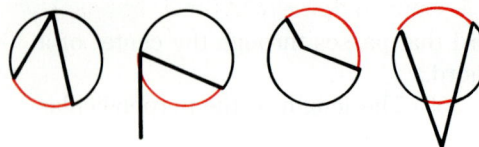

intersecting lines (pp. 18, 388) Two lines that meet in one point. Two intersecting lines are coplanar.

intersecting planes (pp. 34, 389) Two planes that meet in one line.

isosceles trapezoid (p. 220) A trapezoid with equal legs.
There are two pairs of *base angles:*
$\angle 1$ and $\angle 2$, $\angle 3$ and $\angle 4$.

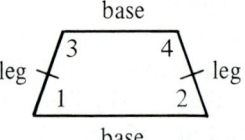

isosceles triangle (p. 97) A triangle with at least two equal sides.
$\angle 1$ and $\angle 2$ are the *base angles.*
$\angle 3$ is the *vertex angle.*

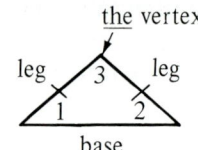

L

lateral area (p. 394) In a right prism or a regular pyramid, the sum of the areas of the lateral faces. In a cylinder or a cone, the area of the lateral surface.

legs of a right triangle (p. 97) The two sides other than the hypotenuse.

legs of a trapezoid (p. 220) The nonparallel sides of a trapezoid.

length of a segment (p. 7) The distance between the endpoints of a segment.

line parallel to a plane (p. 388) A line and a plane which have no point in common.

line perpendicular to a plane (p. 388) A line which is perpendicular to every line in the plane that passes through the point of intersection.

M

measure of an angle (p. 10) The number of degrees of an angle. The number is greater than 0 and less than or equal to 180.

measure of a major arc (p. 355) The difference of 360° and the measure of the minor arc.

measure of a minor arc (p. 355) The measure of its central angle.

measure of a semicircle (p. 355) The measure of a semicircle is 180°.

median of a trapezoid (p. 220) The segment joining the midpoints of the legs of a trapezoid.

median of a triangle (p. 158) A segment that joins a vertex of a triangle to the midpoint of the opposite side.

midpoint of an arc (p. 359) If $\overset{\frown}{CM} = \overset{\frown}{MD}$, then M is the midpoint of $\overset{\frown}{CD}$.

midpoint of a segment (p. 7) Point M is the midpoint of $\overline{AB}$ if M lies on $\overline{AB}$ and $AM = MB$.

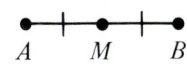

N

noncollinear points (p. 33) Points that do not all lie on one line.
number line (p. 7) A line which matches points with the real numbers.

O

obtuse angle (p. 15) An angle whose measure is between 90° and 180°.
obtuse triangle (p. 97) A triangle with one obtuse angle.
origin of the coordinate plane (p. 468) The point in which the coordinate axes intersect. Point (0, 0).

P

parallel line and plane (p. 388) A line and a plane that have no point in common.
parallel lines (p. 66) Two coplanar lines that have no point in common.

parallel lines cut by a transversal (p. 66) The following angles are formed:
  alternate interior angles: $\angle 3$ and $\angle 6$, $\angle 4$ and $\angle 5$
  corresponding angles: $\angle 1$ and $\angle 5$, $\angle 2$ and $\angle 6$, $\angle 3$ and $\angle 7$, $\angle 4$ and $\angle 8$
  interior angles: $\angle 3$, $\angle 4$, $\angle 5$, $\angle 6$
  same-side interior angles: $\angle 3$ and $\angle 5$, $\angle 4$ and $\angle 6$

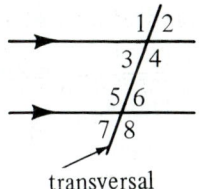

parallel planes (p. 34) Two planes that have no points in common.

parallelogram (p. 198) A quadrilateral with both pairs of opposite sides parallel. Either pair of sides may be considered the *bases*.

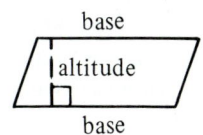

perimeter (p. 240) The distance around a region; the sum of the lengths of the sides of a polygon.
perpendicular bisector of a segment (p. 154) A segment bisector that is perpendicular to the segment.
perpendicular lines (p. 15) Two lines that meet to form four right angles.
perpendicular planes (p. 389) If a plane contains a line that is perpendicular to another plane, the two planes are perpendicular.
pi (p. 264) A Greek letter denoted by π. The quotient $\dfrac{\text{circumference}}{\text{diameter}}$ for any circle.
plotting a point (p. 468) Locating the point in the coordinate plane.

polygon (p. 190) The figure shows one example of a polygon. Parts of a polygon are named as shown. ∠ 1 is an *interior angle*. ∠ 2 is an *exterior angle*.

postulate (p. 29) A statement which is accepted without proof.

proportion (p. 289) An equation which states that two ratios are equal. In the proportion $\frac{4}{6} = \frac{2}{3}$, 6 and 2 are the *means*, and 4 and 3 are the *extremes*.

Q

quadrilateral (p. 190) A polygon with four sides.

R

radius of a circle (p. 346) A segment that joins the center and a point on the circle. Also, the length of such a segment.

ratio (p. 286) If a and b are numbers and $b \neq 0$, the ratio of a to b is the quotient $\frac{a}{b}$. This is sometimes denoted by $a:b$.

ray (p. 6) Part of a line. $\overrightarrow{AB}$ starts at *endpoint A*, goes through point B, and continues indefinitely.

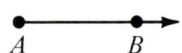

rectangle (p. 198) A parallelogram with four right angles. Any side of a rectangle may be considered a *base*.

rectangular solid (pp. 393, 440) A right prism whose *bases* are rectangular. The segment shown in color is a *diagonal* of the rectangular solid.

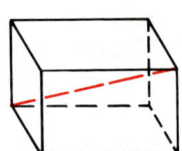

regular polygon (p. 191) A polygon which is both equilateral and equiangular.

regular pyramid (p. 402) The *base* of a regular pyramid is a regular polygon. The *lateral faces* are congruent isosceles triangles. The altitude from the vertex of a face is a *slant height*.

rhombus (p. 198) A parallelogram with four equal sides.

right angle (p. 15) An angle whose measure is 90°.

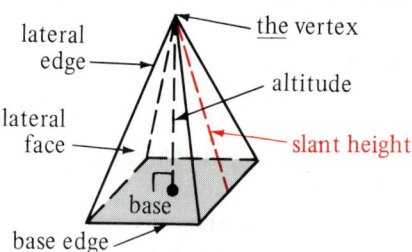

right prism (p. 393) In a right prism, the *bases* are congruent and parallel. The *lateral faces* are rectangular. The *lateral edges* are perpendicular to the bases. Any segment, such as a lateral edge, that is perpendicular to the bases and has an endpoint on each base, is an *altitude* of the right prism.

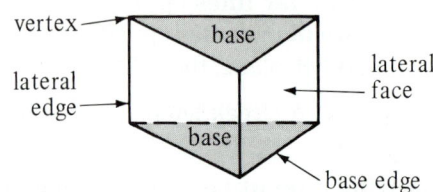

right triangle (p. 97) A triangle with one right angle. The hypotenuse is the longest side of a right triangle.

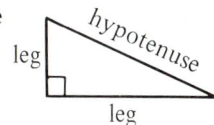

S

scale (p. 296) The ratio of a distance in a drawing to the actual distance represented.

scale factor of two similar polygons (p. 313) The ratio of two corresponding sides.

scalene triangle (p. 97) A triangle with no two sides equal.

secant of a circle (p. 346) A line that contains a chord.

segment (p. 6) Part of a line. A and B are the *endpoints* of $\overline{AB}$.

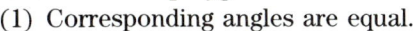

similar polygons (p. 312) Two polygons with the same shape.
  In two similar polygons:
  (1) Corresponding angles are equal.
  (2) Corresponding sides are in proportion.

similar solids (p. 414) Two solids with the same shape.
  In two similar solids:
  (1) Corresponding angles are equal.
  (2) Corresponding segments are in proportion.

skew lines (p. 66) Two lines that are not coplanar and do not intersect.

slope of a line (pp. 486, 487) $\dfrac{\text{rise}}{\text{run}}$. Also, if a line contains points (x_1, y_1) and (x_2, y_2), then the slope of the line is $\dfrac{y_2 - y_1}{x_2 - x_1}$, provided $x_1 \neq x_2$.

sphere (p. 346) A figure in space whose points are the same distance from a particular point. This point is the *center* of the sphere.

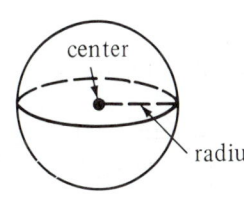

square (p. 198) A parallelogram with four right angles and four equal sides.

straight angle (p. 15) An angle whose measure is 180°.

supplementary angles (p. 47) Two angles whose measures total 180°.

T

tangent to a circle (p. 346) A line or segment, in the plane of a circle, that intersects the circle in exactly one point. This point is called the *point of tangency*.

tangent circles (p. 351) Two coplanar circles that are tangent to a line at one point. $\odot R$ and $\odot S$ are *externally* tangent. $\odot S$ and $\odot T$ are *internally* tangent.

theorem (p. 46) A statement that can be proved.

total area (p. 394) The sum of the lateral area and the area of the base or bases.

trapezoid (p. 198) A quadrilateral with just one pair of opposite sides parallel. The segment joining the midpoints of the legs is the *median* of a trapezoid.

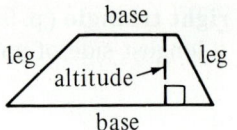

triangle (p. 15) A polygon with three sides. In △ABC, each of points A, B, and C is a *vertex* of the triangle. $\overline{AB}$, $\overline{BC}$, and $\overline{AC}$ are the *sides* of the triangle. ∠1, ∠2, and ∠3 are the *angles* of the triangle. ∠4 is an *exterior angle* of △ABC. ∠1 and ∠2 are *opposite angles* with respect to exterior ∠4.

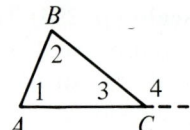

trigonometric ratios (pp. 447, 452)

tangent of ∠A = $\dfrac{\text{leg opposite } \angle A}{\text{leg adjacent to } \angle A}$

sine of ∠A = $\dfrac{\text{leg opposite } \angle A}{\text{hypotenuse}}$

cosine of ∠A = $\dfrac{\text{leg adjacent to } \angle A}{\text{hypotenuse}}$

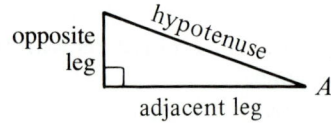

V

vertical angles (p. 18) Two angles formed by intersecting lines. ∠1 and ∠2 are vertical angles.

volume (p. 394) The amount of space in a solid. Volume is measured in cubic units.

Index

A

Acute angle, 15
Acute triangle, 97, 260
Alternate interior angles, 66–68, 72
Altitude
    of cone, 406
    of cylinder, 398
    of parallelogram, 243
    of regular pyramid, 402
    of right prism, 393
    of trapezoid, 251
    of triangle, 158
Angle(s), 10
    acute, 15
    alternate interior, 66–68, 72
    bisector of, 20
        construction of, 25, 149
    central, 355–356
    complementary, 46
    corresponding
        of congruent triangles, 102–103, 144–145
        of parallel lines, 67, 72–73
        of similar polygons, 312–313, 317–318
        of similar solids, 414
    of elevation and depression, 457–458, 475
    equal, 15
        construction of, 77, 151
        ways to prove, 144, 173
    inscribed, 364–365
    interior, 66–67
    measure of, 10–11, 35
    obtuse, 15
    of polygon, 194–195
    right, 15
    same-side interior, 67–68, 72
    sides of, 10
    straight, 15
    supplementary, 47
    of triangle, 15, 92–93
    vertex of, 10
    vertical, 18, 47
Angle sum
    of polygon, 194–195
    of triangle, 92
Answers
    to Selected Exercises, 562–579
    to Self-Tests, 559–561
Applications
    Angles of Elevation and Depression, 457–459
    Art and Geometry, 54–55
    Golden Rectangle, 208
    Indirect Proof, 334–335
    Latitude and Longitude, 23–24
    Locus, 482–485
    Mechanical Drawing, 431
    Mirrors and Billiards, 272–273
    Rigidity of Triangles, 113–114
    Shapes in Nature, 505
    Surface Area and Volume, 413
    Triangle Inequality, 179
    Using Circles to Set Up Schedules, 377
    Using a Compass, 302–303
Arc(s) of circle, 355–356
    bisector of, 359
    of chord, 359
    equal, 355–356, 359–360
    intercepted, 364–365, 369–370
    major, 355
    measure of, 355
    midpoint of, 359
    minor, 355–356
Area(s), 238
    of circle, 268–269
    of cone, 406, 443–444
    of cylinder, 398–399
    of equilateral triangle, 322
    of parallelogram, 243–244
    of rectangle, 238–240
    of regular polygon, 481
    of regular pyramid, 402–403, 443
    of right prism, 394
    of similar polygons, 329–330

Area(s) (*continued*)
    of similar solids, 414–415
    of sphere, 410
    of square, 239
    of trapezoid, 250–251
    of triangle, 247, 292, 481
Axes, coordinate, 468

B

Base(s)
    of cone, 406
    of cylinder, 398
    of isosceles triangle, 97, 168
    of parallelogram, 243
    of rectangle, 238
    of regular pyramid, 402
    of right prism, 393
    of trapezoid, 220
    of triangle, 247
Bisector(s)
    of angle, 20
        construction of, 25, 149
    of angles of triangle, 164
    of arc, 359
    perpendicular, of segment, 154–155
        construction of, 154
    perpendicular, of sides of triangle, 163
    of segment, 154

C

Calculator Activities, 71, 292, 322, 332, 409, 436, 475, 481
Career Notes
    Actuary, 316
    Aircraft Assembly Technician, 451
    Astronomer, 32
    Auto Mechanic, 262
    Carpenter, 500
    Marine Dietician, 59
    Physical Therapist, 202
    Plumber, 172
    Realtor, 125
Center
    of circle, 263, 346
    of sphere, 346
Central angle(s) of circle, 355–356

Chapter reviews. *See* Reviews.
Chapter tests. *See* Tests.
Charts, 180–181
Chord(s) of circle, 346
    arc of, 359
    diameter perpendicular to, 359–360
    equal, 359–360
    intersecting within a circle, 369–370, 373
Circle(s), 263, 346
    arc(s) of, 355–356
    area of, 268–269
    center of, 263, 346
    central angle of, 355–356
    chord of, 346
    circumference of, 263–265
    circumscribed, 163–164, 361
    diameter of, 263–264, 346
    externally tangent, 351
    inscribed, 164, 361
    inscribed angle of, 364–365
    internally tangent, 351
    major arc of, 355
    minor arc of, 355
    radius of, 263, 346
    secant of, 346
    semicircle of, 355
    tangent(s) to, 346, 350–351
Circumference of circle, 263–265
Circumscribed circle, 163–164, 361
    construction of, 163
Circumscribed polygon, 164, 361
Collinear points, 6, 33–34
Compass, 25
Complementary angles, 46, 134–135
Computer Activities
    Best Box, 508–509
    Complementary and Supplementary
        Angles, 134–135
    Random Numbers and Pi, 276–277
    Ten-Speed Bicycles, 378–379
Conclusions, 50
Conditional, 50
Cone, 406, 443–444
Congruence, methods for proving, 127
Congruent triangles, 102–133, 144–145
Consecutive sides of polygon, 190–191
Consecutive vertices of polygon, 190–191

Constructions, 25
- angles, equal, 77
- bisector of angle, 25
- circumscribed circle, 163
- inscribed circle, 164
- justifying, 149
- list of, 522
- parallel lines, 78
- perpendicular bisector of segment, 154
- perpendicular lines, 26
- 60° angle, 174
- tangent to circle, 351

Consumer Applications
- Calling Long Distance, 392
- Consumer Price Index, 148
- Cost of Driving, 246
- Do You Get What You Pay For?, 120
- Energy Usage, 76
- Life Insurance, 349
- Time is Money, 14

Converse, 50–51
Convex polygon, 190
- angle sums of, 194–195

Coordinate, 7
Coordinate axes, 468
Coordinate geometry, 468–474, 476–480, 486–504
Coordinate plane, 468
Coordinates, 468
Coplanar lines, 66, 388
Coplanar points, 33–34
Corollary, 93
Corresponding angles of parallel lines, 67, 72–73
Corresponding parts
- of congruent triangles, 102–103, 144–145
- of similar polygons, 312–313, 317–318
- of similar solids, 414–415

Cosine ratio, 452–453
Counterexamples, 418–419
Cumulative reviews. See Reviews.
Cylinder, 398–399

D

Diagonal(s)
- of parallelogram, 203–204
- of polygon, 190–191
- of rectangle, 209–210
- of rectangular solid, 440–441
- of rhombus, 209–210

Diagrams, 81, 180–181
Diameter of circle, 263–264, 346
- perpendicular to chord, 359–360

Distance
- between points, 7, 34
- in coordinate plane, 472
- from point to line, 207, 359

E

Edge
- of regular pyramid, 402
- of right prism, 393
- of solid, 405

End-of-course reviews. See Reviews.
Endpoint(s)
- of ray, 6
- of segment, 6, 154–155

Enrichment materials. See:
- Applications
- Calculator Activities
- Career Notes
- Computer Activities
- Consumer Applications
- Experiments
- Explorations
- Problem Solving Strategies
- Puzzle Problems
- Reading Geometry

Equal angles, 15
- construction of, 77, 151
- ways to prove, 144, 168

Equal arcs, 355–356, 359–360
Equal chords, 359–360
Equal segments, 97
- ways to prove, 144–145, 174

Equality, postulates of, 29–30
Equation of line, 496–498
Equiangular polygon, 191
Equiangular triangle, 97, 167–168, 173–174
Equilateral polygon, 191
Equilateral triangle, 97, 167–168, 173–174
- area of, 292, 322

Experiments, 22, 120, 132, 162, 178, 193, 333, 349, 392, 405, 409

Explorations, 167, 203, 209, 214, 215, 220, 243, 250, 255, 317, 350, 359, 364, 369, 414, 476–477, 482, 488, 491–492, 544–545
Exterior angle
    of polygon, 195
    of triangle, 93
Exterior angle sum of polygon, 195

F
Face
    of regular pyramid, 402
    of right prism, 393
    of solid, 405
45°-45°-90° triangle, 432

G
Golden rectangle, 208
Graph
    in coordinate plane, 468
    of equation, 496–498
    on number line, 7

H
Hexagon, 190
Hypotenuse, 97, 210
Hypothesis, 50

I
If . . . then statement, 50–51, 56–57
Indirect proof, 334–335
Inscribed angle, 364–365
Inscribed circle, 164, 361
    construction of, 164
Inscribed polygon, 163, 361
Intercepted arc, 364–365, 369–370
Interior angles of two parallel lines, 66–68, 72–73
Interior angle sum of polygon, 194–195
Intersecting line and plane, 388
Intersecting lines, 18, 388
Intersecting planes, 34, 389
Isosceles trapezoid, 220–221
Isosceles triangle, 97, 167–168, 173
    parts of, 97, 168

L
Lateral area
    of cone, 406, 444
    of cylinder, 398–399
    of regular pyramid, 402–403, 443
    of right prism, 394
    of similar solids, 414–415
Lateral edge, 393, 402
Lateral face, 393, 402–403
Legs
    of isosceles triangle, 97, 168
    of right triangle, 97
    of trapezoid, 220
Length of segment, 7, 34
Line(s), 6, 388
    auxiliary, 336–337
    coplanar, 66, 388
    equation of, 496–498
    intersecting, 18, 388
    intersecting plane, 388
    parallel, 66–73, 326, 388, 491
        construction of, 78
    parallel to plane, 388
    perpendicular, 15, 72, 491–492
        construction of, 26, 149, 151
    skew, 66, 388
    slope of, 486–488
Locus, 482–483

M
Major arc, 355
Maps and scale drawings, 296–297, 299
Measure
    of angle, 10–11, 35
    of arc, 355
Median
    of trapezoid, 220–222, 501
    of triangle, 157, 158–159
Midpoint
    of arc, 359
    of segment, 7, 154, 476–478
Midpoint formula, 477
Minor arc(s), 355–356
Mixed reviews. *See* Reviews.
Möbius band, 392

N

Noncollinear points, 6, 33–34
Number line, 7

O

Obtuse angle, 15
Obtuse triangle, 97, 260
Opposite angles of a triangle, 93
Origin of coordinate plane, 468

P

Parallel lines, 66–73, 326, 388, 491
    construction of, 78
Parallel planes, 34, 389
Parallelogram, 198, 203–204, 214–216
    area of, 243–244
    properties of, 204, 231
Patterns, 506–507
Pentagon, 190, 193, 194
Perimeter
    of polygon, 98, 240
    of similar polygons, 330
Perpendicular bisector(s)
    of segment, 154–155
        construction of, 154
    of sides of triangle, 163
Perpendicular lines, 15, 72, 491–492
    construction of, 26, 149, 151
Perpendicular planes, 389
Pi (π), 264, 276–277
Point(s), 6
    collinear, 6, 33–34
    coordinate of, 7
    in coordinate plane, 468
    coplanar, 33–34
    distance between, 7, 34, 472
    noncollinear, 6, 33–34
Plane(s), 33
    intersecting, 34, 389
    intersecting line, 388
    parallel, 34, 389
    parallel to line, 388
    perpendicular, 389
Polygon(s), 190–191
    angle of, 194–195
    circumscribed, 164, 361
    consecutive sides of, 190–191
    consecutive vertices of, 190–191
    convex, 190
    diagonal of, 190–191
    equiangular, 191
    equilateral, 191
    inscribed, 163, 361
    perimeter of, 98, 240
    regular, 191, 481
    side of, 190–191
    similar, 312–313, 329–330
    vertex of, 190–191
Postulate(s), 29
    AA, 317–318
    ASA, 121
    of equality, 29–30
    list of, 518
    Protractor, 35
    Ruler, 34
    SAS, 115–116
    SSS, 107–108
Prism, right, 393–395
Problem Solving Strategies
    Asking Questions, 82–83
    Auxiliary Lines, 336–337
    Counterexamples, 418–419
    Diagrams and Charts, 180–181
    Patterns, 506–507
    Working Backwards, 274–275
Proof, 60–61
    indirect, 334–335
    using coordinate geometry, 501
    See also Strategies of proof.
Proportion(s), 289–290, 293
Protractor, 10–11
Puzzle Problems, 17, 101, 112, 132, 157, 162, 224, 254, 261, 295, 324, 354, 368, 376, 401, 405, 442, 446, 456, 459
Pyramid, regular, 402–403, 443
Pythagorean Theorem, 255–256, 428
    converse of, 259–260
    use of, 432–444

Q

Quadrant, 468
Quadrilateral, 190, 194
    See also Parallelogram, Rectangle, Rhombus, Square, Trapezoid.

R

Radius
    of circle, 263, 346
    of cone, 406
    of cylinder, 398
    of sphere, 346
Ratio, 286–287, 378–379
Ray, 6
Reading Geometry
    Diagrams, 81
    How to Read Your Geometry Textbook, xi–xii
    Independent Study, 153
    Symbols, 133
    Three-Dimensional Figures, 417
    Words and Meanings, 229
Rectangle, 198, 209–210
    area of, 238–240
    properties of, 231
Rectangular solid, 393–395, 508–509
    diagonal of, 440–441
Reflections, 537–539
Regular polygon, 191
    area of, 481
Regular pyramid, 402–403, 443
Reviews
    chapter, 39–40, 85–86, 137–138, 183–184, 231–232, 279–280, 305–306, 339–340, 381–382, 421–422, 461–462, 511–512
    cumulative, 88, 186, 282, 342, 424, 514
    end-of-course, 523–530
    mixed, 42, 140, 234, 308, 384, 464
    skills
        Algebraic Expressions and Equations, 278
        Algebraic Fractions, 460, 510
        Arithmetic Operations, 38
        Coordinates in the Plane, 84
        Decimals and Percents, 182
        Distributive Property, 338
        Fractional and Decimal Equations, 304
        Quotients of Square Roots, 380
        Squares and Square Roots, 230
        Whole Numbers and Fractions, 136
        Working with Polynomials, 420
Rhombus, 198, 209–210
    properties of, 231

Right angle, 15
Right prism, 393–395
Right triangle(s), 97, 210, 255–260, 428
    45°-45°-90°, 432, 437
    hypotenuse of, 97, 210
    legs of, 97
    list of common lengths, 428
    30°-60°-90°, 432–433, 437
    using, 440–453
Rotations, 540–545

S

Same-side interior angles, 67–68, 72–73
Scale drawings, 296–297, 299
Scalene triangle, 97
Secant(s) of circle, 346, 369–370
Segment(s), 6
    bisector of, 154
    divided proportionally, 323, 325–326
    endpoints of, 6, 155
    equal, 97
        ways to prove, 144, 174
    length of, 7, 34
    midpoint of, 7, 154, 476–478
    perpendicular bisector of, 154–155
        construction of, 154
Self-Tests. See Tests.
Semicircle, 355
Side(s)
    of angle, 10
    corresponding, 102–103, 144, 312, 414
    of polygon, 190–191
    of triangle, 97
Similar polygons, 312–313, 329–330
Similar solids, 414–415
Similar triangles, 317–333
Sine ratio, 452
Skew lines, 66, 388
Skills reviews. See Reviews.
Slant height
    of cone, 406
    of pyramid, 402
Slope of line, 475, 486–488
    of parallel lines, 491
    of perpendicular lines, 491–492

Solid(s)
    models of, 417
    rectangular, 393–395
    regular pyramids, 402–403
    right circular cones, 406
    similar, 414–415
    spheres, 410
Sphere, 346
    area and volume of, 410
Square, 198, 209–210
    area of, 239
    properties of, 231
Square roots, table of, 516
Squares, table of, 515
Straight angle, 15
Strategies of proof
    proving a quadrilateral is a parallelogram, 216
    proving two angles equal, 144, 168
    proving two lines parallel, 73
    proving two segments equal, 144, 174
    proving two triangles congruent, 127
Supplementary angles, 47, 134–135
Symbols, 133
    list of, x
Symmetry, 531–533, 543–546

T

Table
    of square roots, 516
    of squares, 515
    of trigonometric ratios, 517
Tangent(s) to circle, 346, 350–351
    angle formed by chord and, 365
    common, 351
    construction of, 351
Tangent ratio, 447–448, 452
Tessellations, 543–546
Tests
    chapter, 41, 87, 139, 185, 233, 281, 307, 341, 383, 423, 463, 513
    self-tests
        answers to, 559–561
        Ch. 1, 14, 22, 37
        Ch. 2, 53, 80
        Ch. 3, 101, 112, 125, 132
        Ch. 4, 152, 166, 178
        Ch. 5, 197, 207, 219, 228
        Ch. 6, 246, 254, 262, 271
        Ch. 7, 295, 301
        Ch. 8, 322, 333
        Ch. 9, 358, 368, 376
        Ch. 10, 401, 412
        Ch. 11, 439, 446, 456
        Ch. 12, 480, 495, 504
Theorem(s), 46
    AAS, 126
    HL, 126–127
    list of, 519–522
    Midpoints, 225–226, 502
    Pythagorean, 255–256
        converse of, 259–260
30°-60°-90° triangle, 432–433, 437
Three-dimensional figures, 417
Total area
    of cone, 406
    of cylinder, 398
    of regular pyramid, 402
    of right prism, 394
    of similar solids, 414
Transformational geometry
    reflections, 537–539
    rotations, 540–542
    symmetry, 531–533
    tessellations, 543–546
    translations, 534–536
Translations, 534–536, 543
Transversal, 66–68, 72, 326
Trapezoid, 198, 220–222
    area of, 250–251
    isosceles, 220–221
    median of, 220–221, 501
    parts of, 220–221, 250
Triangle(s), 15
    acute, 97, 260
    altitudes of, 158–159
    angle bisectors of, 164
    angle sum of, 92
    angles of, 15, 92–93
    area of, 247, 292, 481
    circumscribed, 164
    congruent, 102–132, 144–145
    corresponding parts of, 102–103, 144, 317
    equiangular, 97, 167, 173–174

Triangle(s) *(continued)*
    equilateral, 97, 167, 173–174
    exterior angle of, 93
    inscribed, 163
    isosceles, 97, 167–168, 173
    medians of, 158–159, 162
    obtuse, 97, 260
    opposite angles of, 93
    perpendicular bisectors of sides, 163
    right, 97, 210, 255–260, 428, 432
    scalene, 97
    segment joining midpoints of two sides
        of, 225–226, 502
    side of, 97
    sides divided proportionally, 323,
        325–326
    similar, 317–330
    vertex of, 15
Triangle Inequality, 179
Triangle Proportionality, 325
Trigonometry, 447–458, 475
    finding areas with, 481
    table of, 517

V
Vertex,
    of angle, 10
    of cone, 406
    of polygon, 190–191
    of regular pyramid, 402
    of right prism, 393
    of solid, 405
    of triangle, 15
Vertical angles, 18, 47
Volume(s), 394
    of cone, 406, 443–444
    of cylinder, 398–399
    of regular pyramid, 402–403, 443
    of right prism, 394–395
    of similar solids, 414–415
    of sphere, 410

X
x-axis, 468
x-coordinate, 468

Y
y-axis, 468
y-coordinate, 468

Answers to Self-Tests

CHAPTER 1
Page 14 **1.** True **2.** False **3.** True
4. True **5.** point R **6.** $\angle PQS$ or $\angle SQP$
7. $\angle PQR$ or $\angle RQP$ **8.** 180°

Page 22 **1.** right **2.** acute **3.** obtuse
4. straight **5.** $\angle BED$, $\angle AEF$; $\angle BEA$, $\angle DEF$ **6.** $\angle ABF$, $\angle FBC$

Page 37 **1.** 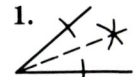 **2.** The Substitution Postulate
3. The Addition Postulate **4.** 2; 5 **5.** W
6. WXYZ **7.** WX

CHAPTER 2
Page 53 **1.** 70° **2.** 142° **3.** $\angle SAL$
4. $\angle SAK$ **5.** Hyp.: you read the newspaper every day; Concl.: you are well informed
6. If you are well informed, then you read the newspaper every day.

Page 80 **1.** A postulate is accepted without proof; a theorem is proved. **2.** $\angle 2$, $\angle 7$; $\angle 3$, $\angle 6$ **3.** $\angle 2$, $\angle 3$; $\angle 6$, $\angle 7$
4. $\angle 1 = \angle 3 = \angle 8 = 130°$; $\angle 2 = \angle 4 = \angle 5 = \angle 7 = 50°$ **5.** $x = 70$; $y = 50$
6. Show that corresponding angles are equal, alternate interior angles are equal, same-side interior angles are supplementary, or both lines are perpendicular to a third line in the same plane.
7. 1. $l \parallel m$ (Given) 2. $\angle 2 = \angle 3$ (If 2 $\parallel$ lines are cut by a trans., alt. int. angles are equal.) 3. $\angle 1 = \angle 2$ (Given) 4. $\angle 1 = \angle 3$ (Subst. Post.)
8.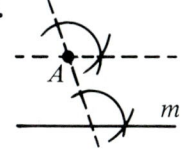

CHAPTER 3
Page 101 **1.** 40 **2.** 120 **3.** 70 **4.** 74
5. d or c **6.** a **7.** c **8.** b

Page 112 **1.** BOJ **2.** $\angle O$ **3.** EP
4. 1. Given 2. From algebra 3. Given: D is the midpt. of $\overline{XY}$. 4. SSS Post.

Page 125
1. 1. $\angle 1 = \angle 2$ (Given: $\overrightarrow{TV}$ bisects $\angle ETO$.)
2. $TE = TO$ (Given) 3. $TV = TV$ (From algebra) 4. $\triangle TEV \cong \triangle TOV$ (SAS Post.)
2. 1. $\angle 1 = \angle 2$; $\angle 3 = \angle 4$ (Given: $\overleftrightarrow{TV}$ bisects $\angle ETO$ and $\angle EVO$.) 2. $TV = TV$ (From algebra) 3. $\triangle TEV \cong \triangle TOV$ (ASA Post.)

Page 132 **1.** SAS, ASA, AAS
2. 1. $\angle 1$ and $\angle 2$ are right angles. (Given: $\overline{RT} \perp \overline{SV}$) 2. $RS = TV$; $RM = TM$ (Given) 3. $\triangle RMS \cong \triangle TMV$ (HL Thm.)

CHAPTER 4
Page 152
1. 1. $RM = TM$; $UM = SM$ (Given)
2. $\angle RMU = \angle TMS$ (Vertical angles are equal.) 3. $\triangle RMU \cong \triangle TMS$ (SAS Post.)
4. $RU = TS$ (Corr. parts of $\cong$ $\triangle$ are $=$.)
2. 1. $\angle X = \angle Y = 90°$ (Given) 2. $AX = AY$ (Given) 3. $AB = AB$ (From algebra)
4. $\triangle AXB \cong \triangle AYB$ (HL Thm.) 5. $\angle 1 = \angle 2$ (Corr. parts of $\cong$ $\triangle$ are $=$.) **3.** SSS

Page 166
1. 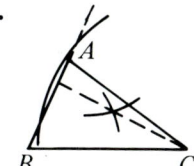 **2.**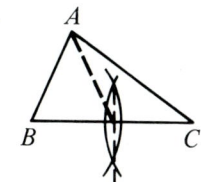

3. 4.

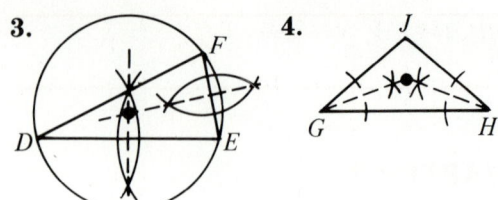

Page 178 **1.** If 2 sides of a triangle are equal, then the angles opposite those sides are equal. **2.** 60 **3.** 40 **4.** 44 **5.** 5 **6.** 4 **7.** 12

CHAPTER 5
Page 197
1. For example:
2. For example:
3. For example:

4. 120° **5.** 60°

Page 207 **1.** A quadrilateral with both pairs of opposite sides parallel **2.** A quadrilateral with just one pair of opposite sides parallel **3.** A parallelogram with four equal sides **4.** A parallelogram with four right angles **5.** A parallelogram with four right angles and four equal sides **6.** CD **7.** $\overline{AD}$ **8.** BX **9.** 5 **10.** BCD **11.** 3 **12.** BCD or DAB

Page 219 **1.** $BD = 12$; $BE = 6$ **2.** $\angle 1 = \angle 4 = 25°$; $\angle 2 = 90°$; $\angle 3 = 65°$ **3.** 28 **4.** 1. $WZ = XY$; $\angle 1 = \angle 2$ (Given) 2. $\overline{WZ} \parallel \overline{XY}$ (If 2 lines are cut by a trans. so that alt. int. angles are equal, the lines are $\parallel$.) 3. $WXYZ$ is a $\square$. (If a quad. has 1 pair of opp. sides both $\parallel$ and equal, the quad. is a $\square$.)

Page 228 **1.** 9 **2.** $\angle G = \angle O = 60°$; $\angle A = 120°$ **3.** $PTKL$, $SPKT$, $PKAT$ **4.** 10 **5.** 9

CHAPTER 6
Page 246 **1. a.** 26 **b.** 30 **2. a.** 5 cm **b.** 20 cm **3.** 48 **4.** 60

Page 254 **1.** 20 **2.** 12 **3.** 84 **4.** 91 **5.** 26

Page 262 **1.** 10 **2.** 4 **3.** 10 **4.** 50 **5.** obtuse **6.** right **7.** acute **8.** right

Page 271 **1.** $\frac{C}{d}$ **2.** 3.14 **3.** $d = 10$; $C = 10\pi$; $A = 25\pi$ **4.** $r = 8$; $C = 16\pi$; $A = 64\pi$ **5.** $r = 10$; $d = 20$; $A = 100\pi$ **6.** $r = 6$; $d = 12$; $C = 12\pi$

CHAPTER 7
Page 295 **1.** $\frac{2}{3}$ **2.** $\frac{5}{3}$ **3.** 5:8 **4.** 1:2:3 **5.** 14 **6.** 15 **7.** 9 **8.** 2 **9.** $4.45 **10.** $79.20

Page 301 **1.** about 3 cm; 1 cm: 3 km or $\frac{1}{300{,}000}$ **2.** 12 km **3.** 8.4 km **4.** Suggested scale: 1 cm : 5 m or $\frac{1}{5000}$

CHAPTER 8
Page 322 **1.** JKP **2.** KP; PJ **3.** 4:8 or 1:2 **4.** J **5.** not similar **6.** similar **7.** not similar **8.** similar

Page 333 **1.** $x = 6$; $y = 15$ **2.** $x = 9$; $y = 2\frac{1}{4}$ **3.** 1:3; 1:9 **4.** 3:7, 9:49 **5.** 5:6; 5:6

CHAPTER 9
Page 358 **1.** $\overline{OX}$ or $\overline{OY}$ **2.** $\overline{XY}$ **3.** $\overleftrightarrow{ST}$ **4.** $\overleftrightarrow{PQ}$ or $\overrightarrow{PQ}$ **5.** $\overline{ST}$ **6.**
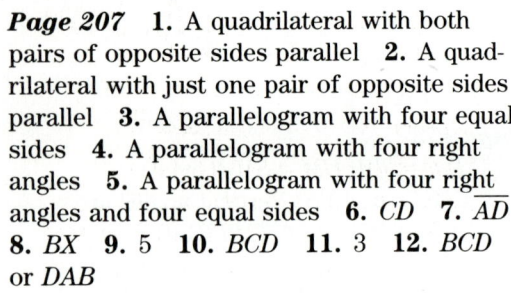
7. 2 **8.** tangent **9.** 80° **10.** 280°

Page 368 **1.** 6 **2.** 32° **3.** 12 **4.** 40°
5. 76° **6.** 156°; 204° **7.** 30°
8. For example:

 (Note that a rhombus inscribed in a circle must be a square.)

Page 376 **1.** 25° **2.** 17° **3.** 110° **4.** 50°
5. 126° **6.** 120° **7.** 10 **8.** 6

CHAPTER 10

Page 401 **1.** $DSTE, ETUF$ **2.** $CDSR, FETU$ **3.** $\overleftrightarrow{CF}, \overleftrightarrow{DE}, \overleftrightarrow{CR},$ or $\overrightarrow{DS}$ **4.** 184
5. 160 **6.** 64π **7.** 192π **8.** 256π

Page 412 **1.**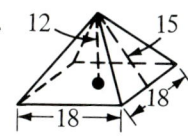

2. 540 **3.** 864 **4.** 1296 **5.** 5 **6.** 15π
7. 24π **8.** 12π **9.** To the nearest integer, 201 **10.** To the nearest integer, 268

CHAPTER 11

Page 439 **1.** $x = 12; y = 9$ **2.** $x = 2; y = 2\sqrt{2}$ **3.** $x = 5; y = 5\sqrt{3}$ **4.** 18
5. $27\sqrt{3}$ **6.** $27\sqrt{3}$

Page 446 **1.** 12 cm **2.** 4 **3.** 16π
4. 260

Page 456 **1.** $x = 14.3$ **2.** $\frac{8}{15}$ **3.** $\frac{8}{17}$
4. $\frac{8}{17}$ **5.** $\frac{15}{17}$ **6.** $y = 16.4; z = 11.5$

CHAPTER 12

Page 480 **1.** $(5, -4)$ **2.** 6 **3.** 10 **4.** 3
5. 13 **6.** $\left(-3, -2\frac{1}{2}\right)$ **7.** $(1, 2)$ **8.** $(1, -1)$

Page 495 **1.** horizontal **2.** vertical **3.** $\frac{4}{3}$
4. -2 **5.** $\frac{3}{7}$ **6.** $\frac{2}{5}$

Page 504 **1.** no **2.** $(0, 8); (8, 0); (3, 5)$
3. $(0, -4); (3, 0); \left(5, 2\frac{2}{3}\right)$

4. For example:

x	y
3	4
4	2
5	0
6	−2

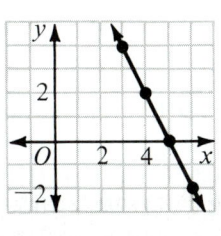

5. Slope of $\overline{AB} = 0 =$ slope of $\overline{DC}$.
Slope of $\overline{AD} = 4 =$ slope of $\overline{BC}$.
Therefore, $\overline{AB} \parallel \overline{DC}$ and $\overline{AD} \parallel \overline{BC}$.
By definition, $ABCD$ is a parallelogram.

Answers to Selected Exercises

CHAPTER 1

Exercises, page 5 **1.** the bottom line **3.** The spaces are the same; for any diam.; the dist. will be $\frac{1}{\pi}$ m.

Written Exercises, pages 8–9 **1. a.** 7 **b.** yes **3.** -3 **5.** -1 **7.** 4 **9.** 2 **11.** 5 **13.** 13 **15.** 17 **17.** true **19.** false **21.** 4 **23.** 16; -10 **25.** 41; 24 **27.** 6; $a + 3$

Written Exercises, page 13 **1.** $\overrightarrow{IA}, \overrightarrow{IH}$ **3.** HIO **5.** DIO; 180 **7.** 30° **9.** 100° **11.** 60° **17.** $\angle 1 = 120°$; $\angle 2 = 65°$; $\angle 3 = 90°$; $\angle 4 = 35°$ **19.** LAS **21.** BAS

Consumer Applications, page 14
toaster: 3 h 7 min; television: 40 h; paperback: 42 min

Written Exercises, pages 16–17 **1.** acute **3.** obtuse **5.** straight **7.** $\angle$s 3, 5, 7, 10 **9.** $\angle$s 2, 3; 5, 6; 6, 7; 8, 10 **11.** $\angle CRA$, $\angle DRA$, $\angle DRB$, $\angle CRB$ **13.** $\angle VQT$, $\angle TQR$, $\angle UQS$ **15. a.** 65° **b.** 65° **c.** 25° **17.** 60°

Puzzle Problems, page 17 NINE

Written Exercises, page 19 **1.** $\angle DOE$ **3.** $\angle FOD$ **5.** 40° **7.** 70° **9.** 65° **11.** 25° **13.** 35° **15.** $x = 20$; $y = 9$

Written Exercises, page 21 **1.** 20; 10 **3. a.** 45° **b.** 65° **5.** 4 **7.** 9 (including $\angle LFA$) **9.** $\overline{FE}$ **11.** Answers will vary; the $\angle$s will be bisected. **13. c.** $\angle 3 = \angle 4$ **d.** $\angle 2 = \angle 3$ (Vert. $\angle$s are =.), and $\angle 1 = \angle 4$ (Vert. $\angle$s are =.); since $\angle 1 = \angle 2$, $\angle 3 = \angle 4$ (Subst. Post.)

Applications, page 24 **1. a.** Arctic **b.** Indian **c.** South Atlantic **d.** North Pacific **e.** North Atlantic **f.** South Pacific **3.** Africa

Written Exercises, pages 27–28 **1.** See Const. 1, p. 25. **3.** See Const. 1, p. 25. **5.** See Const. 2, p. 26. **7.** See Const. 3, p. 26. **9.** See Const. 3, p. 26. **11.** See Const. 1, p. 25.

13–15.
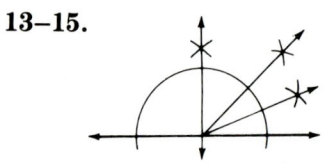

17.
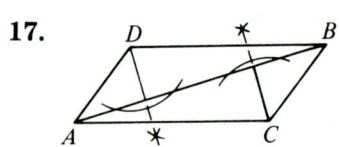

19.
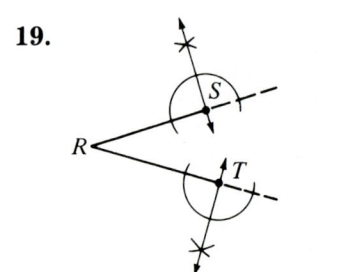

21.
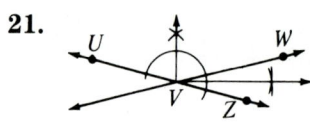

Written Exercises, pages 31–32 **1.** div. **3.** mult. **5.** add. **7.** add. **9.** add. **11.** add. **13.** 2; 3; 5 **15.** $\angle 1 = \angle 2$ since $\overrightarrow{AK}$ bisects $\angle RAC$, and $\angle 3 = \angle 4$ since $\angle 3 = 90° - \angle 1 = 90° - \angle 2 = \angle 4$. **17.** $AB = PQ$ or $BM = QN$

Written Exercises, pages 36–37
1. $EFGH$ **3.** $DCGH$ **5.** BF **7.** D **13.** Through any 2 pts. there is exactly 1 line. **15.** more than one plane **17.** one plane **19.** True **21.** True

Skills Review, page 38 **1.** 9 **3.** -12
5. -21 **7.** -11 **9.** 8.1 **11.** -16.1
13. -4 **15.** 11 **17.** -14 **19.** -7
21. -3.4 **23.** 14.4 **25.** 60 **27.** -91
29. 0 **31.** -88 **33.** 89.6 **35.** -73.5
37. -7 **39.** 13 **41.** 4 **43.** -3
45. -23.1 **47.** 90 **49.** $>$ **51.** $=$ **53.** $<$

Chapter Review Exercises, pages 39–40
1. For example: $\overleftrightarrow{AC}$ **2.** $\overrightarrow{BC}$ or $\overrightarrow{BE}$ **3.** 4
4. -6 **5.** 11 **6.** Q **7.** 180 **8.** $\overline{BN}$ and $\overline{BQ}$ (or $\overline{BK}$) **9.** $\angle KQR$ **10.** 3 or SQN
11. right **12.** acute **13.** obtuse
14. straight **15.** 5 **16.** RQB **17.** 45°

18–19.

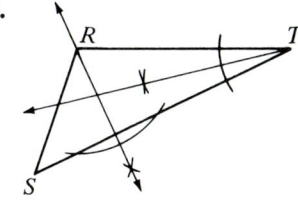

20. div. **21.** add. **22.** subst. **23.** A
24. S **25.** $SLQA$ **26.** KL
27. exactly one

Mixed Review, pages 42–43
Algebra Problems: **1.** $x = 60$ **3.** $k = 6$
5. $x = 20$ **7.** $a = 4$ **9.** $q = -1$ **11.** $j = 3$
13. $x = 4$ **15.** $t = -4$ **17.** $w = -4$
19. $g = \frac{1}{2}$ **21.** no sol. **23.** $h = \frac{9}{2}$ or $4\frac{1}{2}$
25. $k = 3$ **27.** $h = 0$ **29.** All real numbers are solutions.
Geometry Problems: **1.** See Const. 3, p. 26.
3. 1; Through any 2 pts. there is exactly 1 line. **5.** ACE; 180 **7.** True **9.** False
11. Neither is longer; $RS = XY$. **13.** $x = 6$; $y = 40$ **15.** add. **17.** sometimes

CHAPTER 2

Written Exercises, pages 48–49 **1.** 40°; 130° **3.** 72°; 162° **5.** 42°; 138° **7.** 45°; 45°
9. $\angle 4$ **11.** $\angle 2, \angle 8$ **13.** If 2 $\angle$s are supplements of $=$ $\angle$s, then the 2 $\angle$s are $=$.
15. See Const. 2, p. 26 **17.** subtr.
19. add. **21.** 45°

Written Exercises, pages 52–53
1. *Hyp.*: The sum of two angles is 90°; *Conclu.*: The angles are complements.
3. *Hyp.*: $5x + 7 = 27$; *Conclu.*: $x = 4$
5. *Hyp.*: It's a nice day; *Conclu.*: We can go to the beach Saturday. **7.** If two angles are comp., their sum is 90°. **9.** If $x = 4$, then $5x + 7 = 27$. **11.** If we go to the beach Saturday, then it will be a nice day. **13.** If X is a spider, then X is an insect. **15.** If two angles are vert. angles, then they are equal. **17.** If X is a turtle, then X is a reptile. **19.** If a person lives in a glass house, then he or she should not throw stones.
21. a. False **b.** If two angles are vert. angles, then they are equal. **c.** True
23. a. True **b.** If the sides of a figure are all the same length, then it is a square.
c. False **25. a.** False **b.** Every square is a rectangle. **c.** True **27. a.** False
b. Every number divisible by ten is also divisible by five. **c.** True

Applications, page 55 **1.** Art should resemble that shown in text.
3. Art may vary.

Written Exercises, pages 58–59
1. If X is a spider, then X has hairy legs.
3. a. Daddy Long-Legs has hairy legs. (Type 1) **b.** no concl. (Type 3) **c.** no concl. (Type 4) **d.** An octopus is not a spider. (Type 2) **5. a.** Type 1 **b.** correct
7. a. Type 4 **b.** not correct
9. a. Type 2 **b.** correct **11.** Nothing can be concluded.

Written Exercises, pages 62–65
1. 1. Given 2. Vert. $\angle$s are $=$. 3. Subst. Post. **3.** 1. Given 2. Vert. $\angle$s are $=$.
3. Subst. Post. 4. straight 5. Subst. Post.
6. $\angle 3$ and $\angle 2$ are supplements.
5. 1. Def. of $\angle$ bisector 2. Vert. $\angle$s are $=$.
3. 4 (Vert. $\angle$s are $=$.) 4. Subst. Post.
5. $\overleftrightarrow{XY}$ bisects $\angle DOB$.
7. 1. *Given*: $\overrightarrow{RO}$ bisects $\angle YRS$. 2. *Given*: $\overrightarrow{SO}$ bisects $\angle YSR$. 3. Given 4. Subst. Post.
5. Add. Post. 6. Def of supp. $\angle$s 7. Supplements of $=$ $\angle$s are $=$.

9. 1. ∠3 and ∠4 are supp. (Def. of supp. ∠s) 2. ∠5 and ∠6 are supp. (Def. of supp. ∠s) 3. ∠3 = ∠5 (Given) 4. ∠4 = ∠6 (Supplements of = ∠s are = .)
11. 1. ∠3 and ∠6 are supp. (Given) 2. ∠3 and ∠2 are supp. (Def. of supp. ∠s) 3. ∠2 = ∠6 (Supplements of = ∠s are = .)

Written Exercises, pages 69–71 **1.** (3)
3. (1) **5.** (4) **7.** (2) **9.** ∠1 = ∠3 = ∠5 = ∠7 = 110°, ∠2 = ∠4 = ∠6 = ∠8 = 70° **11.** AD, BC, AC; AB, CD, AC
13. corresponding; = **15.** alt. int.; =
17. alt. int.; = **19.** corresponding; =
21. ∠1, ∠5, ∠8, ∠9, ∠12, ∠13, ∠16
23. $x = 48$; $y = 132$ **25.** $x = 75$; $y = 105$
27. $x = 50$; $y = 70$ **29.** 1. Given 2. s.-s. int. ∠s are supp. 3. Given 4. If 2 ∥ lines are cut by a trans., then s.-s. int. ∠s are supp. 5. Supplements of the same ∠ are = .

Calculator Activities, page 71 **1.** 64.360
3. 153.237 **5.** 109.201 **7.** 63.822 **9.** 4.945
11. 2750.019 **13.** 2800.600 **15.** 41.508
17. 28.889 **19.** 13.242 **21.** 384.333
23. 0.014 **25.** 0.2 **27.** 0.96875
29. 0.671875 **31.** 0.5556 **33.** 0.4444
35. 0.8333

Written Exercises, pages 74–76 **1.** yes
3. yes **5.** no

7.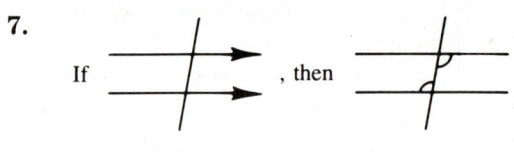

9.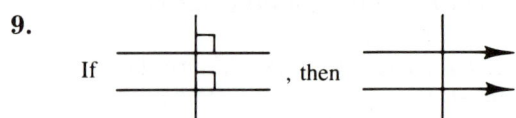

11. $\overleftrightarrow{BC}$ and $\overleftrightarrow{DF}$, $\overleftrightarrow{AB}$ and $\overleftrightarrow{CD}$ **13.** 120
15. $e \parallel f$ **17.** $e \parallel f$ **19.** $c \parallel d$ **21.** $c \parallel d$
23. $c \parallel d$ **25.** 1. Given 2. supp. ∠s 3. the same ∠ are = 4. If 2 lines and a trans. form = corr. ∠s, then the lines are ∥.
27. It is possible for skew lines, or even ⊥ lines, to each be ⊥ to the same line.

Consumer Applications, page 76
1. oven **2.** about 54¢

Written Exercises, pages 79–80 **1.** See Const. 4, p. 77. **3.** See Const. 4, p. 77.
5. See Const. 5, p. 78.

7. 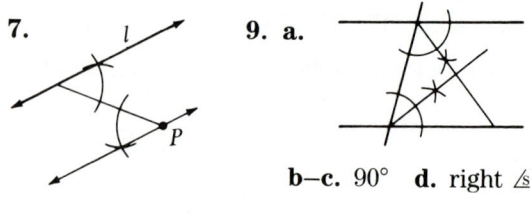 **9. a.**

b–c. 90° **d.** right ∠s

11.

13.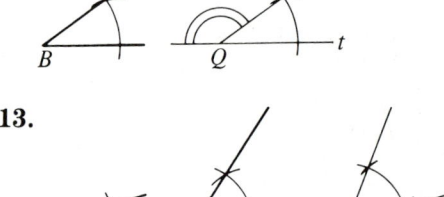

Reading Geometry, page 81 **1.** True
3. False **5.** False **7.** False

Problem Solving Strategies, pages 82–83 **1. b.** Gail's gross income for the entire day **c.** 33 herbs; $33.00 **d.** $49.50 **3. b.** ∠EHB = 90°

Skills Review, page 84 **1.** (6, 2)
3. (−3, −2) **5.** (5, −6) **7.** (−6, 0)
9. (−1, 5) **11.** L **13.** V **15.** M **17.** Z
19. S **33.** (2, 4) **35.** (3, 2) **37.** (5, −1)
39. (6, −4) **41.** (1, −6) **43.** (0, −7)

Chapter Review Exercises, pages 85–86
1. 60 **2.** 150 **3.** ∠4 **4.** Complements of = ∠s are = . **5.** If an angle is acute, then it has a measure less than 90°. **6.** *Hyp.*: An angle is acute; *Conclu.*: It has a measure less than 90°. **7.** If an angle has a measure less than 90°, then it is acute. **8. a.** no concl. **b.** no concl. **c.** Cinnamon has sharp claws. **d.** Cleo is not a cat.
9. 1. ∠1 and ∠2 are supp.; ∠3 and ∠4 are supp. (Def. of supp. ∠s) 2. ∠1 = ∠4 (Given) 3. ∠2 = ∠3 (Supplements of = ∠s

are =.) **4.** ∠3 = ∠5 (Vert. ⩘ are =.)
5. ∠2 = ∠5 (Subst. Post.) **10.** corr. ⩘
11. alt. int. ⩘ **12.** s.-s. int. ⩘ **13.** corr. ⩘
14. $x = 58$; $y = 62$ **15.** $x = 125$; $y = 55$
16. $x = 55$; $y = 60$ **17.** $a \parallel b$ **18.** $a \parallel b$
19. $c \parallel d$ **20.** $c \parallel d$ **21.** See Const. 5, p. 78.

Cumulative Review, page 88 **1.** 4 **3.** 4
5. Z **7.** False **9.** True **11.** False
13. True **15.** If two angles are equal, then they are vert. angles; not true.

CHAPTER 3

Written Exercises, pages 94–96
1. ∠1 = 80°, ∠2 = 100° **3.** ∠5 = 50°, ∠6 = 130° **5.** ∠1 = 30°, ∠2 = 80°, ∠3 = 70° **7.** 90° **9.** 100° **11.** ∠1 = 30°, ∠2 = 50° **13.** ∠1 = 45°, ∠2 = 85° **15.** ∠R = 50°, ∠S = 60°, ∠T = 70° **17.** 30
19. They must also be equal, since each is 180° − (the equal sum of 2 ⩘). **21.** 130
23. ∠D = 40°, ∠E = 55°, ∠F = 85°
25. 1. The ∠ sum of a △ is 180° **2.** Subtr. Post. **3.** straight **4.** Subtr. Post.
5. Subst. Post. **27.** 540 **29.** 90

Written Exercises, pages 99–100
1. isosceles **3.** scalene **5.** acute
7. right **9. a.** Any △ whose ⩘ are all less than 90° with no sides =. **b.** Any △ whose ⩘ are all less than 90° with 2 sides =.
11. a. Any right △ with no sides =.
b. Any right △ with = legs. **13.** ∠C = 60°, ∠CAX = 30°, ∠AXC = 90° **15.** 5 cm, 6 cm, and 7 cm **17.** 10 cm, 10 cm, and 16 cm **19. a-b.** Constructions may vary.
c. The sides opp. the = ⩘ are =.

Written Exercises, pages 104–106
1. a. ∠P **b.** ∠T **c.** ∠I **d.** PI **e.** EJ
f. JM **3.** ∠R = ∠X; ∠S = ∠Y; ∠T = ∠Z; RS = XY; RT = XZ; ST = YZ
5. △ABC ≅ △REW **7.** not ≅ **9.** no
11. no **13.** yes **15.** no **17.** △MNP ≅ △MNQ; ∠PMN = ∠QMN; ∠P = ∠Q; ∠PNM = ∠QNM; MP = MQ; PN = QN; MN = MN

Written Exercises, pages 109–111
1. 1. $\overline{CD}$ **2.** *Given:* X is the mdpt. of $\overline{AB}$.
3. Given **4.** SSS Post. **3. 1.** Given
2. From algebra **3.** SSS Post. **5.** 9
7. $\frac{17}{3}$ or $5\frac{2}{3}$ **9.** ∠B = 60°; ∠R = 100°; ∠S = 60°; ∠T = 20° **11.** ∠A = y°; ∠B = 180° − (x + y)°; ∠S = 180° − (x + y)°; ∠T = x° **13.** See exercise instructions.
15. c. (3, 2) and (3, −2)

Applications, page 114 Models 3 and 6 are rigid figures.

Written Exercises, pages 117–119
1. Yes; yes; the lengths of each pair of sides are =. **3.** yes **5.** ∠QXP; ∠SXR **7.** yes; yes **9.** SSS Post. **11.** SAS Post.
13. 1. *Given:* $\overrightarrow{ZW}$ bisects ∠XZY **2.** From algebra **3.** Given **4.** SAS Post.

15.

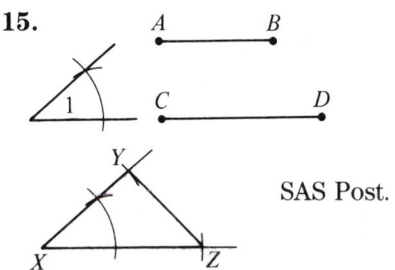

SAS Post.

17. 1. $\overline{RS} \parallel \overline{VT}$ (Given) **2.** ∠1 = ∠2 (If 2 ∥ lines are cut by a trans., then alt. int. ⩘ are =.) **3.** VS = VS (From algebra)
4. RS = VT (Given) **5.** △VRS ≅ △STV (SAS Post.)

Written Exercises, pages 122–124
1. JXP **3.** CA; PJ
5. 1. Given **2.** equals 8 **3.** ASA Post.

7.

Answers to Selected Exercises **569**

9. a. ∠NE **b.** ASA Post. **11. 1.** CB = DB; ∠ABC = ∠ABD (Given) **2.** AB = AB (From algebra) **3.** △ACB ≅ △ADB (SAS Post.) **15. e.** ASA Post.

Written Exercises, pages 128–131
1. HL **3.** ASA **5.** HL (or AAS) **7.** AAS
9. SSS **11.** none **13.** SAS, ASA, AAS
15. 1. Given **2.** From algebra **3.** Given
4. HL Thm. **17. 1.** Def. of supp. ⦞
2. Given **3.** Supplements of = ⦞ are =.
4. Given **5.** From algebra **6.** ASA Post.
19. x = 110, y = 30, z = 12 **21.** x = 40, y = 13, z = 20 **23.** ∠X, $\overline{YM}$ **25.** SAS Post. **27.** no **29.** yes; SAS Post.

Puzzle Problems, page 132 △AZB ≅ △AXB since AB = AB, ∠ZAB = ∠XAB, and ∠ABZ = ∠ABX (ASA Post.). Thus BX = ZX = 12 m.

Reading Geometry, page 133 **1.** No; ∠ABC has vertex B and ∠BAC has vertex A. **3.** Use tick marks to indicate = sides and arcs to indicate = ⦞.

Computer Activities, page 135 **1.** 90 − A
3. a. 55° **b.** 37.5° **c.** 60° **d.** 45°
5. a. 90° **b.** 135° **c.** 45° **d.** 20°

Skills Review, page 136 *Across* **a.** 686
h. 2617 **o.** 361 **q.** 1226 **s.** 2125
u. 6590 **x.** 749 **z.** 2004 *Down* **a.** 6314
c. 615 **e.** 21,960 **g.** 5432 **i.** 625
k. 76,014 **m.** 1936 **s.** 27 **1.** $1\frac{1}{2}$ **3.** $11\frac{1}{6}$
5. $\frac{1}{6}$ **7.** $7\frac{7}{10}$ **9.** $\frac{5}{9}$ **11.** $\frac{5}{6}$ **13.** 32 **15.** 20
17. $5\frac{3}{8}$ **19.** 176 **21.** $\frac{2}{3}$ **23.** $8\frac{1}{6}$

Chapter Review Exercises, pages 137–138 **1.** 80 **2.** 70 **3.** 35
4. 132 **5.** 15 **6.** 40 **7.** 14 cm **8.** 22 cm
9. scalene **10.** △SFZ **11.** △GFZ
12. △SFG **13.** R **14.** RD **15.** RDI
16. SSS **17.** ASA **18.** HL **19.** SAS
20. 1. ∠1 = ∠2 (Given: $\overrightarrow{GH}$ bisects ∠RGS.) **2.** ∠R = ∠S (Given) **3.** GH = GH (From algebra) **4.** △RHG ≅ △SHG (AAS Thm.)

Mixed Review, pages 140–141 **1.** If two lines are perpendicular, then they form right angles. **3. 1.** Given **2.** Supplements of = ⦞ are =. **3.** Given **4.** ASA Post.
5. Def. of mdpt. **7.** add. **9.** 4
11. a. yes **b.** yes **c.** yes **13.** If 2 lines and a trans. form = corr. ⦞, then the lines are ∥. **15.** x = 38, y = 8 **17.** True
19. False **21.** AAS
23.

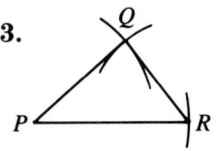 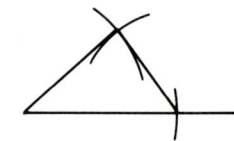

25. 90° **27.** False

CHAPTER 4

Written Exercises, pages 146–148
1. ∠H = ∠U **3.** ∠J = ∠W **5.** IJ = VW
7. △MEN ≅ △OLN **9.** △OEL ≅ △MLE
11. 1. Given **2.** *Given:* $\overrightarrow{CD}$ bisects ∠ACB.
3. From algebra **4.** SAS Post.
5. Corr. parts of ≅ ⦝ are =.
13. 1. Given **2.** From algebra **3.** AAS Thm. **4.** Corr. parts of ≅ ⦝ are =.
15. 1. AX = BX; ∠A = ∠B (Given)
2. ∠1 = ∠2 (Vert. ⦞ are =.)
3. △AXC ≅ △BXD (ASA Post.) **4.** CX = DX (Corr. parts of ≅ ⦝ are =.) **17.** RT = RQ; ST = SQ (Given) **2.** SR = SR (From algebra) **3.** △RTS ≅ △RQS (SSS Post.)
4. ∠T = ∠Q (Corr. parts of ≅ ⦝ are =.)

Written Exercises, pages 150–152
1. f. ASA Post. **3. g.** SSS Post.
5. f. SAS Post. **7. a.** SSS **b.** Corr. parts of ≅ ⦝ are =. **c.** construction; algebra; SAS **d.** Corr. parts of ≅ ⦝ are =. **e.** 90
9. AY = OC, AX = OB, and XY = BC by const.; thus △AXY ≅ △OBC by the SSS Post.; ∠A = ∠O since corr. parts of ≅ ⦝ are =.
11. a. isosceles, since DZ = EZ by const.

c. Yes, $DZ = EZ$, $\angle DZV = \angle EZV$ (by const.) and $ZV = ZV$, so $\triangle DVZ \cong \triangle EVZ$ by the SAS Post.

Reading Geometry, page 153 **1.** Lesson 4–2 **3.** equidistant **5.** inscribed

Written Exercises, pages 156–157
1. midpoint **3.** $\perp$ bisector **5.** Any pt. on the $\perp$ bisector of a seg. is equidistant from the endpts. of the seg. **7.** See Const. 6, p. 154. **9.** See Const. 3, p. 26. **11.** See Const. 6, p. 154. **13.** the midpt. of the hypotenuse **15.** the vertex of the $\angle$ opp. the base

Written Exercises, pages 160–161
1. 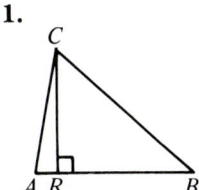 **3.**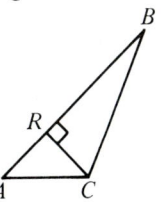

5. 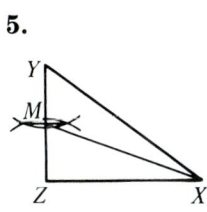 **7.**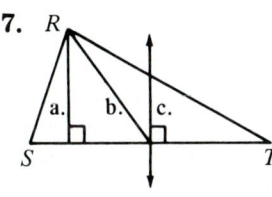

9.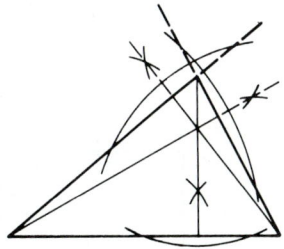

11.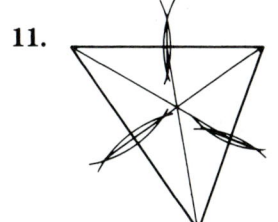

13. d. Actual measurements will vary; in each case, $\frac{2}{3}(AM) = AP$, $\frac{2}{3}(BN) = BP$, and $\frac{2}{3}(CK) = CP$. **15.** The first and third columns are the same; $\frac{2}{3}$. **17. c.** yes

Written Exercises, pages 165–166
1.  **3.**

5. See Const. 7, p. 163. **7.** See Const. 7, p. 163. **9.** See Const. 8, p. 164. **11.** Const. the $\perp$ bisectors of $\overline{XZ}$ and $\overline{YZ}$; label their intersection O and use $\overline{OX}$ as radius.
13. See Const. 7, p. 163 and Const. 8, p. 164.

Written Exercises, pages 169–171
1. $\angle 2 = 65°$, $\angle 3 = \angle 4 = 115°$, $\angle 5 = 50°$
3. $\angle 1 = \angle 2 = 64°$, $\angle 3 = 116°$, $\angle 5 = 52°$
5. $\angle 2 = 68°$, $\angle 3 = \angle 4 = 112°$, $\angle 5 = 44°$
7. $\angle 1 = \angle 2 = 69°$, $\angle 3 = \angle 4 = 111°$
9. $\angle 2 = j°$, $\angle 3 = \angle 4 = 180° - j°$, $\angle 5 = 180° - 2j°$ **11.** 55 **13.** 9 **15.** 6 **17.** 25 **19.** 180 **21.** 1. Given 2. If 2 sides of a $\triangle$ are $=$, then the $\angle$s opp. them are $=$. 3. Subst. Post. 4. The sum of the $\angle$s in a $\triangle$ is 180°. 5. Subst. Post. 6. Div. Post. 7. Subst. Post.

Written Exercises, pages 175–177 **1.** 6
3. 6 **5.** 9 **7.** 10 **9.** 8 **11.** 1. Given 2. If 2 $\angle$s of a $\triangle$ are $=$, then the sides opp. those $\angle$s are $=$. 3. Subst. Post.
13. Bisect the 60° angle from Ex. 12.

15. 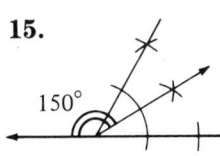 or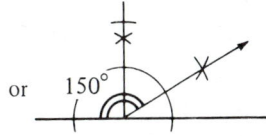

17. other constructions possible

19. 11 **21.** 1. $\angle J = \angle N$ (Given) 2. $PJ = PN$ (If 2 $\angle$s of a $\triangle$ are $=$, then the sides

opp. those ∠s are =.) 3. *JK* = *NM* (Given)
4. △*JPK* ≅ △*NPM* (SAS Post.) 5. ∠5 =
∠6 (Corr. parts of ≅ △ are =.)

Applications, page 179 **1.** yes **3.** no
5. no

Problem Solving Strategies,
pages 180–181 **1.** 15 handshakes
2. Fang

Skills Review, page 182 **1.** 7.569
3. 8.479 **5.** 88.93 **7.** 2.457 **9.** 157.92
11. 1.86 **13.** 58.40 **15.** 11.5817 **17.** 90
19. 205 **21.** 0.071 ≈ 0.07 **23.** 66.666 ≈
66.67 **25.** 75% **27.** 60% **29.** 87.5%
31. 50% **33.** 900.00 **35.** 384 **37.** 20%

Chapter Review Exercises,
pages 183–184 **1.** 1. Given 2. Vert. ∠s
are =. 3. SAS Post. 4. Corr. parts of ≅ △
are =. 5. If 2 lines and a trans. form = alt.
int. ∠s, then the lines are ∥. **2.** 1. Const.
2. Const. 3. SSS Post. 4. Corr. parts of ≅ △
are =. **3.** See Const. 6, p. 154. **4.** on *k*
5. *Z* is equidistant from *C* and *D* **6.** $\overline{RY}$
7. $\overline{TX}$ **8.** Use Const. 3, p. 26, to const. the
⊥ from *M* to $\overline{NO}$. **9.** See Const. 7, p. 163.
10. See Const. 8, p. 164. **11.** 1. Draw the
bisector of ∠*K*, $\overline{KL}$. (Every ∠ has a bisector.) 2. *KX* = *KY* (Given) 3. ∠*XKL* =
∠*YKL* (Def. of ∠ bisector) 4. *KL* = *KL*
(From algebra) 5. △*XKL* ≅ △*YKL*
(SAS Post.) 6. ∠*X* = ∠*Y* (Corr. parts
of ≅ △ are =.) **12.** 9 **13.** 7 **14.** 6

Cumulative Review, page 186 **1.** 1; 3
3. isosceles **5.** obtuse **7.** *J* **9.** HL
11. ASA **13.** AAS **15.** False **17.** True
19. True **21.**

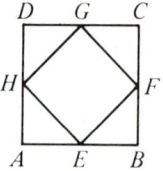

CHAPTER 5

Written Exercises, pages 192–193
1. yes **3.** yes **5.** no **7.** acceptable
9. not acceptable **11.** 3 **13.** equiangular
15. equilateral

17. For example: ◇

19. For example: ⬡

21. For example: ⬡

23. 3 **25.** 5 **27.** *n* − 3 **29.** 5 **31.** 14

Written Exercises, pages 196–197
1. 2 × 180°; 360° **3.** 9 × 180°; 360°
5. 12; 360° **7.** 60°, 120° **9. a.** pentagon
b. 540°; 108° **c.** 360°; 72° **11. a.** octagon
b. 1080°; 135° **c.** 360°; 45° **13.** 6; 60°
15. 18; 160° **17.** No; 3 × 120° = 360°, but
no whole number times 135° is 360°.
19. a. 360 **b.** 180 **c.** supplementary
d. s.-s. int. ∠s are supp.

Written Exercises, pages 200–202
1. a. 1, 3, 4, 6, 8, 10, 11 **b.** 2, 5, 7, 12
c. 3, 6, 10 **d.** 4, 6, 11 **e.** 6 **f.** 6
3. False **5.** False **7.** APTS, STQB,
PDVU, UVCQ, APQB, PDCQ, ADCB
9. APTS, UVCQ **11.** ∠*B*, ∠*PUV*, ∠*UVC*,
∠*UQC*, ∠*QTS*, ∠*AST*, ∠*APT*
13. 1. *Given: ABCD* is a rhombus.
2. From algebra 3. SSS Post. 4. Corr.
parts of ≅ △ are =. 5. If 2 lines and a
trans. form = alt. int. ∠s, then the lines are
∥. 6. Def. of ▱ **15.** A square is also a
rectangle, so it has all the properties of a
rectangle.

17.
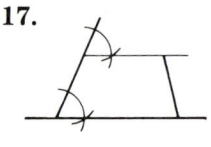

19. rhombus
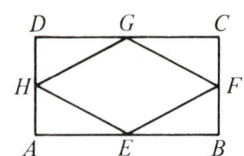

21. square

Written Exercises, pages 205–207 **1.** 6
3. 115° **5.** 10 **7.** 4 **9.** 110° **11.** yes; no
13. 4 **15.** 70° **17.** 70° **19.** 86° **21.** 55°
23. 16 **25.** 36 **27.** 3 **29.** 5
31. 1. ABCD is a ▱. (Given) 2. Draw $\overline{AC}$.
(2 pts. determine a line.) 3. △ ABC ≅
△ CDA (Ex. 30) 4. BC = DA; AB = CD
(Corr. parts of ≅ △ are =.) **33.** 1. Given
2. Def. of ▱ 3. If 2 ∥ lines are cut by a
trans., then alt. int. ∠s are =. 4. Opp. sides
of a ▱ are =. 5. ASA Post. 6. Corr. parts
of ≅ △ are =.

Written Exercises, pages 211–213
1. b, d **3.** b, d **5.** a, b, c, d **7.** c, d
9. c, d **11.** ∠1 = ∠2 = ∠5 = 25°, ∠3 =
∠4 = 65°, ∠6 = 90° **13.** SO = 4; TO = 3
15. 34 **17.** TO = FO = 7 **19.** ∠1 =
∠4 = 15°, ∠2 = ∠3 = 75°, ∠5 = 30°
21. ∠1 = ∠2 = ∠3 = ∠4 = 45°, ∠5 = 90°
23. 5 **25.** 1. Given 2. Opp. sides of a ▱
are =. 3. Def. of rectangle 4. From
algebra 5. SAS Post. 6. Corr. parts of ≅
△ are =.

Written Exercises, pages 216–219 **1.** If
a quad. has diagonals that bisect each other,
then the quad. is a ▱. **3.** If a quad. has
one pair of opp. sides that are both ∥
and =, then the quad. is a ▱. **5.** 1. Given
2. If 2 ∠s of a △ are =, then the sides opp.
those ∠s are =. 3. Given 4. Subst.
Post. 5. Given 6. If a quad. has both
pairs of opp. sides =, then the quad. is a ▱.
7. 1. The diagonals of a ▱ bisect each
other. 2. Mult. Post. 3. *Given:* P is the
mdpt. of $\overline{AO}$ and Q is the mdpt. of $\overline{CO}$.
4. Subst. Post. 5. The diagonals of a ▱ bisect each other. 6. If a quad. has diagonals
that bisect each other, then the quad. is a
▱. **9.** ▱; if a quad. has diagonals that bisect each other, then the quad. is a ▱.
11. square; the sides are = (see Ex. 10) and
⊥. **13.** The jaws are extensions of opp.
sides of the rectangle formed by the four
bolts on the handle.

Written Exercises, pages 223–224
1. ∠Q = 50°, ∠R = ∠S = 130° **3.** ∠P =
∠Q = 80°, ∠R = 100° **5.** $\overline{SP}$; $\overline{RQ}$ **7.** 13
9. 12 **11.** 13 **13.** 5 **15.** 4 **17.** 4.7
19. 4 **21.** 6 **23.** No, the figure would
then be a ▱ by Thm. 10, p. 214.

Puzzle Problems, page 224 The winning horse was Slowpoke, owned by Tic.

Written Exercises, pages 226–228
1. $\overline{XY}$ **3.** 10 **5.** 58 **7.** ST = 4; RT = 5;
RS = 6; p(△ ABC) = 30; p(△ RST) = 15
9. AB = 8; BC = 12; AC = 10; p(△ ABC) =
30; p(△ RST) = 15 **11.** BC = 12; AC = 8;
ST = 7; p(△ ABC) = 34; p(△ RST) = 17
13. 16 **15.** y = 5 **17.** y = 13 **19.** 44
21. a. PRZQ **b.** 50

Reading Geometry, page 229 **2.** Yes; if
2 sides of a △ are =, then the ∠s opp. those
sides are =. **4.** They are =. **5.** No; the
second pair of sides must also be ∥ or the
first pair of sides must be the same length,
in order for the figure to be a ▱.

Skills Review, page 230 **1.** −8 **3.** 7;
−7 **5.** −13 **7.** 10; −10 **9.** 15 **11.** 7n
13. 3√7 **15.** 20 **17.** 8√2 **19.** 7√2
21. 14 **23.** 10√7 **25.** 18 **27.** 16 **29.** 6
31. 5√3 **33.** 4 **35.** 2

Chapter Review Exercises,
pages 231–232 **1.** yes **2.** yes **3.** no
4. no **5.** 540° **6.** 360° **7.** 135° **8.** False
9. True **10.** False **11.** ∠PRQ = 30°;
∠PQR = 132° **12.** OS = 7; PR = 24
13. 10 **14.** ∠RKX = 50°; ∠KXN = 50°
15. ∠FSG = 30°; ∠LOF = 90° **16.** If a
quad. has diagonals that bisect each other,
then the quad. is a ▱. **17.** If a quad. has
one pair of opp. sides that are both ∥ and
=, then the quad. is a ▱. **18.** If a quad.
has both pairs of opp. sides =, then the
quad. is a ▱. **19.** 11 **20.** ∠B = 60°;
∠X = 120°; ∠Y = 120° **21.** 6 **22.** 16

Mixed Review, pages 234–235 **1.** 44
3. trapezoid **5.** Yes; opp. sides of a ▱
are =. **7.** yes **9.** yes **11.** The altitudes
drawn to the base lie on a single line.

13.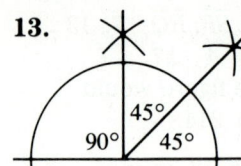

15. If a quad. has equal diagonals, then it is a rectangle; false.
17. never   19. 17
21. −11

23. Sketches may vary. For example,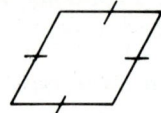

25. 13   27. Use Const. 4 on page 77.
29. (1) Midpoints Thm. (2) Def. of ▱
(3) Opp. ⦞ of a ▱ are =.

CHAPTER 6

Written Exercises, pages 241–242   1. 6 sq. units   3. 84 sq. units   5. 49 sq. units   7. $p = 16; A = 12$ sq. units   9. $p = 14.4; A = 12.8$ sq. units   11. $p = 26\frac{4}{5}; A = 44\frac{2}{5}$ sq. units   13. $p = 4; A = 1$ sq. unit   15. $p = 12; A = 9$ sq. units   17. $p = 20; A = 25$ sq. units   19. $s = 9; A = 81$ 21. $s = 7; p = 28$ 23. $p = 8k; A = 4k^2$ 25. a. 3600 cm² b. 0.36 m²   27. a. 15,000 cm² b. 1.5 m²   29. 35 cm   31. 260 m²   33. a. 72 cm² b. 24 cm²   c. 108 cm²   35. $1296

Written Exercises, page 245   1. 45 sq. units   3. 28 sq. units   5. 56 sq. units   7. 84 sq. units   9. 30.4 sq. units   11. $p = 4k + 14; A = k^2 + 5k$ 13. $p = 16k; A = 5k^2 + 20k$

Written Exercises, page 249   1. 14   3. 32 m²   5. 56   7. 8   9. 20 m²   11. 24   13. $\frac{3t^2 - 5t}{2}$

Written Exercises, pages 252–254   1. d   3. a   5. 96   7. 125   9. 87.5   11. 28 cm²   13. 54 cm²   15. 88 cm²   17. 49   19. 66.5   21. 8   23. 3   25. 20.4 squares

Puzzle Problems, page 254

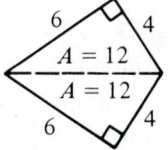

Written Exercises, pages 257–258   1. $x^2 = 3^2 + 4^2; x = 5$ 3. $x^2 = 6^2 + 8^2; x = 10$ 5. $17^2 = x^2 + 15^2; x = 8$ 7. 10   9. 9   11. 10   13. 6.4   15. 7.8   17. 5.2   19. 7.1   21. 7.2   23. $x \approx 7.2; A \approx 86.4$ 25. $x \approx 6.4; A = 20$ 27. $x \approx 3.9; A \approx 66.3$

Written Exercises, page 261   1. yes   3. no   5. yes   7. no   9. yes   11. no   13. right   15. acute   17. right   19. obtuse   21. no   23. $\overline{AC}$

Puzzle Problems, page 261   Luis is winning; George is last.

Written Exercises, pages 266–267   1. 14   3. 10.2 cm   5. 8 cm   7. 10.8   9. 10π   11. 12π   13. 33π m   15. 0.8π km   17. 36.8π km   19. 1   21. 12.5 m   23. 26.5 m   25. 94.2 m   27. 13.2 km   29. 40.8 m   31. 50 cm   33. 3.2 cm   35. 1.5 km   37. 6 cm   39. 1074 cm

Written Exercises, pages 269–271   1. 25π   3. 9π   5. 1.44π cm²   7. 12.25π cm²   9. $\frac{1225}{36}\pi$ 11. 5   13. 9 cm   15. 8 cm   17. 153.9   19. 4.5 cm²   21. 226.9 cm²   23. 576π   25. 30.25π   27. 7π cm   29. 25π − 50   31. 15π

Applications, page 273   1. ∠1 = ∠2 = ∠6 = 60°, ∠3 = ∠4 = 30°, ∠5 = 120°   3. If 2 lines and a trans. form supp. s.-s. int. ⦞, then the lines are ∥.   5. yes   7. Opp. sides of the table are ∥, so alt. int. ⦞ are =.   9. If 2 lines and a trans. form = alt. int. ⦞, then the lines are ∥.

Problem Solving Strategies, pages 274–275   1. A began with $4.50; B began with $7.50   3. a. $3.60   b. 24 sixths of a mile   c. 4 miles   5. 72 days

Computer Activities, pages 276–277   1. Answers will vary.   3. Answers will vary.   5. 3.14

Skills Review, page 278   1. $\frac{86}{n}$ 3. $13 + 3x$ 5. $\frac{1}{2}c - 18$ 7. $7(a + b)$

9. $n - 27 = 15$ **11.** $\frac{1}{5}n - 13 = 27$
13. $3n > 16$ **15.** $n + 43 < 75$ **17.** $b = 24$
19. $n = 50$ **21.** $m = -4$ **23.** $x = -13$
25. $t = -24$ **27.** $a = 0$ **29.** $k = 11$
31. $k = -24$

Chapter Review Exercises, pages 279–280 **1.** $p = 14; A = 10$
2. $p = 100; A = 600$ **3.** $h = 7; p = 28$
4. $p = 22.4; A = 28.8$ **5.** 50 **6.** 42 **7.** 45
8. 36 **9.** 44 **10.** 96 **11.** 50 **12.** 34
13. 32 **14.** 4 **15.** 10 **16.** 9 **17.** 3
18. obtuse **19.** right **20.** acute **21.** 12π
22. 25π **23.** 50.2 **24.** 75.4 **25.** 9 **26.** 9

Cumulative Review, page 282
1. always **3.** sometimes **5.** sometimes
7. sometimes **9.** $\perp$ **11.** $TS; RS$
13. $MF; ME$ **15.** 78 **17.** 40 **19.** 48 m²
21. no

CHAPTER 7

Written Exercises, pages 287–288 **1.** $\frac{1}{2}$
3. $\frac{3}{4}$ **5.** 3:4 **7.** 1:2:3 **9.** $\frac{2}{3}$ **11.** $\frac{2}{5}$
13. $\frac{4}{5}$ **15.** $\frac{5}{9}$ **17.** $\frac{2}{1}$ **19.** $\frac{1}{3}$ **21.** $\frac{3}{2}$ **23.** $\frac{3}{5}$
25. $\frac{5}{8}$ **27.** $\frac{5}{8}$ **29.** $\frac{3}{2}$ **31.** $\frac{1}{1}$ **33. a.** $\frac{3}{5}$
b. $\frac{3}{8}$ **35. a.** $\frac{1}{3}$ **b.** $\frac{1}{4}$ **37.** $\frac{3}{10}$ **39.** $\frac{1}{1500}$

Written Exercises, pages 291–292
1. $3a$ **3.** $3r$ **5.** $\frac{3}{4}$ **7.** $\frac{5}{3}$ **9.** $\frac{12}{4}$ or $\frac{3}{1}$
11. $8f$ **13.** $\frac{4}{3}$ **15.** $\frac{7}{3}$ **17.** $\frac{7}{5}$ **19.** $\frac{-2}{5}$
21. 3 **23.** 30 **25.** 6 **27.** 8 **29.** 4
31. 3 **33.** $\frac{6}{5}$ **35.** 3 **37.** -1

Calculator Activities, page 292 **1.** 24
3. 15.6 **5.** 81.3 **7.** 95.5

Written Exercises, pages 294–295
1. $85 **3.** $9.75 **5.** 5 books **7.** $16.50
9. $240 **11.** 1000 bricks

Puzzle Problems, page 295 Ron is the photographer; Bobbie won the track race; Jo won the math contest.

Written Exercises, page 298 **1.** 2080 m
3. 2960 m **5.** 4880 m **7.** 3120 m **9.** $l = 4$ m; $w = 3.75$ m; $p = 15.5$ m; $A = 15$ m²
11. $l = 4.75$ m; $w = 3.5$ m; $p = 16.5$ m; $A \approx 16.6$ m² **13.** about 21 km **15.** about 38 km **17. a.** Concord-Manchester: 20 mm; Manchester-Portsmouth: 48 mm; Portsmouth-Concord: 56 mm **b.** 70 km
c. 25 km

Written Exercises, pages 299–301
1. Using a scale of 1 cm:5 m, the drawing should be 10 cm × 4.2 cm. **3.** Drawing should be 4.8 cm × 6 cm.

5.
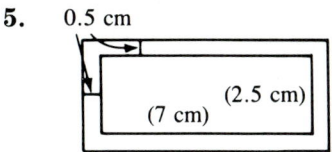

7.
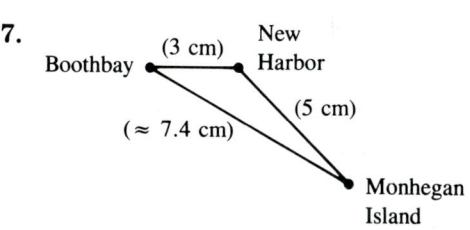

Distance between Boothbay and Monhegan is about 30 km.

9. a. 20 km **b.** 22 km

Applications, page 303 **1.** 90° **2.** 315°
3. 215° **4.** 277°

Skills Review, page 304 **1.** $a = 90$
3. $b = -34$ **5.** $y = 14$ **7.** $n = \frac{6}{7}$ **9.** $b = 2$
11. $c = \frac{3}{2}$ **13.** $a = 5\frac{1}{2}$ **15.** $k = -30$
17. 120 cm tall

Chapter Review Exercises, pages 305–306 **1.** $\frac{3}{5}$ **2.** $\frac{7}{2}$ **3.** $\frac{8}{3}$ **4.** 3:2
5. $b:c$ **6.** 8:5:4 **7.** 20:1 **8.** 1:100
9. 1:9 **10.** $\frac{9}{5}$ **11.** $\frac{1}{2}$ **12.** $\frac{3}{4}$ **13.** $\frac{3}{4}$ **14.** $\frac{4}{5}$

15. $2m$ **16.** $\frac{7}{5}$ **17.** $\frac{y}{4}$ **18.** 6 **19.** 15
20. 20 **21.** $2.25 **22.** $11.00
23. 5 cm; $\frac{1}{1,700,000}$ **24.** 68 km **25.** 51 km
26.

```
                        Courthouse
                       |
                       | (1 cm)
  Business             |
  District •_____|
           (3 cm)   High School

       scale 1 cm : 1 km
```

Mixed Review, pages 308–309 **1.** If a triangle has a 60° angle, then it is equilateral; false. **3.** Each leg is 13 cm. **5.** 105° **7.** 104°, 104°, 152° **9.** 9.5 **11.** $t = 3$ **13.** SSS
15. 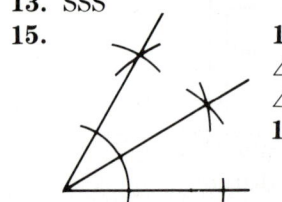 **17.** $\angle 1 = 75°$; $\angle 2 = \angle 3 = 60°$; $\angle 4 = 135°$ **19.** 168 sq. units
21. 120 sq. units **23.** 5 oranges **25.** 2240 km **27.** 2:5 **29.** HL Thm.

CHAPTER 8

Written Exercises, pages 314–316
1. yes **3.** no **5.** yes **7.** The ⚟ are ≅, so corr. ⚟ are equal and corr. sides are in proportion. **9.** Corr. ⚟ are equal and, since all sides in each △ are equal, they are in proportion. **11.** trapezoid *MLKN*; *LK*; *KN* **13.** yes **15. a.** $\frac{2+4+2+4}{3+6+3+6} = \frac{12}{18}$ **b.** yes **17.** $x = 5$; $y = 30$; $z = 24$ **19.** $x = 5$; $y = 12$; $z = 20$ **21.** 20 and 22.5 **23.** *F* is 8 units from *E* in the horizontal direction, *F* at (14, 13) or (−2, 13) **25.** yes

Written Exercises, pages 319–321
1. no **3.** yes **5.** no **7.** yes **9.** yes
11. a. *ADE* **b.** *DE*; *EA* **13. a.** *THG*
b. 12; 10 **c.** $x = 8$; $y = 15$ **15. a.** *DOH*
b. 6 **c.** 8 **17.** 1. Given 2. Opp. sides of a ▱ are ∥. 3. If 2 ∥ lines are cut by a trans., alt. int. ⚟ are =. 4. AA Post.

5. Corr. sides of ∼ ⚟ are in proportion.
19. *BOF* **21.** *FAB*; *FEC*

Calculator Activities, page 322
1. 6; 9; 12; 15; 18; 21

Written Exercises, pages 323–324
1. $\frac{3}{6} = \frac{x}{8} = \frac{y}{12}$; $x = 4$; $y = 6$
3. $\frac{x}{x+6} = \frac{3}{7} = \frac{y}{8}$; $x = 4\frac{1}{2}$; $y = 3\frac{3}{7}$
5. $x = 18$, $y = 21\frac{2}{3}$ **7.** $x = 9$; $y = 10\frac{2}{3}$
9. $x = 9$; $y = 6\frac{6}{7}$ **11.** 7

Puzzle Problems, page 324
The center figure

Written Exercises, pages 327–328 **1.** $\frac{n}{j}$
3. $\frac{y}{x+y}$ **5.** *YC* **7.** *AC* **9.** $x = 12$; $y = 9$
11. $x = 20$; $y = 36$ **13.** $x = 8$; $y = 10$
15. $x = 18$ **17.** $x = 13\frac{1}{3}$ **19.** 1. Given
2. exactly one line 3. If a line intersects a △ and is ∥ to one side, then it divides the other two sides proportionally. 4. Subst. Post. **21.** Lot *A* has the greater area by 1800 m². **23.** 9

Written Exercises, pages 331–332
1. *p*: 1:2; *A*: 1:4 **3.** *p*: 2:3; *A*: 4:9
5. scale factor: 3:5; *A*: 9:25 **7.** scale factor: 2:3; *p*: 2:3 **9.** scale factor: 3:7; *p*: 3:7
11. a. 1:2 **b.** 1:2 **c.** 1:4 **13. a.** 3:5
b. 3:5 **c.** 9:25 **15.** *p*: 3:5; *A*: 9:25
17. a. 20 **b.** 32 **19. a.** $\frac{1}{2}$ **b.** $\frac{1}{4}$
21. 25.2 cm² and 44.8 cm²

Calculator Activities, page 332
1. $d = 12.4$ cm; $C = 38.9$ cm; $A = 120.7$ cm²
3. $r = 15.9$ cm; $d = 31.8$ cm; $A = 793.8$ cm²

Applications, page 335 **1.** Assume $a \parallel b$. Then, since ∠1 and 2 are corr. ⚟, ∠1 = ∠2. This contradicts the given info. that ∠1 ≠ ∠2. Thus, we conclude $a \not\parallel b$.

Problem Solving Strategies, pages 336–337 **1.** $6\sqrt{5}$ **5.** $\overline{OT}$ and $\overline{OV}$
7. $\overline{CX}$ (or $\overline{OC}$ and $\overline{OX}$) **9.** altitude $\overline{DE}$

Skills Review, page 338 **1.** $8x + 33$
3. $a - 9$ **5.** $2 - 9n$ **7.** $1 + 0.4m$ **9.** $8a + 61.3$ **11.** $w = 6$ **13.** $a = -7$ **15.** $b = 2$
17. $m = -4$ **19.** $y = -48$ **21.** $d = 32$
23. $a = 1$ **25.** $1.45 **27.** $4(3n + 5) = 128$; $n = 9$

Chapter Review Exercises, pages 339–340 **1.** JIG **2.** JG; IG **3.** I
4. FLAP **5.** 2:3 **6.** 27 **7.** V; U **8.** VUT
9. 32 **10.** 24 **11.** 21 **12.** $\frac{r}{r+s}$
13. $\frac{c}{c+d}$ **14.** 18 **15.** 4 **16.** 6:5
17. 6:5 **18.** 36:25

Cumulative Review, page 342 **1.** $\frac{1}{7}$
3. $\frac{1}{20}$ **5.** 4 **7.** 11 **9.** $40 **11.** not ~
13. ~ **15.** ROS **17.** SE **19.** A **21.** 2
23. 4:1

CHAPTER 9

Written Exercises, pages 347–348
1. ⊙A **3.** $\overline{RB}, \overline{CD}$ **5.** $\overline{BC}$ **7.** $\overleftrightarrow{BC}$
9. none **11.** 38 **13.** 15 **15.** 13 **17.** $4\frac{1}{4}$
19. J, K, E, F, G, C, D, T **21.** $\overline{EG}$
23. $\overleftrightarrow{JK}$ **25.** A chord that passes through the center of the sphere **27.** possible
29. possible **31.** 10

Written Exercises, pages 353–354
1. internal:2; external:2 **3.** internal:none; external:2 **5.** internal:none; external: none **7.** ext. tangent:Ex. 2; int. tangent: Ex. 6 **9.** True **11.** True **13.** Draw $\overrightarrow{PX}$; use Const. 2, p. 26, to const. the ⊥ to $\overrightarrow{PX}$ through X. **19.** 6 **21.** 3 **23.** 12

Written Exercises, pages 357–358
1. 110° **3.** 58° **5.** $42\frac{1}{2}°$ **7.** 120° **9.** 45°
11. 49 **13.** $\widehat{XP}, \widehat{PY}, \widehat{PQ}, \widehat{YQ}, \widehat{QX}$
15. $\widehat{XPY}, \widehat{XQY}$ **17.** 140 **19.** 21 **21.** 56
23. 90 **25.** 90

Written Exercises, pages 362–363
1. inscribed **3.** circumscribed **5.** neither
7. 10 **9.** 5 **11.** 8 **13.** yes **15.** no

17. See text.
21. 8 **23.** 10
25.

19.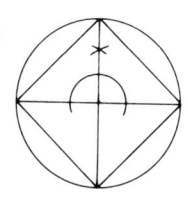

Written Exercises, pages 366–367
1. 25 **3.** 185 **5.** 56 **7.** 184 **9.** 90
11. 60 **13.** 64 **15.** 48; 48 **17.** 88; 44
19. yes; yes **21.** 51° **23.** 37° **25.** 94°
27. 125 **29. a.** 72; 72 **b.** ∠1 = ∠2 = ∠3 = 36°; ∠4 = 108°

Written Exercises, pages 371–372
1. 40° **3.** 38° **5.** 80° **7.** 35° **9.** 59°
11. 45° **13.** 39 **15.** 180 **17.** 84 **19.** 58
21. 27 **23.** 40

Written Exercises, pages 374–376 **1.** 2
3. 4 **5.** $8\frac{3}{4}$ **7.** $2\frac{3}{4}$ **9.** 6 **11.** 0.9
13. $\widehat{XC}; \widehat{XC}$ **15.** PD; PX **17.** 6 **19.** 6

Puzzle Problems, page 376 The construction does not work when the chords are ∥, because the ⊥ bisectors are the same line.

Computer Activities, page 379 **1.** 1.43, 1.67, 2.00, 2.35, 2.86, 1.86, 2.17, 2.60, 3.06, 3.71
3. Change the following lines:
  80 IF G < = 5 THEN R = 36/B
  100 IF G > 5 THEN R = 50/B
5. 744 wheel turns per mile **7. a.** 2600 turns **b.** 1000 turns **c.** 2600 turns
d. 2860 turns

Skills Review, page 380 **1.** $\frac{1}{9}$ **3.** $1\frac{5}{6}$
5. $3\frac{1}{4}$ **7.** $\frac{\sqrt{3}}{10}$ **9.** $\frac{\sqrt{5}}{4}$ **11.** $\frac{6x}{5}$ **13.** $\frac{c\sqrt{2}}{10}$
15. $\frac{3n\sqrt{2}}{7}$ **17.** $\frac{5x\sqrt{3}}{6}$ **19.** $\frac{d\sqrt{5}}{4}$ **21.** $\frac{\sqrt{21}}{7}$
23. $\frac{\sqrt{10}}{4}$ **25.** $\frac{\sqrt{35}}{5}$ **27.** $\frac{2\sqrt{11}}{11}$ **29.** $\frac{\sqrt{6}}{2}$
31. $5\sqrt{2}$ **33.** $2\sqrt{15}$ **35.** $2\sqrt{14}$

Chapter Review Exercises, pages 381–382 **1.** $\overline{OB}$ or $\overline{OC}$ **2.** $\overline{CB}$
3. $\overleftrightarrow{BC}$ **4.** $\overline{BE}$ **5.** 2 **6.** 2 **7.** ⊙R
8. ⊙S **9.** 180 **10.** 50 **11.** 48 **12.** 54

13. 47 **14.** 7 **15.** 70 **16.** 5 **17.** 4
18. 50 **19.** 132 **20.** 7 **21.** 10 **22.** 35
23. 80 **24.** 6 **25.** 45 **26.** 30 **27.** 55

Mixed Review, pages 384–385 **1.** right
3. 10 sides **5.** 1. $WXYZ$ is a $\square$, $\overline{WY}$ and $\overline{XZ}$ intersect at P. (Given) 2. $WX = YZ$ (Opp. sides of a $\square$ are =.) 3. $XP = ZP$; $WP = YP$ (Diagonals of a $\square$ bisect each other.) 4. $\triangle WPX \cong \triangle YPZ$ (SSS Post.)
7. obtuse **9.** GHF **11.** $C = 16\pi$ cm; $A = 64\pi$ cm^2 **13.** 9 **15.** $13\frac{1}{2}$ **17.** See Const. 10, p. 351. **19.** 68.4 sq. units
21. 119 sq. units **23.** $2.45 **25.** 1. $\angle QRT = \angle SRT$; $\angle QTR = \angle STR$ (Given: $\overline{RT}$ bisects $\angle QRS$; $\overline{TR}$ bisects $\angle QTS$.) 2. $RT = RT$ (From algebra) 3. $\triangle QRT \cong \triangle SRT$ (ASA Post.) 4. $QT = ST$ (Corr. parts of $\cong$ $\triangle$ are =.) **27.** none **29.** 10 **31.** 5

CHAPTER 10

Written Exercises, pages 390–391
1, 3. See text examples on p. 388. **5.** $\overleftrightarrow{PQ}$, $\overleftrightarrow{SR}$, $\overleftrightarrow{DC}$ **7.** $\overleftrightarrow{PQ}$, $\overleftrightarrow{SR}$, $\overleftrightarrow{QB}$, $\overleftrightarrow{RC}$ **9.** planes $PQRS$, $SDCR$, $ABCD$, $ABQP$ **11.** $\overleftrightarrow{PQ}$ and $\overleftrightarrow{SD}$, or $\overleftrightarrow{SR}$ and $\overleftrightarrow{PA}$ **13.** $\parallel$ **15.** True
17. False **19.** True **21.** False **23.** False

25. **27.**

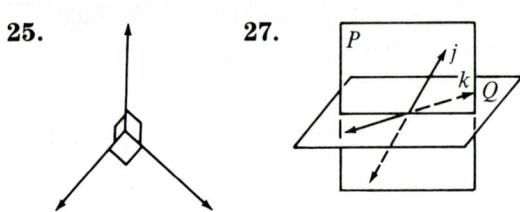

29. **31.** one
33. one

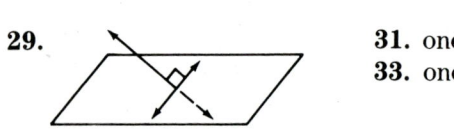

Written Exercises, pages 396–397
1. 4 cm **3.** 8 cm **5.** 24 cm **7.** 20 cm^2
9. 184 cm^2 **11.** 120 cm^3 **13.** 50 cm^2
15. L.A. = 48; T.A. = 72; $V = 36$ **17.** $h = 3$; L.A. = 84; T.A. = 164 **19.** $B = 4$; L.A. = 16;

T.A. = 24; $V = 8$ **21.** $B = 25$; L.A. = 100; T.A. = 150; $V = 125$ **23. a.** 4 **b.** 8
25. 34.4% **27.** 88,750 L

Written Exercises, pages 399–400 **1.** 2
3. 5 **5.** 4π **7.** 28π **9.** $C = 10\pi$; $B = 25\pi$; L.A. = 20π; T.A. = 70π; $V = 50\pi$ **11.** $C = 2\pi$; $B = \pi$; L.A. = 12π; T.A. = 14π; $V = 6\pi$
13. $r = 3$; $B = 9\pi$; L.A. = 12π; T.A. = 30π; $V = 18\pi$ **15.** $h = 4$; $C = 6\pi$; $B = 9\pi$; L.A. = 24π; T.A. = 42π **17.** 226 **19.** shorter can

Puzzle Problems, page 401 $\square$ is opposite $\bullet$; x is opposite $\bigcirc$; $\diamondsuit$ is opposite $+$

Written Exercises, page 404 **1.** True
3. False **5.** True **7.** False **9.** True
17. 64 **19.** 144 **21.** L.A. = 240 cm^2; T.A. = 384 cm^2; $V = 384$ cm^3 **23.** 120 m^2
25. 72 cm^3

Puzzle Problems, page 405
Sam is 1.75 m tall.

Written Exercises, pages 407–408
1. 54π **3.** 30π **5.** 20π **7.** 180π **9.** 48π
11. $C = 24\pi$; $B = 144\pi$; L.A. = 180π; T.A. = 324π; $V = 432\pi$ **13.** $r = 6$; $B = 36\pi$; L.A. = 60π; T.A. = 96π; $V = 96\pi$ **15.** $C = 10\pi$; $B = 25\pi$; $h = 12$; $l = 13$; L.A. = 65π; T.A. = 90π **17.** 5810 g **19.** $\frac{\pi}{4}$

Calculator Activities, page 409
1. ≈ 290 **3.** ≈ 10 **5.** ≈ 10

Written Exercises, pages 411–412
1. $A = 144\pi$; $V = 288\pi$ **3.** $A = 256\pi$; $V = \frac{2048}{3}\pi$ **5.** $A \approx 1017$; $V \approx 3052$
7. $A \approx 452$ cm^2; $V \approx 904$ cm^3 **9.** $d = \frac{1}{4}$; $A = \frac{1}{16}\pi$; $V = \frac{1}{384}\pi$ **11.** $r = 5$; $d = 10$; $V = \frac{500}{3}\pi$ **13.** $d = 3$; $A = 9\pi$; $V = \frac{9}{2}\pi$ **15.** $r = 7$; $d = 14$; $A = 196\pi$ **17.** $A = 4\pi r^2 = 4\pi n^2$; L.A. = $2\pi rh = 2\pi n \cdot 2n = 4\pi n^2$

Applications, page 413 small cube: $\frac{T.A.}{V} = \frac{3}{1}$; larger cube: $\frac{T.A.}{V} = \frac{1}{10}$

Written Exercises, pages 415–416
1. 9:16; 27:64 **3.** 1:36; 1:216 **5.** 2:5; 8:125 **7.** 1:4; 1:16 **9.** 10 cm
11. 1000 cm³ **13.** 1:9 **15.** 53,333L
17. a. $3\frac{1}{3}$ **b.** 1500 cm²; 50,000; $33\frac{1}{3}$

Problem Solving Strategies, pages 418–419

1. 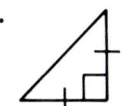   **3.**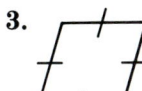

5. 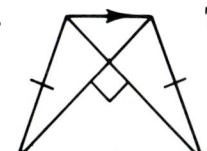   **7.**    **9.**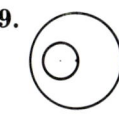

11. The altitudes of any obtuse △ meet outside the △. **13.** Let square A have sides of length 1 and square B have sides of length 2. Then the area of A is 1, the area of B is 4, and their ratio is 1:4. **15.** For example, 7 > 5 and 7 ≠ 6. **17.** For example, $3^2 = 9 = (-3)^2$ and $3 \neq -3$. **19.** For example, a 30° angle and a 60° angle are comp. **21.** $(-5)^2 = 25 > 16 = 4^2$ and $-5 < 4$ **23.** ∠ 1 and ∠ 2 could be the two equal angles of an isosceles △.

Skills Review, page 420 **1.** $3a^5 - a^4b - a^3b^2$ **3.** $r^2 + 8r + 16$ **5.** $100 - 60x + 9x^2$ **7.** $b^2 - 26b + 169$ **9.** $x^2 - 15x + 54$ **11.** $-30hk - 24hk^2 + 6k^5$ **13.** $a^2 - 144b^2$ **15.** $24k^2 + 13hk - 2h^2$ **17.** $9m^2 - 4n^2$ **19.** $5ax(3a - 2x)$ **21.** $(y - 10)(y - 10)$ or $(y - 10)^2$ **23.** $(a + 7)(a - 5)$ **25.** $(k + 7)(k + 7)$ or $(k + 7)^2$ **27.** $(c - 10)(c - 3)$ **29.** $3ab^2(9a^2b + 2c)$ **31.** $(12 + n)(12 + n)$ or $(12 + n)^2$ **33.** $(h + 8)(h - 2)$ **35.** $(n - 9)(n - 3)$ **37.** $(k - 11)(k + 2)$ **39.** $(r + 7)(r + 4)$ **41.** $(d - 19)(d - 1)$

Chapter Review Exercises, page 422
1. True **2.** False **3.** False **4.** False **5.** 90 **6.** 228 **7.** 408 **8.** 540 **9.** 12π
10. 96π **11.** 168π **12.** 288π **13.** 576
14. 720 **15.** 1296 **16.** 1728 **17.** 10π
18. 65π **19.** 90π **20.** 100π **21.** 100π
22. $\frac{256\pi}{3}$ **23.** 4:25; 8:125 **24.** 7:9; 343:729 **25.** 4:3; 16:9

Cumulative Review, page 424 **1.** ⊙R
3. 90 **5.** 47 **7.** RS **9.** 49 **11.** True
13. True **15.** False **17.** True **19.** 56π
21. 24π

CHAPTER 11

Written Exercises, pages 429–430
1. 13 **3.** 6 **5.** 15 **7.** $x = 8; y = \sqrt{37}$
9. $x = 9; y = 12$ **11.** $x = 12; y = 5$
13. $x = 4; y = 4$ **15.** $x = 15; y = 8$
17. 108 sq. units **19.** Yes; $6^2 + 8^2 = 36 + 64 = 100 = 10^2$, so △ACB is a right △.
21. 25π **23.** $p = 20; q = 15$ **25.** yes
27. no **29.** 12 **31.** Yes; $12^2 + 9^2 = 144 + 81 = 225$ and $15^2 = 225$. **33.** 128 sq. units **35.** $25\pi - 48$ sq. units

Applications, page 431
1.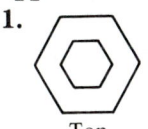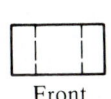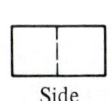
Top Front Side

3.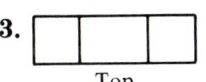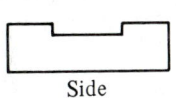
Top Front Side

Written Exercises, pages 434–436
1. $x = 8; y = 8\sqrt{2}$ **3.** $x = 7; y = 7\sqrt{2}$
5. $x = 12; y = 6\sqrt{3}$ **7.** $x = 10; y = 10\sqrt{3}$
9. $x = 8; y = 8\sqrt{3}$ **11.** 90 **13.** 30
15. $3\sqrt{3}$ **17.** $6\sqrt{2}$ **19.** $3\sqrt{2}$ **21.** 8
23. $5\sqrt{2}$ **25.** 50 **27.** $5\sqrt{2}$ **29.** Each side has length $6\sqrt{2}$. **31.** $BC = 6$; $AC = 6\sqrt{3}$; $CD = 3\sqrt{3}$; $AD = 9$; $DB = 3$
33. $AB = 8; BC = 4; CD = 2\sqrt{3}; AD = 6$; $DB = 2$ **35.** $AB = 28; BC = 14; AC = 14\sqrt{3}$; $AD = 21; DB = 7$

Calculator Activities, page 436
1. ≈ 1.4 g **3.** ≈ 1.3 g

Written Exercises, pages 437–439 **1.** 2
3. $50\sqrt{3}$ **5.** $30\sqrt{3}$ **7.** $x = 2\sqrt{3}; A = 4\sqrt{3}$
9. $x = 8; A = 64\sqrt{3}$ **11.** 30° **13.** $4\sqrt{3}$
15. 60° **17.** 7 **19.** $\frac{49}{2}\sqrt{3}$ **21.** 90°
23. 4 **25.** $32\sqrt{3}$ **27.** 5:3 **29.** 25:9
31. 32 **33.** 55 **35.** $\sqrt{3}$ cm **37.** $\sqrt{3}$

Written Exercises, page 442 **1.** $x = 2\sqrt{5}; d = 3\sqrt{5}$ **3.** $x = \sqrt{5}; d = \sqrt{14}$
5. $x = \sqrt{29}; d = \sqrt{65}$ **7.** $c = 8; x = 15$
9. $x^2 = a^2 + b^2 = 1^2 + 1^2 = 2$, so $x = \sqrt{2}$
11. 30.1 cm **13.** no

Puzzle Problems, page 442

a. 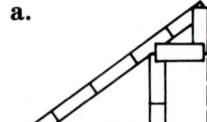 **b.**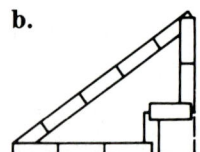

Written Exercises, pages 445–446
1. $l = 5$; L.A. = 80; T.A. = 144; $V = 64$
3. $l = 10$; L.A. = 240; T.A. = 384; $V = 384$
5. $h = 8$; L.A. = 240; T.A. = 384; $V = 384$
7. $l = 10$; L.A. = 80π; T.A. = 144π; $V = 128\pi$
9. $l = 17$; L.A. = 255π; T.A. = 480π; $V = 600\pi$ **11.** $r = 15$; L.A. = 255π; T.A. = 480π; $V = 600\pi$ **13.** $2\sqrt{3}$
15. $16\sqrt{3}$ **17.** 480 cm²
19. $\frac{256\sqrt{161}}{3}$ cm³ **21.** 18 **23.** $\frac{4000}{3}\pi$

Puzzle Problems, page 446 10

Written Exercises, pages 449–451
1. 0.0875 **3.** 2.7475 **5.** 30° **7.** $\tan A = \frac{5}{12}$; $\tan B = \frac{12}{5}$ **9.** $\tan A = 1$; $\tan B = 1$
11. 2.7 **13.** 2.5 **15.** 13.7 **17.** 0.8 m
19. a. $\frac{3}{4}$ **b.** 0.75 **c.** 37° **21. a.** $\frac{2}{3}$ **b.** $0.\overline{6}$
c. 34° **23.** 0.4000

Written Exercises, pages 454–455
1. 0.0872 **3.** 1.0000 **5.** 75° **7.** 65°
9. $\frac{5}{13}$ **11.** $\frac{5}{12}$ **13.** $\frac{5}{13}$ **15.** 67°
17. $\frac{24}{25}$ **19.** $\frac{24}{25}$ **21.** $\frac{24}{7}$ **23.** $\frac{1}{2}$ **25.** 30°

27. 0.9 **29.** 5.7 **31.** 6.7 **33.** 10.0
35. about 3.8 m **37.** 35° **39.** 76°

Applications, pages 458–459 **1. a.** 15°
b. 15° **c.** less than 15° **2. a.** ∠2; angle of elev. **b.** ∠3; angle of dep. **c.** ∠5; angle of dep. **d.** ∠8; angle of elev.
7. $\tan 40° = \frac{16}{x}$ **9.** 92 m **11.** 79 m

Puzzle Problems, page 459

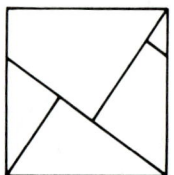

Skills Review, page 460 **1.** $\frac{-a}{4}$ **3.** $\frac{bc^2}{-3}$
5. $\frac{2}{a+b}$ **7.** $\frac{2}{a-b}$ **9.** $a - 8$ **11.** $\frac{d-5}{13}$
13. $\frac{5}{2}$ **15.** $\frac{3}{b}$ **17.** $\frac{x-2}{3}$ **19.** $\frac{y^2}{x+5}$
21. $\frac{x+2}{4}$ **23.** $\frac{m}{2n^2}$ **25.** $\frac{b}{2}$ **27.** $\frac{4}{7}$
29. $\frac{3(1+d)}{c+2}$

Chapter Review Exercises, pages 461–462 **1.** $x = 10; y = 15$
2. $x = 12, y = 15$ **3.** $x = 2; y = 4\sqrt{2}$
4. $x = 18, y = 9\sqrt{3}$ **5.** $x = 4; y = 4\sqrt{2}$
6. $x = 5, y = 5\sqrt{3}$ **7.** $18\sqrt{3}$ **8.** 18
9. $25\sqrt{3}$ **10.** 10 **11.** $\sqrt{109}$ **12.** 240
13. 384 **14.** 8 **15.** 384 **16.** 15
17. 135π **18.** 216π **19.** 324π
20. $\frac{3}{4}$ **21.** $\frac{4}{5}$ **22.** $\frac{4}{5}$ **23.** $\frac{4}{3}$
24. 1.5 **25.** 4.7 **26.** 5.6

Mixed Review, pages 464–465 **1.** True
3. True **5.** False **7.** 1. Given 2. Def. of isos. trap. 3. Base ⦞ of an isos. trap. are =. 4. Given 5. Def. of ⊥ lines
6. △AED ≅ △BFC (AAS Thm.)
7. ∠ADE = ∠BCF (Corr. parts of ≅ △ are =.) **9.** The △ have the same height and equal bases. **11.** 50; 25 **13.** $x = 2$; $y = 5$ **15.** $x = 17; y = 7\sqrt{3}$ **17.** 2940π cubic units **21.** 24 **23.** $3\sqrt{2}$ **25.** 20

CHAPTER 12

Written Exercises, pages 470–471
1. IV; II 3. III 5. y 7. b. C 9. b. C
11. $J(2, 2)$; $K(5, 5)$ 13. $J(-5, 5)$; $K(5, 2)$
15. $J(-3, -1)$; $K(5, -4)$ 17. 5 19. 3
21. 6 23. trapezoid 25. rectangle
27. $T(-6, 2)$ 29. $T(8, 0)$ 31. $(-3, 1)$, $(3, -1)$, or $(3, 9)$

Written Exercises, pages 473–474
1. $\sqrt{29}$ 3. $\sqrt{61}$ 5. $\sqrt{5}$ 7. $3\sqrt{2}$ 9. 5
11. $\sqrt{61}$ 13. 7.1 15. 4.5 17. $p = 12$; $A = 6$ 19. $p = 30$; $A = 30$ 21. $p = 20$; $A = 25$ 23. $p = 34$; $A = 56$

Calculator Activities, page 475
1. a. tangent b. sine 3. a. sine b. cosine 5. 0.2079; 0.9781; 0.2126; 2.1 m; 10.2 m

Written Exercises, pages 479–480
1. $(5, 3)$ 3. $(-1, 4)$ 5. $(4, -2)$
7. $(0, -3)$ 9. $(8, 3)$ 11. $(8, 6)$ 13. $(5, 3)$
15. $(2, -5)$ 17. $(5, 4\frac{1}{2})$ 19. $(3, 2.85)$
21. $(\frac{3}{4}, -5)$ 23. $(10, 5)$ 25. $(3, -1)$

Calculator Activities, page 481
1. 28.19 3. 259.81 5. 50.23

Applications, pages 484–485
1.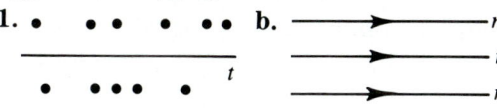
c. Lines m and n, $\parallel$ to t, each 3 cm from t.
3. A plane $\parallel$ to planes R and S, midway between them 5. Any pt. on the $\perp$ bisector of a seg. is equidistant from the endpts. of the seg. 7. At the intersection of the diagonals of the rectangular ceiling 9. 2 planes, each $\parallel$ to plane P, and 1 m from it.
11.
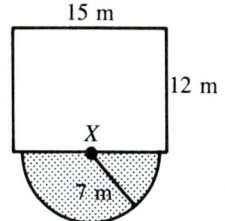

13. At the intersection of the $\perp$ bisectors of the sides of the $\triangle$ formed by the 3 towns 15. an infinite number 17. none

Written Exercises, pages 489–490
1. a. a b. c c. d d. b 3. $\frac{1}{3}$ 5. $-\frac{7}{3}$
7. 6 9. -7; -3; $\frac{7}{3}$ 11. 3; -4; $\frac{3}{-4}$
13. -5; 2; $\frac{-5}{2}$ 15. $\overline{CD}$ 17. $\overline{AB}$, $\overline{CD}$, and $\overline{AD}$ 19. For example: $(1, 5)$, $(2, 6)$, and $(-1, 3)$ 21. For example: $(-1, -4)$, $(2, -4)$, and $(6, -4)$

Written Exercises, pages 493–495
1. a. $-\frac{5}{7}$ b. $\frac{7}{5}$ 3. a. not defined b. 0
5. $\parallel$ 7. $\perp$ 9. 0; 0; yes; no 11. -2; -2; yes; no 13. $\overleftrightarrow{EF}$: $\frac{2}{7}$; $\overleftrightarrow{FG}$: $-\frac{7}{2}$ 15. $\overline{RT} \perp \overline{SV}$
17. $\square$ 19. a. $(1, 3)$ b. $(6, 3)$ c. 0; 0; 0
d. Their slopes are =. e. 8; 2; 5; yes

Written Exercises, pages 499–500
1. For example: $(0, -1)$, $(3, -3)$, $(-3, 1)$, $(-6, 3)$

3.
x	y
0	6
6	0
2	4

5.
x	y
0	0
0	0
3	9

7.
x	y
0	-4
6	0
12	4

9. [graph]

11. [graph]

13. [graph]

Answers to Selected Exercises **581**

15.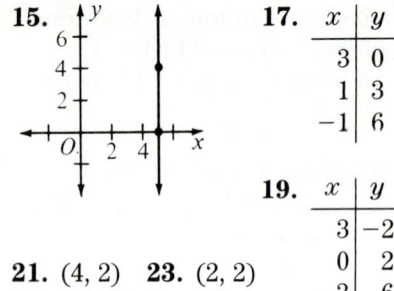

17.

x	y
3	0
1	3
-1	6

19.

x	y
3	-2
0	2
-3	6

21. (4, 2) **23.** (2, 2) **25.** 8 **27.** 3

Written Exercises, pages 502–504
1. a. 5 **b.** 4 **c.** 4; 41 **d.** 5; 4; 41 **e.** $\sqrt{41}$ **3. a.** (2, 3) **b.** See text. **c.** (0, 3); (2, 0) **d.** 2; 13 **e.** 2; 3; 13 **f.** 3; 2; 13 **g.** 13 **5.** T has coord. (a, a). Slope of $\overline{RT} = \frac{a-0}{a-0} = \frac{a}{a} = 1$ and slope of $\overline{VS} = \frac{a-0}{0-a} = \frac{a}{-a} = -1$. Since (slope of $\overline{RT}$)·(slope of $\overline{VS}$) = 1(-1) = -1, $\overline{RT} \perp \overline{VS}$.
7. S must have coord. $(4a, 0)$. Let A and B be the mdpts. of $\overline{RT}$ and $\overline{ST}$, respectively. A has coord. $(\frac{2a}{2}, \frac{2b}{2})$, or (a, b) and B has coord. $(\frac{2a+4a}{2}, \frac{2b}{2})$, or $(3a, b)$. $RB = \sqrt{(3a)^2 + b^2} = \sqrt{9a^2 + b^2}$ and $AS = \sqrt{(-3a)^2 + b^2} = \sqrt{9a^2 + b^2}$. Thus $AS = RB$.

Problem Solving Strategies, pages 506–507
1. 204 squares; $30 = 16 + 9 + 4 + 1$ **2.** 36π; $9 \times 4\pi$ **3.** $4 + 3 + 2 + 1 = 10$; 45 lines

Computer Activities, page 509
2. a. 600 cm² **b.** 10 cm × 10 cm × 10 cm **3. a.** H = 11.08; TA = 601.54 **b.** H = 9.07; TA = 601.44 **c.** no **4. a.** TA = 952.4592 **b.** 12.6 cm × 12.6 cm × 12.6 cm **5. a.** They are all =. **b.** cube

Skills Review, page 510 **1.** $\frac{4x}{7}$ **3.** $-xy$ **5.** $\frac{6b^3 + 2b}{a}$ **7.** $\frac{1}{x+y}$ **9.** $\frac{m}{n}$ **11.** $b-1$ **13.** $\frac{7a}{10}$ **15.** $\frac{5a}{6}$ **17.** $\frac{7}{2a}$ **19.** $\frac{7-5a^2}{ab}$ **21.** $\frac{m}{12}$ **23.** $\frac{14+x}{2x}$ **25.** $\frac{-2}{3n}$ **27.** $\frac{4x}{3}$

Chapter Review Exercises, pages 511–512 **1.** (3, 0) **2.** (-2, 0) **3.** x **4.** (3, -3) **5.** III **6.** 10 **7.** 7 **8.** 5 **9.** 10 **10.** (4, 0) **11.** (3, -4) **12.** (4, 0) **13.** $(-1, \frac{5}{2})$ **14.** $\frac{1}{2}$ **15.** $\frac{3}{-2}$ **16.** $\frac{-1}{2}$ **17.** 0 **18.** $\frac{3}{4}$ **19.** not defined **20.** True **21.** False **22.** False **23.** True **24.** no **25.** no **26.** 7 **27.** Slope of $\overleftrightarrow{AB} = \frac{6-4}{6-0} = \frac{2}{6} = \frac{1}{3}$; slope of $\overleftrightarrow{CD} = \frac{3-2}{4-1} = \frac{1}{3}$; since their slopes are =, $\overleftrightarrow{AB} \parallel \overleftrightarrow{CD}$. **28.** $CD = \sqrt{1^2 + 3^2} = \sqrt{10}$; $AB = \sqrt{6^2 + 2^2} = \sqrt{40} = 2\sqrt{10}$; $\frac{1}{2} \cdot 2\sqrt{10} = \sqrt{10}$, so $CD = \frac{1}{2} AB$.

Cumulative Review, page 514 **1.** 3 **3.** $3\sqrt{2}$ **5.** 12 **7.** 5 **9. a.** 16 **b.** 1280 **11.** $\frac{4}{5}$ **13.** $\frac{4}{3}$ **15.** 0.9063 **17.** 4.7 **19.** (-2, 3) **21.** (0, -2) **23.** $-\frac{1}{3}$ **25.** (Slope of $\overline{AB}$) = $-\frac{1}{3}$ = (slope of $\overline{CD}$) and (slope of $\overline{AD}$) = 4 = (slope of $\overline{BC}$), so $\overline{AB} \parallel \overline{CD}$ and $\overline{AD} \parallel \overline{BC}$. Thus $ABCD$ is a ▱.

REVIEW EXERCISES

Chapter 1, page 523 **1. a.** 30° **b.** acute **3. a.** 90° **b.** right **5.** vertical **7.** XOY **9.** Subst. Post. **11.** KSCA **13.** R, U, F, and S are not coplanar points.

Chapter 2, pages 523–524 **1.** 2 **3.** If 2 ⚓ are complements of = ⚓, then the 2 ⚓ are =. **5.** corresponding **7.** alternate interior **9.** $c \parallel d$ **11.** $a \parallel b$

Chapter 3, page 524 **1.** 75 **3.** 30 **5. a.** LR **b.** NT **c.** E **7.** ASA **9.** 1. $\overline{AB} \perp \overline{XY}$ (Given) 2. △ABY and △ABX are rt. ⚓. (Def. of rt. △) 3. $AX = AY$ (Given) 4. $AB = AB$ (From algebra) 5. △ABX ≅ △ABY (HL Theorem)

Chapter 4, page 525 **1.** Corr. parts of ≅ ⚓ are =. **3.** See Const. 7 **5.** 6 **7.** 1. $\angle A = \angle X$; $AB = XY$; $AC = XZ$

(Given) 2. $\triangle ABC \cong \triangle XYZ$ (SAS Post.)
3. $BC = YZ$ (Corr. parts of $\cong$ $\triangle$s are =.)

Chapter 5, pages 525–526 **1.** False
3. True **5.** True **7.** True **9.** True
11. True **13.** False **15.** 1. X, Y, and Z
are the midpoints of the sides of $\triangle ABC$.
(Given) 2. $\overline{ZY} \parallel \overline{AX}$; $\overline{YX} \parallel \overline{ZA}$ (The Midpoints
Thm.) 3. $AXYZ$ is a $\square$. (Def. of a $\square$)

Chapter 6, page 526 **1.** $p = 20$ m; $A =$
21 m² **3.** 40 **5.** 63 **7.** 15 **9.** right
11. acute **13.** 3π **15.** 113.0 **17.** 50.2

Chapter 7, page 527 **1.** $\frac{5}{3}$ **3.** $\frac{3}{4}$ **5.** 9:10
7. 5:1 **9.** $\frac{3}{2}$ **11.** $\frac{3}{5}$ **13.** 6 **15.** 2
17. \$4.45 **19.** 62.5 km

Chapter 8, pages 527–528 **1.** no
3. yes **5.** $x = 20$, $y = 16$ **7.** $x = 7.5$, $y =$
15 **9.** $\frac{3}{4}$ **11.** 1. $ABCD$ is a trapezoid.
(Given) 2. $\overline{DC} \parallel \overline{AB}$ (Def. of trapezoid) 3. $\angle$
$CAB = \angle DCA$; $\angle DBA = \angle CDB$ (If 2 $\parallel$
lines are cut by a trans., corr. $\angle$s are =.)
4. $\triangle DOC \sim \triangle BOA$ (AA Post.)
5. $\frac{AO}{CO} = \frac{BO}{DO}$ (Corr. sides of $\sim$ $\triangle$s are
in proportion.)

Chapter 9, page 528 **1.** 150° **3.** 90°
5. 35° **7.** 50; 130 **9.** 6 **11.** 50 **13.** 15
15. Use Construction 2 to construct a line
$\perp$ to the line passing through the center of
the constructed $\odot$ and any point on the $\odot$.

Chapter 10, pages 528–529 **1.** intersecting, parallel, skew **3.** no **5.** 220
7. 600 **9. a.** 65π **b.** 90π **c.** 100π
11. 56.5 **13.** 36π

Chapter 11, pages 529–530 **1.** $x = 8$,
$y = 17$ **3.** $x = 90$, $y = 2$ **5.** 60 **7.** $9\sqrt{3}$
cm² **9.** 60 **11. a.** 6 **b.** 10 **c.** 96π
d. 96π **13.** 4.3

Chapter 12, page 530 **1.** II **3.** 3; $\sqrt{5}$
5. 0; $\frac{4}{3}$ **7. a.** $\frac{3}{8}$ **b.** $-\frac{8}{3}$
9.
x	y
0	4
6	0
3	2
−3	6

11. $BC = 5$;
$AC^2 = 4^2 + 3^2 = 25$;
$AC = \sqrt{25} = 5$; $BC = AC$
So $\triangle ABC$ is an isosceles
triangle.

APPENDIX: Transformational Geometry

Exercises, page 533 **1.** yes **3.** yes
5. no **7.** 2 **9.** 2 **11.** an infinite number
13. 8

Exercises, page 539 **1.** $(-7, 1)$
3. $(4, -6)$ **5.** $(9, 2)$ **7.** A, H, I, M, O, T, U,
V, W, X, Y **9.** H, I, O, X

Exercises, page 542 **1.** reflection
3. translation **5.** translation **7.** reflection

Exercises, pages 545–546 **1. a.** yes
b. yes **c.** yes (60°, 120°, 180°, 240°, 300°, 360°)
3. **5.**

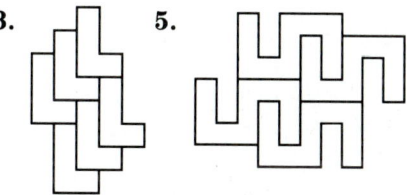

7. yes **9.** no **11.** yes **13.** yes **15.** yes

CREDITS

Book design and production: TextArt, Inc.
Cover design: Sylvia Steiner
Illustrations: Eva Burg Vagreti
Technical art: Network Graphics and ANCO / Boston

PHOTOGRAPHY

Chapter openers

pages 2–3:   Harald Sund / Image Bank
pages 44–45:   Peter Gavin / Photo Researchers
pages 90–91:   John Maher / EKM Nepenthe
pages 142–143: Chris Jones / The Stock Market
pages 188–189: Frank Siteman / EKM Nepenthe
pages 236–237: Grant Heilman
pages 284–285: L.L.T. Rhodes / Earth Scenes
pages 310–311: Lou Jones / Image Bank
pages 344–345: Georg Gerster / Photo Researchers
pages 386–387: Tim Jewett / EKM Nepenthe
pages 426–427: Hazel Hankin / Stock Boston
pages 466–467: Tom Ballard / EKM Nepenthe

Other photographs

page 23:   Shirley Richards / Photo Researchers
page 32:   Alan Carey / The Image Works
page 54:   Art Resource, N.Y.
page 55:   Michael Holford Photographs
page 57:   Mark Boulton / Photo Researchers
page 59:   Tyrone Hall
page 114: Daniel S. Brody / Stock Boston
page 125: Ken Lax
page 172: Glyn Cloyd / Taurus Photos
page 202: Robert Frerck / Odyssey Productions
page 208: Art Resource, N.Y.
page 258: The Granger Collection
page 262: Ken Lax
page 294: Michal Heron / Woodfin Camp & Assoc.
page 300: George Whiteley / Photo Researchers
page 302: George Bellerose / Stock Boston
page 316: Craig Hammel / The Stock Market
page 334: Ellis Herwig / Stock Boston
page 377: Bob Daemmrich
page 378: Courtesy of Panasonic Co.
page 402: Robert Frerck / Odyssey Productions
page 451: Carl Wolinsky / Stock Boston
page 482: Pierre Berger / Photo Researchers
page 500: Donald C. Dietz / Stock Boston
page 505: Richard B. Hoit / Photo Researchers
           Walter G. Hodsdon
           Joel E. Arem, Ph. D., F.G.A.
page 531: Richard Parker / Photo Researchers
           Courtesy Museo Nacional De Artes E Industrias Populares, Mexico City
           Jerome Wexler / Photo Researchers
page 537: J.L. Stage / Image Bank